P9-AZX-976

Chevrolet Camaro Automotive Repair Manual

by John B Raffa
and John H Haynes

Member of the Guild of Motoring Writers

Models covered

Chevrolet Camaro, Berlinetta and Z28
1982 through 1992

ISBN 1 56392 060 3

© **Haynes North America, Inc. 1984, 1987, 1989, 1991, 1993**
With permission from J. H. Haynes & Co. Ltd.

All rights reserved. No part of this book may be reproduced or transmitted in any form or by any means, electronic or mechanical, including photocopying, recording or by any information storage or retrieval system, without permission in writing from the copyright holder.

ABCDE
FGHIJ
KLMNO
PQ 2

Printed in the USA *(4W9 - 866)*

Haynes Publishing Group
Sparkford Nr Yeovil
Somerset BA22 7JJ England

Haynes North America, Inc.
861 Lawrence Drive
Newbury Park
California 91320 USA

Library of Congress
Catalog card number
93-77332

Acknowledgments

We are grateful for the help and cooperation of the Chevrolet Motor Division of the General Motors Corporation for their assistance with technical information, certain illustrations and vehicle photos, and the Champion Spark Plug Company who supplied the illustrations showing the various spark plug conditions.

About this manual

Its purpose

The purpose of this manual is to help you get the best value from your vehicle. It can do so in several ways. It can help you decide what work must be done, even if you choose to have it done by a dealer service department or a repair shop; it provides information and procedures for routine maintenance and servicing; and it offers diagnostic and repair procedures to follow when trouble occurs.

It is hoped that you will use the manual to tackle the work yourself. For many simpler jobs, doing it yourself may be quicker than arranging an appointment to get the vehicle into a shop and making the trips to leave it and pick it up. More importantly, a lot of money can be saved by avoiding the expense the shop must pass on to you to cover its labor and overhead costs. An added benefit is the sense of satisfaction and accomplishment that you feel after having done the job yourself.

Using the manual

The manual is divided into Chapters. Each Chapter is divided into numbered Sections, which are headed in bold type between horizontal lines. Each Section consists of consecutively numbered paragraphs.

The two types of illustrations used (figures and photographs), are referenced by a number preceding their caption. Figure reference numbers denote Chapter and numerical sequence within the Chapter; (i.e. Fig. 3.4 means Chapter 3, figure number 4). Figure captions are followed by a Section number which ties the figure to a specific portion of the text. All photographs apply to the Chapter in which they appear and the reference number pinpoints the pertinent Section and paragraph; i.e., 3.2 means Section 3, paragraph 2.

Procedures, once described in the text, are not normally repeated. When it is necessary to refer to another Chapter, the reference will be given as Chapter and Section number i.e. Chapter 1/16). Cross references given without use of the word 'Chapter' apply to Sections and/or paragraphs in the same Chapter. For example, 'see Section 8' means in the same Chapter.

Reference to the left or right side of the vehicle is based on the assumption that one is sitting in the driver's seat, facing forward.

Even though extreme care has been taken during the preparation of this manual, neither the publisher nor the author can accept responsibility for any errors in, or omissions from, the information given.

Introduction to the Chevrolet Camaro

For the years covered in this manual, the Camaro is available in three models: Sport Coupe, Berlinetta and Z-28.

Engines available include the 151-cubic-inch in-line four-cylinder (standard in the Sport Coupe and designated L4 in this manual); a V6 of 173-cubic-inch displacement (standard in the Berlinetta and optional in the Sport Coupe); a standard V8 of 305-cubic-inch displacement (standard in the Z-28 and optional in the Sport Coupe and Berlinetta); and a high output (HO) version of the 305-cubic-inch V8 (optional in the Z-28 only).

Four and five-speed manual transmissions and three and four-speed automatic transmissions are available as standard or optional equipment, depending on year, model and engine.

Front suspension is independent, featuring modified McPherson struts, while the rear suspension is via torque arm with coil springs and lower control arms.

Front brakes on all models are power assisted sliding caliper discs; standard rear brakes are drum-type, with discs optionally available on some models.

Contents

	Page	
Acknowledgments	2	
About this manual	2	
Introduction to the Chevrolet Camaro	2	
General dimensions	7	
Vehicle identification numbers	7	
Buying parts	9	
Maintenance techniques, tools and working facilities	9	
Booster battery (jump) starting	16	
Jacking and towing	16	
Automotive chemicals and lubricants	17	
Safety first!	18	
Troubleshooting	19	
Chapter 1 Tune-up and routine maintenance	26	1
Chapter 2 Part A L4 (four-cylinder) engine	61	2A
Chapter 2 Part B V6 engine	74	2B
Chapter 2 Part C V8 engine	94	2C
Chapter 2 Part D General engine overhaul procedures	103	2D
Chapter 3 Cooling, heating and air-conditioning systems	125	3
Chapter 4 Fuel and exhaust systems	132	4
Chapter 5 Engine electrical systems	158	5
Chapter 6 Emissions control systems	169	6
Chapter 7 Part A Manual transmission	185	7A
Chapter 7 Part B Automatic transmission	189	7B
Chapter 8 Clutch and driveline	194	8
Chapter 9 Brakes	204	9
Chapter 10 Chassis electrical system	226	10
Chapter 11 Steering and suspension systems	265	11
Chapter 12 Body	281	12
Chapter 13 Supplement: Revisions and information on 1985 and later models	294	13
Conversion factors	338	
Index	339	

1984 Camaro Z28

1984 Camaro Z28

1984 Camaro Berlinetta

General dimensions

Overall length	187.8 inches
Wheelbase	101.0 inches
Overall width	72.8 inches
Tread (front)	60.7 inches
Tread (rear)	61.6 inches
Overall height	
Except Z-28	50.0 inches
Z-28	50.3 inches
Minimum ground clearance	
Except Z-28	4.8 inches
Z-28	5.1 inches

Vehicle identification numbers

Modifications are a continuing and unpublicized process in vehicle manufacturing. Since spare parts manuals and lists are compiled on a numerical basis, the individual vehicle numbers are essential to correctly identify the component required.

Vehicle identification number (VIN)

This very important identification number is located on a plate attached to the top left corner of the dashboard and can easily be seen while looking through the windshield from the outside of the vehicle (photo). The VIN also appears on the Vehicle Certificate of Title and Registration. It contains valuable information such as where and when the vehicle was manufactured, the model year and the body style.

Body identification plate

This metal plate is located on the top left side of the radiator support (photo). Like the VIN, it contains valuable information concerning the production of the vehicle, as well as information about the way in which the vehicle is equipped. This plate is especially useful for matching the color and type of paint during repair work.

Engine identification numbers

The ID number on the L4 engine is found at the left rear side of the engine, on the casting to the rear of the exhaust manifold.

The ID number on the V6 engine is found either below the left cylinder

The Vehicle Identification Number (VIN) is located on the driver's side of the dashboard and is visible through the windshield

The body identification plate is located at the top right side of the radiator support (A), and the Emissions Control Information label is located on the fan shroud (B)

head or below the right cylinder head, to the rear of the timing cover.

The ID number on the V8 engine is found just below the right cylinder head, to the rear of the timing cover.

Automatic transmission numbers

Automatic transmission ID numbers may be stamped in a variety of locations on pads on either side of the transmission (refer to the accompanying illustration).

Manual transmission numbers

Manual transmission ID numbers are stamped on the driver's side, adjacent to the rear of the cover (four-speed 83 mm and five-speed 70 mm), or on the driver's side, below the side cover (four-speed 76 mm).

Rear axle numbers

The rear axle ID number is located on the right axle tube, adjacent to the carrier.

Alternator numbers

The alternator ID number is located on top of the drive end frame.

Starter numbers

The starter ID number is stamped on the outer case, toward the rear.

Battery numbers

The battery ID number is located on the cell cover segment on top of the battery.

Emissions Control Information label

The Emissions Control Information label is attached to the radiator fan shroud.

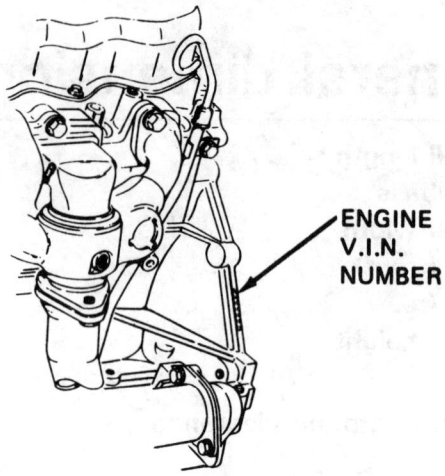

Location of the engine ID number on L4 engines

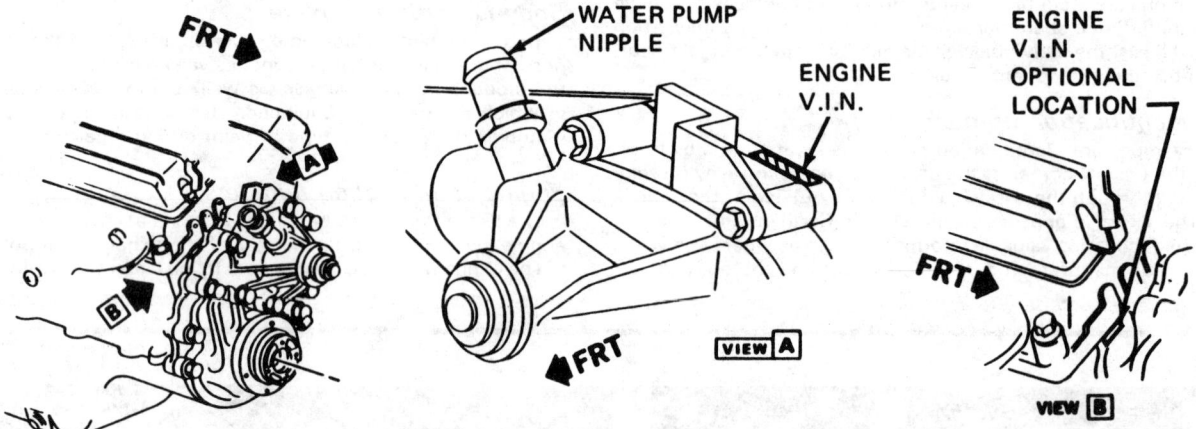

Location of the engine ID number on V6 engines

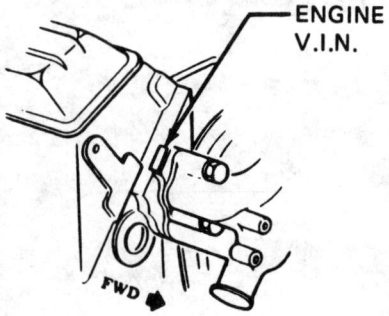

Location of the engine ID number on V8 engines

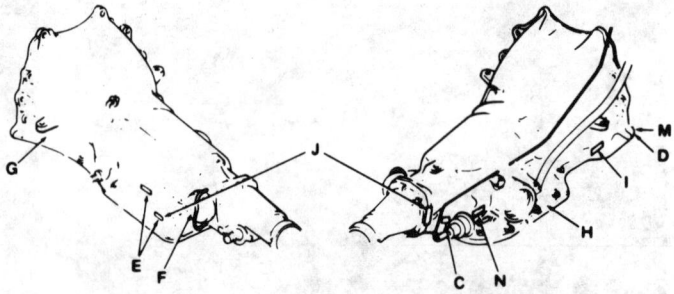

TYPICAL TRANSMISSION – I D LOCATIONS

C	THM 200 AND THM 200-4R I D TAG LOCATION	H	THM 350C STAMPED I D LOCATION
D	THM 200 AND THM 200-4R VIN LOCATION	I	THM 350C VIN LOCATION
E	THM 200 AND THM 200-4R VIN OPTIONAL LOCATIONS	J	THM 350C OPTIONAL VIN LOCATIONS
F	THM 250C I D STAMP LOCATION	M	THM 700-R4 STAMPED I D LOCATION
G	THM 250C VIN LOCATION	N	THM 700-R4 VIN LOCATION

Location of the ID numbers on automatic transmissions

Buying parts

Replacement parts are available from many sources, which generally fall into one of two categories – authorized dealer parts departments and independent retail auto parts stores. Our advice concerning these parts is as follows:

Retail auto parts stores: Good auto parts stores will stock frequently needed components which wear out relatively fast, such as clutch components, exhaust systems, brake parts, tune-up parts, etc. These stores often supply new or reconditioned parts on an exchange basis, which can save a considerable amount of money. Discount auto parts stores are often very good places to buy materials and parts needed for general vehicle maintenance such as oil, grease, filters, spark plugs, belts, touch-up paint, bulbs, etc. They also usually sell tools and general accessories, have convenient hours, charge lower prices and can often be found not far from home.

Authorized dealer parts department: This is the best source for parts which are unique to the vehicle and not generally available elsewhere (such as major engine parts, transmission parts, trim pieces, etc.).

Warranty information: If the vehicle is still covered under warranty, be sure that any replacement parts purchased – regardless of the source – do not invalidate the warranty!

To be sure of obtaining the correct parts, have engine and chassis numbers available and, if possible, take the old parts along for positive identification.

Maintenance techniques, tools and working facilities

Maintenance techniques

There are a number of techniques involved in maintenance and repair that will be referred to throughout this manual. Application of these techniques will enable the home mechanic to be more efficient, better organized and capable of performing the various tasks properly, which will ensure that the repair job is thorough and complete.

Fasteners

Fasteners are nuts, bolts, studs and screws used to hold two or more parts together. There are a few things to keep in mind when working with fasteners. Almost all of them use a locking device of some type, either a lock washer, locknut, locking tab or thread adhesive. All threaded fasteners should be clean and straight, with undamaged threads and undamaged corners on the hex head where the wrench fits. Develop the habit of replacing all damaged nuts and bolts with new ones. Special locknuts with nylon or fiber inserts can only be used once. If they are removed, they lose their locking ability and must be replaced with new ones.

Rusted nuts and bolts should be treated with a penetrating fluid to ease removal and prevent breakage. Some mechanics use turpentine in a spout-type oil can, which works quite well. After applying the rust penetrant, let it "work" for a few minutes before trying to loosen the nut or bolt. Badly rusted fasteners may have to be chiseled or sawed off or removed with a special nut breaker, available at tool stores.

If a bolt or stud breaks off in an assembly, it can be drilled and removed with a special tool commonly available for this purpose. Most automotive machine shops can perform this task, as well as other repair procedures (such as repair of threaded holes that have been stripped out).

Flat washers and lock washers, when removed from an assembly, should always be replaced exactly as removed. Replace any damaged washers with new ones. Always use a flat washer between a lock washer and any soft metal surface (such as aluminum), thin sheet metal or plastic.

Fastener sizes

For a number of reasons, automobile manufacturers are making wider and wider use of metric fasteners. Therefore, it is important to be able to tell the difference between standard (sometimes called U.S., English or SAE) and metric hardware, since they cannot be interchanged.

All bolts, whether standard or metric, are sized according to diameter, thread pitch and length. For example, a standard 1/2 — 13 x 1 bolt is 1/2 inch in diameter, has 13 threads per inch and is 1 inch long. An M12 — 1.75 x 25 metric bolt is 12 mm in diameter, has a thread pitch of 1.75 mm (the distance between threads) and is 25 mm long. The two bolts are nearly identical, and easily confused, but they are not interchangeable.

In addition to the differences in diameter, thread pitch and length, metric and standard bolts can also be distinguished by examining the bolt heads. To begin with, the distance across the flats on a standard bolt head is measured in inches, while the same dimension on a metric bolt is measured in millimeters (the same is true for nuts). As a result, a standard wrench should not be used on a metric bolt and a metric wrench should not be used on a standard bolt. Also, standard bolts have slashes radiating out from the center of the head to denote the grade or strength of the bolt (which is an indication of the amount of torque that can be applied to it). The greater the number of slashes, the greater the strength of the bolt (grades 0 through 5 are commonly used on automobiles). Metric bolts have a property class (grade) number, rather than a slash, molded into their heads to indicate bolt strength. In this case, the higher the number, the stronger the bolt (property class numbers 8.8, 9.8 and 10.9 are commonly used on automobiles).

Strength markings can also be used to distinguish standard hex nuts from metric hex nuts. Standard nuts have dots stamped into one side,

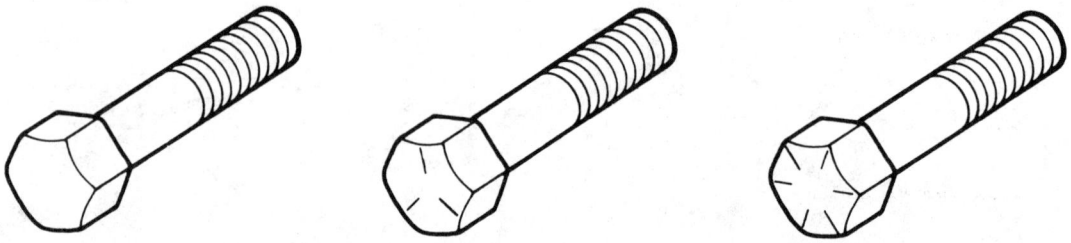

Grade 1 or 2 Grade 5 Grade 8

Bolt strength markings (top — standard/SAE; bottom — metric)

Grade	Identification
Hex Nut Grade 5	3 Dots
Hex Nut Grade 8	6 Dots

Standard hex nut strength markings

Grade	Identification
Hex Nut Property Class 9	Arabic 9
Hex Nut Property Class 10	Arabic 10

Metric hex nut strength markings

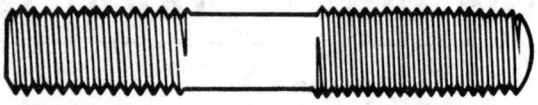

CLASS 10.9 CLASS 9.8 CLASS 8.8

Metric stud strength markings

while metric nuts are marked with a number. The greater the number of dots, or the higher the number, the greater the strength of the nut.

Metric studs are also marked on their ends according to property class (grade). Larger studs are numbered (the same as metric bolts), while smaller studs carry a geometric code to denote grade.

It should be noted that many fasteners, especially Grades 0 through 2, have no distinguishing marks on them. When such is the case, the only way to determine whether it is standard or metric is to measure the thread pitch or compare it to a known fastener of the same size.

Since fasteners of the same size (both standard and metric) may have different strength ratings, be sure to reinstall any bolts, studs or nuts removed from your vehicle in their original locations. Also, when replacing a fastener with a new one, make sure that the new one has a strength rating equal to or greater than the original.

Tightening sequences and procedures

Most threaded fasteners should be tightened to a specific torque value (torque is basically a twisting force). Over-tightening the fastener can weaken it and cause it to break, while under-tightening can cause it to eventually come loose. Bolts, screws and studs, depending on the material they are made of and their thread diameters, have specific torque values (many of which are noted in the Specifications at the beginning of each Chapter). Be sure to follow the torque recommendations closely. For fasteners not assigned a specific torque, a general torque value chart is presented here as a guide. As was previously mentioned, the size and grade of a fastener determine the amount of torque that can safely be applied to it. The figures listed here are approximate for Grade 2 and Grade 3 fasteners (higher grades can tolerate higher torque values).

Metric thread sizes	Ft-lb	Nm
M-6	6 to 9	9 to 12
M-8	14 to 21	19 to 28
M-10	28 to 40	38 to 54
M-12	50 to 71	68 to 96
M-14	80 to 140	109 to 154
Pipe thread sizes		
1/8	5 to 8	7 to 10
1/4	12 to 18	17 to 24
3/8	22 to 33	30 to 44
1/2	25 to 35	34 to 47
U.S. thread sizes		
1/4 — 20	6 to 9	9 to 12
5/16 — 18	12 to 18	17 to 24
5/16 — 24	14 to 20	19 to 27
3/8 — 16	22 to 32	30 to 43
3/8 — 24	27 to 38	37 to 51
7/16 — 14	40 to 55	55 to 74
7/16 — 20	40 to 60	55 to 81
1/2 — 13	55 to 80	75 to 108

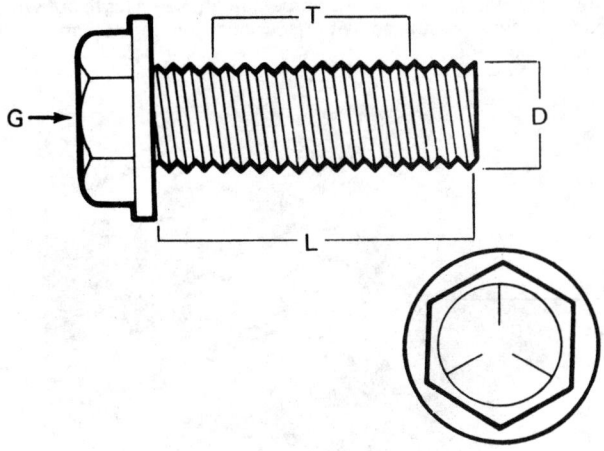

Standard (SAE) bolt dimensions/grade marks

G Grade marks (bolt strength)
L Length (in inches)
T Thread pitch (number of threads per inch)
D Nominal diameter (in inches)

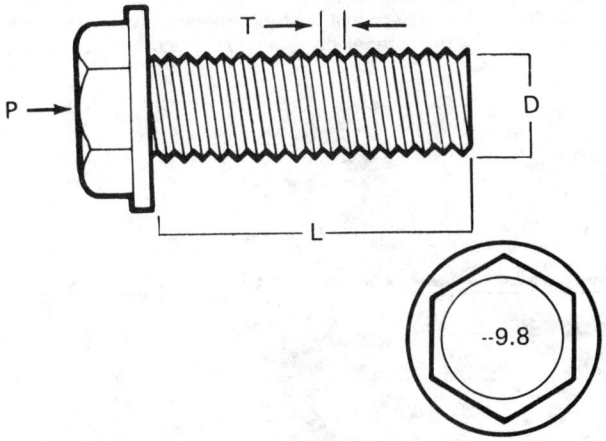

Metric bolt dimensions/grade marks

P Property class (bolt strength)
L Length (in millimeters)
T Thread pitch (distance between threads in millimeters)
D Diameter

Fasteners laid out in a pattern (i.e. cylinder head bolts, oil pan bolts, differential cover bolts, etc.) must be loosened or tightened in a sequence to avoid warping the component. This sequence will normally be shown in the appropriate Chapter. If a specific pattern is not given, the following procedures can be used to prevent warping. Initially, the bolts or nuts should be assembled finger-tight only. Next, they should be tightened one full turn each, in a crisscross or diagonal pattern. After each one has been tightened one full turn, return to the first one and tighten them all one-half turn, following the same pattern. Finally, tighten each of them one-quarter turn at a time until each fastener has been tightened to the proper torque. To loosen and remove the fasteners, the procedure would be reversed.

Component disassembly

Component disassembly should be done with care and purpose to help ensure that the parts go back together properly. Always keep track of the sequence in which parts are removed. Make note of special characteristics or marks on parts that can be installed more than one way (such as a grooved thrust washer on a shaft). It is a good idea to lay the disassembled parts out on a clean surface in the order that they were removed. It may also be helpful to make sketches or take instant photos of components before removal.

When removing fasteners from a component, keep track of their locations. Sometimes threading a bolt back in a part, or putting the washers and nut back on a stud, can prevent mix-ups later. If nuts and bolts cannot be returned to their original locations, they should be kept in a compartmented box or a series of small boxes. A cupcake or muffin tin is ideal for this purpose, since each cavity can hold the bolts and nuts from a particular area (i.e. oil pan bolts, valve cover bolts, engine mount bolts, etc.). A pan of this type is especially helpful when working on assemblies with very small parts, such as the carburetor, alternator, valve train or interior dash and trim pieces. The cavities can be marked with paint or tape to identify the contents.

Whenever wiring looms, harnesses or connectors are separated, it's a good idea to identify the two halves with numbered pieces of masking tape so they can be easily reconnected.

Gasket sealing surfaces

Throughout any vehicle, gaskets are used to seal the mating surfaces between two parts and keep lubricants, fluids, vacuum or pressure contained in an assembly.

Many times these gaskets are coated with a liquid or paste-type gasket sealing compound before assembly. Age, heat and pressure can sometimes cause the two parts to stick together so tightly that they are very difficult to separate. Often, the assembly can be loosened by striking it with a soft-faced hammer near the mating surfaces. A regular hammer can be used if a block of wood is placed between the hammer and the part. Do not hammer on cast parts or parts that could be easily damaged. With any particularly stubborn part, always recheck to make sure that every fastener has been removed.

Avoid using a screwdriver or bar to pry apart an assembly, as they can easily mar the gasket sealing surfaces of the parts (which must remain smooth). If prying is absolutely necessary, use an old broom handle, but keep in mind that extra clean-up will be necessary if the wood splinters.

After the parts are separated, the old gasket must be carefully scraped off and the gasket surfaces cleaned. Stubborn gasket material can be soaked with rust penetrant or treated with a special chemical to soften it so it can be easily scraped off. A scraper can be fashioned from a piece of copper tubing by flattening and sharpening one end. Copper is recommended because it is usually softer than the surfaces to be scraped, which reduces the chance of gouging the part. Some gaskets can be removed with a wire brush, but regardless of the method used, the mating surfaces must be left clean and smooth. If for some reason the gasket surface is gouged, then a gasket sealer thick enough to fill scratches will have to be used during reassembly of the components. For most applications, a non-drying (or semi-drying) gasket sealer should be used.

Hose removal tips

Caution: *If the vehicle is equipped with air conditioning, do not disconnect any of the A/C hoses without first having the system depressurized by a dealer service department or an air conditioning specialist.*

Hose removal precautions closely parallel gasket removal precautions. Avoid scratching or gouging the surface that the hose mates against or the connection may leak. This is especially true for radiator hoses. Because of various chemical reactions, the rubber in hoses can bond itself to the metal spigot that the hose fits over. To remove a hose, first loosen the hose clamps that secure it to the spigot. Then, with slip-joint pliers, grab the hose at the clamp and rotate it around the spigot. Work it back and forth until it is completely free, then pull it off. Silicone or other lubricants will ease removal if they can be applied between the hose and the outside of the spigot. Apply the same lubricant to the inside of the hose and the outside of the spigot to simplify installation.

As a last resort (and if the hose is to be replaced with a new one anyway), the rubber can be slit with a knife and the hose peeled from the spigot. If this must be done, be careful that the metal connection is not damaged.

If a hose clamp is broken or damaged, do not reuse it. Wire-type clamps usually weaken with age, so it is a good idea to replace them with screw-type clamps whenever a hose is removed.

Tools

A selection of good tools is a basic requirement for anyone who plans to maintain and repair his or her own vehicle. For the owner who has few tools, if any, the initial investment might seem high, but when compared to the spiraling costs of professional auto maintenance and repair, it is a wise one.

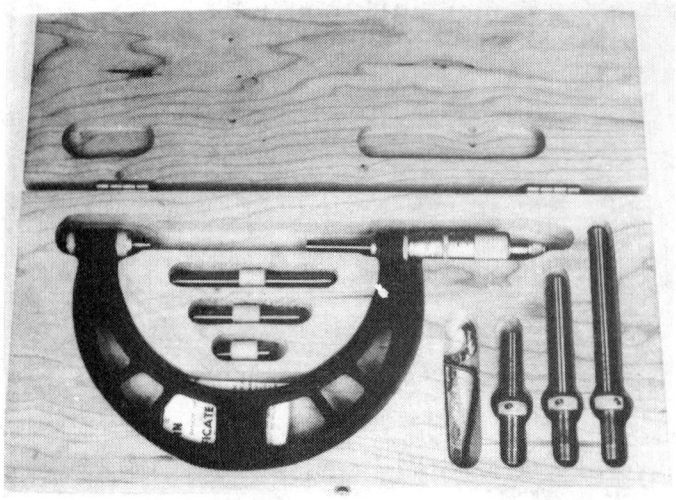

Micrometer set

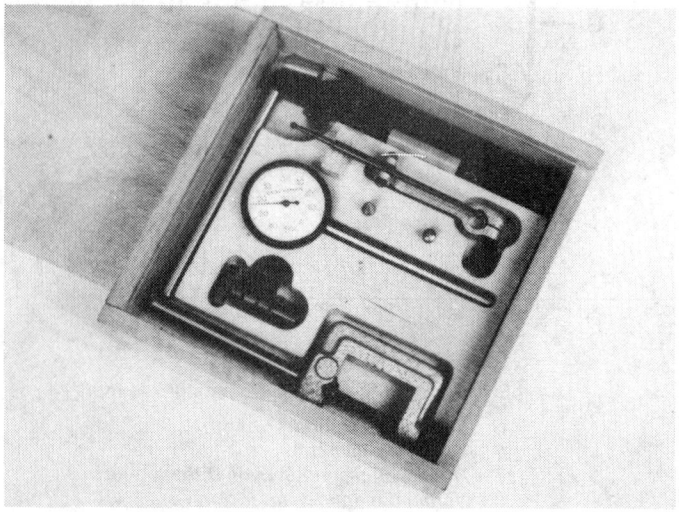

Dial indicator set

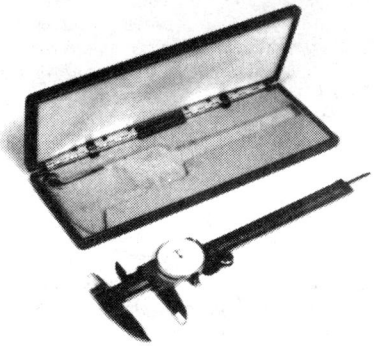

Dial caliper

Hydraulic lifter removal tool

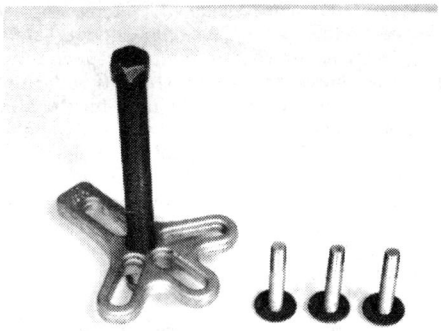

Universal-type puller

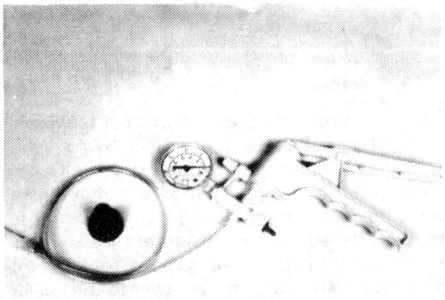

Hand-operated vacuum pump

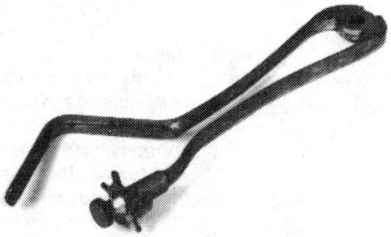

Piston ring groove cleaning tool

Piston ring compressor

Piston ring removal/installation tool

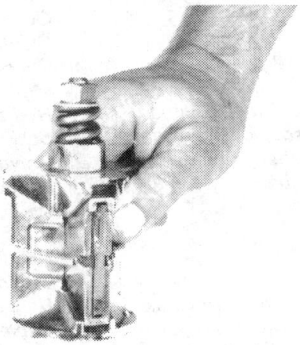

Cylinder ridge reamer

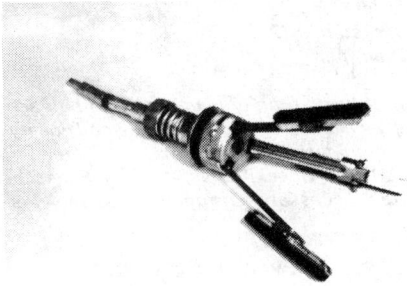

Cylinder surfacing hone

Cylinder bore gauge

Brake shoe spring tool

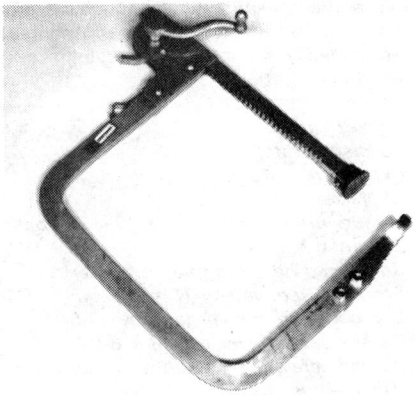

Valve spring compressor

To help the owner decide which tools are needed to perform the tasks detailed in this manual, the following tool lists are offered: *Maintenance and minor repair, Repair and overhaul* and *Special.* The newcomer to practical mechanics should start off with the *Maintenance and minor repair tool kit*, which is adequate for the simpler jobs performed on a vehicle. Then, as confidence and experience grow, the owner can tackle more difficult tasks, buying additional tools as they are needed. Eventually the basic kit will be expanded into the *Repair and overhaul tool set*. Over a period of time, the experienced do-it-yourselfer will assemble a tool set complete enough for most repair and overhaul procedures and will add tools from the *Special* category when it is felt that the expense is justified by the frequency of use.

Maintenance and minor repair tool kit

The tools in this list should be considered the minimum required for performance of routine maintenance, servicing and minor repair work. We recommend the purchase of combination wrenches (box-end and open-end combined in one wrench); while more expensive than open-ended ones, they offer the advantages of both types of wrench.

Combination wrench set (1/4 in to 1 in or 6 mm to 19 mm)
Adjustable wrench — 8 in
Spark plug wrench (with rubber insert)
Spark plug gap adjusting tool
Feeler gauge set
Brake bleeder wrench
Standard screwdriver (5/16 in x 6 in)
Phillips screwdriver (No. 2 x 6 in)
Combination pliers — 6 in
Hacksaw and assortment of blades
Tire pressure gauge
Grease gun
Oil can
Fine emery cloth
Wire brush
Battery post and cable cleaning tool
Oil filter wrench
Funnel (medium size)
Safety goggles
Jackstands (2)
Drain pan

Note: *If basic tune-ups are going to be part of routine maintenance, it will be necessary to purchase a good quality stroboscopic timing light and combination tachometer/dwell meter. Although they are included in the list of Special tools, it is mentioned here because they are absolutely necessary for tuning most vehicles properly.*

Repair and overhaul tool set

These tools are essential for anyone who plans to perform major repairs and are in addition to those in the *Maintenance and minor repair tool kit*. Included is a comprehensive set of sockets which, though expensive, are invaluable because of their versatility (especially when various extensions and drives are available). We recommend the 1/2-inch drive over the 3/8-inch drive. Although the larger drive is bulky and more expensive, it has the capacity of accepting a very wide range of large sockets (ideally, the mechanic would have a 3/8-inch drive set and a 1/2-inch drive set).

Socket set(s)
Reversible ratchet
Extension — 10 in
Universal joint
Torque wrench (same size drive as sockets)
Ball peen hammer — 8 oz
Soft-faced hammer (plastic/rubber)
Standard screwdriver (1/4 in x 6 in)
Standard screwdriver (stubby — 5/16 in)
Phillips screwdriver (No. 3 x 8 in)
Phillips screwdriver (stubby — No. 2)
Pliers — vise grip
Pliers — lineman's
Pliers — needle nose
Pliers — snap-ring (internal and external)
Cold chisel — 1/2 in
Scriber
Scraper (made from flattened copper tubing)
Center punch
Pin punches (1/16, 1/8, 3/16 in)
Steel rule/straightedge — 12 in
Allen wrench set (1/8 to 3/8 in or 4 mm to 10 mm)
A selection of files
Wire brush (large)
Jackstands (second set)
Jack (scissor or hydraulic type)

Note: *Another tool which is often useful is an electric drill motor (with a chuck capacity of 3/8-inch) and a set of good-quality drill bits.*

Special tools

The tools in this list include those which are not used regularly, are expensive to buy, or which need to be used in accordance with their manufacturer's instructions. Unless these tools will be used frequently, it is not very economical to purchase many of them. A consideration would be to split the cost and use between yourself and a friend or friends. In addition, most of these tools can be obtained from a tool rental shop on a temporary basis.

This list primarily contains only those tools and instruments widely available to the public, and not those special tools produced by the vehicle manufacturer for distribution to dealer service departments. Occasionally, references to the manufacturer's special tools are included in the text of this manual. Generally, an alternative method of doing the job without the special tool is offered. However, sometimes there is no alternative to their use. Where this is the case, and the tool cannot be purchased or borrowed, the work should be turned over to the dealer service department or an automotive repair shop.

Valve spring compressor
Piston ring groove cleaning tool
Piston ring compressor
Piston ring installation tool
Cylinder compression gauge
Cylinder ridge reamer
Cylinder surfacing hone
Cylinder bore gauge
Micrometer(s) and/or dial calipers
Hydraulic lifter removal tool
Balljoint separator
Universal-type puller
Impact screwdriver
Dial indicator set
Stroboscopic timing light (inductive pick-up)
Hand-operated vacuum/pressure pump
Tachometer/dwell meter
Universal electrical multimeter
Cable hoist
Brake spring removal and installation tools
Floor jack

Buying tools

For the do-it-yourselfer who is just starting to get involved in vehicle maintenance and repair, there are a number of options available when purchasing tools. If maintenance and minor repair is the extent of the work to be done, the purchase of individual tools is satisfactory. If, on the other hand, extensive work is planned, it would be a good idea to purchase a modest tool set from one of the large retail chain stores. A set can usually be bought at a substantial savings over the individual tool prices (and they often come with a tool box). As additional tools are needed, add-on sets, individual tools and a larger tool box can be purchased to expand the tool selection. Building a tool set gradually allows the cost of the tools to be spread over a longer period of time and gives the mechanic the freedom to choose only those tools that will actually be used.

Tool stores will often be the only source of some of the special tools that are needed, but regardless of where tools are bought, try to avoid cheap ones (especially when buying screwdrivers and sockets) because they won't last very long. The expense involved in replacing cheap tools will eventually be greater than the initial cost of quality tools.

Care and maintenance of tools

Good tools are expensive, so it makes sense to treat them with respect. Keep them clean and in usable condition and store them properly when not in use. Always wipe off any dirt, grease or metal chips before putting them away. Never leave tools lying around in the work area. Upon completion of a job, always check closely under the hood for tools that may have been left there (so they don't get lost during a test drive).

Some tools, such as screwdrivers, pliers, wrenches and sockets, can be hung on a panel mounted on the garage or workshop wall, while others should be kept in a tool box or tray. Measuring instruments, gauges, meters, etc. must be carefully stored where they cannot be damaged by weather or impact from other tools.

When tools are used with care and stored properly, they will last a very long time. Even with the best of care, tools will wear out if used frequently. When a tool is damaged or worn out, replace it; subsequent jobs will be safer and more enjoyable if you do.

Working facilities

Not to be overlooked when discussing tools is the workshop. If anything more than routine maintenance is to be carried out, some sort of suitable work area is essential.

It is understood, and appreciated, that many home mechanics do not have a good workshop or garage available and end up removing an engine or doing major repairs outside. It is recommended, however, that the overhaul or repair be completed under the cover of a roof.

A clean, flat workbench or table of comfortable working height is an absolute necessity. The workbench should be equipped with a vise that has a jaw opening of at least four inches.

As mentioned previously, some clean, dry storage space is also required for tools, as well as the lubricants, fluids, cleaning solvents, etc. which soon become necessary.

Sometimes waste oil and fluids, drained from the engine or cooling system during normal maintenance or repairs, present a disposal problem. To avoid pouring them on the ground or into a sewage system, simply pour the used fluids into large containers, seal them with caps and take them to an authorized disposal site or recycling center. Plastic jugs (such as old antifreeze containers) are ideal for this purpose.

Always keep a supply of old newspapers and clean rags available. Old towels are excellent for mopping up spills. Many mechanics use rolls of paper towels for most work because they are readily available and disposable. To help keep the area under the vehicle clean, a large cardboard box can be cut open and flattened to protect the garage or shop floor.

Whenever working over a painted surface (such as when leaning over a fender to service something under the hood), always cover it with an old blanket or bedspread to protect the finish. Vinyl covered pads, made especially for this purpose, are available at auto parts stores.

Booster battery (jump) starting

Certain precautions must be observed when using a booster battery to 'jump start' a vehicle.

a) Before connecting the booster battery, make sure that the ignition switch is in the Off position.
b) Turn off the lights, heater and other electrical loads.
c) The eyes should be shielded; safety goggles are a good idea.
d) Make sure the booster battery is the same voltage as the dead one in the vehicle.
e) The two vehicles must not touch each other.
f) Make sure the transmission is in Neutral (manual transmission) or Park (automatic transmission).

Connect the red jumper cable to the *positive* (+) terminals of each battery.

Connect one end of the black jumper cable to the *negative* (–) terminal of the booster battery. The other end of this cable should be connected to a good ground on the vehicle to be started, such as a bolt or bracket on the engine block.

Start the engine using the booster battery, then, with the engine running at idle speed, disconnect the jumper cables in the reverse order of connection.

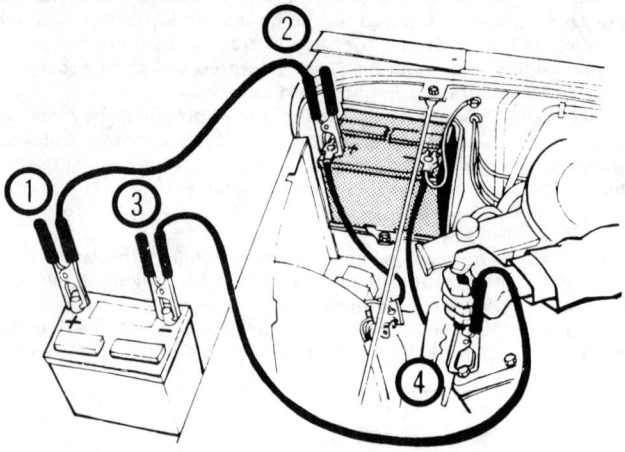

Make the booster battery cable connections in the numerical order shown (note that the negative cable of the booster battery in not attached to the negative terminal of the dead battery)

Jacking and towing

Jacking

The jack supplied with the vehicle should only be used for raising the vehicle when changing a tire or placing jackstands under the frame. **Caution:** *Never work under the vehicle or start the engine while this jack is being used as the only means of support.*

The vehicle should be on level ground with the wheels blocked and the transmission in Park (automatic) or Reverse (manual). Pry off the hub cap (if equipped) using the tapered end of the lug wrench. Loosen the wheel nuts one-half turn and leave them in place until the wheel is raised off the ground.

Place the jack under the side of the vehicle in the indicated position and place the jack lever in the 'up' position. Raise the jack until the jack head groove fits into the rocker flange notch. Operate the jack with a slow, smooth motion, using your hand or foot to pump the handle until the wheel is raised off the ground. Remove the wheel nuts, pull off the wheel and replace it with the spare. (If you have a stowaway spare, refer to the instructions accompanying the supplied inflator.)

With the beveled side in, replace the wheel nuts and tighten them until snug. Place the jack lever in the 'down' position and lower the vehicle. Remove the jack and tighten the nuts in a crisscross sequence by turning the wrench clockwise. Replace the hub cap (if equipped) by placing it into position and using the heel of your hand or a rubber mallet to seat it.

Towing

The vehicle can be towed with all four wheels on the ground, provided that speeds do not exceed 35 mph and the distance is not over 50 miles, otherwise transmission damage can result.

Towing equipment specifically designed for this purpose should be used and should be attached to the main structural members of the vehicle and not the bumper or brackets.

Safety is a major consideration when towing and all applicable state and local laws must be obeyed. A safety chain system must be used for all towing.

While towing, the parking brake should be released and the transmission should be in Neutral. The steering must be unlocked (ignition switch in the Off position). Remember that power steering and power brakes will not work with the engine off.

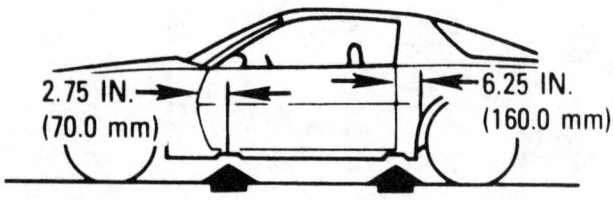

Location of rocker panel flange notches used for jack placement (left side shown, right side similar)

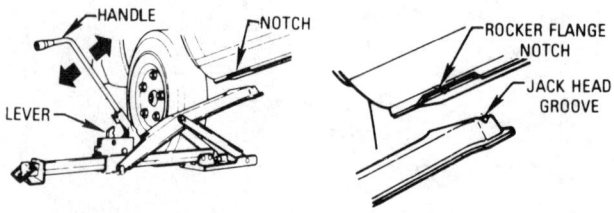

Placement and operation of jack for tire changing (left) and details of matching jack head groove-to-rocker panel flange notch (right)

Automotive chemicals and lubricants

A number of automotive chemicals and lubricants are available for use during vehicle maintenance and repair. They include a wide variety of products ranging from cleaning solvents and degreasers to lubricants and protective sprays for rubber, plastic and vinyl.

Contact point/spark plug cleaner is a solvent used to clean oily film and dirt from points, grime from electrical connectors and oil deposits from spark plugs. It is oil free and leaves no residue. It can also be used to remove gum and varnish from carburetor jets and other orifices.

Carburetor cleaner is similar to contact point/spark plug cleaner but it is a stronger solvent and may leave a slight oily residue. It is not recommended for cleaning electrical components or connections.

Brake system cleaner is used to remove grease or brake fluid from brake system components where clean surfaces are absolutely necessary and petroleum-based solvents cannot be used. It also leaves no residue.

Silicone-based lubricants are used to protect rubber parts such as hoses, weatherstripping and grommets, and are used as lubricants for hinges and locks.

Multi-purpose grease is an all-purpose lubricant used wherever grease is more practical than a liquid lubricant such as oil. Some multi-purpose grease is white and specially formulated to be more resistant to water than ordinary grease.

Bearing grease/wheel bearing grease is a heavy grease used where increased loads and friction are encountered (i.e. wheel bearings, universal joints, etc.).

High-temperature wheel bearing grease is designed to withstand the extreme temperatures encountered by wheel bearings in disc-brake equipped vehicles. It usually contains molybdenum disulfide, which is a 'dry' type lubricant.

Gear oil (sometimes called gear lube) is a specially designed oil used in differentials, manual transmissions and transfer cases, as well as other areas where high-friction, high-temperature lubrication is required. It is available in a number of viscosities (weights) for various applications.

Motor oil, of course, is the lubricant specially formulated for use in engines. It normally contains a wide variety of additives to prevent corrosion and reduce foaming and wear. Motor oil comes in various weights (viscosity ratings) of from 5 to 80. The recommended weight of the oil depends on the seasonal temperature and the demands on the engine. Light oil is used in cold climates and under light load conditions; heavy oil is used in hot climates and where high loads are encountered. Multi-viscosity oils are designed to have characteristics of both light and heavy oils and are available in a number of weights from 5W-20 to 20W-50.

Oil additives range from viscosity index improvers to slick chemical treatments that purportedly reduce friction. It should be noted that most oil manufacturers caution against using additives with their oils.

Gas additives perform several functions, depending on their chemical makeup. They usually contain solvents that help dissolve gum and varnish that build up on carburetor and intake parts. They also serve to break down carbon deposits that form on the inside surfaces of the combustion chambers. Some additives contain upper cylinder lubricants for valves and piston rings.

Brake fluid is a specially formulated hydraulic fluid that can withstand the heat and pressure encountered in brake systems. Care must be taken that this fluid does not come in contact with painted surfaces or plastics. An opened container should always be resealed to prevent contamination by water or dirt.

Undercoating is a petroleum-based, tar-like substance that is designed to protect metal surfaces on the underside of a vehicle from corrosion. It also acts as a sound-deadening agent by insulating the bottom of the vehicle.

Weatherstrip cement is used to bond weatherstripping around doors, windows and trunk lids. It is sometimes used to attach trim pieces as well.

Degreasers are heavy-duty solvents used to remove grease and grime that may accumulate on engine and chassis components. They can be sprayed or brushed on and, depending on the type, are rinsed off with either water or solvent.

Solvents are used alone or in combination with degreasers to clean parts and assemblies during repair and overhaul. The home mechanic should use only solvents that are non-flammable and that do not produce irritating fumes.

Gasket sealing compounds may be used in conjunction with gaskets, to improve their sealing capabilities, or alone, to seal metal-to-metal joints. Many gasket sealers can withstand extreme heat, some are impervious to gasoline and lubricants, while others are capable of filling and sealing large cavities. Depending on the intended use, gasket sealers either dry hard or stay relatively soft and pliable. They are usually applied by hand, with a brush, or are sprayed on the gasket sealing surfaces.

Thread cement is an adhesive locking compound that prevents threaded fasteners from loosening because of vibration. It is available in a variety of types for different applications.

Moisture dispersants are usually sprays that can be used to dry out electrical components such as the distributor, fuse block and wiring connectors. Some types can also be used as treatment for rubber and as a lubricant for hinges, cables and locks.

Waxes and polishes are used to help protect painted and plated surfaces from the weather. Different types of paint may require the use of different types of wax polish. Some polishes utilize a chemical or abrasive cleaner to help remove the top layer of oxidized (dull) paint on older vehicles. In recent years many non-wax polishes that contain a wide variety of chemicals such as polymers and silicones have been introduced. These non-wax polishes are usually easier to apply and last longer than conventional waxes and polishes.

Safety first!

Regardless of how enthusiastic you may be about getting on with the job at hand, take the time to ensure that your safety is not jeopardized. A moment's lack of attention can result in an accident, as can failure to observe certain simple safety precautions. The possibility of an accident will always exist, and the following points should not be considered a comprehensive list of all dangers. Rather, they are intended to make you aware of the risks and to encourage a safety conscious approach to all work you carry out on your vehicle.

Essential DOs and DON'Ts

DON'T rely on a jack when working under the vehicle. Always use approved jackstands to support the weight of the vehicle and place them under the recommended lift or support points.
DON'T attempt to loosen extremely tight fasteners (i.e. wheel lug nuts) while the vehicle is on a jack — it may fall.
DON'T start the engine without first making sure that the transmission is in Neutral (or Park where applicable) and the parking brake is set.
DON'T remove the radiator cap from a hot cooling system — let it cool or cover it with a cloth and release the pressure gradually.
DON'T attempt to drain the engine oil until you are sure it has cooled to the point that it will not burn you.
DON'T touch any part of the engine or exhaust system until it has cooled sufficiently to avoid burns.
DON'T siphon toxic liquids such as gasoline, antifreeze and brake fluid by mouth, or allow them to remain on your skin.
DON'T inhale brake lining dust — it is potentially hazardous (see *Asbestos* below)
DON'T allow spilled oil or grease to remain on the floor — wipe it up before someone slips on it.
DON'T use loose fitting wrenches or other tools which may slip and cause injury.
DON'T push on wrenches when loosening or tightening nuts or bolts. Always try to pull the wrench toward you. If the situation calls for pushing the wrench away, push with an open hand to avoid scraped knuckles if the wrench should slip.
DON'T attempt to lift a heavy component alone — get someone to help you.
DON'T rush or take unsafe shortcuts to finish a job.
DON'T allow children or animals in or around the vehicle while you are working on it.
DO wear eye protection when using power tools such as a drill, sander, bench grinder, etc. and when working under a vehicle.
DO keep loose clothing and long hair well out of the way of moving parts.
DO make sure that any hoist used has a safe working load rating adequate for the job.
DO get someone to check on you periodically when working alone on a vehicle.
DO carry out work in a logical sequence and make sure that everything is correctly assembled and tightened.
DO keep chemicals and fluids tightly capped and out of the reach of children and pets.
DO remember that your vehicle's safety affects that of yourself and others. If in doubt on any point, get professional advice.

Asbestos

Certain friction, insulating, sealing, and other products — such as brake linings, brake bands, clutch linings, torque converters, gaskets, etc. — contain asbestos. *Extreme care must be taken to avoid inhalation of dust from such products since it is hazardous to health.* If in doubt, assume that they *do* contain asbestos.

Fire

Remember at all times that gasoline is highly flammable. Never smoke or have any kind of open flame around when working on a vehicle. But the risk does not end there. A spark caused by an electrical short circuit, by two metal surfaces contacting each other, or even by static electricity built up in your body under certain conditions, can ignite gasoline vapors, which in a confined space are highly explosive. Do not, under any circumstances, use gasoline for cleaning parts. Use an approved safety solvent.

Always disconnect the battery ground (–) cable *at the battery* before working on any part of the fuel system or electrical system. Never risk spilling fuel on a hot engine or exhaust component.

It is strongly recommended that a fire extinguisher suitable for use on fuel and electrical fires be kept handy in the garage or workshop at all times. Never try to extinguish a fuel or electrical fire with water.

Fumes

Certain fumes are highly toxic and can quickly cause unconsciousness and even death if inhaled to any extent. Gasoline vapor falls into this category, as do the vapors from some cleaning solvents. Any draining or pouring of such volatile fluids should be done in a well ventilated area.

When using cleaning fluids and solvents, read the instructions on the container carefully. Never use materials from unmarked containers.

Never run the engine in an enclosed space, such as a garage. Exhaust fumes contain carbon monoxide, which is extremely poisonous. If you need to run the engine, always do so in the open air, or at least have the rear of the vehicle outside the work area.

If you are fortunate enough to have the use of an inspection pit, never drain or pour gasoline and never run the engine while the vehicle is over the pit. The fumes, being heavier than air, will concentrate in the pit with possibly lethal results.

The battery

Never create a spark or allow a bare light bulb near the battery. The battery normally gives off a certain amount of hydrogen gas, which is highly explosive.

Always disconnect the battery ground (–) cable *at the battery* before working on the fuel or electrical systems.

If possible, loosen the filler caps or cover when charging the battery from an external source. Do not charge at an excessive rate or the battery may burst.

Take care when adding water and when carrying a battery. The electrolyte, even when diluted, is very corrosive and should not be allowed to contact clothing or skin.

Always wear eye protection when cleaning the battery to prevent the caustic deposits from entering your eyes.

Household current

When using an electric power tool, inspection light, etc., which operates on household current, always make sure that the tool is correctly connected to its plug and that, where necessary, it is properly grounded. Do not use such items in damp conditions and, again, do not create a spark or apply excessive heat in the vicinity of fuel or fuel vapor.

Secondary ignition system voltage

A severe electric shock can result from touching certain parts of the ignition system (such as the spark plug wires) when the engine is running or being cranked, particularly if components are damp or the insulation is defective. In the case of an electronic ignition system, the secondary system voltage is much higher and could prove fatal.

Troubleshooting

Contents

Symptom	Section
Engine	
Engine backfires	13
Engine 'diesels' (continues to run) after switching off	15
Engine hard to start when cold	4
Engine hard to start when hot	5
Engine lacks power	12
Engine 'lopes' while idling or idles erratically	8
Engine misses at idle speed	9
Engine misses throughout driving speed range	10
Engine rotates but will not start	2
Engine stalls	11
Engine starts but stops immediately	7
Engine will not rotate when attempting to start	1
Pinging or knocking engine sounds during acceleration or uphill	14
Starter motor noisy or excessively rough in engagement	6
Starter motor operates without rotating engine	3
Engine electrical system	
Battery will not hold a charge	16
'Check engine' light comes on	19
Ignition light fails to come on when key is turned on	18
Ignition light fails to go out	17
Fuel system	
Excessive fuel consumption	20
Fuel leakage and/or fuel odor	21
'Rapping' noise in vicinity of fuel pump below 70°F (1982 V8 engines with four-barrel carburetor only)	22
Cooling system	
Coolant loss	27
External coolant leakage	25
Internal coolant leakage	26
Overcooling	24
Overheating	23
Poor coolant circulation	28
Clutch	
Clutch slips (engine speed increases with no increase in vehicle speed)	30
Clutch pedal stays on floor when disengaged	34
Fails to release (pedal pressed to the floor — shift lever does not move freely in and out of Reverse)	29
Grabbing (chattering) as clutch is engaged	31
Squeal or rumble with clutch fully disengaged (pedal depressed)	33
Squeal or rumble with clutch fully engaged (pedal released)	32
Manual transmission	
Buzzing noise in shifter assembly at high speeds (1982 models only)	40
Difficulty in engaging gears	39
Noisy in all gears	36
Noisy in Neutral with engine running	35
Noisy in one particular gear	37
Oil leakage	42
Slips out of high gear	38
Throttle sensitive 'boom' vibration in vicinity of the transmission at approximately 4000 rpm (1982 models with L4 engines only)	41
Automatic transmission	
Engine will start in gears other than Park or Neutral	45
Fluid leakage	47
General shift mechanism problems	43
Transmission slips, shifts rough, is noisy or has no drive in forward or reverse gears	46
Transmission will not downshift with accelerator pedal pressed to the floor	44
Driveshaft	
Knock or clunk when the transmission is under initial load (just after transmission is put into gear)	49
Metallic grating sound consistent with vehicle speed	50
Oil leak at front of driveshaft	48
Vibration	51
Rear axle	
Noise (same when in drive as when vehicle is coasting)	52
Oil leakage	54
Vibration	53
Brakes	
Brake pedal feels spongy when depressed	59
Brake pedal pulsates during brake application	62
Excessive brake pedal travel	58
Excessive effort required to stop vehicle	60
Noise (high-pitched squeal without the brakes applied)	56
Pedal travels to the floor with little resistance	61
Rattle in parking brake cable at left rear of vehicle (1982 models only)	63
Rear brake 'moan' at low speeds under light brake application (1982 models only)	57
Vehicle pulls to one side during braking	55
Suspension and steering systems	
Excessive pitching and/or rolling around corners or during braking	66
Excessive play in steering	68
Excessive tire wear (not specific to one area)	70
Excessive tire wear on inside edge	72
Excessive tire wear on outside edge	71
Excessively stiff steering	67
Lack of power assistance	69
Shimmy, shake or vibration	65
Tire tread worn in one place	73
Vehicle pulls to one side	64

This section provides an easy-reference guide to the more common problems which may occur during the operation of your vehicle. These problems and possible causes are grouped under various components or systems i.e. Engine, Cooling system, etc., and also refer to the Chapter and/or Section which deals with the problem.

Remember that successful troubleshooting is not a mysterious 'black art' practiced only by professional mechanics; it's simply the result of a bit of knowledge combined with an intelligent, systematic approach to the problem. Always work by a process of elimination, starting with the simplest solution and working through to the most complex — and never overlook the obvious. Anyone can forget to fill the gas tank or leave the lights on overnight, so don't assume that you are above such oversights.

Finally, always get clear in your mind why a problem has occurred and take steps to ensure that it doesn't happen again. If the electrical system fails because of a poor connection, check all other connections in the system to make sure that they don't fail as well; if a particular fuse continues to blow, find out why — don't just go on replacing fuses. Remember, failure of a small component can often be indicative of potential failure or incorrect functioning of a more important component or system.

Engine

1 Engine will not rotate when attempting to start

1 Battery terminal connections loose or corroded. Check the cable terminals at the battery; tighten the cable or remove corrosion as necessary.
2 Battery discharged or faulty. If the cable connections are clean and tight on the battery posts, turn the key to the On position and switch on the headlights and/or windshield wipers. If they fail to function, the battery is discharged.
3 Automatic transmission not completely engaged in Park or clutch not completely depressed.
4 Broken, loose or disconnected wiring in the starting circuit. Inspect all wiring and connectors at the battery, starter solenoid and ignition switch.
5 Starter motor pinion jammed in flywheel ring gear. If manual transmission, place transmission in gear and rock the vehicle to manually turn the engine. Remove starter and inspect pinion and flywheel at earliest convenience.
6 Starter solenoid faulty (Chapter 5).
7 Starter motor faulty (Chapter 5).
8 Ignition switch faulty (Chapter 10).

2 Engine rotates but will not start

1 Fuel tank empty.
2 Battery discharged (engine rotates slowly). Check the operation of electrical components as described in previous Section.
3 Battery terminal connections loose or corroded. See previous Section.
4 Carburetor flooded and/or fuel level in carburetor incorrect. This will usually be accompanied by a strong fuel odor from under the hood. Wait a few minutes, depress the accelerator pedal all the way to the floor and attempt to start the engine.
5 Choke control inoperative (Chapter 1).
6 Fuel not reaching carburetor. With ignition switch in Off position, open hood, remove the top plate of air cleaner assembly and observe the top of the carburetor (manually move choke plate back if necessary). Have an assistant depress accelerator pedal and check that fuel spurts into carburetor. If not, check fuel filter (Chapter 1), fuel lines and fuel pump (Chapter 4).
7 Fuel injector or fuel pump faulty (fuel-injected vehicles) (Chapter 4).
8 Excessive moisture on, or damage to, ignition components (Chapter 5).
9 Worn, faulty or incorrectly gapped spark plugs (Chapter 1).
10 Spark plug fouling (on V6 engines only) after repeated short trips, leading to a no-start condition. A General Motors dealer technical service bulletin concerning this problem has been issued. Take the vehicle to your dealer and inform him of the problem.
11 Broken, loose or disconnected wiring in the starting circuit (see previous Section).
12 Distributor loose, causing ignition timing to change. Turn the distributor as necessary to start the engine, then set ignition timing as soon as possible (Chapter 1).
13 Broken, loose or disconnected wires at the ignition coil or faulty coil (Chapter 5).

3 Starter motor operates without rotating engine

1 Starter pinion sticking. Remove the starter (Chapter 5) and inspect.
2 Starter pinion or flywheel teeth worn or broken. Remove the cover at the rear of the engine and inspect.

4 Engine hard to start when cold

1 Battery discharged or low. Check as described in Section 1.
2 Choke control inoperative or out of adjustment (Chapter 4).
3 Carburetor flooded (see Section 2).
4 Fuel supply not reaching the carburetor (see Section 2).
5 Carburetor/fuel-injection system in need of overhaul (Chapter 4).
6 Distributor rotor carbon tracked and/or mechanical advance mechanism rusted (Chapter 5).

5 Engine hard to start when hot

1 Choke sticking in the closed position (Chapter 1).
2 Carburetor flooded (see Section 2).
3 Air filter clogged (Chapter 1).
4 Fuel not reaching the carburetor (see Section 2).

6 Starter motor noisy or excessively rough in engagement

1 Pinion or flywheel gear teeth worn or broken. Remove the cover at the rear of the engine (if so equipped) and inspect.
2 Starter motor mounting bolts loose or missing.

7 Engine starts but stops immediately

1 Loose or faulty electrical connections at distributor, coil or alternator.
2 Insufficient fuel reaching the carburetor/fuel injector(s). Disconnect the fuel line at the carburetor/fuel injector(s) and remove the filter (Chapter 1). Place a container under the disconnected fuel line. Observe the flow of fuel from the line. If little or none at all, check for blockage in the lines and/or replace the fuel pump (Chapter 4).
3 Vacuum leak at the gasket surfaces of the intake manifold and/or carburetor/fuel injection unit(s). Make sure that all mounting bolts (nuts) are tightened securely and that all vacuum hoses connected to the carburetor/fuel injection unit(s) and manifold are positioned properly and in good condition.

8 Engine lopes while idling or idles erratically

1 Vacuum leakage. Check mounting bolts (nuts) at the carburetor/fuel injection unit(s) and intake manifold for tightness. Make sure that all vacuum hoses are connected and in good condition. Use a stethoscope or a length of fuel hose held against your ear to listen for vacuum leaks while the engine is running. A hissing sound will be heard. A soapy water solution will also detect leaks. Check the carburetor/fuel injector and intake manifold gasket surfaces.
2 Leaking EGR valve or plugged PCV valve (see Chapters 1 and 6).
3 Air filter clogged (Chapter 1).
4 Fuel pump not delivering sufficient fuel to the carburetor/fuel injector (see Section 7).
5 Carburetor out of adjustment (Chapter 4).
6 Leaking head gasket. If this is suspected, take the vehicle to a

repair shop or dealer where the engine can be pressure checked.
7 Timing chain and/or gears worn (Chapter 2).
8 Camshaft lobes worn (Chapter 2).
9 Spark plug fouling (on V6 engines only) after repeated short trips. A General Motors dealer technical service bulletin concerning this problem has been issued. Take the vehicle to your dealer annd inform him of the problem.

9 Engine misses at idle speed

1 Spark plugs worn or not gapped properly (Chapter 1).
2 Spark plug fouling (on V6 engines only) after repeated short trips (see Section 8).
3 Faulty spark plug wires (Chapter 1).
4 Choke not operating properly (Chapter 1).

10 Engine misses throughout driving speed range

1 Fuel filter clogged and/or impurities in the fuel system (Chapter 1). Also check fuel output at the carburetor/fuel injector (see Section 7).
2 Faulty or incorrectly gapped spark plugs (Chapter 1).
3 Incorrect ignition timing (Chapter 1).
4 Check for cracked distributor cap, disconnected distributor wires and damaged distributor components (Chapter 1).
5 Leaking spark plug wires (Chapter 1).
6 Faulty emissions system components (Chapter 6).
7 Low or uneven cylinder compression pressures. Remove spark plugs and test compression with gauge (Chapter 1).
8 Weak or faulty ignition system (Chapter 5).
9 Vacuum leaks at carburetor/fuel injection unit(s), intake manifold or vacuum hoses (see Section 8).
10 Voltage spike during operation of power accessories (power windows, power door locks, etc.). A General Motors dealer technical service bulletin concerning this problem has been issued. Take the vehicle to your dealer and inform him of the problem.

11 Engine stalls

1 Idle speed incorrect (Chapter 1).
2 Fuel filter clogged and/or water and impurities in the fuel system (Chapter 1).
3 Choke improperly adjusted or sticking (Chapter 1).
4 Distributor components damp or damaged (Chapter 5).
5 Faulty emissions system components (Chapter 6).
6 Faulty or incorrectly gapped spark plugs (Chapter 1). Also check spark plug wires (Chapter 1).
7 Vacuum leak at the carburetor/fuel injection unit(s), intake manifold or vacuum hoses. Check as described in Section 8.

12 Engine lacks power

1 Incorrect ignition timing (Chapter 1).
2 Excessive play in distributor shaft. At the same time, check for worn rotor, faulty distributor cap, wires, etc. (Chapters 1 and 5).
3 Faulty or incorrectly gapped spark plugs (Chapter 1).
4 Spark plug fouling (on V6 engines only) after repeated short trips (see Section 8).
5 Carburetor/fuel injection unit not adjusted properly or excessively worn (Chapter 4).
6 Faulty coil (Chapter 5).
7 Brakes binding (Chapter 1).
8 Automatic transmission fluid level incorrect (Chapter 1).
9 Clutch slipping (Chapter 8).
10 Fuel filter clogged and/or impurities in the fuel system (Chapter 1).
11 Emissions control system not functioning properly (Chapter 6).
12 Use of sub-standard fuel. Fill tank with proper octane fuel.
13 Low or uneven cylinder compression pressures. Test with compression tester, which will detect leaking valves and/or blown head gasket (Chapter 1).

13 Engine backfires

1 Emissions system not functioning properly (Chapter 6).
2 Ignition timing incorrect (Chapter 1).
3 Faulty secondary ignition system (cracked spark plug insulator, faulty plug wires, distributor cap and/or rotor) (Chapters 1 and 5).
4 Carburetor/fuel injection unit in need of adjustment or worn excessively (Chapter 4).
5 Vacuum leak at carburetor/fuel injection unit(s), intake manifold or vacuum hoses. Check as described in Section 8.
6 Valve clearances incorrectly set, and/or valves sticking (Chapter 2).

14 Pinging or knocking engine sounds during acceleration or uphill

1 Incorrect grade of fuel. Fill tank with fuel of the proper octane rating.
2 Ignition timing incorrect (Chapter 1).
3 Carburetor/fuel injection unit in need of adjustment (Chapter 4).
4 Improper spark plugs. Check plug type against Emissions Control Information label located in engine compartment. Also check plugs and wires for damage (Chapter 1).
5 Worn or damaged distributor components (Chapter 5).
6 Faulty emissions system (Chapter 6).
7 Vacuum leak. Check as described in Section 8.

15 Engine 'diesels' (continues to run) after switching off

1 Idle speed too high (Chapter 1).
2 Electrical solenoid at side of carburetor not functioning properly (not all models, see Chapter 4).
3 Ignition timing incorrectly adjusted (Chapter 1).
4 Thermo-controlled air cleaner heat valve not operating properly (Chapter 6).
5 Excessive engine operating temperature. Probable causes of this are malfunctioning thermostat, clogged radiator, faulty water pump (Chapter 3).

Engine electrical system

16 Battery will not hold a charge

1 Alternator drivebelt defective or not adjusted properly (Chapter 1).
2 Electrolyte level low or battery discharged (Chapter 1).
3 Battery terminals loose or corroded (Chapter 1).
4 Alternator not charging properly (Chapter 5).
5 Loose, broken or faulty wiring in the charging circuit (Chapter 5).
6 Short in vehicle wiring causing a continual drain on battery.
7 Battery defective internally.

17 Ignition light fails to go out

1 Fault in alternator or charging circuit (Chapter 5).
2 Alternator drivebelt defective or not properly adjusted (Chapter 1).

18 Ignition light fails to come on when key is turned on

1 Warning light bulb defective (Chapter 10).
2 Alternator faulty (Chapter 5).
3 Fault in the printed circuit, dash wiring or bulb holder (Chapter 10).

19 'Check engine' light comes on

See Chapter 6.

Fuel system

20 Excessive fuel consumption

1 Dirty or clogged air filter element (Chapter 1).
2 Incorrectly set ignition timing (Chapter 1).
3 Choke sticking or improperly adjusted (Chapter 1).
4 Emissions system not functioning properly (not all vehicles, see Chapter 6).
5 Carburetor idle speed and/or mixture not adjusted properly (Chapter 1).
6 Carburetor/fuel injection internal parts excessively worn or damaged (Chapter 4).
7 Low tire pressure or incorrect tire size (Chapter 1).

21 Fuel leakage and/or fuel odor

1 Leak in a fuel feed or vent line (Chapter 4).
2 Tank overfilled. Fill only to automatic shut-off.
3 Emissions system filter clogged (Chapter 1).
4 Vapor leaks from system lines (Chapter 4).
5 Carburetor/fuel injection internal parts excessively worn or out of adjustment (Chapter 4).

22 'Rapping' noise in vicinity of fuel pump below 70°F (1982 V8 engines with four-barrel carburetor only)

A General Motors dealer technical service bulletin concerning this problem has been issued. Take the vehicle to your dealer and inform him of the problem.

Engine cooling system

23 Overheating

1 Insufficient coolant in system (Chapter 1).
2 Water pump drivebelt defective or not adjusted properly (Chapter 1).
3 Radiator core blocked or radiator grille dirty and restricted (Chapter 3).
4 Thermostat faulty (Chapter 3).
5 Fan blades broken or cracked (Chapter 3).
6 Radiator cap not maintaining proper pressure. Have cap pressure tested by gas station or repair shop.
7 Ignition timing incorrect (Chapter 1).

24 Overcooling

1 Thermostat faulty (Chapter 3).
2 Inaccurate temperature gauge (Chapter 10)

25 External coolant leakage

1 Deteriorated or damaged hoses. Loosen clamps at hose connections (Chapter 1).
2 Water pump seals defective. If this is the case, water will drip from the 'weep' hole in the water pump body (Chapter 1).
3 Leakage from radiator core or header tank. This will require the radiator to be professionally repaired (see Chapter 3 for removal procedures).
4 Engine drain plugs or water jacket core plugs leaking (see Chapter 2).

26 Internal coolant leakage

Note: *Internal coolant leaks can usually be detected by examining the oil. Check the dipstick and inside of the rocker arm cover(s) for water deposits and an oil consistency like that of a milkshake.*

1 Leaking cylinder head gasket. Have the cooling system pressure-tested.
2 Cracked cylinder bore or cylinder head. Dismantle engine and inspect (Chapter 2).

27 Coolant loss

1 Too much coolant in system (Chapter 1).
2 Coolant boiling away due to overheating (see Section 15).
3 Internal or external leakage (see Sections 25 and 26).
4 Faulty radiator cap. Have the cap pressure-tested.

28 Poor coolant circulation

1 Inoperative water pump. A quick test is to pinch the top radiator hose closed with your hand while the engine is idling, then let it loose. You should feel the surge of coolant if the pump is working properly (Chapter 1).
2 Restriction in cooling system. Drain, flush and refill the system (Chapter 1). If necessary, remove the radiator (Chapter 3) and have it reverse-flushed.
3 Water pump drivebelt defective or not adjusted properly (Chapter 1).
4 Thermostat sticking (Chapter 3).

Clutch

29 Fails to release (pedal pressed to the floor — shift lever does not move freely in and out of Reverse)

1 Improper linkage free play adjustment (Chapter 8).
2 Clutch fork off ball stud. Look under the vehicle, on the left side of transmission.
3 Clutch plate warped or damaged (Chapter 8).

30 Clutch slips (engine speed increases with no increase in vehicle speed)

1 Linkage out of adjustment (Chapter 8).
2 Clutch plate oil soaked or lining worn. Remove clutch (Chapter 8) and inspect.
3 Clutch plate not seated. It may take 30 or 40 normal starts for a new one to seat.

31 Grabbing (chattering) as clutch is engaged

1 Oil on clutch plate lining. Remove (Chapter 8) and inspect. Correct any leakage source.
2 Worn or loose engine or transmission mounts. These units move slightly when clutch is released. Inspect mounts and bolts.
3 Worn splines on clutch plate hub. Remove clutch components (Chapter 8) and inspect.
4 Warped pressure plate or flywheel. Remove clutch components and inspect.

32 Squeal or rumble with clutch fully engaged (pedal released)

1 Improper adjustment; no free play (Chapter 1).
2 Release bearing binding on transmission bearing retainer. Remove clutch components (Chapter 8) and check bearing. Remove any burrs or nicks, clean and relubricate before reinstallation.
3 Weak iinkage return spring. Replace the spring.

33 Squeal or rumble with clutch fully disengaged (pedal depressed)

1 Worn, defective or broken release bearing (Chapter 8).
2 Worn or broken pressure plate springs (or diaphragm fingers) (Chapter 8).

34 Clutch pedal stays on floor when disengaged

1 Bind in linkage or release bearing. Inspect linkage or remove clutch components as necessary.
2 Linkage springs being over-extended. Adjust linkage for proper free play. Make sure proper pedal stop (bumper) is installed.

Manual transmission

35 Noisy in Neutral with engine running

1 Input shaft bearing worn.
2 Damaged main drive gear bearing.
3 Worn countershaft bearings.
4 Worn or damaged countershaft end play shims.

36 Noisy in all gears

1 Any of the above causes, and/or:
2 Insufficient lubricant (see checking procedures in Chapter 1).

37 Noisy in one particular gear

1 Worn, damaged or chipped gear teeth for that particular gear.
2 Worn or damaged synchronizer for that particular gear.

38 Slips out of high gear

1 Transmission loose on clutch housing (Chapter 7).
2 Shift rods interfering with engine mounts or clutch lever (Chapter 7).
3 Shift rods not working freely (Chapter 7).
4 Damaged mainshaft pilot bearing.
5 Dirt between transmission case and engine or misalignment of transmission (Chapter 7).
6 Worn or improperly adjusted linkage (Chapter 7).

39 Difficulty in engaging gears

1 Clutch not releasing completely (see clutch adjustment in Chapter 7).
2 Loose, damaged or out-of-adjustment shift linkage. Make a thorough inspection, replacing parts as necessary (Chapter 7).

40 Buzzing noise in shifter assembly at high speeds (1982 models only)

A General Motors dealer technical service bulletin concerning this problem has been issued. Take the vehicle to your dealer and inform him of the problem.

41 Throttle sensitive 'boom' vibration in vicinity of the transmission at approximately 4000 rpm (1982 models with L4 engines only)

A General Motors dealer technical service bulletin concerning this problem has been issued. Take the vehicle to your dealer and inform him of the problem.

42 Oil leakage

1 Excessive amount of lubricant in transmission (see Chapter 1 for correct checking procedures). Drain lubricant as required.
2 Side cover loose or gasket damaged.
3 Rear oil seal or speedometer oil seal in need of replacement (Chapter 7).

Automatic transmission

Note: *Due to the complexity of the automatic transmission, it is difficult for the home mechanic to properly diagnose and service this component. For problems other than the following, the vehicle should be taken to a dealer or reputable mechanic.*

43 General shift mechanism problems

1 Chapter 7 deals with checking and adjusting the shift linkage on automatic transmissions. Common problems which may be attributed to poorly adjusted linkage are:

Engine starting in gears other than Park or Neutral.
Indicator on shifter pointing to a gear other than the one actually being used.
Vehicle moves when in Park.

2 Refer to Chapter 7 to adjust the linkage.

44 Transmission will not downshift with accelerator pedal pressed to the floor

Chapter 7 deals with adjusting the throttle valve (TV) cable to enable the transmission to downshift properly.

45 Engine will start in gears other than Park or Neutral

Chapter 7 deals with adjusting the Neutral start switch used with automatic transmissions.

46 Transmission slips, shifts rough, is noisy or has no drive in forward or reverse gears

1 There are many probable causes for the above problems, but the home mechanic should be concerned with only one possibility — fluid level.
2 Before taking the vehicle to a repair shop, check the level and condition of the fluid as described in Chapter 1. Correct fluid level as necessary or change the fluid and filter if needed. If the problem persists, have a professional diagnose the probable cause.

47 Fluid leakage

1 Automatic transmission fluid is a deep red color. Fluid leaks should not be confused with engine oil, which can easily be blown by air flow to the transmission.
2 To pinpoint a leak, first remove all built-up dirt and grime from around the transmission. Degreasing agents and/or steam cleaning will achieve this. With the underside clean, drive the vehicle at low speeds so air flow will not blow the leak far from its source. Raise the vehicle and determine where the leak is coming from. Common areas of leakage are:

a) Pan: Tighten mounting bolts and/or replace pan gasket as necessary (see Chapters 1 and 7).
b) Rear extension: Tighten bolts and/or replace oil seal as necessary (Chapter 7).

c) Filler pipe: Replace the rubber seal where pipe enters transmission case.
d) Transmission oil lines: Tighten connectors where lines enter transmission case and/or replace lines.
e) Vent pipe: Transmission over-filled and/or water in fluid (see checking procedures, Chapter 1).
f) Speedometer connector: Replace the O-ring where speedometer cable enters transmission case (Chapter 7).

Driveshaft

48 Oil leak at front of driveshaft

Defective transmission rear oil seal. See Chapter 7 for replacement procedures. While this is done, check the splined yoke for burrs or a rough condition which may be damaging the seal. They can be removed with crocus cloth or a fine whetstone.

49 Knock or clunk when the transmission is under initial load (just after transmission is put into gear)

1 Loose or disconnected rear suspension components. Check all mounting bolts and bushings (Chapter 11).
2 Loose driveshaft bolts. Inspect all bolts and nuts and tighten them to the specified torque (Chapter 8).
3 Worn or damaged universal joint bearings. Check for wear (Chapter 8).

50 Metallic grating sound consistent with vehicle speed

Pronounced wear in the universal joint bearings. Check as described in Chapter 8.

51 Vibration

Note: *Before assuming that the driveshaft is at fault, make sure the tires are perfectly balanced and perform the following test.*

1 Install a tachometer inside the vehicle to monitor engine speed as the vehicle is driven. Drive the vehicle and note the engine speed at which the vibration (roughness) is most pronounced. Now shift the transmission to a different gear and bring the engine speed to the same point.
2 If the vibration occurs at the same engine speed (rpm) regardless of which gear the transmission is in, the driveshaft is NOT at fault since the driveshaft speed varies.
3 If the vibration decreases or is eliminated when the transmission is in a different gear at the same engine speed, refer to the following probable causes.
4 Bent or dented driveshaft. Inspect and replace as necessary (Chapter 8).
5 Undercoating or build-up dirt, etc. on the driveshaft. Clean the shaft thoroughly and recheck.
6 Worn universal joint bearings. Remove and inspect (Chapter 8).
7 Driveshaft and/or companion flange out of balance. Check for missing weights on the shaft. Remove driveshaft (Chapter 8) and reinstall 180° from original position, then retest. Have driveshaft professionally balanced if problem persists.

Rear axle

52 Noise (same when in drive as when vehicle is coasting)

1 Road noise. No corrective procedures available.
2 Tire noise. Inspect tires and check tire pressures (Chapter 1).
3 Front wheel bearings loose, worn or damaged (Chapter 1).

53 Vibration

See probable causes under *Driveshaft*. Proceed under the guidelines listed for the driveshaft. If the problem persists, check the rear axleshaft bearings by raising the rear of the vehicle and spinning the wheels by hand. Listen for evidence of rough (noisy) bearings. Remove and inspect (Chapter 8).

54 Oil leakage

1 Pinion seal damaged (Chapter 8).
2 Axleshaft oil seals damaged (Chapter 8).
3 Differential cover leaking. Tighten mounting bolts or replace the gasket as required (Chapter 8).

Brakes

Note: *Before assuming that a brake problem exists, make sure that the tires are in good condition and inflated properly (see Chapter 1), that the front end alignment is correct and that the vehicle is not loaded with weight in an unequal manner.*

55 Vehicle pulls to one side during braking

1 Defective, damaged or oil contaminated disc brake pads on one side. Inspect as described in Chapter 9.
2 Excessive wear of brake pad material or disc on one side. Inspect and correct as necessary.
3 Loose or disconnected front suspension components. Inspect and tighten all bolts to the specified torque (Chapter 11).
4 Defective caliper assembly. Remove caliper and inspect for stuck piston or other damage (Chapter 9).

56 Noise (high-pitched squeal without the brakes applied)

Disc brake pads worn out. The noise comes from the wear sensor rubbing against the disc (does not apply to all vehicles). Replace pads with new ones immediately (Chapter 9).

57 Rear brake 'moan' at low speeds under light brake application (1982 models only)

A General Motors dealer technical service bulletin concerning this problem has been issued. Take the vehicle to your dealer and inform him of the problem.

58 Excessive brake pedal travel

1 Partial brake system failure. Inspect entire system (Chapter 9) and correct as required.
2 Insufficient fluid in master cylinder. Check (Chapter 1), add fluid and bleed system if necessary (Chapter 9).
3 Rear brakes not adjusting properly. Make a series of starts and stops while the vehicle is in Reverse. If this does not correct the situation, remove drums and inspect self-adjusters (Chapter 9).

59 Brake pedal feels spongy when depressed

1 Air in hydraulic lines. Bleed the brake system (Chapter 9).
2 Faulty flexible hoses. Inspect all system hoses and lines. Replace parts as necessary.
3 Master cylinder mounting bolts/nuts loose.
4 Master cylinder defective (Chapter 9).

60 Excessive effort required to stop vehicle

1 Power break booster not operating properly (Chapter 9).
2 Excessively worn linings or pads. Inspect and replace if necessary (Chapter 9).
3 One or more caliper pistons or wheel cylinders seized or sticking. Inspect and rebuild as required (Chapter 9).
4 Brake linings or pads contaminated with oil or grease. Inspect and replace as required (Chapter 9).
5 New pads or shoes installed and not yet seated. It will take a while for the new material to seat against the drum (or rotor).

61 Pedal travels to the floor with little resistance

Little or no fluid in the master cylinder reservoir caused by leaking wheel cylinder(s), leaking caliper piston(s), loose, damaged or disconnected brake lines. Inspect entire system and correct as necessary.

62 Brake pedal pulsates during brake application

1 Wheel bearings not adjusted properly or in need of replacement (Chapter 1).
2 Caliper not sliding properly due to improper installation or obstructions. Remove and inspect (Chapter 9).
3 Rotor defective. Remove the rotor (Chapter 9) and check for excessive lateral runout and parallelism. Have the rotor resurfaced or replace it with a new one.

63 Rattle in parking brake cable at left rear of vehicle (1982 models only)

A General Motors dealer technical service bulletin concerning this problem has been issued. Take the vehicle to your dealer and inform him of the problem.

Suspension and steering systems

64 Vehicle pulls to one side

1 Tire pressures uneven (Chapter 1).
2 Defective tire (Chapter 1).
3 Excessive wear in suspension or steering components (Chapter 11).
4 Front end in need of alignment.
5 Front brakes dragging. Inspect brakes as described in Chapter 9.

65 Shimmy, shake or vibration

1 Tire or wheel out-of-balance or out-of-round. Have professionally balanced.
2 Loose, worn or out-of-adjustment wheel bearings (Chapters 1 and 8).
3 Shock absorbers and/or suspension components worn or damaged (Chapter 11).

66 Excessive pitching and/or rolling around corners or during braking

1 Defective shock absorbers. Replace as a set (Chapter 11).
2 Broken or weak springs and/or suspension components. Inspect as described in Chapter 11.

67 Excessively stiff steering

1 Lack of fluid in power steering fluid reservoir (Chapter 1).
2 Incorrect tire pressures (Chapter 1).
3 Lack of lubrication at steering joints (Chapter 1).
4 Front end out of alignment.
5 See also section titled *Lack of power assistance.*

68 Excessive play in steering

1 Loose front wheel bearings (Chapter 1).
2 Excessive wear in suspension or steering components (Chapter 11).
3 Steering gearbox out of adjustment (Chapter 11).

69 Lack of power assistance

1 Steering pump drivebelt faulty or not adjusted properly (Chapter 1).
2 Fluid level low (Chapter 1).
3 Hoses or lines restricted. Inspect and replace parts as necessary.
4 Air in power steering system. Bleed system (Chapter 11).

70 Excessive tire wear (not specific to one area)

1 Incorrect tire pressures (Chapter 1).
2 Tires out of balance. Have professionally balanced.
3 Wheels damaged. Inspect and replace as necessary.
4 Suspension or steering components excessively worn (Chapter 11).

71 Excessive tire wear on outside edge

1 Inflation pressures incorrect (Chapter 1).
2 Excessive speed in turns.
3 Front end alignment incorrect (excessive toe-in). Have professionally aligned.
4 Suspension arm bent or twisted (Chapter 11).

72 Excessive tire wear on inside edge

1 Inflation pressures incorrect (Chapter 1).
2 Front end alignment incorrect (toe-out). Have professionally aligned.
3 Loose or damaged steering components (Chapter 11).

73 Tire tread worn in one place

1 Tires out of balance.
2 Damaged or buckled wheel. Inspect and replace if necessary.
3 Defective tire (Chapter 1).

Chapter 1 Tune-up and routine maintenance

Refer to Chapter 13 for specifications and information related to 1985 and later models

Contents

Air filter and PCV filter replacement 29
Automatic transmission fluid change 28
Battery check and maintenance 5
Brake check 13
Carburetor choke check 14
Carburetor/throttle body injection (TBI) mounting torque check 21
Chassis lubrication 10
Check engine light . . . . . Chapter 6
Clutch pedal free play check and adjustment 24
Compression check 37
Cooling system check 7
Cooling system servicing (draining, flushing and refilling) 26
Drivebelt check and adjustment 6
Early Fuel Evaporation (EFE) system check 22
Engine idle speed check and adjustment 15
Engine oil and filter change 16
Evaporative Emissions Control System (EECS) filter replacement 33
Exhaust Gas Recirculation (EGR) valve check 32
Exhaust system check 11
Fluid level checks 4
Fuel filter replacement 18
Fuel system check 17
Ignition timing check and adjustment 34
Introduction and routine maintenance schedule 1
Oxygen sensor replacement 30
Positive Crankcase Ventilation (PCV) valve replacement 31
Rear axle oil change 25
Spark plug replacement 35
Spark plug wires, distributor cap and rotor check and replacement 36
Suspension and steering check 12
Thermo-controlled Air Cleaner (THERMAC) check 20
Throttle linkage check 19
Tire and tire pressure checks 3
Tire rotation 23
Tune-up sequence 2
Underhood hose check and replacement 8
Wheel bearing check and repack 27
Windshield wiper blade inspection and replacement 9

Specifications

Recommended lubricants, fluids and capacities

Engine oil type	Grade SF, SF/CC or SF/CD
Engine oil viscosity	
40° to 100° plus F (4° to 38° plus C)	SAE 30
20° to 100° plus F (-7° to 38° plus C)	SAE 20W-20, 20W-40, 20W-50
10° to 100° plus F (-14° to 38° plus C)	SAE 15W-40
0° to 100° plus F (-18° to 38° plus C)	SAE 10W-30, 10W-40
0° to 60° F (-18° to 17° C)	SAE 10W
Below -20° to 60° F (below -30° to 17° C)	SAE 5W-30 (may be used up to 100° F/38°C in L4 and V6 engines)
Below -20° to 20° F (below -30° to -7° C)	SAE 5W-20 (not to be used for continuous high speed driving)
Engine oil capacity (without filter)	
V8	4 qts
V6	4 qts
L4	3 qts
Coolant type	Ethylene glycol
Cooling system capacity	
V8 (all)	15 qts
V6 (all)	12.5 qts
L4 without A/C	12.8 qts
L4 with A/C	13.0 qts
Brake fluid type	DOT-3
Manual transmission oil type*	
4-speed	
1982 and 1983	GL-5 gear lubricant, SAE 80W or 80W-90
1984	Dexron II or equivalent ATF
5-speed	
1982	SAE 5W-30
1983	GL-5 gear lubricant, SAE 80W or 80W-90
1984	Dexron II or equivalent ATF
Manual transmission oil capacity	
4-speed	2.4 qts
5-speed	3.5 qts
Automatic transmission	
Fluid type	Dexron II or equivalent
Fluid capacity*	
200-C	3.5 qts
700-R4	4.7 qts
Power steering fluid type	GM no. 1050017 or equivalent
Front wheel bearing grease	GM no. 1051344 or equivalent
Windshield washer solvent	GM Optikleen or equivalent

***Note:** *Capacities are approximate for refill after draining for routine maintenance; actual capacity should be measured on the transmission dipstick as discussed in Section 28.*

*** Note:** *Manual transmission fluid recommendations varied on these particular models. The following represents the best possible recommendation; however, refer to the owner's manual or dealer service department for information specific to your vehicle.*

Radiator cap opening pressure	15 psi
Thermostat rating	195° F
Engine idle speed	See Emission Control Information label in engine compartment
Ignition system	
Distributor direction of rotation	Clockwise (all)
Firing order	
V8	1-8-4-3-6-5-7-2
V6	1-2-3-4-5-6
L4	1-3-4-2
Spark plug type and gap	
L4	
1982 to 1983	AC R44TSX (or equivalent) @ 0.060 inch
1984 on	AC R43TS6 (or equivalent) @ 0.060 inch
V6	
1982 to 1984	AC R43TS (or equivalent) @ 0.045 inch
1985 to 1988	AC R42TS (or equivalent) @ 0.045 inch
1989 on	AC R43TSK (or equivalent) @ 0.045 inch
V8	
1982 to 1985	
Carbureted models	AC R45TS (or equivalent) @ 0.045 inch
Fuel-injected models	AC R43TS (or equivalent) @ 0.045 inch
1986 to 1987	
Carbureted models	AC R45TS (or equivalent) @ 0.035 inch
Fuel-injected models	
with cast iron heads	AC R43TS (or equivalent) @ 0.035 inch
with aluminum heads	AC FR3LS (or equivalent) @ 0.035 inch
1988 on	
with cast iron heads	AC R45TS (or equivalent) @ 0.035 inch
with aluminum heads	AC FR5LS (or equivalent) @ 0.035 inch
Ignition timing	See Emission Control Information label in engine compartment
Engine compression pressure	
L4	140 psi at 160 rpm
V6	N/A
V8	N/A
Clutch pedal free play *(non-hydraulic clutch)*	0.845 to 0.905 in
Torque specifications	
Oil pan drain plug	**Ft-lbs**
V8	20
V6	20
L4	25
Spark plugs	
V8	22
V6	7 to 15
L4	15
Carburetor/TBI mounting nuts/bolts	
V8	
Carburetor (long studs/bolts)	7
Carburetor (short studs/bolts)	11
TBI mounting bolts	15
V6	11
L4	15
Distributor hold-down bolt	20
Manual transmission fill plug	
4-speed	15
5-speed	20
Automatic transmission pan bolts	18
Rear axle filler/inspection plug	20
Rear axle cover bolts	30
Brake caliper mounting bolts	35
Wheel lug nuts	
Standard wheels	70
Cast aluminum wheels	80

1 Introduction and routine maintenance schedule

This Chapter was designed to help the home mechanic maintain his (or her) vehicle for peak performance, economy, safety and long life.

On the following pages you will find a maintenance schedule along with Sections which deal specifically with each item on the schedule. Included are visual checks, adjustments and item replacements.

Servicing your vehicle using the time/mileage maintenance schedule and the sequenced Sections will give you a planned program of maintenance. Keep in mind that it is a full plan, and maintaining only a few items at the specified intervals will not give you the same results.

You will find as you service your vehicle that many of the procedures can, and should, be grouped together, due to the nature of the job at hand. Examples of this are as follows:

If the vehicle is fully raised for a chassis lubrication, for example, this is the ideal time for the following checks: manual transmission oil, rear axle oil, exhaust system, suspension, steering and the fuel system.

If the tires and wheels are removed, as during a routine tire rotation, go ahead and check the brakes and wheel bearings at the same time.

If you must borrow or rent a torque wrench, it is a good idea to service the spark plugs, repack (or replace) the wheel bearings and check the carburetor mounting bolt torque all in the same day to save time and money.

The first step of this or any maintenance plan is to prepare yourself before the actual work begins. Read through the appropriate Sections for all work that is to be performed before you begin. Gather together all necessary parts and tools. If it appears that you could have a problem during a particular job, don't hesitate to seek advice from your local parts man or dealer service department.

Routine maintenance intervals

The following recommendations are given with the assumption that the vehicle owner will be doing the maintenance or service work (as opposed to having a dealer service department do the work). The following are factory maintenance recommendations; however, subject to the preference of the individual owner, in the interest of keeping his or her vehicle in peak condition at all times and with the vehicle's ultimate resale in mind, many of these operations may be performed more often. We encourage such owner initiative.

When the vehicle is new, it should be serviced initially by a factory authorized dealer service department to protect the factory warranty. In most cases the initial maintenance check is done at no cost to the owner.

Every 250 miles or weekly, whichever comes first

Check the engine oil level (Sec 4)
Check the engine coolant level (Sec 4)
Check the windshield washer fluid level (Sec 4)
Check the tires and tire pressures (Sec 3)

Every 6000 miles or 6 months, whichever comes first

Check the automatic transmission fluid level (Sec 4)
Check the power steering fluid level (Sec 4)
Change the engine oil and oil filter (Sec 16)
Lubricate the chassis components (Sec 10)
Check the cooling system (Sec 7)
Check and replace (if necessary) the underhood hoses (Sec 8)
Check the exhaust system (Sec 11)
Check the steering and suspension components (Sec 12)
Check and adjust (if necessary) the engine drivebelts (Sec 12)
Check the brake master cylinder fluid level (Sec 4)
Check the manual transmission oil level (Sec 4)
Check the rear axle oil level (Sec 4)
Check the operation of the choke (at this interval, then every 12000 miles or 12 months, whichever comes first) (Sec 14)
Check and adjust (if necessary) the engine idle speed (at this interval, then every 12000 miles or 12 months, whichever comes first) (Sec 15)
Check the disc brake pads (Sec 13)
Check the brake system (Sec 13)
Check and service the battery (Sec 5)
Check and replace (if necessary) the windshield wiper blades (Sec 9)

Every 12000 miles or 12 months, whichever comes first

Check the drum brake linings (Sec 13)
Check the parking brake (Sec 13)
Check the clutch pedal free play (Sec 24)
Rotate the tires (Sec 23)
Check the Thermo-controlled Air Cleaner (THERMAC) for proper operation (Sec 20)
Check the EFE system (Sec 22)
Check the fuel system components (Sec 17)
Replace the fuel filter (Sec 18)
Check the operation of the EGR valve (Sec 32)
Change the rear axle oil (if vehicle is used to pull a trailer) (Sec 25)
Check the throttle linkage (Sec 19)

Every 18000 miles or 18 months, whichever comes first

Check and repack the front wheel bearings (perform this procedure whenever brake pads are replaced, regardless of maintenance interval) (Sec 27)
Change the automatic transmission fluid and filter (if driven mainly in heavy city traffic in hot-climate regions, in hilly or mountainous areas, or for frequent trailer pulling) (Sec 28)
Check the torque of the carburetor or TBI mounting bolts (Sec 21)
Drain, flush and refill the cooling system (Sec 26)

Every 24000 miles or 24 months, whichever comes first

Replace the oxygen sensor (if equipped (Sec 30)
Check the EGR system (Sec 32)
Replace the PCV valve (Sec 31)
Replace the air filter and PCV filter (Sec 29)
Replace the spark plugs (Sec 35)
Check the spark plug wires, distributor cap and rotor (Sec 36)
Check and adjust (if necessary) the ignition timing (Sec 34)
Check the EECS emissions system and replace the canister filter (Sec 33)
Change the automatic transmission fluid (if driven under severe conditions, see 18000 mile interval) (Sec 28)
Change the rear axle oil (if driven under severe conditions, see 12000 mile interval) (Sec 25)
Lubricate the clutch cross-shaft (Sec 10)
Check the engine compression (Sec 37)

2 Tune-up sequence

The term 'tune-up' is loosely used for any general operation that puts the engine back in its proper running condition. A tune-up is not a specific operation, but rather a combination of individual operations, such as replacing the spark plugs, adjusting the idle speed, setting the ignition timing, etc.

If, from the time the vehicle is new, the routine maintenance schedule (Section 1) is followed closely and frequent checks are made of fluid levels and high wear items, as suggested throughout this manual, the engine will be kept in relatively good running condition and the need for additional tune-ups will be minimized.

More likely than not, however, there will be times when the engine is running poorly due to lack of regular maintenance. This is even more likely if a used vehicle (which has not received regular and frequent maintenance checks) is purchased. In such cases, an engine tune-up will be needed outside of the regular routine maintenance intervals.

The following series of operations are those most often needed to bring a generally poor running engine back into a proper state of tune.

Minor tune-up

Clean, inspect and test battery (Sec 5)
Check all engine-related fluids (Sec 4)
Check engine compression (Sec 37)
Check and adjust drivebelts (Sec 6)
Replace spark plugs (Sec 35)
Inspect distributor cap and rotor (Sec 36)
Inspect spark plug and coil wires (Sec 36)
Check and adjust idle speed (Sec 15)
Check and adjust timing (Sec 34)
Check and adjust fuel/air mixture (see underhood VECI label)
Replace fuel filter (Sec 18)
Check PCV valve (Sec 31)
Check cooling system (Sec 7)

Major tune-up

(the above operations and those listed below)
Check EGR system (Chapter 6)
Check ignition system (Chapter 5)
Check charging system (Chapter 5)
Check fuel system (Sec 17)

3 Tire and tire pressure checks

1 Periodically inspecting the tires may not only prevent you from being stranded with a flat tire, but can also give you clues as to possible problems with the steering and suspension systems before major damage occurs.
2 Proper tire inflation adds miles to the lifespan of the tires, allows the vehicle to achieve maximum miles per gallon figures and contributes to the overall quality of the ride.
3 When inspecting the tires, first check the wear of the tread. Irregularities in the tread pattern (cupping, flat spots, more wear on one side than the other) are indications of front end alignment and/or balance problems. If any of these conditions are noted, take the vehicle to a reputable repair shop to correct the problem.
4 Also check the tread area for cuts and punctures. Many times a nail or tack will embed itself into the tire tread and yet the tire will hold its air pressure for a short time. In most cases, a repair shop or gas station can repair the punctured tire.
5 It is also important to check the sidewalls of the tires, both inside and outside. Check for deteriorated rubber, cuts, and punctures. Also inspect the inboard side of the tire for signs of brake fluid leakage, indicating that a thorough brake inspection is needed immediately.
6 Incorrect tire pressure cannot be determined merely by looking at the tire. This is especially true for radial tires. A tire pressure gauge must be used (photo). If you do not already have a reliable gauge, it is a good idea to purchase one and keep it in the glovebox. Built-in pressure gauges at gas stations are often unreliable.
7 Always check tire inflation when the tires are cold. Cold, in this case, means the vehicle has not been driven more than one mile after sitting for three hours or more. It is normal for the pressure to increase four to eight pounds or more when the tires are hot.
8 Unscrew the valve cap protruding from the wheel or hubcap and press the gauge firmly onto the valve stem. Observe the reading on the gauge and compare the figure to the recommended tire pressure listed on the tire placard. The tire placard is usually attached to the driver's door.
9 Check all tires and add air as necessary to bring them up to the recommended pressure levels. Do not forget the spare tire. Be sure to reinstall the valve caps (which will keep dirt and moisture out of the valve stem mechanism).

4 Fluid level checks

1 There are a number of components on a vehicle which rely on the use of fluids to perform their job. During normal operation of the vehicle, these fluids are used up and must be replenished before damage occurs. See *Recommended lubricants and fluids* at the front of this Chapter for the specific fluid to be used when addition is required. When checking fluid levels, it is important to have the vehicle on a level surface.

Engine oil

2 The engine oil level is checked with a dipstick which is located at the side of the engine block. The dipstick travels through a tube and into the oil pan to the bottom of the engine.
3 The oil level should be checked preferably before the vehicle has been driven, or about 15 minutes after the engine has been shut off. If the oil is checked immediately after driving the vehicle, some of the oil will remain in the upper engine components, producing an inaccurate reading on the dipstick.
4 Pull the dipstick from the tube (photo) and wipe all the oil from the end with a clean rag. Insert the clean dipstick all the way back into the oil pan and pull it out again. Observe the oil at the end of the dipstick (photo). At its highest point, the level should be between the Add and Full marks.
5 It takes approximately one quart of oil to raise the level from the Add mark to the Full mark on the dipstick. Do not allow the level to drop below the Add mark as engine damage due to oil starvation may occur. On the other hand, do not overfill the engine by adding oil above the Full mark since it may result in oil-fouled spark plugs, oil leaks or oil seal failures.
6 Oil is added to the engine after removing a twist-off cap located on the rocker arm cover or through a raised tube near the front of the engine. The cap should be marked 'Engine oil' or 'Oil.' An oil can spout or funnel will reduce spills as the oil is poured in.
7 Checking the oil level can also be an important preventative maintenance step. If you find the oil level dropping abnormally, it is an indication of oil leakage or internal engine wear which should be corrected. If there are water droplets in the oil, or if it is milky looking, component failure is indicated and the engine should be checked immediately. The condition of the oil can also be checked

3.6 The use of an accurate tire pressure gauge is essential for long tire life

4.4A After wiping off the engine oil dipstick, make sure it is reinserted all the way before withdrawing it for the oil level check

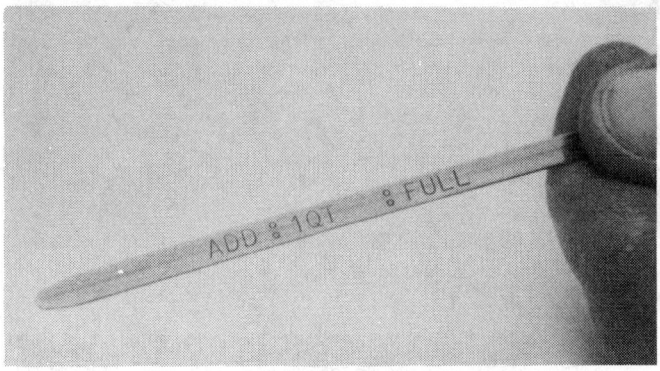

4.4B The oil level should appear between the Add and Full marks; do not overfill the crankcase

along with the level. With the dipstick removed from the engine, take your thumb and index finger and wipe the oil up the dipstick, looking for small dirt or metal particles which will cling to the dipstick. This is an indication that the oil should be drained and fresh oil added (Section 16).

Engine coolant

8 All vehicles covered by this manual are equipped with a pressurized coolant recovery system which makes coolant level checks very easy. A clear or white coolant reservoir attached to the inner fender panel is connected by a hose to the radiator cap. As the engine heats up during operation, coolant is forced from the radiator, through the connecting tube and into the reservoir. As the engine cools, the coolant is automatically drawn back into the radiator to keep the level correct.

9 The coolant level should be checked when the engine is hot. Merely observe the level of fluid in the reservoir, which should be at or near the Full Hot mark on the side of the reservoir (photo). If the system is completely cool, also check the level in the radiator by removing the cap (photo).

10 **Caution:** *Under no circumstances should either the radiator cap or the coolant recovery reservoir cap be removed when the system is hot, because escaping steam and scalding liquid could cause serious personal injury. In the case of the radiator cap, wait until the system has cooled completely, then wrap a thick cloth around the cap and turn it to the first stop. If any steam escapes, wait until the system has cooled further, then remove the cap. The coolant recovery cap may be removed carefully after it is apparent that no further 'boiling' is occurring in the recovery tank.*

11 If only a small amount of coolant is required to bring the system up to the proper level, regular water can be used. However, to maintain the proper antifreeze/water mixture in the system, both should be mixed together to replenish a low level. High-quality antifreeze offering protection to −20°F should be mixed with water in the proportion specified on the container. Do not allow antifreeze to come in contact with your skin or painted surfaces of the vehicle. Flush contacted areas immediately with plenty of water.

12 Coolant should be added to the reservoir until it reaches the Full Cold mark.

13 As the coolant level is checked, note the condition of the coolant. It should be relatively clear. If it is brown or a rust color, the system should be drained, flushed and refilled (Section 26).

14 If the cooling system requires repeated additions to maintain the proper level, have the radiator cap checked for proper sealing ability. Also check for leaks in the system (cracked hoses, loose hose connections, leaking gaskets, etc.).

Windshield washer fluid

15 Fluid for the windshield washer system is located in a plastic reservoir (photo). The level in the reservoir should be maintained at the Full mark, except during periods when freezing temperatures are expected, at which times the fluid level should be maintained no higher than 3/4 full to allow for expansion should the fluid freeze. The use of an additive such as Optikleen will help lower the freezing point of the fluid and will result in better cleaning of the windshield surface. *Do not use antifreeze because it will cause damage to the vehicle's paint.*

16 Also, to help prevent icing in cold weather, warm the windshield with the defroster before using the washer.

Battery electrolyte

17 All vehicles with which this manual is concerned are equipped with a 'freedom' battery which is permanently sealed (except for vent holes) and has no filler caps (photo). Water does not have to be added to these batteries at any time.

Brake fluid

18 The master cylinder is mounted directly on the firewall (manual

4.9A The engine coolant level should appear near the Full Hot mark with the engine at normal operating temperature

4.9B Heed the warning on the radiator cap and never remove it while the engine is hot (arrows align with the vent tube)

4.15 The windshield washer fluid level should be kept an inch or so below the fill line when freezing temperatures are expected

4.17 This type of battery never requires adding water, but normal maintenance should be performed

brake models) or on the front of the power booster unit (power brake models) in the engine compartment.

19 The master cylinder reservoir incorporates two windows, which allow checking the brake fluid level without removal of the reservoir cover. The level should be maintained at 1/4-inch below the lowest edge of each reservoir opening (photo).

20 If a low level is indicated, be sure to wipe the top of the reservoir cover with a clean rag, to prevent contamination of the brake system, before lifting the cover.

21 When adding fluid, pour it carefully into the reservoir, taking care not to spill any onto surrounding painted surfaces. Be sure the specified fluid is used, since mixing different types of brake fluid can cause damage to the system. See *Recommended lubricants and fluids* or your owner's manual.

22 At this time the fluid and master cylinder can be inspected for contamination. Normally, the brake system will not need periodic draining and refilling, but if rust deposits, dirt particles or water droplets are seen in the fluid, the system should be dismantled, drained and refilled with fresh fluid.

23 After filling the reservoir to the proper level, make sure the lid is properly seated to prevent fluid leakage and/or system pressure loss.

24 The brake fluid in the master cylinder will drop slightly as the brake shoes or pads at each wheel wear down during normal operation. If the master cylinder requires repeated replenishing to keep it at the proper level, this is an indication of leakage in the brake system, which should be corrected immediately. Check all brake lines and connections, along with the wheel cylinders and booster (see Section 13 for more information).

4.19 It is essential that dirt and grease be wiped off the master cylinder reservoir cover before it is removed

25 If upon checking the master cylinder fluid level you discover one or both reservoirs empty or nearly empty, the brake system should be bled (Chapter 9).

Manual transmission oil

26 Manual shift transmissions do not have a dipstick. The fluid level is checked with the engine cold by removing a plug from the side of the transmission case. Locate the plug and use a rag to clean the plug and the area around it, then remove it with a wrench.

27 If oil immediately starts leaking out, thread the plug back into the transmission because the level is all right. If there is no leakage, completely remove the plug and place your little finger inside the hole. The oil level should be just at the bottom of the plug hole.

28 If the transmission needs more oil, use a syringe to squeeze the appropriate lubricant into the plug hole until the level is correct.

29 Thread the plug back into the transmission and tighten it securely.

30 Drive the vehicle a short distance, then check to make sure the plug is not leaking.

Automatic transmission fluid

31 The level of the automatic transmission fluid should be carefully maintained. Low fluid level can lead to slipping or loss of drive, while overfilling can cause foaming and loss of fluid.

32 With the parking brake set, start the engine, then move the shift lever through all the gear ranges, ending in Park. The fluid level must be checked with the vehicle level and the engine running at idle. **Note:** *Incorrect fluid level readings will result if the vehicle has just been driven at high speeds for an extended period, in hot weather in city traffic, or if it's been pulling a trailer.* If any of these conditions apply, wait until the fluid has cooled (about 30 minutes).

33 Locate the dipstick at the rear of the engine compartment on the passenger's side and pull it out of the filler tube (photo).

34 Carefully touch the end of the dipstick to determine if the fluid is cool (about room temperature), or warm or hot (uncomfortable to the touch).

35 Wipe the fluid from the dipstick and push it back into the filler tube until the cap seats.

36 Pull the dipstick out again and note the fluid level.

37 If the fluid felt cool, the level should be 1/8 to 3/8-inch below the Add mark (photo). The two dimples below the Add mark indicate this range.

38 If the fluid felt warm, the level should be close to the Add mark (just above or below it).

39 If the fluid felt hot, the level should be at the Full mark.

40 Add just enough of the recommended fluid to fill the transmission to the proper level. It takes about one pint to raise the level from the Add mark to the Full mark with a hot transmission, so add the fluid

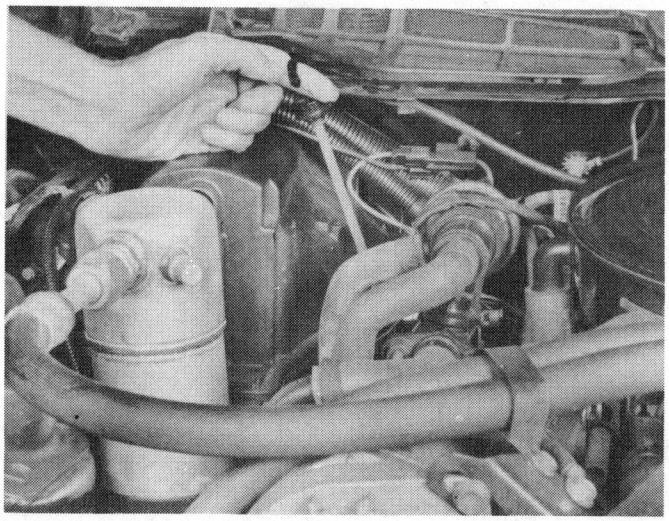

4.33 Do not confuse the automatic transmission fluid level dipstick (shown here) with the engine oil level dipstick

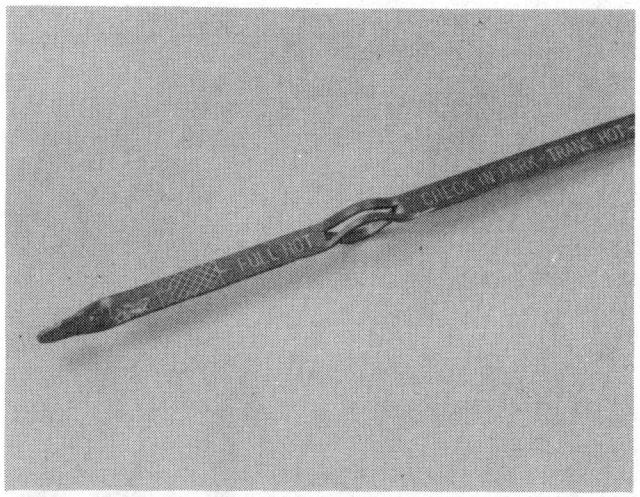

4.37 Follow the directions stamped on the dipstick to get an accurate reading (use high quality transmission fluid, as a few cents saved on 'bargain' brands could lead to a high repair bill!)

a little at a time and keep checking the level until it is correct.
41 The condition of the fluid should also be checked along with the level. If the fluid at the end of the dipstick is a dark reddish-brown color, or if the fluid has a burnt smell, the transmission fluid should be changed. If you are in doubt about the condition of the fluid, purchase some new fluid and compare the two for color and smell.

Rear axle oil

42 Like the manual transmission, the rear axle has an inspection and fill plug which must be removed to check the oil level.
43 Remove the plug which is located either in the removable cover plate or on the side of the differential carrier (photo). Use your little finger to reach inside the rear axle housing to feel the oil level. It should be at the bottom of the plug hole.
44 If this is not the case, add the proper lubricant to the rear axle carrier through the plug hole. A syringe or a small funnel can be used for this.
45 Make certain the correct lubricant is used, as regular and Positraction rear axles require different lubricants. You can determine which type of axle you have by reading the stamped number on the axle tube (refer to Chapter 8).
46 Tighten the plug securely and check for leaks after the first few miles of driving.

Power steering fluid

47 Unlike manual steering, the power steering system relies on fluid which may, over a period of time, require replenishing.
48 The reservoir for the power steering pump will be located near the front of the engine and can be mounted on either the left or right side.
49 For the check, the front wheels should be pointed straight ahead and the engine should be off.
50 Use a clean rag to wipe off the reservoir cap and the area around the cap. This will help prevent any foreign matter from entering the reservoir during the check.
51 Make sure the engine is at normal operating temperature.
52 Remove the dipstick (photo), wipe it off with a clean rag, reinsert it, then withdraw it and read the fluid level (photo). The level should be between the Add and Full Hot marks.
53 If additional fluid is required, pour the specified type directly into the reservoir using a funnel to prevent spills.
54 If the reservoir requires frequent fluid additions, all power steering hoses, hose connections, the power steering pump and the steering box should be carefully checked for leaks.

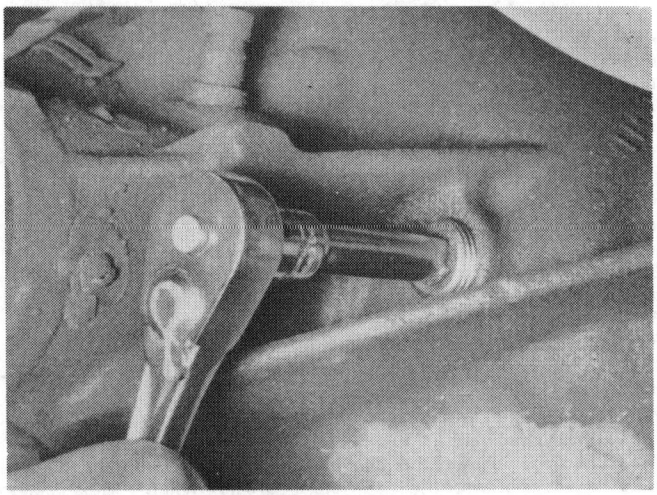

4.43 This rear axle oil filler/check plug is on the differential housing, though on some models the plug may be in the rear cover plate

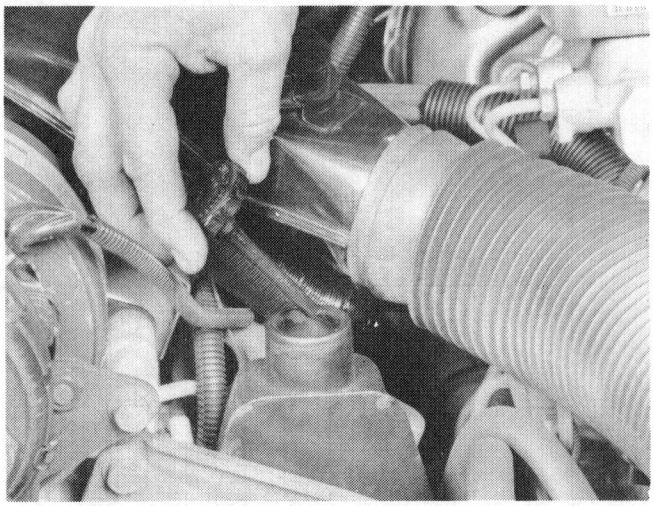

4.52A Make certain no foreign matter gets into the power steering fluid reservoir when the check is made

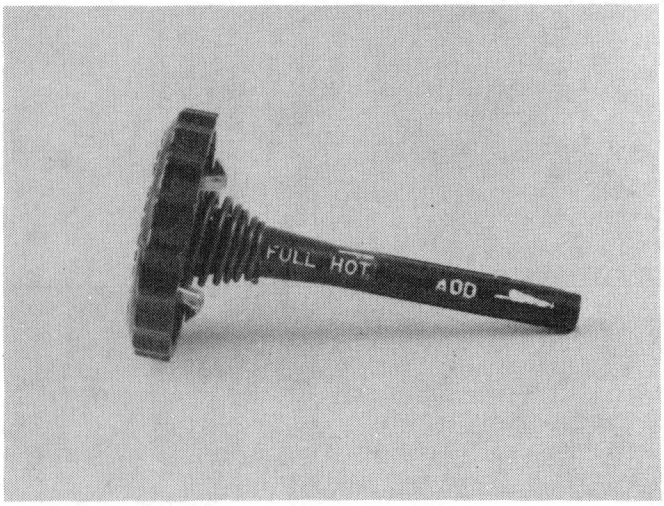

4.52B Checking of the power steering fluid level is done with the engine at normal operating temperature

5 Battery check and maintenance

1 A sealed 'freedom' battery is standard equipment on all vehicles with which this manual is concerned. Although this type of battery has many advantages over the older, capped cell type and never requires the addition of water, it should nevertheless be routinely maintained according to the procedures which follow. **Warning:** *Hydrogen gas in small quantities is present in the area of the two small side vents on sealed batteries, so keep lighted tobacco and open flames or sparks away from them.*
2 The external condition of the battery should be monitored periodically for damage such as a cracked case or cover.
3 Check the tightness of the battery cable clamps to ensure good electrical connections and check the entire length of each cable for cracks and frayed conductors (photo).
4 If corrosion (visible as white, fluffy deposits) is evident, remove the cables from the terminals, clean them with a battery brush and reinstall the cables. Corrosion can be kept to a minimum by applying a layer of petroleum jelly or grease to the terminals and cable clamps after they are assembled.
5 Make sure that the rubber protector (if so equipped) over the positive terminal is not torn or missing. It should completely cover the terminal.
6 Make sure that the battery carrier is in good condition and that the hold-down clamp bolts are tight. If the battery is removed from the carrier, make sure that no parts remain in the bottom of the carrier when the battery is reinstalled (photo). When reinstalling the hold-down clamp bolts, do not overtighten them.
7 Corrosion on the hold-down components, battery case and surrounding areas may be removed with a solution of water and baking soda, but take care to prevent any solution from coming in contact with

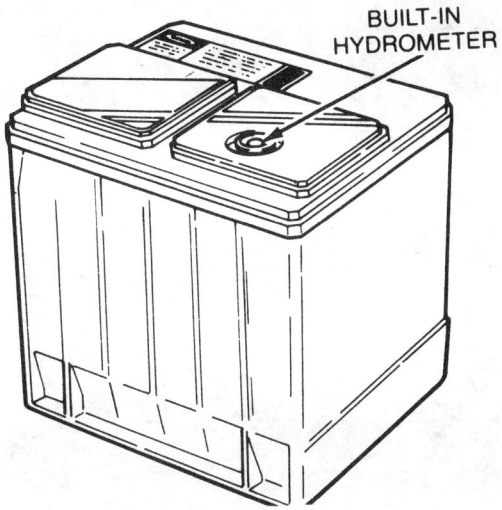

Fig. 1.1 Original equipment batteries have a built-in hydrometer to indicate battery condition (Sec 5)

your eyes, skin or clothes, as it contains acid. Protective gloves should be worn. Thoroughly wash all cleaned areas with plain water.

8 Any metal parts of the vehicle damaged by corrosion should be covered with a zinc-based primer, then painted after the affected areas have been cleaned and dried.

9 Further information on the battery, charging and jump-starting can be found in Chapter 5.

6 Drivebelt check and adjustment

1 The drivebelts, or V-belts as they are sometimes called, are located at the front of the engine and play an important role in the overall operation of the vehicle and its components. Due to their function and material make-up, the belts are prone to failure after a period of time and should be inspected and adjusted periodically to prevent major engine damage.

2 The number of belts used on a particular vehicle depends on the accessories installed. Drivebelts are used to turn the generator/alternator, air injection smog pump, power steering pump, water pump, fan and air-conditioning compressor. Depending on the pulley arrangement, a single belt may be used to drive more than one of these components.

3 With the engine off, open the hood and locate the various belts

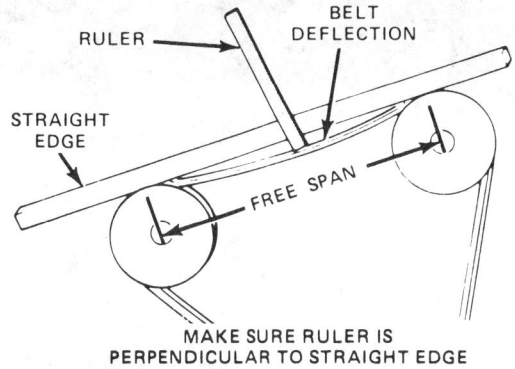

Fig. 1.2 Drivebelt tension can be checked with a straightedge and ruler (Sec 6)

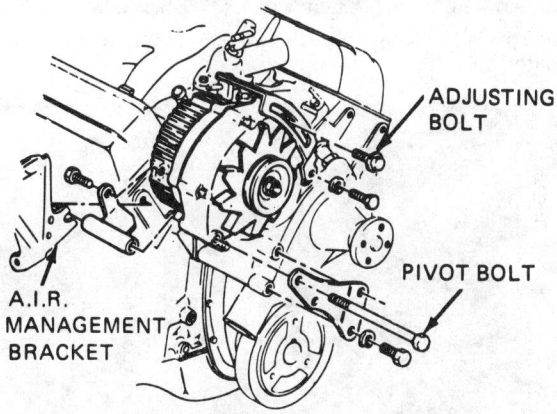

Fig. 1.3 Typical alternator drivebelt adjusting hardware (V6 shown) (Sec 6)

at the front of the engine. Using your fingers (and a flashlight, if necessary), move along the belts checking for cracks and separation of the belt plies. Also check for fraying and glazing, which gives the belt a shiny appearance. Both sides of the belt should be inspected, which means you will have to twist the belt to check the underside.

4 The tension of each belt is checked by pushing on the belt at a distance halfway between the pulleys. Push firmly with your thumb and see how much the belt moves down (deflects). A rule of thumb is that if the distance from pulley center-to-pulley center is be-

5.3 Unlike older model batteries, the cable terminals are on the side of 'freedom' type batteries

5.6 Use a socket extension to reach the battery hold-down bolt

6.6A This is the location of the adjusting bolt on the alternator

6.6B You'll reach the power steering pump adjusting bolts a lot easier if you remove the air cleaner snorkel first

6.6C Loosening the power steering pump adjusting bolt

6.8 A breaker bar is used to tension the air-conditioning drivebelt while the adjusting bolt is tightened

tween 7 and 11 inches, the belt should deflect 1/4-inch. If the belt is longer and travels between pulleys spaced 12 to 16 inches apart, the belt should deflect 1/2-inch.

5 If it is necessary to adjust the belt tension, either to make the belt tighter or looser, it is done by moving the belt-driven accessory on the bracket.

6 For each component there will be an adjustment or strap bolt and a pivot bolt (photos). Both bolts must be loosened slightly to enable you to move the component.

7 After the two bolts have been loosened, move the component away from the engine (to tighten the belt) or toward the engine (to loosen the belt). Hold the accessory in position and check the belt tension. If it is correct, tighten the two bolts until just snug, then recheck the tension. If it is all right, tighten the bolts.

8 It will often be necessary to use some sort of pry bar to move the accessory while the belt is adjusted (photo). If this must be done to gain the proper leverage, be very careful not to damage the component being moved or the part being pried against.

7 Cooling system check

1 Many major engine failures can be attributed to a faulty cooling system. If the vehicle is equipped with an automatic transmission, the cooling system also plays an important role in prolonging its life.

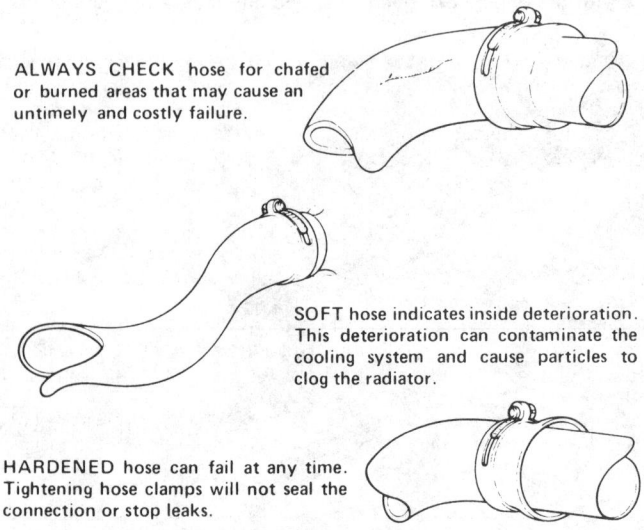

Fig. 1.4 Simple checks can detect radiator hose defects (Sec 7)

7.4 Although this radiator hose appears to be in good condition, it should still be periodically checked for cracks (more easily revealed when squeezed)

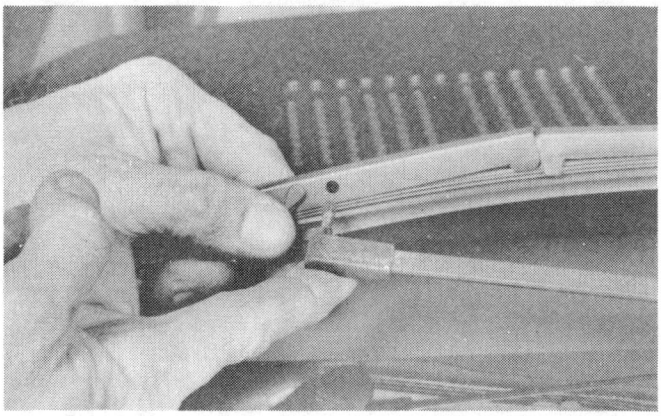

9.4 You don't even have to worry about broken fingernails with this convenient windshield wiper arm assembly quick-release lever!

2 The cooling system should be checked with the engine cold. Do this before the vehicle is driven for the day or after it has been shut off for at least three hours.
3 Remove the radiator cap and thoroughly clean the cap (inside and out) with clean water. Also clean the filler neck on the radiator. All traces of corrosion should be removed.
4 Carefully check the upper and lower radiator hoses along with the smaller diameter heater hoses. Inspect each hose along its entire length, replacing any hose which is cracked, swollen or shows signs of deterioration. Cracks may become more apparent if the hose is squeezed (photo).
5 Also make sure that all hose connections are tight. A leak in the cooling system will usually show up as white or rust colored deposits on the areas adjoining the leak.
6 Use compressed air or a soft brush to remove bugs, leaves, etc. from the front of the radiator or air-conditioning condenser (photo). Be careful not to damage the delicate cooling fins or cut yourself on them.
7 Finally, have the cap and system pressure tested. If you do not have a pressure tester, most gas stations and repair shops will do this for a minimal change.

8 Underhood hose check and replacement

Caution: *Replacement of air-conditioner hoses must be left to a dealer or air-conditioning specialist who can depressurize the system and perform the work safely.*

7.6 'Worm's-eye' view of radiator reveals that it isn't in dire need of debugging yet, but it's getting there!

1 The high temperatures present under the hood can cause deterioration of the numerous rubber and plastic hoses.
2 Periodic inspection should be made for cracks, loose clamps and leaks because some of the hoses are part of the emissions control systems and can affect the engine's performance.
3 Remove the air cleaner if necessary and trace the entire length of each hose. Squeeze each hose to check for cracks and look for swelling, discoloration and leaks.
4 If the vehicle has considerable mileage or if one or more of the hoses is suspect, it is a good idea to replace all of the hoses at one time.
5 Measure the length and inside diameter of each hose and obtain and cut the replacement to size. Since original equipment hose clamps are often good for only one or two uses, it is a good idea to replace them with screw-type clamps.
6 Replace each hose one at a time to eliminate the possibility of confusion. Hoses attached to the heater and radiator contain coolant, so newspapers or rags should be kept handy to catch the spills when they are disconnected.
7 After installation, run the engine until it reaches operating temperature, shut it off and check for leaks. After the engine has cooled, retighten all of the screw-type clamps.

9 Windshield wiper blade inspection and replacement

1 The windshield wiper and blade assembly should be inspected periodically for damage, loose components and cracked or worn blade elements.
2 Road film can build up on the wiper blades and affect their efficiency so they should be washed regularly with a mild detergent solution.
3 The action of the wiping mechanism can loosen the bolts, nuts and fasteners so they should be checked and tightened, as necessary, at the same time the wiper blades are checked.
4 If the wiper blade elements are cracked, worn or warped, they should be replaced with new ones. The wiper blade assemblies are equipped with press-type release levers. Push the lever up or down and slide the old wiper blade assembly off the wiper arm pin (photo). Install the new blade by positioning it over the wiper arm pin and pressing on it until it snaps into place.
5 The wiper blade element is locked in place at either end of the wiper blade assembly by a spring-loaded retainer and metal tabs. To remove the element, use a screwdriver to slide the blade under the element near the tabs and rotate the screwdriver, then slide the element up, out of the tabs.
6 To install, slide the element into the retaining tabs, lining up the slot in the element with the tabs, and snap the element into place.
7 Snap the blade assembly into place on the wiper arm pin.

10 Chassis lubrication

1 A grease gun and a cartridge filled with the proper grease (see *Recommended lubricants and fluids*) are usually the only equipment necessary to lubricate the chassis components. In some chassis locations, plugs may be installed rather than grease fittings, in which case grease fittings will have to be purchased and installed.

2 Refer to the accompanying illustration which shows where the various grease fittings are located. Look under the vehicle to find these components and determine if grease fittings or solid plugs are installed. If there are plugs, remove them with a wrench and buy grease fittings which will thread into the component. A GM dealer or auto parts store will be able to supply replacement fittings. Straight, as well as angled, fittings are available.

3 For easier access under the vehicle, raise it with a jack and place jackstands under the frame. Make sure the vehicle is securely supported by the stands.

4 Before proceeding, force a little of the grease out of the nozzle to remove any dirt from the end of the gun. Wipe the nozzle clean with a rag.

5 With the grease gun, plenty of clean rags and the diagram, crawl under the vehicle and begin lubricating the components.

6 Wipe the grease fitting nipple clean and push the nozzle firmly over the fitting nipple. Squeeze the trigger on the grease gun to force grease into the component (photo). **Note:** *The lower control arm balljoints (one for each front wheel) should be lubricated until the rubber reservoir is firm to the touch. Do not pump too much grease into these fittings as it could rupture the reservoir.* For all other suspension and steering fittings, continue pumping grease into the nipple until grease seeps out of the joint between the two components. If the grease seeps out around the grease gun nozzle, the nipple is clogged or the nozzle is not seated on the fitting nipple. Resecure the gun nozzle to the fitting and try again. If necessary, replace the fitting.

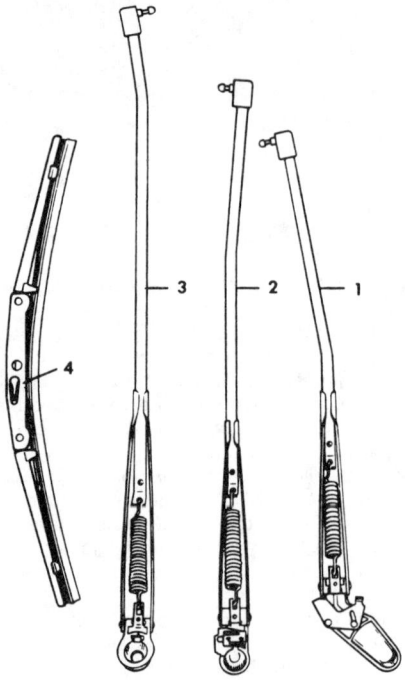

Fig. 1.5 Wiper blade assemblies (Sec 9)

1 Right wiper arm
2 Left wiper arm
3 Rear wiper arm
4 Location of blade release lever

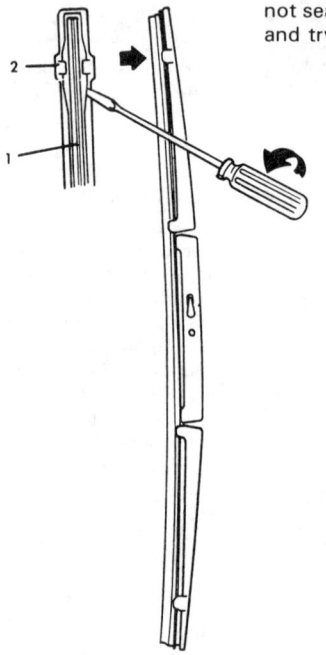

Fig. 1.6 Wiper blade element installation details (Sec 9)

1 Element 2 Housing tabs

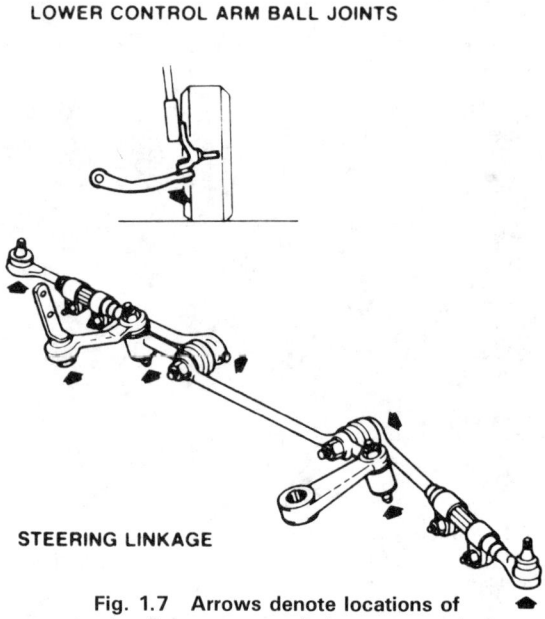

Fig. 1.7 Arrows denote locations of steering linkage grease fittings on a typical vehicle (Sec 10)

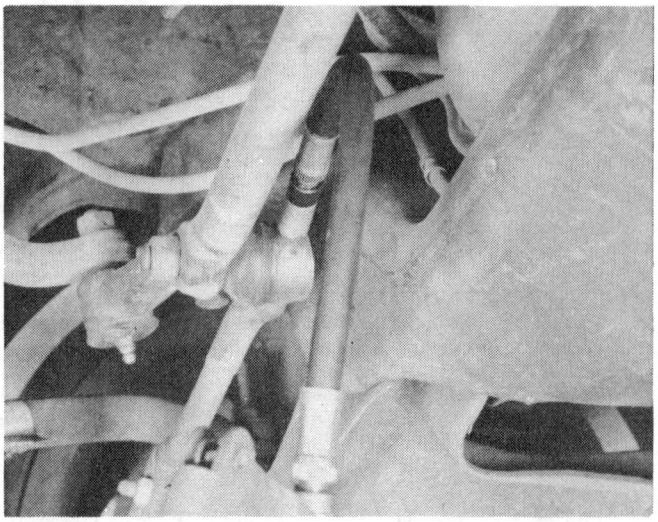

10.6 Hand-operated grease guns work just fine when lubricating steering and suspension components

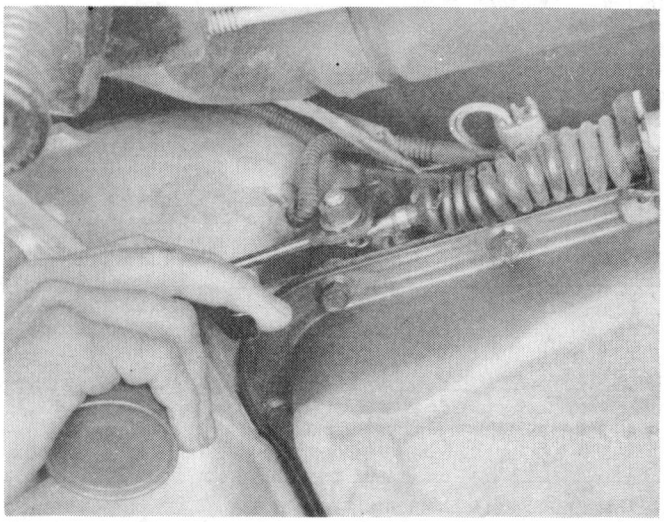

10.8 Lubricate the automatic transmission shift linkage with a few drops of oil from an oil can

7 Wipe the excess grease from the components and the grease fitting. Follow the procedures for the remaining fittings.
8 While you are under the vehicle, clean and lubricate the parking brake cable, along with the cable guides and levers. This can be done by smearing some of the chassis grease onto the cable and its related parts with your fingers. Place a few drops of light engine oil on the transmission shift linkage rods and swivels (photo).
9 Lower the vehicle to the ground for the remaining body lubrication process.
10 Open the hood and smear a little chassis grease on the hood latch mechanism (photo). If the hood has an inside release, have an assistant pull the release knob from inside the vehicle as you lubricate the cable at the latch.
11 Lubricate all the hinges (door, hood, hatch) with a few drops of light engine oil to keep them in proper working order.
12 Finally, the key lock cylinders can be lubricated with spray-on graphite which is available at auto parts stores.

11 Exhaust system check

1 With the engine cold (at least three hours after the vehicle has been driven), check the complete exhaust system from its starting point at the engine to the end of the tailpipe. This should be done on a hoist where unrestricted access is available.
2 Check the pipes and connections for signs of leakage and/or corrosion indicating a potential failure. Make sure that all brackets and hangers are in good condition and tight.
3 At the same time, inspect the underside of the body for holes, corrosion, open seams, etc. which may allow exhaust gases to enter the passenger compartment. Seal all body openings with silicone or body putty.
4 Rattles and other noises can often be traced to the exhaust system, especially the mounts and hangers. Try to move the pipes, muffler and catalytic converter (if so equipped). If the components can come in contact with the body or driveline parts, secure the exhaust system with new mounts.
5 This is also an ideal time to check the running condition of the engine by inspecting inside the very end of the tailpipe. The exhaust deposits here are an indication of engine state-of-tune. If the pipe is black and sooty or coated with white deposits, the engine is in need of a tune-up (including a thorough carburetor inspection and adjustment).

12 Suspension and steering check

1 Whenever the front of the vehicle is raised for service, it is a good idea to visually check the suspension and steering components for wear.
2 Indications of a fault in these systems are excessive play in the steering wheel before the front wheels react, excessive sway around corners, body movement over rough roads or binding at some point as the steering wheel is turned.
3 Before the vehicle is raised for inspection, test the shock absorbers by pushing down to rock the vehicle at each corner. If you push down and the vehicle does not come back to a level position within one or two bounces, the shocks/struts are worn and must be replaced. As this is done, check for squeaks and strange noises coming from the suspension components. Information on suspension components can be found in Chapter 11.
4 Now raise the front end of the vehicle and support it firmly on jackstands placed under the frame rails. Because of the work to be done, make sure the vehicle cannot fall from the stands.
5 Grab the top and bottom of the front tire with your hands and rock the tire/wheel on the spindle. If there is movement of more than 0.005 inch, the wheel bearings should be serviced (see Section 27).
6 Crawl under the vehicle and check for loose bolts, broken or disconnected parts and deteriorated rubber bushings on all suspension and steering components. Look for grease or fluid leaking from around the steering box. Check the power steering hoses and connections for leaks. Check the balljoints for wear.
7 Have an assistant turn the steering wheel from side-to-side and check the steering components for free movement, chafing and binding. If the steering does not react with the movement of the steering wheel, try to determine where the slack is located.

13 Brake check

Note: *For detailed photographs of the brake system, refer to Chapter 9.*

1 The brakes should be inspected every time the wheels are removed or whenever a defect is suspected. Indications of a potential brake system defect are: the vehicle pulls to one side when the brake pedal is depressed; noises coming from the brakes when they are applied; excessive brake pedal travel; pulsating pedal; and leakage of fluid, usually seen on the inside of the tire or wheel.

Disc brakes

2 Both front and rear disc brakes (if the rear is so equipped) can be visually checked without removing any parts except the wheels.
3 Raise the vehicle and place it securely on jackstands. Remove the wheels (see *Jacking and towing* at the front of the manual, if necessary).
4 The disc brake calipers, which contain the pads, are now visible. There is an outer pad and an inner pad in each caliper. All pads should be inspected.
5 The inner pads on the front wheels are equipped with a wear sensor. This is a small, bent piece of metal which is visible from the inboard side of the brake caliper. When the pads wear to the danger limit,

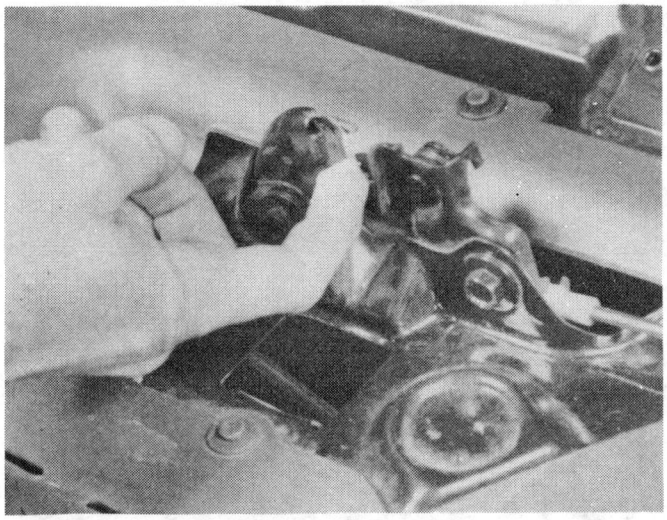

10.10 Multi-purpose grease is used to lubricate the hood latch mechanism

13.6 Disc pad wear inspection hole is visible at the center of caliper

the metal sensor rubs against the rotor and makes a screeching sound.
6 Check the pad thickness by looking at each end of the caliper and through the inspection hole in the caliper body (photo). If the wear sensor clip is very close to the rotor, or if the lining material is 1/16-inch or less in thickness, the pads should be replaced. Keep in mind that the lining material is riveted or bonded to a metal backing shoe and the metal portion is not included in this measurement.
7 Since it will be difficult, if not impossible, to measure the exact thickness of the remaining lining material, remove the pads for further inspection or replacement if you are in doubt as to the quality of the pad.
8 Before installing the wheels, check for leakage around the brake hose connections leading to the caliper and damage (cracking, splitting, etc.) to the brake hose. Replace the hose or fittings as necessary, referring to Chapter 9.
9 Also check the condition of the rotor. Look for scoring, gouging and burnt spots. If these conditions exist, the hub/rotor assembly should be removed for servicing (Chapter 9).

Drum brakes — rear (if so equipped)

10 Using a scribe or chalk, mark the drum, hub and backing plate so they can be reinstalled in the same positions.
11 Pull the brake drum off the axle and brake assembly. If this proves difficult, make sure the parking brake is released, then squirt some penetrating oil around the center hub area. Allow the oil to soak in and again try to pull the drum off. Then, if the drum cannot be pulled off, the brake shoes will have to be adjusted in. This is done by first removing the lanced knock-out in the backing plate with a hammer and chisel. With the lanced area punched in, pull the lever off the sprocket and then use a small screwdriver to turn the adjuster wheel, which will move the shoes away from the drum.
12 With the drum removed, carefully brush away any accumulations of dirt and dust. **Warning:** *Do not blow the dust out with compressed air. Make an effort not to inhale the dust because it contains asbestos and is harmful to your health.*
13 Note the thickness of the lining material on both the front and rear brake shoes. If the material has worn away to within 1/16-inch of the recessed rivets or metal backing, the shoes should be replaced. If the linings look worn, but you are unable to determine their exact thickness, compare them with a new set at an auto parts store. The shoes should also be replaced if they are cracked, glazed (shiny surface) or contaminated with brake fluid.
14 Check to see that all the brake assembly springs are connected and in good condition.
15 Check the brake components for signs of fluid leakage. With your finger, carefully pry back the rubber cups on the wheel cylinder located at the top of the brake shoes. Any leakage is an indication that the wheel cylinders should be overhauled immediately (Chapter 9). Also check the hoses and connections for signs of leakage.
16 Wipe the inside of the drum with a clean rag and denatured alcohol. Again, be careful not to breathe the dangerous asbestos dust.

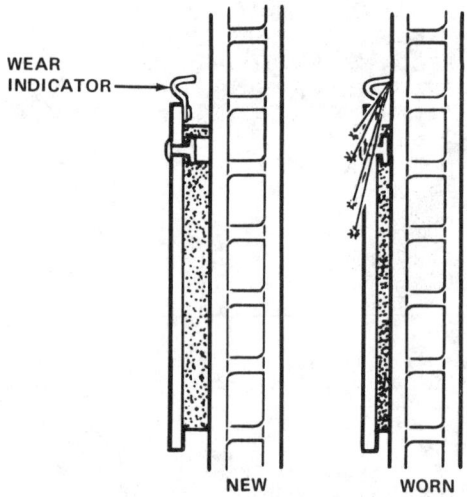

Fig. 1.8 Position of the wear indicator relative to the face of the disc brake rotor on a new pad (left) and a worn pad (right) (Sec 13)

17 Check the inside of the drum for cracks, scores, deep scratches and hard spots which will appear as small discolored areas (photo). If imperfections cannot be removed with fine emery cloth, the drum must be taken to a machine shop for resurfacing.
18 After the inspection process, if all parts are found to be in good condition, reinstall the brake drum (using a metal plug if the lanced knock-out was removed). Install the wheel and lower the vehicle to the ground.

Parking brake

19 The easiest way to check the operation of the parking brake is to park the vehicle on a steep hill with the parking brake set and the transmission in Neutral. If the parking brake cannot prevent the vehicle from rolling, it is in need of adjustment (see Chapter 9).

14 Carburetor choke check

1 The choke only operates when the engine is cold, so this check should be performed before the vehicle has been started for the day.
2 Open the hood and remove the top plate of the air cleaner assembly. It is usually held in place by a wing nut. If any vacuum hoses must be disconnected, make sure you tag them to ensure reinstallation in their original positions. Place the top plate and wing nut aside, out of the way of moving engine components.
3 Look at the top of the carburetor at the center of the air cleaner housing. You will notice a flat plate at the carburetor opening (photo).
4 Have an assistant press the accelerator pedal to the floor. The plate should close completely. Start the engine while you observe the plate at the carburetor. **Caution:** *Do not position your face directly over the carburetor, because the engine could backfire and cause serious burns.* When the engine starts, the choke plate should open slightly.
5 Allow the engine to continue running at an idle speed. As the engine warms up to operating temperature, the plate should slowly open, allowing more air to enter through the top of the carburetor.
6 After a few minutes, the choke plate should be all the way open to the vertical position.
7 You will notice that the engine speed corresponds with the plate opening. With the plate completely closed, the engine should run at a fast idle. As the plate opens, the engine speed will decrease.
8 If a malfunction is detected during the above checks, refer to Chapter 4 for specific information related to adjusting and servicing the choke components.

15 Engine idle speed check and adjustment

1 The engine idle speed is adjustable on some models and should

14.3 With the air cleaner top plate removed, the choke plate is easily checked

be checked at the scheduled maintenance interval.

2 On those vehicles with provisions for idle speed adjustment, the specifications for such adjustments are shown on the vehicle Emissions Control Information label; however, the adjustments must be made using calibrated test equipment. The adjustments should therefore be made by a dealer or automotive repair facility.

16 Engine oil and filter change

1 Frequent oil changes may be the best form of preventative maintenance available for the home mechanic. When engine oil ages, it gets diluted and contaminated, which ultimately leads to premature

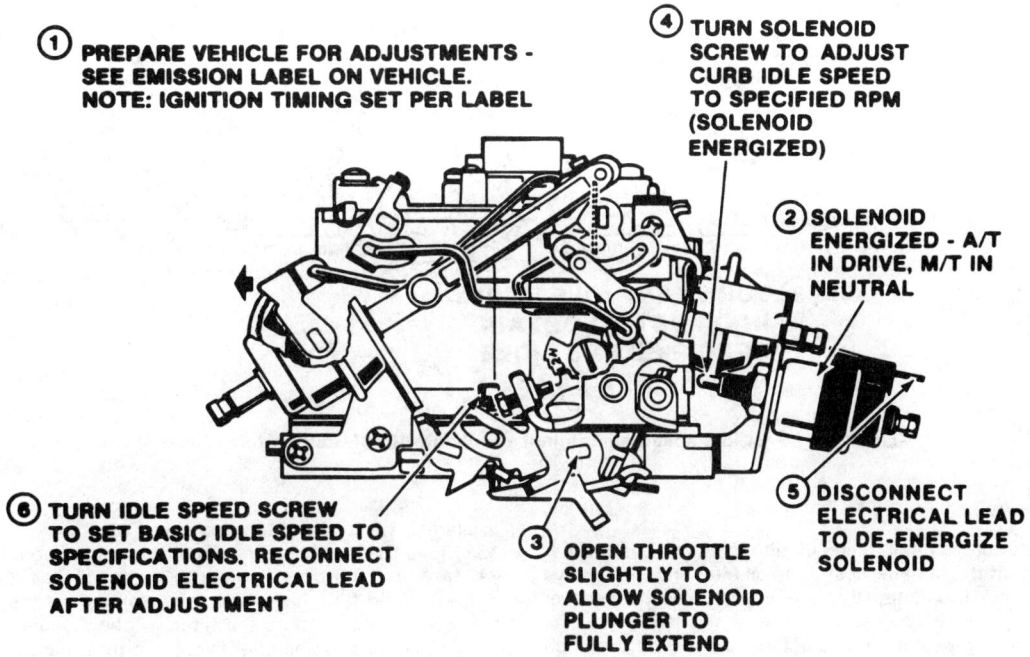

Fig. 1.9 Idle speed adjustment procedure for vehicles without air-conditioning (E2SE carburetor) (Sec 15)

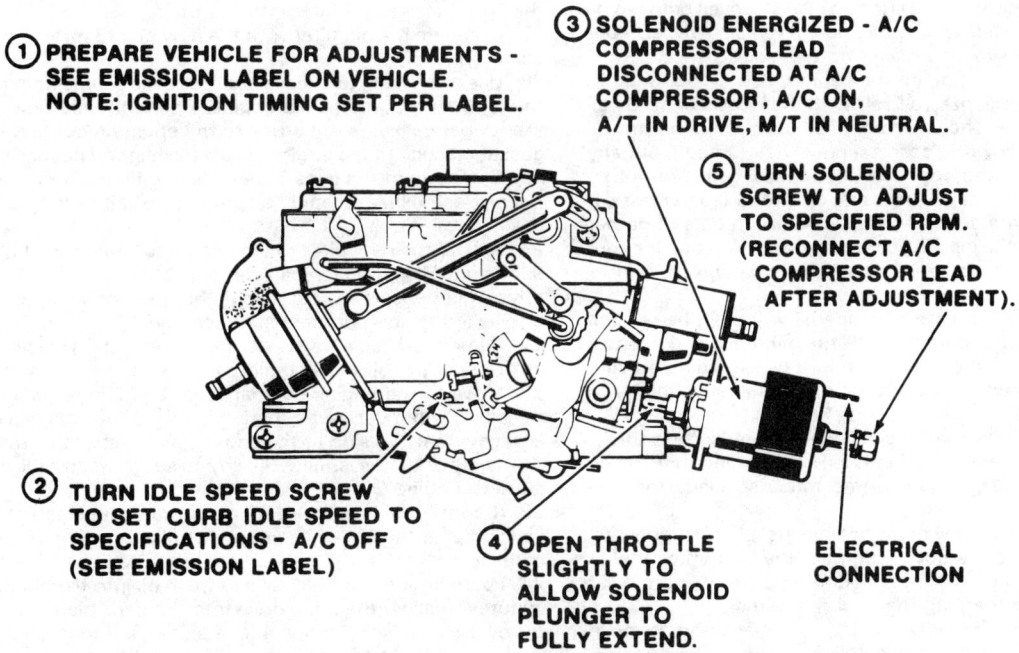

Fig. 1.10 Idle speed adjustment procedure for vehicles with air-conditioning (E2SE) carburetor) (Sec 15)

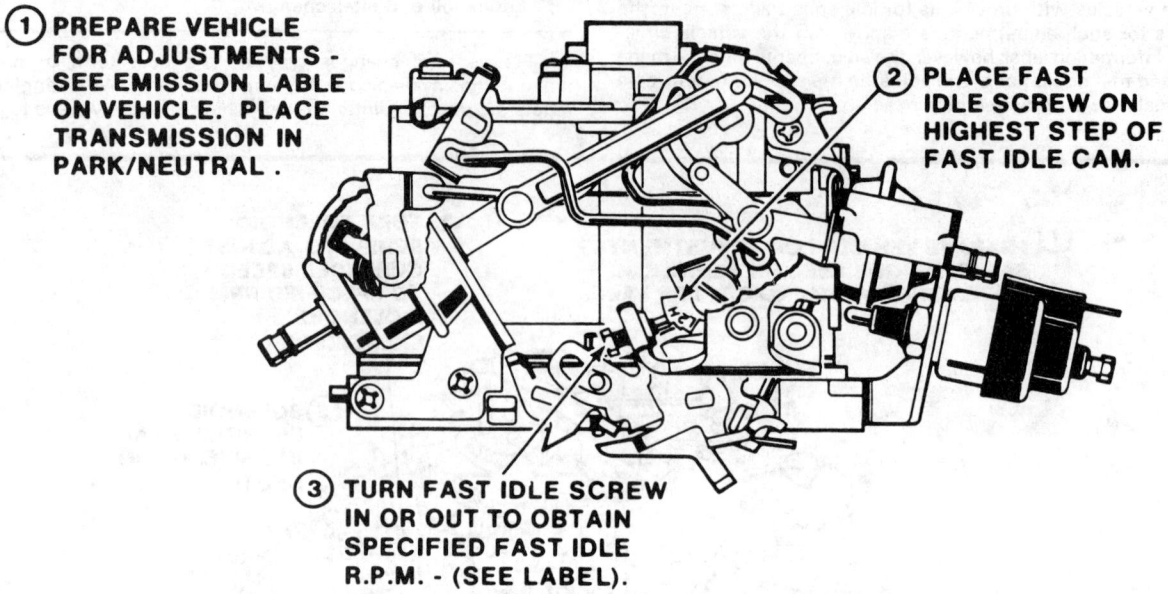

Fig. 1.11 Fast idle speed adjustment procedure (E2SE carburetor) (Sec 15)

engine wear.

2 Although some sources recommend oil filter changes every other oil change, we feel that the minimal cost of an oil filter and the relative ease with which it is installed dictate that a new filter be used whenever the oil is changed.

3 The tools necessary for a normal oil and filter change are a wrench to fit the drain plug at the bottom of the oil pan, an oil filter wrench to remove the old filter, a container with at least a six-quart capacity to drain the old oil into and a funnel or oil can spout to help pour fresh oil into the engine.

4 In addition, you should have plenty of clean rags and newspapers handy to mop up any spills. Access to the underside of the vehicle is greatly improved if the vehicle can be lifted on a hoist, driven onto ramps or supported by jackstands. *Do not work under a vehicle which is supported only a bumper, hydraulic or scissors-type jack.*

5 If this is your first oil change on the vehicle, it is a good idea to crawl underneath and familiarize yourself with the locations of the oil drain plug and the oil filter. The engine and exhaust components will be warm during the actual work, so it is a good idea to figure out any potential problems before the engine and its accessories are hot.

6 Allow the engine to warm up to normal operating temperature. If the new oil or any tools are needed, use this warm-up time to gather everything necessary for the job. The correct type of oil to buy for your application can be found in *Recommended lubricants and fluids* near the front of this manual.

7 With the engine oil warm (warm engine oil will drain better and more built-up sludge will be removed with the oil), raise and support the vehicle. Make sure it is firmly supported. If jackstands are used, they should be placed toward the front of the frame rails which run the length of the vehicle.

8 Move all necessary tools, rags and newspapers under the vehicle. Position the drain pan under the drain plug. Keep in mind that the oil will initially flow from the pan with some force, so place the pan accordingly.

9 Being careful not to touch any of the hot exhaust pipe components, use the wrench to remove the drain plug near the bottom of the oil pan. Depending on how hot the oil has become, you may want to wear gloves while unscrewing the plug the final few turns.

10 Allow the old oil to drain into the pan. It may be necessary to move the pan farther under the engine as the oil flow slows to a trickle.

11 After all the oil has drained, wipe off the drain plug with a clean rag. Small metal particles may cling to the plug and would immediately contaminate the new oil.

12 Clean the area around the drain plug opening and reinstall the plug. Tighten the plug securely with the wrench. If a torque wrench is available, use it to tighten the plug.

13 Move the drain pan into position under the oil filter.

14 Now use the filter wrench to loosen the oil filter (photo). Chain or metal band-type filter wrenches may distort the filter canister, but this is of no concern as the filter will be discarded anyway.

15 Sometimes the oil filter is on so tight it cannot be loosened, or it is positioned in an area which is inaccessible with a filter wrench. As a last resort, you can punch a metal bar or long screwdriver directly through the bottom of the canister and use it as a T-bar to turn the filter. If so, be prepared for oil to spurt out of the canister as it is punctured.

16 Completely unscrew the old filter. Be careful, it is full of oil. Empty the oil oil inside the filter into the drain pan.

17 Compare the old filter with the new one to make sure they are the same type.

18 Use a clean rag to remove all oil, dirt and sludge from the area where the oil filter mounts to the engine. Check the old filter to make sure the rubber gasket is not stuck to the engine mounting surface. If the gasket is stuck to the engine (use a flashlight if necessary), remove it.

19 Open one of the cans of new oil and fill the new filter about half-full of fresh oil. Also, apply a light coat of oil to the rubber gasket of the new oil filter.

20 Attach the new filter to the engine following the tightening directions printed on the filter canister or packing box. Most filter manufacturers recommend against using a filter wrench due to the possibility of overtightening and damage to the seal.

21 Remove all tools, rags, etc. from under the vehicle, being careful not to spill the oil in the drain pan, then lower the vehicle.

22 Move to the engine compartment and locate the oil filler cap on the engine. In most cases there will be a screw-off cap on the rocker arm cover (at the side of the engine) or a cap at the end of a fill tube at the front of the engine. In any case, the cap will most likely be labeled 'Engine Oil' or 'Oil.'

23 If an oil can spout is used, push the spout into the top of the oil can and pour the fresh oil through the filler opening. A funnel may also be used.

24 Pour about three (3) quarts of fresh oil into the engine. Wait a few minutes to allow the oil to drain into the pan, then check the level on the oil dipstick (see Section 4 if necessary). If the oil level is at or near the lower Add mark, start the engine and allow the new oil to circulate.

25 Run the engine for only about a minute and then shut it off. Immediately look under the vehicle and check for leaks at the oil pan drain plug and around the oil filter. If either is leaking, tighten with a bit more force.

26 With the new oil circulated and the filter now completely full,

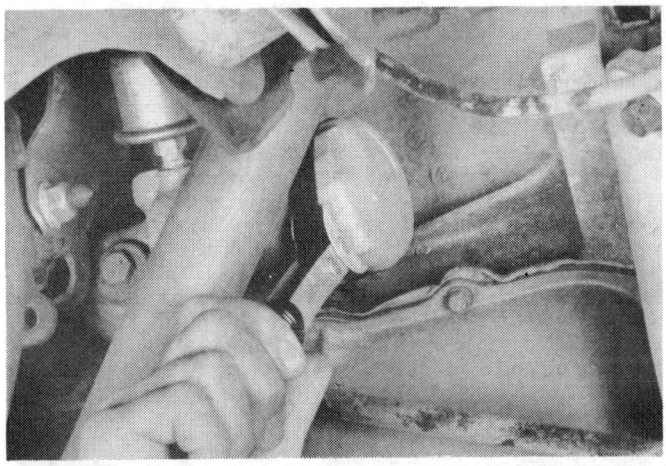

16.14 A strap-type oil filter wrench works well in hard-to-reach locations

recheck the level on the dipstick and add enough oil to bring the level to the Full mark on the dipstick.
27 During the first few trips after an oil change, make it a point to check frequently for leaks and proper oil level.
28 The old oil drained from the engine cannot be reused in its present state and should be disposed of. Oil reclamation centers, auto repair shops and gas stations will normally accept the oil, which can be refined and used again. After the oil has cooled, it can be drained into a suitable container (capped plastic jugs, topped bottles, milk cartons, etc.) for transport to one of these disposal sites.

17 Fuel system check

Caution: *There are certain precautions to take when inspecting or servicing the fuel system components. Work in a well-ventilated area and do not allow open flames (cigarettes, appliance pilot lights, etc.) to get near the work area. Mop up spills immediately and do not store fuel-soaked rags where they could ignite.*

1 If your vehicle is equipped with fuel injection (TBI), refer to the fuel injection pressure relief procedure (Chapter 4) before servicing any component of the fuel system.
2 The fuel system is under a small amount of pressure, so if any fuel lines are disconnected for servicing, be prepared to catch the fuel as it spurts out. Plug all disconnected fuel lines immediately after disconnection to prevent the tank from emptying itself.
3 The fuel system is most easily checked with the vehicle raised on a hoist so the components underneath the vehicle are readily visible and accessible.
4 If the smell of gasoline is noticed while driving or after the vehicle has been in the sun, the system should be thoroughly inspected immediately.
5 Remove the gas filler cap and check for damage, corrosion and an unbroken sealing imprint on the gasket (photo). Replace the cap with a new one, if necessary.
6 With the vehicle raised, inspect the gas tank and filler neck for punctures, cracks and other damage. The connection between the filler neck and the tank is especially critical. Sometimes a rubber filler neck will leak due to loose clamps or deteriorated rubber, problems a home mechanic can usually rectify. **Caution:** *Do not, under any circumstances, try to repair a fuel tank yourself (except rubber components) unless you have had considerable experience. A welding torch or any open flame can easily cause the fuel vapors to explode if the proper precautions are not taken.*
7 Carefully check all rubber hoses and metal lines leading away from the fuel tank. Check for loose connections, deteriorated hoses, crimped lines and other damage. Follow the lines up to the front of the vehicle, carefully inspecting them all the way. Repair or replace damaged sections as necessary.
8 If a fuel odor is still evident after the inspection, refer to Section 33.

17.5 With today's sophisticated emissions systems, it is essential that seals in the gas tank cap be checked regularly

18.3 Tag all hoses as they are removed from the air cleaner assembly to prevent 'spaghetti soup' later on!

18 Fuel filter replacement

Carburetor equipped models

1 On these models, the fuel filter is located inside the fuel inlet at the carburetor. It is made of pleated paper and cannot be cleaned or reused.
2 This job should be done with the engine cold (after sitting at least three hours). The necessary tools include open-end wrenches to fit the fuel line nuts. Flare nut wrenches (which wrap around the nut) should be used, if available. In addition, you will have to obtain the replacement filter (make sure it is for your specific vehicle and engine) and some clean rags.
3 Remove the air cleaner assembly. If vacuum hoses must be disconnected, be sure to note their positions and/or tag them to ensure that they are reinstalled correctly (photo).
4 Now follow the fuel line from the fuel pump to the point where it enters the carburetor. The fuel pump is located low on the engine, at the right front on V8 engines and at the left front on V6's. In most cases the fuel line will be metal all the way from the fuel pump to the carburetor.
5 Place some rags under the fuel inlet fittings to catch spilled fuel as the fittings are disconnected.
6 With the proper size wrench, hold the nut immediately next to the

carburetor body. Now loosen the nut fitting and the end of the metal fuel line. A flare nut wrench on this fitting will help prevent slipping and possible damage. However, an open-end wrench should do the job. Make sure the larger nut next to the carburetor is held securely while the fuel line is disconnected.

7 After the fuel line is disconnected, move it aside for better access to the inlet filter nut. Do not crimp the fuel line.

8 Now unscrew the fuel inlet filter nut which was previously held steady. As this fitting is drawn away from the carburetor body, be careful not to lose the thin washer-type gasket or the spring located behind the fuel filter. Also, pay close attention to how the filter is installed.

9 Compare the old filter with the new one to make sure they are the same length and design.

10 Reinstall the spring in the carburetor body.

11 Place the new filter in position behind the spring. It will have a rubber gasket and check valve at one end, which should point *away* from the carburetor.

12 Install a new washer-type gasket on the fuel inlet filter nut (a gasket is usually supplied with the new filter) and tighten the nut in the carburetor. Make sure it is not cross-threaded. Tighten it securely (if a torque wrench is available, tighten the nut to 18 ft-lbs). Do not overtighten it, as the hole can strip easily, causing fuel leaks.

13 Hold the fuel inlet nut securely with a wrench while the fuel line is connected. Again, be careful not to cross-thread the connector. Tighten the fitting securely.

14 Plug the vacuum hose which leads to the air cleaner snorkel motor so the engine can be run.

15 Start the engine and check carefully for leaks. If the fuel line connector leaks, disconnect it using the above procedures and check for stripped or damaged threads. If the fuel line connector has stripped threads, remove the entire line and have a repair shop install a new fitting. If the threads look all right, purchase some thread sealing tape and wrap the connector threads with it. Now reinstall and tighten it securely. Inlet repair kits are available at most auto parts stores to overcome leaking at the fuel inlet filter nut.

16 Reinstall the air cleaner assembly, connecting the hoses in their original positions.

Fuel-injected models

Caution: *Refer to the Fuel injection pressure relief procedure in Chapter 4 before performing this procedure.*

17 Fuel-injected engines employ a stainless steel in-line fuel filter. On

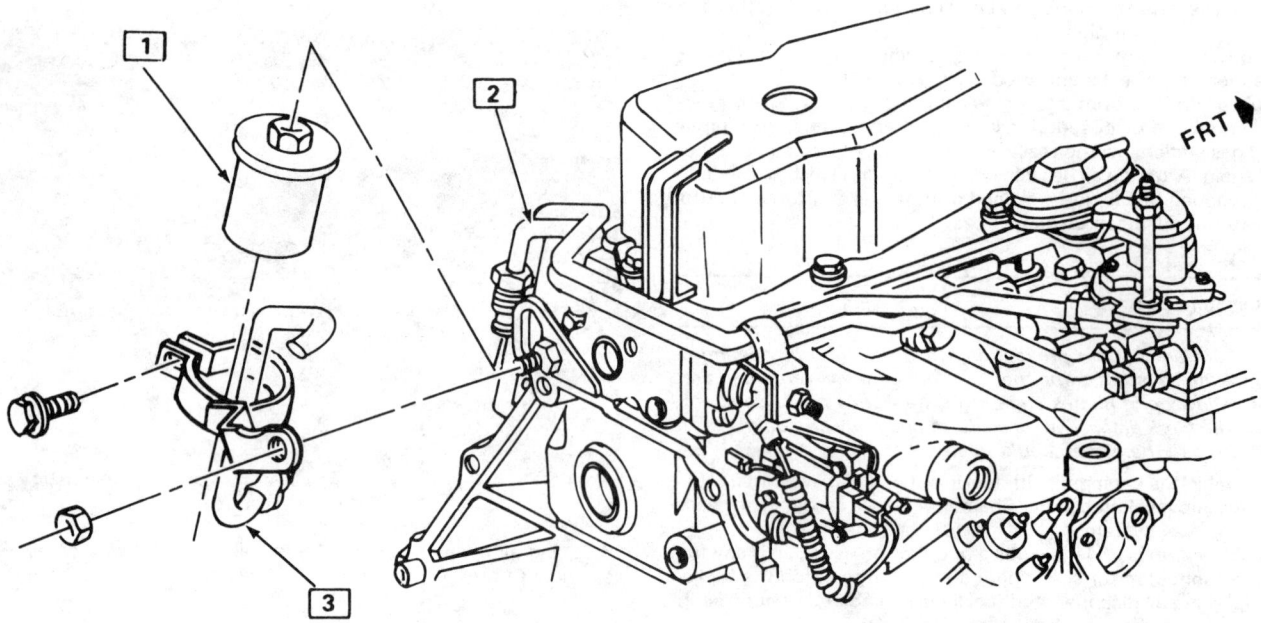

Fig. 1.12 Location of the fuel filter and related parts on an L4 engine (Sec 18)

1 Filter assembly
2 Fuel feed line
3 Fuel return line

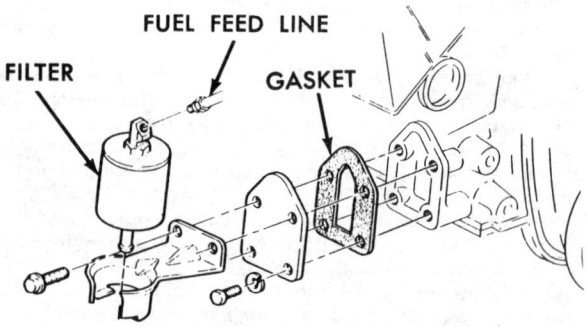

Fig. 1.13 Location of the fuel filter and related parts on a V8 engine with fuel injection (Sec 18)

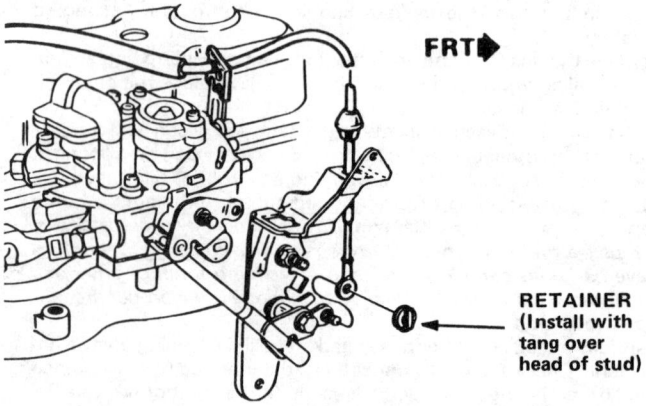

Fig. 1.14 Details of the accelerator control cable assembly on an L4 engine (Sec 19)

four-cylinder engines it is located at the left rear of the engine, clamped to the cylinder head. On V8 fuel-injected models, the filter is bracketed in the position normally occupied by the fuel pump on carbureted models, at the lower right front of the engine.

18 With the engine cold, place a container under the fuel filter.

19 Remove the bolts attaching the fuel filter bracket to the engine.

20 Using wrenches of the proper size, remove the line from the top of the filter.

21 Unclamp and remove the fuel line from the bottom of the filter and remove the filter.

22 Install the new filter by reversing the removal procedure. Do not overtighten the fitting at the top of the fuel filter. If a torque wrench is available, tighten the fitting to 18 ft-lbs.

19 Throttle linkage check

1 The throttle linkage is a cable type and, although there are no adjustments for the linkage itself, periodic maintenance is necessary to assure its proper function.

2 Remove the air cleaner (refer to Chapter 2) so the entire linkage is visible.

3 Check the entire length of the cable to make sure that it is not binding.

4 Check all nylon bushings for wear, replacing them with new ones as necessary.

5 Lubricate the cable mechanisms with engine oil at the pivot points.

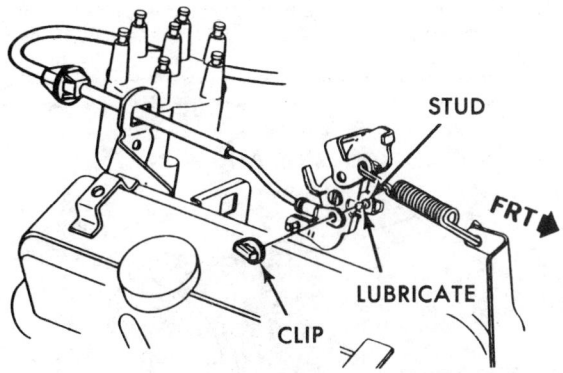

Fig. 1.15 Details of the accelerator control cable assembly on a V6 engine (Sec 19)

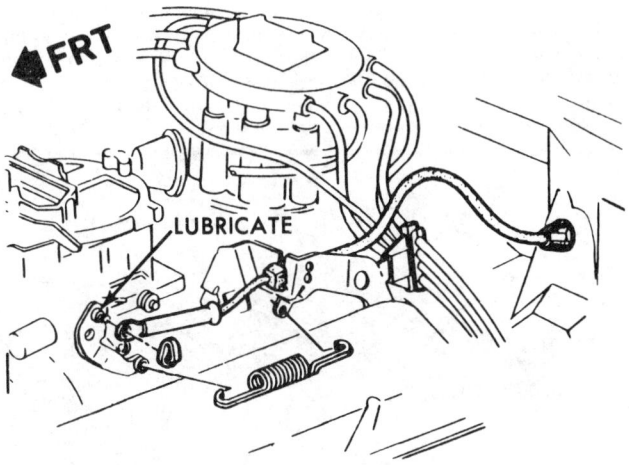

Fig. 1.16 Details of the accelerator control cable assembly on carburetor equipped V8 engines (Sec 19)

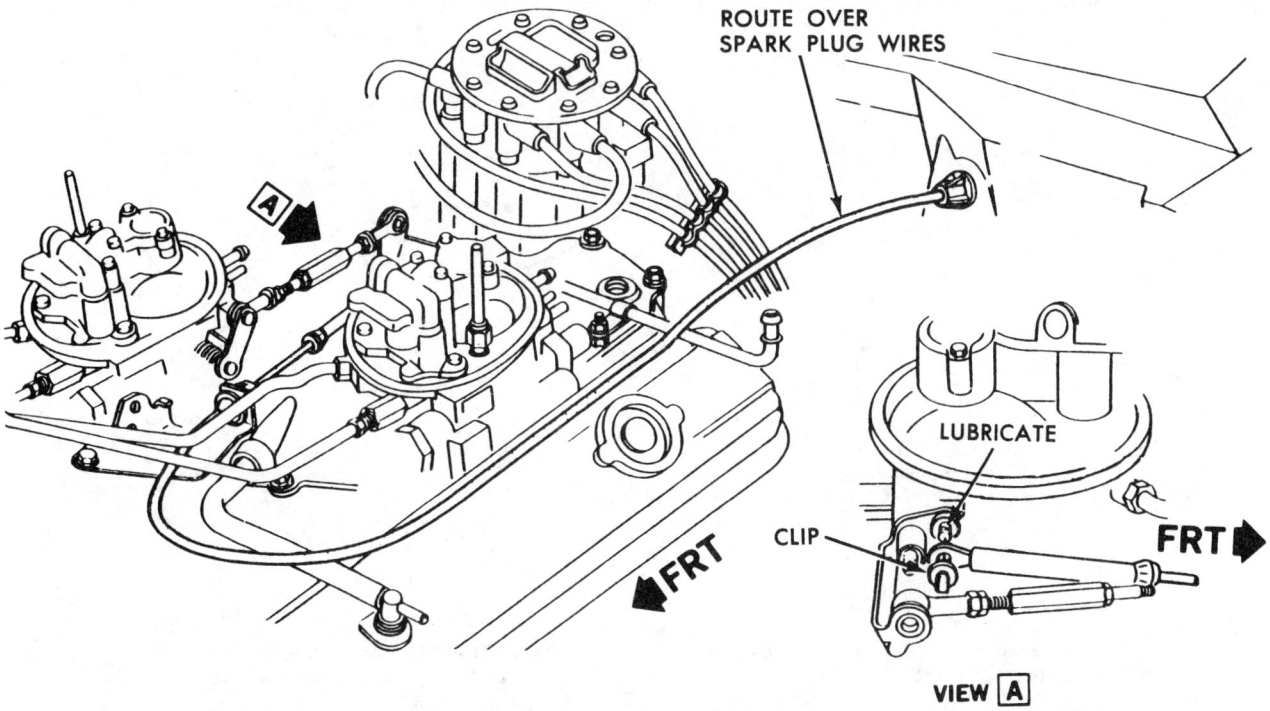

Fig. 1.17 Details of the accelerator control cable assembly on fuel-injected V8 engines (Sec 19)

20 Thermo-controlled Air Cleaner (THERMAC) check

1 All engines are equipped with a thermostatically controlled air cleaner which draws air to the carburetor from different locations, depending upon engine temperature.
2 This is a simple visual check; however, if access is limited, a small mirror may have to be used.
3 Open the hood and locate the baffle inside the air cleaner assembly. It will be located inside the long snorkel of the metal air cleaner housing. Make sure that the flexible air hose(s) are securely attached and undamaged.
4 If there is a flexible air duct attached to the end of the snorkel, leading to an area behind the grille, disconnect it at the snorkel. This will enable you to look through the end of the snorkel and see the baffle inside.
5 The check should be done when the engine and outside air are cold. Start the engine and look through the snorkel at the baffle, which should move to a closed position. With the baffle closed, air cannot enter through the end of the snorkel, but instead enters the air cleaner through the flexible duct attached to the exhaust manifold.
6 As the engine warms up to operating temperature, the baffle should open to allow air through the snorkel end. Depending on ambient temperature, this may take 10 to 15 minutes. To speed up this check you can reconnect the snorkel air duct, drive the vehicle and then check to see if the baffle is completely open.
7 If the thermo-controlled air cleaner is not operating properly, see Chapter 6 for more information.

21 Carburetor/throttle body injection (TBI) mounting torque check

1 The carburetor/TBI is attached to the top of the intake manifold by four nuts or bolts. These fasteners can sometimes work loose from vibration and temperature changes during normal engine operation and cause a vacuum leak.
2 To properly tighten the mounting nuts/bolts, a torque wrench is necessary. If you do not own one, they can usually be rented on a daily basis.
3 Remove the air cleaner assembly, tagging each hose to be disconnected with a piece of numbered tape to make reassembly easier.
4 Locate the mounting nuts/bolts at the base of the carburetor/TBI. Decide what special tools or adapters will be necessary, if any, to tighten the fasteners with a socket and the torque wrench (photo).

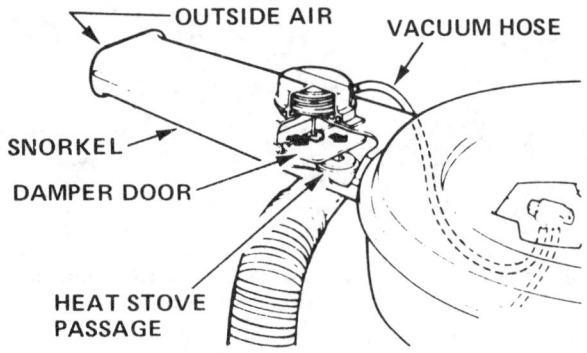

Fig 1.18 THERMAC assembly shown with the snorkel passage open (Sec 20)

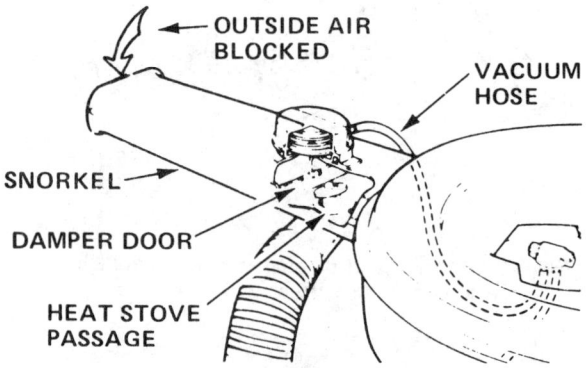

Fig. 1.19 THERMAC assembly shown with the snorkel passage closed (Sec 20)

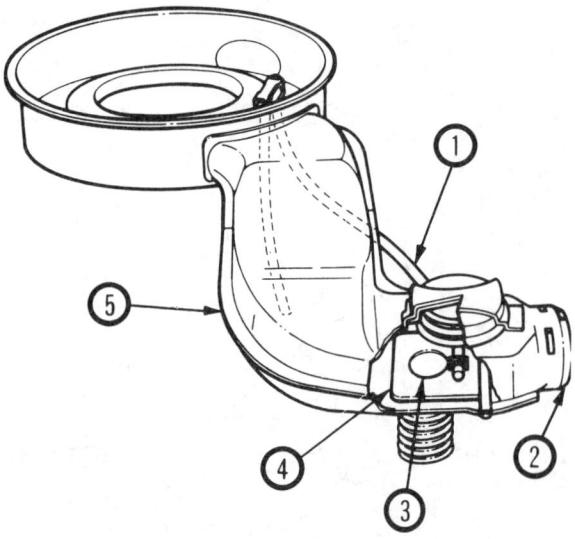

Fig. 1.20 THERMAC assembly (fuel-injected engine) shown with the snorkel passage open (Sec 20)

1 *Vacuum hose*
2 *Outside air passage (open)*
3 *Heat stove passage*
4 *Damper door*
5 *Snorkel*

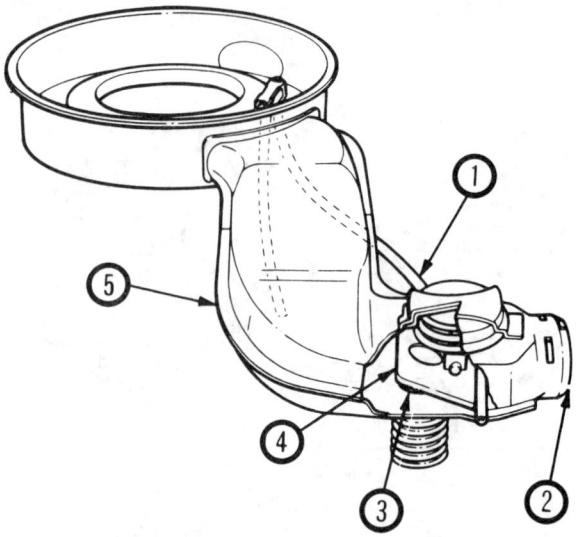

Fig. 1.21 THERMAC assembly (fuel-injected engine) shown with the snorkel passage closed (Sec 20)

1 *Vacuum hose*
2 *Outside air passage (blocked)*
3 *Heat stove passage*
4 *Damper door*
5 *Snorkel*

5 Tighten the nuts/bolts to the specified torque. Do not overtighten them, as the threads could strip.
6 If you suspect that a vacuum leak exists at the bottom of the carburetor/TBI, obtain a length of hose about the diameter of fuel hose. Start the engine and place one end of the hose next to your ear as you probe around the base with the other end. You will hear a hissing sound if a leak exists.
7 If, after the nuts/bolts are properly tightened, a vacuum leak still exists, the carburetor/TBI must be removed and a new gasket installed. See Chapter 4 for more information.
8 After tightening the fasteners, reinstall the air cleaner and return all hoses to their original positions.

22 Early Fuel Evaporation (EFE) system check

1 The EFE system is designed to recirculate exhaust gases to help preheat the engine's induction system, thereby improving cold engine driveability and, by reducing the time that the choke is closed, reducing exhaust emissions levels. There are two basic types of EFE systems. The first is a valve in the exhaust system, between the exhaust manifold and the exhaust pipe, and is called a vacuum servo type; the second type is an electrically heated unit located between the carburetor base and the intake manifold. The procedures which follow in this Section are concerned only with the first type. If you have the electrically heated type, refer to Chapter 6 for more information.

21.4 A torque check of the carburetor (or fuel injection unit) mounting bolts will help eliminate vacuum leaks in the induction system

2 Locate the EFE actuator, which is bolted to a bracket on the right side of the engine. It will have an actuating rod attached to it which will lead down to the valve inside the exhaust pipe. In some cases the entire mechanism, including actuator, will be located at the exhaust pipe.
3 With the engine cold, have an assistant start it. Observe the movement of the actuator rod. It should immediately be drawn into the diaphragm, closing the valve.
4 If the valve does not close, it could be seized. Shut off the engine and apply penetrating oil to the shaft at the pivot points in the valve assembly, allow it to work, then restart the engine and observe the actuator rod. If it still does not close, disconnect the vacuum hose at the actuator.
5 Hold your thumb over the disconnected hose. If vacuum is felt, test the actuator and valve as follows.
6 Attach a piece of hose to the actuator vacuum outlet, apply vacuum to the hose, then quickly place your thumb over it. If the actuator holds vacuum, it is operating correctly and the valve is defective. Refer to Chapter 6 and replace the valve assembly.
7 If the actuator does not hold vacuum, the actuator diaphragm is defective. Refer to Chapter 6 and replace the actuator.
8 If no vacuum was felt at the disconnected vacuum source hose in Step 5, either the hose is crimped or plugged, or the thermal vacuum switch (TVS) or EFE solenoid is not functioning properly.
9 Check the hose for cracks and restrictions and replace it as necessary. Refer to Chapter 6 to test the TVS. Because of its interrelationship with the ECM, the solenoid must be tested by a dealer service department.
10 With the hose again connected to the EFE actuator, continue to warm up the engine until it reaches normal operating temperature.
11 Make sure that the actuating rod has moved the valve to the open position.
12 If the valve does not open, remove the hose from the actuator. If the valve opens, there is no air bleed for the valve actuator diaphragm, the electrical solenoid plunger or TVS is stuck in the cold mode or the engine is not reaching proper operating temperature. Refer to Chapter 6 to test the TVS or replace the actuator diaphragm. If the engine is not reaching proper operating temperature, check the thermostat (refer to Chapter 3). If the valve does not open with the hose removed from the actuator, it may be stuck closed by corrosion.

23 Tire rotation

1 The tires should be rotated at the specified intervals and whenever uneven wear is noticed. Since the vehicle will be raised and the tires

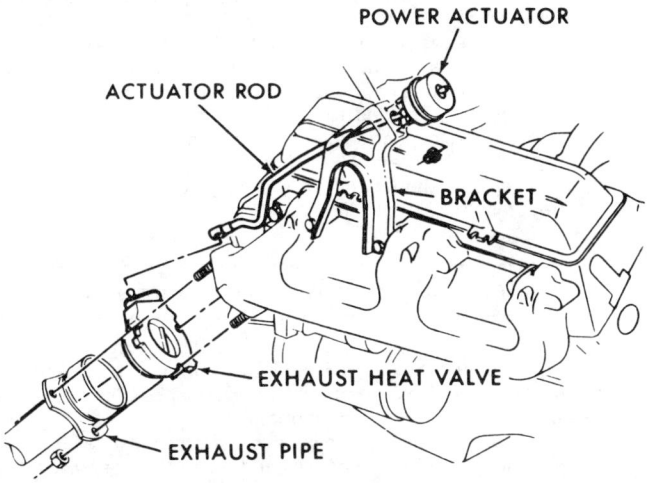

Fig. 1.22 Components of a typical valve-actuated EFE system (Sec 22)

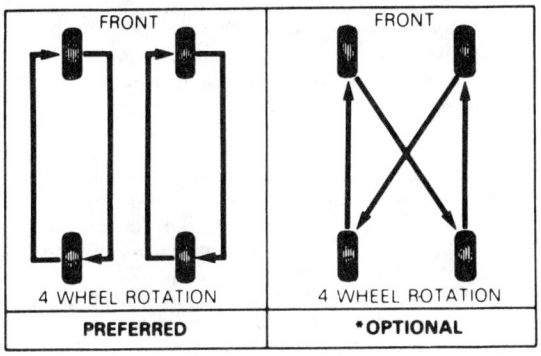

DO NOT INCLUDE "TEMPORARY USE ONLY" SPARE TIRE IN ROTATION

* THE OPTIONAL 'X' ROTATION PATTERN FOR RADIALS IS ACCEPTABLE WHEN REQUIRED FOR MORE UNIFORM TIRE WEAR

Fig. 1.23 Tire rotation diagram (Sec 23)

removed anyway, this is a good time to check the brakes (Section 13) and/or repack the wheel bearings (Section 27). Read over these Sections if this is to be done at the same time.

2 Refer to the accompanying illustration of the 'preferred' and 'optional' tire rotation patterns. Do not include 'Temporary Use Only' spare tires in the rotation sequence. The 'optional' X-rotation procedure is acceptable when required for more uniform tire wear.

3 Refer to the information in *Jacking and towing* at the front of this manual for the proper procedures to follow when raising the vehicle and changing a tire; however, if the brakes are to be checked, do not apply the parking brake as stated. Make sure the tires are blocked to prevent the vehicle from rolling.

4 Preferably, the entire vehicle should be raised at the same time. This can be done on a hoist or by jacking up each corner and then lowering the vehicle onto jackstands placed under the frame rails. Always use four jackstands and make sure the vehicle is firmly supported.

5 After rotation, check and adjust the the tire pressures as necessary and be sure to check the lug nut tightness.

24 Clutch pedal free play check and adjustment

Note: *The information which follows applies to 1982 and 1983 vehicles only (non-hydraulic assisted clutch assemblies). A hydraulic clutch operating mechanism is used on all 1984 vehicles. This system locates the clutch pedal and provides automatic clutch adjustment, so no adjustment of the clutch linkage or pedal position is required.*

1 If your vehicle is equipped with a manual transmission, it is important to have the clutch free play properly adjusted. Basically, free play is the distance the clutch pedal moves before all play in the linkage is removed and the clutch begins to disengage. It is measured at the pedal pad. Slowly depress the pedal and determine how far it moves before resistance is felt.

2 There is only one linkage adjustment to compensate for all normal clutch wear.

3 To check for correct adjustment, apply the parking brake, block the front wheels and start the engine. Hold the clutch pedal approximately 1/2-inch from the floor and move the shift lever between First and Reverse gears several times. If the shift is not smooth, clutch adjustment is necessary.

4 Raise the vehicle to permit access to the clutch linkage underneath it. Place the vehicle securely on jackstands.

5 Disconnect the return spring at the clutch fork.

6 Rotate the clutch lever and shaft assembly until the clutch pedal is firmly against the rubber bumper on the dashboard brace. If in doubt about this, have an assistant check it from inside the vehicle.

7 Push the outer end of the clutch fork to the rear until the release bearing lightly contacts the pressure plate fingers.

8 Keep the fork in this position while the pushrod is removed from the operating hole in the lever assembly. Now place the rod in the gauge hole of the lever assembly (refer to the accompanying illustration for application to your particular vehicle model).

9 With the pushrod in the gauge hole, loosen the locking nut and turn the rod to increase its length. Increase the length of the rod until all lash is removed from the system.

10 Remove the rod from the gauge hole and return it to the lower operating hole in the shaft. Install the retainer and then carefully tighten the nut, being sure the length of the rod is not changed.

11 Connect the return spring and check the pedal free play as described previously. Lower the vehicle, then check for correct operation and free play. **Note:** *A clutch that has been slipping prior to free play adjustment may still slip right after the adjustment has been made due to previous heat damage. If so, allow the clutch to cool for at least 12 hours, then proceed to check the clutch as follows:*

12 While driving the vehicle in High gear at 20 to 25 mph, depress the clutch pedal to the floor, then engage the clutch quickly by snapping your foot off the pedal and press the accelerator to the full open throttle position.

13 Engine speed should drop noticeably, then increase as vehicle speed increases. If the clutch is not working properly, the engine speed will increase. **Caution:** *Do not repeat Step 12 more than once or the clutch may overheat.*

14 If the need for further clutch service is indicated, refer to Chapter 8.

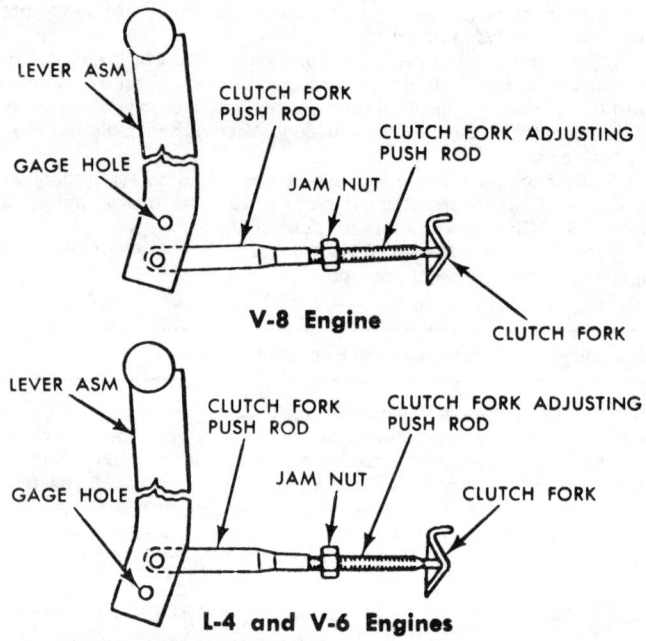

Fig. 1.24 Components involved in adjustment of clutch pedal free play (Sec 24)

25 Rear axle oil change

1 To change the oil in the rear axle it is necessary to remove the cover plate on the differential housing. Because of this, purchase a new gasket at the same time the gear lubricant is bought.

2 Move a drain pan, rags, newspapers and wrenches under the rear of the vehicle. With the drain pan under the differential cover, loosen each of the inspection plate bolts.

3 Remove the bolts on the lower half of the plate, but use the upper bolts to keep the cover loosely attached to the differential. Allow the oil to drain into the pan, then completely remove the cover.

4 Using a lint-free rag, clean the inside of the cover and the accessible areas of the differential housing. As this is done, check for chipped gears and metal particles in the oil, indicating that the differential should be thoroughly inspected and repaired (see Chapter 8 for more information).

5 Thoroughly clean the gasket mating surface on the cover and the differential housing. Use a gasket scraper or putty knife to remove all traces of the old gasket.

6 Apply a thin film of gasket sealant to the cover flange and then press a new gasket into position on the cover. Make sure the bolt holes align properly.

7 Place the cover on the differential housing and install the bolts. Tighten the bolts a little at a time, working across the cover in a diagonal fashion until all bolts are tight. If a torque wrench is available, tighten the bolts to the specified torque.

8 Remove the inspection plug on the side of the differential housing (or inspection cover) and fill the housing with the proper lubricant until the level is at the bottom of the plug hole.

9 Securely install the plug.

26 Cooling system servicing (draining, flushing and refilling)

1 Periodically, the cooling system should be drained, flushed and refilled to replenish the antifreeze mixture and prevent formation of rust and corrosion, which can impair the performance of the cooling system and ultimately cause engine damage.

2 At the same time the cooling system is serviced, all hoses and the radiator cap should be inspected and replaced if defective (see Section 8).

3 Since antifreeze is a corrosive and poisonous solution, be careful not to spill any of the coolant mixture on the vehicle's paint or your skin. If this happens, rinse immediately with plenty of clean water. Also,

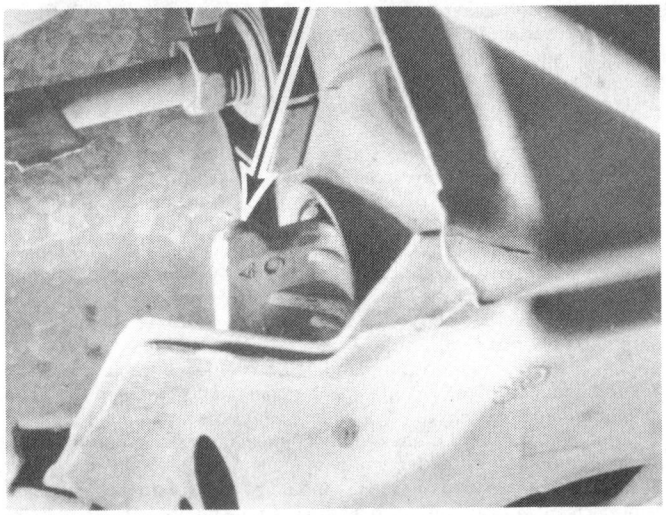

26.6 This type of radiator drain plug (arrow) is easy to get a grip on, saving a lot of 'grated' knuckles

27.6 Removing the wheel bearing dust cap

27.7A Straightening the ends of the cotter pin

27.7B Removing the cotter pin

consult your local authorities about the dumping of antifreeze before draining the cooling system. In many areas, reclamation centers have been set up to collect automobile oil and drained antifreeze/water mixtures rather than allowing them to be added to the sewage system.

4 With the engine cold, remove the radiator cap.

5 Move a large container under the radiator to catch the coolant as it is drained.

6 Drain the radiator. Most models are equipped with a drain plug at the bottom (photo). If this drain has excessive corrosion and cannot be turned easily, or if the radiator is not equipped with a drain, disconnect the lower radiator hose to allow the coolant to drain. Be careful that none of the solution is splashed on your skin or into your eyes.

7 If accessible, remove the two engine drain plugs. There is one plug on each side of the engine about halfway back, on the lower edge near the oil pan rail. These will allow the coolant to drain from the engine itself.

8 Disconnect the hose from the coolant reservoir and remove the reservoir. Flush it out with clean water.

9 Place a garden hose in the radiator filler neck at the top of the radiator and flush the system until the water runs clear at all drain points.

10 In severe cases of contamination or clogging of the radiator, remove it (see Chapter 3) and reverse flush it. This involves simply inserting the hose in the bottom radiator outlet to allow the clear water to run against the normal flow, draining through the top. A radiator repair shop should be consulted if further cleaning or repair is necessary.

11 When the coolant is regularly drained and the system refilled with the correct antifreeze/water mixture, there should be no need to use chemical cleaners or descalers.

12 To refill the system, reconnect the radiator hoses and install the drain plugs securely in the engine. Special thread-sealing tape (available at auto parts stores) should be used on the drain plugs. Install the reservoir and the overflow hose where applicable.

13 Fill the radiator to the base of the filler neck and then add more coolant to the reservoir until it reaches the mark.

14 Run the engine until normal operating temperature is reached and, with the engine idling, add coolant up to the Full Hot level. Install the radiator cap so that the arrows are in alignment with the overflow hose. Install the reservoir cap.

15 Always refill the system with a mixture of high quality antifreeze and water in the proportion called for on the antifreeze container or in your owner's manual. Chapter 3 also contains information on antifreeze mixtures.

16 Keep a close watch on the coolant level and the various cooling system hoses during the first few miles of driving. Tighten the hose clamps and/or add more coolant as necessary.

27 Wheel bearing check and repack

1 In most cases, the front wheel bearings will not need servicing until the brake pads are changed. However, these bearings should be checked whenever the front wheels are raised for any reason.

2 With the vehicle securely supported on jackstands, spin the wheel

and check for noise, rolling resistance and free play. Now grab the top of the tire with one hand and the bottom of the tire with the other. Move the tire in-and-out on the spindle. If it moves more than 0.005-inch, the bearings should be checked and then repacked with grease or replaced if necessary.

3 To remove the bearings for replacement or repacking, begin by removing the hub cap and wheel.

4 Using an Allen wrench of the proper size, remove the two bolts which secure the disc brake caliper to the support (see Chapter 9).

5 Fabricate a wood block (1-1/16-inch by 1-1/16-inch by 2 inches in length) which will be slid between the brake pads to keep them separated. Carefully slide the caliper off the disc and insert the wood block between the pads. Use wire to hang the caliper assembly out of the way. Be careful not to kink or damage the brake hose.

6 Pry the dust cap out of the hub using a screwdriver or hammer and chisel (photo). The cap is located at the center of the hub.

7 Use needle-nose pliers or a screwdriver to straighten the bent ends of the cotter pin and then pull the cotter pin out of the locking nut (photos). Discard the cotter pin and use a new one during reassembly.

8 Remove the spindle nut and washer from the end of the spindle (photos).

9 Pull the hub assembly out slightly and then push it back into its original position. This should force the outer bearing off the spindle enough so that it can be removed with your fingers (photo). remove the outer bearing, noting how it is installed on the end of the spindle.

10 Now the hub assembly can be pulled off the spindle (photo).

11 On the rear side of the hub, use a screwdriver to pry out the inner bearing lip seal (photo). As this is done, note the direction in which the seal is installed.

12 The inner bearing can now be removed from the hub, again noting how it is installed.

13 Use solvent to remove all traces of the old grease from the bearings, hub and spindle. A small brush may prove useful; however, make sure no bristles from the brush embed themselves inside the bearing rollers. Allow the parts to air dry.

14 Carefully inspect the bearings for cracks, heat discoloration, bent rollers, etc. Check the bearing races inside the hub for cracks, scoring and uneven surfaces. If the bearing races are defective, the hubs should be taken to a machine shop with the facilities to remove the old races and press new ones in.

15 Use an approved high-temperature front wheel bearing grease to pack the bearings. Work the grease completely into the bearings, forcing it between the rollers, cone and cage (photo).

16 Apply a thin coat of grease to the spindle at the outer bearing seat, inner bearing seat, shoulder and seal seat (photo).

17 Put a small quantity of grease inboard of each bearing race inside the hub. Using your finger, form a dam at these points to provide extra grease availability and to keep thinned grease from flowing out of the bearing.

18 Place the grease-packed inner bearing into the rear of the hub and

27.8A Removing the spindle nut

27.8B Removing the spindle washer

27.9 Removing the outer wheel bearing

27.10 Removing the hub assembly from the spindle

put a little more grease outboard of the bearing.
19 Place a new seal over the inner bearing and tap the seal with a hammer until it is flush with the hub (photo).
20 Carefully place the hub assembly onto the spindle and push the grease-packed outer bearing into position (photo).
21 Install the washer and spindle nut. Tighten the nut only slightly (12 ft-lbs of torque) (photos).
22 Spin the hub in a forward direction to seat the bearings and remove any grease or burrs which could cause excessive bearing play later.
23 Put a little grease outboard of the outer bearing to provide extra grease availability.
24 Now check to see that the tightness of the spindle nut is still 12 ft-lbs.
25 Loosen the spindle nut until it is just loose, no more.
26 Using your hand (not a wrench of any kind), tighten the nut until it is snug. Install a new cotter pin through the hole in the spindle and spindle nut (photo). If the nut slits do not line up, loosen the nut slightly until they do. From the hand-tight position, the nut should not be loosened more than one-half flat to install the cotter pin.
27 Bend the ends of the new cotter pin until they are flat against the nut (photo). Cut off any extra length which could interfere with the dust cap.
28 Install the dust cap, tapping it into place with a rubber mallet.
29 Place the brake caliper near the rotor and carefully remove the wood spacer. Slide the caliper over the rotor. Tighten the caliper mount-

27.11 Removing the inner bearing lip seal

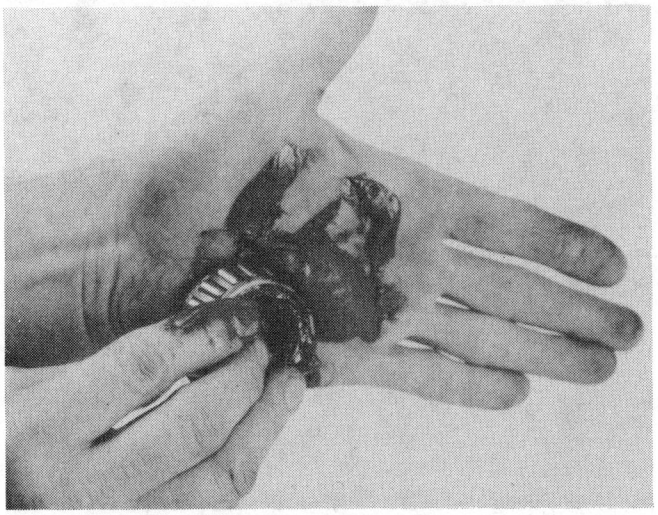

27.15 Packing a bearing with special high-temperature wheel bearing grease (work it well into the rollers)

27.16 Applying a thin coat of grease to the spindle surfaces

27.19 Installing a new seal over the inner bearing using a soft-faced hammer (tap around the entire circumference)

27.20 Installing the outer bearing

27.21A Installing the spindle washer (make sure the tab lines up with the groove)

27.21B Tighten the spindle nut only slightly (no more than 12 ft-lbs)

27.26 Installing the cotter pin after lining up the hole in the spindle with a groove in the spindle nut

27.27 Bending the ends of the cotter pin to secure it

28.7 The first step in automatic transmission oil pan removal is the removal of the keeper at the rear of the selector cable assembly

28.8 Removing the transmission selector cable-to-transmission selector lever clip

ing bolts to 35 ft-lbs. Chapter 9 contains details on the disc brake caliper assembly.
30 Install the tire/wheel assembly on the hub and tighten the lug nuts.
31 Grab the top and bottom of the tire and check the bearings in the same manner as described at the beginning of this Section.
32 Lower the vehicle to the ground and tighten the lug nuts completely. Install the hub cap, using a rubber mallet to seat it.

28 Automatic transmission fluid change

1 At the specified time intervals, the transmission fluid should be changed and the filter replaced with a new one. Since there is no drain plug, the transmission oil pan must be removed from the bottom of the transmission to drain the fluid.
2 Before draining, purchase the specified transmission fluid (see *Recommended lubricants and fluids*) and a new filter. The necessary gaskets should be included with the filter; if not, purchase an oil pan gasket and a strainer-to-valve body gasket.
3 Other tools necessary for this job include jackstands to support the vehicle in a raised position, a wrench to remove the oil pan bolts, a standard screwdriver, a drain pan capable of holding at least eight pints, newspapers and clean rags.
4 The fluid should be drained immediately after the vehicle has been driven. This will remove any built-up sediment better than if the fluid were cold. Because of this, it may be wise to wear protective gloves (fluid temperature can exceed 350°F in a hot transmission).
5 After it has been driven to warm up the fluid, raise the vehicle and place it on jackstands for access underneath. Make sure it is firmly supported by the four stands placed under the frame rails.
6 Move the necessary equipment under the vehicle, being careful not to touch any of the hot exhaust components.
7 Remove the keeper at the rear of the transmission selector cable assembly (photo).
8 Remove the clip that retains the transmission selector cable to the transmission selector lever (photo).
9 Remove the cable from the selector lever.
10 Remove the cable housing from the transmission selector cable bracket.
11 Remove the selector cable bracket. To avoid confusion when reinstalling the transmission pan, note that the bolts retaining the cable bracket are a different size than those retaining the pan.
12 Place the drain pan under the transmission oil pan and remove the oil pan bolts along the rear and sides of the pan. Loosen, but do not remove, the bolts at the front of the pan.
13 Carefully pry the pan down at the rear, allowing the hot fluid to drain into the container. If necessary, use a screwdriver to break the gasket seal at the rear of the pan; however, do not damage the pan or transmission in the process.
14 Support the pan and remove the remaining bolts at the front. Lower the pan and drain the remaining fluid into the container (photo). As

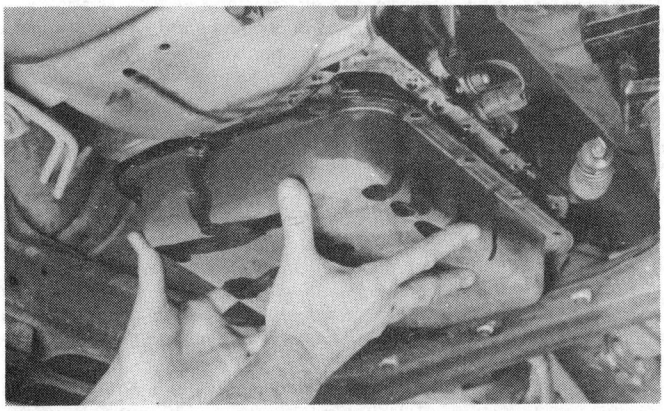

28.14 Easy does it! Carefully support the pan as it is removed and you won't get an unwanted oil bath, as there is still some fluid in the pan

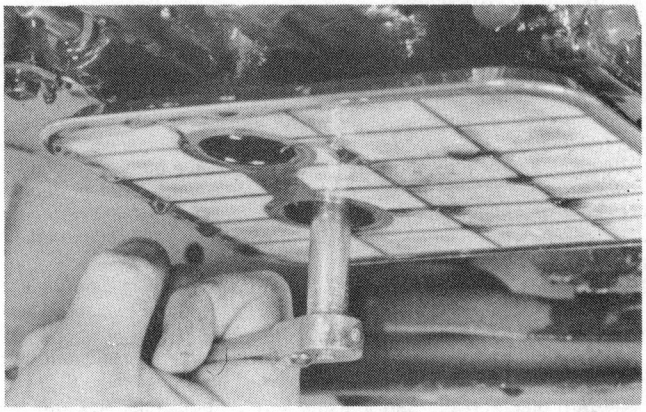

28.16 The filter/strainer is held in place with two bolts located in the recesses

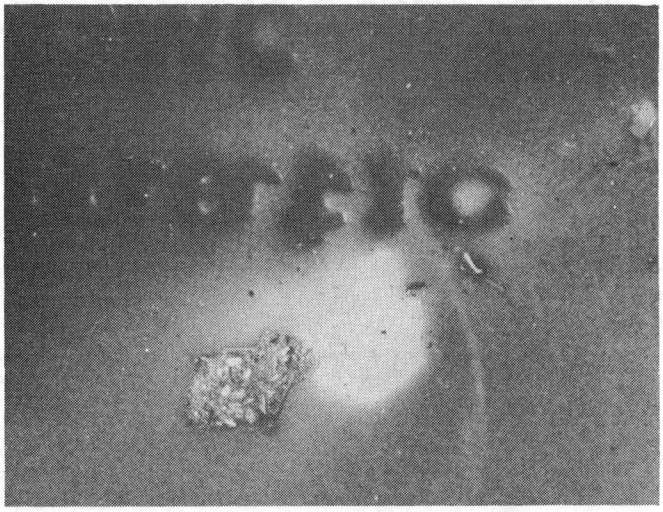

28.17 Finding some metal particles in the bottom of the pan is normal (if excessive metal is found, you should consult an expert concerning its source and the remedy)

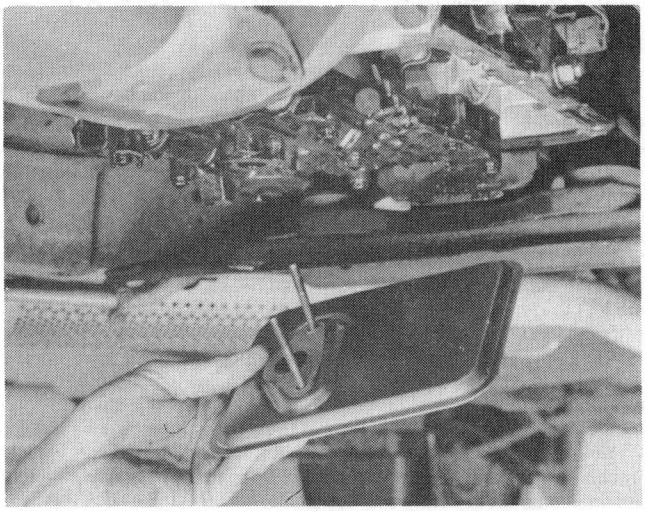

28.20 Be sure to install a new gasket (shown mounted over the bolts) when attaching the filter to the valve body

29.2 Removing the wing nut atop the air cleaner assembly

29.4 Removing the air filter element

this is done, check the fluid for metal particles, which may be an indication of internal failure.

15 Now visible on the bottom of the transmission is the filter/strainer held in place by two screws.

16 Remove the two screws, the filter and the gasket (photo).

17 Thoroughly clean the transmission oil pan with solvent. Inspect it for metal particles and foreign matter (photo). Dry it with compressed air if available. It is important that all remaining gasket material be removed from the oil pan mounting flange. Use a gasket scraper or putty knife for this.

18 Clean the filter mounting surface on the valve body. Again, this surface should be smooth and free of any leftover gasket material.

19 Clean the screen assembly with solvent, then dry it thoroughly (use compressed air, if available). Paper or felt-type filters should be replaced with new ones.

20 Place the cleaned or new filter into position, with a new gasket between it and the transmission valve body. Install the two mounting screws and tighten them securely (photo).

21 Apply a light bead of gasket sealant around the oil pan mounting surface, with the sealant to the inside of the bolt holes. Press the new gasket into place on the pan, making sure all bolt holes line up.

22 Lift the pan up to the bottom of the transmission and install the mounting bolts. Tighten the bolts in a diagonal fashion, working around the pan. Using a torque wrench, tighten the bolts to the specified torque.

23 Lower the vehicle off the jackstands.

24 Open the hood and remove the transmission fluid dipstick from the guide tube.

25 Since fluid capacities vary between the various transmission types, it is best to add a little fluid at a time, continually checking the level with the dipstick. Allow the fluid time to drain into the pan. Add fluid until the level just registers on the end of the dipstick. In most cases, a good starting point will be four to five pints added to the transmission through the filler tube (use a funnel to prevent spills).

26 With the selector lever in Park, apply the parking brake and start the engine without depressing the accelerator pedal (if possible). Do not race the engine at a high speed; run it at slow idle only.

27 Depress the brake pedal and shift the transmission through each gear. Place the selector back into Park and check the level on the dipstick (with the engine still idling). Look under the vehicle for leaks around the transmission oil pan mating surface.

28 Add more fluid through the dipstick tube until the level on the dipstick is 1/4-inch below the Add mark on the dipstick. Do not allow the fluid level to go above this point, as the transmission would then be overfull, necessitating the removal of the pan to drain the excess fluid.

29 Push the dipstick firmly back into the tube and drive the vehicle to reach normal operating temperature (15 miles of highway driving or its equivalent in the city). Park on a level surface and check the fluid level on the dipstick with the engine idling and the transmission in Park. The level should now be at the Full mark on the dipstick. If not, add more fluid as necessary to bring the level up to this point. Again, do not overfill.

29 Air filter and PCV filter replacement

1 At the specified intervals, the air filter and PCV filter should be replaced with new ones. A thorough program of preventative maintenance would call for the two filters to be inspected between changes.

2 The air filter is located inside the air cleaner housing on the top of the engine. The filter is generally replaced by removing the wing nut at the top of the air cleaner assembly and lifting off the top plate (photo). If vacuum hoses are connected to the plate, note their positions and disconnect them.

3 While the top plate is off, be careful not to drop anything down into the carburetor.

4 Lift the air filter element out of the housing (photo).

5 To check the filter, hold it up to strong sunlight or place a flashlight or droplight on the inside of the filter. If you can see light coming through the paper element, the filter is all right. Check all the way around the filter.

6 Wipe out the inside of the air cleaner housing with a clean rag.

7 Place the old filter (if in good condition) or the new filter (if the specified interval has elapsed) back into the air cleaner housing. Make sure it seats properly in the bottom of the housing.

8 Connect any disconnected vacuum hoses to the top plate and reinstall the plate.

9 The PCV filter is also located inside the air cleaner housing. Remove the top plate and air filter as described previously, then locate the PCV filter on the side of the housing.

10 Remove the retaining clip from the outside of the housing, then remove the PCV filter (photos).

11 Install a new PCV filter, then reinstall the retaining clip, air filter, top plate and any hoses that were disconnected.

30 Oxygen sensor replacement

Note: *Special care must be taken when handling the sensitive oxygen sensor:*

a) The oxygen sensor has a permanently attached pigtail and connector, which should not be removed from the sensor. Damage or removal of the pigtail or connector can adversely affect its operation.
b) Grease, dirt and other contaminants should be kept away from the electrical connector and the louvered end of the sensor.
c) Do not use cleaning solvents of any kind on the oxygen sensor.
d) Do not drop or roughly handle the sensor.
e) The silicone boot must be installed in the correct position to prevent the boot from being melted and to allow the sensor to operate properly.

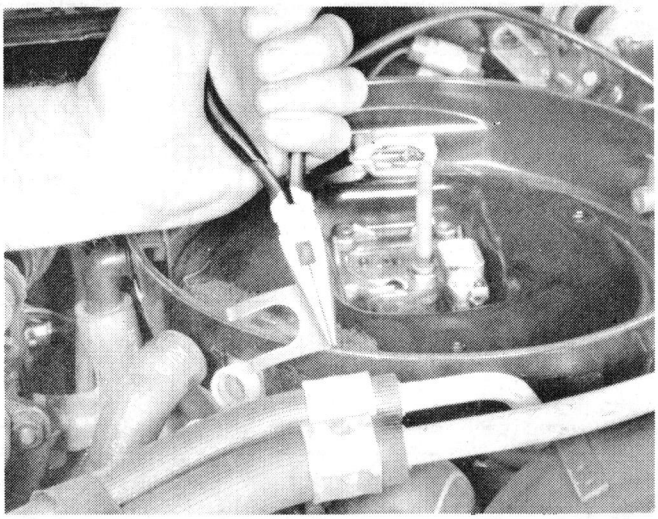

29.10A Removing the PCV filter retaining clip

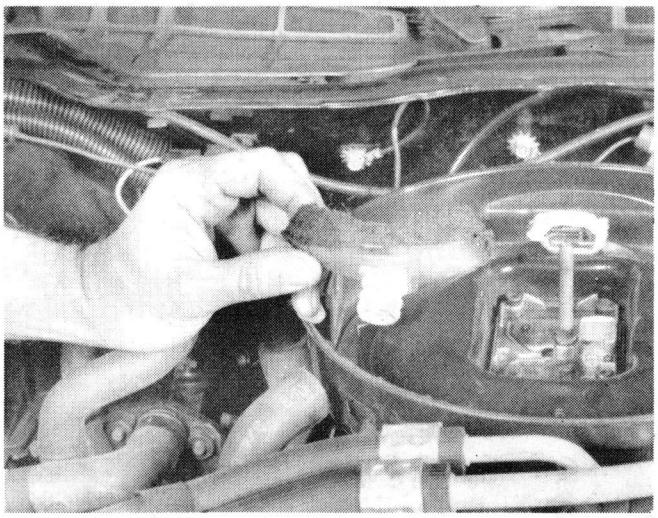

29.10B Removing the PCV filter

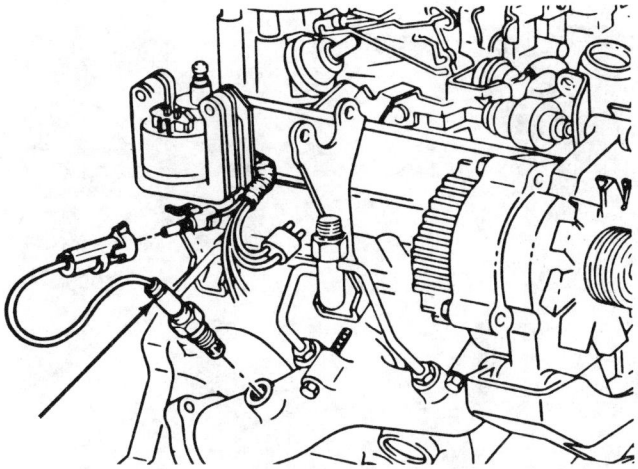

Fig. 1.25 Arrow indicates the oxygen sensor assembly (V6 engine shown, others similar) (Sec 30)

1 The sensor is located in the exhaust manifold or exhaust pipe and is accessible from under the hood.
2 Since the oxygen sensor may be difficult to remove with the engine cold, begin by operating the engine until it has warmed to at least 120°F (48°C).
3 Disconnect the electrical connector from the oxygen sensor.
4 Note the position of the silicone boot and carefully back out the oxygen sensor from the exhaust manifold. Be advised that excessive force may damage the threads.
5 A special anti-seize compound must be used on the threads of the oxygen sensor to aid in future removal. New or service sensors will have this compound already applied, but if for any reason an oxygen sensor is removed and then reinstalled, the threads must be coated before reinstallation.
6 Install the sensor and tighten it to 30 ft-lbs.
7 Connect the electrical connector.

31 Positive Crankcase Ventilation (PCV) valve replacement

1 The PCV valve is located in the rocker arm cover or intake manifold. A hose connected to the valve runs to either the carburetor or intake manifold.
2 When purchasing a replacement PCV valve, make sure it is for your particular vehicle, model year and engine size.
3 Pull the valve (with hose attached) from the rubber grommet in the rocker arm cover or manifold.
4 Loosen the retaining clamp and pull the PCV valve from the end of the hose, noting its installed position and direction.
5 Compare the old valve with the new one to make sure they are the same.
6 Push the new valve into the end of the hose until it is seated and reinstall the clamp.
7 Inspect the rubber grommet for damage and replace it with a new one, if faulty.
8 Push the PCV valve and hose securely into position.
9 More information on the PCV system can be found in Chapter 6.

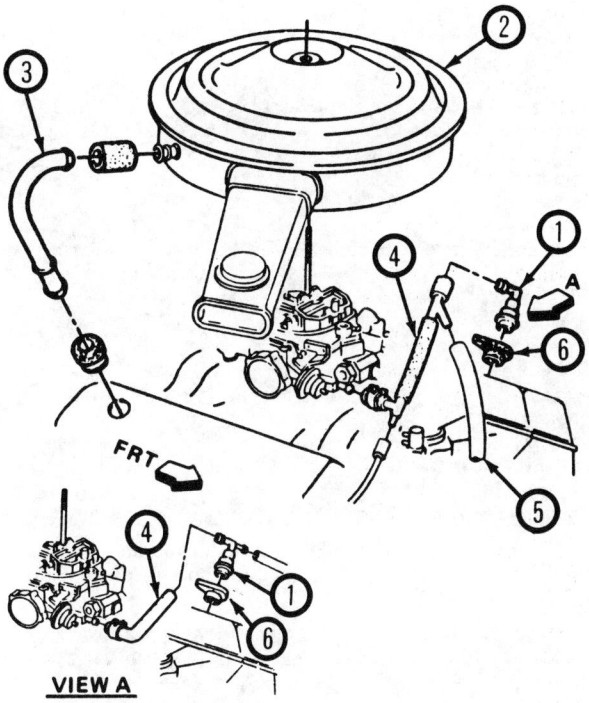

Fig. 1.26 Details of a typical PCV system (Sec 31)

1	*PCV valve*	*4*	*PCV valve hose*
2	*Air cleaner*	*5*	*Hose to canister*
3	*Crankcase vent tube*	*6*	*Grommet*

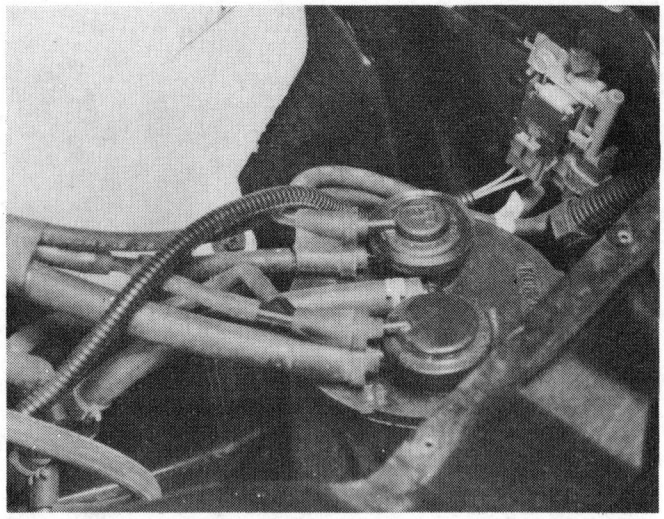

33.3 The EECS canister is located in the front of the engine compartment, just behind the headlight assembly

32 Exhaust Gas Recirculation (EGR) valve check

1 On GM vehicles, the EGR valve is located on the intake manfold, adjacent to the carburetor or TBI unit. Most of the time when a problem develops in this emissions system, it is due to a stuck or corroded EGR valve.
2 With the engine cold to prevent burns, reach under the EGR valve and manually push on the diaphragm. Using moderate pressure, you should be able to press the diaphragm up and down within the housing.
3 If the diaphragm does not move or moves only with much effort, replace the EGR valve with a new one. If in doubt about the quality of the valve, compare the free movement of your EGR valve with a new valve.
4 Refer to Chapter 6 for more information on the EGR system.

33 Evaporative Emissions Control System (EECS) filter replacement

1 The function of the Evaporative Emissions Control System is to draw fuel vapors from the tank and carburetor, store them in a charcoal canister and then burn them during normal engine operation.
2 The filter at the bottom of the charcoal canister should be replaced at the specified intervals. If, however, a fuel odor is detected,

33.4 If the hoses on the EECS canister top are very tight (and they probably are!), this is a good method of 'persuading' them off without damage

33.6 With the canister retaining bolts removed and the canister inverted, the filter is easily removed

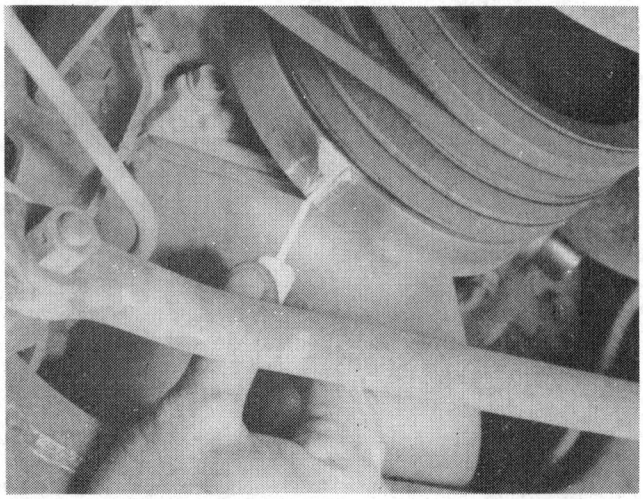

34.3 Marking the crankshaft balancer notch with white paint

34.5 Aiming the timing light at the timing mark

the canister, filter and system hoses should immediately be inspected.
3 To replace the filter, locate the canister at the front of the engine compartment. It will have between three and six hoses running out of the top (photo).
4 Mark the hoses with tape to simplify reinstallation, then disconnect them from the canister (photo).
5 Remove the two bolts which secure the bottom of the canister to the body.
6 Turn the canister upside-down and pull the old filter from the bottom of the canister (photo).
7 Push the new filter into the bottom of the canister, making sure it is seated all the way around.
8 Place the canister back into position and tighten the two mounting bolts. Connect the various hoses if disconnected.
9 The EECS is explained in more detail in Chapter 6.

34 Ignition timing check and adjustment

Note: *It is imperative that the procedures included on the Vehicle Emissions Control Information label be followed when adjusting the ignition timing. The label will include all information concerning preliminary steps to be performed before adjusting the timing, as well as the timing specifications.*

On fuel-injected V8 engines (TBI 400), be sure to put the Electronic Spark Timing (EST) in bypass mode by disconnecting the single wire connector. It is a tan wire with a black stripe that comes out of the wiring harness conduit near the rear of the right-hand rocker arm cover. Do not disconnect the four-wire connector to the distributor.

1 Locate the VECI label under the hood and read through and perform all preliminary instructions concerning ignition timing.
2 Locate the timing mark pointer plate located beside the crankshaft pulley. The O mark represents top dead center (TCD). The pointer plate will be marked in either one or two-degree increments and should have the proper timing mark for your particular engine noted. If not, count back from the O mark the correct number of degrees BTDC, as noted on the VECI label, and mark the plate.
3 Locate the notch on the crankshaft balancer or pulley and mark it with chalk or a dab or paint so it will be visible under the timing light (photo).
4 With the ignition off, connect the pick-up lead of the timing light to the number one spark plug. Use either a jumper lead between the wire and plug or an inductive-type pick-up. *Do not pierce the wire or attempt to insert a wire between the boot and the wire.* Connect the

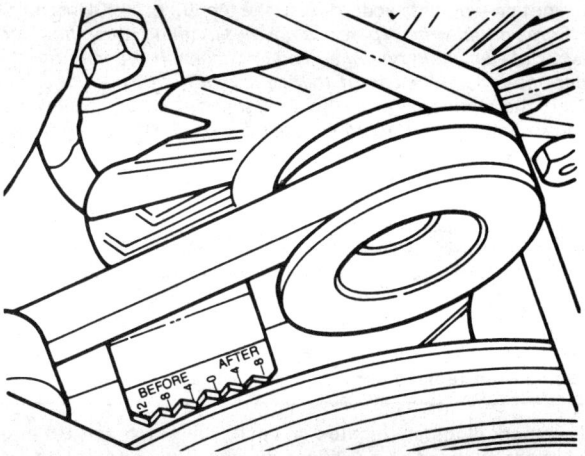

Fig. 1.27 Location of a typical timing mark plate (Sec 34)

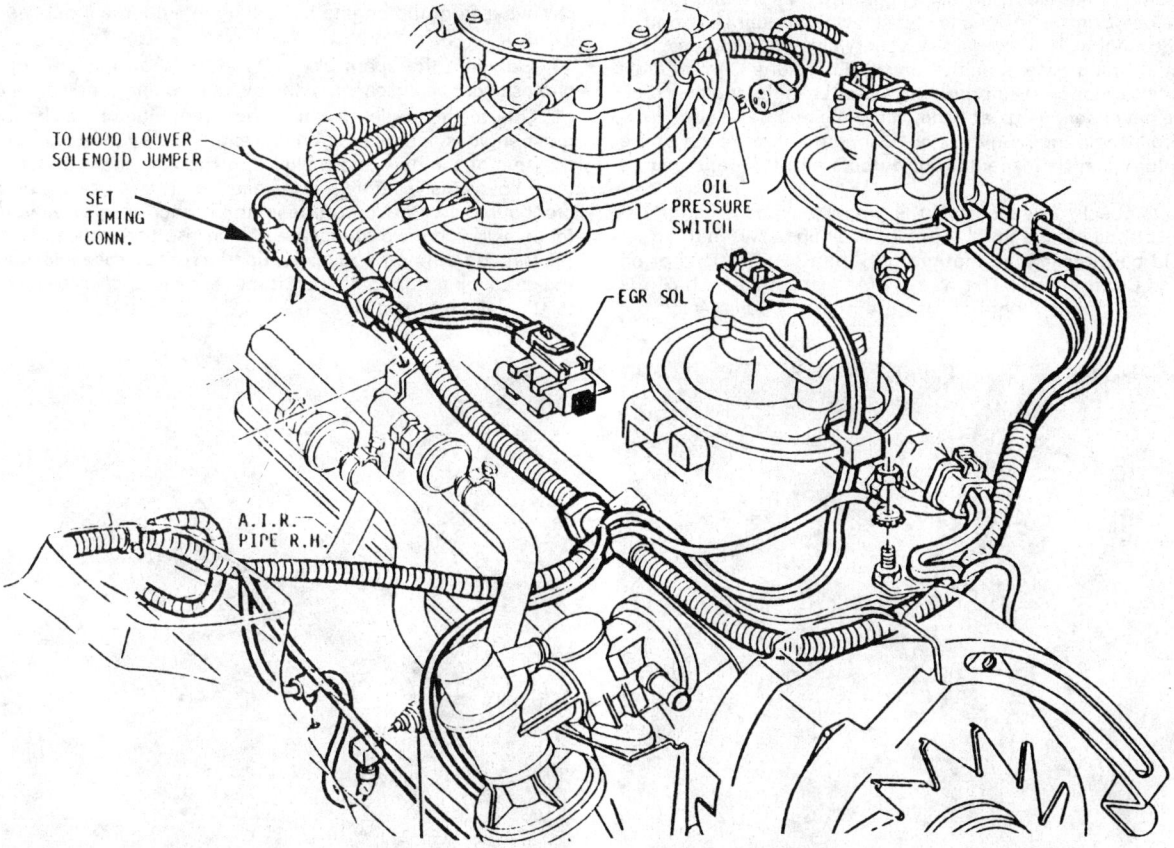

Fig. 1.28 Arrow indicates the wire to be disconnected to put the EST in bypass mode on fuel-injected V8 (TBI 400) engines (Sec 34)

timing light power leads according to the manufacturer's instructions.
Note: *Some engines incorporate a magnetic timing probe hole for use with special electronic timing equipment. Consult the manufacturer's instructions for proper use of this equipment.*
5 Start the engine, aim the timing light at the timing mark by the crankshaft pulley and note which timing mark the notch on the pulley is lining up with (photo).
6 If the notch is not lining up with the correct mark, loosen the distributor hold-down bolt and rotate the distributor until the notch is lined up with the correct timing mark.
7 Retighten the hold-down bolt and recheck the timing.
8 Turn off the engine and disconnect the timing light. Reconnect the number one spark plug wire, if removed.

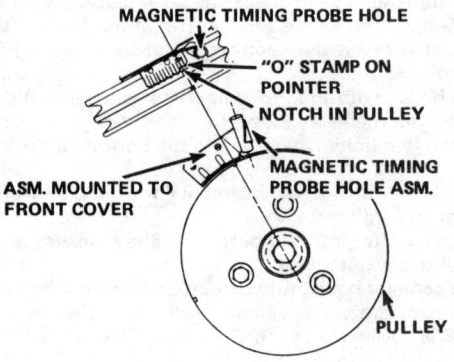

Fig. 1.29 Location of the magnetic timing probe hole incorporated on some engines (Sec 34)

35 Spark plug replacement

1 The spark plugs are located on each side of V6 and V8 engines and on the left side of four-cylinder engines. They may or may not be easily accessible for removal. If the vehicle is equipped with air-conditioning or power steering, some of the plugs may be tricky to remove. Special extension or swivel tools may be necessary. Make a survey under the hood to determine if special tools will be needed.
2 In most cases, the tools necessary for a spark plug replacement job include a plug wrench or spark plug socket which fits onto a ratchet wrench (this special socket will be padded inside to protect the plug) and a feeler gauge to check and adjust the spark plug gaps. Also, a special spark plug wire removal tool is available for separating the wires from the spark plugs. To ease installation, obtain a piece of 3/16-inch inside diameter rubber hose, 8 to 12 inches in length.
3 The best procedure to follow when replacing the spark plugs is to purchase the new spark plugs beforehand, adjust them to the proper gap and then replace each plug one at a time. When buying the new spark plugs, it is important to obtain the correct plugs for your specific engine. This information can be found in the Specifications at the front of this Chapter, but should be checked against the information found on the Emissions Control Information label located under the hood or in the owner's manual. If differences exist between these sources, purchase the spark plug type specified on the Emissions Control label because the information was printed for your specific engine.
4 With the new spark plugs at hand, allow the engine to cool completely before attempting plug removal. During this time, each of the new spark plugs can be inspected for defects and the gaps can be checked.
5 The gap is checked by inserting the proper thickness gauge between the electrodes at the tip of the plug. The gap between the electrodes should be the same as that given in the Specifications or on the Emissions Control label. The wire should just touch each of the electrodes. If the gap is incorrect, use the notched adjuster on the feeler gauge body to bend the curved side electrode slightly until the proper gap is achieved. If the side electrode is not exactly over the center electrode, use the notched adjuster to align the two. Also at this time check for cracks in the porcelain insulator, indicating the spark plug should not be used.
6 Cover the fenders of the vehicle to prevent damage to the paint.
7 With the engine cool, remove the spark plug wire from one spark plug. Do this by grabbing the boot at the end of the wire, not the wire itself. Sometimes it is necessary to use a twisting motion while the boot and plug wire are pulled free. Using a plug wire removal tool is the easiest and safest method.
8 If compressed air is available, use it to blow any dirt or foreign material away from the spark plug area. A common bicycle pump will also work. The idea here is to eliminate the possibility of material falling into the cylinder as the spark plug is removed.
9 Now place the spark plug wrench or socket over the plug and remove it from the engine by turning in a counterclockwise direction (photo).
10 Compare the spark plug with those shown in the accompanying photos to get an indication of the overall running condition of the engine.
11 Due to the angle at which the spark plugs must be installed on most engines, installation will be simplified by inserting the plug wire terminal of the new spark plug into the 3/16-inch rubber hose, mentioned previously, before it is installed in the cylinder head (photo). This procedure serves two purposes: the rubber hose gives you flexibility for establishing the proper angle of plug insertion in the head and, should the threads be improperly lined up, the rubber hose will merely slip on the spark plug terminal when it meets resistance, preventing damage to the cylinder head threads.

35.9 Removing a spark plug (U-joint sockets will be necessary to remove most plugs)

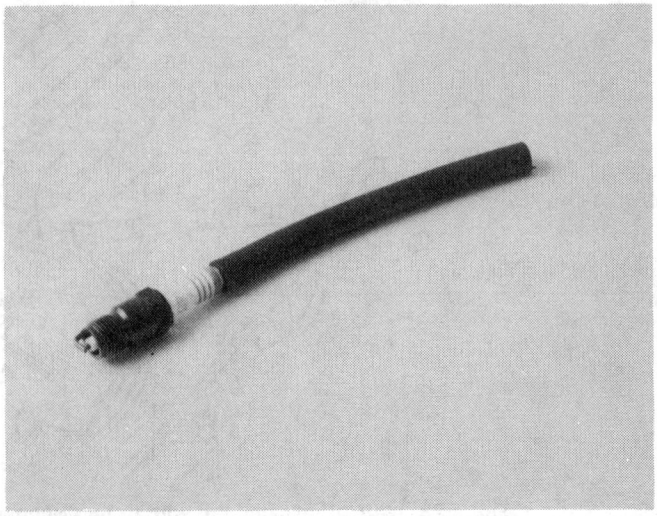

35.11 The trick set-up for installing plugs is a length of 3/16-inch ID rubber hose

For a COLOR version of this spark plug diagnosis page, please see the inside rear cover of this manual

CARBON DEPOSITS

Symptoms: Dry sooty deposits indicate a rich mixture or weak ignition. Causes misfiring, hard starting and hesitation.

Recommendation: Check for a clogged air cleaner, high float level, sticky choke and worn ignition points. Use a spark plug with a longer core nose for greater anti-fouling protection.

OIL DEPOSITS

Symptoms: Oily coating caused by poor oil control. Oil is leaking past worn valve guides or piston rings into the combustion chamber. Causes hard starting, misfiring and hesition.

Recommendation: Correct the mechanical condition with necessary repairs and install new plugs.

TOO HOT

Symptoms: Blistered, white insulator, eroded electrode and absence of deposits. Results in shortened plug life.

Recommendation: Check for the correct plug heat range, over-advanced ignition timing, lean fuel mixture, intake manifold vacuum leaks and sticking valves. Check the coolant level and make sure the radiator is not clogged.

PREIGNITION

Symptoms: Melted electrodes. Insulators are white, but may be dirty due to misfiring or flying debris in the combustion chamber. Can lead to engine damage.

Recommendation: Check for the correct plug heat range, over-advanced ignition timing, lean fuel mixture, clogged cooling system and lack of lubrication.

HIGH SPEED GLAZING

Symptoms: Insulator has yellowish, glazed appearance. Indicates that combustion chamber temperatures have risen suddenly during hard acceleration. Normal deposits melt to form a conductive coating. Causes misfiring at high speeds.

Recommendation: Install new plugs. Consider using a colder plug if driving habits warrant.

GAP BRIDGING

Symptoms: Combustion deposits lodge between the electrodes. Heavy deposits accumulate and bridge the electrode gap. The plug ceases to fire, resulting in a dead cylinder.

Recommendation: Locate the faulty plug and remove the deposits from between the electrodes.

NORMAL

Symptoms: Brown to grayish-tan color and slight electrode wear. Correct heat range for engine and operating conditions.

Recommendation: When new spark plugs are installed, replace with plugs of the same heat range.

ASH DEPOSITS

Symptoms: Light brown deposits encrusted on the side or center electrodes or both. Derived from oil and/or fuel additives. Excessive amounts may mask the spark, causing misfiring and hesitation during acceleration.

Recommendation: If excessive deposits accumulate over a short time or low mileage, install new valve guide seals to prevent seepage of oil into the combustion chambers. Also try changing gasoline brands.

WORN

Symptoms: Rounded electrodes with a small amount of deposits on the firing end. Normal color. Causes hard starting in damp or cold weather and poor fuel economy.

Recommendation: Replace with new plugs of the same heat range.

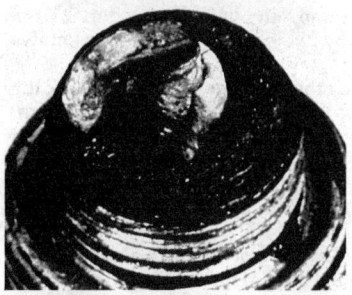

DETONATION

Symptoms: Insulators may be cracked or chipped. Improper gap setting techniques can also result in a fractured insulator tip. Can lead to piston damage.

Recommendation: Make sure the fuel anti-knock values meet engine requirements. Use care when setting the gaps on new plugs. Avoid lugging the engine.

SPLASHED DEPOSITS

Symptoms: After long periods of misfiring, deposits can loosen when normal combustion temperature is restored by an overdue tune-up. At high speeds, deposits flake off the piston and are thrown against the hot insulator, causing misfiring.

Recommendation: Replace the plugs with new ones or clean and reinstall the originals.

MECHANICAL DAMAGE

Symptoms: May be caused by a foreign object in the combustion chamber or the piston striking an incorrect reach (too long) plug. Causes a dead cylinder and could result in piston damage.

Recommendation: Remove the foreign object from the engine and/or install the correct reach plug.

12 After installing the plug to the limit of the hose grip, tighten it with the socket. It is a good idea to use a torque wrench for this to ensure that the plug is seated correctly. The correct torque figure is included in the Specifications.
13 Before pushing the spark plug wire onto the end of the plug, inspect it following the procedures outlined in Section 36.
14 Attach the plug wire to the new spark plug, again using a twisting motion on the boot until it is firmly seated on the spark plug. Make sure the wire is routed away from the exhaust manifold.
15 Follow the above procedure for the remaining spark plugs, replacing them one at a time to prevent mixing up the spark plug wires.

36 Spark plug wires, distributor cap and rotor check and replacement

1 Begin this procedure by making a visual check of the spark plug wires while the engine is running. In a darkened garage (make sure there is ventilation) start the engine and observe each plug wire. Be careful not to come into contact with any moving engine parts. If there is a break in the wire, you will see arcing or a small spark at the damaged area. If arcing is noticed, make a note to obtain new wires, then allow the engine to cool and check the distributor cap and rotor.
2 Disconnect the negative cable from the battery. At the distributor, disconnect the ECM connector and the coil connector (coil-in-cap models) or coil wire (models with separately mounted coil).
3 Remove the distributor cap by placing a screwdriver on the slotted head of each latch. Press down on the latch and turn it 90° to release the hooked end at the bottom (photo). On some engines, due to restricted working room, a stubby screwdriver will work best. With all latches disengaged, separate the cap from the distributor with the spark plug wires still attached.
4 Inspect the cap for cracks and other damage. Closely examine the contacts on the inside of the cap for excessive corrosion (photo). Slight scoring is normal. Deposits on the contacts may be removed with a small file.
5 If the inspection reveals damage to the cap, make a note to obtain a replacement for your particular engine, then examine the rotor.
6 The rotor is visible, with the cap removed, at the top of the distributor shaft. It is held in place by two screws. Remove the screws and the rotor (photo).
7 Inspect the rotor for cracks and other damage. Carefully check the condition of the metal contact at the top of the rotor for excessive burning and pitting (photo). On coil-in-cap models, also check the top of the rotor for carbon tracks (photo). This is a sign of moisture contamination due to a leaking seal between the distributor cap and the coil. The rotor and seal should be replaced with new ones if carbon tracks are visible.
8 If it is determined that a new rotor is required, make a note to that effect. If the rotor and cap are in good condition, reinstall them at this time. Be sure to apply a small dab of silicone lubricant to the contacts inside the cap before installing it. Note that the rotor has two raised pegs on the bottom and that it has a wide slot and a narrow slot. Make sure that the slots are correctly aligned and that the pegs are firmly seated when the rotor is installed.
9 If the cap must be replaced, do not reinstall it. Leave it off the distributor with the wires still connected.
10 If the spark plug wires are being replaced, now is the time to obtain a new set, along with a new cap and rotor as determined in the checks above. Purchase a wire set for your particular engine, pre-cut to the proper size, with the rubber boots already installed.
11 If the spark plug wires passed the check in Step 1, they should be checked further as follows.
12 Examine the wires one at a time to avoid mixing them up.
13 Disconnect the plug wire from the spark plug. A removal tool can be used for this, or you can grab the rubber boot, twist slightly and then pull the wire free. Do not pull on the wire itself, only on the rubber boot.
14 Inspect inside the boot for corrosion, which will look like a white crusty powder. Some models use a conductive white silicone lubricant which should not be mistaken for corrosion.
15 Now push the wire and boot back onto the end of the spark plug. It should be a tight fit on the plug end. If not, remove the wire and use pliers to carefully crimp the metal connector inside the wire boot until the fit is snug.
16 Now, using a clean rag, clean the entire length of the wire. Remove all built-up dirt and grease. As this is done, check for burns, cracks and any other form of damage. Bend the wires in several places to ensure that the conductive wire inside has not hardened.
17 Next, the wires should be checked at the distributor cap in the same manner. On four-cylinder engines, remove the wire from the cap by pulling on the boot, again examining the wires one at a time, and reinstalling each one after examination. Apply new silicone lubricant before reinstallation. On V6 and V8 engines, the distributor boots are connected to a circular retaining ring attached to the distributor cap. Release the locking tabs, turn the ring upside down and check all wire boots at the same time (photos).
18 If the wires appear to be in good condition, reinstall the retaining ring (V6 and V8 engines) and make sure that all wires are secure at both ends. If the cap and rotor are also in good condition, the check is finished. Reconnect the wires at the distributor and at the battery.
19 If it was determined in Steps 12 through 17 that new wires are required, obtain them at this time, along with a new cap and rotor if so determined in the checks above.
20 If a new cap is being installed on a coil-in-cap type distributor, the coil and cover from the cap being replaced should be transferred to the new cap.
21 Remove the three coil cover attaching screws and lift off the cover.
22 Remove the coil attaching screws, disconnect the leads and separate the coil from the distributor.
23 Attach the coil to the new cap by reversing Steps 21 and 22. Use a new seal between the coil and cap and be sure to lubricate the seal

36.3 Releasing the distributor cap hold-down latches (the spark plug wire ring has previously been removed for visibility)

36.4 Close-up shows deposit build-up on cap terminal

36.6 Removing the rotor

36.7A Close-up of metal contact on this rotor reveals normal wear pattern

36.7B Carbon tracking on the rotor is caused by a leaking seal between the distributor cap and the ignition coil (always replace both the seal and the rotor when this condition is present)

36.17A Removing the spark plug wire retaining ring clip from the top of the distributor cap (the cap is still installed on the distributor in this photo; normally, the cap would have been removed previously)

36.17B Removing the plug wire retainer from the cap

37.4 Using a special gauge to check cylinder compression

with multi-purpose grease.

24 Attach the rotor to the distributor. Make sure that the carbon brush is properly installed in the cap, as a wide gap between the carbon brush and the rotor will cause rotor burn-through and/or damage to the distributor cap.

25 If new wires are being installed, replace them one at a time. **Note:** *It is important to replace the wires one at a time, noting the routing as each wire is removed and installed, to maintain the correct firing order and to prevent short-circuiting.*

26 Attach the cap to the distributor, reconnecting all wires disconnected in Step 2, then reconnect the battery cable.

37 Compression check

1 A compression check will tell you what mechanical condition the engine is in. Specifically, it can tell you if the compression is down due to leakage caused by worn piston rings, defective valves and seats or a blown head gasket.

2 Begin by cleaning the area around the spark plugs before you remove them. This will keep dirt from falling into the cylinders while you are performing the compression test.

3 Remove the coil high-tension lead from the distributor and ground it on the engine block. Block the throttle and choke valves wide open.

4 With the compression gauge in the number one cylinder's spark plug hole, crank the engine over at least four compression strokes and observe the gauge (the compression should build up quickly in a healthy engine) (photo). Low compression on the first stroke, followed by gradually increasing pressure on successive strokes, indicates worn piston rings. A low compression reading on the first stroke, which does not build up during successive strokes, indicates leaking valves or a blown head gasket (a cracked head could also be the cause). Record the highest gauge reading obtained.

5 Repeat the procedure for the remaining cylinders and compare the results to the Specifications. Compression readings approximately 10 percent above or below the specified amount can be considered normal.

6 Pour a couple of teaspoons of engine oil (a squirt can works great for this) into each cylinder, through the spark plug hole, and repeat the test.

7 If the compression increases after the oil is added, the piston rings are definitely worn. If the compression does not increase significantly, the leakage is occurring at the valves or head gasket. Leakage past the valves may be caused by burned valve seats or faces, warped, cracked or bent valves or valves that require adjustment

8 If two adjacent cylinders have equally low compression, there is a strong possibility that the head gasket between them is blown. The appearance of coolant in the combustion chambers or the crankcase would verify this condition.

9 If the compression is higher than normal, the combustion chambers are probably coated with carbon deposits. If that is the case, the cylinder head(s) should be removed and decarbonized.

10 If compression is way down or varies greatly between cylinders, it would be a good idea to have a leak-down test performed by a reputable automotive repair shop. This test will pinpoint exactly where the leakage is occurring and how severe it is.

Chapter 2 Part A L4 (four-cylinder) engine

Refer to Chapter 13 for specifications and information related to 1985 and later models

Contents

Air filter replacement . . . . . Chapter 1
Camshaft — removal and installation 14
Check engine light . . . . . Chapter 6
Compression check . . . . . Chapter 1
Crankshaft pulley hub and oil seal — removal and installation . 12
Cylinder head — removal and installation 7
Drivebelt check and adjustment . . . . . Chapter 1
Engine — removal and installation 17
Engine mounts — replacement with engine in vehicle 16
Engine oil and filter change . . . . . Chapter 1
Engine oil level check . . . . . Chapter 1
Engine overhaul — general information . . . . . Chapter 2D
Engine removal — methods and precautions . . . . . Chapter 2D
Exhaust manifold — removal and installation 6
Flywheel/driveplate and rear main bearing oil seal — removal and installation 15
General information 1
Hydraulic lifters — removal, inspection and installation 8
Intake manifold — removal and installation 5
Oil pan — removal and installation 10
Oil pump — removal and installation 11
Oil pump driveshaft — removal and installation 9
Pushrod cover — removal and installation 3
Repair operations possible with the engine in the vehicle . . . . . Chapter 2D
Rocker arm cover — removal and installation 2
Spark plug replacement . . . . . Chapter 1
Timing gear cover — removal and installation 13
Valve train components — replacement (cylinder head installed) 4
Water pump — removal and installation . . . . . Chapter 3

Specifications

Torque specifications

	Ft-lbs	M-kg
Oil pan bolts	6	0.8
Oil pan drain plug	25	3.4
Oil screen support nut	37	5.1
Oil pump-to-block bolts	22	3.0
Oil pump cover bolts	10	1.4
Pushrod cover-to-block bolts	7	1.0
Crank pulley hub bolt	160	21.2
Flywheel/driveplate-to-crankshaft bolts	44	6.0
EFI assembly-to-manifold nuts/bolts	15	2.0
Intake manifold-to-cylinder head bolts	29	4.0
Exhaust manifold-to-cylinder head bolts	44	6.0
Fuel pump-to-block bolts	18	2.5
Distributor hold-down bolt	22	3.0
EGR valve-to-manifold bolts	10	1.4
Water outlet housing bolts	20	2.7
Thermostat housing bolts	20	2.7
Water pump bracket-to-block bolts	25	3.4
Timing cover-to-block bolts	7	1.0
Fan and pulley-to-water pump bolts	18	2.4
Cylinder head bolts	85	11.5
Rocker arm bolts	20	2.7
Rocker arm cover bolts	6	0.8
Camshaft thrust plate-to-block bolts	7	1.0
Oil pump driveshaft retainer plate bolts	10	1.4
Engine mount nuts	41	5.6

Cylinder numbering and direction distributor rotor rotates (arrow)

Note: *Refer to Chapter 2, Part D, for additional specifications.*

1 General information

The forward Sections in this Part of Chapter 2 are devoted to 'in-vehicle' repair procedures for the L4 engine. The latter Sections in this Part of Chapter 2 involve the removal and installation procedures for the L4 engine. All information concerning engine block and cylinder head servicing can be found in Part D of this Chapter.

The repair procedures included in this Part are based on the assumption that the engine is still installed in the vehicle. Therefore, if this information is being used during a complete engine overhaul — with the engine already out of the vehicle and on a stand — many of the steps included here will not apply.

The Specifications included in this Part of Chapter 2 apply only to the engine and procedures found here. For Specifications regarding engines other than the L4, see Part B or C, whichever applies. Part D of Chapter 2 contains the Specifications necessary for engine block and cylinder head rebuilding.

2 Rocker arm cover — removal and installation

1 Remove the air cleaner assembly, tagging each hose to be disconnected with a piece of numbered tape to simplify installation.
2 Disconnect the throttle cable from the fuel injection assembly, making careful note of the exact locations of the cable components and hardware to ensure correct reinstallation (refer to the illustration in Chapter 1, Section 19, if necessary).
3 Remove the PCV valve from the rocker arm cover.
4 Remove the spark plug wires from the spark plugs (refer to the removal technique described in Chapter 1), then remove the wires and retaining clips from the rocker arm cover. Be sure to label each wire before removal to ensure that all wires are reinstalled correctly.
5 Loosen the fuel injection mounting nuts and bolts to provide clearance for removal of the EGR valve, then remove the EGR valve.
6 Remove the rocker arm cover bolts (photo).
7 Remove the rocker arm cover. **Note:** *If the cover sticks to the cylinder head, use a block of wood and a hammer to dislodge it. If the cover still will not come loose, pry on it carefully, but do not distort the sealing flange surface.*
8 Prior to installation of the cover, clean all dirt, oil and old gasket material from the sealing surfaces of the cover and cylinder head with a scraper and degreaser.
9 Apply a continuous 3/16-inch (5 mm) diameter bead of RTV-type sealant to the sealing flange of the cover. Be sure to apply the sealant *inboard* of the bolt holes.
10 Place the rocker arm cover on the cylinder head while the sealant is still wet and install the mounting bolts. Tighten the bolts a little at a time to the specified torque.
11 Complete the installation by reversing the removal procedure.

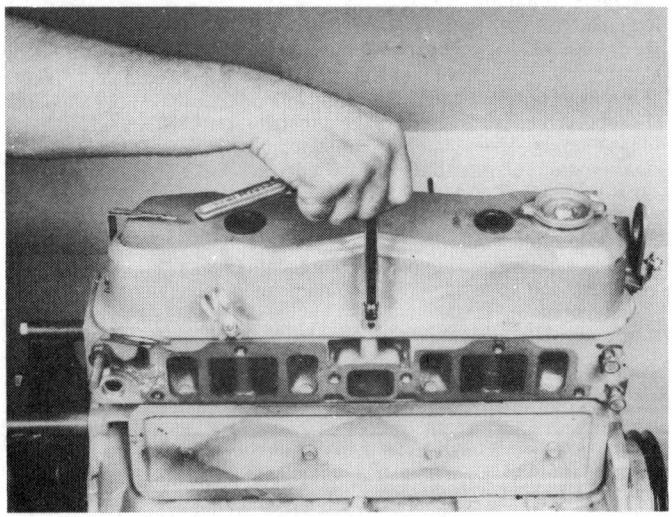

2.6 Removing the rocker arm cover bolts

3 Pushrod cover — removal and installation

1 Remove the intake manifold as described in Section 5.
2 Remove the distributor (refer to Chapter 5).
3 Remove the pushrod cover bolts and lift off the cover. If the gasket seal is difficult to break, tap on the cover gently with a rubber hammer. Do not pry on the cover.
4 Using a scraper and degreaser, clean the sealing surfaces on the cover and engine block to remove all oil and old gasket materal.
5 Prior to installation of the cover, apply a continuous 3/16-inch (5 mm) bead of RTV-type sealant to the mounting flange of the pushrod cover.
6 With the sealant still wet, place the cover in position on the block and install the cover bolts. Tighten the bolts gradually, following a crisscross pattern, to the specified torque.
7 Install the distributor and intake manifold.

4 Valve train components — replacement (cylinder head installed)

1 Remove the rocker arm cover as described in Section 2.
2 If only the pushrod is to be replaced, loosen the rocker bolt enough to allow the rocker arm to be rotated away from the pushrod. Pull the pushrod out of the hole in the cylinder head. If the rocker arm is to

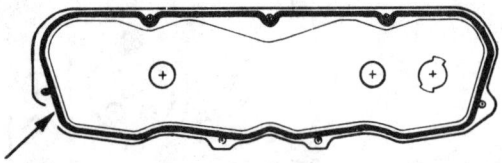

Fig. 2A.1 Apply a continuous 3/16-inch bead of RTV-type sealant (arrow) to the rocker arm cover flange (make sure it is applied to the inside of the bolt holes as shown) (Sec 2)

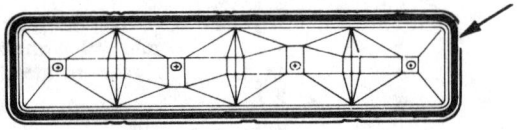

Fig. 2A.2 Apply a continuous 3/16-inch bead of RTV-type sealant (arrow) to the pushrod cover flange (Sec 3)

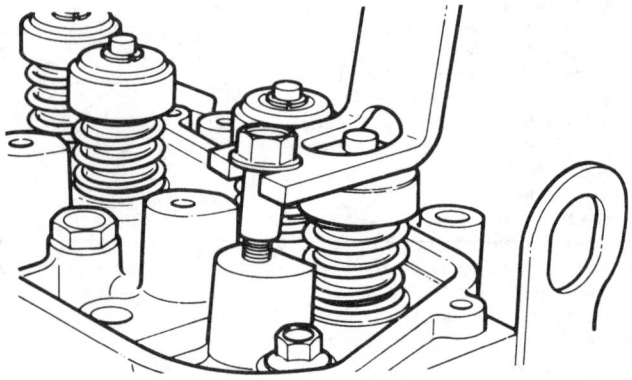

Fig. 2A.3 A special tool can be used to compress the valve springs and remove the spring keepers (Sec 4)

be removed, remove the rocker bolt and pivot and lift off the rocker arm.
3 If the valve spring is to be removed, remove the spark plug from the affected cylinder.
4 There are two methods that will allow the valve to remain in place while the valve spring is removed. If you have access to compressed air, install an air hose adapter (GM part number J-22794) in the spark plug hole. When air pressure is applied to the adapter, the valves will be held in place by the pressure.
5 If you do not have access to compressed air, bring the piston of the affected cylinder to top dead center (TDC) on the compression stroke. Feed a long piece of 1/4-inch nylon cord in through the spark plug hole until it fills the combustion chamber. Be sure to leave the end of the cord hanging out of the spark plug hole so it can be removed easily. Rotate the crankshaft with a wrench (in the normal direction of rotation) until slight resistance is felt.
6 Install the rocker arm bolt (without the rocker arm).
7 Insert the slotted end of a valve spring compressing tool under the bolt head and compress the spring just enough to remove the spring keepers, then release the pressure on the tool.
8 Remove the retainer, cup shield, O-ring seal, spring, spring damper (if so equipped) and valve stem oil seal (if so equipped).
9 Inspection procedures for valve train components are covered in Section 8 and in Chapter 2, Part D.
10 Installation of the valve train components is the reverse of the removal procedure. *Always use new valve stem oil seals whenever the spring keepers have been disturbed.* Prior to installing the rocker arms, coat the bearing surfaces of the arms and rocker arm pivots with moly-based grease or engine assembly lube. The L4 engine valve mechanisms require no special valve lash adjustment. Simply tighten the rocker arm bolts to the specified torque.

5 Intake manifold — removal and installation

1 Disconnect the cable from the negative battery terminal.
2 Remove the air cleaner assembly, tagging each hose to be disconnected with a piece of numbered tape to simplify reinstallation.
3 Remove the PCV valve and hose.
4 Drain the cooling system (refer to Chapter 1).
5 Label and disconnect the fuel line, vacuum lines and electrical leads from the fuel injection assembly. When disconnecting the fuel line, be prepared to catch some fuel, then plug the fuel line to prevent contamination.
6 Disconnect the fuel injection throttle linkage, making careful note of how it is installed.
7 Disconnect the cruise control linkage (if so equipped).
8 Disconnect the coil wire, remove the two retaining screws and one bolt, then remove the coil.
9 Remove the coolant hoses from the manifold.
10 Remove the manifold retaining bolts and separate the manifold from the cylinder head. *Do not pry between the manifold and head, as damage to the gasket sealing surfaces may result.*
11 Remove the manifold gasket.
12 If the intake manifold is to be replaced with another, transfer all components still attached to the old manifold to the new one.
13 Before installing the manifold, clean the cylinder head and manifold gasket surfaces. All gasket material and sealing compound must be removed prior to installation.
14 Apply a thin bead of RTV-type sealant to the intake manifold and cylinder head mating surfaces. Make certain that the sealant will not spread into the air or coolant passages when the manifold is installed.
15 Place a new intake manifold gasket on the manifold, place the manifold in position against the cylinder head and install the mounting bolts finger tight.
16 Tighten the mounting bolts a little at a time in the sequence shown in the accompanying illustration until they are all at the specified torque.
17 Install the remaining components in the reverse order of removal.
18 Fill the radiator with coolant, start the engine and check for leaks. Adjust the ignition timing and idle speed as necessary (Chapter 1).

6 Exhaust manifold — removal and installation

1 If the vehicle is equipped with air-conditioning, carefully examine the routing of the hoses and the mounting of the compressor. You may be able to remove the exhaust manifold without disconnecting the system. If you are in doubt, take the vehicle to a GM dealer or automotive repair shop to have the system depressurized. **Caution:** *Do not, under any circumstances, disconnect any air-conditioning system lines while the system is under pressure.*
2 Remove the cable from the negative battery terminal.
3 Remove the air cleaner assembly, tagging each hose to be disconnected with a piece of numbered tape to simplify reinstallation.
4 Remove the fuel injection pre-heat tube.
5 Remove the engine oil dipstick and dipstick tube.
6 Remove the oxygen sensor (refer to Chapter 1).
7 Label the four spark plug wires, then disconnect them and secure them to the side out of the way.
8 Disconnect the exhaust pipe from the exhaust manifold. You may have to apply penetrating oil to the fastener threads as they are usual-

2A

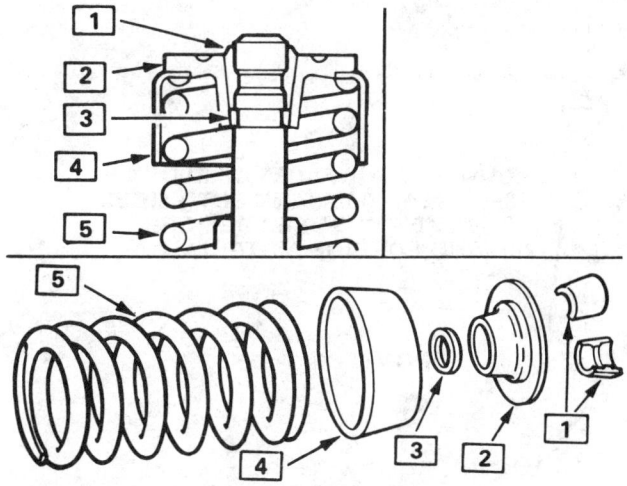

Fig. 2A.4 Valve train parts (typical) — exploded view (Sec 4)

1 Keepers
2 Retainer
3 Oil seal
4 Spring shield
5 Spring

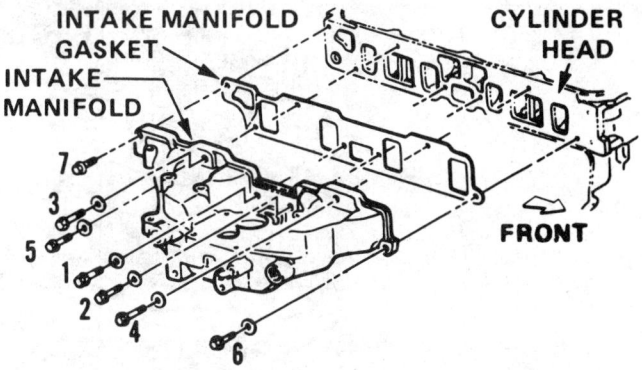

Fig. 2A.5 Tighten the intake manifold mounting bolts a little at a time in the order shown until they are all at the specified torque (Sec 5)

ly corroded. The exhaust pipe can be hung from the frame with a piece of wire.

9 Remove the exhaust manifold end bolts first, then remove the center bolts and separate the exhaust manifold from the engine.

10 Remove the exhaust manifold gasket.

11 Before installing the manifold, clean the gasket mating surfaces on the cylinder head and manifold. All leftover gasket material and carbon deposits must be removed.

12 Place a new exhaust manifold gasket into position on the cylinder head, then place the manifold in position and install the mounting bolts finger tight.

13 Tighten the mounting bolts a little at a time in the sequence shown in the accompanying illustration until all of the bolts are at the specified torque.

14 Install the remaining components in the reverse order of removal, using new gaskets wherever one has been removed.

15 Start the engine and check for exhaust leaks between the manifold and cylinder head and between the manifold and exhaust pipe.

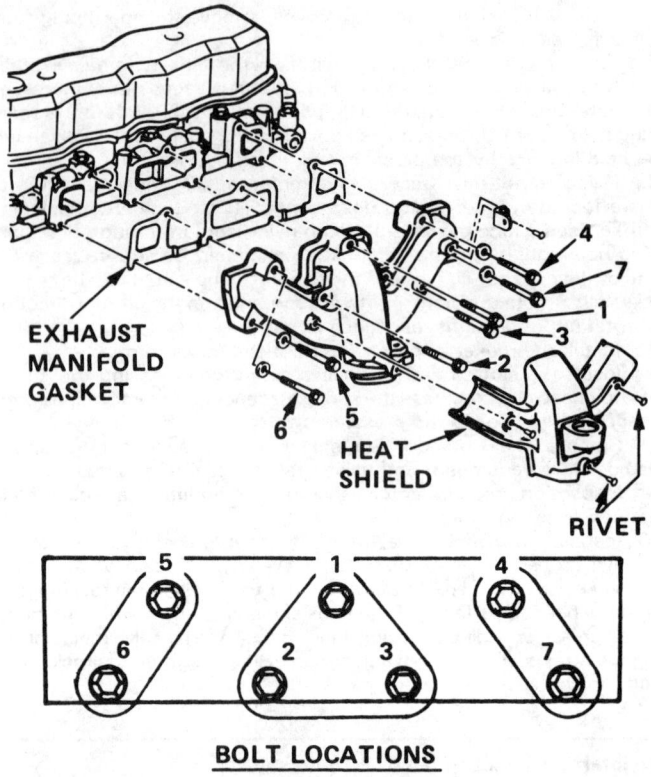

Fig. 2A.6 Recommended tightening sequence for the exhaust manifold bolts (Sec 6)

7 Cylinder head — removal and installation

Removal

1 Remove the intake manifold as described in Section 5.

2 Remove the exhaust manifold as described in Section 6.

3 Remove the bolts that secure the alternator bracket to the cylinder head.

4 If so equipped, unbolt the air-conditioner compressor and swing it out of the way for clearance. **Caution:** *Do not disconnect any of the air-conditioning lines unless the system has been depressurized by a dealer or repair shop, because personal injury may occur.* Disconnection of the lines should not be necessary in this case.

5 Disconnect all electrical and vacuum lines from the cylinder head. Be sure to label the lines to simplify reinstallation.

6 Remove the upper radiator hose.

7 Disconnect the spark plug wires and remove the spark plugs. Be sure to label the plug wires to simplify reinstallation.

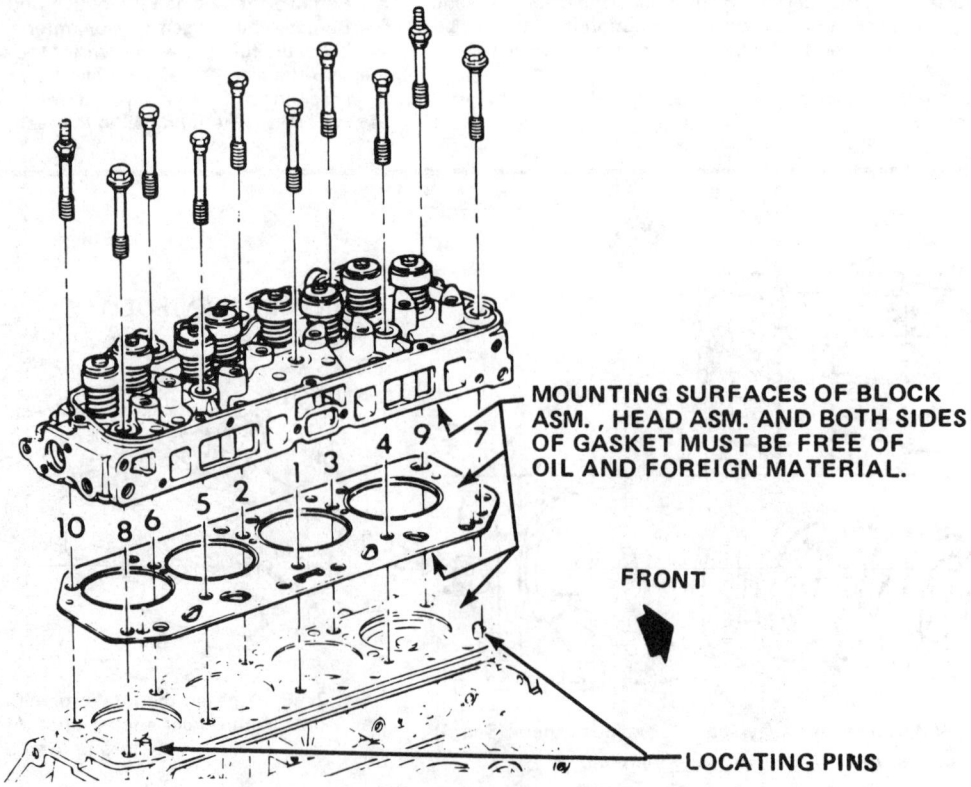

Fig. 2A.7 Cylinder head bolt positions and bolt tightening sequence (Sec 7)

8 Remove the rocker arm cover. To break the gasket seal it may be necessary to strike the cover with your hand or a rubber hammer. Do not pry between the sealing surfaces. Refer to Section 2 if necessary.
9 When disassembling the valve mechanisms, keep all of the components separate so they can be reinstalled in their original positions. A cardboard box or rack, numbered to correspond to the engine cylinders, can be used for this purpose.
10 Remove each of the rocker arm nuts and separate the rocker arms and pivots from the cylinder head (photo).
11 Remove the pushrods (photo).
12 If the ignition coil is mounted separately from the distributor, disconnect the wires and remove the coil.
13 Loosen each of the cylinder head mounting bolts one turn at a time until they can be removed (photo). Note the length and position of each bolt to ensure correct reinstallation.
14 Lift the head off of the engine. If it is stuck to the engine block, do not attempt to pry it free, as you could damage the sealing surfaces. Instead, use a hammer and block of wood to tap the head and break the gasket seal. Place the head on a block of wood to prevent damage to the gasket surface.
15 Remove the cylinder head gasket.
16 Refer to Chapter 2D for cylinder head disassembly and valve service procedures.

Installation

17 If a new cylinder head is being installed, transfer all external parts from the old cylinder head to the new one.
18 If not already done, thoroughly clean the gasket surfaces on the cylinder head and the engine block. Do not gouge or otherwise damage the gasket surfaces.
19 To get the proper torque readings, the threads of the head bolts must be clean. This also applies to the threaded holes in the engine block. Run a tap through the holes to ensure that they are clean.
20 Place the gasket in position over the engine block dowel pins.
21 Carefully lower the cylinder head onto the engine, over the dowel pins and the gasket.
22 Coat the threads of each cylinder head bolt and the point at which the head and the bolt meet with a sealing compound and install the bolts finger tight (photo). Do not tighten any of the bolts at this time.
23 Tighten each of the bolts a little at a time in the sequence shown in the accompanying illustration. Continue tightening in this sequence until the proper torque reading is obtained. As a final check, work around the head in a front-to-rear sequence to make sure none of the bolts have been left out of the sequence.
24 The remaining installation steps are the reverse of removal.

8 Hydraulic lifters — removal, inspection and installation

1 A noisy valve lifter can be isolated when the engine is idling. Place a length of hose or tubing near the position of each valve while listening at the other end of the tube. Another method is to remove the rocker arm cover and, with the engine idling, place a finger on each of the

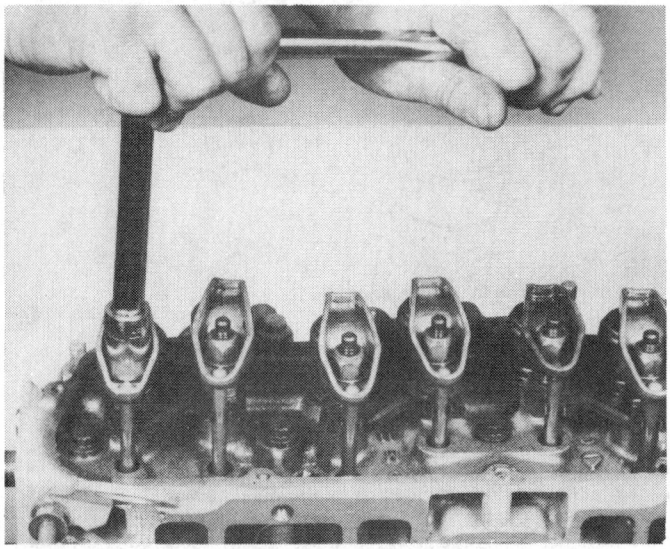

7.10 Removing the rocker arm nuts

7.11 Removing the pushrods

7.13 Loosening the cylinder head mounting bolts

7.22 The cylinder head mounting bolts should be coated with sealant (arrows) before installation

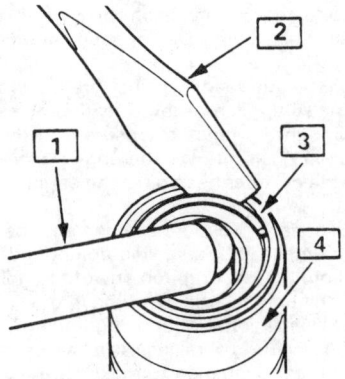

Fig. 2A.8 Removing the pushrod seat retainer from the lifter (Sec 8)

1 Pushrod
2 Screwdriver
3 Retainer ring
4 Lifter body

8.8 Removing the hydraulic lifters with a special tool

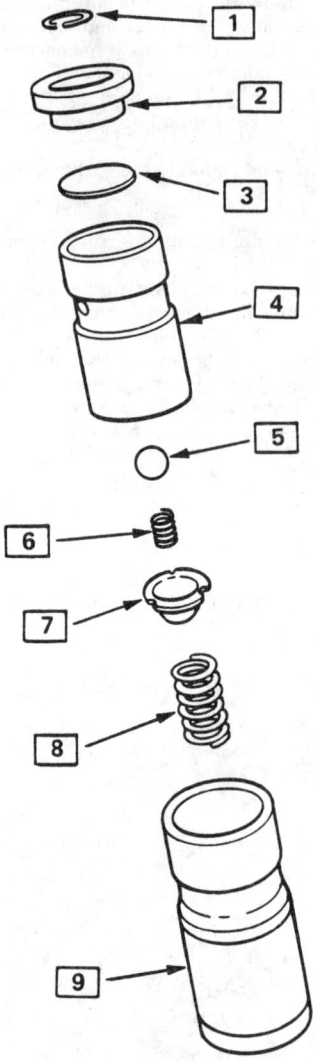

Fig. 2A.9 Hydraulic valve lifter — exploded view (Sec 8)

1 Retainer spring
2 Pushrod seat
3 Rocker feed metering valve
4 Plunger
5 Ball check valve
6 Ball check valve spring (not on all lifters)
7 Ball check valve retainer
8 Plunger spring
9 Lifter body

valve spring retainers, one at a time. If a valve lifter is defective, it will be evident from the shock felt at the retainer as the valve seats.

2 The most likely cause of a noisy valve lifter is a piece of dirt trapped between the plunger and the lifter body.

3 Remove the rocker arm cover as described in Section 2.

4 Remove the intake manifold as described in Section 5.

5 Remove the pushrod cover as described in Section 3.

6 Loosen the rocker arm bolt and rotate the rocker arm away from the pushrod.

7 Remove the pushrod.

8 To remove the lifter, a special hydraulic lifter removal tool should be used (photo) or a scribe can be positioned at the top of the lifter and used to force the lifter up and out of the bore. *Do not use pliers or other tools on the outside of the lifter body, because they will damage the finished surface and render the lifter useless.*

9 The lifters should be kept separate for reinstallation in their original positions.

10 To dismantle a valve lifter, hold the plunger down with a pushrod and then extract the retainer spring with a small screwdriver.

11 Remove the pushrod seat and the metering valve.

12 Remove the plunger, ball check valve and plunger spring. Remove the ball check valve and spring by prying them out with a small screwdriver.

13 Clean the lifter components with solvent and dry them with compressed air. Examine the internal components for wear and check the ball carefully for flat spots. **Note:** *Refer to Chapter 2, Part D, for additional lifter (and camshaft) inspection procedures.*

14 If the lifters are worn, they must be replaced with new ones and the camshaft must be replaced as well (see Chapter 2, Part D). If the lifters were contaminated with dirt, they can be reinstalled — they *may* operate normally.

15 Reassembly should be done in the following manner:

a) Place the ball check valve on the small hole in the bottom of the plunger.
b) Insert the ball check spring into the seat in the valve retainer and place the retainer over the ball so that the spring rests on the ball. Using a small screwdriver, carefully press the retainer into position in the plunger.
c) Place the plunger spring over the ball retainer, invert the lifter body and slide it over the spring and plunger. Make sure the oil holes in the body and plunger line up.
d) Fill the assembly with 10-weight oil. Place the metering valve and pushrod seat in position, press down on the seat and install the retainer spring.

16 When installing the lifters, make sure they are replaced in their original bores and coat them with moly-based grease or engine assembly lube.

17 The remaining installation steps are the reverse of removal.

9 Oil pump driveshaft — removal and installation

1 Remove the air cleaner, marking the hoses as they are disconnected

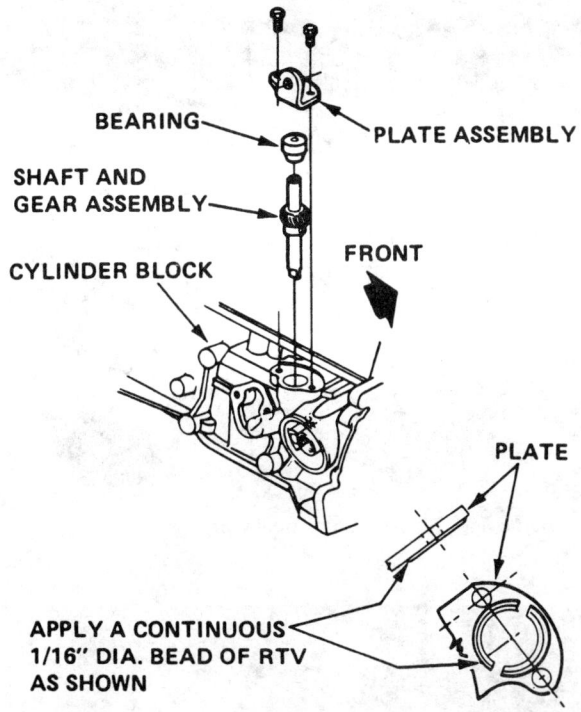

Fig. 2A.10 Oil pump driveshaft assembly (Sec 9)

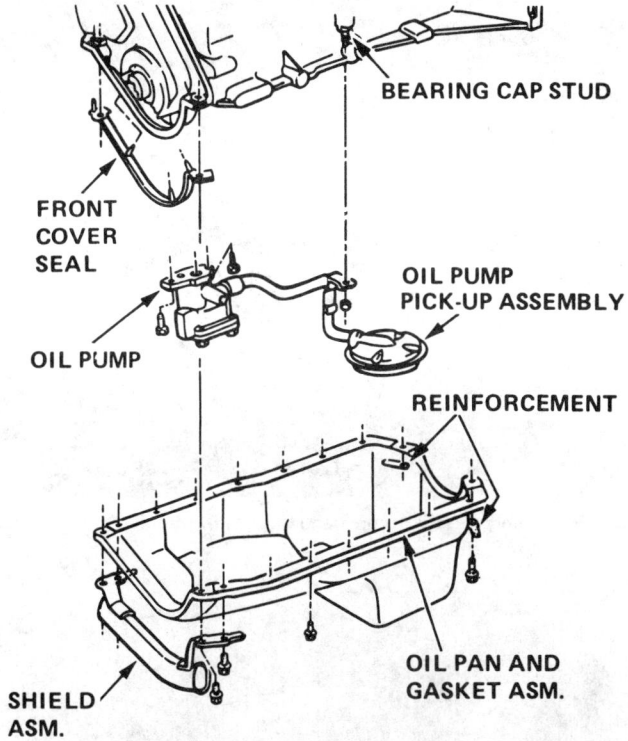

Fig. 2A.11 Oil pan and oil pump details (Sec 10)

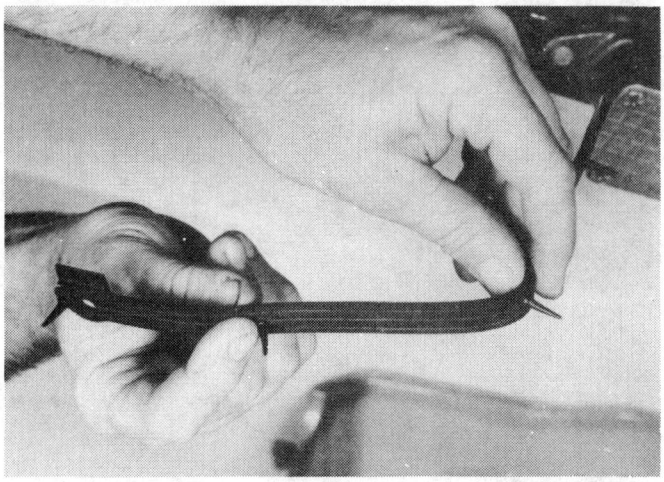

10.9 During installation, the rubber tips on the front oil pan gasket should be pressed into the holes in the timing gear cover

10.10 All four sections of the oil pan gasket in position on the engine block

to simplify reinstallation.
2 Remove the alternator (refer to Chapter 5).
3 Remove the oil pump driveshaft retainer plate bolts.
4 Remove the bushing.
5 Remove the shaft and gear assembly.
6 Thoroughly clean the sealing surfaces on the engine block and retainer plate.
7 Inspect the gear teeth to see if they are chipped or cracked.
8 Install the oil pump driveshaft in the block and turn it until it engages with the camshaft drive gear in the oil pump body.
9 Apply a 1/16-inch (1.5 mm) diameter bead of RTV-type sealant to the retainer plate so that it completely seals around the oil pump driveshaft hole in the block, as shown in the accompanying illustration.
10 Install the retainer plate mounting bolts and tighten them to the specified torque.
11 Complete the installation by reversing the removal procedure.

10 Oil pan — removal and installation

1 Disconnect the cable from the negative battery terminal.
2 Raise the vehicle, place it securely on jackstands and drain the engine oil (refer to Chapter 1 if necessary).
3 Disconnect the exhaust pipe at the exhaust manifold and loosen the hanger bracket.
4 Remove the through-bolts from the front motor mounts.
5 Using a hoist or block-and-tackle, raise the engine enough so the oil pan can be removed without interference from the steering and suspension components.
6 Remove the oil pan bolts and separate the oil pan from the block.
7 Clean the pan with solvent and remove all old sealant and gasket material from the block and pan sealing surfaces.
8 Install the rear pan oil seal in the groove in the rear main bearing

11.2 Removing the oil pump flange mounting bolts

12.3 Removing the crankshaft pulley bolt

12.4 Marking the position of the pulley in relation to the hub

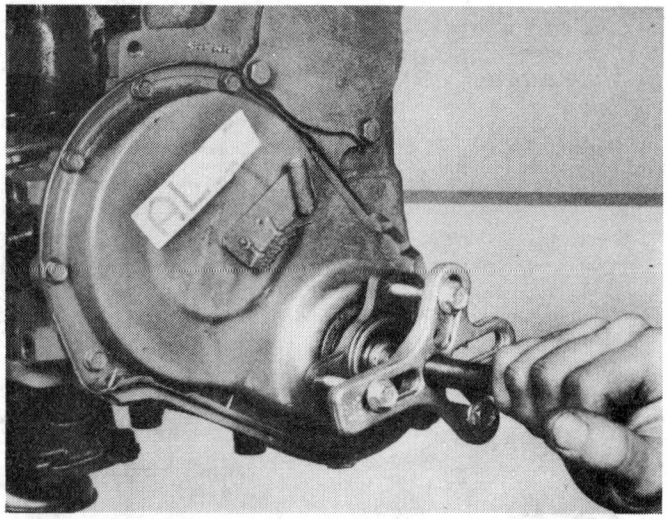

12.5 Using a puller to remove the hub from the crankshaft

12.9 Using the pulley hub bolt to press the hub onto the crankshaft

13.3 Removing the timing gear cover mounting bolts

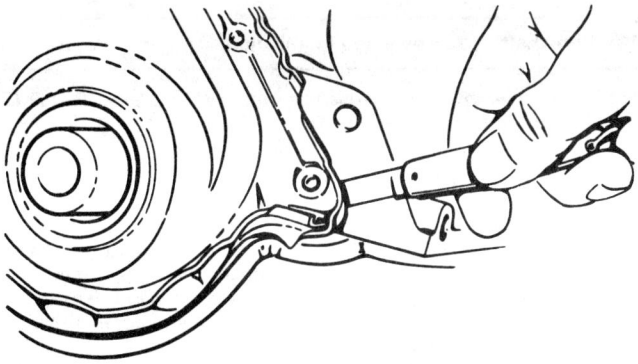

Fig. 2A.12 Cutting the oil pan front gasket flush with the engine block (Sec 13)

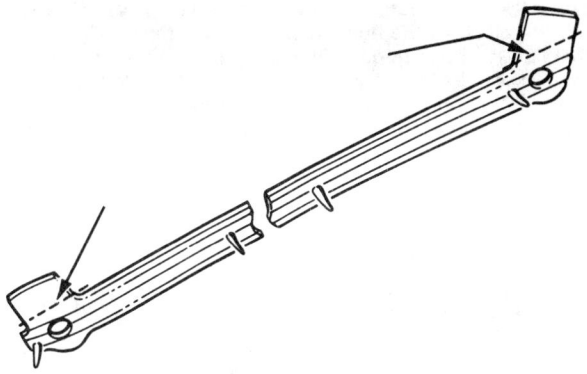

Fig. 2A.13 Cut the oil pan front gasket as shown before installing it (Sec 13)

cap and apply a small quantity of RTV-type sealant to the depressions where the seal meets the block.
9 Install the front pan oil seal on the timing cover, pressing the tips into the holes in the cover (photo).
10 Using a light coat of grease as a retainer, install the pan side gaskets (photo). Apply a 1/8-inch diameter by 1/4-inch long bead of RTV-type sealant at the parting lines of the front seal and side gaskets.
11 Attach the oil pan to the block. The bolts that secure the pan to the timing cover should be installed last. They are installed at an angle into holes which will line up as the rest of the pan bolts are tightened.
12 After all bolts are installed, tighten them to the specified torque (use a criss-cross pattern and work up to the final torque in three or four steps).
13 The remaining steps are the reverse of the removal procedure.

11 Oil pump — removal and installation

1 Remove the oil pan (refer to Section 10).
2 Remove the two oil pump flange mounting bolts and the nut from the main bearing cap bolt (photo).
3 Lift out the oil pump and screen as an assembly.
4 To install the pump, align the shaft so it mates with the oil pump driveshaft tang, then install the pump on the block, over the oil pump driveshaft lower bushing. No gasket is used. The oil pump should slide easily into place — if not, remove it and relocate the slot.
5 Install the mounting bolts and nut and tighten them to the specified torque.
6 Reinstall the oil pan (refer to Section 10).

12 Crankshaft pulley hub and oil seal — removal and installation

1 Remove the cable from the negative battery terminal.

13.9 Using a block of wood to install the oil seal

2 Loosen the accessory drivebelt tension adjusting bolts, as necessary, and remove the drivebelts. Tag each belt as it is removed to simplify reinstallation.
3 With the parking brake applied and the shifter in Park (automatic) or in gear (manual) to prevent the engine from turning over, remove the crank pulley bolt (photo). A breaker bar will probably be necessary, since the bolt is very tight.
4 Mark the position of the pulley in relation to the hub (photo). Remove the bolts and separate the pulley from the hub.
5 Using a puller, remove the hub from the crankshaft (photo).
6 Carefully pry the oil seal out of the front cover with a large screwdriver. Be careful not to distort the cover.
7 Install the new seal with the helical lip toward the *rear* of the engine. Drive the seal into place using a seal installation tool or a large socket. If there is enough room, a block of wood and hammer can also be used.
8 Apply a thin layer of clean multi-purpose grease to the seal contact surface of the hub.
9 Position the pulley hub on the crankshaft and slide it through the seal until it bottoms against the crankshaft gear. Note that the slot in the hub must be aligned with the Woodruff key in the end of the crankshaft. The hub-to-crankshaft bolt can also be used to press the hub into position (photo).
10 Install the crank pulley on the hub, noting the alignment marks made during removal. The pulley-to-hub bolts should be coated with Drylock 299, or equivalent, whenever they are removed and installed.
11 Tighten the hub-to-crankshaft and pulley-to-hub bolts to the specified torque.
12 The remaining installation steps are the reverse of removal. Tighten the drivebelts to the proper tension (refer to Chapter 1).

13 Timing gear cover — removal and installation

1 Remove the crankshaft pulley hub as described in Section 12.
2 Remove the oil pan-to-timing gear cover bolts.
3 Remove the cover-to-block bolts (photo).
4 Using a sharp knife, cut the oil pan front gasket flush with the engine block at both sides.
5 Remove the cover and the attached portion of the oil pan gasket.
6 Remove the cover gasket.
7 Using a scraper and degreaser, remove all dirt and old gasket material from the sealing surfaces of the timing gear cover, engine block and oil pan.
8 Replace the front oil seal by carefully prying it out of the timing gear cover with a large screwdriver. Do not distort the cover.
9 Install the new seal with the helical lip toward the *inside* of the cover. Drive the seal into place using a seal installation tool or a large socket and hammer. A block of wood will also work (photo).
10 Prior to installing the cover, install a new front oil pan gasket. Cut the ends off of the gasket, as shown in the accompanying illustration, and attach it to the cover by pressing the rubber tips into the holes provided.
11 Apply a thin coat of RTV-type gasket sealant to the timing gear cover gasket and place it in position on the cover.
12 Apply a bead of RTV-type sealant to the joint between the oil pan and engine block.

13 Insert the hub through the cover seal and place the cover in position on the block as the hub slides onto the crankshaft.
14 Install the oil pan-to-cover bolts and partially tighten them.
15 Install the bolts that secure the cover to the block, then tighten all of the mounting bolts to the specified torque.
16 Complete the installation by reversing the removal procedure.

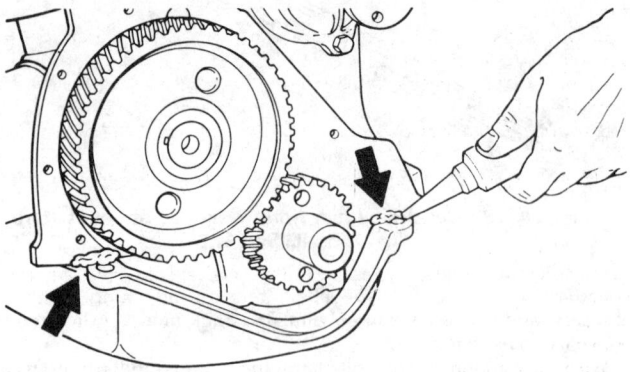

Fig. 2A.14 Apply RTV-type sealant at the joints (arrows) between the oil pan and the engine block (Sec 13)

14.16 Removing the camshaft thrust plate screws

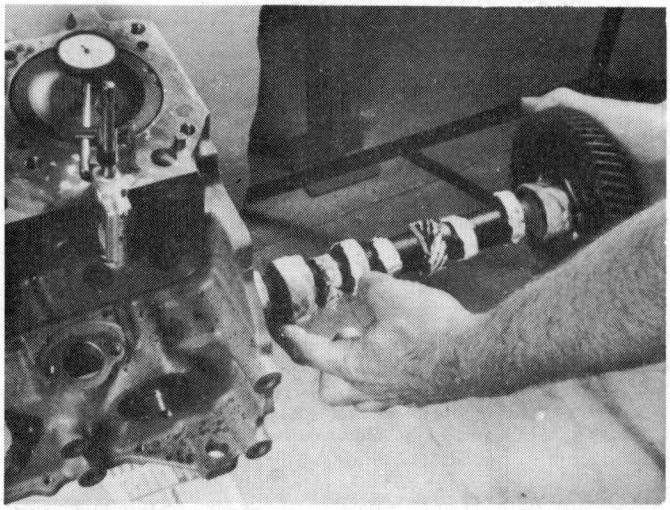

14.21 The camshaft lobes and journals should be lubricated prior to installing the camshaft in the block

14 Camshaft — removal and installation

Note: *Before removing the camshaft, refer to Chapter 2, Part D (Section 15), and measure the lobe lift.*

1 Remove the cable from the negative battery terminal.
2 Drain the oil from the crankcase (refer to Chapter 1).
3 Drain the coolant from the radiator (refer to Chapter 1).
4 Remove the radiator (refer to Chapter 3).
5 If equipped with air-conditioning, remove the condenser (refer to Chapter 3). **Caution:** *The air-conditioning system must be evacuated by an air-conditioning system technician before the condenser can be removed. Under no circumstances should this be attempted by the home mechanic, as personal injury may result.*
6 Remove the fan and water pump pulley (refer to Chapter 3).
7 Remove the rocker arm cover (refer to Section 2).
8 Loosen the rocker arm bolts and pivot the rocker arms away from the pushrods.
9 Remove the oil pump driveshaft assembly (refer to Section 9).
10 Remove the spark plugs (refer to Chapter 1).
11 Remove the distributor (refer to Chapter 5).
12 Remove the pushrod cover and gasket (refer to Section 3), then remove the pushrods.
13 Remove the valve lifters (refer to Section 8).
14 Remove the crankshaft pulley hub (refer to Section 12).
15 Remove the timing gear cover (refer to Section 13).

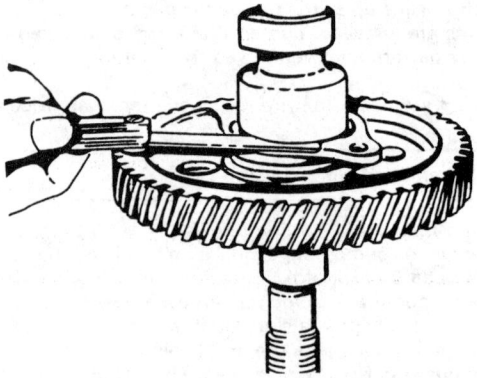

Fig. 2A.15 Using a feeler gauge to check the thrust plate end clearance (Sec 14)

14.23 The camshaft and crankshaft gears must be positioned so that the timing marks line up

16 Working through the holes in the camshaft gear, remove the two camshaft thrust plate screws (photo).
17 Supporting the camshaft carefully to prevent damage to the bearings, remove the camshaft and gear assembly by pulling it out through the front of the engine.
18 Refer to Chapter 2, Part D, for the camshaft inspection procedures.
19 If the gear must be removed from the camshaft, it must be pressed off. If you do not have access to a press, take it to your dealer or an automotive machine shop. The thrust plate must be positioned so that the Woodruff key in the shaft does not damage it when the shaft is pressed off.
20 If the gear has been removed, it must be pressed back on prior to installation of the camshaft.

a) Support the camshaft in an arbor press by placing press plate adaptors behind the front journal.
b) Place the gear spacer ring and the thrust plate over the end of the shaft.
c) Install the Woodruff key in the shaft keyway.
d) Position the camshaft gear and press it onto the shaft until it bottoms against the gear spacer ring.
e) Use a feeler gauge to check the end clearance of the thrust plate. It should be 0.0015 to 0.0050-inch. If the clearance is less than 0.0015-inch, the spacer ring should be replaced. If the clearance is more than 0.0050-inch the thrust plate should be replaced.

21 Prior to installing the camshaft, coat each of the lobes and journals with engine assembly lube or moly-based grease (photo).
22 Slide the camshaft into the engine block. Again, be extra careful not to damage the bearings.
23 Position the camshaft and crankshaft gears so that the valve timing marks line up (photo). With the shafts in this position, the engine is in the number four (4) cylinder firing position.
24 Install the camshaft thrust plate mounting screws and tighten them to the specified torque.
25 Complete the installation by reversing the removal procedure. When installing the distributor, be sure to rotate the engine 360° to the number one (1) cylinder firing position (see Chapter 5).
26 Have the air-conditioning system recharged (if so equipped).

15 Flywheel/driveplate and rear main bearing oil seal — removal and installation

1 The rear main oil seal can be replaced without removal of the oil pan or crankshaft.
2 Refer to Chapter 7, follow all precautionary notes and remove the transmission.
3 If equipped with a manual transmission, remove the clutch pressure plate and disc.
4 Remove the flywheel/driveplate mounting bolts and separate it from the crankshaft.
5 Using a screwdriver or pry bar, carefully remove the oil seal from the block.
6 Using solvent, thoroughly clean the block-to-seal mating surfaces, then dry them with compressed air.
7 Apply a light coat of engine oil to the outside surface of the new seal.
8 Using your fingertips, press the new seal evenly into position in the block.
9 Install the flywhee/driveplate and tighten the bolts to the specified torque.
10 If equipped with a manual transmission, reinstall the clutch disc and pressure plate and tighten the pressure plate bolts to the specified torque.
11 Reinstall the transmission as described in Chapter 7.

16 Engine mounts — replacement with engine in vehicle

1 If the rubber mounts have become hard, split or separated from the metal backing, they must be replaced. This operation may be carried out with the engine/transmission still in the vehicle. See Section 10 for the proper way to raise the engine while it is still in place.

Front mounts

2 Remove the through-bolt and nut.

2A

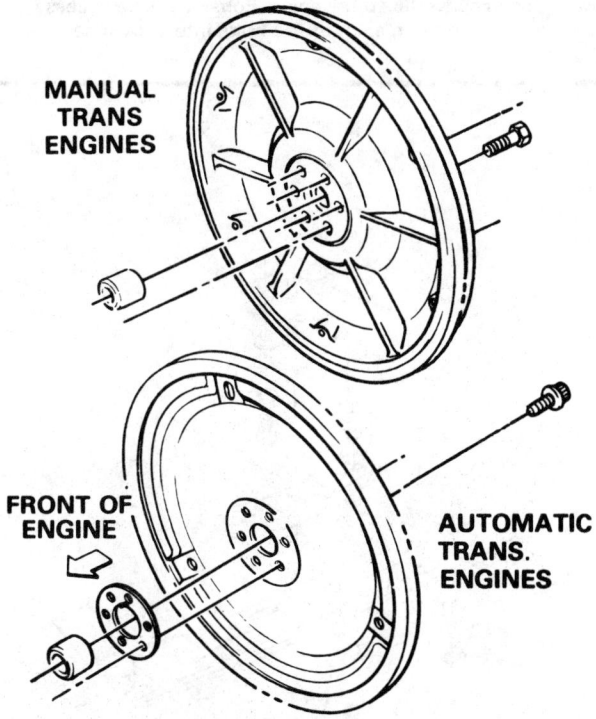

Fig. 2A.16 Manual transmission flywheel and automatic transmission driveplate installation details (Sec 15)

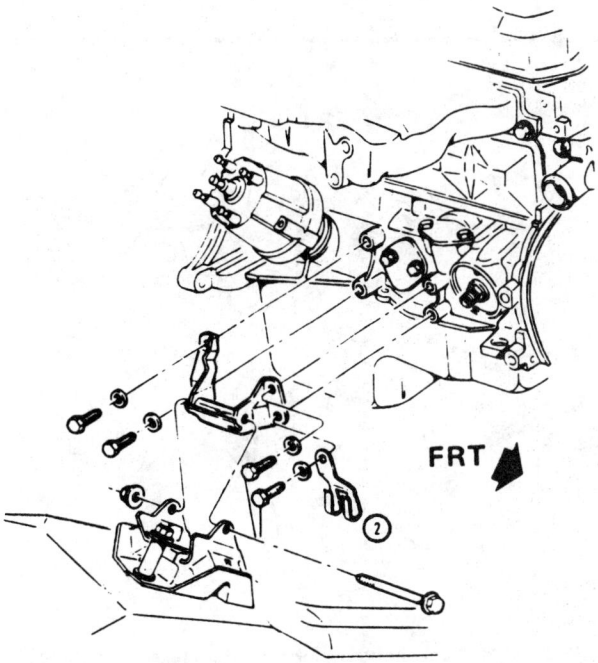

Fig. 2A.17 Right-hand motor mount installation details (Sec 16)

3 Raise the engine slightly using a hoist or jack with a wood block under the oil pan, then remove the mount and frame bracket assembly from the crossmember.
4 Position the new mount, install the through-bolt and nut, then tighten all the bolts to the specified torque.

Rear mount

5 Remove the crossmember-to-mount bolts, then raise the transmission slightly with a jack.
6 Remove the mount-to-transmission bolts, followed by the mount.
7 Install the new mount, lower the transmission and align the crossmember-to-mount bolts.
8 Tighten all the bolts to the specified torque.

17 Engine — removal and installation

Note: *On some models, the upper bellhousing-to-engine bolts are difficult to get at, in some cases necessitating cutting of the interior floor pan to expose the bolts. If this is the case with your vehicle and you do not wish to cut the floor pan, read through the procedure in this Section to familiarize yourself with the details of the four-cylinder engine removal, then refer to the engine removal section in Chapter 2, Part B, and remove your engine and transmission as a unit, noting any differences between the V6 and four-cylinder applications and adjusting the procedure accordingly.*

Removal

1 Disconnect the cable from the negative battery terminal.
2 Remove the clip retaining the underhood light assembly wire harness to the hood.
3 Disconnect the underhood light electrical connector and unbolt the assembly from the hood.
4 Scribe very light lines on the underside of the hood, around the hood mounting bracket, so the hood can be installed in the same position.
5 Apply white paint around the bracket-to-hood bolts so they can be installed in exactly the same positions during installation.
6 Remove the bracket-to-hood bolts.
7 With the help of an assistant, separate the hood from the vehicle.
8 Drain the radiator by turning the petcock handle located at the lower right-hand side of the radiator.
9 Remove the hose from the radiator coolant recovery tank.
10 Remove the upper radiator hose. Take care not to damage the plastic fittings when detaching the hose.
11 If so equipped, remove the automatic transmission cooler lines from the radiator.
12 Remove the bolts securing the top of the radiator fan shroud to the lower half and remove the upper shroud.
13 If equipped with an automatic transmission, remove the radiator from the vehicle. This step is not necessary with manual transmission-equipped vehicles, though it may be desirable to gain more working room for pulling the engine.
14 Mark the drivebelts to simplify installation.
15 Unbolt and remove the fan from the fan pulley.
16 If so equipped, remove the air-conditioning compressor and brackets and lay them to one side. **Caution:** *Do not disconnect any of the air-conditioning lines unless the system has been depressurized by a dealer service department or air-conditioning technician, as personal injury may occur.* Disconnection of the lines should not be necessary in this case.
17 Disconnect the engine electrical connector at the firewall.
18 Disconnect the EFI inlet and return fuel lines at the flexible hoses, then plug the lines to prevent loss of fuel.
19 Disconnect the vacuum brake hose from the vacuum booster.
20 Remove the ground strap from the rear of the cylinder head, then disconnect the wires at the oil pressure and coolant temperature sending units.
21 From inside the vehicle, remove the right-hand hush panel, then disconnect the ECM harness from the main connector of the ECM unit.
22 Remove the right-hand splash shield from the right fender and feed the ECM harness through the opening from inside the vehicle.
23 Remove the heater hoses from the heater core.
24 Disconnect the hoses at the air cleaner. Carefully label them as they are removed to simplify installation.
25 Remove the air cleaner assembly.
26 Remove the canister hose from the EFI.
27 Remove the throttle cable from the EFI.
28 Raise the vehicle and place it securely on jackstands, then drain the engine oil.
29 Disconnect the electrical connectors at the transmission.
30 Unbolt and remove the flywheel/driveplate dust cover.
31 On vehicles equipped with an automatic transmission, remove the now-exposed converter-to-driveplate bolts. It will be necessary to turn the crankshaft to bring each of the bolts into view (use a wrench on

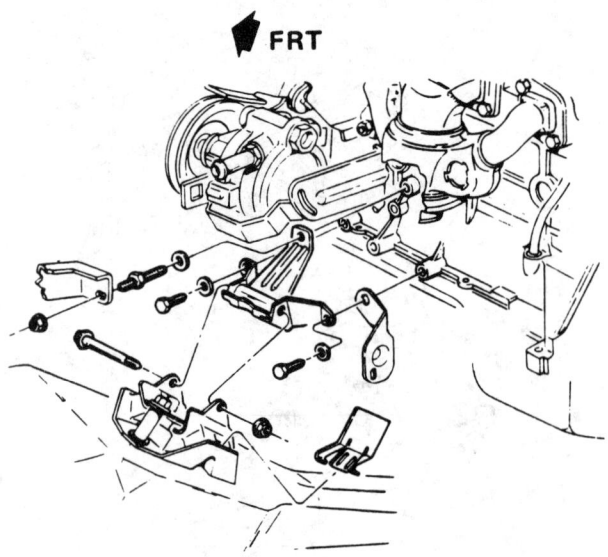

Fig. 2A.18 Left-hand motor mount installation details (Sec 16)

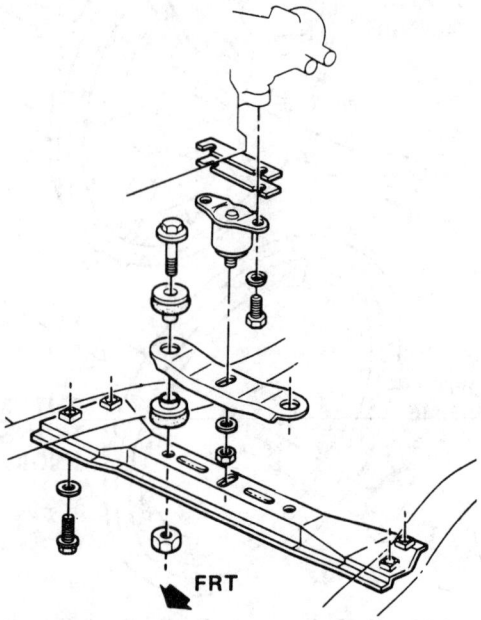

Fig. 2A.19 Transmission mount and crossmember installation details (Sec 16)

the large bolt at the front of the crankshaft). Mark the relative position of the converter and driveplate with a scribe so it can be reinstalled in the same position. Engage a long screwdriver in the teeth of the driveplate to prevent movement as the bolts are loosened.
32 Unbolt and remove the exhaust pipe from the exhaust manifold.
33 Unbolt and remove the exhaust pipe support from the bellhousing.
34 Disconnect the catalytic converter at the tailpipe joint, then remove the converter and tailpipe from the vehicle.
35 Disconnect and label the wires at the starter motor.
36 Unbolt and remove the starter motor.
37 If equipped with a manual transmission, remove the clutch fork return spring.
38 Remove the bellhousing-to-engine bolts (refer to the note at the beginning of this Section).
39 Remove the motor mount through-bolt on each side of the engine.
40 Make sure that all the wires and hoses have been disconnected from under the vehicle.
41 Remove the jackstands and lower the vehicle.
42 Move back into the engine compartment and make one last check to ensure that all wires and hoses have been disconnected from the engine and that all accessories have enough clearance.
43 Attach the hoist lifting chains to the lifting 'eyes' mounted on the engine. There is one bracket at the front of the engine and one at the rear. Make sure the chain is looped properly through the engine brackets and secured with strong nuts and bolts through the chain links. The hook on the hoist should be over the center of the engine with the lengths of chain equal, to lift the engine straight up.
44 Raise the engine hoist until all slack is out of the chains. Do not lift any further at this time.
45 Support the transmission using a jack with wood blocks as cushions.
46 Raise the engine slightly and then pull it forward to clear the input shaft (manual transmission). Where an automatic transmission is involved, keep the torque converter pushed well to the rear to retain engagement of the converter tangs with the oil pump inside the transmission.
47 Carefully lift the engine straight up and out of the engine compartment, continually checking clearances around it.
48 The transmission should remain supported by the floor jack or wood blocks while the engine is out of place.
49 Attach the engine to an engine stand.
50 Refer to Chapter 2, Part D, for further disassembly and rebuilding procedures.

Installation

51 Lift the engine off the engine stand with a hoist. The chains should be positioned just as they were during removal, with the engine sitting level.
52 Lower the engine into place inside the engine compartment, closely watching clearances. On manual transmission-equipped vehicles, carefully guide the engine onto the transmission input shaft. The two components should be at the same angle, with the shaft sliding easily into the engine.
53 Install the engine mount through-bolts and the bellhousing bolts. Tighten them to the specified torque.
54 Install the remaining engine components in the reverse order of removal.
55 Fill the cooling system with the specified coolant mixture (Chapter 1).
56 Fill the engine with the correct grade of engine oil (Chapter 1).
57 Check the transmission fluid level, adding fluid as necessary.
58 Connect the positive battery cable, followed by the negative cable. If sparks or arcing occur as the negative cable is connected to the battery, make sure that all electrical accessories are turned off (check the dome light first). If arcing still occurs, make sure that all electrical wiring is properly connected to the engine and transmission.
59 Refer to Chapter 2, Part D, for the recommended engine start-up sequence.

Chapter 2 Part B V6 engine

Refer to Chapter 13 for specifications and information related to 1985 and later models

Contents

Air filter — replacement . . . . . Chapter 1
Camshaft — removal and installation 16
Check engine light . . . . . Chapter 6
Crankcase front cover — removal and installation 13
Compression check . . . . . Chapter 1
Cylinder heads — removal and installation 8
Drivebelt check and adjustment . . . . . Chapter 1
Engine — removal and installation 18
Engine mounts — replacement with engine in vehicle 17
Engine oil and filter change . . . . . Chapter 1
Engine overhaul — general information . . . . . Chapter 2D
Engine removal — methods and precautions . . . . . Chapter 2D
Engine repairs possible with the engine in the vehicle . . . . . Chapter 2D
Exhaust manifolds — removal and installation 7
Front cover oil seal — replacement 14
General information 1
Hydraulic lifters — removal, inspection and installation 5
Intake manifold — removal and installation 4
Oil pan — removal and installation 9
Oil pump — removal and installation 10
Rear main bearing oil seal — replacement (engine in vehicle) . 11
Rocker arm covers — removal and installation 2
Spark plug replacement . . . . . Chapter 1
Timing chain and sprockets — inspection, removal and installation 15
Valve lash — adjustment 6
Valve train components — replacement (cylinder head installed) 3
Vibration damper — removal and installation 12
Water pump — removal and installation . . . . . Chapter 3

Specifications

Torque specifications

	Ft-lbs	M-kg
Camshaft sprocket bolts	18	2.5
Clutch cover-to-flywheel bolts	15	2.0
Rear camshaft cover bolts	7	1.0
Cylinder head bolts	68	9.4
Connecting rod cap nuts	37	5.1
Crankshaft pulley bolts	25	3.4
Crankshaft pulley hub bolt	75	10.3
Distributor hold-down bolt	25	3.4
Driveplate-to-torque converter bolts	27	3.7
EGR valve mounting bolts	15	2.0
Engine mounting bracket bolts	80	11.0
Engine strut bracket	35	4.8
Exhaust manifold mounting bolts	25	3.4
Flywheel mounting bolts	50	7.0
Front cover mounting bolts (small)	15	2.0
Front cover mounting bolts (large)	25	3.4
Fuel pump mounting bolts	15	2.0
Intake manifold mounting bolts	23	3.2
Main bearing cap bolts	70	9.6
Oil filter bolt	15	2.0
Oil filter connector bolt	29	3.9
Oil pan mounting bolts (small)	7	1.0
Oil pan mounting bolts (large)	18	2.5
Oil pump mounting bolt	30	4.1
Oil pump cover bolts	8	1.1
Oil pressure switch	5	0.7
Oil drain plug	18	2.5
Rear lifting bracket bolt	25	3.4
Rocker arm cover bolts	8	1.1
Rocker arm studs	45	6.2
Spark plugs	12	1.6
Starter motor mounting bolts	32	4.4
Strut bracket assembly nut and bolt	35	4.8
Timing chain tensioner bolts	15	2.0
Water outlet housing bolts	25	3.4
Water pump mounting bolts (small)	7	1.0
Water pump mounting bolts (medium)	15	2.0
Water pump mounting bolts (large)	25	3.4
Water pump pulley bolt	15	2.0

Note: *Refer to Chapter 2, Part D for additional specifications.*

Cylinder numbering and direction distributor rotor rotates (arrow)

Fig. 2B.1 V6 engine (partially disassembled) — right side view

1 *Distributor*
2 *Transmission fluid level dipstick*
3 *Ignition coil*
4 *AIR management valve*
5 *AIR system check valve*
6 *Intake manifold*
7 *Exhaust manifold*
8 *Strut*
9 *Starter motor*
10 *Starter solenoid*
11 *Transmission dust cover*
12 *Transmission oil cooler lines*

Fig. 2B.2 V6 engine (partially disassembled) — left side view

1 *AIR system deceleration valve*
2 *EGR valve*
3 *PCV valve*
4 *Rocker arm cover*
5 *Exhaust manifold insulator*
6 *Cylinder head*
7 *Bellhousing bolts*
8 *Oil filter*
9 *Oil pan*
10 *Fuel pump*
11 *Crankcase front cover*

Fig. 2B.3 V6 engine (partially disassembled) — front view

1 Intake manifold
2 Oil level dipstick
3 Left cylinder head
4 Air-conditioning compressor mounting bracket
5 Power steering pump mounting bracket
6 Water pump
7 Fuel pump
8 Left motor mount
9 Crankshaft pulley
10 AIR system pump bracket
11 Right motor mount
12 Alternator mounting bracket
13 Right cylinder head
14 Thermostat housing

1 General information

The forward Sections in this Part of Chapter 2 are devoted to 'in-vehicle' repair procedures for the V6 engine. The latter Sections in this Part of Chapter 2 involve the removal and installation procedures for the V6 engine. All information concerning engine block and cylinder head servicing can be found in Part D of this Chapter.

The repair procedures included in this Part are based on the assumption that the engine is still installed in the vehicle. Therefore, if this information is being used during a complete engine overhaul — with the engine already out of the vehicle and on a stand — many of the steps included here will not apply.

The Specifications included in this Part of Chapter 2 apply only to the engine and procedures found here. For Specifications regarding engines other than the V6, see Part A or C, whichever applies. Part D of Chapter 2 contains the Specifications necessary for engine block and cylinder head rebuilding procedures.

2 Rocker arm covers — removal and installation

Note: *Refer to the illustrations in Section 18.*

Right side

1 Disconnect the cable from the negative battery terminal.
2 Remove the air cleaner assembly, tagging each hose to be disconnected with a piece of numbered tape to simplify reinstallation.
3 Remove the bolts retaining the air management valve/coil bracket (photo).
4 Disconnect the wires and hoses that would interfere with the removal of the rocker arm covers, tagging them as they are disconnected.
5 Disconnect the carburetor controls at the carburetor, then remove them from the support bracket.
6 Remove the rocker arm cover bolts (photo).
7 Remove the rocker arm cover. **Note:** *If the cover sticks to the cylinder head, use a block of wood and a rubber hammer to dislodge it. If the cover still will not come loose, pry on it carefully, but do not distort the sealing flange surface.*
8 Before installing the cover, clean all dirt, oil and old gasket material from the sealing surfaces of the cover and cylinder head with a scraper and degreaser.
9 Apply a continuous 3/16-inch (5 mm) diameter bead of RTV-type sealant to the flange of the cover. Be sure to apply the sealant inboard of the bolt holes.
10 Place the rocker arm cover on the cylinder head while the sealant is still wet and install the mounting bolts. Tighten the bolts a little at a time to the specified torque.
11 Complete the installation by reversing the removal procedure.

Left side

12 Disconnect the cable from the negative battery terminal.
13 Disconnect the hoses at the PCV valve and label them.
14 Remove the air cleaner assembly, tagging each hose to be disconnected with a piece of numbered tape to simplify reinstallation.
15 Disconnect all other wires and hoses that would interfere with the removal of the rocker arm cover, tagging them as they are disconnected.
16 Remove the bolts retaining the solenoid/hose bracket to the front of the cover and set the assembly aside.
17 Remove the rocker arm cover bolts.
18 Disconnect the fuel line at the carburetor, plugging the fitting at the carburetor and the disconnected fuel line to prevent leakage and contamination.
19 Refer to Paragraphs 7 through 11 in this Section.

3 Valve train components — replacement (cylinder head installed)

1 Remove the rocker arm cover(s) as described in Section 2.
2 If only the pushrod is to be replaced, loosen the rocker nut enough to allow the rocker arm to be rotated away from the pushrod. Pull the pushrod out of the hole in the cylinder head.
3 If the rocker arm is to be removed, remove the rocker arm nut and pivot and lift off the rocker arm.
4 If the valve spring is to be removed, remove the spark plug from the affected cylinder.
5 There are two methods that will allow the valve to remain in place while the valve spring is removed. If you have access to compressed air, install an air hose adapter (GM part number J-22794) in the spark plug hole. When air pressure is applied to the adapter, the valves will be held in place by the pressure.
6 If you do not have access to compressed air, bring the piston of the affected cylinder to top dead center (TDC) on the compression stroke. Feed a long piece of 1/4-inch nylon cord in through the spark plug hole until it fills the combustion chamber. Be sure to leave the end of the cord hanging out of the spark plug hole so it can be removed easily. Rotate the crankshaft with a wrench (in the normal direction of rotation) until slight resistance is felt.
7 Reinstall the rocker arm nut (without the rocker arm).
8 Insert the slotted end of a valve spring compression tool under the nut and compress the spring just enough to remove the spring keepers, then release the pressure on the tool.
9 Remove the retainer, cup shield, O-ring seal, spring, spring damper (if so equipped) and valve stem oil seal (if so equipped).
10 If a rocker arm stud requires replacement, first determine whether it is a screw-in or press-in type. Screw-in studs are retained by a nut atop the stud boss, while press-in studs have no retaining nut.
11 Screw-in studs may be replaced by simply removing the damaged one and replacing it with a new one. Be sure to reinstall the pushrod guide (if so equipped) under the stud nut and tighten the nut to the

2.3 Removing the AIR management valve/coil bracket (V6)

2.6 Removing the rocker arm cover bolts (V6)

specified torque.
12 Press-in studs that have damaged threads or are loose in the cylinder head should be replaced with new studs that are available in 0.003-inch and 0.013-inch oversizes. Press-in stud replacement procedures require the use of special tools and should be performed by an automotive machine shop. If such services are required, remove the cylinder head(s) by referring to Section 8.
13 Inspection procedures for the hydraulic lifters are detailed in Section 5. Procedures regarding other valve train components are detailed in Chapter 2, Part D.
14 Installation of the valve train components is the reverse of the removal procedure. Always use new valve stem oil seals whenever the spring keepers have been disturbed. Before installing the rocker arms, coat the bearing surfaces of the arms and pivots with moly-based grease or engine assembly lube. Be sure to adjust the valve lash as detailed in Section 6.

4 Intake manifold — removal and installation

1 If the vehicle is equipped with air-conditioning, carefully examine the routing of the hoses and the mounting of the compressor. You may be able to remove the intake manifold without disconnecting the system. If you are in doubt, take the vehicle to a dealer or refrigeration specialist to have the system depressurized. *Do not, under any circumstances, disconnect the hoses while the system is under pressure.*
2 Disconnect the cable from the negative battery terminal.
3 Drain the coolant from the radiator (Chapter 1).
4 Remove the air cleaner assembly, tagging each hose to be disconnected with a piece of numbered tape to simplify reinstallation.
5 Label and disconnect all electrical wires and vacuum hoses at the carburetor.
6 Disconnect the fuel line at the carburetor. Be prepared to catch some fuel, then plug the fuel line to prevent contamination.
7 Disconnect the throttle cable, making careful note of how it was installed.
8 Disconnect the spark plug wires at the spark plugs, referring to the removal technique described in Chapter 1.
9 Disconnect the wires at the coil, again using numbered pieces of tape to label them.
10 Remove the distributor cap and the attached spark plug wires (refer to Chapter 5).
11 Remove the distributor (refer to Chapter 5).
12 Remove the power brake fitting from the rear of the manifold (photo).
13 Remove the power brake tube mounting bracket from the manifold boss (photo).
14 Move the power brake tube/hose aside so it will not interfere with the removal of the manifold.
15 Remove the solenoid/hose bracket from the front of the left rocker arm cover.
16 Remove the mounting bolts from the left rocker arm cover, then remove the cover.
17 Remove the AIR management valve/coil mounting bracket assembly bolts from the right cylinder head, disconnect the AIR management hose and remove the assembly.
18 Remove the bolts from the right rocker arm cover, then remove the cover.
19 Remove the upper radiator hose from the manifold.
20 Disconnect the heater hose at the manifold.
21 Disconnect the vacuum lines from the TVS switch at the front of the manifold (photo).
22 Remove the electrical connector from the coolant switch at the front of the manifold.
23 Make sure that all wires, vacuum hoses and coolant hoses that would interfere with manifold removal have been disconnected.
24 If the manifold is to be replaced with a new one, the external components remaining on the manifold must be removed for transfer to the new manifold. These components may be removed either before or after the manifold has been separated from the engine. These components include:

The carburetor choke assembly and carburetor studs or bolts (refer to Chapter 4 for details)
The coolant switch
The EGR valve (use a new gasket when installing)
The emissions system TVS valve

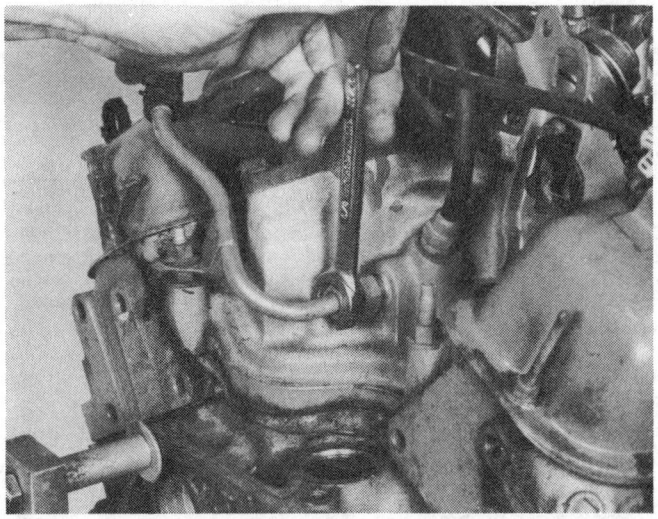

4.12 Removing the power brake tube fitting from the rear of the manifold (V6)

4.13 Removing the power brake tube fitting bracket from the manifold (V6)

4.21 Disconnecting the vacuum lines from the TVS switch (V6)

25 Remove the manifold mounting bolts (photo).
26 Separate the manifold from the engine. Do not pry between the mating surfaces because it could damage them. Tap the manifold with a hammer and wooden block to loosen it if necessary.
27 If a new manifold is being installed, transfer the external components from the old manifold to the new one.
28 Before installing the manifold, place clean, lint-free rags in the engine cavity and clean the engine block, cylinder head and manifold gasket surfaces. All gasket material and sealant must be removed prior to installation. Remove all dirt and gasket remnants from the engine cavity.
29 Clean the gasket sealing surfaces with degreaser, then apply a 3/16-inch (5 mm) diameter bead of RTV-type sealant to the engine block gasket surfaces only.
30 Install the new intake gaskets on the cylinder heads. Notice that the gaskets are marked *Right* and *Left*. Be sure to use the correct gasket on each cylinder head.
31 Hold the gaskets in place by extending the bead of RTV up about 1/4-inch onto the gasket ends. The new gaskets will have to be cut, as illustrated, so they can be installed behind the pushrods.
32 Carefully lower the intake manifold into position, making sure that you do not disturb the gaskets.
33 Install the intake manifold mounting bolts and tighten them following the sequence illustrated. Tighten the bolts a little at a time until they are all at the specified torque.
34 Install the remaining components in the reverse order of removal.
35 Fill the radiator with coolant, start the engine and check for leaks. Adjust the ignition timing and idle speed as necessary (refer to Chapter 1).

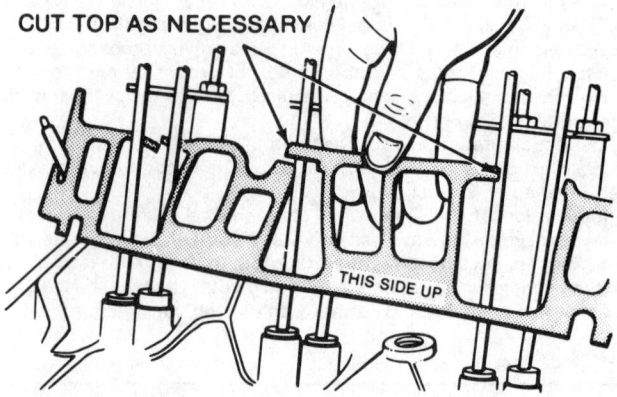

Fig. 2B.4 The intake manifold gasket must be cut as shown before installing it so the top can be positioned behind the pushrods (Sec 4)

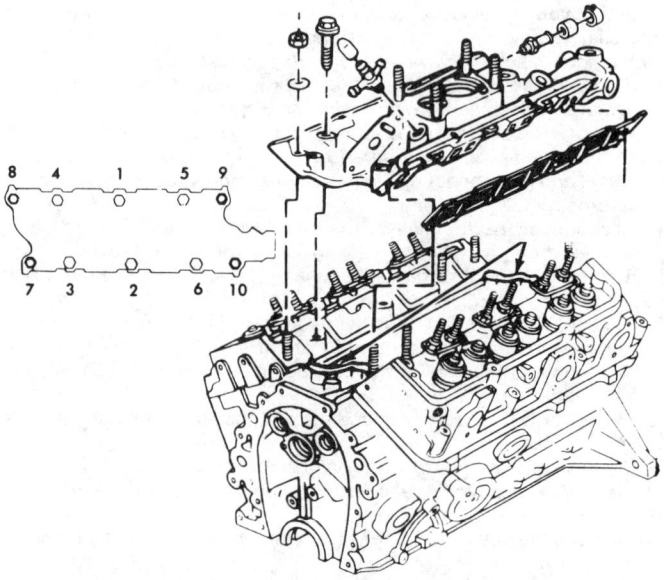

Fig. 2B.5 Recommended tightening sequence for the intake manifold mounting bolts (Sec 4)

5 Hydraulic lifters — removal, inspection and installation

1 A noisy hydraulic lifter can be isolated when the engine is idling. Place a length of hose or tubing near the position of each valve while listening at the other end of the tube. Another method is to remove the rocker arm cover and, with the engine idling, place a finger on each of the valve spring retainers, one at a time. If a valve lifter is defective, it will be evident from the shock felt at the retainer as the valve seats.
2 Assuming that adjustment is correct, the most likely cause of a noisy valve lifter is a piece of dirt trapped between the plunger and the lifter body.
3 Remove the rocker arm covers as described in Section 2.
4 Remove the intake manifold as described in Section 4.
5 Loosen the rocker arm nut and rotate the rocker arm away from the pushrod.
6 Remove the pushrod.
7 To remove the lifter, a special hydraulic lifter removal tool should be used, or a scribe can be positioned at the top of the lifter and used

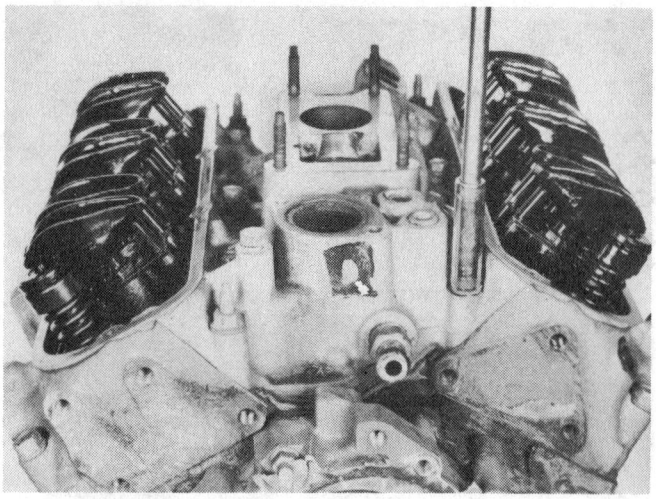

4.25 Removing the manifold mounting bolts (V6)

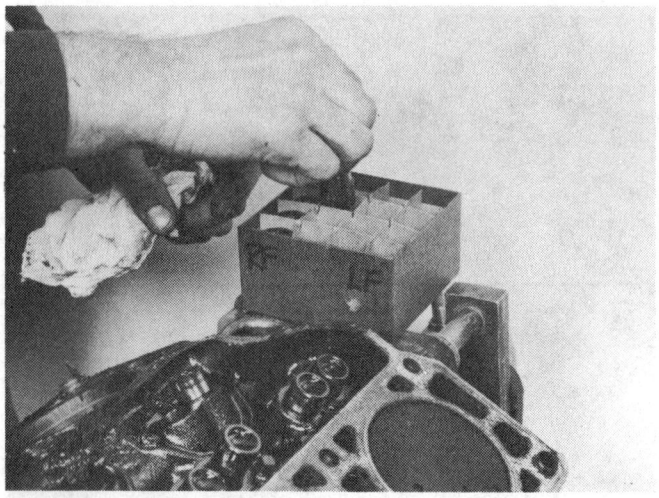

5.8 As the lifters are removed from the engine block, they should be stored separately to ensure reinstallation in their original positions (V6)

to force the lifter up. *Do not use pliers or other tools on the outside of the lifter body, as they will damage the finished surface and render the lifter useless.*

8 The lifters should be kept separate for reinstallation in their original positions (photo).

9 To dismantle a valve lifter, hold the plunger down with a pushrod and then extract the retainer spring with a small screwdriver.

10 Remove the pushrod seat and the metering valve.

11 Remove the plunger, ball check valve and plunger spring. Remove the ball check valve and spring by prying them out with a small screwdriver.

12 Clean the lifter components with solvent and dry them with compressed air. Examine the internal components for wear and check the ball carefully for flat spots. Refer to Chapter 2, Part D, for additional lifter (and camshaft) inspection procedures.

13 If the lifters are worn, they must be replaced with new ones and the camshaft must be replaced as well (see Chapter 2, Part D). If the lifters were contaminated with dirt, they can be reinstalled — they *may* operate normally.

14 Reassembly should be done in the following manner:

a) Place the ball check valve on the small hole in the bottom of the plunger.
b) Insert the ball check valve spring into the seat in the retainer and place the retainer over the ball so that the spring rests on the ball. Using a small screwdriver, carefully press the retainer into position in the plunger.
c) Place the plunger spring over the ball retainer, invert the lifter body and slide it over the spring and plunger. Make sure the oil holes in the body and plunger line up.
d) Fill the assembly with 10-weight oil. Place the metering valve and pushrod seat in position, press down on the seat and install the retainer spring.

15 When installing the lifters, make sure they are replaced in their original bores and coat them with moly-based grease or engine assembly lube.

16 The remaining installation steps are the reverse of removal.

6 Valve lash — adjustment

1 Disconnect the cable from the negative battery terminal.

2 If the rocker arm covers are still on the engine, refer to Section 2 and remove them.

3 If the valve train components have been serviced just prior to this procedure, make sure that the components are completely reassembled.

4 Rotate the crankshaft until the number one piston is at top dead center (TDC) on the compression stroke. To make sure that you do not mix up the TDC positions of the number one and four pistons, check the position of the rotor in the distributor to see which terminal it is pointing at. Another method is to place your fingers on the number one rocker arms as the timing marks line up at the crankshaft pulley. If the rocker arms are not moving, the number one piston is at TDC. If they move as the timing marks line up, the number four piston is at TDC.

5 Back off the rocker arm nut until play is felt at the pushrod, then turn it back in just until all play is removed. This can be determined by rotating the pushrod while tightening the nut. Just when drag is felt at the pushrod, all lash has been removed. Now turn the nut in an additional 3/4 turns.

6 Adjust the number one, five and six cylinder intake valves and the number one, two and three cylinder exhaust valves, with the crankshaft in this position, using the method just described.

7 Rotate the crankshaft until the number four piston is at TDC on the compression stroke and adjust the number two, three and four cylinder intake valves and the number four, five and six cylinder exhaust valves.

8 Refer to Section 2 and install the rocker arm covers.

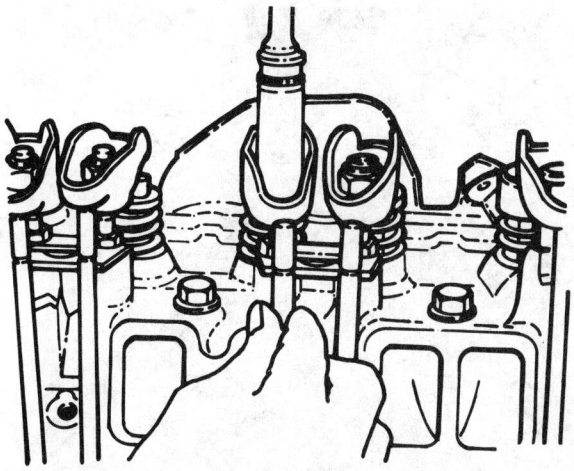

Fig. 2B.6 Determining at what point drag is felt on the pushrod by turning it as the nut is tightened (Sec 6)

7 Exhaust manifolds — removal and installation

Right side

1 Remove the cable from the negative battery terminal.

2 Raise the front of the vehicle and support it securely on jackstands. Block the rear wheels to keep the vehicle from rolling.

3 Remove the bolts attaching the exhaust pipe to the exhaust manifold, then separate the pipe from the manifold.

4 Remove the jackstands and lower the vehicle.

5 Disconnect the oxygen sensor pigtail electrical connector.

6 Disconnect the air management hose at the check valve.

7 Disconnect the spark plug wires from the spark plugs, labeling them as they are disconnected to simplify installation.

8 Remove the exhaust manifold mounting bolts and separate the manifold from the engine (photo).

9 Installation is the reverse of the removal procedure. Before installing the manifold, be sure to thoroughly clean the mating surfaces on the manifold and cylinder head.

Left side

10 Disconnect the cable from the negative battery terminal.

11 Raise the front of the vehicle and place it securely on jackstands. Block the rear wheels to keep the vehicle from rolling.

12 Remove the bolts retaining the exhaust pipe to the manifold, then disconnect the pipe from the manifold.

13 Remove the four bolts and one nut accessible at the rear of the manifold.

14 Remove the jackstands and lower the vehicle.

15 Remove the air cleaner assembly, labeling all hoses.

7.8 Removing the right exhaust manifold mounting bolts (V6)

16 Disconnect the hoses leading to the air management valve.
17 Disconnect and label any wires that will interfere with the removal of the manifold.
18 On some models it may be necessary to remove the power steering pump bracket from the cylinder head. If so, loosen the pump adjusting bracket bolt and remove the pump drivebelt from the pulley first. After removing the bracket from the cylinder head, place the steering pump assembly aside, out of the way.
19 Remove the remaining manifold bolts and separate the manifold and heat shield from the engine.
20 Installation is the reverse of the removal procedure. Be sure to thoroughly clean the cylinder head and manifold surfaces before installing the manifold.

8 Cylinder heads — removal and installation

Left side

1 Remove the intake manifold (refer to Section 4).
2 Raise the vehicle and place it securely on jackstands.
3 Locate the engine block drain plugs to the rear of the motor mounts (the plug on the left side is just above the oil filter). Remove the plugs and drain the block (photo).
4 Disconnect the exhaust pipe from the exhaust manifold.
5 Remove the oil dipstick tube assembly from the side of the engine.
6 Remove the jackstands and lower the vehicle.
7 **Note:** *Steps 8 through 11 are to be followed if the head is to be replaced with a new one. These Steps may be performed either before or after the head has been removed. In the accompanying illustrations, the procedures were performed before the head was removed.*
8 Remove the exhaust manifold (refer to Section 7).
9 Remove the power steering pump bracket from the side of the cylinder head (photo).
10 Remove the air-conditioner compressor bracket from the front of the cylinder head (photo).
11 Remove the coolant temperature sending unit from the front of the cylinder head (photo).
12 Loosen the rocker arm nuts enough to allow removal of the pushrods, then remove the pushrods (photos).
13 Loosen the head bolts in a sequence *opposite* to the one used for tightening them (see accompanying illustration).
14 Remove the cylinder head. To break the gasket seal, insert a bar into one of the exhaust ports, then carefully lift on the tool (photo).
15 If a new cylinder head is being installed, attach the components previously removed from the old head. Before installing the cylinder head, the gasket surfaces of both the head and the engine block must be clean and free of nicks and scratches. Also, the threads in the block and on the head bolts must be completely clean, as any dirt remaining in the threads will affect bolt torque.
16 Place the gasket in position over the locating dowels, with the note 'This Side Up' visible.

8.3 Arrow indicates the location of the left-hand engine drain plug (V6)

8.9 Removing the power steering pump mounting bracket (V6)

8.10 Removing the air-conditioner compressor bracket (V6)

8.11 Removing the coolant temperature sending unit (V6)

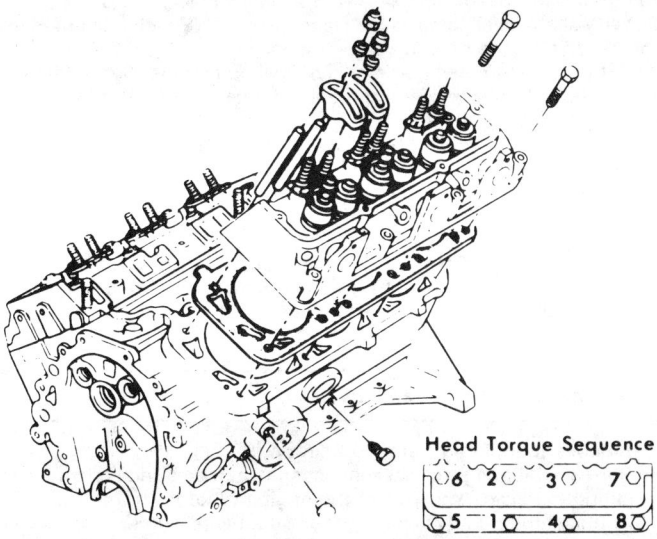

Fig. 2B.7 Recommended tightening sequence for the cylinder head mounting bolts (Sec 8)

17 Position the cylinder head over the gasket.
18 Coat the cylinder head bolts with an appropriate sealer (GM part number 1052080, or equivalent) and install the bolts.
19 Tighten the bolts in the proper sequence (see the accompanying illustration) to the specified torque. Work up to the final torque in three steps.
20 Install the pushrods, making sure the lower ends are in the lifter seats, place the rocker arm ends over the pushrods and loosely install the rocker arm nuts.
21 The remaining installation steps are the reverse of those for removal. Before installing the rocker arm covers, adjust the valve lash (refer to Section 6).

Right side

22 Remove the intake manifold (refer to Section 4).
23 Raise the vehicle and place it securely on jackstands.
24 Locate the engine block drain plugs to the rear of the motor mounts (the plug on the left side is just above the oil filter), remove the plugs and drain the block.
25 Disconnect the exhaust pipe from the exhaust manifold.
26 Remove the jackstands and lower the vehicle.
27 Remove the alternator from the alternator bracket, then remove the bracket from the head (photo).
28 Remove the lifting 'eye' from the rear of the head (necessary only if the head is to be replaced with a new one) (photo).
29 Remove the exhaust manifold (if the head is to be replaced with a new one).

8.12A Loosening the rocker arm adjusting nuts (V6)

8.12B When removing the pushrods, be sure to store them separately to ensure reinstallation in their original positions (V6)

8.14 Using a wrench handle to break the gasket seal on the right-hand cylinder head (V6)

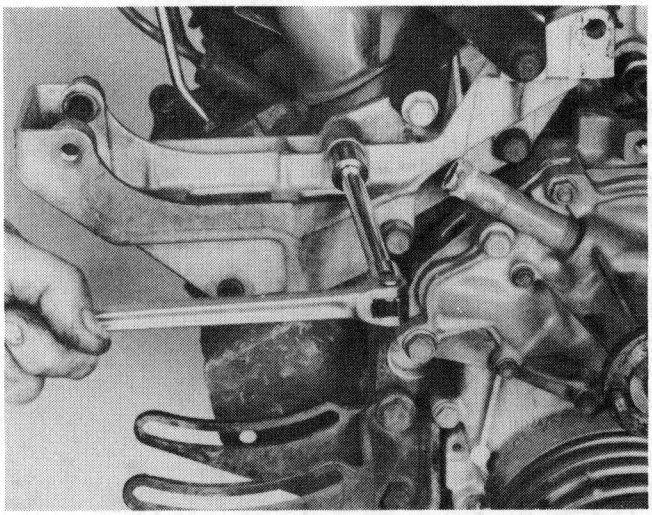

8.27 Removing the alternator bracket from the right-side cylinder head (V6)

30 Loosen the rocker arm nuts sufficiently to allow removal of the pushrods, then remove the pushrods.
31 Loosen the head bolts in a sequence *opposite* to the one used for tightening them (see accompanying illustration).
32 Remove the cylinder head. To break the gasket seal, insert a bar into one of the exhaust ports, then carefully lift on the tool.
33 To install the head, refer to Steps 15 through 21.

9 Oil pan — removal and installation

1 Disconnect the cable from the negative battery terminal.
2 Remove the air cleaner assembly, tagging each hose to be disconnected with a piece of numbered tape to simplify reinstallation.
3 Remove the distributor cap (refer to Chapter 5) to prevent damage when the engine is raised.
4 Unbolt and remove the top half of the fan shroud.
5 Raise the vehicle and support it on jackstands.
6 Drain the engine oil.
7 Disconnect the exhaust crossover pipe at the exhaust manifold flanges. Lower the exhaust pipes and suspend them from the frame with wire.
8 If equipped with an automatic transmission, remove the converter shroud.
9 If equipped with a manual transmission, remove the flywheel cover.
10 Remove the starter (refer to Chapter 5).
11 Remove the through-bolt at each engine mount.
12 At this time the engine must be raised slightly to allow the oil pan to clear the crossmember. The preferred method requires an engine hoist or 'cherry picker'. Hook up the lifting chains as described in Section 18.
13 An alternative method can be used if you are careful. Place a floor jack and a block of wood under the oil pan. The wood block should spread the load across the oil pan to prevent damage or collapse. The oil pump pick-up and screen are very close to the oil pan bottom, so any collapse of the pan may damage the pick-up or prevent the oil pump from drawing oil properly.
14 With either method, raise the engine slowly until wood blocks can be placed between the frame front crossmember and the engine block. The blocks should be approximately three inches thick. Check clearances all around the engine as it is raised. Pay particular attention to the distributor and the cooling fan.
15 Lower the engine onto the wood blocks. Make sure it is firmly supported. If a hoist is being used, keep the lifting chains secured to the engine.
16 Remove the oil pan bolts. Note the different sizes used and their locations.
17 Remove the oil pan.
18 Before installing the pan, make sure that the sealing surfaces on the pan, block and front cover are clean and free of oil. If the old pan is being reinstalled, make sure that all sealant has been removed from the pan sealing flange and from the blind attaching holes.
19 With all the sealing surfaces clean, place a 1/8-inch bead of RTV-type sealant on the oil pan sealing flange.
20 Lift the pan into position and install all bolts finger tight. There is no specific order for torquing the bolts; however, it is a good idea to tighten the end bolts first.
21 Lower the engine onto the mounts and install the through-bolts. Tighten the bolts to the specified torque.
22 Follow the removal steps in reverse order. Fill the crankcase with the correct grade and quantity of oil, start the engine and check for leaks.
23 Most of the above steps will not be required if the engine has been removed from the vehicle. The pan can simply by unbolted and removed, cleaned and installed as indicated.

10 Oil pump — removal and installation

1 Remove the oil pan (refer to Section 9).
2 Remove the pump-to-rear main bearing cap bolt and separate the pump and extension shaft from the engine (photo).
3 To install the pump, move it into position and align the top end of the hexagonal extension shaft with the hexagonal socket in the lower end of the distributor drive gear. The distributor drives the oil pump, so it is essential that this alignment is correct.
4 Install the oil pump-to-rear main bearing cap bolt and tighten it to the specified torque.
5 Reinstall the oil pan.

11 Rear main bearing oil seal — replacement (engine in vehicle)

1982 and 1983 engines

Note: *Special tools, as noted in the Steps which follow, are required for this procedure. They are available from your dealer or may in some cases be rented from an auto parts store or tool rental shop.*

1 Although the crankshaft must be removed to install a new seal, the upper portion of the seal can be repaired with the crankshaft in place.
2 Remove the oil pan and oil pump (Sections 9 and 10).
3 Using GM tool J-29114-2, drive the old seal gently back into the groove, packing it tight. It will pack in to a depth of 1/4-to-3/4-inch.
4 Repeat the procedure on the other end of the seal.
5 Measure the amount that the seal was driven up into the groove on one side and add 1/16-inch. Remove the old seal from the main bearing cap. Use the main bearing cap as a fixture and cut off a piece of the old seal to the predetermined length. Repeat this process for the other side.
6 Install the guide tool J-29114-1 on the block.
7 Using the packing tool J-29114-2, work the short pieces of the

8.28 Removing the lifting 'eye' from the right-side cylinder head (V6)

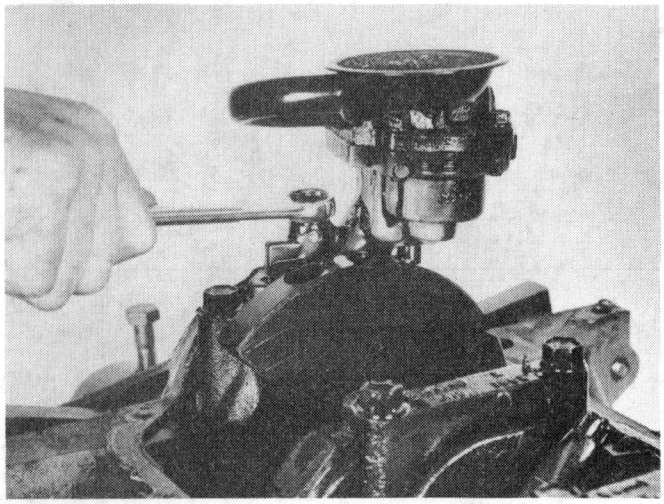

10.2 Removing the oil pump-to-rear main bearing cap bolt (V6)

previously cut seal into the guide tool J-29114-1 and pack them into the block groove on each side. The guide and packing tools have been machined to provide a built-in stop. Use of oil on the seal pieces will ease installation.

8 Remove the guide tool.

9 Install a new seal in the main bearing cap.

10 Apply a thin, even coat of anaerobic-type gasket sealant to the areas of the rear main bearing cap indicated in the illustration in Chapter 2, Part D (Section 21). **Caution:** *Do not get any sealant on the bearing or seal faces.*

11 Tighten the rear main bearing cap bolts to the specified torque.

12 Install the oil pump and oil pan.

1984 engines

13 Always service both halves of the rear main oil seal. While replacement of this seal is much easier with the engine removed from the vehicle, the job can be done with the engine in place.

14 Remove the oil pan and oil pump as described previously in this Chapter.

15 Remove the rear main bearing cap from the engine.

16 Using a screwdriver, pry the lower half of the oil seal from the bearing cap.

17 To remove the upper half of the seal, use a small hammer and a brass pin punch to roll the seal around the crankshaft journal. Tap one end of the seal with the hammer and punch (be careful not to strike the crankshaft) until the other end of the seal protrudes enough to pull the seal out with pliers.

18 Remove all sealant and foreign material from the main bearing cap. Do not use an abrasive cleaner for this.

19 Inspect the components for nicks, scratches and burrs at all sealing surfaces. Remove any defects with a fine file or deburring tool.

20 Apply a very thin coat of RTV-type gasket sealant to the outer surface of the upper seal as shown in the accompanying illustration. Do not get any sealant on the seal lips.

21 Included in the purchase of the rear main oil seal should be a small plastic installation tool; if not, a tool may be fashioned from an old feeler gauge blade (refer to the accompanying illustration).

22 With the upper half of the seal positioned so that the seal lip faces

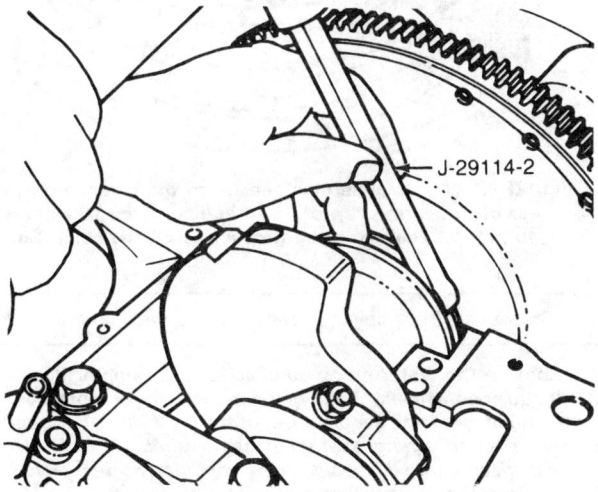

Fig. 2B.8 Using a packing tool to pack the old rear main bearing oil seal into the groove — 1982 and 1983 V6 engines (Sec 11)

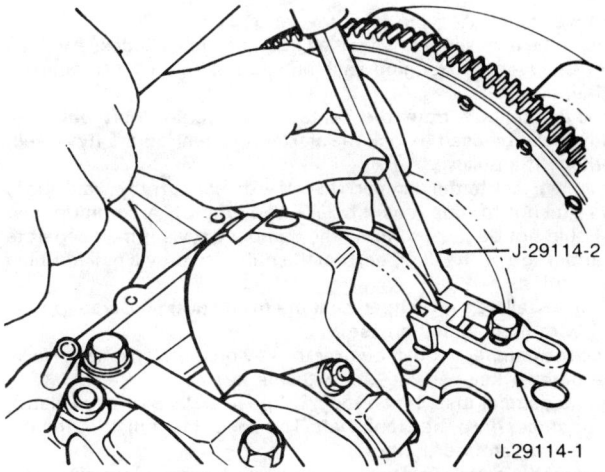

Fig. 2B.9 Using guide tools to install short pieces of seal cut from the old main bearing seal in the upper seal groove — 1982 and 1983 V6 engines (Sec 11)

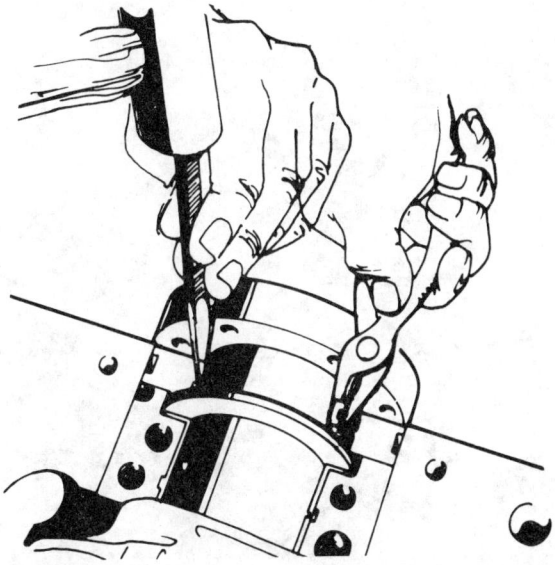

Fig. 2B.10 Removing the crankshaft rear main oil seal upper half from a 1984 engine (Sec 11)

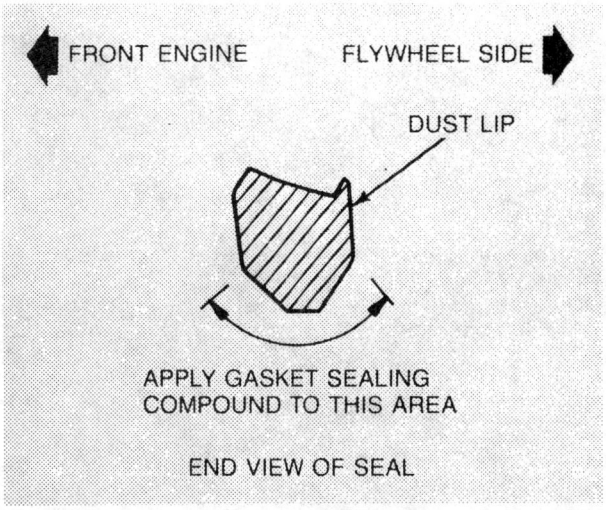

Fig. 2B.11 A very thin coat of RTV-type gasket sealant should be applied to the area shown on 1984 rear main bearing seals (avoid getting sealant on the seal lips) (Sec 11)

toward the *front* of the engine and the small dust lip faces toward the flywheel, install the seal by rolling it around the crankshaft using the installation tool as a 'shoehorn' for protection.
23 Apply sealing compound as described in Step 20 to the other half of the seal and install it in the bearing cap.
24 Apply a 1/32-inch bead of anaerobic sealant to the cap between the rear main oil seal end and oil pan rear seal groove. Be sure to keep the sealant off the rear main oil seal and bearing and out of the drain slot.
25 Just before installing the cap, apply a light coat of moly-based grease or engine assembly lube to the crankshaft surface that will contact the seal.
26 Install the rear main bearing cap and tighten the bolts to the specified torque.
27 Install the oil pump and oil pan.

12 Vibration damper — removal and installation

1 Remove the hose from the retaining strap atop the radiator shroud by releasing the end of the strap from the underside of the shroud with pliers. Move the hose to the side, out of the way.
2 Remove the bolts and separate the upper radiator shroud from the radiator.
3 Remove the cooling fan from the fan pulley.
4 Loosen the accessory drivebelt adjusting bolts as necessary, then remove the drivebelts, tagging each one as it is removed to simplify reinstallation.
5 Remove the bolt from the center of the crankshaft pulley (a screwdriver can be used to lock the starter ring gear on the flywheel), then remove the pulley.
6 Attach a puller to the damper. Draw the damper off the crankshaft, being careful not to drop it as it breaks free (photo). A common gear puller should not be used to draw the damper off, as it may separate the outer portion of the damper from the hub. Use only a puller which bolts to the hub.
7 Before installing the damper, coat the front cover seal area (on the damper) with moly-based grease.
8 Place the damper in position over the key on the crankshaft. Make sure the damper keyway lines up with the key.
9 Using a damper installation tool (GM no. J-23523 or equivalent), push the damper onto the crankshaft. The special tool distributes the pressure evenly around the hub.
10 Remove the installation tool and install the damper retaining bolt. Tighten the bolt to the specified torque.
11 Follow the removal procedure in the reverse order for the remaining components.
12 Adjust the drivebelts (refer to Chapter 1).

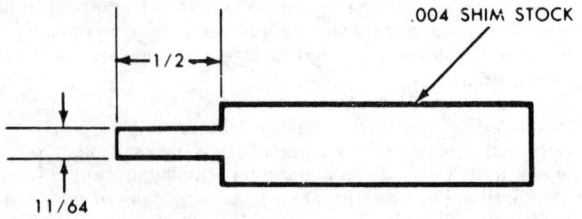

Fig. 2B.12 Protective tool used for installing rear main bearing oil seal in 1984 engines (Sec 11)

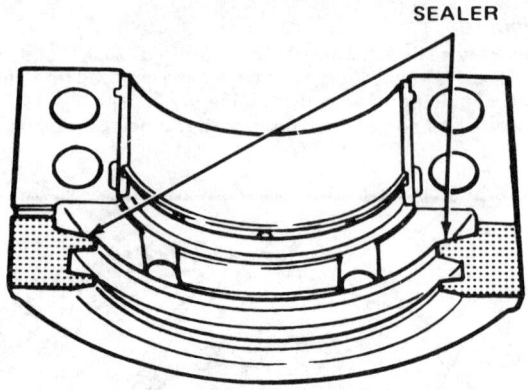

Fig. 2B.13 Apply anaerobic sealant to the areas shown on the rear main bearing cap of 1984 engines prior to installation (do not get sealant in the grooves or on the seal) (Sec 11)

13 Crankcase front cover — removal and installation

1 Remove the water pump as described in Chapter 3.
2 If equipped with air-conditioning, remove the compressor from the mounting bracket and secure it out of the way. *Do not disconnect any of the air-conditioning system hoses without having the system depressurized by a GM dealer or air-conditioning technician.*
3 Remove the compressor mounting bracket.
4 Remove the vibration damper as described in Section 12.
5 Disconnect the lower radiator hose at the front cover.
6 Remove the front cover mounting bolts and separate the cover from the engine (photo).

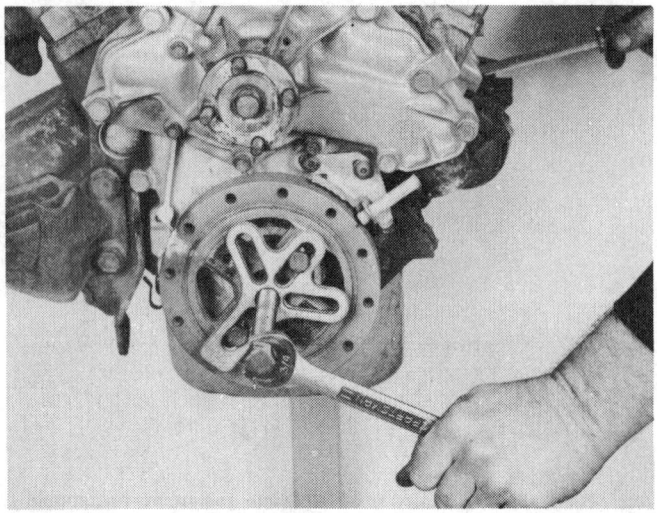

12.6 Using a puller to remove the vibration damper (V6)

13.6 Locations of the crankcase front cover mounting bolts (V6)

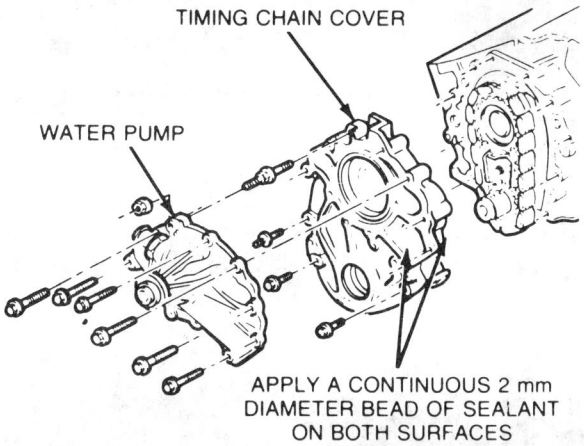

Fig. 2B.14 Details of the water pump and front cover orientation (Sec 13)

7 Clean all oil, dirt and old gasket material from the sealing surfaces of the front cover and engine block. Replace the front cover oil seal as described in Section 14.
8 Apply a continuous 3/32-inch (2 mm) bead of anaerobic sealant to both mating surfaces of the front cover (except the cover-to-block mating surface where the cover engages the oil pan lip). Apply RTV-type sealant to the cover-to-oil pan area. Also apply anaerobic sealant to the areas surrounding the coolant passages.
9 Place the front cover in position on the engine block and install the mounting bolts. Tighten the bolts to the specified torque.
10 The remaining installation procedures are the reverse of removal.

14 Front cover oil seal — replacement

With front cover installed on engine

1 With the vibration damper removed (Section 12), pry the old seal out of the crankcase front cover with a large screwdriver. Be very careful not to damage the surface of the crankshaft.
2 Place the new seal in position with the open end of the seal (seal lip) toward the *inside* of the cover.
3 Drive the seal into the cover until it is seated. GM tool J-23042 is available for this purpose. These tools are designed to exert even pressure around the entire circumference of the seal as it is hammered into place. A section of large-diameter pipe or a large socket could also be used.
4 Be careful not to distort the front cover.

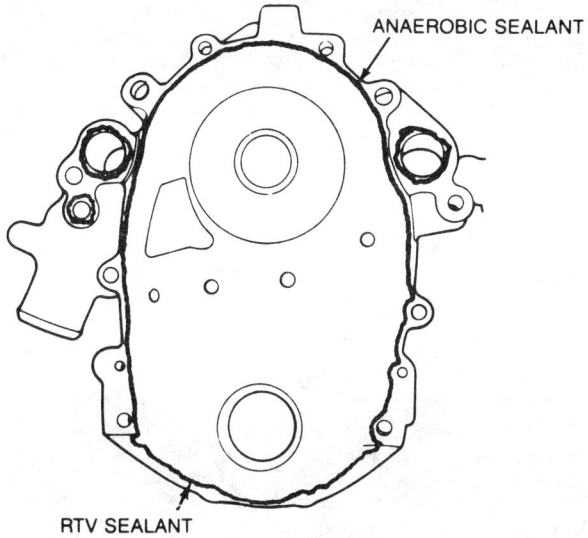

Fig. 2B.15 Apply sealant to the front cover as shown prior to installation (Sec 13)

With front cover removed from engine

5 This method is preferred, as the cover can be supported while the old seal is removed and the new one is installed.
6 Remove the crankcase front cover (refer to Section 13).
7 Using a large screwdriver, pry the old seal out of the bore from the front of the cover. Alternatively, support the cover and drive the seal out from the rear (photo). Be careful not to damage the cover.
8 With the front of the cover facing up, place the new seal in position with the open end of the seal toward the *inside* of the cover.
9 Using a wooden block and hammer, drive the new seal into the cover until it is completely seated (photo).
10 If the cover was removed, install it by reversing the removal procedure.

14.7 Driving the seal out of the front cover (V6)

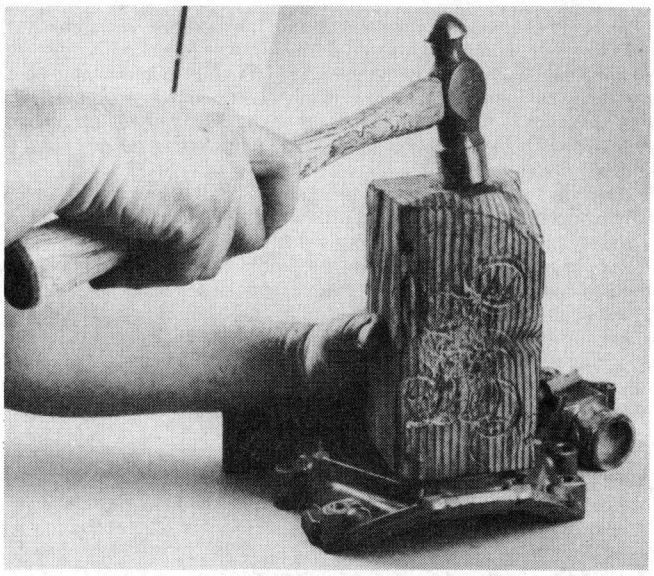

14.9 Installing the front cover oil seal with a wood block and hammer (V6)

15.13 Lubricating the thrust surface of the camshaft sprocket (V6)

16.11 Removing the camshaft from the engine block (be very careful not to damage the cam bearings as the shaft is withdrawn) (V6)

15 Timing chain and sprockets — inspection, removal and installation

1 Disconnect the cable from the negative battery terminal.
2 Remove the vibration damper (refer to Section 12).
3 Remove the crankcase front cover (refer to Section 13).
4 Before removing the chain and sprockets, visually inspect the teeth on the sprockets for signs of wear and the chain for looseness. Also check the condition of the timing chain tensioners.
5 If either or both sprockets show any signs of wear (edges on the teeth of the camshaft sprocket not 'square,' bright or blue areas on the teeth of either sprocket, chipping, pitting, etc.), they should be replaced with new ones. Wear in these areas is very common.
6 If the timing chain has not recently been replaced, or if the engine has over 25000 miles on it, it is almost certainly in need of replacement. Failure to replace a worn timing chain may result in erratic engine performance, loss of power and lowered gas mileage.
7 If any one component requires replacement, all related components, including the tensioners, should be replaced as well.
8 If it is determined that the timing components require replacement, proceed as follows.
9 Turn the engine over until the marks on the camshaft and crankshaft are in exact alignment. At this point the number one and four pistons will be at top dead center with the number four piston in the firing position (verify by checking the position of the rotor in the distributor). *Do not attempt to remove either sprocket or the timing chain until this is done and do not turn the crankshaft or camshaft after the sprockets/chain are removed.*
10 Remove the three camshaft sprocket retaining bolts and lift the camshaft sprocket and timing chain off the front of the engine. It may be necessary to tap the sprocket with a soft-faced hammer to dislodge it.
11 If it is necessary to remove the crankshaft sprocket, it can be withdrawn from the crankshaft with a special puller.
12 Attach the crankshaft sprocket to the crankshaft using a bolt and washer from the puller set.
13 Lubricate the thrust (rear) surface of the camshaft sprocket with moly-based grease or engine assembly lube (photo). Install the timing chain over the camshaft sprocket with slack in the chain hanging down over the crankshaft sprocket.
14 With the timing marks aligned, slip the chain over the crankshaft sprocket and then draw the camshaft sprocket into place with the three retaining bolts. Do not hammer or attempt to drive the camshaft sprocket into place, as it could dislodge the welch plug at the rear of the engine.
15 With the chain and both sprockets in place, check again to ensure that the timing marks on the two sprockets are properly aligned. If not, remove the camshaft sprocket and move it until the marks align.
16 Lubricate the chain with engine oil and install the remaining components in the reverse order of removal.

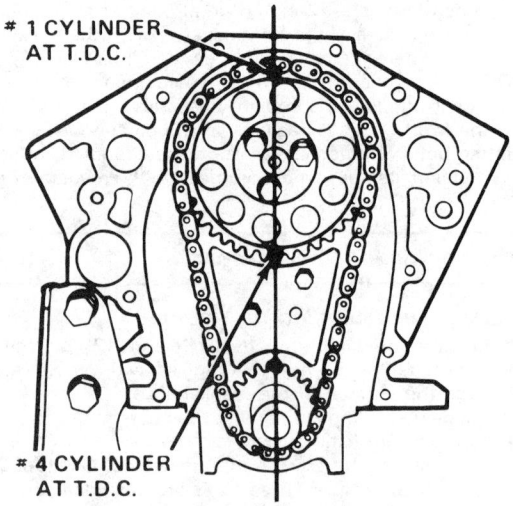

Fig. 2B.16 Proper alignment of the camshaft and crankshaft timing marks (Sec 15)

16 Camshaft — removal and installation

Note: *Before removing the camshaft, refer to Chapter 2, Part D (Section 15), and measure the lobe lift.*

1 Remove the cable from the negative battery terminal.
2 Drain the oil from the crankcase (refer to Chapter 1).
3 Drain the coolant from the radiator (refer to Chapter 1).
4 Remove the radiator (refer to Chapter 3).
5 If equipped with air-conditioning, remove the condenser (refer to Chapter 3). **Caution:** *The air-conditioning system must be evacuated by an air-conditioning technician before the condenser can be removed. Under no circumstances should this be attempted by the home mechanic, as personal injury may result.*
6 Remove the valve lifters (refer to Section 5).
7 Remove the crankcase front cover (refer to Section 13).
8 Remove the fuel pump and pushrod (refer to Chapter 4).
9 Remove the timing chain and sprocket (refer to Section 15).
10 Install a long bolt in one of the camshaft bolt holes to be used as a handle to pull on and support the camshaft.
11 Carefully draw the camshaft out of the engine block. Do this very slowly to avoid damage to the camshaft bearings as the lobes pass over the bearing surfaces. Always support the camshaft with one hand

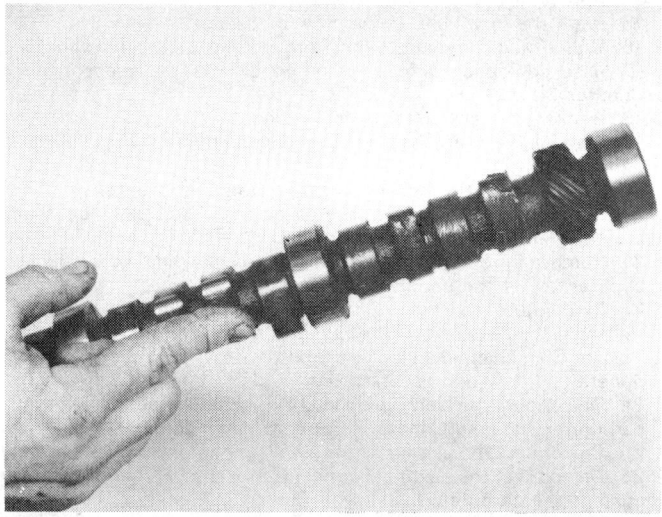

16.13 Be sure to apply engine assembly lube or moly-based grease to the cam lobes and bearing journals before installing the camshaft (V6)

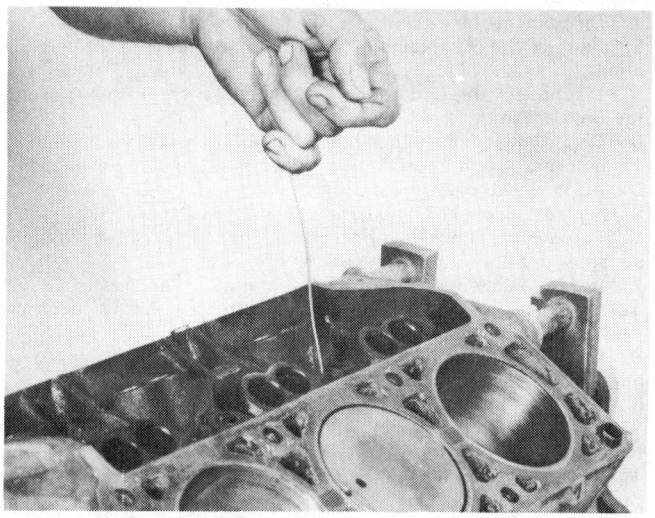

16.14 A length of wire is used to support the camshaft, preventing damage to the bearings from the cam lobes as it is passed through the bearings during installation (V6)

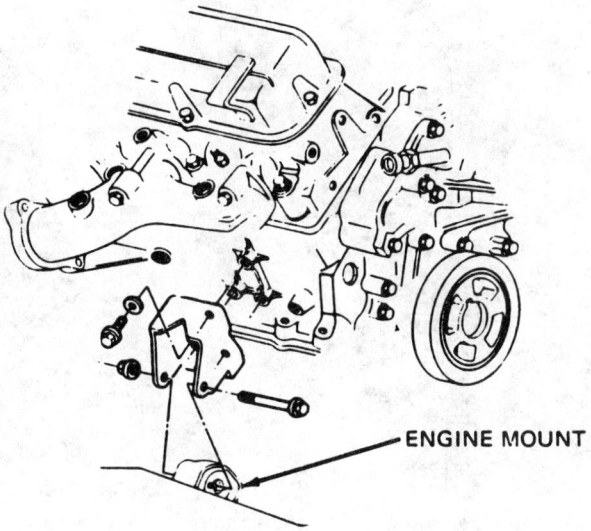

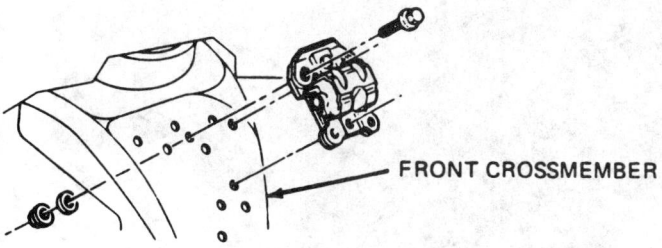

Fig. 2B.17 Front engine mount installation details (Sec 17)

near the engine block (photo).
12 Refer to Chapter 2, Part D, for the camshaft inspection procedures.
13 Prior to installing the camshaft, coat each of the lobes and journals with engine assembly lube or moly-based grease (photo).
14 Slide the camshaft into the engine block, again taking extra care not to damage the bearings (photo).
15 Install the camshaft sprocket and timing chain as described in Section 15.
16 Install the remaining components in the reverse order of removal by referring to the appropriate Chapter or Section.
17 Adjust the valve lash (refer to Section 6).
18 Have the air-conditioning system (if so equipped) recharged.

17 Engine mounts — replacement with engine in vehicle

1 If the mounts have become hard, split or separated from the metal backing, they must be replaced. This operation may be carried out with the engine/transmission still in the vehicle. See Section 9 for the proper way to raise the engine while it is still in place.

Front mount

2 Remove the through-bolt and nut.
3 Raise the engine slightly using a hoist or jack with a wood block under the oil pan, then remove the mount and frame bracket assembly from the crossmember.
4 Position the new mount, install the through-bolt and nut, then tighten all the bolts to the specified torque.

Rear mount

5 Remove the crossmember-to-mount bolts, then raise the transmission slightly with a jack.
6 Remove the mount-to-transmission bolts, followed by the mount.
7 Install the new mount, lower the transmission and align the crossmember-to-mount bolts.
8 Tighten all the bolts to the specified torque.

18 Engine — removal and installation

Note: *Due to the inaccessibility of the upper bellhousing-to-engine bolts in vehicles equipped with a V6 engine, it is recommended that the transmission and engine be removed as a unit, as detailed in this Section. To unbolt the transmission from the engine while both are in the vehicle, in order to remove the engine separately, the floor pan in the interior must be cut away to reveal the upper bellhousing bolts. If this method of removal is pursued, read through the procedure in this Section to familiarize yourself with the V6 engine/transmission removal. Refer to the engine removal section in Chapter 2, Part C, and note the differences between the V6 and V8 and perform the engine removal procedure accordingly — after the floor pan has been cut to reveal the upper bellhousing bolts.*

Removal

1 Disconnect the cable from the negative battery terminal.
2 Remove the clip retaining the underhood light wire harness to the hood.
3 Disconnect the underhood light electrical connector and unbolt the bulb assembly from the hood.
4 Scribe lines on the underside of the hood, around the hood mounting bracket, so the hood can be installed in the same position.
5 Apply white paint around the bracket-to-hood bolts so they can be aligned quickly and accurately during installation (photo).
6 Remove the bracket-to-hood bolts and, along with an assistant, separate the hood from the vehicle.
7 If the vehicle is equipped with a standard transmission, remove the transmission linkage (four-speed) or shift control lever (five-speed). Refer to Chapter 7 for either procedure.
8 Drain the radiator by opening the petcock located at the lower right-hand side of the radiator.
9 Remove the coolant recovery tank hose from the radiator.
10 Remove the upper radiator hose from the radiator. Take care not to damage the plastic fitting when breaking the hose loose.
11 If so equipped, remove the automatic transmission cooler lines at the radiator (photo).
12 Remove the bolts securing the top of the radiator fan shroud to the lower half and remove the upper shroud (photo).
13 Remove the lower radiator hose from the radiator.
14 Remove the radiator from the vehicle.
15 Mark the drivebelts to simplify installation.
16 Unbolt and remove the fan from the fan pulley (photo).
17 If so equipped, remove the air-conditioner compressor (refer to Chapter 3).
18 Remove the alternator (refer to Chapter 5).
19 Remove the heater hoses from the water pump and intake manifold outlets.
20 Remove the AIR system air pump (refer to Chapter 6).
21 Disconnect the hoses at the air cleaner. Carefully label them as they are removed to simplify installation.
22 Remove the air cleaner assembly, then disconnect the vacuum brake booster line at the booster.
23 Disconnect the linkage at the carburetor (including the cruise control cable, if so equipped). The cables may be removed from the linkage bracket with channel lock pliers (photo). Label the linkage/cable components as they are disconnected.
24 Disconnect the electronic metering valve wire connector from the carburetor (refer to Chapter 4 for carburetor details).
25 Disconnect the electric choke wire connector from the carburetor.
26 Disconnect the electrical connector for the EFE heater wire at the base of the carburetor.
27 Disconnect the wiring harness ground, located just ahead of the thermostat housing.
28 Disconnect the wire connector at the oil pressure sending unit

18.5 Removing the bracket-to-hood bolts (note white paint around bolt head to ensure correct installation) (V6)

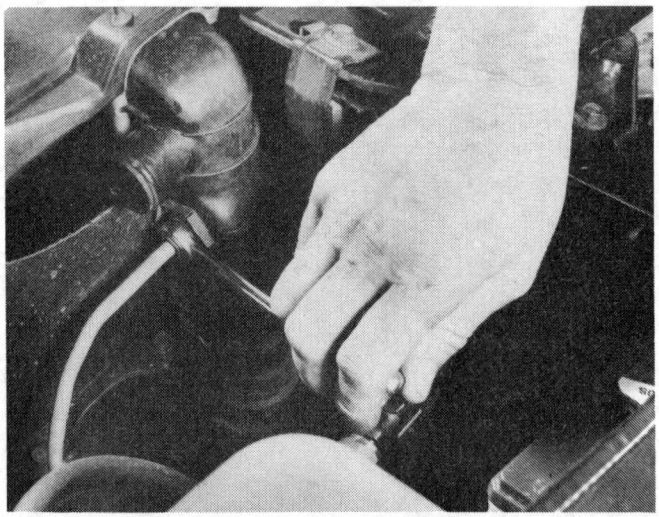

18.11 When removing the automatic transmission cooler lines from the radiator, use a flare nut wrench (V6)

18.12 Removing the retaining bolts from the upper radiator shroud (V6)

18.16 Unbolting the fan from the fan pulley (V6)

mounted atop the oil filter.
29 Remove the power steering pump mounting bolts. Lay the pump aside, out of the way, with the hoses still attached.
30 Unbolt the solenoid valve hose bracket from the left-hand rocker arm cover so it will not be damaged when the engine is lifted out of the vehicle (photo). Lay the assembly aside.
31 Remove the belts from the water pump pulley, then unbolt and remove the pulley from the front of the water pump.
32 Remove the oxygen sensor from the right side exhaust manifold so it will not be damaged as other components are removed (photo). Because it is a very sensitive component, be sure to protect it.
33 Raise the vehicle and place it securely on jackstands.
34 Remove the hoses from the fuel pump, then plug the openings in the pump and hoses to prevent contamination and loss of fuel (photo).
35 Remove the bolts securing the exhaust pipes to the exhaust manifolds on both sides of the engine (photo).
36 Pull the exhaust pipes off the exhaust manifolds.
37 If equipped with a manual transmission, disconnect the clutch fork return spring.
38 Remove the driveshaft (refer to Chapter 8).
39 Remove the bolts retaining the torque arm to the transmission bracket (refer to Chapter 11 for an illustration if necessary). **Caution:** *Rear spring force will cause the torque arm to move toward the floor pan when the arm is disconnected from the transmission bracket. Carefully place a wooden block between the torque arm and floor pan*

18.23 Removing the carburetor linkage from the linkage bracket (V6)

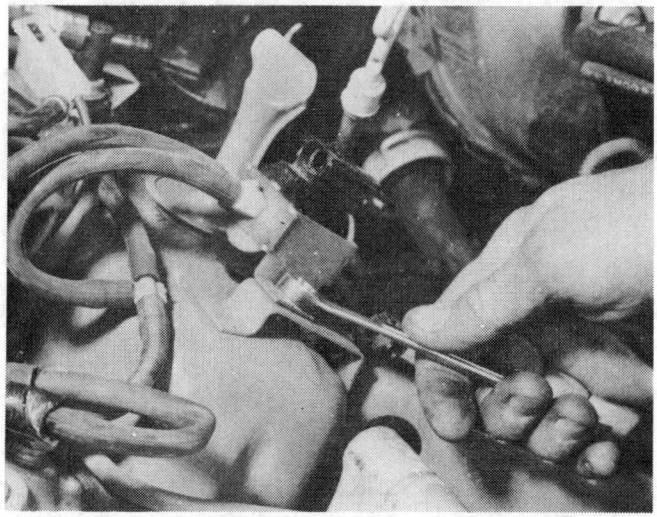

18.30 Removing the solenoid/hose bracket from the left-side rocker arm cover (V6)

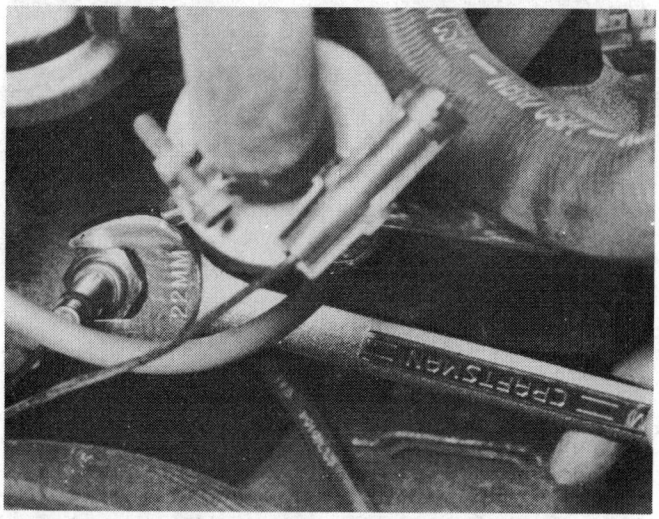

18.32 Removing the oxygen sensor from the right-side exhaust manifold (V6)

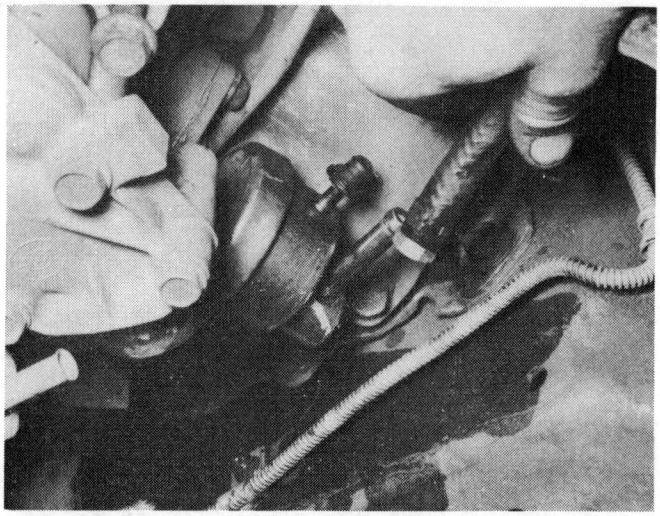

18.34 Remove the hoses from the fuel pump and plug the pump fittings (V6)

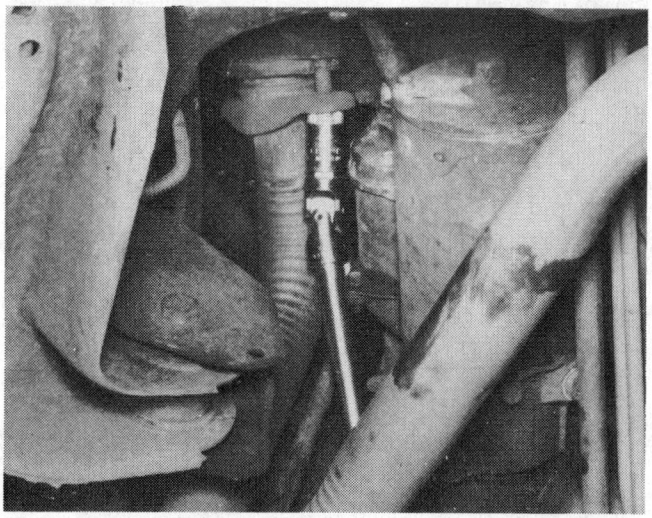

18.35 Removing the right-side exhaust pipe from the exhaust manifold (V6)

when disconnecting the arm to avoid possible injury to the hands and fingers and possible damage to the floor pan.

40 Loosen the rear torque arm bolts and pull the front of the arm free of the bracket, heeding the caution in Step 39.

41 Remove the catalytic converter/exhaust pipe bracket at the rear of the transmission.

42 Disconnect the speedometer cable from the transmission.

43 If so equipped, remove the automatic transmission throttle valve cable from the bracket on the transmission, then remove the bracket and shift linkage.

44 If so equipped, disconnect the TCC electrical connector at the transmission (automatic transmissions only) (photo).

45 Remove the TCC wiring bracket from the transmission.

46 Disconnect the electrical wires at the starter, labeling them to ensure proper installation.

47 Position a movable jack (floor jack or transmission jack) under the transmission oil pan using a block of wood as an insulator. Raise the transmission slightly.

48 Attach the hoist lifting chains to the lifting 'eyes' on the engine. There is one bracket at the front of the engine and one at the rear, diagonally opposite each other. Make sure the chain is looped properly through the engine brackets and secured with bolts and nuts through the chain links. The hook on the lifting hoist should be at the center of the engine. Position the chains so the engine/transmission unit will be at a steep angle with the front higher than the rear.

49 Raise the hoist until all slack is removed from the chains. Do not lift any further at this time.

50 Remove the transmission-to-crossmember bolts and the crossmember-to-frame bolts. Raise the transmission slightly and slide the crossmember to the rear until it can be removed. If equipped with an automatic transmission, insert a plug into the rear to prevent fluid loss when the engine/transmission unit is tilted while being removed. A plastic bag secured with tape will generally work.

51 Remove the engine mount through-bolts (photo).

52 Make sure that all wires and hoses, both under the vehicle and in the engine compartment, have been disconnected.

53 Remove the jack from under the transmission.

54 Start pulling the engine out. Pay special attention to the clearance at the distributor and the AIR system assembly. When the distributor is clear of the firewall, if not previously done, detach the wire connector at the rear of the distributor, remove the ground strap at the rear of the left cylinder head and remove the two wiring harness clamps at the bellhousing.

55 Continue to lift the engine at an angle until it is high enough to move forward without hitting the body (photo). Work slowly and be on the lookout for hoses and wires that are still connected.

56 Move to an appropriate work area and support the engine and transmission while they are separated.

Separating the engine and automatic transmission

57 Remove the transmission cooler lines from the transmission.

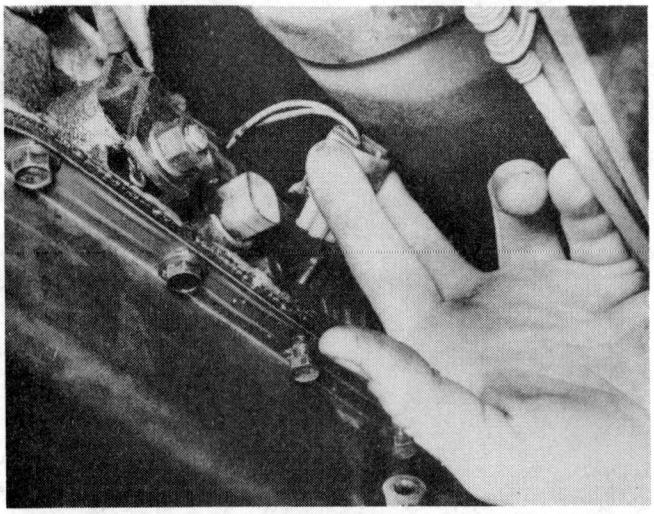

18.44 Disconnecting the TCC connector at the transmission (automatic transmission models only) (V6)

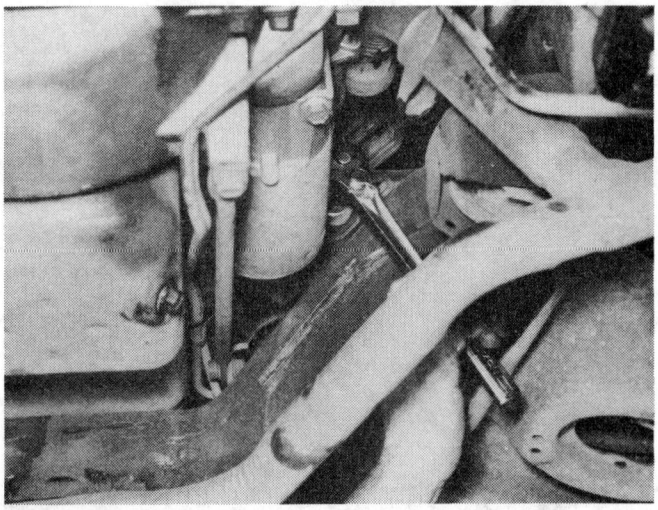

18.51 Removing the engine mount through-bolt on the right side (V6)

18.55 Lifting the engine from the vehicle (note assistant checking clearance at firewall) (V6)

18.61 Mark the relationship of the driveplate to the torque converter before removing the bolts (V6)

58 Remove the starter motor.
59 Remove the strut rods that run between the front of the engine and the transmission dust cover.
60 Remove the dust cover.
61 Using white paint, mark the relationship of the driveplate to the converter (photo).
62 Remove the driveplate-to-converter bolts. Use a large screwdriver to turn the driveplate and expose the bolts.
63 When the bolts have all been removed, push the converter toward the rear of the transmission to protect the transmission input seal.
64 Loosen the bellhousing bolts a little at a time.
65 Remove the transmission fluid dipstick and housing by pulling the housing out of the boss in the transmission, then unbolt the throttle valve cable housing from the transmission.
66 Remove the bellhousing bolts and separate the engine and transmission. Make sure that the converter does not slide forward during the separation procedure.
67 Mount the engine on an engine stand.
68 Refer to the appropriate Sections of this Chapter and Chapter 2, Part D, for further engine disassembly and rebuilding procedures.

Separating the engine and manual transmission

69 Remove the bolts securing the dust cover to the bottom of the bellhousing and remove the dust cover.
70 Loosen the bolts securing the bellhousing to the engine a little at a time.
71 When all bellhousing-to-transmission bolts have been loosened, remove the bolts.
72 Support the weight of the transmission and withdraw it from the engine in a straight line so the clutch disc is not damaged while the input shaft is still engaged in the clutch splines.
73 Refer to Steps 67 and 68.

Installation

74 To rejoin the automatic transmission to the engine, reverse Steps 57 through 66. Be sure to align the mating marks on the torque converter and driveplate and tighten all bolts to the specified torque.
75 To rejoin the manual transmission to the engine, reverse Steps 69 through 72. Be sure to tighten all bolts to the specified torque.
76 Attach the lifting chains to the engine/transmission assembly just like they were during removal.
77 Tilt and lower the engine/transmission unit into the engine compartment, guiding the engine mounts onto the frame mounts, and at the same time raising the transmission into position.
78 Install the engine mount through-bolts and the rear transmission crossmember. Tighten all bolts to the specified torque.
79 Install the remaining components in the reverse order of removal.
80 If equipped with a manual transmission, adjust the clutch as described in Chapter 1.
81 If equipped with a manual four-speed transmission, adjust the shift linkage as described in Chapter 7A.
82 If equipped with an automatic transmission, adjust the throttle valve and neutral safety and back-up light switch as described in Chapter 7B.
83 If so equipped, adjust the cruise control as described in Chapter 10.
84 Fill the cooling system with the specified antifreeze and water mixture (Chapter 1).
85 Fill the engine with the correct grade of engine oil (Chapter 1).
86 Check the fluid level in the transmission and add fluid as necessary.
87 Connect the positive battery cable, followed by the negative cable. If sparks or arcing occur as the negative cable is connected to the battery, make sure that all electrical accessories are turned off (check interior dome lights first). If arcing still occurs, make sure that all electrical wires are properly connected to the engine and transmission.
88 Refer to Chapter 2, Part D, for the engine start-up procedure.

Chapter 2 Part C V8 engine

Refer to Chapter 13 for specifications and information related to 1985 and later models

Contents

Air filter — replacement . . . . . Chapter 1
Camshaft — removal and installation 16
Check engine light . . . . . Chapter 6
Compression check . . . . . Chapter 1
Crankcase front cover — removal and installation 13
Cylinder heads — removal and installation 8
Drivebelt check and adjustment . . . . . Chapter 1
Engine — removal and installation 18
Engine mounts — replacement with engine in vehicle 17
Engine oil and filter change . . . . . Chapter 1
Engine overhaul — general information . . . . . Chapter 2D
Engine removal — methods and precautions . . . . . Chapter 2D
Exhaust manifolds — removal and installation 7
Front cover oil seal — replacement 14
General information 1
Hydraulic lifters — removal, inspection and installation 5
Intake manifold — removal and installation 4
Oil pan — removal and installation 9
Oil pump — removal and installation 10
Rear main bearing oil seal — replacement (engine in vehicle) . 11
Repair operations possible with the engine in the vehicle . . . . . Chapter 2D
Rocker arm covers — removal and installation 2
Spark plug replacement . . . . . Chapter 1
Timing chain and sprockets — removal and installation 15
Valve lash — adjustment 6
Valve train components — replacement (cylinder head installed) 3
Vibration damper — removal and installation 12
Water pump — removal and installation . . . . . Chapter 3

Specifications

Torque specifications

Cylinder head bolts	65	9.0
Cylinder head bolts	65	9.0
Camshaft sprocket bolts	20	2.8
Rear camshaft cover bolts	7	1.0
Clutch pressure plate bolts	35	4.8
Crankcase front cover bolts	7	1.0
Distributor clamp bolt	20	2.8
Exhaust manifold bolts	20	2.8
Driveplate-to-torque converter bolts	27	3.7
Flywheel-to-crankshaft bolts	60	8.3
Flywheel housing bolts	30	4.2
Flywheel housing cover bolts	7	1.0
Fuel pump mounting bolts	15	2.1
Intake manifold bolts	30	4.2
Intake manifold cover-to-intake manifold (fuel-injected models)	20	2.8
Oil filter bypass valve	7	1.0
Oil pan-to-crankcase bolts		
5/16 – 18	14	1.9
1/4 – 20	7	1.0
Rocker arm cover bolts	4	0.6
Spark plugs	22	3.0
Vibration damper bolt	60	8.3
Crankshaft pulley bolts	25	3.4
Engine mounting bracket bolts	80	11.0
Engine strut bracket bolts	35	4.8

Note: *Refer to Chapter 2, Part D, for additional specifications.*

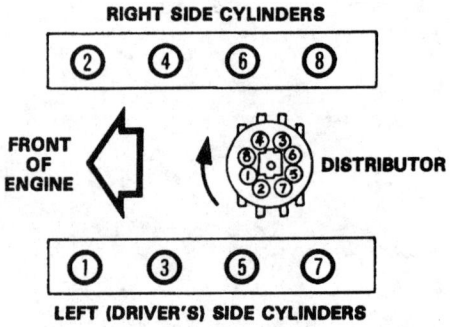

Cylinder numbers and distributor spark plug wire terminal locations

1 General information

The forward Sections in this Part of Chapter 2 are devoted to 'in-vehicle' repair procedures for the V8 engine. The latter Sections in this Part of Chapter 2 involve the removal and installation procedures for the V8 engine. All information concerning engine block and cylinder head servicing can be found in Part D of this Chapter.

The repair procedures included in this Part are based on the assumption that the engine is still installed in the vehicle. Therefore, if this information is being used during a complete engine overhaul — with the engine already out of the vehicle and on a stand — many of the steps included here will not apply.

The Specifications included in this Part of Chapter 2 apply only to the engine and procedures found here. For Specifications regarding engines other than the V8, see Part A or B, whichever applies. Part D of Chapter 2 contains the Specifications necessary for engine block and cylinder head rebuilding procedures.

2 Rocker arm covers — removal and installation

Right side

1 Disconnect the cable from the negative battery terminal.
2 Remove the air cleaner assembly, tagging each hose to be disconnected with a piece of numbered tape to simplify reinstallation.
3 Disconnect the AIR management hoses at the manifold.
4 Disconnect the wires and hoses that would interfere with the rocker arm cover removal. Label all hoses and wires to simplify reinstallation.
5 Remove the EGR solenoid.
6 Unbolt the AIR management valve bracket and move it aside.
7 Remove the AIR management tubes.
8 Remove the rocker arm cover bolts.
9 Remove the cover. **Note:** *If the cover adheres to the cylinder head, use a block of wood and a rubber hammer to dislodge it. If the cover still will not come loose, pry on it carefully, but do not distort the sealing flange surface.*
10 Prior to installation of the cover, clean all dirt, oil and old gasket material from the sealing surfaces of the cover and cylinder head with a scraper and degreaser.
11 Apply a continuous 3/16-inch (5 mm) diameter bead of RTV-type sealant to the sealing flange of the cover. Be sure to apply the sealant inboard of the bolt holes.
12 Place the rocker arm cover on the cylinder head while the sealant is still wet and install the mounting bolts. Tighten the bolts a little at a time to the specified torque.
13 **Note:** *On later model engines that have been leaking oil at rocker arm covers sealed with RTV, rather than reinstalling the covers with RTV, use a new gasket, reinforcement bar and bolt unit as depicted in the accompanying illustration. These parts are available from your dealer or auto supply store.*
14 The remaining installation steps are the reverse of removal.

Left side

15 Disconnect the cable from the negative battery terminal.
16 Remove the air cleaner assembly, tagging each hose to be disconnected with a piece of numbered tape to simplify reinstallation.
17 Remove the power brake booster line.
18 Disconnect the AIR hoses.
19 Disconnect the PCV valve hose.
20 Unbolt the wire harness and move it aside.
21 Remove the fuel return line from the TBI (if so equipped).
22 Remove the rocker arm cover bolts.
23 Follow Steps 9 through 14.

3 Valve train components — replacement (cylinder head installed)

1 Remove the rocker arm cover(s) as described in Section 2.
2 If only the pushrod is to be replaced, loosen the rocker nut enough to allow the rocker arm to be rotated away from the pushrod. Pull the pushrod out of the hole in the cylinder head.
3 If the rocker arm is to be removed, remove the rocker arm nut and pivot and lift off the rocker arm.
4 If the valve spring is to be removed, remove the spark plug from the affected cylinder.
5 There are two methods that will allow the valve to remain in place while the valve spring is removed. If you have access to compressed air, install an air hose adapter (GM part number J-23590) in the spark plug hole. When air pressure is applied to the adapter, the valves will be held in place by the pressure.
6 If you do not have access to compressed air, bring the piston of the affected cylinder to top dead center (TDC) on the compression stroke. Feed a long piece of 1/4-inch nylon cord in through the spark plug hole until it fills the combustion chamber. Be sure to leave the end of the cord hanging out of the plug hole so it can be removed easily. Rotate the crankshaft with a wrench (in the normal direction of rotation) until slight resistance is felt.
7 Reinstall the rocker arm nut (without the rocker arm).
8 Insert the slotted end of a valve spring compressing tool under the nut and compress the spring just enough to remove the spring keepers.
9 Remove the retainer, cup shield, O-ring seal, spring, spring damper (if so equipped) and valve stem oil seal (if so equipped).
10 If a rocker arm stud requires replacement, first determine whether it is a screw-in or press-in type. Screw-in studs are retained by a nut atop the stud boss, while press-in studs have no retaining nut.
11 Screw-in studs may be replaced by simply removing the damaged one and replacing it with a new one. Be sure to reinstall the pushrod guide (if so equipped) under the stud nut and tighten the nut to the specified torque.
12 Press-in studs that have damaged threads or are loose in the cylinder head should be replaced with new studs that are available in 0.003-inch and 0.013-inch oversizes. Press-in stud replacement procedures require the use of special tools and should be performed by an automotive machine shop. If such services are required, remove the cylinder head(s) by referring to Section 8.
13 Inspection procedures for the hydraulic lifters are detailed in Section 5. Procedures regarding other valve train components are detailed in Chapter 2, Part D.
14 Installation of the valve train components is the reverse of the

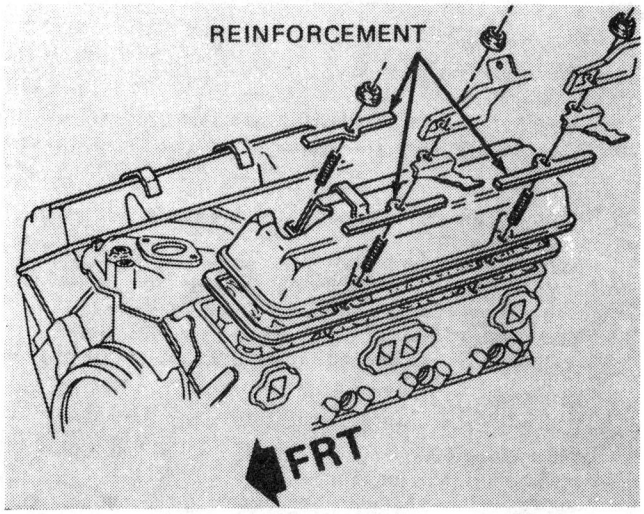

Fig. 2C.1 Reinforcement bars are used along with gaskets to remedy leaking RTV-sealed rocker arm covers (Sec 2)

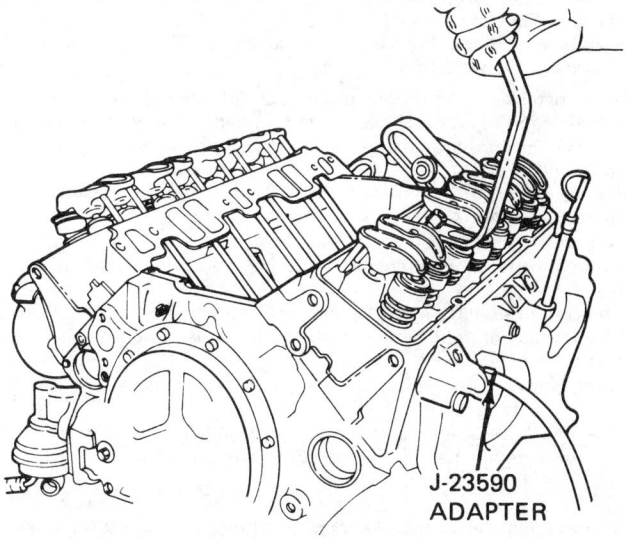

Fig. 2C.2 Using compressed air to keep the valve in place while the valve spring is removed (Sec 3)

removal procedure. Always use new valve stem oil seals whenever the spring keepers have been disturbed. Before installing the rocker arms, coat the bearing surfaces of the arms and pivots with moly-based grease or engine assembly lube. Be sure to adjust the valve lash as detailed in Section 6.

4 Intake manifold — removal and installation

1 If the vehicle is equipped with air-conditioning, carefully examine the routing of the hoses and the mounting of the compressor. You may be able to remove the intake manifold without disconnecting the system. If you are in doubt, take the vehicle to a dealer or refrigeration specialist to have the system depressurized. *Do not, under any circumstances, disconnect the hoses while the system is under pressure.*
2 Disconnect the cable from the negative battery terminal.
3 Drain the coolant from the radiator (refer to Chapter 1).
4 Remove the air cleaner assembly, tagging each hose to be disconnected with a piece of numbered tape to simplify reinstallation.

Carburetor-equipped models

5 Label and disconnect all electrical wires and vacuum hoses at the carburetor.
6 Disconnect the fuel line at the carburetor. Be prepared to catch some fuel, then plug the fuel line to prevent contamination.
7 Remove the upper radiator hose from the manifold.
8 Disconnect the heater hose at the manifold.
9 Disconnect the carburetor linkage, making careful note of how it was installed.
10 Disconnect the spark plug wires at the spark plugs on the right-hand side, referring to the removal technique described in Chapter 1.
11 Remove the distributor cap and spark plug wires.
12 Remove the distributor (refer to Chapter 5).
13 Remove the cruise control servo and bracket (if so equipped).
14 Remove the alternator upper mounting bracket (refer to Chapter 5).
15 Remove the EGR solenoid and bracket from the rear of the manifold.
16 Disconnect the vacuum line leading to the power brake booster (if so equipped).
17 Remove the manifold mounting bolts.
18 Make sure that all electrical wires, vacuum hoses and coolant hoses that would interfere with manifold removal have been disconnected.
19 Separate the manifold from the engine. Do not pry between the mating surfaces as it could damage them. Tap the manifold with a hammer and wooden block, if necessary.
20 If the intake manifold is to be replaced with another one, transfer all components that have not yet been removed. These include:

Carburetor and carburetor mounting bolts or studs (see Chapter 4 for procedures)
Thermostat and housing (use a new gasket or RTV, as applicable)
All necessary switches and fittings

21 Before installing the manifold, place clean lint-free rags in the engine cavity and clean the engine block, cylinder head and manifold gasket surfaces. All gasket material and sealing compound must be removed prior to installation.
22 Remove all dirt and gasket remnants from the engine cavity.
23 Clean the gasket sealing surfaces with degreaser.
24 Position the gaskets on the cylinder heads and apply a 3/16-inch (5 mm) bead of RTV-type sealant to the front and rear ridges of the engine block.
25 Extend the bead 1/2-inch (13 mm) up each cylinder head to seal and retain the manifold side gaskets. Use sealant at the coolant passages.
26 Carefully lower the intake manifold into position, making sure you do not disturb the gaskets.
27 Install the intake manifold mounting bolts and tighten them following the sequence illustrated. Tighten the bolts a little at a time until they are all at the specified torque.
28 Install the remaining components in the reverse order of removal.
29 Fill the radiator with coolant, start the engine and check for leaks. Adjust the ignition timing and idle speed as necessary (refer to Chapter 1).

Fuel-injected models

30 Disconnect the fuel inlet line at the front TBI unit.

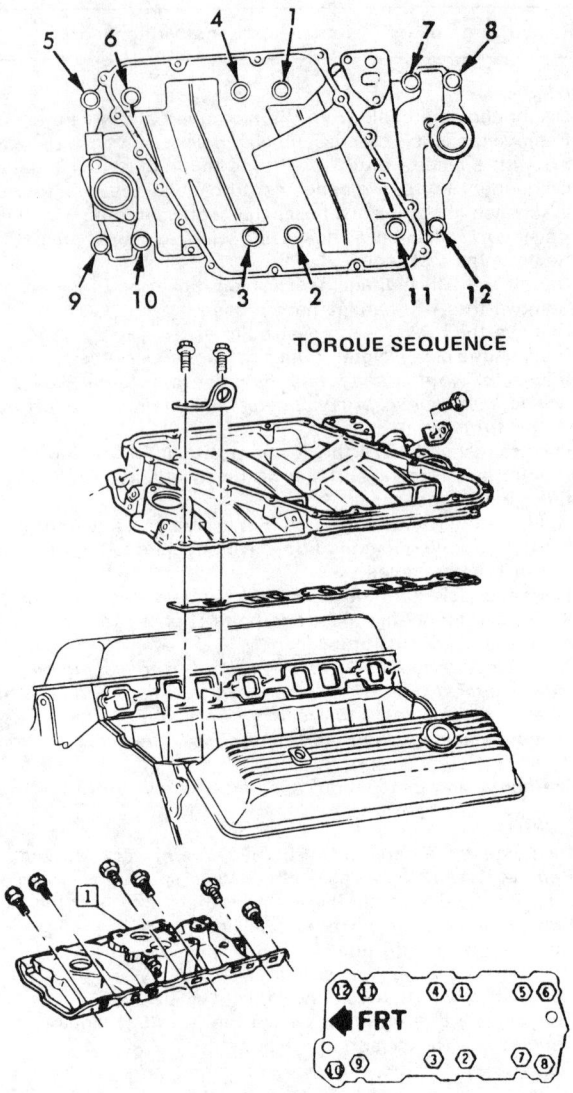

Fig. 2C.3 Recommended bolt tightening sequence for installing fuel injection (TBI 400 — top) and carburetor intake manifolds (Sec 4)

31 Remove the EGR solenoid from the rear of the manifold.
32 Unbolt and remove the alternator adjusting bracket from the front of the manifold.
33 Disconnect the electrical wires from the idle air control (IAC) assemblies (one on each TBI unit).
34 Disconnect the electrical wires from the injectors (one on each TBI unit).
35 Disconnect the electrical wire at the throttle position sensor (TPS) on the rear TBI unit.
36 Disconnect the power brake booster lines at the manifold.
37 Disconnect the fuel return line at the rear TBI unit.
38 Remove the air-conditioning strut (if so equipped).
39 Disconnect the throttle cable from the rear TBI unit.
40 Disconnect the automatic transmission detent cable (if so equipped).
41 Disconnect the cruise control cable (if so equipped).
42 Remove the PCV valve.
43 Label and disconnect any vacuum hoses not already disconnected.
44 Remove the fuel balance tube.
45 Remove the nuts and bolts retaining the intake manifold cover to the intake manifold.
46 Make sure that all electrical wires and vacuum and coolant hoses have been disconnected.
47 Carefully remove the manifold cover with the TBI units attached.
48 Remove the distributor (refer to Chapter 5).

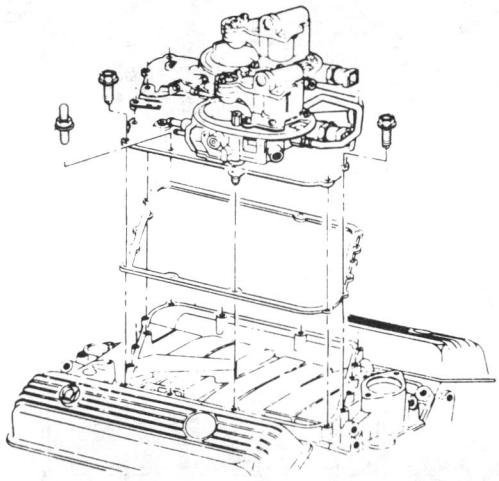

Fig. 2C.4 Exploded view of the TBI 400 injector plate and manifold assemblies (Sec 4)

49 Remove the radiator hose at the thermostat housing.
50 Disconnect the heater hose at the manifold.
51 Remove the manifold-to-cylinder head bolts and separate the intake manifold from the engine.
52 If the manifold and/or manifold cover is to be replaced with a new one, transfer all components still attached to the old unit to the new one.
53 Follow the procedures outlined in Steps 21 through 29.

5 Hydraulic lifters — removal, inspection and installation

1 A noisy valve lifter can be isolated when the engine is idling. Place a length of hose or tubing near the position of each valve while listening at the other end of the tube. Another method is to remove the rocker arm cover and, with the engine idling, place a finger on each of the valve spring retainers, one at a time. If a valve lifter is defective, it will be evident from the shock felt at the retainer as the valve seats.
2 Assuming that adjustment is correct, the most likely cause of a noisy valve lifter is a piece of dirt trapped between the plunger and the lifter body.
3 Remove the rocker arm covers as described in Section 2.
4 Remove the intake manifold as described in Section 4.
5 Loosen the rocker arm nut and rotate the rocker arm away from the pushrod.
6 Remove the pushrod.
7 To remove the lifter, a special hydraulic lifter removal tool should be used, or a scribe can be positioned at the top of the lifter and used to force the lifter up. *Do not use pliers or other tools on the outside of the lifter body as they will damage the finished surface and render the lifter useless.*
8 The lifters should be kept separate for reinstallation in their original positions.
9 To dismantle a valve lifter, hold the plunger down with a pushrod and then extract the retainer spring with a small screwdriver.
10 Remove the pushrod seat and the metering valve.
11 Remove the plunger, ball check valve and plunger spring. Remove the ball check valve and spring by prying them out with a small screwdriver.
12 Clean the lifter components with solvent and dry them with compressed air. Examine the internal components for wear and check the ball carefully for flat spots. Refer to Chapter 2, Part D, for additional lifter (and camshaft) inspection procedures.
13 If the lifters are worn, they must be replaced with new ones and the camshaft must be replaced as well (see Chapter 2, Part D). If the lifters were contaminated with dirt, they can be reinstalled — they *may* operate normally.
14 Reassembly should be done in the following manner:

a) Place the ball check valve on the small hole in the bottom of the plunger.
b) Insert the ball check valve spring into the seat in the retainer and place the retainer over the ball so that the spring rests on the ball. Using a small screwdriver, carefully press the retainer into position in the plunger.
c) Place the plunger spring over the ball retainer, invert the lifter body and slide it over the spring and plunger. Make sure the oil holes in the body and plunger line up.
d) Fill the assembly with 10-weight oil. Place the metering valve and pushrod seat in position, press down on the seat and install the retainer spring.

15 When installing the lifters, make sure they are replaced in their original bores and coat them with moly-based grease or engine assembly lube.
16 The remaining installation steps are the reverse of removal.

6 Valve lash — adjustment

Note: *For illustrations, refer to Chapter 2, Part B.*

1 Disconnect the cable from the negative battery terminal.
2 If the rocker arm covers are still on the engine, refer to Section 2 and remove them.
3 If the valve train components have been serviced just prior to this procedure, make sure that the components are completely reassembled.
4 Rotate the crankshaft until the timing mark on the vibration damper aligns with the zero (0) on the timing indicator. If the No. 1 cylinder valves are moving, the engine is in the No. 6 cylinder firing position and the crankshaft must be rotated an additional 360°. If the No. 1 cylinder valves are not moving, the piston is at top dead center (TDC), which is correct.
5 Back off the rocker arm stud adjusting nut on the No. 1 intake and exhaust valves until there is play in the pushrods; tighten the nut to *just* eliminate the play, then tighten the nut one (1) turn. **Note:** *You can determine the point where all lash has just been removed by rotating the pushrod while tightening the nut. At the point where drag at the pushrod is first felt, all lash has been removed.*
6 With the engine in the No. 1 cylinder firing position, as determined in paragraph 4, also adjust the exhaust valves of cylinders 3, 4 and 8 and the intake valves of cylinders 2, 5 and 7.
7 Rotate the crankshaft 360° and align the timing mark once more, then repeat the procedure for exhaust valves 2, 5, 6 and 7, and intake valves 3, 4, 6 and 8.
8 Refer to Section 2 and install the rocker arm covers.

7 Exhaust manifolds — removal and installation

1 If the vehicle is equipped with air- conditioning, carefully examine the routing of the hoses and the mounting of the compressor. Depending on the system used, you may be able to remove the exhaust manifolds without disconnecting the system. If you are in doubt, take the vehicle to a dealer or air-conditioning technician to have the system depressurized. *Do not, under any circumstances, disconnect the hoses while the system is under pressure.*
2 Disconnect the negative battery cable from the battery.
3 Disconnect the carburetor hot-air pipe routed from the right exhaust manifold to the air cleaner.
4 Mark each spark plug wire with a strip of numbered tape (numbered 1—8) and then remove the plug wires from the spark plugs. Remember to pull on the rubber boot and not the wire itself.
5 Remove the spark plug heat shields. These four shields are held in place with bolts.
6 Disconnect the rubber AIR hoses where they meet the manifold check valves. The single metal pipe and the AIR manifold can be removed with the exhaust manifold.
7 If equipped with power steering and/or air-conditioning, remove the brackets attached to the left-side manifold.
8 Disconnect the exhaust pipes from the manifold flanges. The exhaust pipes can be hung from the frame with wire.
9 If equipped with bolt locking strips, use a screwdriver to pry the metal locking strips away from the bolt heads. This will enable you to put a standard wrench on the bolts.
10 Remove the exhaust manifold end bolts first, followed by the center

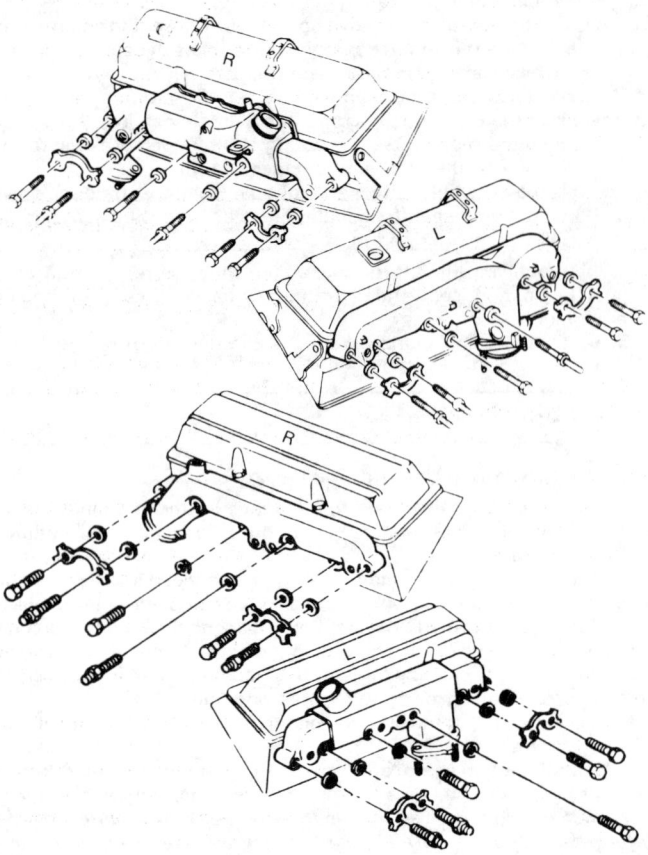

Fig. 2C.5 Exhaust manifold installation details for carbureted (top) and fuel-injected engines (Sec 7)

bolts. Lift the manifold and AIR injection tubes away from the engine.
11 If you are replacing the exhaust manifolds with new ones, the carburetor hot-air stove and the AIR injection tube system must be transfered to the new manifolds.
12 Before installing the exhaust manifolds, clean the mating surfaces on the cylinder heads and manifolds. All old gasket material must be removed.
13 Place new gaskets into position on the cylinder heads, using a thin film of sealer to hold them in place.
14 Install the manifold and tighten the center bolts first, followed by the end bolts. On some models it is easier to install the left rear spark plug heat shield first due to inaccessibility after the exhaust manifold is in place.
15 Install the remaining components in the reverse order of removal. Use a new gasket or packing at the exhaust manifold-to-exhaust pipe flange connections.
16 Start the engine and check for exhaust leaks where the manifolds meet the cylinder heads and where the manifolds join the exhaust pipes.

8 Cylinder heads — removal and installation

1 Locate the engine block drain plugs at the center of the block on each side, remove the plugs and drain the coolant.
2 Remove the intake manifold (refer to Section 4).
3 Remove the exhaust manifolds (refer to Section 7).
4 Remove the lower power steering pump adjusting bracket from the left-side cylinder head (if not previously removed). Remove the lower alternator mounting bolt from the right-side cylinder head.
5 Remove the rocker arm covers (refer to Section 2).
6 Loosen the rocker arm nuts enough to allow removal of the pushrods.
7 Remove the pushrods and store them separately so they can be reinstalled in their original positions. A cardboard box with numbered

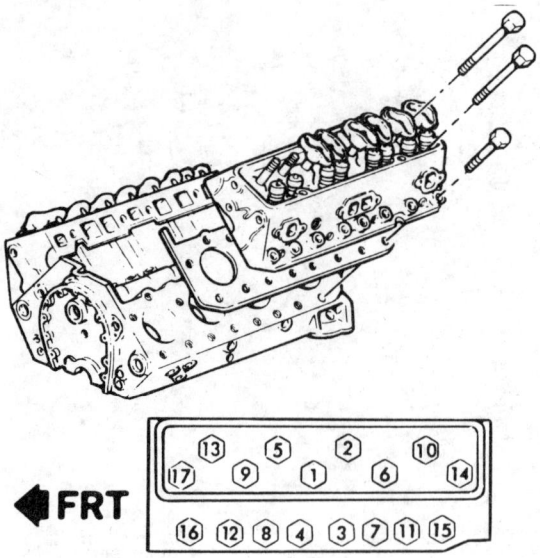

Fig. 2C.6 Recommended tightening sequence for cylinder head bolts (Sec 8)

holes can be used effectively.
8 Remove the head bolts in a sequence *opposite* to the tightening sequence (refer to the accompanying illustration).
9 Remove the cylinder head(s). To break the gasket seal, insert a bar into one of the exhaust ports, then carefully lift on the tool.
10 If a new cylinder head is being installed, transfer the external components from the old head to the new head.
11 Before installing the head(s), the gasket surfaces on both the head and engine block must be clean and free of nicks and scratches. Also, the threads in the block and on the head bolts must be completely clean, as any dirt remaining in the threads will affect the bolt torque.
12 When installing the cylinder heads on engines using a plain steel gasket, coat both sides of the new gasket with a good quality sealer made expressly for head gasket installation. To ensure an even coat of the proper thickness, use a paint roller to spread the sealer. The application of too much sealer may hold the gasket away from the head or block.
13 Do not use sealer on engines requiring a sandwich-type steel/asbestos gasket.
14 Place the gasket in position over the dowel pins with the bead up.
15 Position the cylinder head over the gasket.
16 Coat the cylinder head bolt threads with an appropriate sealer (GM no. 1052080 or equivalent) and install the bolts.
17 Tighten the bolts in the proper sequence (see the accompanying illustration) to the specified torque. Work up the the final torque in three steps.
18 Install the pushrods, making sure the lower ends are in the lifter seats, place the rocker arm ends over the pushrods and loosely install the rocker arm nuts.
19 The remaining installation steps are the reverse of removal. Before installing the rocker arm covers, adjust the valve lash (refer to Section 6).

9 Oil pan — removal and installation

1 Disconnect the cable from the negative battery terminal.
2 Remove the air cleaner assembly, tagging each hose to be disconnected with a piece of numbered tape to simplify reinstallation.
3 Remove the distributor cap (refer to Chapter 5) to prevent damage when the engine is raised.
4 Unbolt and remove the top half of the fan shroud.
5 Raise the vehicle and support it on jackstands.
6 Drain the engine oil.
7 Disconnect the exhaust crossover pipe at the exhaust manifold flanges. Lower the exhaust pipes and suspend them from the frame with wire.
8 If equipped with an automatic transmission, remove the converter shroud.

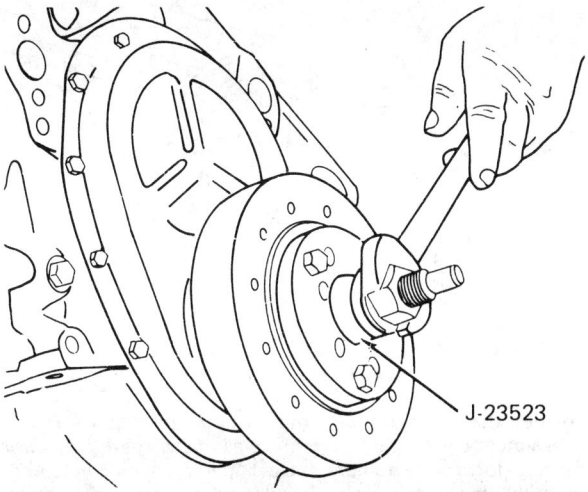

Fig. 2C.7 Installing the vibration damper using tool no. J-23523 (this tool or its equivalent is necessary to evenly distribute pressure around the hub) (Sec 12)

9 If equipped with a manual transmission, remove the flywheel cover.
10 Remove the starter (refer to Chapter 5).
11 Turn the bolt at the center of the vibration damper to rotate the crankshaft until the timing mark is facing down, at the 6 o'clock position. This will move the forward crankshaft throw to the top, providing clearance at the front of the oil pan.
12 Remove the through-bolt at each engine mount.
13 At this time the engine must be raised slightly to allow the oil pan to clear the crossmember. The preferred method requires an engine hoist or 'cherry picker.' Hook up the lifting chains as described in Section 18.
14 An alternative method can be used if you are careful. Place a floor jack and a block of wood under the oil pan. The wood block should spread the load across the oil pan to prevent damage or collapse of the oil pan metal. The oil pump pick-up and screen are very close to the oil pan bottom, so any collapse of the pan may damage the pick-up or prevent the oil pump from drawing oil properly.
15 With either method, raise the engine slowly until wood blocks can be placed between the frame front crossmember and the engine block. The blocks should be approximately three (3) inches thick. Check clearances all around the engine as it is raised. Pay particular attention to the distributor and the cooling fan.
16 Lower the engine onto the wood blocks. Make sure it is firmly supported. If a hoist is being used, keep the lifting chains secured to the engine.
17 Remove the oil pan bolts. Note the different sizes used and their locations.
18 Remove the oil pan.
19 Before installing the pan, make sure that the sealing surfaces on the pan, block and front cover are clean. If the original oil pan is being reinstalled, make sure that all old gasket material has been removed.
20 Always use new gaskets and seals when installing the pan.
21 Lift the pan into position and install all bolts finger tight. There is no specific order for torquing the bolts; however, it is a good idea to tighten the end bolts first.
22 Lower the engine onto the mounts and install the through-bolts.
23 Follow the removal steps in reverse order. Fill the crankcase with the correct grade and quantity of oil, start the engine and check for leaks.
24 Most of the above steps will not be required if the engine has been removed from the vehicle. The pan can simply be unbolted and removed, cleaned and installed as indicated.

10 Oil pump — removal and installation

1 Remove the oil pan (refer to Section 9).
2 Remove the pump-to-rear main bearing cap bolt and separate the pump and extension shaft from the engine.
3 To install the pump, move it into position and align the slot on the top end of the extension shaft with the drive tang on the lower end of the distributor driveshaft. The distributor drives the oil pump, so it is essential that this alignment is correct.
4 Install the oil pump-to-main bearing cap bolt and tighten it to the specified torque.
5 Reinstall the oil pan (refer to Section 9).

11 Rear main bearing oil seal — replacement (engine in vehicle)

Note: *Refer to Chapter 2, Part B, for illustrations. The illustrations pertaining to the 1984 V6 engine are also valid for all 1982 through 1984 V8 engines.*

1 Always replace both halves of the rear main oil seal. While replacement of the seal is much easier with the engine removed, the job can be done with the engine in place.
2 Remove the oil pan and oil pump as described previously in this Chapter.
3 Remove the rear main bearing cap from the engine.
4 Using a screwdriver, pry the lower half of the oil seal out of the bearing cap.
5 To remove the upper half of the seal, use a small hammer and a brass pin punch to roll the seal around the crankshaft journal. Tap one end of the seal with the hammer and punch (be careful not to strike the crankshaft) until the other end of the seal protrudes enough to pull the seal out with pliers.
6 Remove all sealant and foreign material from the block. *Do not use abrasives.*
7 Inspect the components for nicks, scratches and burrs on all sealing surfaces.
8 Coat the lips of the new seal with engine assembly lube or moly-based grease. Do not get any on the seal mating ends.
9 Included with the new rear main oil seal should be a small plastic installation tool.
10 Position the narrow end of the installation tool between the crankshaft and the seal seat. The idea is to keep the new seal from being damaged by the sharp edge of the seal seat.
11 Raise the new upper half of the seal into position with the seal lips facing toward the *front* of the engine. Push the seal into the seat, using the installation tool as a 'shoehorn' to keep the seal from contacting the sharp edge.
12 Roll the seal around the crankshaft, all the time using the tool for protection. When both ends of the seal are flush with the engine block, remove the installation tool (be careful not to withdraw the seal as well).
13 Install the lower half of the oil seal in the bearing cap, again using the installation tool to protect the seal from the sharp edge. Make sure the seal is firmly seated, then withdraw the installation tool.
14 Apply a thin, even film of anaerobic-type sealant to the cap-to-block mating surfaces. *Do not get any on the seal lip.*
15 Install the bearing cap and seal and tighten the bolts to about 10-to-12 ft-lbs only. Now tap the end of the crankshaft forward and backward to line up the thrust bearing surfaces. Tighten the bearing cap bolts to the final torque and install the oil pump and oil pan.

12 Vibration damper — removal and installation

1 Remove the hose from the retaining strap atop the radiator shroud by releasing the end of the strap from the underside of the shroud with pliers.
2 Remove the bolts and separate the upper radiator shroud from the radiator.
3 Remove the cooling fan from the fan pulley.
4 Loosen the accessory drivebelt adjusting bolts, as necessary, and remove the drivebelts, tagging each one as it is removed to simplify reinstallation.
5 Remove the bolt from the center of the crankshaft pulley (a screwdriver can be used to lock the starter ring gear on the flywheel), then remove the pulley.
6 Attach a puller to the damper. Draw the damper off the crankshaft, being careful not to drop it as it breaks free. A common gear puller should not be used to draw the damper off, as it may separate the outer portion from the hub. Use only a puller which bolts to the hub.

7 Before installing the damper, coat the front cover seal area (on the damper) with engine oil or moly-based grease.
8 Place the damper in position over the key on the crankshaft. Make sure the damper keyway lines up with the key.
9 Using a damper installation tool (GM no. J-23523 or its equivalent), push the damper onto the crankshaft. The special tool distributes the pressure evenly around the hub.
10 Remove the installation tool and install the damper retaining bolt. Tighten the bolt to the specified torque.
11 Follow the removal procedure in the reverse order for the remaining components.
12 Adjust the drivebelts (refer to Chapter 1).

13 Crankcase front cover — removal and installation

1 Remove the vibration damper as described in Section 12.
2 Remove the water pump as described in Chapter 3.
3 Remove the crankcase front cover mounting bolts and separate the cover from the engine.
4 Remove the old gasket and discard it.
5 Clean all oil, dirt and old gasket material from the sealing surfaces of the front cover and engine block. Use a sharp knife to remove any excess oil pan gasket material that may be protruding at the oil pan-to-engine block junction.
6 Replace the front cover seal as described in Section 14.
7 Apply a 1/8-inch (3 mm) bead of RTV-type sealant to the joint formed by the oil pan and engine block.
8 Coat the new cover gasket with RTV-type sealant and position it on the engine.
9 Install the cover-to-oil pan gasket, then coat the crankshaft seal with engine assembly lube or moly-based grease and position the cover over the crankshaft end.
10 Loosely install the upper cover-to-block retaining bolts.
11 Tighten the bolts in a crisscross pattern while pressing down on the cover so the dowels in the block are aligned with the holes in the cover.
12 Install the remaining cover bolts and tighten them to the specified torque.
13 The remaining steps are the reverse of removal.

14 Front cover oil seal — replacement

Note: *For illustrations, refer to Chapter 2, Part B.*

With front cover installed on engine

1 With the vibration damper removed (Section 12), pry the old seal out of the crankcase front cover with a large screwdriver. Be very careful not to damage the surface of the crankshaft.
2 Place the new seal into position with the open end of the seal (seal 'lip') toward the *inside* of the cover.
3 Drive the seal into the cover until it is completely seated. GM tool J-23042 is available for this purpose. This tool is designed to exert even pressure around the entire circumference of the seal as it is hammered into place. A section of large-diameter pipe or a large socket could also be used.
4 Be careful not to distort the front cover.

With front cover removed from engine

5 This method is preferred because the cover can be supported as the new seal is driven into place, preventing the possibility of cover distortion.
6 Remove the crankcase front cover as described in Section 13.
7 Pry the old seal out of the bore with a large screwdriver.
8 Support the inside of the cover around the seal area and install the new seal in the same manner as described above.

15 Timing chain and sprockets — removal and installation

1 Disconnect the cable from the negative battery terminal.
2 Remove the vibration damper (refer to Section 12).
3 Remove the crankcase cover (refer to Section 13).

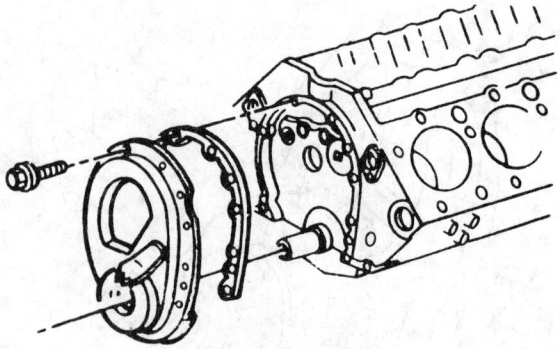

Fig. 2C.8 Exploded view of the crankcase front cover installation (Sec 13)

4 Position the number one piston at top dead center (TDC) on the compression stroke. To do this, remove all of the spark plugs from the engine, then locate the number one cylinder spark plug wire and trace it back to the distributor. Make a mark on the distributor body directly below the terminal where the number one spark plug wire attaches to the distributor cap. Remove the cap and wires from the distributor.
5 Slip a wrench or socket over the large bolt at the front of the crankshaft and slowly turn it in a clockwise direction until the timing mark on the damper is aligned with the zero (0). The rotor should now be pointing at the mark you made on the distributor body. If it is not, turn the crankshaft one more complete revolution (360°) in a clockwise direction. If the rotor is now pointing at the mark on the distributor body, then the number one piston is at TDC on the compression stroke. Make sure the marks on the crankshaft and camshaft are aligned as shown in the accompanying illustration.
6 Remove the camshaft sprocket mounting bolts and lift the camshaft sprocket and timing chain together off the front of the engine. In some cases it may be necessary to tap the sprocket off with a soft-faced hammer.
7 If it is necessary to remove the crankshaft sprocket, it can be withdrawn from the crankshaft with a special puller.
8 Attach the crankshaft sprocket to the crankshaft using a bolt and washer from the puller set.
9 Install the timing chain over the camshaft sprocket with slack in the chain hanging down over the crankshaft sprocket.
10 With the timing marks aligned as shown, slip the chain over the crankshaft sprocket and then draw the camshaft sprocket into place with the mounting bolts. Do not hammer or attempt to drive the camshaft sprocket into place, as it could dislodge the welch plug at the rear of the engine.
11 With the chain and both sprockets in place, check again to ensure that the timing marks on the two sprockets are properly aligned. If not, remove the camshaft sprocket and move it until the marks align.
12 Lubricate the chain with engine oil and install the remaining components in the reverse order of removal.

16 Camshaft — removal and installation

Note: *Before removing the camshaft, refer to Chapter 2, Part D (Section 15), and measure the lobe lift.*

1 Remove the cable from the negative battery terminal.
2 Drain the oil from the crankcase (refer to Chapter 1).
3 Drain the coolant from the radiator (refer to Chapter 1).
4 Remove the radiator (refer to Chapter 3).
5 If equipped with air-conditioning, remove the condenser (refer to Chapter 3). **Caution:** *The air-conditioning system must be evacuated by an air-conditioning technician before the condenser can be removed. Under no circumstances should this be attempted by the home mechanic, because personal injury may result.*
6 Remove the grille.
7 Remove the valve lifters (refer to Section 5).
8 Remove the crankcase front cover (refer to Section 13).
9 Remove the fuel pump and pushrod (refer to Chapter 4).
10 Remove the timing chain and sprocket (refer to Section 15).
11 Install a long bolt in one of the camshaft bolt holes to be used as a handle to pull on and support the camshaft.

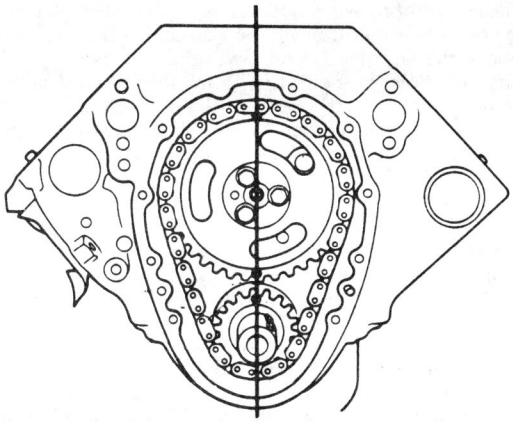

Fig. 2C.9 Proper alignment of the crankshaft and camshaft gear timing marks (Sec 16)

12 Carefully draw the camshaft out of the engine block. Do this very slowly to avoid damage to the camshaft bearings as the lobes pass over the bearing surfaces. Always support the camshaft with one hand near the engine block.
13 Refer to Chapter 2, Part D, for the camshaft inspection procedures.
14 Before installing the camshaft, coat each of the lobes and journals with engine assembly lube or moly-based grease.
15 Slide the camshaft into the engine block, again taking extra care not to damage the bearings.
16 Install the camshaft sprocket and timing chain as described in Section 15.
17 Install the remaining components in the reverse order of removal by referring to the appropriate Chapter or Section.
18 Adjust the valve lash (refer to Section 6).
19 Have the air-conditioning system recharged (if so equipped).

17 Engine mounts — replacement with engine in vehicle

1 If the rubber mounts have become hard, split or separated from the metal backing, they must be replaced. The operation may be carried out with the engine/transmission still in the vehicle. See Section 9 for the proper way to raise the engine while it is still in place.

Front mounts

2 Remove the through-bolt and nut.
3 Raise the engine slightly using a hoist or jack with a wood block under the oil pan, then remove the mount and frame bracket assembly from the crossmember.
4 Position the new mount, install the through-bolt and nut, then tighten all the bolts to the specified torque.

Rear mount

5 Remove the crossmember-to-mount bolts, then raise the transmission slightly with a jack.
6 Remove the mount-to-transmission bolts, followed by the mount.
7 Install the mount, lower the transmission and align the crossmember-to-mount bolts.
8 Tighten all the bolts to the specified torque.

18 Engine — removal and installation

Note: *On some models, the upper bellhousing-to-engine bolts are inaccessible, in some cases requiring cutting the floor pan to expose the bolts. If this is the case with your vehicle and you do not want to cut the floor pan, read through the procedure in this Section to familiarize yourself with the V8 engine removal, then refer to the engine removal Section in Chapter 2, Part B, and remove the engine and transmission as a unit, noting any differences between the V6 and V8 engines and adjusting the procedure accordingly (many of the V6 illustrations also apply to the V8).*

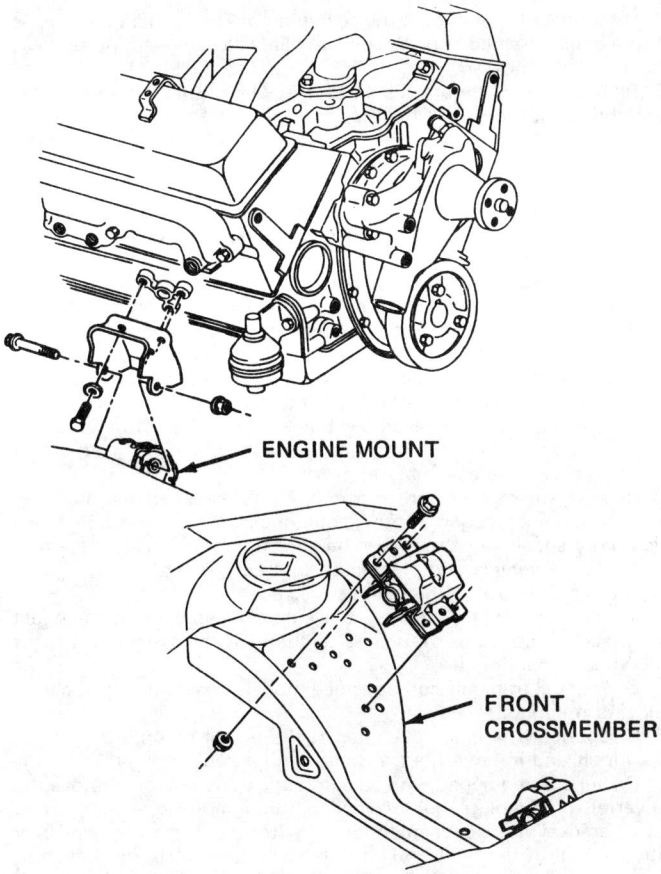

Fig. 2C.10 Front motor mount installation details (Sec 17)

Removal

1 Disconnect the cable from the negative battery terminal.
2 Remove the clip retaining the underhood light wire harness to the hood.
3 Disconnect the underhood light electrical connector and unbolt the bulb assembly from the hood.
4 Scribe lines on the underside of the hood, around the hood mounting bracket, so the hood can be installed in the same position.
5 Apply white paint around the bracket-to-hood bolts so they can be aligned quickly and accurately during installation.
6 Remove the bracket-to-hood bolts.
7 With the help of an assistant, lift the hood and separate it from the vehicle.
8 Drain the radiator by opening the petcock located at the lower right-hand side of the radiator.
9 Remove the coolant recovery tank hose from the radiator.
10 Remove the upper radiator hose from the radiator. Take care not to damage the plastic fitting when breaking the hose loose.
11 If so equipped, remove the automatic transmission cooler lines at the radiator.
12 Remove the bolts securing the top of the radiator fan shroud to the lower half and remove the upper shroud.
13 Remove the lower radiator hose from the radiator.
14 Remove the radiator from the vehicle.
15 Mark the drivebelts to simplify installation.
16 Unbolt and remove the fan from the fan pulley.
17 If so equipped, remove the air-conditioning compressor (refer to Chapter 3).
18 Remove the alternator (refer to Chapter 5).
19 Remove the heater hoses from the water pump and intake manifold outlets.
20 Remove the AIR system air pump (refer to Chapter 6).
21 Disconnect the hoses at the air cleaner. Carefully label them as they are removed to simplify installation.
22 Remove the air cleaner assembly.
23 Disconnect the vacuum brake booster line at the booster.

24 Disconnect the linkage at the carburetor or TBI, including the cruise control cable (if so equipped). Label the linkage/cable components as they are disconnected.
25 Remove the distributor cap with the wires attached and lay the assembly aside (refer to Chapter 5 if necessary).
26 On carburetor-equipped models, disconnect the electronic metering valve wire connector from the carburetor (refer to Chapter 4 for carburetor details).
27 Disconnect the electric choke wire connector from the carburetor.
28 Disconnect the electrical connector for the EFE heater wire at the base of the carburetor.
29 On fuel-injected (TBI) models, remove the fuel hoses at the TBI unit, then plug the openings to prevent fuel loss and contamination.
30 Disconnect the wires at the coolant temperature and oil pressure sending units.
31 Remove the power steering pump mounting bolts and lay the pump aside, with the hoses still attached, out of the way.
32 Unbolt the solenoid valve/hose bracket from the left-hand rocker arm cover so it will not be damaged when the engine is lifted out of the vehicle. Lay the assembly aside.
33 Remove the oxygen sensor from the right-side exhaust manifold so it will not be damaged as other components are removed. Because it is a very sensitive component, be sure to protect it.
34 Raise the vehicle and place it securely on jackstands.
35 Drain the oil from the engine crankcase.
36 Remove the bolts securing the exhaust pipes to the exhaust manifolds on both sides of the engine, then pull the exhaust pipes off the exhaust manifolds.
37 On manual transmission-equipped models, disconnect the clutch fork return spring.
38 Disconnect the electrical wires at the transmission.
39 Unbolt and remove the transmission dust cover.
40 On vehicles equipped with an automatic transmission, remove the converter-to-driveplate bolts, working through the opening gained by the removal of the cover mentioned in Step 39. Turn the engine over with the bolt at the center of the vibration damper to bring each of the bolts into view. Mark the relative position of the converter with a scribe so it can be reinstalled in the same position. Wedge a screwdriver in the teeth of the flywheel to prevent movement as the bolts are loosened.
41 Disconnect and label the wires at the starter motor.
42 On carburetor-equipped models, disconnect the lines at the fuel pump and plug the openings to prevent loss of fuel and contamination.
43 Remove the AIR system/catalytic converter pipe mounting bracket from the transmission.
44 Remove the bellhousing-to-engine bolts (refer to the note at the beginning of this Section).
45 Remove the motor mount through-bolt on each side of the engine.
46 Make sure that all wires and hoses, both under the vehicle and in the engine compartment, have been disconnected.
47 Remove the jackstands and lower the vehicle.
48 Attach the hoist lifting chains to the lifting 'eyes' on the engine. There is one bracket at the front of the engine and one at the rear, diagonally opposite each other. Make sure the chain is looped properly through the engine brackets and secured with strong nuts and bolts through the chain links. The hook on the hoist should be over the center of the engine with the lengths of chain equal to lift the engine straight up.
49 Raise the engine hoist until all slack is out of the chains. Do not lift any further at this time.
50 Support the transmission using a jack with wood blocks as cushions.
51 Raise the engine slightly and then pull it forward to clear the input shaft (manual transmission). Where an automatic transmission is installed, keep the torque converter pushed well to the rear to ensure engagement of the converter tangs with the oil pump inside the transmission.
52 Carefully lift the engine straight up and out of the engine compartment, continually checking clearances around the engine.
53 The transmission should remain supported by the floor jack or wood blocks while the engine is out of place.
54 Attach the engine to an engine stand.
55 Refer to Chapter 2, Part D, for further disassembly and rebuilding procedures.

Installation

56 Lift the engine off the engine stand with a hoist. The chains should be positioned as they were during removal, with the engine sitting level.
57 Lower the engine into place inside the engine compartment, closely watching clearances. On manual transmissions, carefully guide the engine onto the transmission input shaft. The two components should be at the same angle, with the shaft sliding easily into the engine.
58 Install the engine mount through-bolts and the bellhousing bolts. Tighten the bolts to the specified torque.
59 Install the remaining engine components in the reverse order of removal.
60 Fill the cooling system with the specified antifreeze and water mixture (Chapter 1).
61 Fill the engine with the correct grade of engine oil (Chapter 1).
62 Check the transmission fluid level, adding fluid as necessary.
63 Connect the positive battery cable, followed by the negative cable. If sparks or arcing occur as the negative cable is connected to the battery, make sure that all electrical accessories are turned off (check the dome light first). If arcing still occurs, make sure that all electrical wires are properly connected to the engine and transmission.
64 Refer to Chapter 2, Part D, for the engine start-up procedure.

Chapter 2 Part D
General engine overhaul procedures

Refer to Chapter 13 for specifications and information related to 1985 and later models

Contents

Camshaft, lifters and bearings — inspection and bearing replacement . . . 15
Crankshaft — inspection . . . 17
Crankshaft — installation and main bearing oil clearance check . . . 21
Crankshaft — removal . . . 12
Cylinder head — cleaning and inspection . . . 8
Cylinder head — disassembly . . . 7
Cylinder head — reassembly . . . 10
Engine block — cleaning . . . 13
Engine block — inspection . . . 14
Engine overhaul — disassembly sequence . . . 6
Engine overhaul — general information . . . 3
Engine overhaul — reassembly sequence . . . 23
Engine rebuilding alternatives . . . 4
Engine removal — methods and precautions . . . 5
General information . . . 1
Initial start-up and break-in after overhaul . . . 25
Main and connecting rod bearings — inspection . . . 18
Piston/connecting rod assembly — inspection . . . 16
Piston/connecting rod assembly — installation and bearing oil clearance check . . . 22
Piston/connecting rod assembly — removal . . . 11
Piston rings — installation . . . 19
Pre-oiling engine after overhaul (V6 and V8 engines only) . . . 24
Rear main oil seal (V6 and V8 engines only) — installation . . . 20
Repair operations possible with the engine in the vehicle . . . 2
Valves — servicing . . . 9

Specifications

L4 engine

Oil pressure	36 to 41 psi at 2000 rpm
Valves and related components	
Valve face angle	45°
Valve seat angle	46°
Valve stem diameter	
Intake	0.3418 to 0.3425 in (8.6817 to 8.6995 mm)
Exhaust	0.3418 to 0.3425 in (8.6817 to 8.6995 mm)
Stem-to-guide clearance	
Intake and top exhaust	0.001 to 0.0027 in (0.0254 to 0.0686 mm)
Bottom exhaust	0.002 to 0.0037 in (0.0508 to 0.0939 mm)
Valve seat width	
Intake	0.0353 to 0.0747 in (0.897 to 1.897 mm)
Exhaust	0.058 to 0.097 in (1.468 to 2.468 mm)
Valve spring installed height	1.69 in (42.93 mm)
Valve spring pressure and length (intake and exhaust)	
Valve closed	1.66 in at 76 to 86 lbs.
Valve open	1.254 in at 122 to 180 lbs.
Pushrod length	9.754 in (242.316 mm)
Crankshaft and connecting rods	
Crankshaft end play	0.0035 to 0.0085 in (0.0889 to 0.2159 mm)
Connecting rod end play (side clearance)	0.006 to 0.022 in (0.1524 to 0.5588 mm)
Main bearing journal diameter	2.300 in (59.182 mm)
Main bearing oil clearance	0.0005 to 0.0022 in (0.0127 to 0.05588 mm)
Connecting rod bearing journal diameter	2.000 in (50.8 mm)
Connecting rod bearing oil clearance	0.0005 to 0.0026 in (0.0127 to 0.0660 mm)
Crankshaft journal taper/out-of-round limit	0.0005 in (0.0127 mm)
Engine block	
Cylinder bore diameter	4.000 in (101.6 mm)
Out-of-round limit	0.0014 in (0.0356 mm)
Taper limit	0.005 in (0.127 mm)

L4 engine (continued)

Pistons and rings

Compression ring side clearance	0.003 in (0.0762 mm)
Piston diameter	3.9971 to 3.9975 in (101.526 to 101.536 mm)
Piston-to-bore clearance	
Top	0.0025 to 0.0033 in (0.0635 to 0.0838 mm)
Bottom	0.0017 to 0.0041 in (0.043 to 0.104 mm)
Piston pin diameter	0.938 to 0.942 in (23.825 to 23.927 mm)
Pin-to-piston clearance	Loose; 0.003 to 0.005 in (0.076 to 0.127 mm)
Pin-to-rod clearance	Loose
Piston ring end gap	
Top ring	0.015 to 0.025 in (0.381 to 0.635 mm)
Second ring	0.009 to 0.019 in (0.2286 to 0.04826 mm)
Oil ring	0.015 to 0.055 in (0.381 to 1.397 mm)

Camshaft

Lobe lift (intake and exhaust)	0.406 in (10.3124 mm)
Bearing journal diameter	1.869 in (47.4726 mm)
Bearing oil clearance	0.0007 to 0.0027 in (0.0178 to 0.0685 in)
Camshaft end play	0.0015 to 0.005 in (0.0381 to 0.127 mm)

Cylinder numbers	1–2–3–4 (front-to-rear)
Firing order	1–3–4–2

Torque specifications

	Ft-lbs	M-Kg
Rocker arm bolts	20	2.8
Main bearing cap bolts	70	9.6
Connecting rod cap nuts	32	4.4
Cylinder head bolts	85	11.7
Oil pump driveshaft retainer plate bolts	10	1.4

V6 engine

Valves and related components

Valve face angle	45°
Valve seat angle	46°
Valve seat runout	0.002 in (0.05 mm) maximum
Stem-to-guide clearance	0.001 to 0.0028 in (0.026 to 0.068 mm)
Valve seat width	
Intake	0.049 to 0.059 in (1.25 to 1.50 mm)
Exhaust	0.063 to 0.075 in (1.60 to 1.90 mm)
Valve spring installed height	1.57 in (40.0 mm)
Valve spring free length	1.91 in (48.5 mm)
Valve spring pressure and length (intake and exhaust)	
Valve closed	1.57 in at 87.9 lbs.
Valve open	1.18 in at 194.9 lbs.

Crankshaft and connecting rods

Crankshaft end play	0.002 to 0.007 in (0.05 to 0.17 mm)
Connecting rod end play (side clearance)	0.006 to 0.017 in (0.16 to 0.44 mm)
Main bearing journal diameter	
1982 (all)	2.4937 to 2.4946 in (63.340 to 63.364 mm)
1983 and 1984 (1, 2 and 4)	2.4937 to 2.4946 in (63.340 to 63.364 mm)
1983 and 1984 (3 only)	2.4932 to 2.4941 in (63.327 to 63.351 mm)
Main bearing oil clearance	
1982 (all)	0.0017 to 0.0030 in (0.044 to 0.076 mm)
1983 and 1984 (1, 2 and 4)	0.0016 to 0.0032 in (0.041 to 0.081 mm)
1983 and 1984 (3 only)	0.0021 to 0.0033 in (0.054 to 0.084 mm)
Connecting rod bearing journal diameter	1.9983 to 1.9994 in (50.758 to 50.784 mm)
Connecting rod bearing oil clearance	0.0012 to 0.0037 in (0.035 to 0.095 mm)
Crankshaft journal taper/out-of-round limit	0.0002 in (0.005 mm)

Engine block

Cylinder bore diameter	3.504 to 3.507 in (88.992 to 89.070 mm)
Out-of-round limit	0.0008 in (0.02 mm)
Taper limit	0.0008 in (0.02 mm)

Pistons and rings

Compression ring side clearance	
Top ring	0.0019 to 0.0028 in (0.030 to 0.070 mm)
Second ring	0.0016 to 0.0037 in (0.040 to 0.095 mm)
Oil ring side clearance	0.0078 in (0.199 mm)
Piston-to-bore clearance	0.0017 to 0.0027 in (0.043 to 0.069 mm)
Piston pin diameter	0.90526 to 0.90557 in (22.994 to 23.002 mm)
Pin-to-piston clearance	0.00026 to 0.00036 in (0.0065 to 0.0091 mm)
Pin-to-rod clearance	0.00074 to 0.00203 in (0.0187 to 0.0515 mm)
Piston ring end gap	
Top ring	0.0098 to 0.0196 in (0.25 to 0.50 mm)
Second ring	0.0098 to 0.0196 in (0.25 to 0.50 mm)
Oil ring	0.020 to 0.055 in (0.51 to 1.40 mm)

Camshaft		
Lobe lift		
Intake	0.231 in (5.87 mm)	
Exhaust	0.262 in (6.67 mm)	
Bearing journal diameter	1.868 to 1.870 in (47.44 to 47.49 mm)	
Bearing oil clearance	0.001 to 0.004 in (0.026 to 0.101 mm)	
Cylinder numbers		
Right bank	1—3—5 (front-to-rear)	
Left bank	2—4—6 (front-to-rear)	
Firing order	1—2—3—4—5—6	
Torque specifications	**Ft-lbs**	**M-Kg**
Rocker arm studs	45	6.2
Main bearing cap bolts	70	9.6
Connecting rod cap nuts	37	5.1
Cylinder head bolts	68	9.4

V8 engine

Valves and related components	
Valve face angle	45°
Valve seat angle	46°
Valve seat runout	0.002 in maximum
Stem-to-guide clearance	
Standard	0.001 to 0.0027 in
Service limit	0.002 to 0.0037 in
Valve seat width	
Intake	1/32 to 1/16 in
Exhaust	1/16 to 3/32 in
Valve spring installed height	1-23/32 in
Valve spring pressure and length (intake and exhaust)	
Valve closed	1.70 in at 76 to 84 lbs.
Valve open	1.25 in at 194 to 206 lbs.
Valve spring free length	2.03 in
Valve spring damper free length	1.86 in
Crankshaft and connecting rods	
Main journal diameter	
Journal 1	2.4484 to 2.4493 in
Journals 2, 3 and 4	2.4481 to 2.4490 in
Journal 5	2.4479 to 2.4488 in
Journal taper	
Standard	0.0002 in maximum
Service limit	0.0010 in maximum
Journal out-of-round	
Standard	0.0002 in maximum
Service limit	0.0010 in maximum
Main bearing oil clearance	
Journal 1	0.0008 to 0.0020 in
Journals 2, 3 and 4	0.0011 to 0.0023 in
Journal 5	0.0017 to 0.0032 in
Crankshaft end play	0.002 to 0.006 in
Connecting rod journal diameter	2.0986 to 2.0998 in
Connecting rod journal taper	
Standard	0.0005 in maximum
Service limit	0.001 in maximum
Connecting rod journal out-of-round	
Standard	0.0002 in maximum
Service limit	0.001 in maximum
Rod bearing oil clearance	0.0018 to 0.0039 in
Connecting rod end play (side clearance)	0.008 to 0.014 in
Engine block	
Bore diameter	3.7350 to 3.7385 in
Out-of-round	
Standard	0.0010 in maximum
Service limit	0.0020 in maximum
Taper—thrust side	
Standard	0.0005 in maximum
Service limit	0.0015 in maximum
Pistons and rings	
Piston-to-bore clearance	
Standard	0.0007 to 0.0017 in
Service limit	0.0027 in
Piston ring side clearance	
Top ring	
Standard	0.0012 to 0.0032 in

V8 engine (continued)

Service limit	0.0042 in
Second ring	
Standard	0.0012 to 0.0032 in
Service limit	0.0042 in
Oil ring	
Standard	0.002 to 0.007 in
Service limit	0.008 in
Piston ring end gap	
Top ring	
Standard	0.010 to 0.020 in
Service limit	0.030 in
Second ring	
Standard	0.010 to 0.025 in
Service limit	0.035 in
Oil ring	
Standard	0.015 to 0.055 in
Service limit	0.065 in
Piston pin diameter	0.9270 to 0.9273 in
Pin-to-piston clearance	
Standard	0.00025 to 0.00035 in
Service limit	0.001 in maximum
Piston pin interference fit in rod	0.0008 to 0.0016 in
Camshaft	
Lobe lift (intake)	
Carbureted engine (standard)	0.234 in
Carbureted engine (HO)	0.269 in
TBI engine	0.257 in
Lobe lift (exhaust)	
Carbureted engine (standard)	0.257 in
Carbureted engine (HO)	0.276 in
TBI engine	0.269 in
Bearing journal diameter	1.8682 to 1.8692 in
Camshaft end play	0.004 to 0.012 in
Cylinder numbers	
Left bank	1–3–5–7 (front-to-rear)
Right bank	2–4–6–8 (front-to-rear)
Firing order	1–8–4–3–6–5–7–2

Torque specifications	**Ft-lbs**	**M-Kg**
Main bearing cap bolts	70	9.6
Connecting rod cap nuts	45	6.2
Cylinder head bolts	65	9.0
Oil pump bolts	65	9.0
Oil pump cover bolts	80	11.0

1 General information

Included in this portion of Chapter 2 are the general overhaul procedures for the cylinder head(s) and internal engine components. The information ranges from advice concerning preparation for an overhaul and the purchase of replacement parts to detailed, step-by-step procedures covering removal and installation of internal engine components and the inspection of parts.

The following Sections have been written based on the assumption that the engine has been removed from the vehicle. For information concerning in-vehicle engine repair, as well as removal and installation of the external components necessary for the overhaul, see Part A, B or C of Chapter 2 (depending on engine type) and Section 2 of this Part.

The Specifications included here in Part D are only those necessary for the inspection and overhaul procedures which follow. Refer to Part A, B or C for additional Specifications related to the various engines covered in this manual.

2 Repair operations possible with the engine in the vehicle

Many major repair operations can be accomplished without removing the engine from the vehicle.

It is a very good idea to clean the engine compartment and the exterior of the engine with some type of pressure washer before any work is begun. A clean engine will make the job easier and will prevent the possibility of getting dirt into internal areas of the engine.

Remove the hood (Chapter 12) and cover the fenders to provide as much working room as possible and to prevent damage to the painted surfaces.

If oil or coolant leaks develop, indicating a need for gasket or seal replacement, the repairs can generally be made with the engine in the vehicle. The oil pan gasket, the cylinder head gasket(s), intake and exhaust manifold gaskets, timing chain cover gaskets and the front and rear crankshaft oil seals are accessible with the engine in place.

Exterior engine components, such as the water pump, the starter motor, the alternator, the distributor, the fuel pump and the carburetor, as well as the intake and exhaust manifolds, are quite easily removed for repair with the engine in place.

Since the cylinder head(s) can be removed without pulling the engine, valve component servicing can also be accomplished with the engine in the vehicle.

Replacement of, repairs to or inspection of the timing gears or sprockets and chain and the oil pump are all possible with the engine in place.

In extreme cases caused by a lack of necessary equipment, repair or replacement of piston rings, pistons, connecting rods and rod bearings and reconditioning of the cylinder bores is possible with the engine in the vehicle. However, this practice is not recommended because of the cleaning and preparation work that must be done to the components involved.

Detailed removal, inspection, repair and installation procedures for the above mentioned components can be found in the appropriate Part of Chapter 2 or the other Chapters in this manual.

3 Engine overhaul — general information

It is not always easy to determine when, or if, an engine should be completely overhauled, as a number of factors must be considered.

High mileage is not necessarily an indication that an overhaul is needed while low mileage, on the other hand, does not preclude the need for an overhaul. Frequency of servicing is probably the single most important consideration. An engine that has regular (and frequent) oil and filter changes, as well as other required maintenance, will most likely give many thousands of miles of reliable service. Conversely, a neglected engine may require an overhaul very early in its life.

Excessive oil consumption is an indication that piston rings and/or valve guides are in need of attention (make sure that oil leaks are not responsible before deciding that the rings and guides are bad). Have a cylinder compression or leak-down test performed by an experienced tune-up mechanic to determine for certain the extent of the work required.

If the engine is making obvious knocking or rumbling noises, the connecting rod and/or main bearings are probably at fault. Check the oil pressure with a gauge (installed in place of the oil pressure sending unit) and compare it to the Specifications. If it is extremely low, the bearings and/or oil pump are probably worn out.

Loss of power, rough running, excessive valve train noise and high fuel consumption rates may also point to the need for an overhaul (especially if they are all present at the same time). If a complete tune-up does not remedy the situation, major mechanical work is the only solution.

An engine overhaul generally involves restoring the internal parts to the specifications of a new engine. During an overhaul, the piston rings are replaced and the cylinder walls are reconditioned (rebored and/or honed). If a rebore is done, then new pistons are also required. The main and connecting rod bearings are replaced with new ones and, if necessary, the crankshaft may be reground to restore the journals. Generally, the valves are serviced as well since they are usually in less-than-perfect condition at this point. While the engine is being overhauled, other components such as the carburetor, the distributor, the starter and the alternator can be rebuilt also. The end result should be a like-new engine that will give as many trouble-free miles as the original.

Before beginning the engine overhaul, read through the entire procedure to familiarize yourself with the scope and requirements of the job. Overhauling an engine is not that difficult, but it is time consuming. Plan on the vehicle being tied up for a minimum of two weeks, especially if parts must be taken to an automotive machine shop for repair or reconditioning. Check on availability of parts and make sure that any necessary special tools and equipment are obtained in advance. Most work can be done with typical shop hand tools, although a number of precision measuring tools are required for inspecting parts to determine if they must be replaced. Often a reputable automotive machine shop will handle the inspection of parts and offer advice concerning reconditioning and replacement. **Note:** *Always wait until the engine has been completely disassembled and all components, especially the engine block, have been inspected before deciding what service and repair operations must be performed by an automotive machine shop.* Since the block's condition will be the major factor to consider when determining whether to overhaul the original engine or buy a rebuilt one, never purchase parts or have machine work done on other components until the block has been thoroughly inspected. As a general rule, time is the primary cost of an overhaul, so it does not pay to install worn or sub-standard parts.

As a final note, to ensure maximum life and minimum trouble from a rebuilt engine, everything must be assembled with care in a spotlessly clean environment.

4 Engine rebuilding alternatives

The do-it-yourselfer is faced with a number of options when performing an engine overhaul. The decision to replace the engine block, piston/connecting rod assemblies and crankshaft depends on a number of factors, with the number one consideration being the condition of the block. Other considerations are cost, access to machine shop facilities, parts availability, time required to complete the project and experience.

Some of the rebuilding alternatives include:

Individual parts — *If the inspection procedures reveal that the engine block and most engine components are in reusable condition, purchasing individual parts may be the most economical alternative. The block, crankshaft and piston/connecting rod assemblies should all be inspected carefully. Even if the block shows little wear, the cylinder bores should receive a finish hone; a job for an automotive machine shop.*

Master kit *(crankshaft kit)* — *This rebuild package usually consists of a reground crankshaft and a matched set of pistons and connecting rods. The pistons will already be installed on the connecting rods. Piston rings and the necessary bearings may or may not be included in the kit. These kits are commonly available for standard cylinder bores, as well as for engine blocks which have been bored to a regular oversize.*

Short block — *A short block consists of an engine block with a crankshaft and piston/connecting rod assemblies already installed. All new bearings are incorporated and all clearances will be correct. Depending on where the short block is purchased, a guarantee may be included. The existing camshaft, valve train components, cylinder head(s) and external parts can be bolted to the short block with little or no machine shop work necessary.*

Long block — *A long block consists of a short block plus an oil pump, oil pan, cylinder head(s), rocker arm cover(s), camshaft and valve train components, timing gears or sprockets and chain and timing gear/chain cover. All components are installed with new bearings, seals and gaskets incorporated throughout. The installation of manifolds and external parts is all that is necessary. Some form of guarantee is usually included with the purchase.*

Give careful thought to which alternative is best for you and discuss the situation with local automotive machine shops, auto parts dealers or dealership partsmen before ordering or purchasing replacement parts.

5 Engine removal — methods and precautions

If it has been decided that an engine must be removed for overhaul or major repair work, certain preliminary steps should be taken.

Locating a suitable work area is extremely important. A shop is, of course, the most desirable place to work. Adequate work space along with storage space for the vehicle is very important. If a shop or garage is not available, at the very least a flat, level, clean work surface made of concrete or asphalt is required.

Cleaning the engine compartment and engine prior to removal will help keep tools clean and organized.

An engine hoist or A-frame will also be necessary. Make sure that the equipment is rated in excess of the combined weight of the engine and its accessories. Safety is of primary importance, considering the potential hazards involved in lifting the engine out of the vehicle.

If the engine is being removed by a novice, a helper should be available. Advice and aid from someone more experienced would also be helpful. There are many instances when one person cannot simultaneously perform all of the operations required when lifting the engine out of the vehicle.

Plan the operation ahead of time. Arrange for or obtain all of the tools and equipment you will need prior to beginning the job. Some of the equipment necessary to perform engine removal and installation safely and with relative ease are (in addition to an engine hoist) a heavy-duty floor jack, complete sets of wrenches and sockets as described in the front of this manual, wooden blocks and plenty of rags and cleaning solvent for mopping up the inevitable spills. If the hoist is to be rented, make sure that you arrange for it in advance and perform beforehand all of the operations possible without it. This will save you money and time.

Plan for the vehicle to be out of use for a considerable amount of time. A machine shop will be required to perform some of the work which the do-it-yourselfer cannot accomplish due to a lack of special equipment. These shops often have a busy schedule so it would be wise to consult them before removing the engine in order to accurately estimate the amount of time required to rebuild or repair components that may need work.

Always use extreme caution when removing and installing the engine; serious injury can result from careless actions. Plan ahead. Take your time and a job of this nature, although major, can be accomplished successfully.

6 Engine overhaul disassembly sequence

1 It is much easier to disassemble and work on the engine if it is mounted on a portable engine stand. These stands can often be rented for a reasonable fee from an equipment rental yard. Before the engine is mounted on a stand, the flywheel/driveplate should be removed from the engine (refer to Chapter 8).
2 If a stand is not available, it is possible to disassemble the engine with it blocked up on a sturdy workbench or on the floor. Be extra careful not to tip or drop the engine when working without a stand.
3 If you are going to obtain a rebuilt engine, all external components must come off first in order to be transferred to the replacement engine (just as they will if you are doing a complete engine overhaul yourself). These include:

Alternator and brackets
Emissions control components
Distributor, spark plug wires and spark plugs
Thermostat and housing cover
Water pump
Carburetor/fuel injection components
Intake/exhaust manifolds
Oil filter
Fuel pump
Engine mounts
Flywheel/driveplate

Note: *When removing the external components from the engine, pay close attention to details that may be helpful or important during installation. Note the installed position of gaskets, seals, spacers, pins, washers, bolts and other small items.*

4 If you are obtaining a short block (which consists of the engine block, crankshaft, pistons and connecting rods all assembled), then the cylinder heads, oil pan and oil pump will have to be removed also. See *Engine rebuilding alternatives* for additional information regarding the different possibilities to be considered.
5 If you are planning a complete overhaul, the engine must be disassembled and the internal components removed in the following order:

Rocker arm cover(s)
Cylinder head(s) and pushrods
Valve lifters
Timing chain/gear cover
Timing chain/sprockets or gears
Camshaft
Oil pan
Oil pump
Piston/connecting rod assemblies
Crankshaft

6 Before beginning the disassembly and overhaul procedures, make sure the following items are available:

Common hand tools
Small cardboard boxes or plastic bags for storing parts
Gasket scraper
Ridge reamer
Vibration damper puller
Micrometers
Small hole gauges
Telescoping gauges
Dial indicator set
Valve spring compressor
Cylinder surfacing hone
Piston ring groove cleaning tool
Electric drill motor
Tap and die set
Wire brushes
Cleaning solvent

7 Cylinder head — disassembly

Note: *New and rebuilt cylinder heads are commonly available for most engines at dealerships and auto parts stores. Due to the fact that some specialized tools are necessary for the disassembly and inspection procedures, and replacement parts may not be readily available, it may be more practical and economical for the home mechanic to purchase a replacement head (or heads) rather than taking the time to disassemble, inspect and recondition the original head(s).*

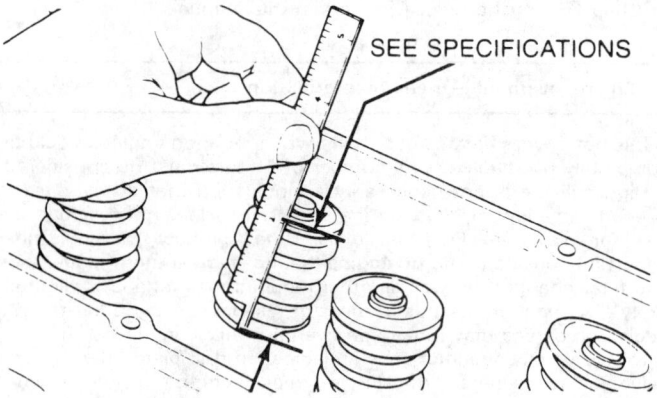

Fig. 2D.1 Measuring the installed valve spring height with a steel rule (V6 engine *intake* valves only) (Sec 7)

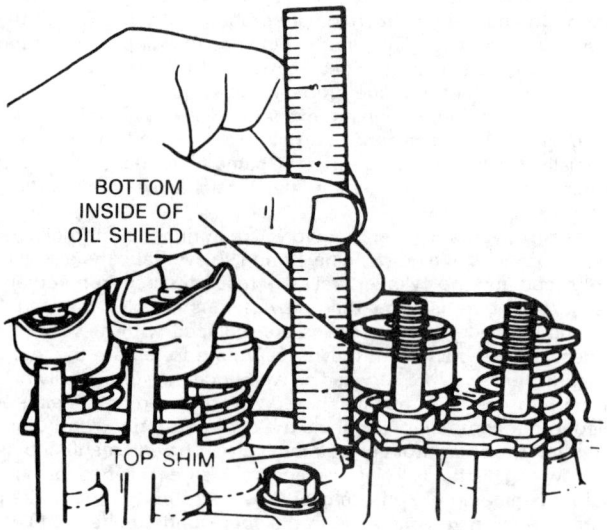

Fig. 2D.2 When checking the valve spring height on V6 engine exhaust valves and all four-cylinder and V8 engine valves, the measurement is made up to the bottom inside surface of the oil shield (Sec 7)

1 Cylinder head disassembly involves removal and disassembly of the intake and exhaust valves and their related components. If they are still in place, remove the nuts or bolts and pivot balls, then separate the rocker arms from the cylinder head. Label the parts or store them separately so they can be reinstalled in their original locations.
2 Before the valves are removed, arrange to label and store them, along with their related components, so they can be kept separate and reinstalled in the same valve guides they are removed from. Also, measure the valve spring installed height (for each valve) and compare it to the Specifications. If it is greater than specified, the valve seats and valve faces need attention.
3 Compress the valve spring on the first valve with a spring compressor and remove the keepers (photo). Carefully release the valve spring compressor and remove the retainer (or rotator), the shield (if so equipped), the springs, the valve guide seal and/or O-ring seal, the spring seat and the valve from the head. If the valve binds in the guide (won't pull through), push it back into the head and deburr the area around the keeper groove with a fine file or whetstone.
4 Repeat the procedure for the remaining valves. Remember to keep together all the parts for each valve so they can be reinstalled in the same locations.

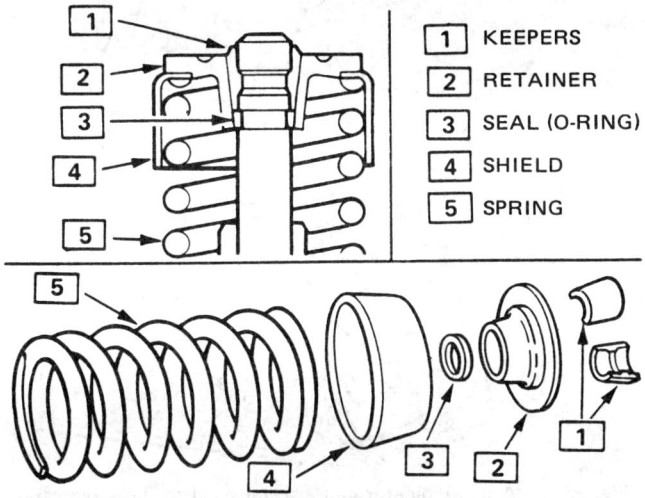

Fig. 2D.3 L4 engine valve components — note the location of the valve stem seals (Sec 7)

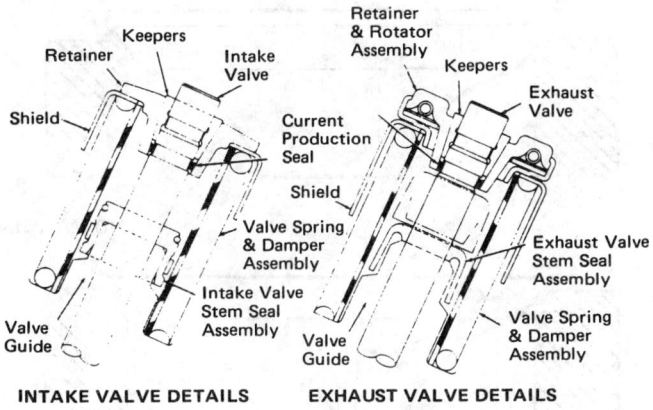

Fig. 2D.5 V8 engine valve components — note that two types of seals are used (Sec 7)

5 Once the valves have been removed and safely stored, the head should be thoroughly cleaned and inspected. If a complete engine overhaul is being done, finish the engine disassembly procedures before beginning the cylinder head cleaning and inspection process.

8 Cylinder head — cleaning and inspection

1 Thorough cleaning of the cylinder head and related valve train components, followed by a detailed inspection, will enable you to decide how much valve service work must be done during the engine overhaul.

Cleaning

2 Scrape away all traces of old gasket material and sealing compound from the head gasket, intake manifold and exhaust manifold sealing surfaces.

3 Remove any built-up scale around the coolant passages.

4 Run a stiff wire brush through the oil holes to remove any deposits that may have formed in them.

5 It is a good idea to run an appropriate size tap into each of the threaded holes to remove any corrosion and thread sealant that may be present. If compressed air is available, use it to clear the holes of debris produced by this operation.

6 Clean the exhaust and intake manifold stud threads in a similar manner with an appropriate size die. Clean the rocker arm pivot bolt or stud threads with a wire brush.

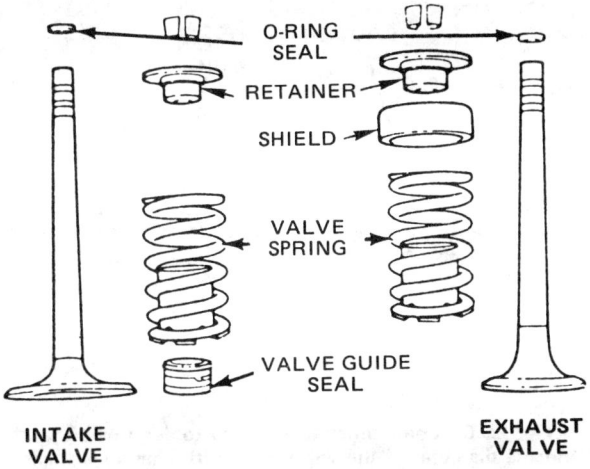

Fig. 2D.4 V6 engine valve components — note the valve stem seal locations (Sec 7)

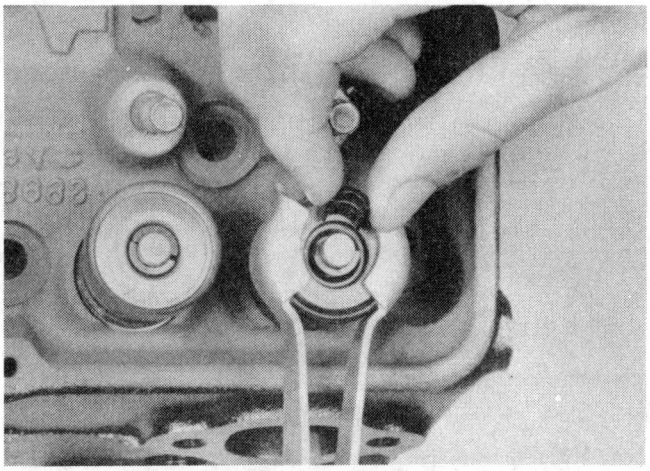

7.3 Use a valve spring compressor to compress the springs, then remove the keepers from the valve stem

7 Next, clean the cylinder head with solvent and dry it thoroughly. Compressed air will speed the drying process and ensure that all holes and recessed areas are clean. **Note:** *Decarbonizing chemicals are available and may prove very useful when cleaning cylinder heads and valve train components. They are very caustic and should be used with caution. Be sure to follow the instructions on the container.*

8 Clean the rocker arms, pivot balls and pushrods with solvent and dry them thoroughly. Compressed air will speed the drying process and can be used to clean out the oil passages.

9 Clean all the valve springs, keepers, retainers, rotators, shields and spring seats with solvent and dry them thoroughly. Do the components from one valve at a time to avoid mixing up the parts.

10 Scrape off any heavy deposits that may have formed on the valves, then use a motorized wire brush to remove deposits from the valve heads and stems. Again, make sure the valves do not get mixed up.

Inspection

Cylinder head

11 Inspect the head very carefully for cracks, evidence of coolant leakage and other damage. If cracks are found, a new cylinder head should be obtained.

12 Using a straightedge and feeler gauges, check the head gasket mating surface for warpage. If the head is warped beyond the limits given in the Specifications, it can be resurfaced at an automotive machine shop.

13 Examine the valve seats in each of the combustion chambers. If they are pitted, cracked or burned, the head will require valve service

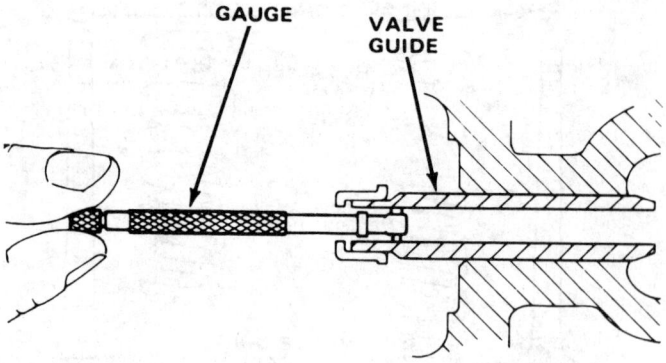

Fig. 2D.6 Use a small hole gauge to determine the inside diameter of the valve guides (the gauge is then measured with a micrometer) (Sec 8)

Fig. 2D.7 A dial indicator can also be used to determine the valve stem-to-guide clearance (Sec 8)

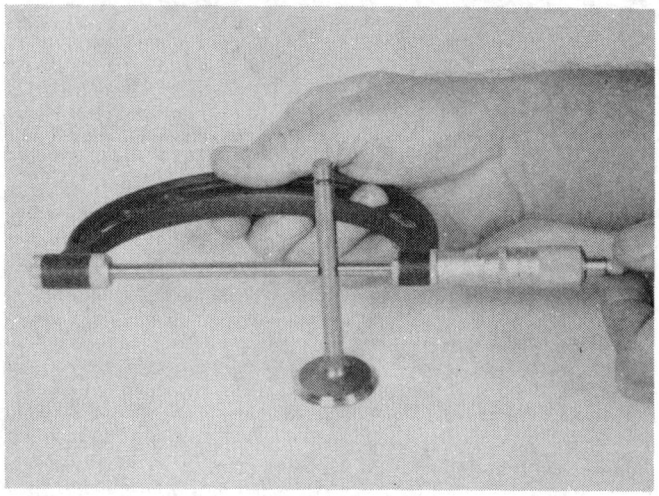

8.20 Measure the valve stem diameter at three points

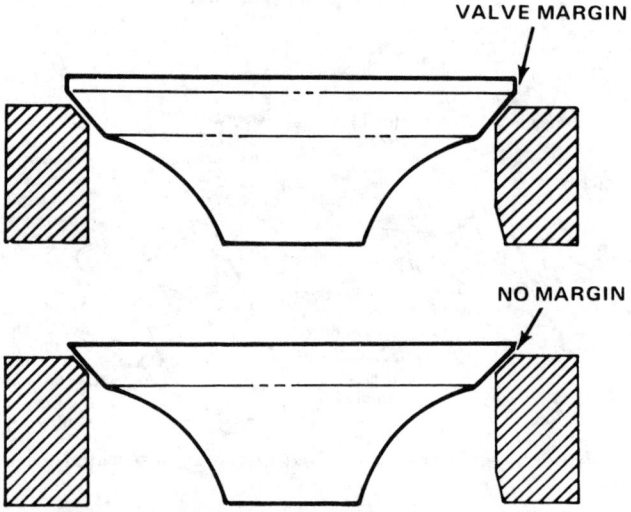

Fig. 2D.8 The margin width on each valve must be as specified (if no margin exists, the valve must be replaced) (Sec 8)

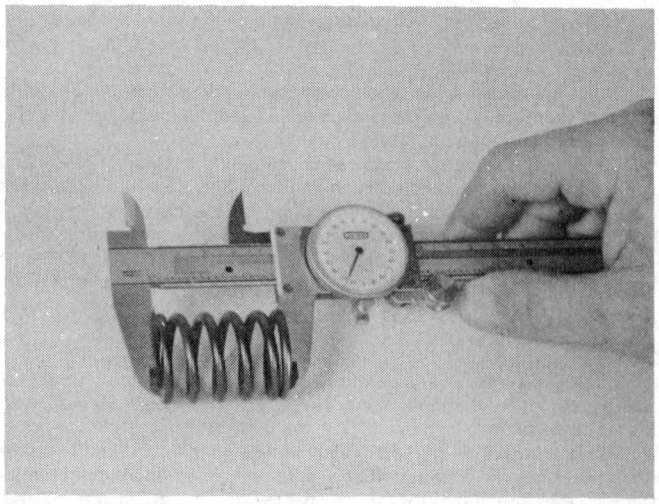

8.21A Measure the free length of each valve spring with a dial or Vernier caliper

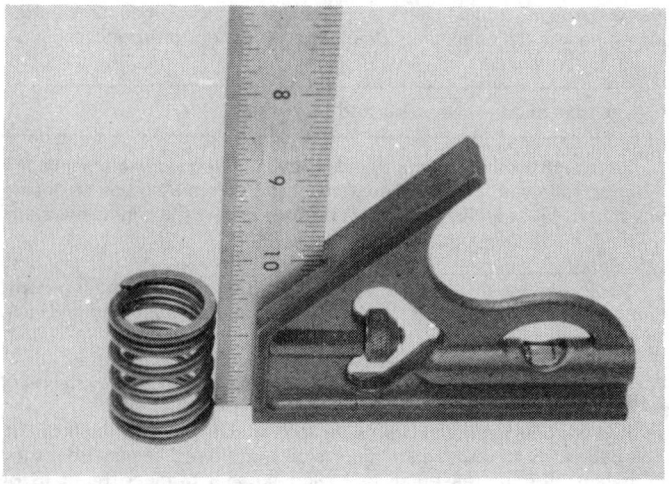

8.21B Check each valve spring for squareness

that is beyond the scope of the home mechanic.

14 Measure the inside diameters of the valve guides (at both ends and the center of each guide) with a small hole gauge and a 0-to-1-inch micrometer. Record the measurements for future reference. These measurements, along with the valve stem diameter measurements, will enable you to compute the valve stem-to-guide clearances. These clearances, when compared to the Specifications, will be one factor that will determine the extent of valve service work required. The guides are measured at the ends and at the center to determine if they are worn in a bell-mouth pattern (more wear at the ends). If they are, guide reconditioning or replacement is necessary. As an alternative, use a dial indicator to measure the lateral movement of each valve stem with the valve in the guide and approximately 1/16-inch off the seat (see the accompanying illustration).

Rocker arm components

15 Check the rocker arm faces (that contact the pushrod ends and valve stems) for pits, wear and rough spots. Check the pivot contact areas as well.

16 Inspect the pushrod ends for scuffing and excessive wear. Roll the pushrod on a flat surface, such as a piece of glass, to determine if it is bent.

17 Any damaged or excessively worn parts must be replaced with new ones.

Valves

18 Carefully inspect each valve face for cracks, pits and burned spots. Check the valve stem and neck for cracks. Rotate the valve and check for any obvious indication that it is bent. Check the end of the stem for pits and excessive wear. The presence of any of these conditions indicates the need for valve service by a properly equipped professional.

19 Measure the width of the valve margin (on each valve) and compare it to the Specifications. Any valve with a margin narrower than specified will have to be replaced with a new one.

20 Measure the valve stem diameter (photo). **Note:** *The exhaust valves used in the GM 151 cu in four-cylinder engine have tapered stems and are approximately 0.001-inch larger at the tip end than at the head end.* By subtracting the stem diameter from the corresponding valve guide diameter, the valve stem-to-guide clearance is obtained. Compare the results to the Specifications. If the stem-to-guide clearance is greater than specified, the guides will have to be reconditioned and new valves may have to be installed, depending on the condition of the old ones.

Valve components

21 Check each valve spring for wear (on the ends) and pits. Measure the free length (photo) and compare it to the Specifications. Any springs that are shorter than specified have sagged and should not be reused. Stand the spring on a flat surface and check it for squareness (photo).

22 Check the spring retainers or rotators and keepers for obvious wear and cracks. Any questionable parts should be replaced with new ones, as extensive damage will occur in the event of failure during engine operation.

23 If the inspection process indicates that the valve components are in generally poor condition and worn beyond the limits specified, which is usually the case in an engine that is being overhauled, reassemble the valves in the cylinder head and refer to Section 9 for valve servicing recommendations.

24 If the inspection turns up no excessively worn parts, and if the valve faces and seats are in good condition, the valve train components can be reinstalled in the cylinder head without major servicing. Refer to the appropriate Section for cylinder head reassembly procedures.

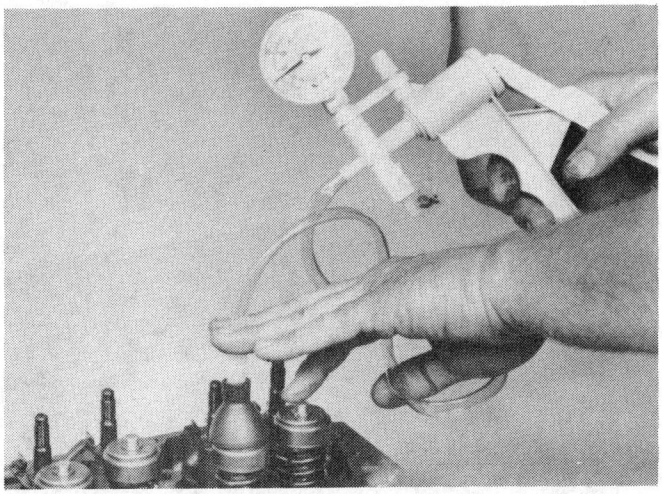

10.6 Checking the valve stem seals for leakage

9 Valves — servicing

1 Because of the complex nature of the job and the special tools and equipment needed, servicing of the valves, the valve seats and the valve guides (commonly known as a 'valve job') is best left to a professional.

2 The home mechanic can remove and disassemble the head, do the initial cleaning and inspection, then reassemble and deliver the head to a dealer service department or a reputable automotive machine shop for the actual valve servicing.

3 The dealer service department, or automotive machine shop, will remove the valves and springs, recondition or replace the valves and valve seats, recondition the valve guides, check and replace the valve springs, spring retainers or rotators and keepers (as necessary), replace the valve seals with new ones, reassemble the valve components and make sure the installed spring height is correct. The cylinder head gasket surface will also be resurfaced if it is warped.

4 After the valve job has been performed by a professional, the head will be in like-new condition. When the head is returned, be sure to clean it again, very thoroughly (before installation on the engine), to remove any metal particles and abrasive grit that may still be present from the valve service or head resurfacing operations. Use compressed air, if available, to blow out all the oil holes and passages.

10 Cylinder head — reassembly

1 Regardless of whether or not the head was sent to an automotive machine shop for valve servicing, make sure it is clean before beginning reassembly.

2 If the head was sent out for valve servicing, the valves and related components will already be in place. Begin the reassembly procedure with Step 6.

3 Lay all the spring seats in position, then lubricate and install new seals (or deflectors) on each of the valve guides (V6 intake valves and all V8 valves). Using a hammer and an appropriate size deep socket, gently tap each seal into place until it is properly seated on the guide. Do not twist or cock the seals during installation or they will not seal properly on the valve stems.

4 Next, install the valves (taking care not to damage the new valve guide seals), the springs, the shields (if so equipped), the retainers or rotators, the O-ring valve stem seals and the keepers. Coat the valve stems with clean moly-based grease or engine assembly lube before slipping them into the guides. Install the O-ring seals in the lower valve stem grooves (one in each groove). Make sure they are not twisted (a light coat of oil will prevent twisting). When compressing the springs with the valve spring compressor, do not let the retainers contact the valve guide seals. Make certain that the keepers are securely locked in their retaining grooves.

5 Double-check the installed valve spring height. If it was correct before disassembly, it should still be within the specified limits.

6 Check the valve stem O-ring seals with a vacuum pump and adapter (photo). A properly installed seal should not leak.

7 Install the rocker arms and tighten the bolts/nuts to the specified torque. Be sure to lubricate the ball pivots with moly-based grease or engine assembly lube.

11 Piston/connecting rod assembly — removal

1 Prior to removal of the piston/connecting rod assemblies, the engine should be positioned upright.

2 Using a ridge reamer, completely remove the ridge at the top of each cylinder (follow the manufacturer's instructions provided with

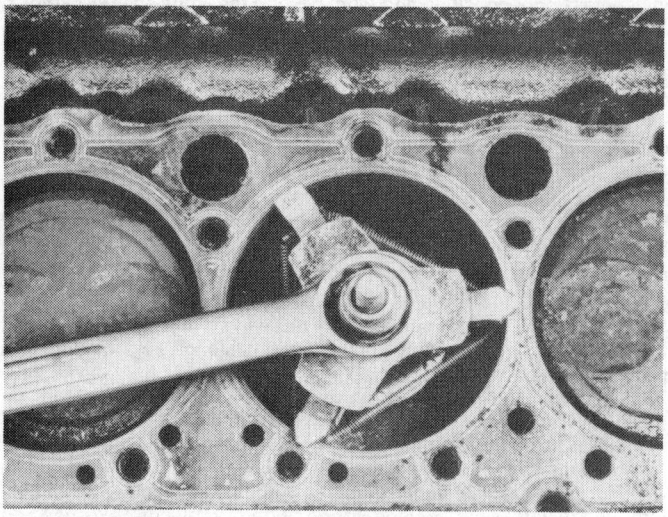

11.2 A special tool is required to remove the ridge from the top of each cylinder (do it *before* removing the piston)

the ridge reaming tool) (photo). *Failure to remove the ridge before attempting to remove the piston/connecting rod assemblies will result in piston breakage.*

3 After all of the cylinder wear ridges have been removed, turn the engine upside-down.

4 Before the connecting rods are removed, check the end play as follows. Mount a dial indicator with its stem in line with the crankshaft and touching the side of the number one cylinder connecting rod cap.

5 Push the connecting rod forward, as far as possible, and zero the dial indicator. Next, push the connecting rod all the way to the rear and check the reading on the dial indicator. The distance that it moves is the end play. If the end play exceeds the service limit, a new connecting rod will be required. Repeat the procedure for the remaining connecting rods.

6 An alternative method is to slip feeler gauges between the connecting rod and the crankshaft throw until the play is removed (photo). The end play is then equal to the thickness of the feeler gauge(s).

7 Check the connecting rods and connecting rod caps for identification marks. If they are not plainly marked, identify each rod and cap using a small punch to make the appropriate number of indentations to indicate the cylinders they are associated with.

8 Loosen each of the connecting rod cap nuts approximately 1/2-turn each. Remove the number one connecting rod cap and bearing insert. Do not drop the bearing insert out of the cap. Slip a short length of plastic or rubber hose over each connecting rod cap bolt (to pro-

11.6 Checking connecting rod end play with a feeler gauge

11.8 To prevent damage to the crankshaft journals and cylinder walls, slip sections of hose over the rod bolts before removing the pistons

12.3 Checking crankshaft end play with a feeler gauge (V8 engine shown)

12.4 Mark the bearing caps with a center punch before removing them

tect the crankshaft journal and cylinder wall when the piston is removed) (photo) and push the connecting rod/piston assembly out through the *top* of the engine. Use a wooden tool to push on the upper bearing insert in the connecting rod. If resistance is felt, double-check to make sure that all of the ridge was removed from the cylinder.
9 Repeat the procedure for the remaining cylinders. After removal, reassemble the connecting rod caps and bearing inserts in their respective connecting rods and install the cap nuts finger tight. Leaving the old bearing inserts in place until reassembly will help prevent the connecting rod bearing surfaces from being accidentally nicked or gouged.

12 Crankshaft — removal

1 Before the crankshaft is removed, check the end play as follows. Mount a dial indicator with the stem in line with the crankshaft and just touching one of the crank throws (see accompanying illustration).
2 Push the crankshaft all the way to the rear and zero the dial indicator. Next, pry the crankshaft to the front as far as possible and check the reading on the dial indicator. The distance that it moves is the end play. If it is greater than specified, check the crankshaft thrust surfaces for wear. If no wear is apparent, new main bearings should correct the end play.
3 If a dial indicator is not available, feeler gauges can be used. Gently pry or push the crankshaft all the way to the front of the engine. Slip feeler gauges between the crankshaft and the front face of the thrust main bearing (photo) to determine the clearance (which is equivalent to crankshaft end play).
4 Loosen each of the main bearing cap bolts 1/4 of a turn at a time, until they can be removed by hand. Check the main bearing caps to see if they are marked as to their locations. They are usually numbered consecutively (beginning with 1) from the front of the engine to the rear. If they are not, mark them with number stamping dies or a center punch (photo). Most main bearing caps have a cast-in arrow, which points to the front of the engine.
5 Gently tap the caps with a soft-faced hammer, then separate them from the engine block. If necessary, use the main bearing cap bolts as levers to remove the caps. Try not to drop the bearing insert if it comes out with the cap.
6 Carefully lift the crankshaft out of the engine. It is a good idea to have an assistant available, since the crankshaft is quite heavy. With the bearing inserts in place in the engine block and in the main bearing caps, return the caps to their respective locations on the engine block and tighten the bolts finger tight.

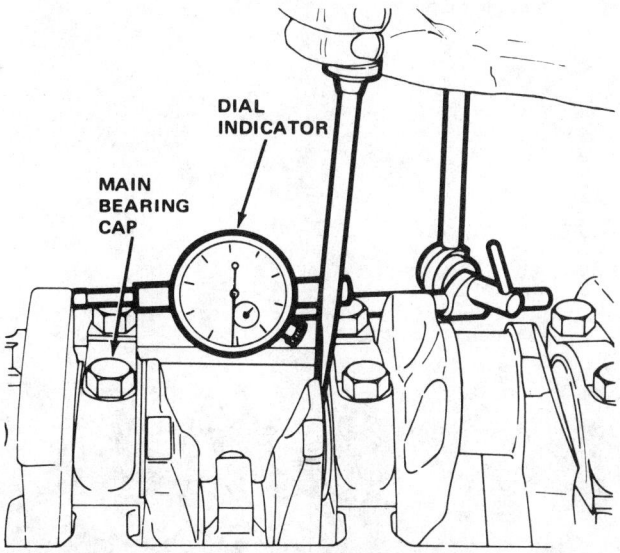

Fig. 2D.9 Checking crankshaft end play with a dial indicator (Sec 12)

13 Engine block — cleaning

1 Remove the soft plugs from the engine block. To do this, knock the plugs into the block (using a hammer and punch), then grasp them with large pliers and pull them back through the holes (photos).
2 Using a gasket scraper, remove all traces of gasket material from the engine block. Be very careful not to nick or gouge the gasket sealing surfaces.
3 Remove the main bearing caps and separate the bearing inserts from the caps and the engine block. Tag the bearings according to which cylinder they removed from (and whether they were in the cap or the block) and set them aside.
4 Using a hex wrench of the appropriate size, remove the threaded oil gallery plugs from the front and back of the block.
5 If the engine is extremely dirty, it should be taken to an automotive machine shop to be steam cleaned or hot tanked. Any bearings left in the block (such as the camshaft bearings) will be damaged by the cleaning process, so plan on having new ones installed while the block is at the machine shop.
6 After the block is returned, clean all oil holes and oil galleries one more time (brushes for cleaning oil holes and galleries are available at most auto parts stores). Flush the passages with warm water until the water runs clear, dry the block thoroughly and wipe all machined surfaces with a light, rust-preventative oil. If you have access to com-

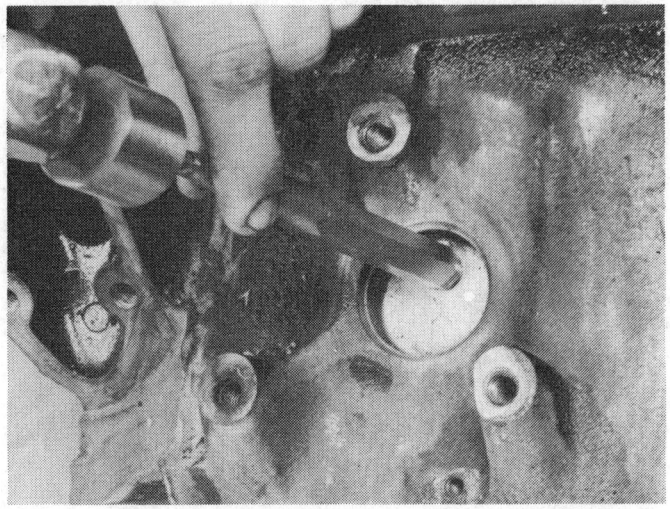

13.1A A hammer and large punch can be used to drive the soft plugs into the block

13.1B Using pliers to remove a soft plug from the block

pressed air, use it to speed the drying process and to blow out all the oil holes and galleries.

7 If the block is not extremely dirty or sludged up, you can do an adequate cleaning job with warm soapy water and a stiff brush. Take plenty of time and do a thorough job. Regardless of the cleaning method used, be very sure to thoroughly clean all oil holes and galleries, dry the block completely and coat all machined surfaces with light oil.

8 The threaded holes in the block must be clean to ensure accurate torque readings during reassembly. Run the proper size tap into each of the holes to remove any rust, corrosion, thread sealant or sludge and to restore any damaged threads. If possible, use compressed air to clear the holes of debris produced by this operation. Now is a good time to thoroughly clean the threads on the head bolts and the main bearing cap bolts as well.

9 Reinstall the main bearing caps and tighten the bolts finger tight.

10 After coating the sealing surfaces of the new soft plugs with a good quality gasket sealer, install them in the engine block (photo). Make sure they are driven in straight and seated properly or leakage could result. Special tools are available for this purpose, but equally good results can be obtained using a large socket (with an outside diameter that will just slip into the soft plug) and a large hammer.

11 If the engine is not going to be reassembled right away, cover it with a large plastic trash bag to keep it clean.

14 Engine block — inspection

1 Thoroughly clean the engine block as described in Section 13 and double-check to make sure that the ridge at the top of each cylinder has been completely removed.

2 Visually check the block for cracks, rust and corrosion. Look for stripped threads in the threaded holes. It is also a good idea to have the block checked for hidden cracks by an automotive machine shop that has the special equipment to do this type of work. If defects are found, have the block repaired, if possible, or replaced.

3 Check the cylinder bores for scuffing and scoring.

4 Using the appropriate precision measuring tools, measure each cylinder's diameter at the top (just under the ridge), center and bottom of the cylinder bore, *parallel* to the crankshaft axis (photos). Next, measure each cylinder's diameter at the same three locations *across* the crankshaft axis. Compare the results to the Specifications. If the cylinder walls are badly scuffed or scored, or if they are out-of-round or tapered beyond the limits given in the Specifications, have the engine block rebored and honed at an automotive machine shop. If a rebore is done, oversize pistons and rings will be required as well.

5 If the cylinders are in reasonably good condition and not worn to

13.10 A large socket on an extension can be used to force the new soft plugs into their bores

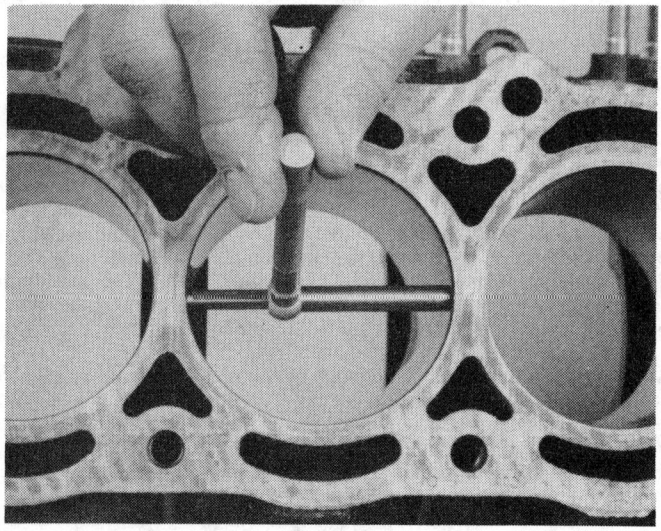

14.4A A telescoping gauge can be used to determine the cylinder bore diameter

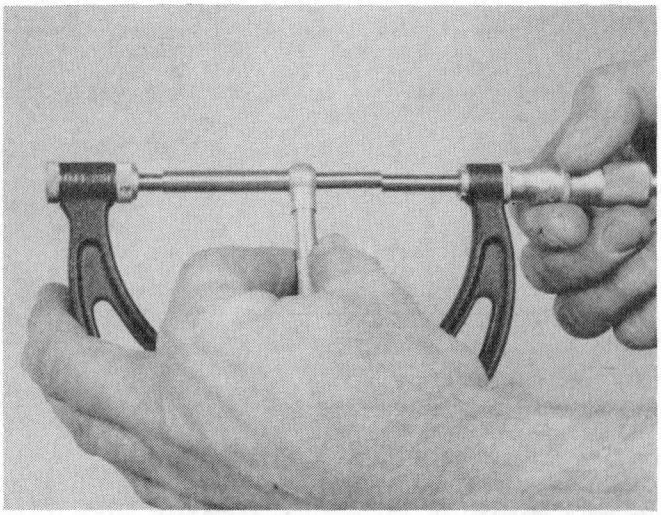

14.4B The gauge is then measured with a micrometer to determine the bore size in inches

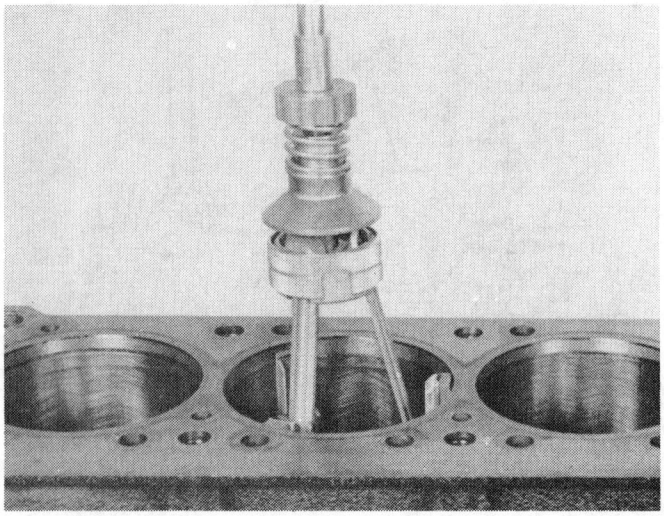

14.7 Honing a cylinder with a surfacing hone

the outside of the limits, and if the piston-to-cylinder clearances can be maintained properly, then they do not have to be rebored; honing is all that is necessary.

6 Before honing the cylinders, install the main bearing caps (without the bearings) and tighten the bolts to the specified torque.

7 To perform the honing operation, you will need the proper size flexible hone (with fine stones), plenty of light oil or honing oil, some rags and an electric drill motor. Mount the hone in the drill motor, compress the stones and slip the hone into the first cylinder (photo). Lubricate the cylinder thoroughly, turn on the drill and move the hone up and down in the cylinder at a pace which will produce a fine cross-hatch pattern on the cylinder walls (with the cross-hatch lines intersecting at approximately a 60° angle). Be sure to use plenty of lubricant and do not take off any more material than is absolutely necessary to produce the desired finish. Do not withdraw the hone from the cylinder while it is running. Instead, shut off the drill and continue moving the hone up and down in the cylinder until it comes to a complete stop, then compress the stones and withdraw the hone. Wipe the oil out of the cylinder and repeat the procedure on the remaining cylinders. Remember, do not remove too much material from the cylinder wall. If you do not have the tools or do not desire to perform the honing operation, most automotive machine shops will do it for a reasonable fee.

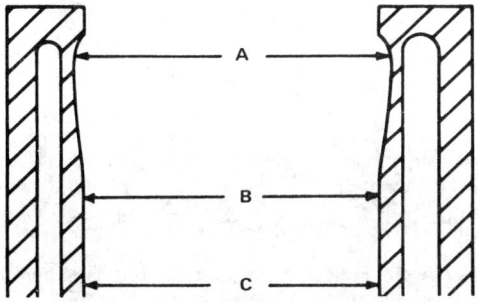

Fig. 2D.10 Measure the diameter of each cylinder just under the wear ridge (A), at the center (B) and at the bottom (C) (Sec 14)

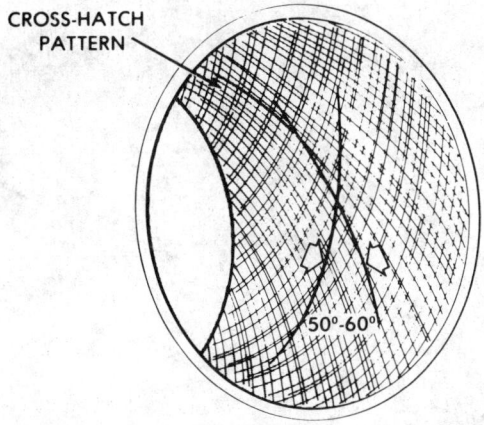

Fig. 2D.11 The cylinder hone should leave a cross-hatch pattern with the lines intersecting at approximately a 60° angle (Sec 14)

8 After the honing job is complete, chamfer the top edges of the cylinder bores with a small file so the rings will not catch when the pistons are installed.

9 Next, the entire engine block must be thoroughly washed again with warm, soapy water to remove all traces of the abrasive grit produced during the honing operation. Be sure to run a brush through all oil holes and galleries and flush them with running water. After rinsing, dry the block and apply a coat of light rust preventative oil to all machined surfaces. Wrap the block in a plastic trash bag to keep it clean and set it aside until reassembly.

15 Camshaft, lifters and bearings — inspection and bearing replacement

Camshaft

1 The most critical camshaft inspection procedure is lobe lift measurement, which must be done *before* the engine is disassembled.

2 Remove the rocker arm cover(s), then remove the nuts/bolts and separate the rocker arms and ball pivots from the cylinder head(s).

3 Beginning with the number one (1) cylinder, mount a dial indicator with the stem resting on the end of, and directly in line with, the exhaust valve pushrod.

4 Rotate the crankshaft very slowly in the direction of rotation until the lifter is on the heel of the cam lobe. At this point the pushrod will be at its lowest position.

5 Zero the dial indicator, then very slowly rotate the crankshaft in the direction of rotation until the pushrod is at its highest position. Note and record the reading on the dial indicator, then compare it to the lobe lift specifications.

6 Repeat the procedure for each of the remaining valves. Note that intake and exhaust valves may have different lobe lift specifications.

7 If the lobe lift measurements are not as specified, a new camshaft should be installed.

8 After the camshaft has been removed from the engine, cleaned with solvent and dried, inspect the bearing journals for uneven wear, pit-

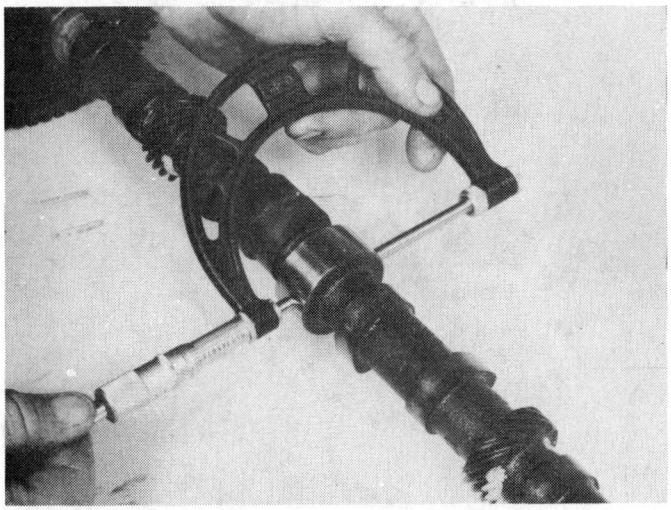

15.9 The camshaft bearing journal diameter is subtracted from the bearing inside diameter to obtain the oil clearance, which must be as specified

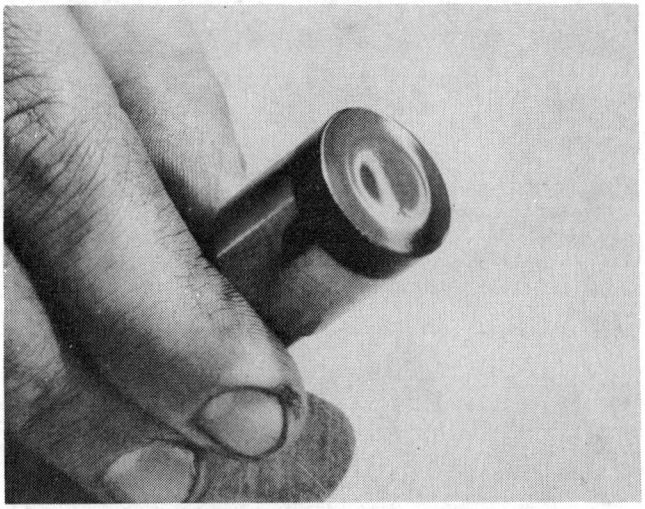

15.12 If the bottom of any of the lifters is worn concave, scratched or galled, they should all be replaced with new ones

ting and evidence of seizure. If the journals are damaged, the bearing inserts in the block are probably damaged as well. Both the camshaft and bearings will have to be replaced with new ones. Measure the inside diameter of each camshaft bearing and record the results (take two measurements, 90° apart, at each bearing).

9 Measure the bearing journals with a micrometer (photo) to determine if they are excessively worn or out-of-round. If they are more than 0.001-inch out-of-round, the camshaft should be replaced with a new one. Subtract the bearing journal diameter(s) from the corresponding bearing inside diameter measurement to obtain the oil clearance. If it is excessive, new bearings must be installed.

10 Check the camshaft lobes for heat discoloration, score marks, chipped areas, pitting and uneven wear. If the lobes are in good condition and if the lobe lift measurements (Steps 1 through 7) were as specified, the camshaft can be reused.

Lifters

11 Clean the lifters with solvent and dry them thoroughly *without mixing them up.*

12 Check each lifter wall, pushrod seat and foot for scuffing, score marks and uneven wear. Each lifter foot (the surface that rides on the cam lobe) must be slightly convex — if they are concave (photo), the lifters and camshaft must be replaced with new ones. If the lifter walls are damaged or worn (which is not very likely), inspect the lifter bores in the engine block as well. If the pushrod seats are worn, check the

Fig. 2D.12 A dial indicator can be mounted as shown here to check the camshaft lobe lift (Sec 15)

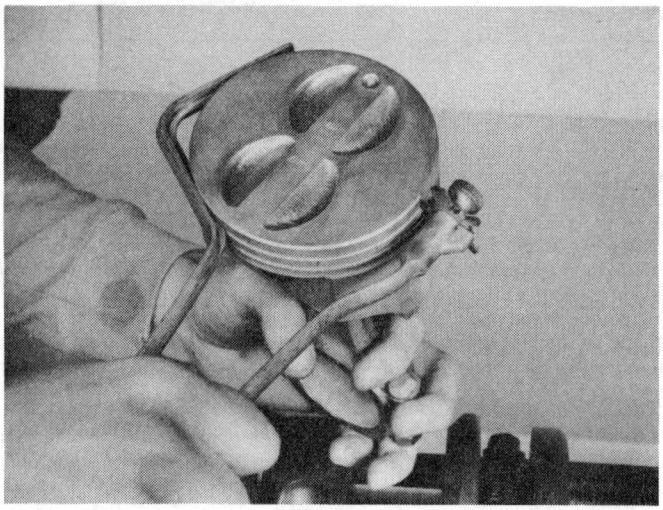

16.4 Cleaning the piston ring grooves with a piston ring groove cleaning tool

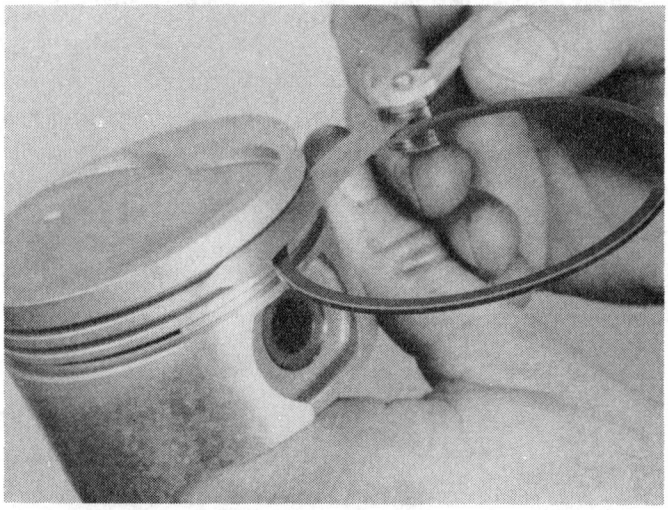

16.10 Checking the piston ring side clearance with a feeler gauge

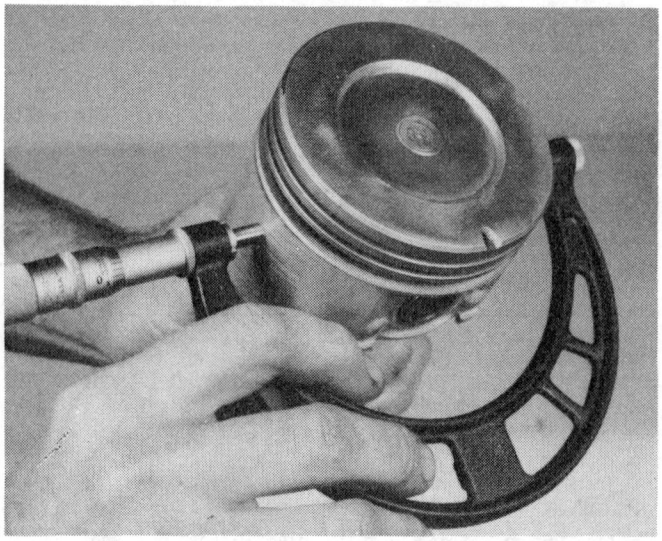

16.11 Measure the piston diameter directly in line with the piston pin hole

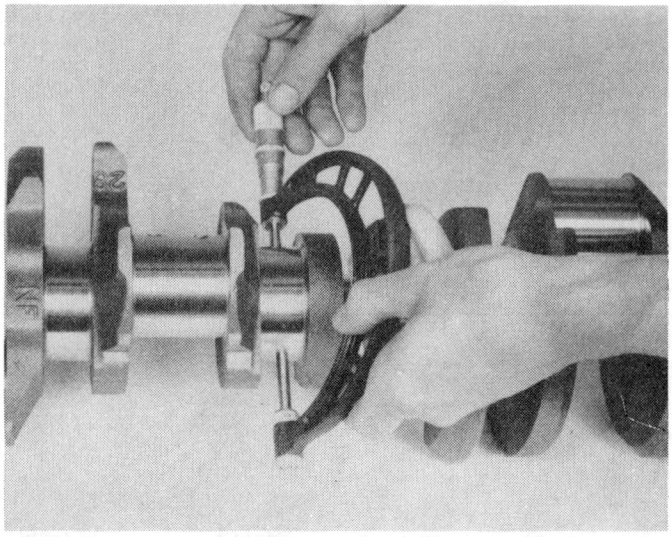

17.2 Measure the diameter of each crankshaft journal at several points to detect taper and out-of-round conditions

pushrod ends.
13 If new lifters are being installed, a new camshaft must also be installed. If a new camshaft is installed, then use new lifters as well. *Never install used lifters unless the original camshaft is used and the lifters can be installed in their original locations.*

Bearing replacement

14 Camshaft bearing replacement requires special tools and expertise that place it outside the scope of the do-it-yourselfer. Take the block to an automotive machine shop to ensure that the job is done correctly.

16 Piston/connecting rod assembly — inspection

1 Before the inspection process can be carried out, the piston/connecting rod assemblies must be cleaned and the original piston rings removed from the pistons. **Note:** *Always use new piston rings when the engine is reassembled.*
2 Using a piston ring installation tool, carefully remove the rings from the pistons. Do not nick or gouge the pistons in the process.
3 Scrape all traces of carbon from the top (or crown) of the piston. A hand-held wire brush or a piece of fine emery cloth can be used once the majority of the deposits have been scraped away. Do not, under any circumstances, use a wire brush mounted in a drill motor to remove deposits from the pistons. The piston material is soft and will be eroded away by the wire brush.
4 Use a piston ring groove cleaning tool to remove any carbon deposits from the ring grooves. If a tool is not available, a piece broken off the old ring will do the job. Be very careful to remove only the carbon deposits. Do not remove any metal and do not nick or scratch the sides of the ring grooves (photo).
5 Once the deposits have been removed, clean the piston/rod assemblies with solvent and dry them thoroughly. Make sure that the oil hole in the big end of the connecting rod and the oil return holes in the back sides of the ring grooves are clear.
6 If the pistons are not damaged or worn excessively, and if the engine block is not rebored, new pistons will not be necessary. Normal piston wear appears as even vertical wear on the piston thrust surfaces and slight looseness of the top ring in its groove. New piston rings, on the other hand, should always be used when an engine is rebuilt.
7 Carefully inspect each piston for cracks around the skirt, at the pin bosses and at the ring lands.
8 Look for scoring and scuffing on the thrust faces of the skirt, holes in the piston crown and burned areas at the edge of the crown. If the skirt is scored or scuffed, the engine may have been suffering from overheating and/or abnormal combustion, which caused excessively high operating temperatures. The cooling and lubrication systems should be checked thoroughly. A hole in the piston crown, an extreme to be sure, is an indication that abnormal combustion (preignition) was occurring. Burned areas at the edge of the piston crown are usually evidence of spark knock (detonation). If any of the above problems exist, the causes must be corrected or the damage will occur again.
9 Corrosion of the piston (evidenced by pitting) indicates that coolant is leaking into the combustion chamber and/or the crankcase. Again, the cause must be corrected or the problem may persist in the rebuilt engine.
10 Measure the piston ring side clearance by laying a new piston ring in each ring groove and slipping a feeler gauge in beside it (photo). Check the clearance at three or four locations around each groove. Be sure to use the correct ring for each groove; they are different. If the side clearance is greater than specified, new pistons and/or rings will have to be used.
11 Check the piston-to-bore clearance by measuring the bore (see Section 14) and the piston diameter (photo). Make sure that the pistons and bores are correctly matched. Measure the piston across the skirt, on the thrust faces (at a 90° angle to the piston pin), directly in line with the center of the pin hole. Subtract the piston diameter from the bore diameter to obtain the clearance. If it is greater than specified, the block will have to be rebored and new pistons and rings installed. Check the piston-to-rod clearance by twisting the piston and rod in opposite directions. Any noticeable play indicates that there is excessive wear, which must be corrected. The piston/connecting rod assemblies should be taken to an automotive machine shop to have new piston pins installed and the pistons and connecting rods rebored.
12 If the pistons must be removed from the connecting rods, such as when new pistons must be installed, or if the piston pins have too much play in them, they should be taken to an automotive machine shop. While they are there, it would be convenient to have the connecting rods checked for bend and twist, as automotive machine shops have special equipment for this purpose. *Unless new pistons or connecting rods must be installed, do not disassemble the pistons from the connecting rods.*
13 Check the connecting rods for cracks and other damage. Temporarily remove the rod caps, lift out the old bearing inserts, wipe the rod and cap bearing surfaces clean and inspect them for nicks, gouges and scratches. After checking the rods, replace the old bearings, slip the caps into place and tighten the nuts finger tight.

17 Crankshaft — inspection

1 Clean the crankshaft with solvent and dry it thoroughly. Be sure to clean the oil holes with a stiff brush and flush them with solvent. Check the main and connecting rod bearing journals for uneven wear, scoring, pitting and cracks. Check the remainder of the crankshaft for cracks and damage.
2 Using an appropriate size micrometer, measure the diameter of the main and connecting rod journals (photo) and compare the results to the Specifications. By measuring the diameter at a number of points around the journal's circumference, you will be able to determine whether or not the journal is worn out-of-round. Take the measurement at each end of the journal, near the crank throw, to determine whether the journal is tapered.
3 If the crankshaft journals are damaged, tapered, out-of-round or worn beyond the limits given in the Specificatioins, have the crankshaft reground by a reputable automotive machine shop. Be sure to use the correct undersize bearing inserts if the crankshaft is reconditioned.
4 Refer to Section 18 and examine the main and rod bearing inserts. If the bearing inserts and journals are all in good condition, do not decide to reuse the bearings until the oil clearances have been checked.

18 Main and connecting rod bearings — inspection

1 Even though the main and connecting rod bearings should be replaced with new ones during the engine overhaul, the old bearings should be retained for close examination, as they may reveal valuable information about the condition of the engine.
2 Bearing failure occurs mainly because of lack of lubrication, the presence of dirt or other foreign particles, overloading the engine and corrosion. Regardless of the cause of bearing failure, it must be corrected before the engine is reassembled to prevent it from happening again.
3 When examining the bearings, remove them from the engine block, the main bearing caps, the connecting rods and the rod caps and lay them out on a clean surface in the same general position as their location in the engine. This will enable you to match any noted bearing problems with the corresponding crankshaft journal.
4 Dirt and other foreign particles get into the engine in a variety of ways. If may be left in the engine during assembly, or it may pass through filters or breathers. It may get into the oil, and from there into the bearings. Metal chips from machining operations and normal engine wear are often present. Abrasives are sometimes left in engine components after reconditioning, especially when parts are not thoroughly cleaned using the proper cleaning methods. Whatever the source, these foreign objects often end up embedded in the soft bearing material and are easily recognized. Large particles will not embed in the bearing and will score or gouge the bearing and shaft. The best prevention for this cause of bearing failure is to clean all parts thoroughly and keep everything spotlessly clean during engine assembly. Frequent and regular engine oil and filter changes are also recommended.
5 Lack of lubrication (or lubrication breakdown) has a number of interrelated causes. Excessive heat (which thins the oil), overloading (which squeezes the oil from the bearing face) and oil leakage or throw-off (from excessive bearing clearances, worn oil pump or high engine

19.3A Use the piston to square up the ring in the cylinder prior to checking the ring end gap

19.3B Measure the ring end gap with a feeler gauge

19.9A Installing the spacer/expander in the oil control ring groove

19.9B *Do not* use a piston ring tool when installing the oil ring side rails

speeds) all contribute to lubrication breakdown. Blocked oil passages, which usually are the result of misaligned oil holes in a bearing shell, will also oil-starve a bearing and destroy it. When lack of lubrication is the cause of bearing failure, the bearing material is wiped or extruded from the steel backing of the bearing. Temperatures may increase to the point where the steel backing turns blue from overheating.

6 Driving habits can have a definite effect on bearing life. Full-throttle, low-speed operation (or 'lugging' the engine) puts very high loads on bearings, which tends to squeeze out the oil film. These loads cause the bearings to flex, which produces fine cracks in the bearing face (fatigue failure). Eventually the bearing material will loosen in pieces and tear away from the steel backing. Short-trip driving leads to corrosion of bearings because insufficient engine heat is produced to drive off the condensed water and corrosive gases. These products collect in the engine oil, forming acid and sludge. As the oil is carried to the engine bearings, the acid attacks and corrodes the bearing material.

7 Incorrect bearing installation during engine assembly will lead to bearing failure as well. Tight-fitting bearings leave insufficient bearing oil clearance and will result in oil starvation. Dirt or foreign particles trapped behind a bearing insert result in high spots on the bearing which lead to failure.

19 Piston rings — installation

1 Before installing the new piston rings, the ring end gaps must be checked. It is assumed that the piston ring side clearance has been checked and verified correct (Section 16).

2 Lay out the piston/connecting rod assemblies and the new ring sets so the ring sets will be matched with the same piston and cylinder during the end gap measurement and engine assembly.

3 Insert the top (number one) ring into the first cylinder and square it up with the cylinder walls by pushing it in with the top of the piston (photo). The ring should be near the bottom of the cylinder at the lower limit of ring travel. To measure the end gap, slip a feeler gauge between the ends of the ring (photo). Compare the measurement to the Specifications.

4 If the gap is larger or smaller than specified, double-check to make sure that you have the correct rings before proceeding.

5 If the gap is too small, it must be enlarged or the ring ends may come in contact with each other during engine operation, which can cause serious damage to the engine. The end gap can be increased by filing the ring ends very carefully with a fine file. Mount the *file* in a vise equipped with soft jaws, slip the ring over the file with the ends contacting the file face and slowly move the ring to remove material from the ends. *When performing this operation, file only from the outside in.*

6 Excess end gap is not critical unless it is greater than 0.040-inch (1 mm). Again, double-check to make sure you have the correct rings for your engine.

7 Repeat the procedure for each ring that will be installed in the first cylinder and for each ring in the remaining cylinders. Remember to keep rings, pistons and cylinders matched up.

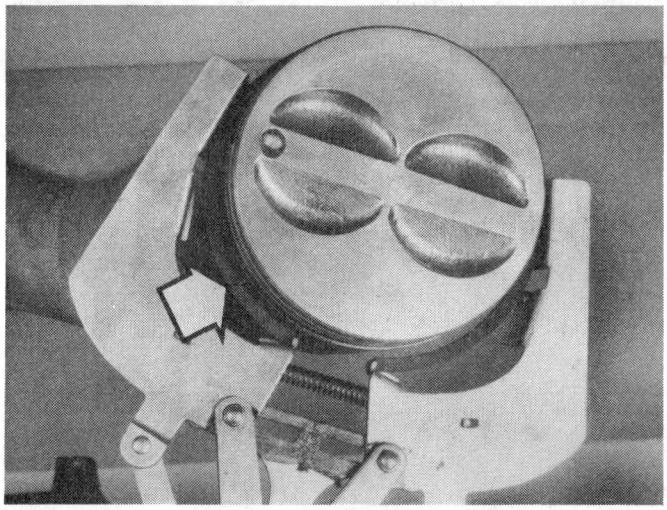

19.12 Installing the compression rings with the special tool — the mark (arrow) must face *up*

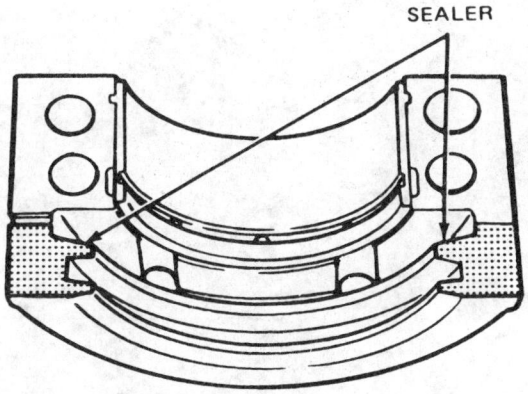

Fig. 2D.13 Apply anaerobic-type gasket sealant to the shaded areas of the rear main bearing cap (V6 engine) (Sec 20)

8 Once the ring end gaps have been checked/corrected, the rings can be installed on the pistons.

9 The oil control ring (lowest one on the piston) is installed first. It is composed of three separate components. Slip the spacer expander into the groove (photo), then install the upper side rail. *Do not use a piston ring installation tool on the oil ring side rails, as they may be damaged.* Instead, place one end of the side rail into the groove between the spacer expander and the ring land, hold it firmly in place and slide a finger around the piston while pushing the rail into the groove (photo). Next, install the lower side rail in the same manner.

10 After the three oil ring components have been installed, check to make sure that both the upper and lower side rails can be turned smoothly in the ring groove.

11 The number two (middle) ring is installed next. It should be stamped with a mark so it can be readily distinguished from the top ring. **Note:** *Always follow the instructions printed on the package that the new rings are in — different manufacturers may require different approaches. Do not mix up the top and middle rings, as they have different cross sections.*

12 Use a piston ring installation tool and *make sure that the identification mark is facing the top of the piston,* then slip the ring into the middle groove on the piston (photo). Do not expand the ring any more than is necessary to slide it over the piston.

13 Finally, install the number one (top) ring in the same manner. Make sure the identifying mark is facing up.

14 Repeat the procedure for the remaining pistons and rings. Be careful not to confuse the number one and number two rings.

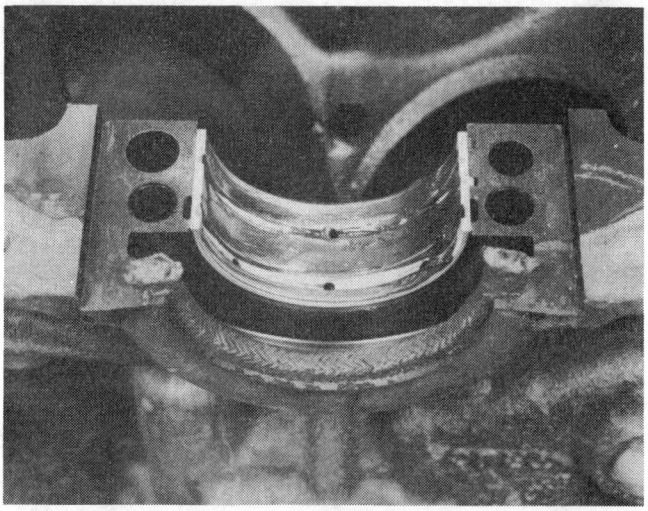

20.1 When correctly installed, the ends of the rope-type seal should extend out of the block

20.2 Seat the seal in the groove, but do not depress it below the bearing surface (the seal must contact the crankshaft journal)

20 Rear main oil seal (V6 and V8 engines only) — installation

Rope-type seal

1 Lay one seal section on edge in the seal groove in the block and push it into place with your thumbs. Both ends of the seal should extend out of the block slightly (photo).

2 Seat it in the groove by rolling a large socket or piece of bar stock along the entire length of the seal (photo). As an alternative, push the seal very carefully into place with a wooden hammer handle.

3 Once you are satisfied that the seal is completely seated in the groove, trim off the excess on the ends with a single-edge razor blade or razor knife (the seal ends must be flush with the block-to-cap mating surfaces) (photos). Make sure that no seal fibers get caught between the block and cap.

4 Repeat the entire procedure to install the other half of the seal in the bearing cap. Apply a thin film of engine assembly lube to the edge of the seal (where it contacts the crankshaft) (photo).

5 During final installation of the crankshaft (after the main bearing oil clearances have been checked with Plastigage) as described in Section 21, apply a thin, even film of anaerobic-type gasket sealant to the areas of the rear main bearing cap indicated in the accompanying illustration (photo). **Caution:** *Do not get any sealant on the bearing or seal faces.*

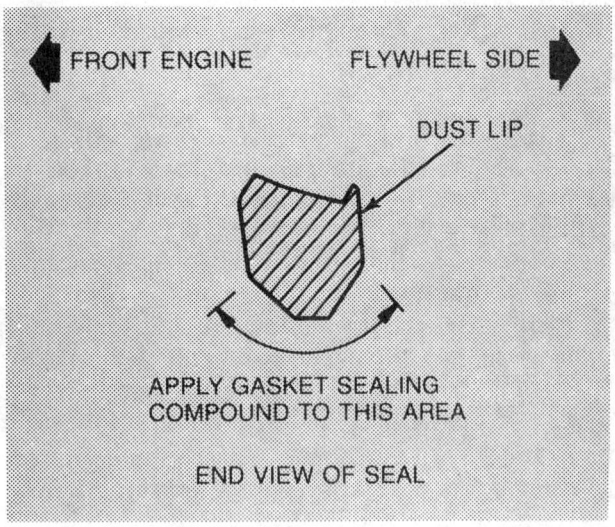

Fig. 2D.14 When installing the rear main oil seal on 1984 V6 engines, apply RTV-type sealant to the area indicated (Sec 20)

Neoprene lip-type seal

6 Inspect the bearing cap and engine block mating surfaces and seal grooves for nicks, burrs and scratches. Remove any defects with a fine file or deburring tool.

7 Install one seal section in the block with the lip facing the *front* of the engine. Leave one end protruding from the block approximately 1/4 to 3/8-inch and make sure it is completely seated. **Note:** *On V6*

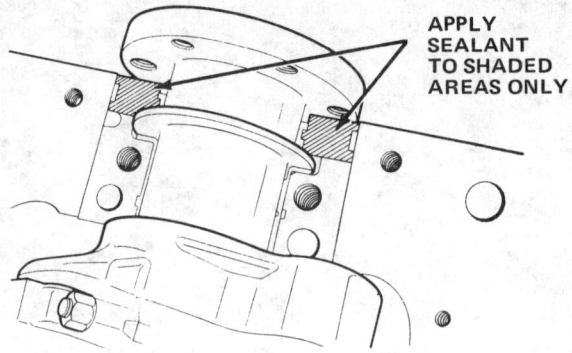

Fig. 2D.15 Apply anaerobic-type gasket sealant to the shaded areas of the block when installing the bearing cap (V8 engine) (Sec 20)

20.3A Trim the ends flush with the block, . . .

20.3B . . . but leave the inner edge (arrow) protruding slightly

20.4 Lubricate the seal with assembly lube or moly-based grease

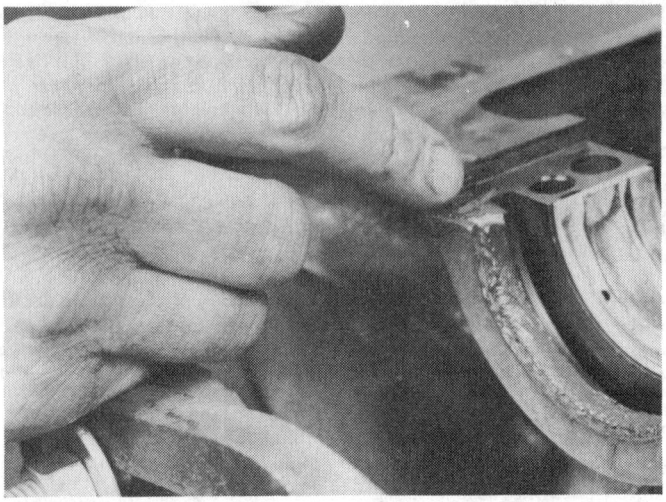

20.5 When applying the sealant, be sure it gets into the corner and onto the vertical cap-to-block mating surface or oil leaks will result

models only, apply a very thin coat of RTV-type gasket sealant to the outer surface of the seal as shown in the accompanying illustration. Do not get any sealant on the seal lips.

8 Repeat the procedure to install the remaining seal half in the rear main bearing cap. In this case, leave the opposite end of the seal protruding from the cap approximately the same distance the block seal is protruding from the block.

9 During final installation of the crankshaft (after the main bearing oil clearances have been checked with Plastigage) as described in Section 21, apply a thin, even film of anaerobic-type gasket sealant to the areas of the cap or block indicated in the accompanying illustration. **Caution:** *Do not get any sealant on the bearing face, crankshaft journal or seal lip.* Also, lubricate the seal lips with moly-based grease or engine assembly lube.

21 Crankshaft — installation and main bearing oil clearance check

1 Crankshaft installation is generally one of the first steps in engine reassembly; it is assumed at this point that the engine block and crankshaft have been cleaned, inspected and repaired or reconditioned. **Note:** *On V6 and V8 engines only, refer to Section 20 and install the rear main oil seal sections before preceeding.*

2 Position the engine with the bottom facing up.

3 Remove the main bearing cap bolts and lift out the caps. Lay them out in the proper order to help ensure that they are installed correctly.

4 If they are still in place, remove the old bearing inserts from the block and the main bearing caps. Wipe the main bearing surfaces of the block and caps with a clean, lint-free cloth (they must be kept spotlessly clean).

5 Clean the back sides of the new main bearing inserts and lay one bearing half in each main bearing saddle in the block. Lay the other bearing half from each bearing set in the corresponding main bearing cap. Make sure the tab on the bearing insert fits into the recess in the block or cap. Also, the oil holes in the block and cap must line up with the oil holes in the bearing insert. *Do not hammer the bearing into place and do not nick or gouge the bearing faces. No lubrication should be used at this time.*

6 The flanged thrust bearing must be installed in the number three (3) cap and saddle on V6 engines and the number five (5-rear) cap and saddle on V8 and four-cylinder engines.

7 Clean the faces of the bearings in the block and the crankshaft main bearing journals with a clean, lint-free cloth. Check or clean the oil holes in the crankshaft, as any dirt here can only go one way — straight through the new bearings.

8 Once you are certain that the crankshaft is clean, carefully lay it in position (an assistant would be very helpful here) in the main bearings with the counterweights lying sideways.

9 Before the crankshaft can be permanently installed, the main bearing oil clearance must be checked.

10 Trim several pieces of the appropriate type of Plastigage (so they are slightly shorter than the width of the main bearings) and place one piece on each crankshaft main bearing journal, parallel with the journal axis. Do not lay them across the oil holes.

11 Clean the faces of the bearings in the caps and install the caps in their respective positions (do not mix them up) with the arrows pointing toward the front of the engine. Do not disturb the Plastigage.

12 Starting with the center main and working out toward the ends, tighten the main bearing cap bolts, in three steps, to the specified torque. *Do not rotate the crankshaft at any time during this operation.*

13 Remove the bolts and carefully lift off the main bearing caps. Keep them in order. Do not disturb the Plastigage or rotate the crankshaft. If any of the main bearing caps are difficult to remove, tap them gently from side-to-side with a soft-faced hammer to loosen them.

14 Compare the width of the crushed Plastigage on each journal to the scale printed on the Plastigage container (photo) to obtain the main bearing oil clearance. Check the Specifications to make sure it is correct.

15 If the clearance is not correct, double-check to make sure you have the right size bearing inserts. Also, make sure that no dirt or oil was between the bearing inserts and the main bearing caps or the block when the clearance was measured.

16 Carefully scrape all traces of the Plastigage material off the main bearing journals and/or the bearing faces. Do not nick or scratch the bearing faces.

17 Carefully lift the crankshaft out of the engine. Clean the bearing faces in the block, then apply a thin, uniform layer of clean, high-quality moly-based grease or engine assembly lube to each of the bearing faces. Be sure to coat the thrust flange faces as well as the journal face of the thrust bearing.

18 Lubricate the rear main oil seal (where it contacts the crankshaft) with moly-based grease or engine assembly lube. Note that on four-cylinder engines the oil seal is installed *after* the crankshaft is in place.

19 If you are working on a V6 or V8 engine, refer to Section 20 and apply anaerobic-type gasket sealant to the rear main bearing cap or block as described there. Make sure the crankshaft journals are clean, then lay it back in place in the block. Clean the faces of the bearings in the caps, then apply a thin, uniform layer of clean, moly-based grease to each of the bearing faces and install the caps in their respective positions with the arrows pointing toward the front of the engine. Install the bolts and tighten them to the specified torque, starting with the center main and working out toward the ends. Work up to the final torque in three steps.

20 Rotate the crankshaft a number of times by hand and check for any obvious binding.

21 The final step is to check the crankshaft end play. This can be done with a feeler gauge or a dial indicator set. Refer to Section 12 for the procedure.

22 Piston/connecting rod assembly — installation and bearing oil clearance check

1 Before installing the piston/connecting rod assemblies, the cylinder walls must be perfectly clean, the top edge of each cylinder must be chamfered, and the crankshaft must be in place.

2 Remove the connecting rod cap from the end of the number one connecting rod. Remove the old bearing inserts and wipe the bearing surfaces of the connecting rod and cap with a clean, lint-free cloth (they must be kept spotlessly clean).

3 Clean the back side of the new upper bearing half, then lay it in place in the connecting rod. Make sure that the tab on the bearing fits into the recess in the rod. Do not hammer the bearing insert into place and be very careful not to nick or gouge the bearing face. *Do not lubricate the bearing at this time.*

4 Clean the back side of the other bearing insert and install it in the rod cap. Again, make sure the tab on the bearing fits into the recess in the cap, and do not apply any lubricant. It is critically important that the mating surfaces of the bearing and connecting rod are perfectly clean and oil-free when they are assembled.

5 Position the piston ring gaps as shown in the accompanying illustrations, then slip a section of plastic or rubber hose over the connecting rod cap bolts.

6 Lubricate the piston and rings with clean engine oil and attach a piston ring compressor to the piston. Leave the skirt protruding about 1/4-inch to guide the piston into the cylinder. The rings must be compressed as far as possible.

7 Rotate the crankshaft until the number one connecting rod journal is as far from the number one cylinder as possible (bottom dead center), and apply a uniform coat of engine oil to the cylinder walls.

8 With the notch on top of the piston facing to the front of the engine, gently place the piston/connecting rod assembly into the number one cylinder bore and rest the bottom edge of the ring compressor on the engine block. Tap the top edge of the ring compressor to make sure it is contacting the block around its entire circumference. **Note:** *On V8 engines, the bearing tab recess in the rod must be opposite the camshaft.*

9 Clean the number one connecting rod journal on the crankshaft and the bearing faces in the rod.

10 Carefully tap on the top of the piston with the end of a wooden hammer handle (photo) while guiding the end of the connecting rod into place on the crankshaft journal. The piston rings may try to pop out of the ring compressor just before entering the cylinder bore, so keep some downward pressure on the ring compressor. Work slowly, and if any resistance is felt as the piston enters the cylinder, stop immediately. Find out what is hanging up and fix it before proceeding. *Do not, for any reason, force the piston into the cylinder, as you will break a ring and/or the piston.*

11 Once the piston/connecting rod assembly is installed, the connecting rod bearing oil clearance must be checked before the rod cap is permanently bolted in place.

12 Cut a piece of the the appropriate type Plastigage slightly shorter than the width of the connecting rod bearing and lay it in place on the number one connecting rod journal, parallel with the journal axis (it must not cross the oil hole in the journal) (photo).

13 Clean the connecting rod cap bearing face, remove the protective hoses from the connecting rod bolts and gently install the rod cap in place. Make sure the mating mark on the cap is on the same side as the mark on the connecting rod. Install the nuts and tighten them to the specified torque, working up to it in three steps. *Do not rotate the crankshaft at any time during this operation.*

14 Remove the rod cap, being very careful not to disturb the Plastigage. Compare the width of the crushed Plastigage to the scale printed on the Plastigage container to obtain the oil clearance (photo). Compare it to the Specifications to make sure the clearance is correct. If the clearance is not correct, double-check to make sure that you have the correct size bearing inserts. Also, recheck the crankshaft connecting rod journal diameter and make sure that no dirt or oil was between the bearing inserts and the connecting rod or cap when the clearance was measured.

15 Carefully scrape all traces of the Plastigage material off the rod journal and/or bearing face (be very careful not to scratch the bearing — use your fingernail or a piece of hardwood). Make sure the bearing faces are perfectly clean, then apply a uniform layer of clean, high-quality moly-based grease or engine assembly lube to both of them. You will have to push the piston into the cylinder to expose the face of the bearing insert in the connecting rod; be sure to slip the protective hoses over the rod bolts first.

16 Slide the connecting rod back into place on the journal, remove the protective hoses from the rod cap bolts, install the rod cap and tighten the nuts to the specified torque. Again, work up to the torque in three steps.

17 Repeat the entire procedure for the remaining piston/connecting rod assemblies. Keep the back sides of the bearing inserts and the inside of the connecting rod and cap perfectly clean when assembling them. Make sure you have the correct piston for the cylinder and that the notch, arrow or F on the piston faces to the front of the engine when the piston is installed. Remember, use plenty of oil to lubricate the piston before installing the ring compressor and cap. Also, when installing the rod caps for the final time, be sure to lubricate the bearing faces adequately.

18 After all the piston/connecting rod assemblies have been properly installed, rotate the crankshaft a number of times by hand and check for any obvious binding.

19 As a final step, the connecting rod end play must be checked. Refer to Section 11 for the procedure to follow. Compare the measured end play to the Specifications to make sure it is correct.

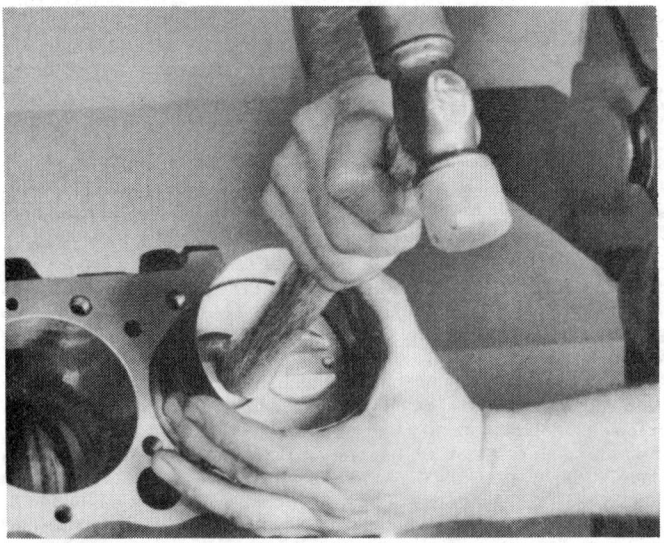

22.10 If resistance is encountered when tapping the piston/connecting rod assembly into the block, *stop immediately* and make sure the rings are fully compressed

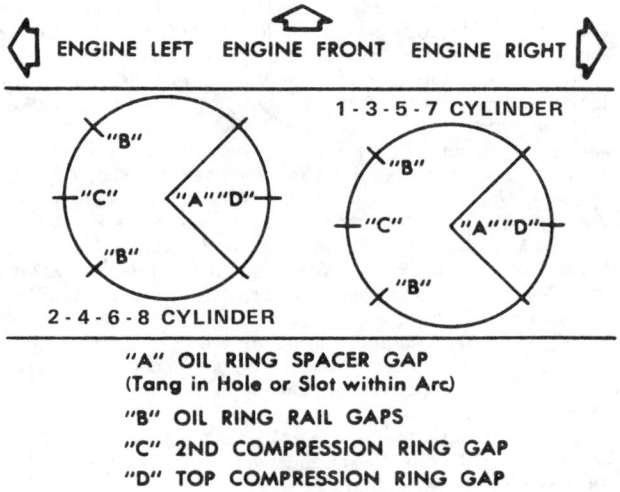

Fig. 2D.16 Before installing the pistons, position the ring end gaps as shown here (Sec 22)

22.12 Position the Plastigage strip on the bearing journal, parallel to the journal axis

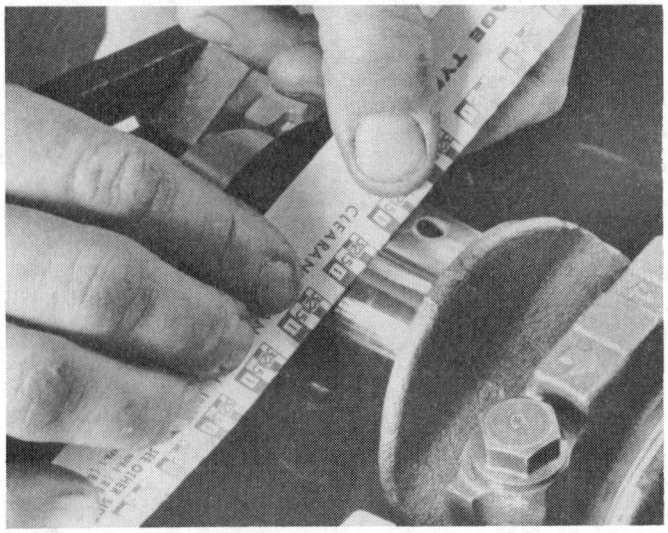

22.14 The crushed Plastigage is compared to the scale printed on the container to obtain the bearing oil clearance

23 Engine overhaul — reassembly sequence

1 Before beginning engine reassembly, make sure you have all the necessary new parts, gaskets and seals as well as the following items on hand:

Common hand tools
A 1/2-inch drive torque wrench
Piston ring installation tool
Piston ring compressor
Short lengths of rubber or plastic hose to fit over connecting rod bolts
Plastigage
Feeler gauges
A fine-tooth file
New, clean engine oil
Engine assembly lube or moly-based grease
RTV-type gasket sealant
Anaerobic-type gasket sealant
Thread locking compound

2 In order to save time and avoid problems, engine reassembly must be done in the following order.

Rear main oil seal (V6 and V8 engines only)
Crankshaft and main bearings
Piston rings
Piston/connecting rod assemblies
Oil pump
Oil pan
Camshaft
Timing chain/sprockets or gears
Timing chain/gear cover
Valve lifters
Cylinder head(s) and pushrods
Intake and exhaust manifolds
Oil filter
Pre-oil the engine (V6 and V8 engines only — Section 24)
Rocker arm cover(s)
Fuel pump
Water pump
Rear main oil seal (four-cylinder engine only)
Flywheel/driveplate
Carburetor/fuel injection components
Thermostat and housing cover
Distributor, spark plug wires and spark plugs
Emissions control components
Alternator

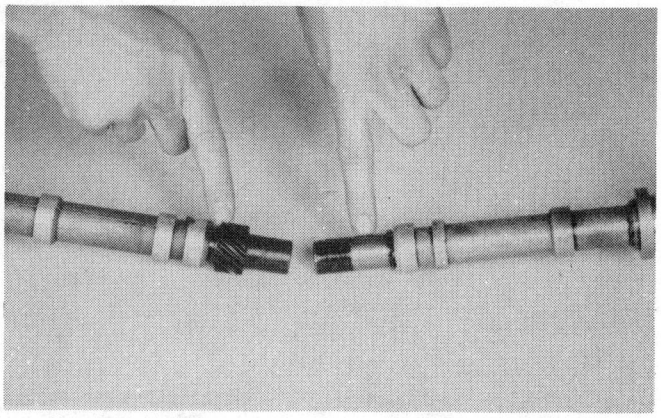

24.3 The pre-oiling tool (right) has the gear and advance weights ground off

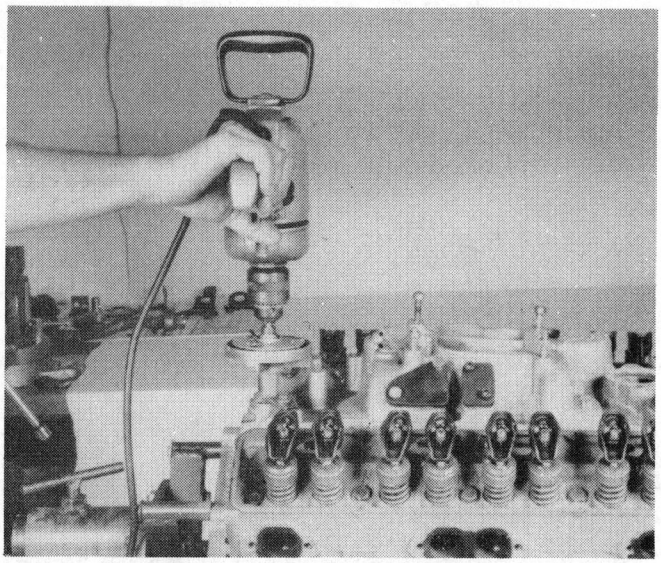

24.4 A drill motor connected to the modified distributor drives the oil pump

24 Pre-oiling engine after overhaul (V6 and V8 engines only)

1 After an overhaul it is a good idea to pre-oil the engine before it is installed and initially started. This will reveal any problems with the lubrication system at a time when corrections can be made easily and without major engine damage. Pre-oiling the engine will also allow the parts to be lubricated thoroughly in a normal fashion, but without the heavy loads associated with the combustion process placed upon them.
2 The engine should be assembled completely with the exception of the distributor and the rocker arm covers.
3 A modified distributor will be needed for this procedure. This pre-oil tool is a distributor body with the bottom gear ground off and the advance weight assembly removed from the top of the shaft (photo).
4 Place the pre-oiler into the distributor shaft access hole at the rear of the intake manifold and make sure the bottom of the shaft mates with the oil pump. Clamp the modified distributor into place just as you would an ordinary distributor. Now attach an electric drill motor to the top of the shaft (photo).
5 With the oil filter installed, all oil ways plugged (oil-pressure sending unit at rear of block) and the crankcase full of oil as shown on the dipstick, rotate the pre-oiler with the drill. Make sure the rotation is in a *clockwise* direction. Soon, oil should start to flow from the rocker arms, signifying that the oil pump and lubrication system are functioning properly. It may take two or three minutes for oil to flow to all of the rocker arms (photo). Allow the oil to circulate through the engine for a few minutes, then shut off the drill motor.
6 Check for oil leaks at the filter and all gasket and seal locations.
7 Remove the pre-oil tool, then install the distributor and rocker arm covers.

24.5 Oil and assembly lube or grease will spurt out of the holes in the rocker arms if the lubrication system is functioning properly

25 Initial start-up and break-in after overhaul

1 Once the engine has been properly installed in the vehicle, double-check the engine oil and coolant levels.
2 With the spark plugs out of the engine and the coil high-tension lead grounded to the engine block, crank the engine over until oil pressure registers on the gauge (if so equipped) or until the oil light goes off.
3 Install the spark plugs, hook up the plug wires and the coil high-tension lead.
4 Make sure the carburetor choke plate is closed, then start the engine. It may take a few moments for the gasoline to reach the carburetor, but the engine should start without a great deal of effort.
5 As soon as the engine starts, it should be set at a fast idle (to ensure proper oil circulation) and allowed to warm up to normal operating temperature. While the engine is warming up, make a thorough check for oil and coolant leaks.
6 Shut the engine off and recheck the engine oil and coolant levels. Also, check the ignition timing and the engine idle speed (refer to Chapter 1) and make any necessary adjustments.
7 Drive the vehicle to an area with minimum traffic, accelerate at full throttle from 30 to 50 mph, then allow the vehicle to slow to 30 mph with the throttle closed. Repeat the procedure 10 or 12 times. This will load the piston rings and cause them to seat properly against the cylinder walls. Check again for oil and coolant leaks.
8 Drive the vehicle gently for the first 500 miles (no sustained high speeds) and keep a constant check on the oil level. It is not unusual for an engine to use oil during the break-in period.
9 At approximately 500 to 600 miles, change the oil and filter, retorque the cylinder head bolts and recheck the valve clearances (if applicable).
10 For the next few hundred miles, drive the vehicle normally. Do not pamper it or abuse it.
11 After 2000 miles, change the oil and filter again and consider the engine fully broken in.

Chapter 3
Cooling, heating and air-conditioning systems

Refer to Chapter 13 for information related to 1985 and later models

Contents

Air-conditioning system – servicing 11
Air-conditioning system compressor – removal and installation 13
Air-conditioning system condenser – removal and installation 12
Antifreeze – general information 2
Coolant level check ..... Chapter 1
Coolant temperature sending unit – check and replacement 8
Cooling system check ..... Chapter 1
Cooling system servicing (draining, flushing and refilling) ..... Chapter 1
General information 1
Heater blower motor – removal and installation 9
Heater core – removal and installation 10
Radiator – removal, servicing and installation 5
Thermostat – check 4
Thermostat – replacement 3
Water pump – check 6
Water pump – removal and installation 7

Specifications

Operating pressure	14 to 17 psi
Thermostat rating	195°
Coolant capacity	See Chapter 1
Torque specifications	**Ft-lbs**
Fan-to-water pump bolts	20
Water pump bolts	15
Water outlet (thermostat housing) bolts	25
Coolant temperature sending unit	20

3

1 General information

The cooling system is very conventional in design, utilizing a cross flow radiator, an engine driven water pump and thermostat controlled coolant flow. The fan is equipped with a clutch, which allows it to draw air through the radiator at lower speeds. At higher speeds the fan is not needed for cooling, so the clutch automatically lowers the fan speed and reduces the engine power required for fan operation.

The water pump is mounted on the front of the engine and is driven by a belt from the pulley mounted on the front of the crankshaft. The belt is also used to drive other components.

The heater utilizes the heat produced by the engine and absorbed by the coolant to warm the interior of the vehicle. It is manually controlled from inside by the driver or passenger.

Air-conditioning is optional equipment. All components of the system are mounted in the engine compartment and the system is driven by a belt from the pulley mounted on the front of the crankshaft. Output of the system is controlled from inside the vehicle.

2 Antifreeze — general information

Caution: *Do not allow antifreeze to come in contact with your skin or painted surfaces of the vehicle. Flush contacted areas immediately with plenty of water. Antifreeze can be fatal to children and pets. They like because it is sweet. Just a few licks can cause death. Wipe up garage floor and drip pan coolant spills immediately. Keep antifreeze containers covered and repair leaks in your cooling system immediately.*

The cooling system should be filled with a water/ethylene glycol-based antifreeze solution, which will prevent freezing down to at least -20°F at all times. It also provides protection against corrosion and increases the coolant boiling point.

The cooling system should be drained, flushed and refilled at least every other year (see Chapter 1). The use of antifreeze solutions for periods of longer than two years is likely to cause damage and encourage the formation of rust and scale in the system.

Before adding antifreeze to the system, check all hose connections and retorque the cylinder head bolts, because antifreeze tends to search out and leak through very minute openings.

The exact mixture of antifreeze-to-water which you should use depends on the relative weather conditions. The mixture should contain at least 50 percent antifreeze, but should never contain more than 70 percent antifreeze.

3 Thermostat — replacement

Caution: *The engine must be completely cool before beginning this procedure. Also, when working in the vicinity of the electric fan, disconnect the negative battery cable from the battery to prevent the fan from coming on accidentally.*

Four-cylinder engine

1 Refer to the **Caution** in Section 2.
2 Disconnect the cable from the negative battery terminal (if not done previously).
3 Drain the cooling system until the level is below the thermostat by opening the petcock at the bottom right side of the radiator. Close the petcock when enough coolant has drained.
4 Remove the upper radiator hose from the water outlet housing.
5 Remove the two housing bolts and separate the housing from the engine.
6 Remove the thermostat from the thermostat housing.
7 Before installing the thermostat, clean the gasket sealing surfaces on the water outlet and thermostat housing.
8 To install the thermostat, place it in the housing, install a new gasket (not RTV), install the water outlet and tighten the water outlet bolts to the specified torque.
9 Install the upper radiator hose.
10 Fill the cooling system with the proper antifreeze/water mixture (refer to Chapter 1).
11 Reconnect the battery cable and start the engine.
12 Run the engine with the radiator cap removed until the upper radiator hose is hot (thermostat open).
13 With the engine idling, add coolant to the radiator until the level reaches the bottom of the filler neck.
14 Install the radiator cap, making sure the arrows on the cap line up with the radiator overflow tube.

V6 and V8 engines

15 Refer to the **Caution** in Section 2.
16 Disconnect the cable from the negative battery terminal.
17 Remove the air cleaner, tagging all hoses as they are removed to simplify installation.
18 Drain the cooling system until the level is below the thermostat (refer to Step 3).
19 Disconnect the upper radiator hose from the thermostat housing.
20 Remove the thermostat housing bolts and separate the housing from the engine (photo). On some models, the alternator mounting bracket will have to be disconnected first, as it is attached to the housing

3.20 Removing the thermostat housing (L4 engine shown)

3.21 Removing the thermostat from the engine (L4 engine shown)

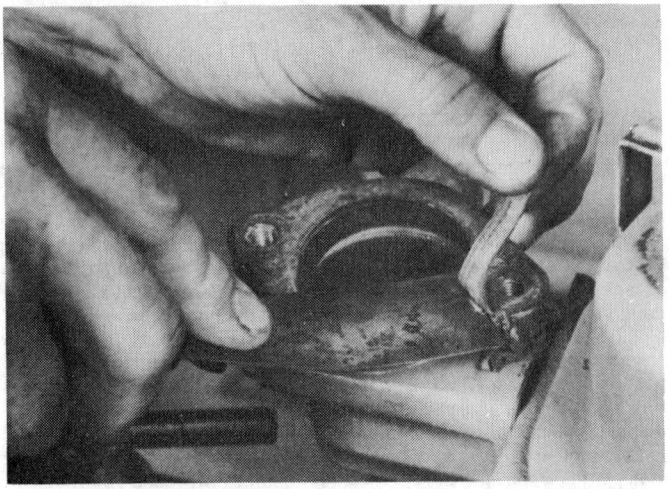

3.22 Using a putty knife to remove old gasket material from the thermostat gasket sealing surface

5.9 Removing the lower radiator shroud bolts (V6 engine shown)

mounting stud. Also, late model vehicles may have a TVS switch installed in the thermostat housing. If so, disconnect each of the vacuum hoses from the switch (noting their installed positions) and then unscrew the switch from the housing.

21 After separating the thermostat housing from the engine, the thermostat will be visible and can be lifted out (photo). Note how the thermostat is positioned in the recess, as it must be replaced in the same position.

22 Before installation, use a gasket scraper or putty knife to carefully remove all traces of the old gasket from the thermostat housing and the engine sealing surface (photo). Do not allow the gasket pieces to drop down into the intake manifold.

23 Apply a 1/8-inch bead of RTV-type sealant to the sealing surface on the engine and place the thermostat into the recess.

24 Immediately place the thermostat housing with sealant and a new gasket into position and tighten the bolts to the specified torque.

25 Where applicable, install the alternator brace and/or the TVS switch and vacuum hoses.

26 Connect the upper radiator hose and tighten the hose clamp securely.

27 Reinstall the air cleaner.

28 To complete the installation, refer to Steps 10 through 14.

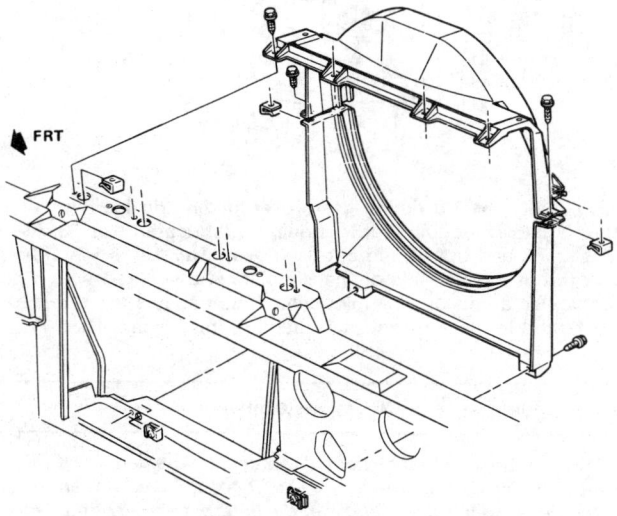

Fig. 3.1 Typical radiator fan shroud mounting details (Sec 5)

4 Thermostat — check

1 The best way to check the operation of the thermostat is with it removed from the engine. In most cases, if the thermostat is suspect, it is more economical to simply buy and install a replacement thermostat, as they are not very costly.

2 To check, first remove the thermostat as described in Section 3.

3 Inspect the thermostat for excessive corrosion and damage. Replace it with a new one if either of these conditions is noted.

4 Place the thermostat in hot water (25 degrees above the temperature stamped on the thermostat). The water temperature should be approximately 220 degrees. When submerged in the water (which should be agitated thoroughly), the valve should open all the way.

5 Next, remove the thermostat using a piece of bent wire and place it in water which is 10 degrees below the temperature on the thermostat, or about 185 degrees. At this temperature, the thermostat valve should close completely.

6 Reinstall the thermostat if it operates properly. If it does not, purchase a new thermostat of the same temperature rating.

5 Radiator — removal, servicing and installation

Caution: *The engine must be completely cool before beginning this procedure. Also, when working in the vicinity of the electric fan, disconnect the negative battery cable from the battery to prevent the fan from coming on accidentally.*

1 Refer to the **Caution** in Section 2.

2 Disconnect the cable from the negative battery terminal.

3 Drain the radiator (refer to Chapter 1, if necessary).

4 Unbolt the fan from the fan pulley hub and remove the fan assembly. On vehicles equipped with a fan clutch, store the clutch in an upright position to prevent seal leakage.

5 Disconnect the upper and lower radiator hoses from the radiator.

6 On vehicles equipped with an automatic transmission, disconnect the fluid cooler lines at the radiator and immediately plug the lines.

7 Remove the clamp securing the coolant reservoir hose to the radiator outlet. On V6 models, remove the hose from the upper radiator shroud retainers.

8 Remove the bolts securing the top of the upper radiator shroud to the metal support.

9 Remove the bolts securing the bottom of the radiator shroud to the chassis, if so equipped (photo).

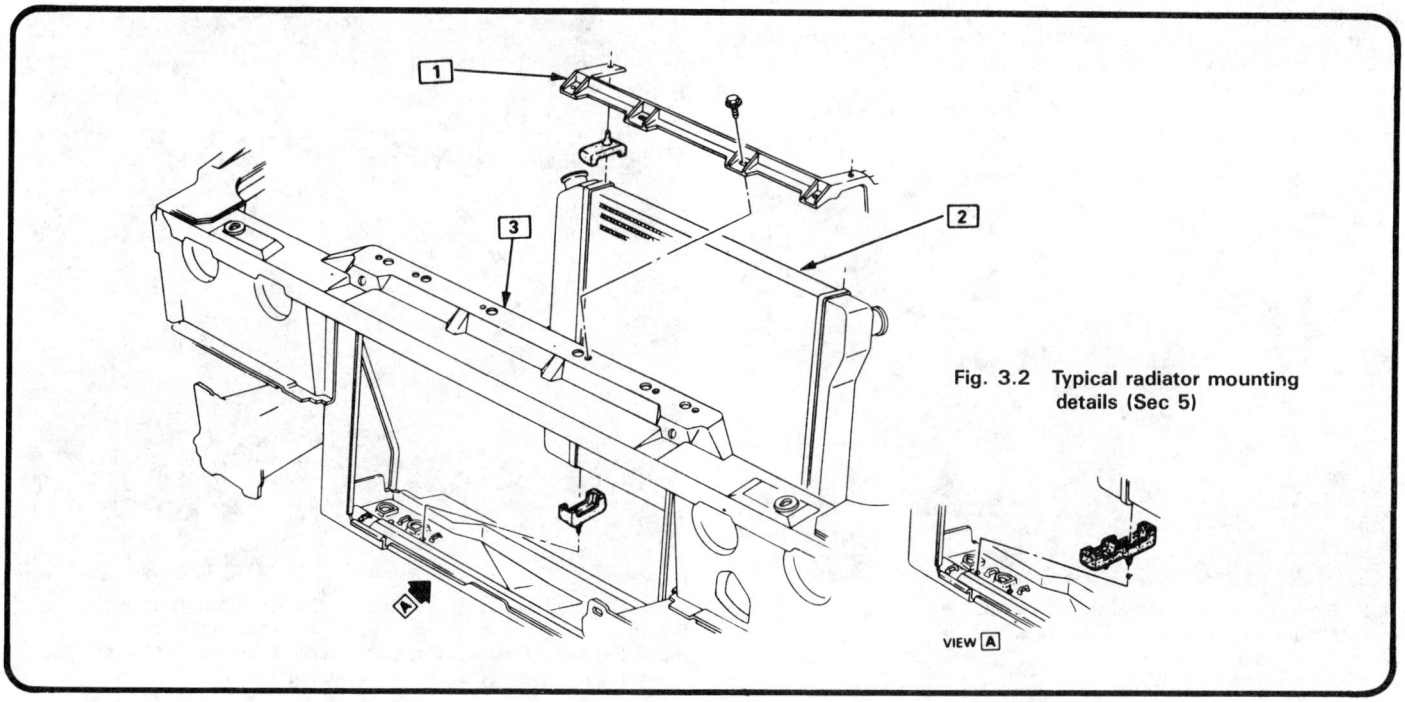

Fig. 3.2 Typical radiator mounting details (Sec 5)

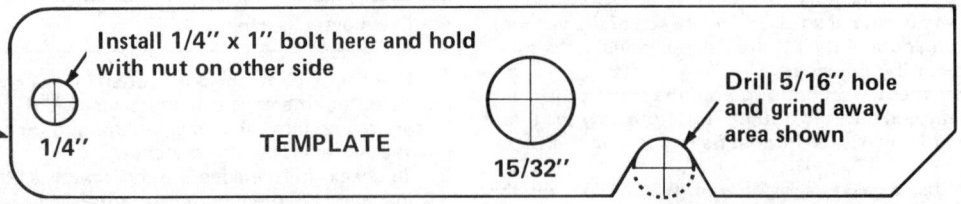

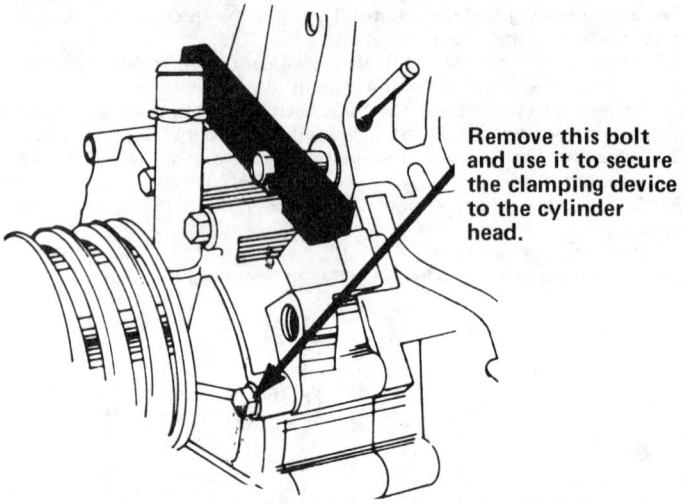

Fig. 3.3 The V6 timing gear cover holding device template (top) is shown actual size. It should be fabricated from 3/8-inch flat stock and drilled and cut as shown. The 1/4 x 1-inch bolt inserted at the left side of the template and secured with a nut serves as a spacer. The lower illustration shows the template bolted in place on the right-hand cylinder head. (Sec 7)

10 Pull straight up on the radiator and shroud assembly and remove it from the vehicle.
11 Carefully examine the radiator for evidence of leaks and damage. It is recommended that any necessary repairs be performed by a reputable radiator repair shop.
12 With the radiator removed, brush accumulations of insects and leaves from the fins and examine and replace, if necessary, any hoses or clamps which have deteriorated.
13 The radiator can be flushed as described in Chapter 1.
14 Check the pressure rating of the radiator cap and have it tested by a service station.
15 If you are installing a new radiator, transfer the fittings from the old unit to the new one.
16 Installation is the reverse of the removal procedure. When setting the radiator in the chassis, make sure the bottom of the radiator is seated correctly in the two bottom cradles secured to the radiator support.
17 After installing the radiator, refill it with the proper coolant mixture (refer to Chapter 1), then start the engine and check for leaks.
18 If equipped with a new automatic transmission, check the transmission fluid level (refer to Chapter 1).

6 Water pump — check

1 A failure in the water pump can cause overheating and serious engine damage (the pump will not circulate coolant through the engine).
2 There are three ways to check the operation of the water pump while it is still installed on the engine. If the pump is defective, it should be replaced with a new or rebuilt unit.
3 With the engine at normal operating temperature, squeeze the upper radiator hose. If the water pump is working properly, a pressure surge will be felt as the hose is released.
4 Water pumps are equipped with 'weep' or vent holes. If a pump seal failure occurs, small amounts of coolant will leak from the weep holes. In most cases it will be necessary to use a flashlight from under the vehicle to see evidence of leakage from this point on the pump body.
5 If the water pump shaft bearings fail, there may be a squealing sound emitted from the front of the engine while it is running. Shaft wear can be felt if the water pump pulley is forced up and down. Do not mistake drivebelt slippage (which also causes a squealing sound) for water pump failure.

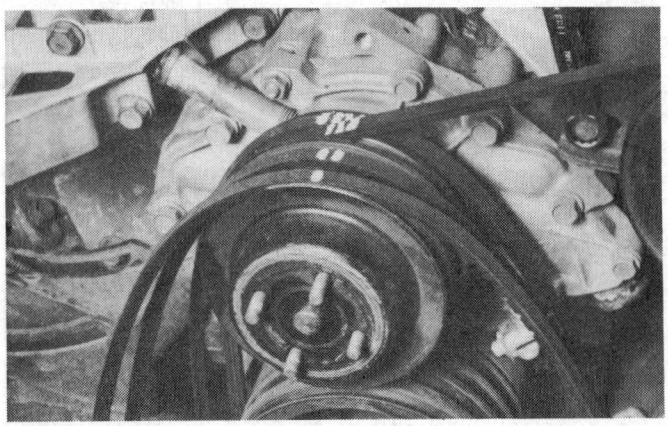

7.4 Mark the drivebelts before removing them to ensure reinstallation on the correct pulleys

7 Water pump — removal and installation

Caution: *The engine must be completely cool before beginning this procedure. Also, when working in the vicinity of the electric fan, disconnect the negative battery cable from the battery to prevent the fan from coming on accidentally.*

7.10 On V6 engines in which stud C goes through both the timing cover and the water pump, remove the water pump mounting bolts (arrows) **after** using bolt A to attach the tool mentioned in the text to the right cylinder head at hole B

1 Refer to the **Caution** in Section 2.
2 Disconnect the cable from the negative battery terminal.
3 Drain the cooling system (refer to Chapter 1, if necessary).
4 Mark the accessory drivebelts with white paint to simplify installation (photo).
5 On V6 and V8 models, remove the upper fan shroud (refer to Section 5, if necessary).
6 Remove the accessory drivebelts (alternator, air-conditioning compressor, AIR system pump, power steering pump, as applicable), by loosening the pivot and adjusting bolts and pushing the accessory toward the engine.
7 On some engines, it may be necessary to remove one or more accessories to gain access to the water pump. Refer to the appropriate Chapter(s) to remove them.
8 Remove the bolts retaining the fan to the fan pulley hub, then separate the fan and pulley hub from the water pump.
9 Disconnect the heater and lower radiator hoses from the water pump.
10 Remove the water pump mounting bolts (photo) and separate the water pump from the engine. **Caution:** *On some V6 engines, the water pump bolts hold the timing cover to the block. Removal of the water pump may break the chemical seal and allow coolant into the oil. To prevent this, before removing the water pump, secure the timing cover to the block with the clamping device shown in the accompanying illustrations.*
11 Prior to installing the water pump, remove all old gasket material and sealant from the gasket sealing surfaces. Clean the threaded holes in the block as well.
12 Installation is the reverse of the removal procedure.
13 If a new water pump is being installed, transfer the heater hose fitting from the old pump to the new one.
14 On L4 engines, use a new gasket or RTV-type sealant when installing the pump.
15 On V6 and V8 engines, use a new gasket when installing the pump.
16 Tighten the bolts to the specified torque after coating the threads with RTV-type sealant to prevent leaks. Follow a crisscross pattern and work up to the final torque in three (3) steps.
17 Adjust all drivebelts (see Chapter 1).
18 Connect the negative battery cable and fill the radiator with a mixture of antifreeze and water. Start the engine and allow it to idle until the upper radiator hose gets hot. Check for leaks. With the engine hot, fill the radiator with more coolant mixture until the level is at the bottom of the filler neck. Install the radiator cap and check the coolant level periodically during the first few miles of driving.

8 Coolant temperature sending unit — check and replacement

1 The coolant temperature indicator system is composed of a light mounted in the instrument panel and a coolant temperature sending unit located at the left rear corner of the cylinder head on L4 engines and at the front of the left cylinder head on V6 and V8 engines. If a temperature gauge is included in the instrument cluster, the temperature sending unit is replaced by a transducer.
2 **Caution:** *Since the ignition key will be in the On position for some of the diagnostic steps, be especially careful to stay clear of the electric fan blades.*
3 If overheating occurs, check the coolant level in the system and then make sure that the wiring between the light or gauge and the sending unit is secure.
4 When the ignition switch is turned on and the starter motor is turning, the indicator light should be on (overheated engine indication). If the light is not on, the bulb may be burned out, the ignition switch may be faulty or the circuit may be open.
5 As soon as the engine starts, the light should go out and remain out unless the engine overheats. Failure of the light to go out may be due to grounded wiring between the light and the sending unit, a defective sending unit or a faulty ignition switch.
6 If the sending unit is to be replaced, it is simply unscrewed from the cylinder head and a replacement installed (photo). *Make sure that the engine is cool before removing the defective sending unit.* There will be some coolant loss, so check the level after the replacement has been installed.

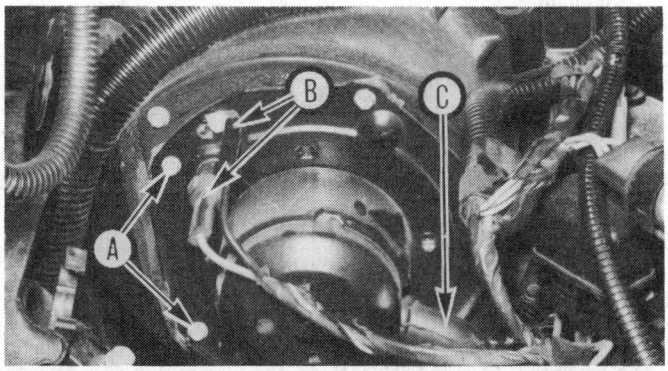

9.2 To remove the heater blower motor, remove the mounting bolts (A — 6 total), the motor and resistor wires (B) and the cooling tube (C)

9 Heater blower motor — removal and installation

1 Disconnect the cable from the negative battery terminal.
2 Working in the engine compartment, disconnect the wires at the lower motor and resistor (photo).
3 Disconnect the blower motor cooling tube.
4 On some models, it may be necessary to disconnect the radio capacitor to allow removal of the blower motor.
5 Remove the blower motor mounting bolts and separate the motor/cage assembly from the case.
6 While holding the cage, remove the cage retaining nut and slide the cage off the motor shaft (photo).
7 Installation is the reverse of the removal procedure.

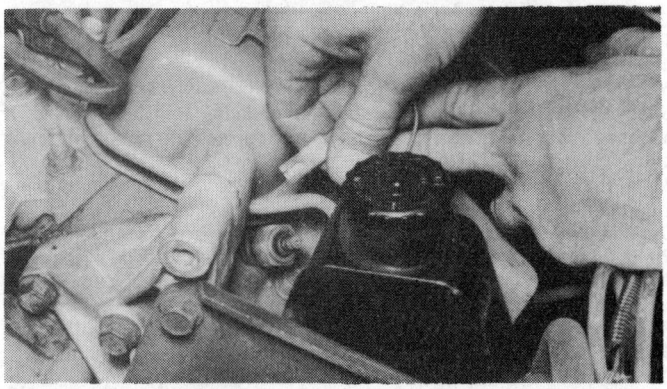
8.6 Removing the wire connector from the coolant temperature sending unit prior to removing it from the cylinder head (V6 engine shown)

9.6 Remove the nut (arrow) to separate the cage from the blower motor shaft

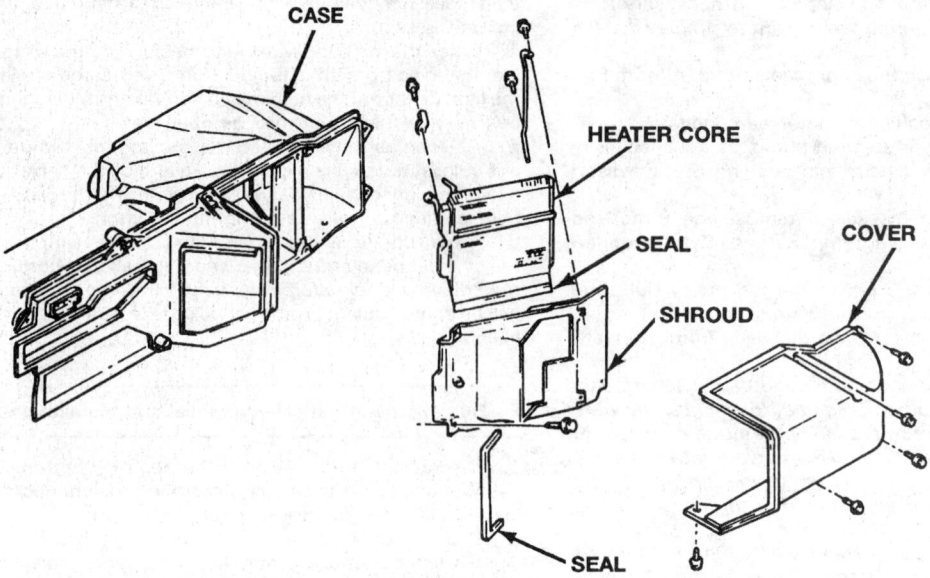

Fig. 3.4 Typical heater module assembly (Sec 10)

10 Heater core – removal and installation

1 Disconnect the cable from the negative battery terminal.
2 Drain the cooling system (see Chapter 1).
3 Remove the heater hoses from the heater core inlet and outlet pipes. Use care in removing the hoses so as not to damage the heater pipes. If necessary, cut the hoses off the pipes and replace the hoses with new ones.
4 Remove the lower hush panels from under the left and right side of the instrument panel.
5 Remove the instrument panel pad. Retaining screws are located in the defroster vents and along the underside of the rear edge of the pad.
6 Remove the retaining nuts to both front speakers and remove the speakers. Also remove the right speaker bracket.
7 Remove the screws retaining the side window defrost ducts, the front carrier braces and carrier shelf and remove the side defrost ducts.
8 Carefully disconnect the Electronic Control Module and set it safely aside.
9 Remove the radio trim plate, the upper console trim, and the glove box assembly.
10 Remove the parking brake handle grip, the screws retaining the console body, and position the console back out of the way.
11 Remove the trim plate under the steering column. Remove the two nuts retaining the steering column and carefully lower and support the steering column out of the way.
12 Remove the two nuts and four screws retaining the instrument panel to the firewall and move the instrument panel back far enough to access the heater module.
13 Remove the screws retaining the heater core cover and remove the cover (photo).
14 Remove the screws retaining the shroud. Remove the shroud and the heater core.
15 Installation is the reverse of removal. Replace the seals with new ones. Refill the cooling system and test for leaks.

11 Air-conditioning system – servicing

1 Regularly inspect the condenser fins (located ahead of the radiator) and brush away leaves and bugs.
2 Clean the evaporator drain tubes.
3 Check the condition of the system hoses. If there is any sign of deterioration or hardening, have them replaced by your dealer.
4 At the recommended intervals, check and adjust the compressor drivebelt as described in Chapter 1.
5 Because of the special tools, equipment and skills required to service air-conditioning systems, and the differences between the various systems that may be installed on vehicles, major air-conditioner servicing procedures cannot be covered in this manual.
6 We will cover component removal, as the home mechanic may realize a substantial savings in repair costs if he removes components himself, takes them to a professional for repair, and/or replaces them with new ones.
7 Problems in the air-conditioning system should be diagnosed, and the system refrigerant evacuated, by an air-conditioning technician before component removal/replacement is attempted.
8 Once the new or reconditioned component has been installed, the system should then be recharged and checked by an air-conditioning technician.
9 Before indiscriminately removing the air-conditioning system components, get more than one estimate of repair costs from reputable air-conditioning service centers. You may find it to be cheaper and less trouble to let the entire operation be performed by someone else.

12 Air-conditioning system condenser – removal and installation

Caution: *Before removing the condenser, the system must be evacuated by an air-conditioning technician. Do not attempt to do this yourself- the refrigerant used in the system can cause serious injuries and respiratory irritation.*

1 After having the system evacuated, disconnect the coupled hose and liquid line fittings.
2 Remove the screws retaining the top radiator shroud.
3 Remove the bolts holding the condenser to the top radiator support.
4 Carefully tilt the radiator to the rear and lift the condenser out of the bottom cradle supports.
5 If the condenser is being replaced with a new one, transfer the brackets and mounts from the old unit to the new one.
6 The installation procedures are the reverse of those for removal. When installing the hose and fittings, use new O-rings and lubricate them with 525 viscosity refrigerant oil.
7 After the condenser is installed, have the system recharged.

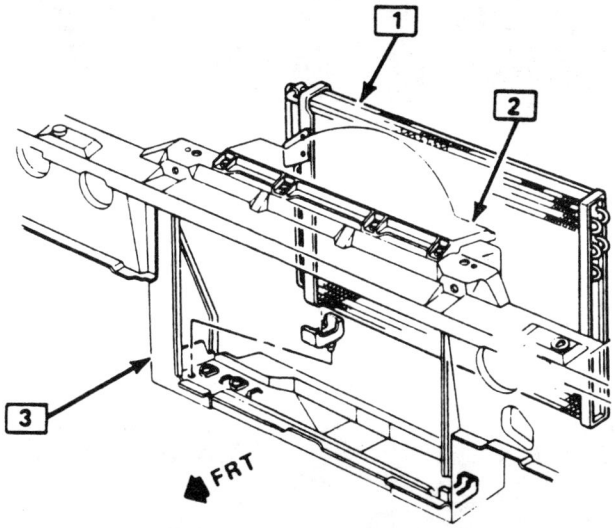

Fig. 3.5 Typical air-conditioning condenser installation details (Sec 12)

1 Condenser
2 Fan shroud
3 Radiator support

13.1A Loosening the compressor pivot bolt

13.1B Removing the compressor adjusting bolt

13.2 Disconnecting the compressor wire

13 Air-conditioning system compressor – removal and installation

Caution: *Before removing the compressor, the system must be evacuated by an air-conditioning technician. Do not attempt to do this yourself – the refrigerant in the system can cause serious injuries and respiratory irritation.*

1 After having the system evacuated, remove the compressor pivot and adjusting bolts (photos).
2 Disconnect the wire (photo).
3 Remove the drivebelt, routing the lower loop behind the vibration damper to gain additional slack, if necessary.
4 Remove the compressor from the mount.
5 Remove the fitting block (coupled hose assembly) bolt at the rear of the compressor.
6 If the compressor is being replaced with a new one, transfer the usable switches from the old unit to the new one.
7 Installation procedures are the reverse of those for removal. When installing the fitting block, use new O-rings and lubricate them with 525 viscosity refrigerant oil.
8 After the compressor is installed, have the system recharged.

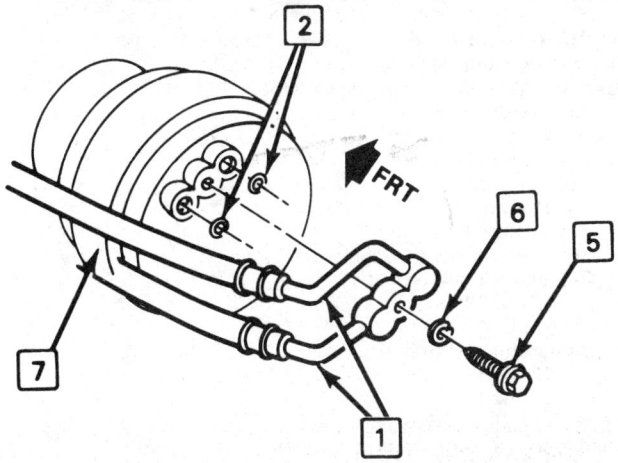

Fig. 3.6 Air-conditioning compressor attachments (Sec 13)

1 Fitting and hose assembly
2 Gasket O-rings
3 Hose fitting assembly attaching bolt
4 Hose fitting assembly washer
5 Compressor

Chapter 4 Fuel and exhaust systems

Refer to Chapter 13 for specifications and information related to 1985 and later models

Contents

Carburetor — removal and installation 7
Carburetor (E2SE) — overhaul 8
Carburetor (E4ME and E4MC) — overhaul 9
Crossfire Injection (CFI) system — general information 18
Electronic Fuel Injection (EFI) system — general information . . 10
Engine idle speed check and adjustment . . . . . Chapter 1
Exhaust system check . . . . . Chapter 1
Exhaust system components — removal and installation 20
Fuel filter replacement . . . . . Chapter 1
Fuel injection pressure relief procedure 11
Fuel line — repair and replacement 4
Fuel pressure regulator — removal and installation 14
Fuel pump — check 2
Fuel pump (carbureted models) — removal and installation . . . 3
Fuel pump (fuel-injected models) — removal and installation . . 12
Fuel system check . . . . . Chapter 1
Fuel tank — removal and installation 5
Fuel tank — repair 6
General information 1
Idle Air Control (IAC) assembly — removal and installation . . . 13
Injector — removal and installation 15
Thermo-controlled Air Cleaner (THERMAC) check . . Chapter 1
Throttle Body Injection (TBI) 300 assembly — overhaul 17
Throttle Body Injection (TBI) 300 assembly — removal and installation 16
Throttle Body Injection (TBI) 400 assemblies — removal and installation 19
Throttle linkage check . . . . . Chapter 1

Specifications

Carburetor specifications — 1982 models

Float adjustment (all)	13/32 in (10.3 mm)
Fast idle cam (choke rod) adjustment	
Carburetor numbers 17082390, 17082490	17°
Carburetor numbers 17082391, 17092491	25°
Primary vacuum break adjustment	
Carburetor numbers 17082390, 17082490	26°
Carburetor numbers 17082391, 17082491	29°
Air valve rod link adjustment (all)	1°
Secondary vacuum break adjustment	
Carburetor numbers 17082390, 17082490	34°
Carburetor numbers 17082391, 17082491	35°
Unloader adjustment (all)	35°

Carburetor specifications — 1983 models

Float adjustment	
Carburetor numbers	13/32 in (10.3 mm)
17083356	
17083357	
17083358	
17083359	
Carburetor numbers	1/8 in (3.2 mm)
17083368	
17083370	
Carburetor numbers	1/8 in (3.2 mm)
17083450	
17083451	
17083452	
17083453	
17083454	
17083455	
17083456	
Carburetor numbers	1/4 in (6.5 mm)
17083630	
17083631	
17083632	
17083633	
17083634	
17083635	
17083636	
Carburetor number	1/8 in (3.2 mm)
17083650	
Air valve spring adjustment	
Carburetor number 17083650	1/2 turn
All others	1 turn
Choke coil lever adjustment (all)	0.085 in (2.16 mm) plug gauge
Fast idle cam (choke rod) adjustment	
Carburetor numbers	22°
17083356	
17083357	
17083358	
17083359	
17083368	
17083370	
All others	28°
Primary vacuum break adjustment	
Carburetor numbers	25°
17083356	
17083357	
17083358	
17083359	
17083368	
17083370	
All others	27°
Air valve rod link adjustment (all)	1°
Secondary vacuum break adjustment (all)	35°
Unloader adjustment	
Carburetor numbers	30°
17083356	
17083357	
17083358	
17083359	
17083368	
17083370	
All others	45°

Carburetor specifications — 1984 models

Float adjustment	
Carburetor numbers	9/32 in (7.0 mm)
17072683	
17074812	
17084356	
17084357	
17084358	
17084359	
Carburetor numbers	1/8 in (3.2 mm)
17084368	
17084370	

Carburetor specifications — 1984 models (continued)

Carburetor numbers	5/32 in (3.6 mm)
17084452	
17084453	
17084455	
17084456	
17084458	
17084532	
17084534	
17084535	
17084537	
17084538	
17084540	
Carburetor number	1/8 in (3.2 mm)
17084542	
Carburetor numbers	9/32 in (7.0 mm)
17084632	
17084633	
17084635	
17084636	
Air valve spring adjustment	
Carburetor numbers	3/4 turn
17084356	
17084357	
17084358	
17084359	
17084368	
17084370	
Carburetor numbers	1 turn
17084430	
17084431	
17084434	
17084435	
All others	1/2 turn
Choke coil lever adjustment	0.085 in (2.16 mm) plug gauge
Fast idle cam (choke rod) adjustment	
Carburetor numbers	22°
17084356	
17084357	
17084358	
17084359	
17084368	
17084370	
Carburetor numbers	15°
17084430	
17084431	
17084434	
17084435	
All others	28°
Primary vacuum break adjustment	
Carburetor numbers	26°
17084430	
17084431	
17084434	
17084435	
All others	25°
Air valve rod link adjustment (all)	1°
Secondary vacuum break adjustment	
Carburetor numbers	30°
17084356	
17084357	
17084358	
17084359	
17084368	
17084370	
Carburetor numbers	38°
17084430	
17084431	
17084434	
17084435	
All others	35°
Unloader adjustment	
Carburetor numbers	30°
17084356	
17084357	

17084358	
17084359	
17084368	
17084370	
Carburetor numbers	42°
17084430	
17084431	
17084434	
17084435	
All others	45°

Torque specifications

	Ft-lbs
Carburetor mounting nuts/bolts	
Long	7
Short	11
TBI mounting bolts	15
Fuel inlet nut-to-carburetor	18
Carburetor fuel line-to-inlet nut	24
300 TBI fuel meter body mounting screws	3
300 TBI fuel feed and return nuts	21
Fuel tank strap-to-chassis bolts	25

1 General information

The fuel system consists of a rear mounted fuel tank, a mechanically operated fuel pump (carburetor-equipped engines) or an electrically operated fuel pump (fuel-injected engines), a carburetor or fuel injection assembly and an air cleaner.

The carburetor may be a two or four-barrel type, depending on the engine. Similarly, the fuel injection unit may be a single or dual throttle body injector.

The exhaust system includes a catalytic converter, muffler, related emissions equipment and associated pipes and hardware.

2 Fuel pump — check

Warning: *Gasoline is extremely flammable, so extra precautions must be taken when working on any part of the fuel system. Do not smoke or allow open flames or bare light bulbs near the work area. Also, do not work in a garage if a natural gas-type appliance with a pilot light is present.*

1 On carburetor-equipped engines the fuel pump is located on the left-front side of the engine (V6 models) or on the right-front side (V8 models).

2 The fuel pump for fuel-injected models is located in the fuel tank.

3 Both types of fuel pumps are sealed and no repairs are possible. However, on carburetor-equipped models, the fuel pump can be inspected and tested on the vehicle as follows.

4 Make sure that there is fuel in the fuel tank.

5 With the engine running, check for leaks at all gasoline line connections between the fuel tank and the carburetor. Tighten any loose connections. Inspect all hoses for flat spots and kinks which would restrict the fuel flow. Air leaks or restrictions on the suction side of the fuel pump will greatly affect the pump's output.

6 Check for leaks at the fuel pump diaphragm flange.

7 Disconnect the high energy ignition (HEI) connector at the distributor, then disconnect the fuel inlet line from the carburetor and place it in a metal container.

8 Crank the engine a few revolutions and make sure that well-defined spurts of fuel are ejected from the open end of the line. If not, the fuel line is clogged or the fuel pump is defective.

9 Disconnect the fuel line at both ends and blow through it with compressed air. If the fuel line is not clogged, replace the fuel pump with a new one.

3 Fuel pump (carbureted models) — removal and installation

Warning: *Gasoline is extremely flammable, so extra precautions must be taken when working on any part of the fuel system. Do not smoke or allow open flames or bare light bulbs near the work area. Also, do not work in a garage if a natural gas-type appliance with a pilot light is present.*

1 Disconnect the cable from the negative battery terminal.

2 Remove the fuel inlet and outlet lines. Use two wrenches to prevent damage to the pump and connections (photo).

3 Remove the fuel pump mounting bolts, the pump and the gasket.

4 If the pushrod is to be removed, first remove the pipe plug or the

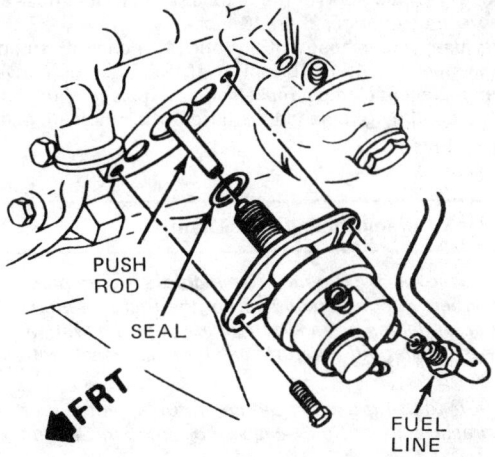

Fig. 4.1 V6 engine fuel pump installation — exploded view (Sec 3)

3.2 Removing the fuel pump (hold the inside fitting with one wrench while turning the outside fitting with another) (V6 engine shown)

pump adapter and gasket, as appropriate.
5 When installing the pump, first install the pushrod using gasket sealant on the pipe plug or gasket (where applicable). Hold the pushrod in position with heavy grease.
6 Install the pump using a new gasket. Use gasket sealant on the screw threads.
7 Connect the fuel lines, start the engine and check for leaks.

4 Fuel line — repair and replacement

Warning: *Gasoline is extremely flammable, so extra precautions must be taken when working on any part of the fuel system. Do not smoke or allow open flames or bare light bulbs near the work area. Also, do not work in a garage if a natural gas-type appliance with a pilot light is present.*

1 If a section of metal fuel line must be replaced, only brazed, seamless steel tubing should be used, since copper or aluminum does not have enough durability to withstand normal operating vibrations.
2 If only one section of a metal fuel line is damaged, it can be cut out and replaced with a piece of rubber hose. Be sure to use only reinforced fuel resistant hose, identified by the word 'Fluroelastomer' on the hose. The inside diameter of the hose should match the outside diameter of the metal line. The rubber hose should be cut four (4) inches longer than the section it's replacing, so there are two (2) inches of overlap between the rubber and metal line at either end of the section. Hose clamps should be used to secure both ends of the repaired section.
3 If a section of metal line longer than six (6) inches is being removed, use a combination of metal tubing and rubber hose so the hose lengths will be no longer than 10 inches.
4 Never use rubber hose within four (4) inches of any part of the exhaust system or within 10 inches of the catalytic converter.
5 When replacing clamps, make sure the replacement clamp is identical to the one being replaced, as different clamps are used depending on location.

5 Fuel tank — removal and installation

Warning: *Gasoline is extremely flammable, so extra precautions must be taken when working on any part of the fuel system. Do not smoke or allow open flames or bare light bulbs near the work area. Also, do not work in a garage if a natural gas-type appliance with a pilot light is present.*

Note: *Before starting this procedure on fuel-injected models, the fuel system pressure must be relieved as outlined in Section 11.*

1 Remove the cable from the negative battery terminal.
2 Remove the filler cap from the fuel tank.
3 Drain all fuel from the tank into a clean container. Since there are no drain plugs on the tank, siphon the fuel through the filler neck or drain the fuel through the fuel line running to the carburetor or fuel injection unit. Do not start the siphoning process with your mouth because serious personal injury could result.
4 Raise the vehicle and support it securely on jackstands.
5 Disconnect the exhaust pipe from the rear of the catalytic converter.
6 Disconnect the rear exhaust pipe hanger and allow the exhaust system to hang over the rear axle assembly.
7 Remove the exhaust pipe and muffler heat shields.
8 Remove the fuel filler neck shield.
9 Remove the rear suspension track bar (refer to Chapter 11).
10 Remove the rear suspension track bar brace.
11 Disconnect the fuel gauge wire (fuel gauge/fuel pump wire on fuel-injected models) **Note:** *The wiring harness on the fuel pump/gauge sending assembly is an integral, permanent part of the sending assembly. Do not pry up on the cover connector. Make the disconnection at the body wiring harness.*
12 Disconnect the tank ventilator hose, the fuel vapor hose and the fuel feed and return hoses from the sender assembly by loosening the hose clamps and pulling the hoses off the fittings.
13 Remove the fuel line retaining bracket on the left side, then remove the brake line clip from the bracket.
14 Position a floor jack under the rear axle assembly and raise it just

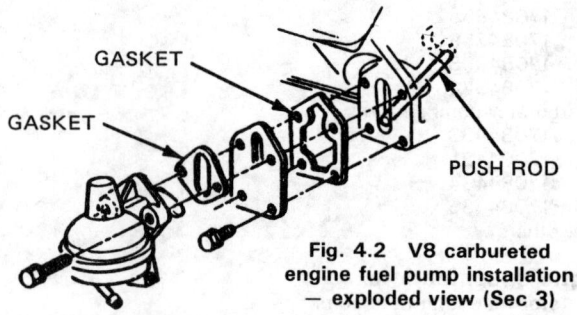

Fig. 4.2 V8 carbureted engine fuel pump installation — exploded view (Sec 3)

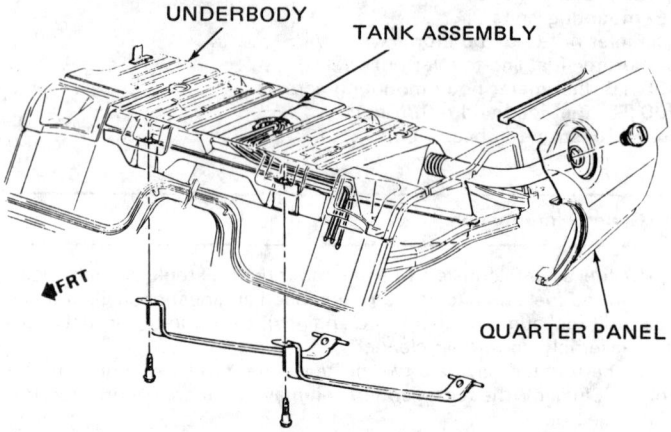

Fig. 4.3 Typical fuel tank installation details (Sec 5)

enough to take the weight off the axle.
15 Disconnect the lower ends of the rear shock absorbers.
16 If so equipped, remove the rear stabilizer bar from the frame.
17 Lower the jack under the axle assembly enough to allow removal of the rear coil springs.
18 Remove the fuel tank supporting strap bolts.
19 Remove the fuel tank by rotating the front of the tank down and sliding the tank to the right. It may be necessary to further lower the jack under the rear axle to remove the tank, but care should be taken to ensure that the brake lines and cables are not stretched or damaged.
20 Installation is the reverse of the removal procedure.
21 When the installation is complete, carefully check all lines, hoses and fittings for leaks.

6 Fuel tank — repair

1 Any repairs to the fuel tank or filler neck should be carried out by a professional who has experience in this critical and potentially dangerous work. Even after cleaning and flushing of the fuel system, explosive fumes can remain and ignite during repair of the tank.
2 If the fuel tank is removed from the vehicle, it should not be placed in an area where sparks or open flames could ignite the fumes coming out of the tank. Be especially careful inside garages where a natural gas-type appliance is located, because the pilot light could cause an explosion.

7 Carburetor — removal and installation

Warning: *Gasoline is extremely flammable, so extra precautions must be taken when working on any part of the fuel system. Do not smoke or allow open flames or bare light bulbs near the work area. Also, do not work in a garage if a natural gas-type appliance with a pilot light is present.*

1 Remove the cable from the negative battery terminal.
2 Remove the air cleaner.
3 Disconnect the fuel and vacuum lines from the carburetor, noting

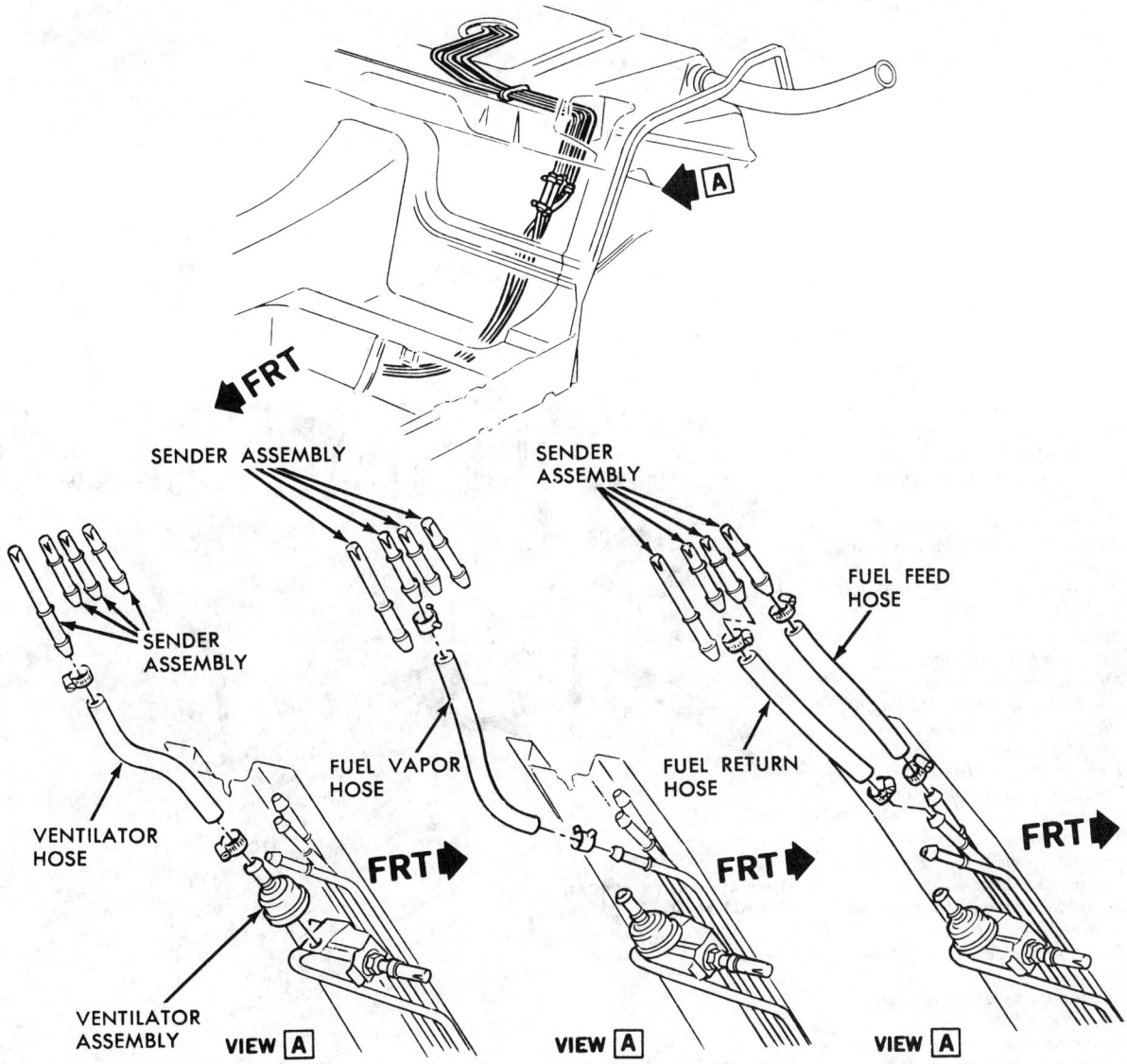

Fig. 4.4 Fuel tank-to-fuel line connections (Sec 5)

their locations.

4 Disconnect the accelerator linkage and cruise control linkage (if so equipped).

5 Disconnect the throttle valve linkage or downshift cable (automatic transmission).

6 Remove all hoses and wires, making very careful note of how they are attached. Tags or coded pieces of tape will help.

7 Remove the carburetor mounting nuts and/or bolts and separate the carburetor from the manifold.

8 Remove the gasket and/or EFE insulator.

9 Installation is the reverse of the removal procedure, but the following points should be noted:

a) By filling the carburetor bowl with fuel, the initial start-up will be easier and less drain on the battery will occur.
b) New gaskets should be used.
c) Idle speed and mixture settings should be checked and adjusted, if necessary.

8 Carburetor (E2SE) — overhaul

Note: *Carburetor overhaul is an involved procedure that requires some experience. The home mechanic without much experience should have the overhaul done by a dealer service department or repair shop. Because of running production changes, some details of the unit overhauled here may not exactly match those of your carburetor, although the home mechanic with previous experience should be able to detect the differences and modify the procedure.*

Disassembly

1 Before disassembling the carburetor, purchase a carburetor rebuild kit for your particular model (the model number will be found on a metal tag at the side of the carburetor). This kit will have all the necessary replacement parts for the overhaul procedure.

2 It will be necessary to have a relatively large, clean workbench to lay out all of the parts as they are removed. Many of the parts are very small and can be lost easily if the work space is cluttered.

3 Carburetor disassembly is illustrated in the following step-by-step photo sequence to make the operation as easy as possible. Work slowly through the procedure and if at any point you feel the reassembly of a certain component may prove confusing, stop and make a rough sketch or apply identification marks. The time to think about reassembling the carburetor is when it is being taken apart. The disassembly photo sequence begins with photo 8.3/1.

4 The final step in disassembly involves the idle mixture needle. It is recessed in the throttle body and sealed with a hardened steel plug. The plug should not be removed unless the needle requires replace-

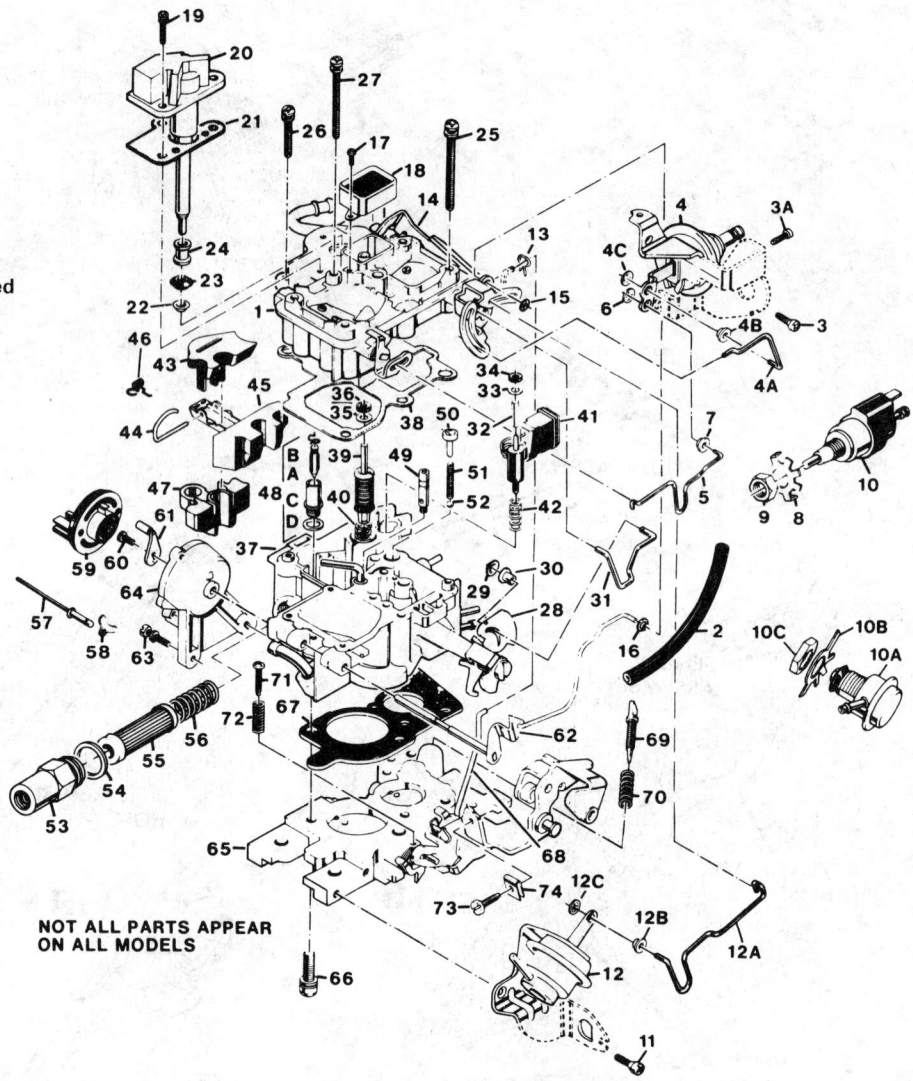

Fig. 4.5 E2SE carubretor — exploded view (Sec 8)

1 *Air horn assembly*
2 *Primary side vacuum break hose*
3 *Primary side vacuum break attaching screw*
3A *Primary side vacuum break attaching screw*
4 *Primary side vacuum break assembly*
4A *Primary side vacuum break line*
4B *Primary side vacuum break link bushing*
4C *Primary side vacuum break link retainer*
5 *Air valve link*
6 *Air valve link retainer*
7 *Air valve link bushing*
8 *Idle stop solenoid nut retainer*
9 *Idle stop solenoid nut*
10 *Idle stop solenoid*
10A *Throttle lever actuator*
10B *Throttle lever actuator nut retainer*
10C *Throttle lever actuator nut*
11 *Secondary side vacuum break attaching screw*
12 *Secondary side vacuum break assembly*
12A *Secondary side vacuum break link*
12B *Secondary side vacuum break link bushing*
12C *Secondary side vacuum break link retainer*
13 *Pump link retainer*
14 *Pump lever*
15 *Intermediate choke shaft link retainer*
16 *Intermediate choke shaft link bushing*
17 *Vent stack attaching screw*
18 *Vent stack*
19 *Mixture control (M/C) solenoid attaching screw assembly*
20 *M/C solenoid*
21 *Solenoid-to-air horn gasket*
22 *M/C solenoid seal retainer*
23 *Solenoid-to-float bowl large screw assembly*
24 *M/C solenoid spacer*
25 *Air horn-to-float bowl large screw assembly*
26 *Air horn-to-float bowl short screw assembly*
27 *Air horn-to-float bowl long screw assembly*
28 *Fast idle cam*
29 *Fast idle cam link retainer*
30 *Fast idle cam link bushing*
31 *Fast idle cam link*
32 *Throttle position sensor actuator plunger*
33 *TPS plunger seal retainer*
34 *TPS plunger seal*
35 *Pump stem seal retainer*
36 *Pump stem seal*
37 *Float bowl assembly*
38 *Air horn-to-float bowl gasket*
39 *Pump assembly*
40 *Pump return spring*
41 *Throttle position sensor (TPS)*
42 *TPS adjusting spring*
43 *Upper float bowl insert*
44 *Float hinge pin*
45 *Float and lever assembly*
46 *Float stabilizing spring*
47 *Lower float bowl insert*
48 *Needle and seat assembly*
A *Float needle*
B *Float needle pull clip*
C *Float needle seat*
D *Float needle seat gasket*
49 *Jet and lean mixture needle assembly*
50 *Pump discharge spring guide*
51 *Pump discharge spring*
52 *Pump discharge ball*
53 *Fuel inlet nut*
54 *Fuel inlet nut gasket*
55 *Fuel inlet filter*
56 *Fuel filter spring*
57 *Choke cover attaching rivet*
58 *Choke cover retainer*
59 *Electric choke cover and stat assembly*
60 *Choke stat lever attaching screw*
61 *Choke stat lever*
62 *Intermediate choke shaft, lever and link assembly*
63 *Choke housing attaching screw assembly*
64 *Choke housing*
65 *Throttle body assembly*
66 *Throttle body-to-float bowl screw assembly*
67 *Throttle body-to-float bowl gasket*
68 *Pump link*
69 *Idle mixture needle*
70 *Idle mixture needle spring*
71 *Throttle stop screw*
72 *Throttle stop screw spring*
73 *Fast idle screw*
74 *Fast idle screw retainer*

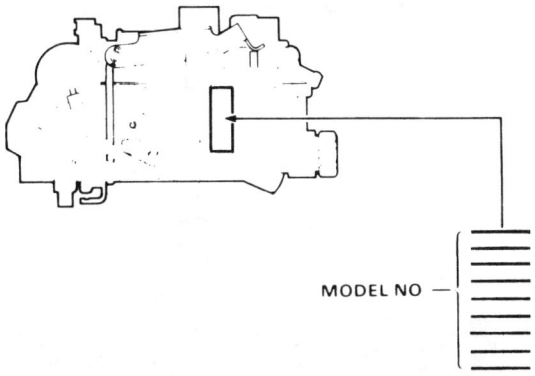

Fig. 4.6 Location of carburetor identification tag (Sec 8)

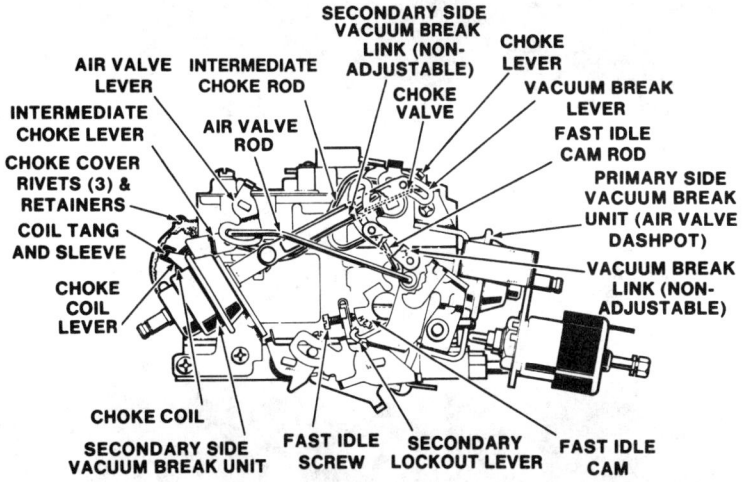

Fig. 4.7 E2SE carburetor choke system (Sec 8)

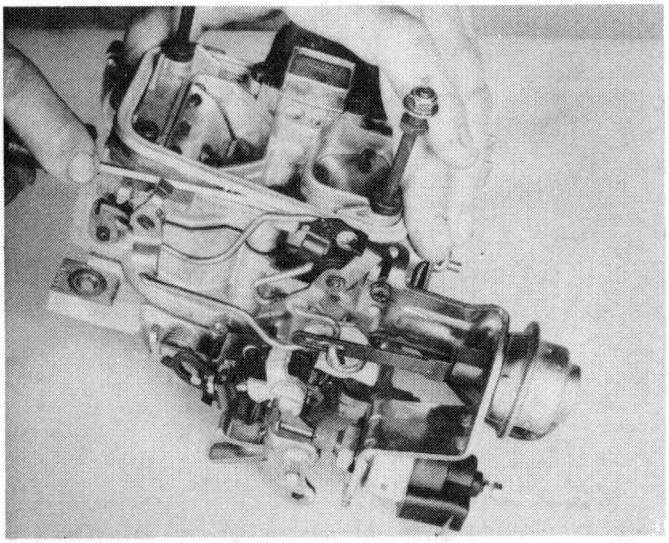

8.3/1 Remove the gasket from the top of the air horn

8.3/2 Remove the fuel inlet nut, fuel filter and spring

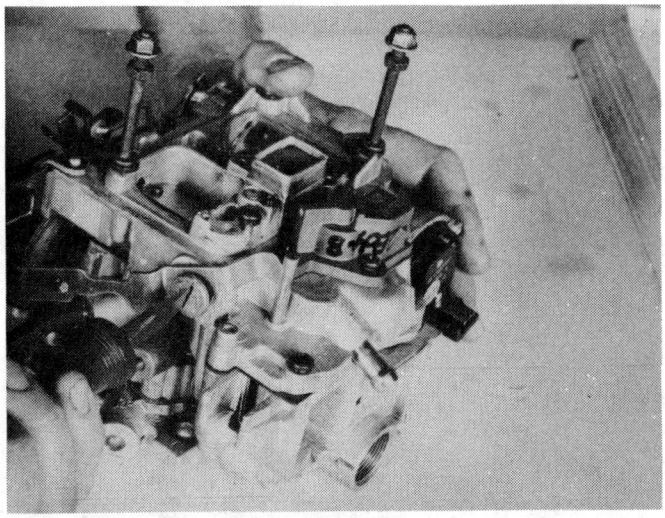

8.3/3 Remove the pump lever attaching screw

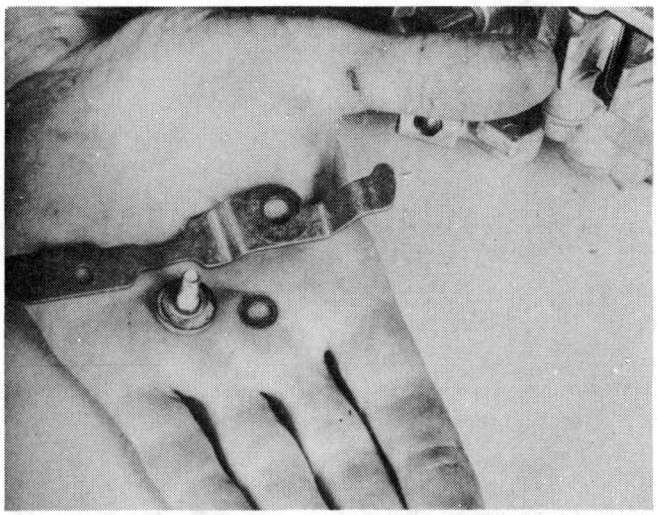

8.3/4 Disconnect the pump rod from the pump lever and remove the pump lever

8.3/5 Disconnect the primary vacuum break diaphragm hose from the throttle body

8.3/6 Remove the screws that retain the idle speed solenoid/vacuum break diaphragm bracket

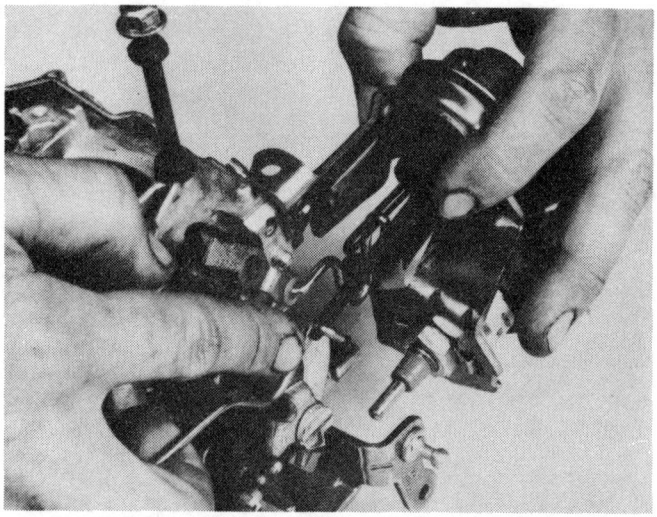

8.3/7 Lift off the idle speed solenoid/vacuum break diaphragm assembly and disconnect the air valve link from the vacuum break plunger (repeat this step for the secondary vacuum break assembly, disconnecting the link from the slot in the choke lever)

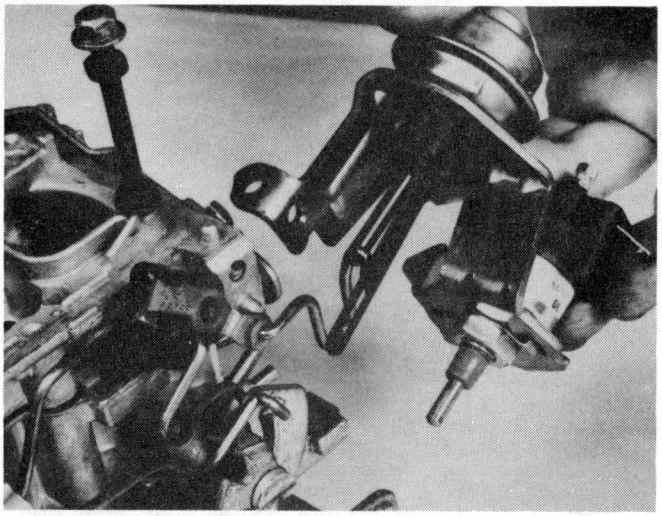

8.3/8 Disconnect the vacuum break and air valve links from the levers (**Note:** *It is not necessary to disconnect the links from the vacuum break plungers unless either the rods or vacuum break units are being replaced)*

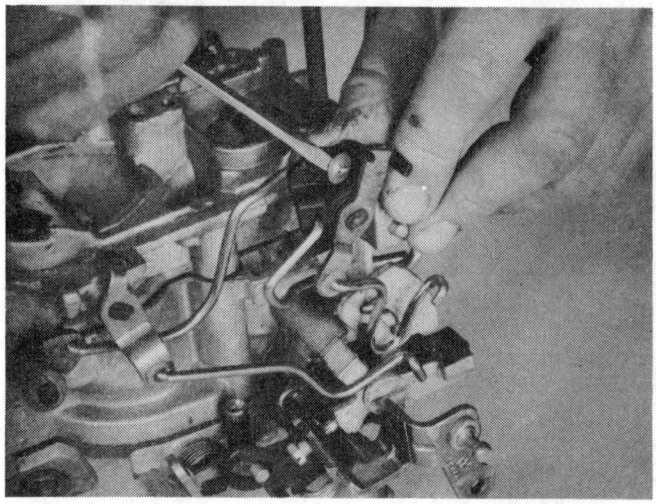

8.3/9 Pry off the clip that retains the intermediate choke link to the choke lever and separate the link from the lever

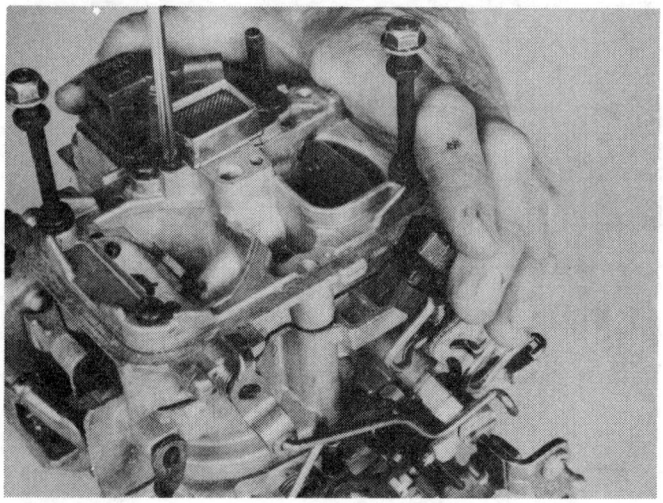

8.3/10 Remove the screws that retain the vent/screen assembly to the air horn and lift off the assembly

8.3/11 Remove the screws that retain the mixture control solenoid and, using a slight twisting motion, lift the solenoid out of the air horn

8.3/12 Remove the screws securing the air horn to the float bowl, noting their various lengths and positions to simplify installation

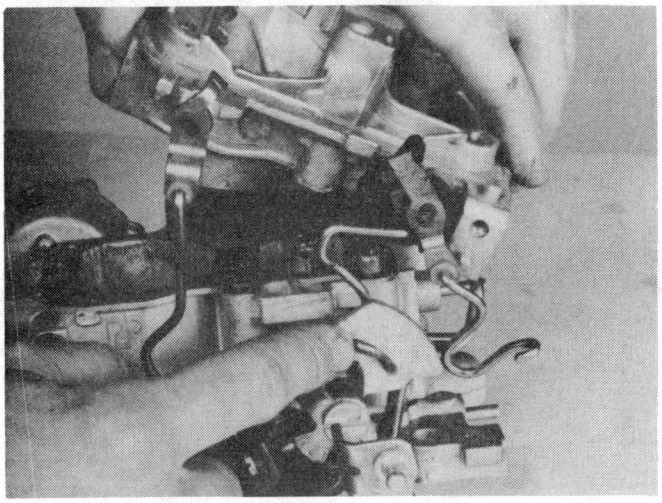

8.3/13 Rotate the fast idle cam up, lift off the air horn and disconnect the fast idle cam link from the fast idle cam

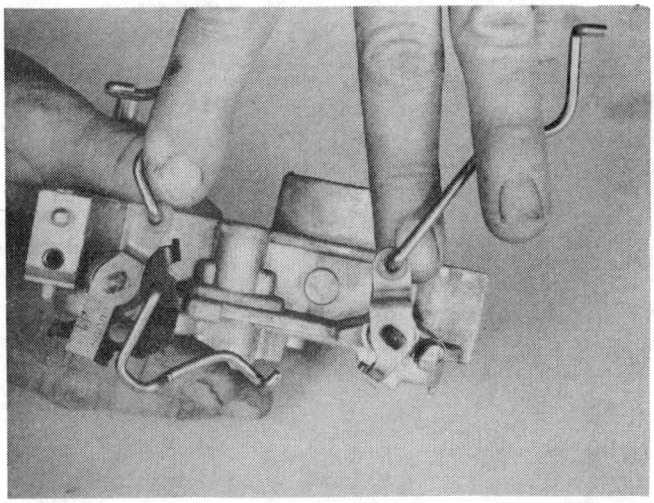

8.3/14 The links attached to the air horn need not be removed unless their replacement or removal is required to service other components

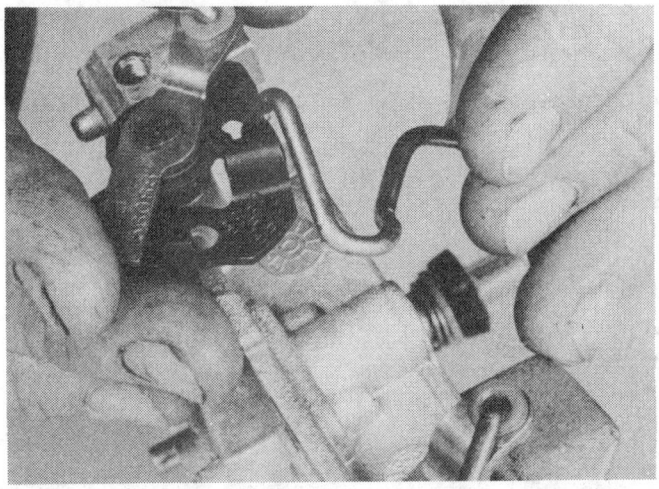

8.3/15 Disengage the fast idle cam link from the choke lever and save the bushing for reassembly

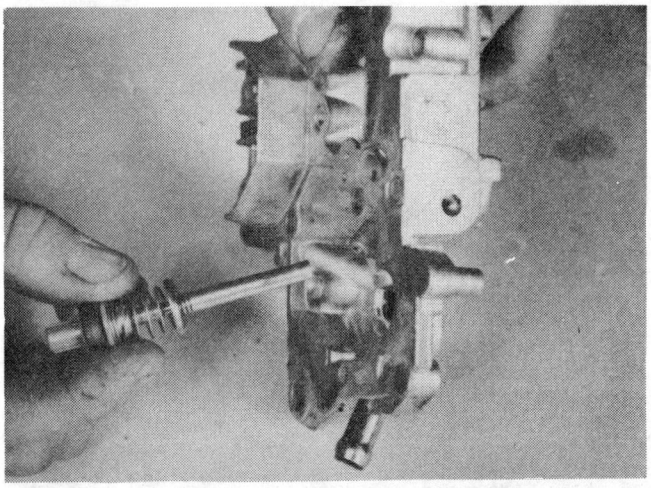

8.3/16 Remove the pump plunger from the air horn or the pump well in the float bowl (**Note:** *For throttle position sensor (TPS) equipped carburetors, refer to the exploded-view drawing at this time and remove the TPS)*

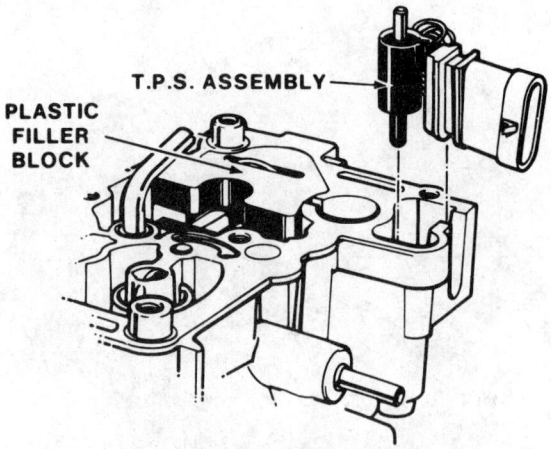

Fig. 4.8 TPS removal on E2SE carburetors (push up from the bottom of the electrical connector and remove the TPS and connector assembly from the float bowl; also remove the spring from the bottom of the bloat bowl) (Sec 8)

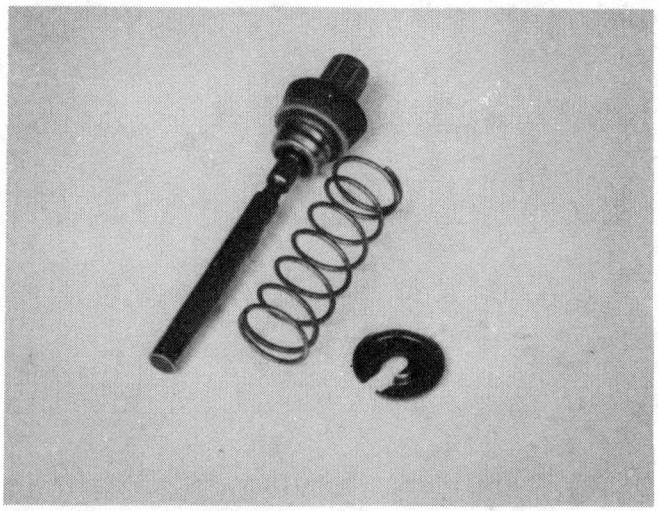

8.3/17 Compress the pump plunger spring and separate the spring retainer clip and spring from the piston

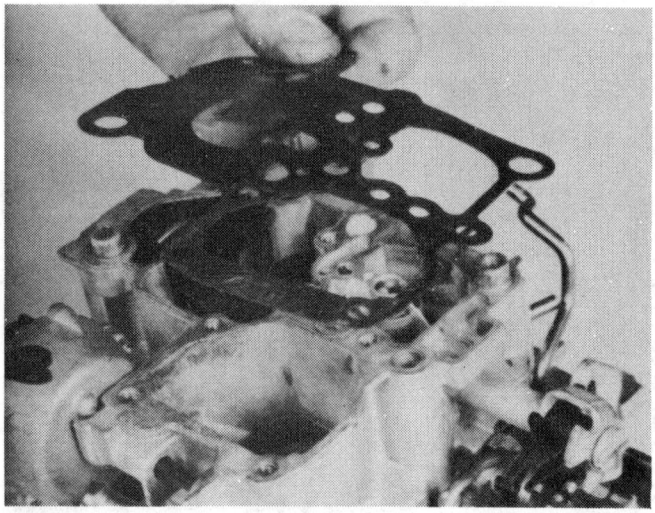

8.3/18 Remove the air horn gasket from the float bowl

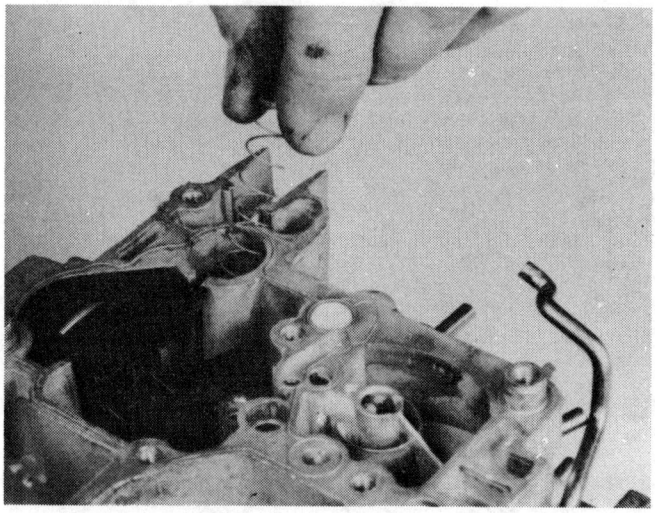

8.3/19 Remove the pump return spring from the pump well

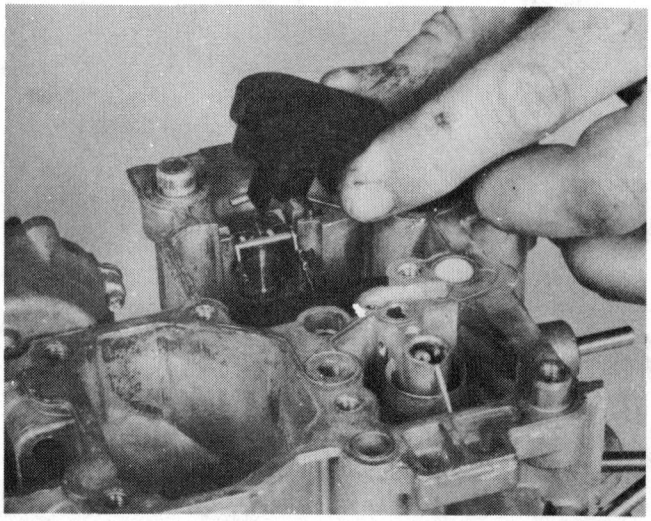

8.3/20 Remove the plastic filler block that covers the float valve

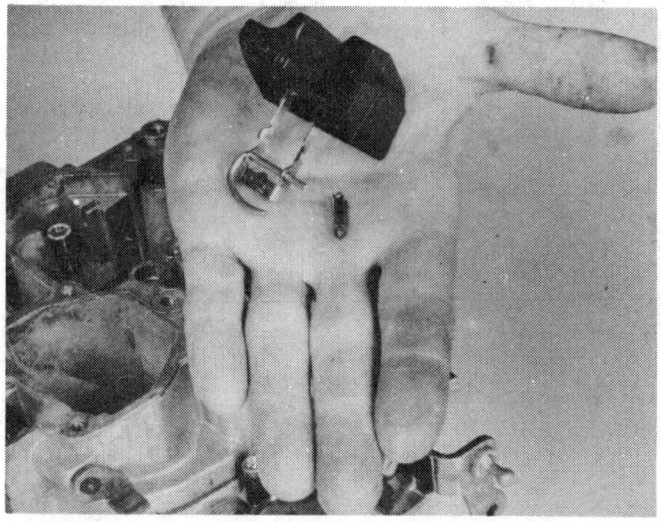

8.3/21 Remove the float and lever assembly, float valve and stabilizing spring (if used) by pulling up on the hinge pin

8.3/22 Remove the float valve seat and gasket (left) and the extended metering jet (right) from the float bowl

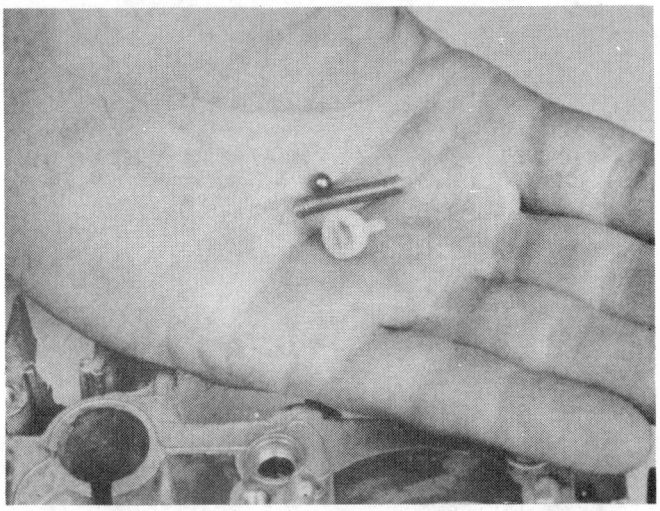

8.3/23 Using the needle-nose pliers, pull out the white plastic retainer and remove the pump discharge spring and check ball (do not pry on the retainer to remove it)

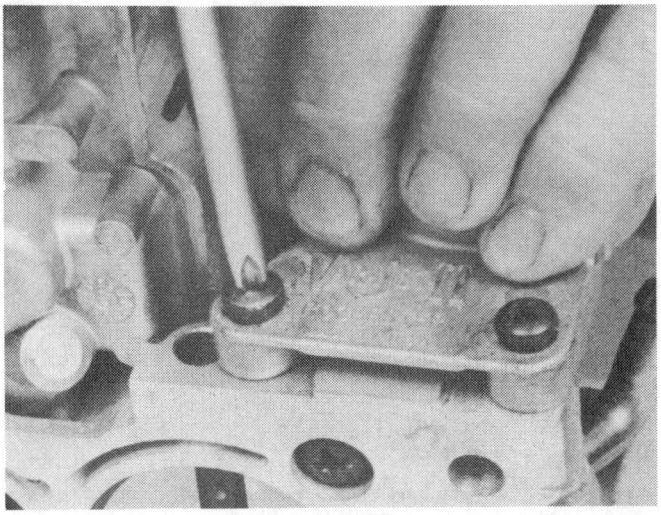

8.3/24 Remove the screws that retain the choke housing to the throttle body

8.3/25 Remove the screws that retain the float bowl to the throttle body

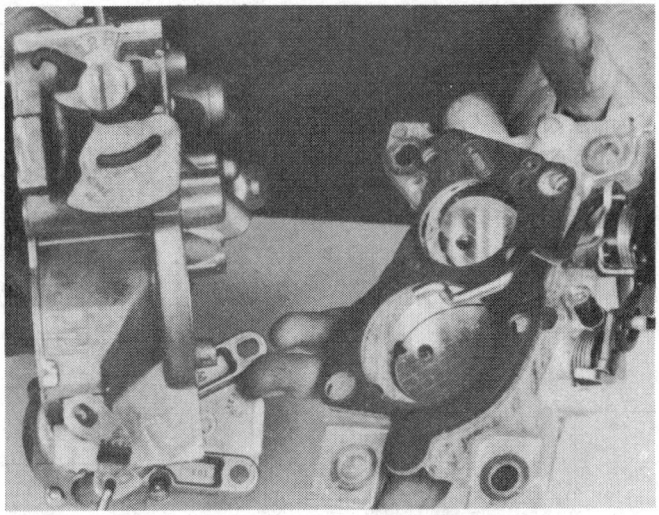

8.3/26 Separate the float bowl from the throttle body

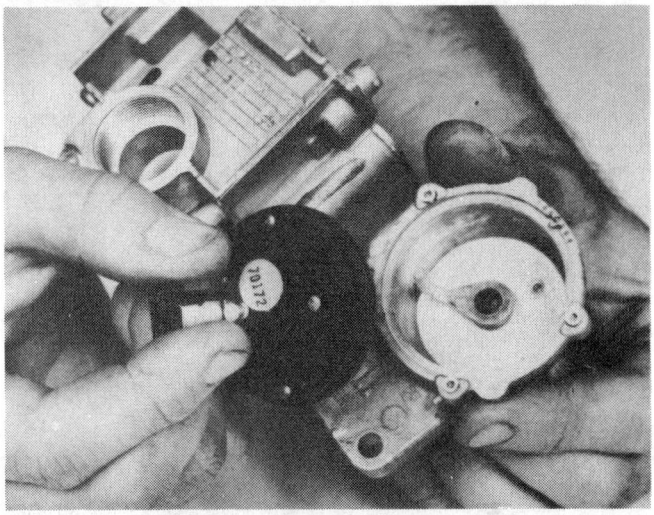

8.3/27 Carefully file the heads off the pop rivets that retain the choke cover to the choke housing, remove the cover and tap out the remainder of the rivets

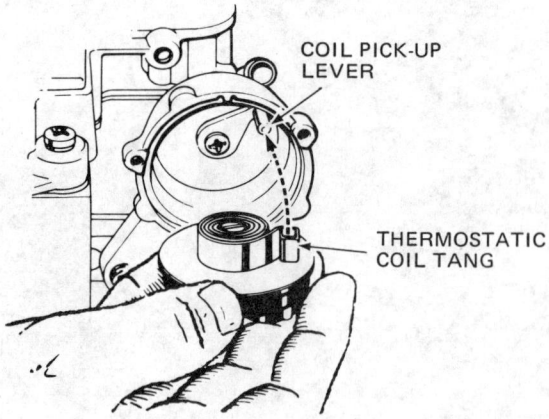

Fig. 4.9 Details of the E2SE choke housing assembly (SEC 8)

8.3/28 Remove the choke coil lever screw and lift out the lever

8.3/29 Remove the intermediate shaft and lever assembly by sliding it out the lever side of the float bowl (**Note:** *For further procedures, refer to Step 4 in the text)*

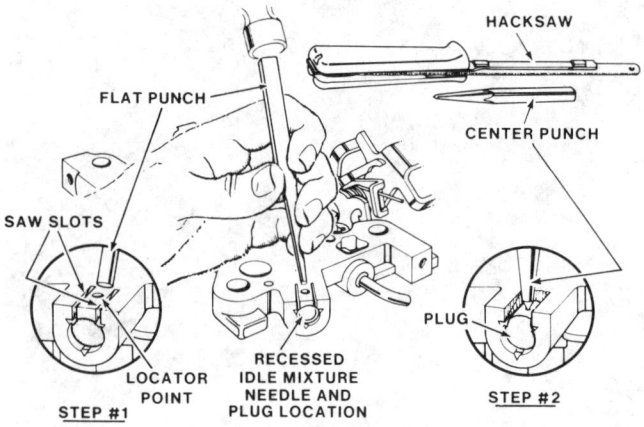

Fig. 4.10 E2SE carburetor idle mixture needle plug removal details (Sec 8)

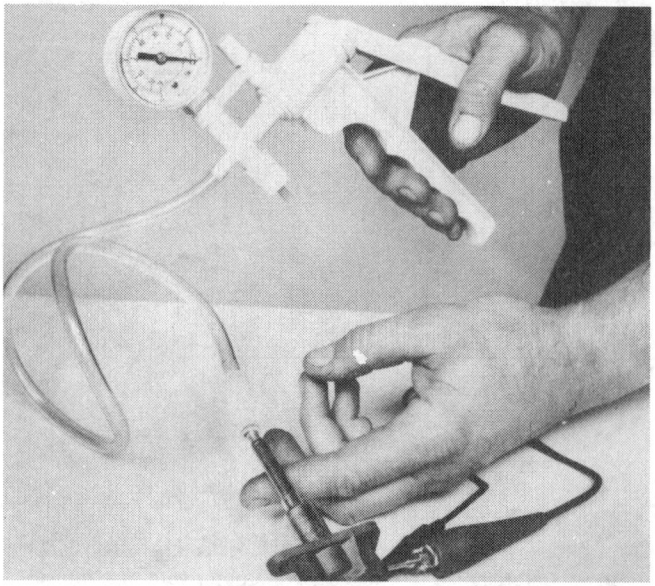

8.22 Attaching a hand vacuum pump to the end of the mixture control solenoid

8.25 Installing the gasket on the bottom of the float bowl

ment or normal cleaning procedures fail to clean the idle mixture passages. If the idle mixture needle must be removed, refer to the accompanying illustration and proceed as follows.

5 Secure the throttle body in a vise so it is inverted with the manifold side up. Use blocks of wood to cushion the throttle body.

6 Locate the idle mixture needle and plug. It should be marked by an indented locator point on the underside of the throttle body. Using a hacksaw, make two parallel cuts in the throttle body on either side of the locator mark. The cuts should be deep enough to touch the steel plug, but should not extend more than 1/8-inch beyond the locator point.

7 Position a flat punch at a point near the ends of the saw marks. Holding it at a 45° angle, drive it into the throttle body until the casting breaks away, exposing the steel plug.

8 Use a center punch to make an indentation in the steel plug. Holding it at a 45° angle, drive the plug from the throttle body casting. **Note:** *If the plug breaks apart, be sure to remove all of the pieces.*

9 Use a 3/16-inch deep socket to remove the idle mixture needle and spring from the throttle body.

Cleaning and inspection

10 Clean the air horn, float bowl, throttle body and related components with clean solvent and blow them out with compressed air. A can of compressed air can be used if an air compressor is not available. *Do not use a piece of wire for cleaning the jets and passages.*

11 The idle speed solenoid, mixture control solenoid, Throttle Position Sensor, electric choke, pump plunger, diaphragm, plastic filler block and other electrical, rubber and plastic parts should *not* be immersed in carburetor cleaner because they will harden, swell or distort.

12 Make sure all fuel passages, jets and other metering components are free of burrs and dirt.

13 Inspect the upper and lower surfaces of the air horn, float bowl and throttle body for damage. Be sure all material has been removed.

14 Inspect all lever holes and plastic bushings for excessive wear and an out-of-round condition and replace them if necessary.

15 Inspect the float valve and seat for dirt, deep wear grooves and scoring and replace it if necessary.

16 Inspect the float valve pull clip for proper installation and adjust it if necessary.

17 Inspect the float, float arms and hinge pin for distortion and binding and correct or replace as necessary.

18 Inspect the rubber cup on the pump plunger for excessive wear and cracks.

19 Check the choke valve and linkage for excessive wear, binding and distortion and correct or replace as necessary.

20 Inspect the choke vacuum diaphragm for leaks and replace if necessary.

21 Check the choke valve for freedom of movement.

22 Check the mixture control solenoid in the following manner.

a) Connect one end of a jumper wire to either end of the solenoid connector and the other end to the positive terminal of the battery.

b) Connect another jumper wire between the other terminal of the solenoid connector and the negative terminal of the battery.

c) Remove the rubber seal and retainer from the end of the solenoid stem and attach a hand vacuum pump to it (photo).

d) With the solenoid fully energized (lean position), apply at least 25 in-Hg of vacuum and time the leak-down rate from 20 to 15 in-Hg. The leak-down rate should not exceed 5 in-Hg in five (5) seconds. If leakage exceeds that amount, replace the solenoid.

e) To check if the solenoid is sticking in the down position, again apply about 25 in-Hg of vacuum to it, then disconnect the jumper lead to the battery and watch the pump gauge reading. It should fall to zero in less than one (1) second.

Reassembly

23 Before reassembling the carburetor, compare all old and new gaskets back-to-back to make sure they match perfectly. Check especially that all the necessary holes are present and in the proper positions in the new gaskets.

24 If the idle mixture needle and spring have been removed, reinstall them by lightly seating the needle, then back it off three (3) turns. This will provide a preliminary idle mixture adjustment. Final idle mixture adjustment must be made on the vehicle. Proper adjustment must be done using special emission sensing equipment, making it impractical for the home mechanic. To have the mixture settings checked or readjusted, take your vehicle to a GM dealer or other qualified mechanic with the proper equipment.

25 Install a new gasket on the bottom of the float bowl (photo).

26 Mount the throttle body on the float bowl so it is properly in-

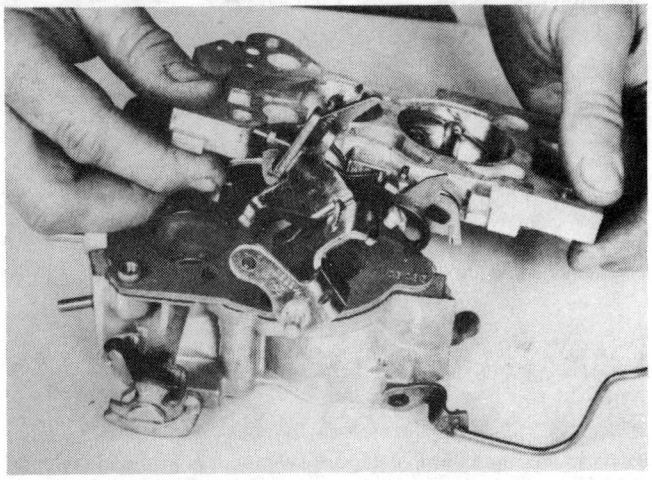

8.26A Mounting the throttle body on the float bowl

8.26B Installing the throttle body-to-float bowl attaching screws

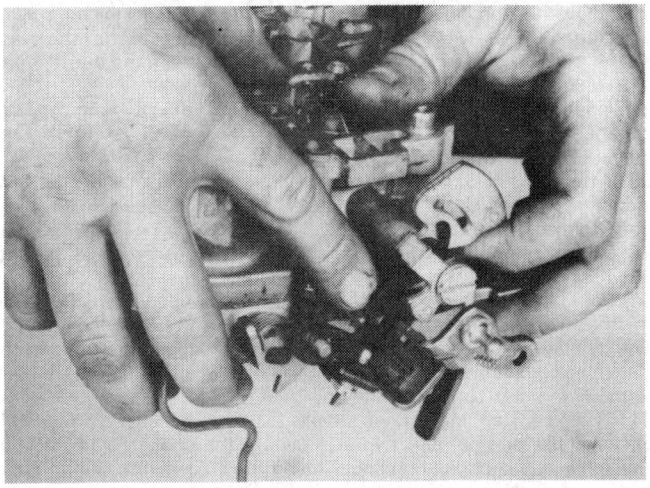

8.27 Checking the engagement of the lockout tang in the secondary lockout lever

8.28A Attaching the choke housing to the throttle body

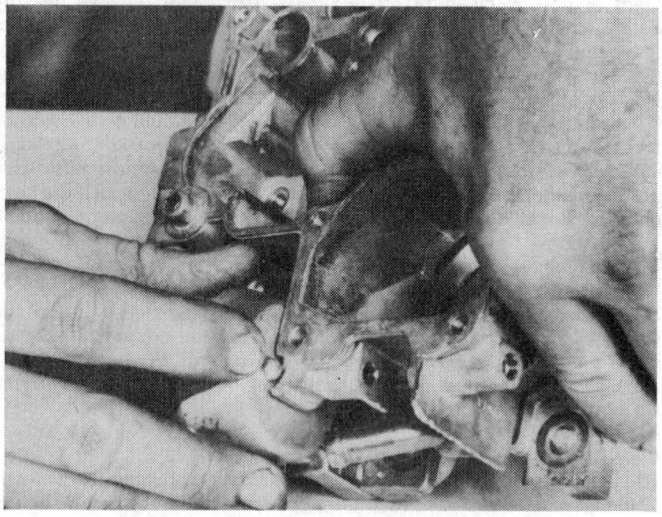

8.28B The lug at the rear of the choke housing should sit in the bowl recess

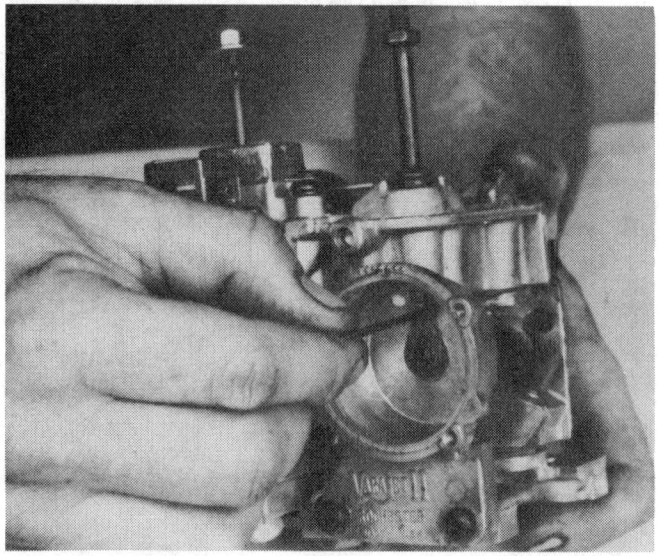

8.30 Install the thermostatic coil lever so it is in the 12 o'clock position when the intermediate choke lever is facing up

8.31 The choke cover is reinstalled with the self-tapping screws supplied in the overhaul kit

stalled over the locating dowels on the bowl (photo), reinstall the screws and tighten them evenly and securely (photo). Be sure that the steps on the fast idle cam face toward the fast idle screw on the throttle lever when installed.

27 Inspect the linkage to make sure that the lockout tang properly engages in the slot of the secondary lockout lever and that the linkage moves freely without binding (photo).

28 Attach the choke housing to the throttle body, making sure the locating lug on the rear of the housing sits in the recess in the float bowl (photos).

29 Install the intermediate choke shaft and lever assembly in the float bowl by pushing it through from the throttle lever side.

30 Position the intermediate choke lever in the up position and install the thermostatic coil lever on the end sticking into the choke housing. The coil lever is properly aligned when the coil pick-up tang is in the 12 o'clock position (photo). Install the screw in the end of the intermediate shaft to secure the coil lever.

31 Three self-tapping screws supplied in the overhaul kit are used in place of the original pop rivets to secure the choke cover and coil assembly to the choke housing. Thread the screws into the housing, making sure they start easily and are properly aligned (photo), then remove them.

32 Place the fast idle screw on the highest step of the fast idle cam, then install the choke cover on the housing, aligning the notch in the cover with the raised casting projection on the housing cover flange (photo). When installing the cover, be sure the coil pick-up tang engages the inside choke lever. **Note:** *The thermostatic coil tang is formed so that it will completely encircle the coil pick-up lever. Make sure the lever is inside of the tang when installing the cover (refer to the accompanying illustration).*

33 With the choke cover in place, install the self-tapping screws and tighten them securely.

34 Install the pump discharge check ball and spring in the passage next to the float chamber, then place a new plastic retainer in the hole so that its end engages the spring and tap it lightly into place until the retainer top is flush with the bowl surface.

35 Install the main metering jet in the bottom of the float chamber (photo).

36 Install the float valve seat assembly and gasket (photo).

37 To make float level adjustments easier, bend the float arm up slightly at the notch before installing the float (photo).

38 Install the float valve onto the float arm by sliding the lever under the pull clip. The correct installation of the pull clip is shown in the accompanying illustrations. Install the float pin in the float lever (photo). Install the float assembly by aligning the valve and seat and the float retaining pin and locating channels in the float bowl.

8.32 Be sure the notch in the choke cover is aligned with the raised casting projection on the housing cover flange

8.35 Installing the main metering jet

8.36 Installing the float valve seat assembly

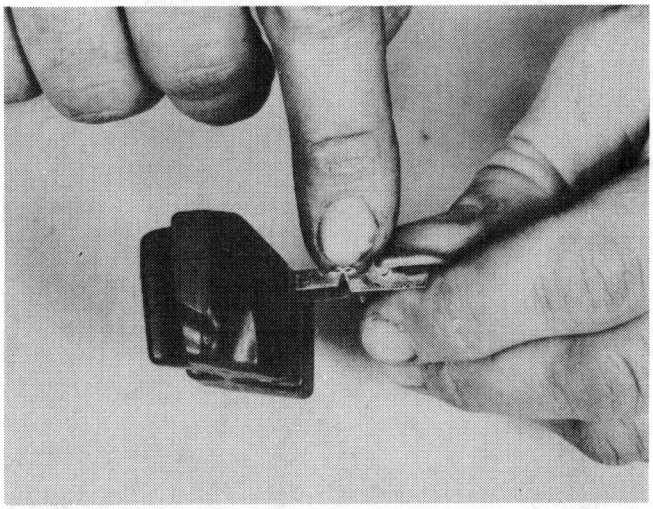

8.37 Prior to installation, bend the float arm up slightly at the point shown

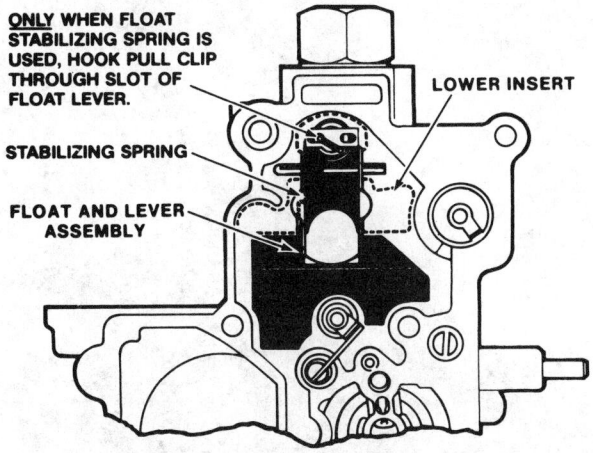

Fig. 4.11 Correct installation of the pull clip on E2SE floats *with* a stabilizing spring (Sec 8)

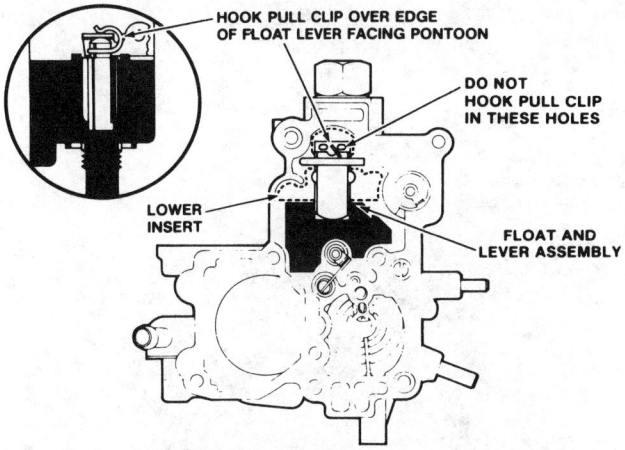

Fig. 4.12 Correct installation of the pull clip on E2SE floats *without* a stabilizing spring (Sec 8)

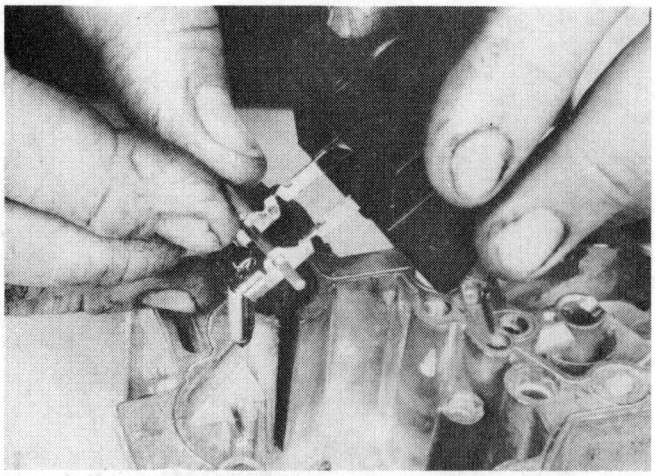

8.38 Installing the float retaining pin in the float lever

39 To adjust the float level, hold the float pin firmly in place, push down on the float arm at the outer end, against the top of the float valve, and see if the top of the float is the specified distance from the float bowl surface (photo). Bend the float arm as necessary to achieve the proper measurement by pushing down on the pontoon. See the Specifications for the proper float measurement for your vehicle. Check the float level visually following adjustment.

40 Install the plastic filler block over the float valve so that it is flush with the float bowl surface (photo).

41 If the carburetor is equipped with a Throttle Position Sensor, install the TPS return spring in the bottom of the well in the float bowl. Then install the TPS and connector assembly by aligning the groove in the electrical connector with the slot in the float bowl. When properly installed, the assembly should sit below the float bowl surface.

42 Install a new air horn gasket on the float bowl (photo).

43 Install the pump return spring in the pump well (photo).

44 Reassemble the pump plunger assembly, lubricate the plunger cap with a thin coat of engine oil and install the pump plunger in the pump well (photo).

45 If used, remove the old pump plunger seal and retainer and the old TPS plunger seal and retainer from the air horn. Install new seals and retainers in both locations and lightly stake both seal retainers in three (3) places other than the original staking locations.

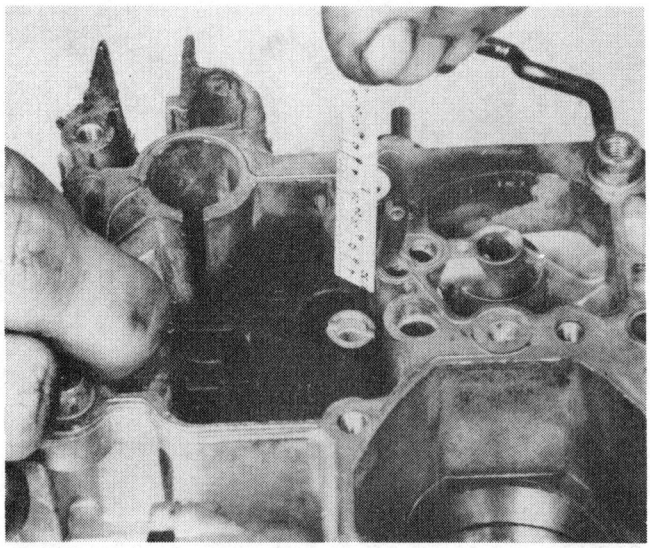

8.39 Measuring the float level

8.40 Installing the plastic float block

8.42 Installing a new air horn gasket on the float bowl

8.43 Installing the pump return spring in the pump well

46 Install the fast idle cam rod in the lower hole of the choke lever.
47 If so equipped, apply a light coat of silicone grease or engine oil to the TPS plunger and push it through the seal in the air horn so that about one-half of the plunger extends above the seal.
48 Before installing the air horn, apply a light coat of silicone grease or engine oil to the pump plunger stem to aid in slipping it through the seal in the air horn.
49 Rotate the fast idle cam to the Up position so it can be engaged with the lower end of the fast idle cam rod (photo). While holding down on the pump plunger assembly, carefully lower the air horn onto the float bowl and guide the pump plunger stem through the seal.
50 Install the air horn retaining screws and washers, making sure the different length screws are inserted into their respective holes, then tighten them in the sequence illustrated.
51 If so equipped, install a new seal in the recess of the float bowl and attach the hot idle compensator valve.
52 Install a new rubber seal on the end of the mixture control solenoid stem until it is up against the boss on the stem (photo).
53 Using a 3/16-inch socket and a hammer (photo), drive the retainer over the mixture control solenoid stem just far enough to retain the rubber seal, while leaving a slight clearance between them for seal expansion.
54 Apply a light coat of engine oil to the rubber seal and, using a new gasket, install the mixture control solenoid in the air horn. Use a slight

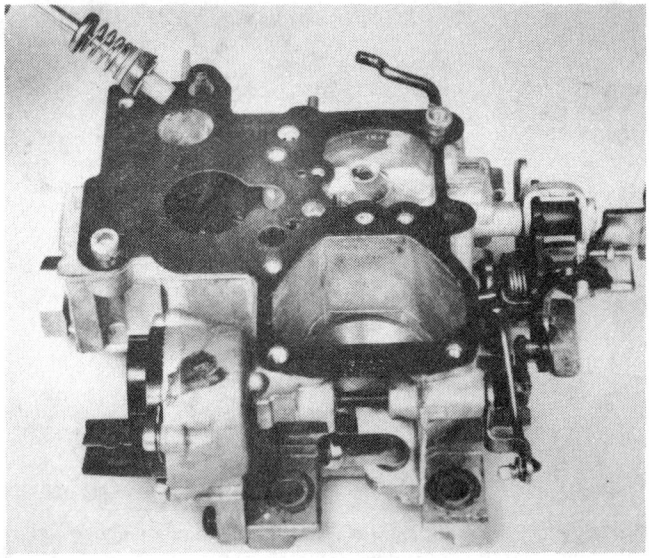

8.44 Installing the pump plunger in the pump well

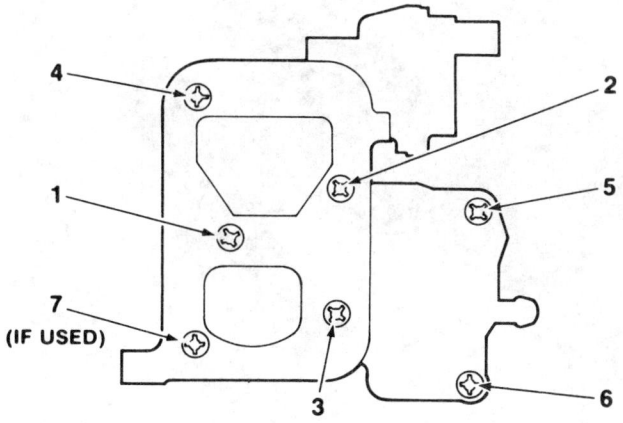

Fig. 4.13 Recommended air horn screw tightening sequence for E2SE carburetors (Sec 8)

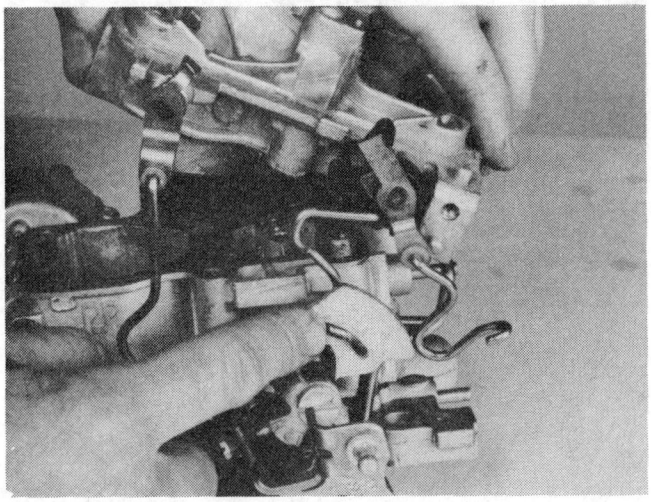

8.49 Engage the fast idle cam link in the fast idle cam prior to installation of the air horn

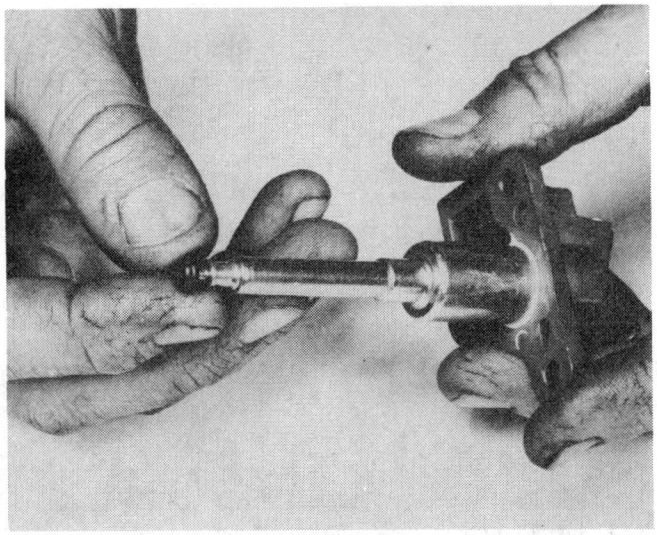

8.52 Attaching a new rubber seal to the end of the mixture control solenoid

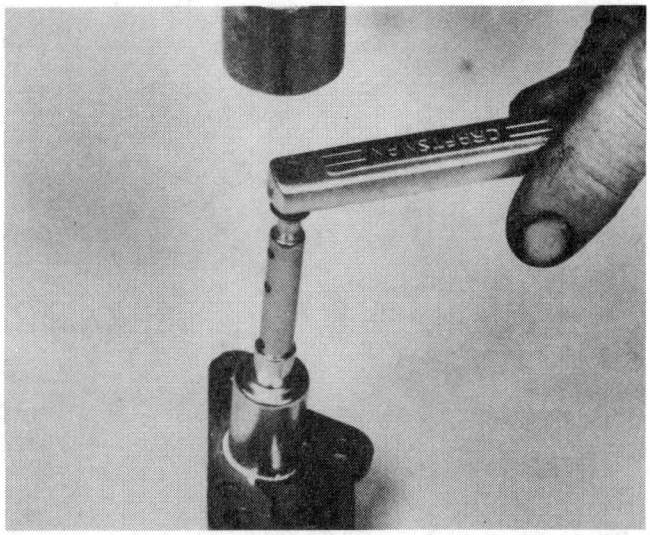

8.53 Using a hammer and hollow tool to tap the seal retainer onto the mixture control solenoid stem

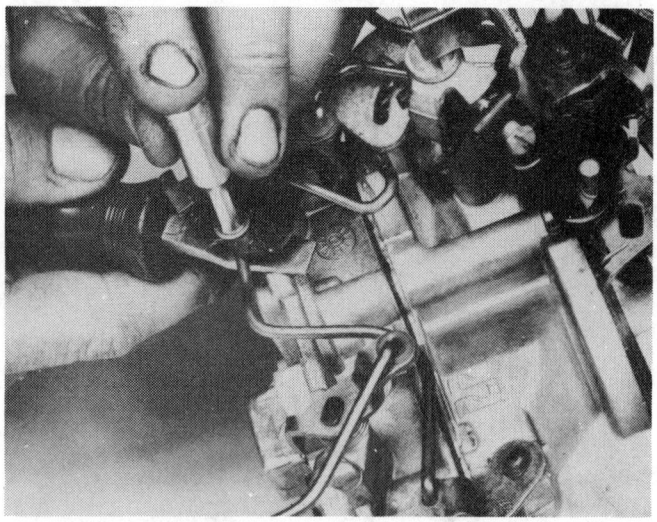

8.56 Attaching a retaining clip to the intermediate choke rod to secure it to the choke lever

twisting motion while installing the solenoid to help the rubber seal slip into the recess.

55 Install the vent/screen assembly on the air horn.

56 Install a plastic bushing in the hole in the choke lever, with the small end facing out, then, with the intermediate choke lever at the 12 o'clock position, install the intermediate choke rod in the bushing. Install a new retaining clip on the end of the rod. Use a broad flat-blade screwdriver and a 3/16-inch socket as shown in the photo. Make sure the clip is not seated tightly against the bushing and that the linkage moves freely.

57 Reattach the primary and secondary vacuum break links and install the vacuum break and idle speed solenoid assemblies (photo).

58 Engage the pump rod with the pump rod lever (photo), install a new retaining clip on the pump rod and install the pump lever on the air horn with the washer between the lever and the air horn (photo).

59 Reconnect the vacuum break hoses.

60 Install the fuel filter with the hole facing toward the inlet nut.

61 Place a new gasket on the inlet nut and install and tighten it securely. Take care not to over tighten the nut, as it could damage the gasket, leading to a fuel leak.

62 Install a new gasket on the top of the air horn (photo).

63 Check that all linkage hook-ups have been made and that they do not bind.

64 For external linkage adjustment procedures, refer to the accompanying illustrations.

8.57 Installing the idle speed solenoid/vacuum break diaphragm assembly

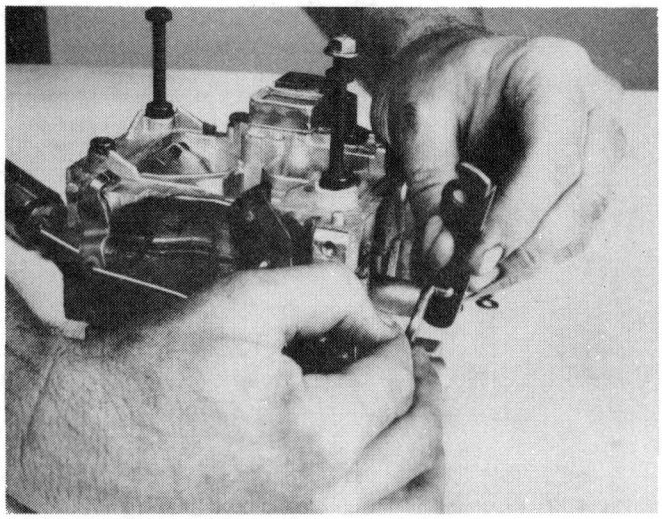

8.58A Engaging the pump rod with the pump rod lever

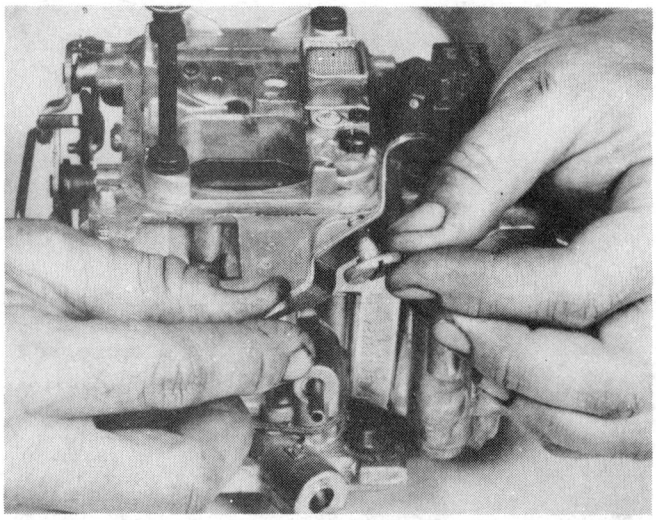

8.58B Inserting the pump rod mounting screw through the pump rod prior to installation

8.62 Installing a new gasket on top of the air horn

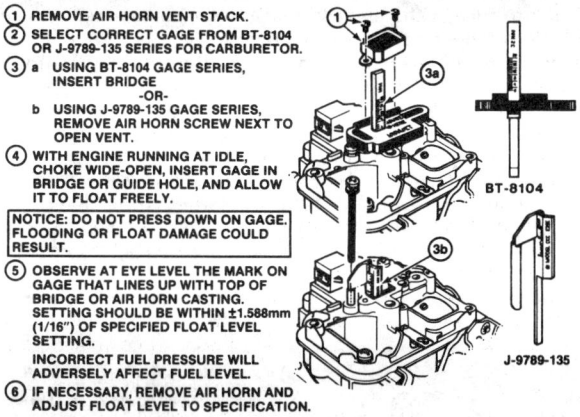

Fig. 4.14 E2SE carburetor external float check procedure (Sec 8)

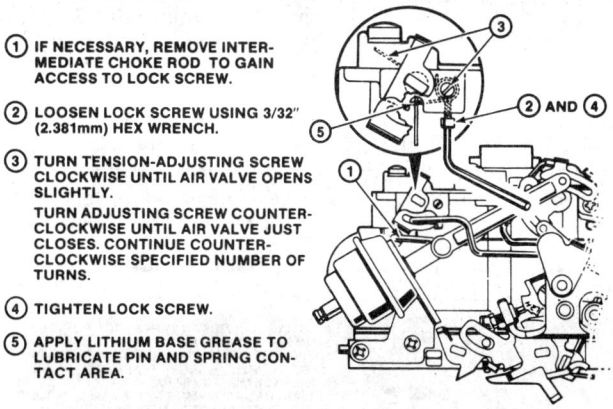

Fig. 4.15 E2SE carburetor air valve spring adjustment procedure (Sec 8)

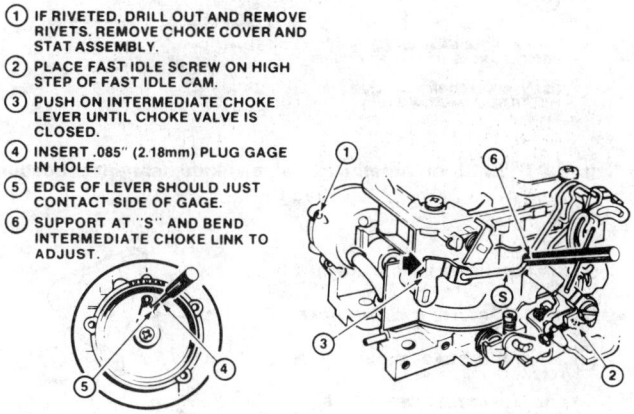

Fig. 4.16 E2SE carburetor choke coil lever adjustment procedure (typical) (Sec 8)

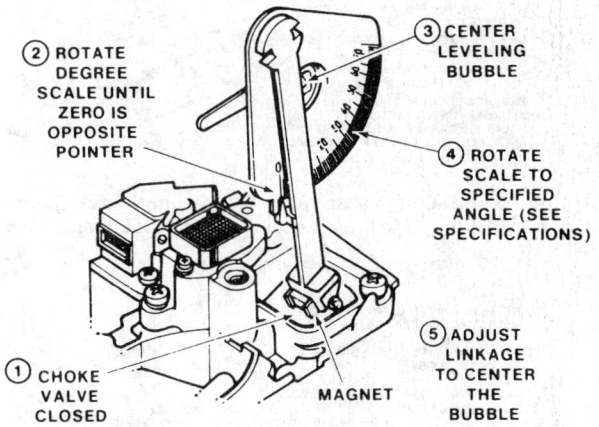

Fig. 4.17 Measuring the E2SE carburetor choke valve angle with an angle gauge (Sec 8)

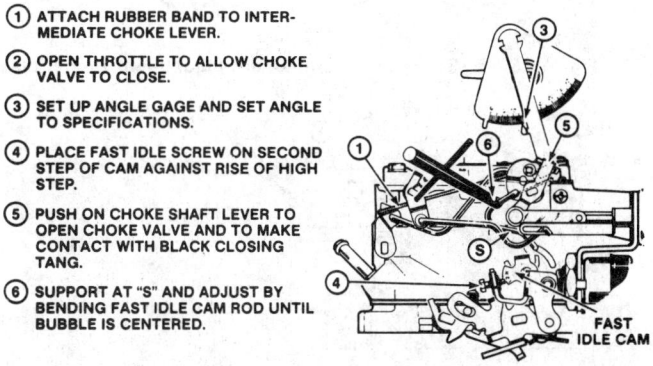

Fig. 4.18 E2SE carburetor choke rod fast idle cam adjustment procedure (Sec 8)

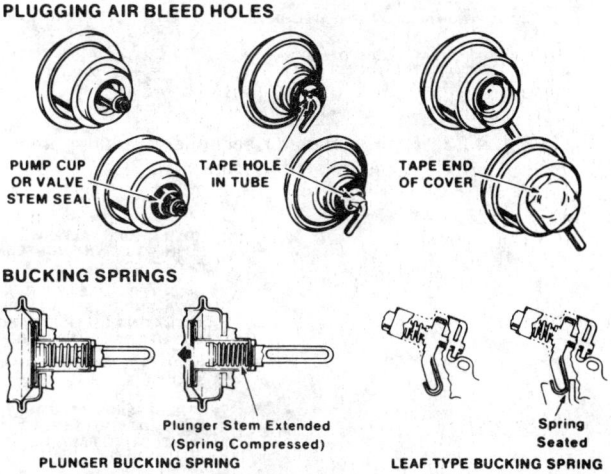

Fig. 4.19 E2SE carburetor vacuum break adjustment details (Sec 8)

9 Carburetor (E4ME and E4MC) — overhaul

Because of the complexity of these carburetors and the special tools required to overhaul and adjust them, it is recommended that major service work be left to a dealer service department or a carburetor repair specialist.

10 Electronic Fuel Injection (EFI) system — general information

All vehicles equipped with a four-cylinder engine employ EFI in place of the conventional carburetor-fed intake system. The specific fuel injection unit used is the Model 300 Throttle Body Injection (TBI). Fuel injection provides optimum mixture ratios at all stages of combustion. Combined with its immediate response characteristics, EFI permits the engine to run on the leanest possible fuel/air mixture, which greatly reduces exhaust gas emissions.

The EFI system is controlled directly by the vehicle's Electronic Control Module (ECM), which automatically adjusts the air/fuel mixture in accordance with engine load and performance. The main component of the EFI system is the Throttle Body Injector (TBI), which is mounted on the intake manifold just like a carburetor. The TBI is made up of two major assemblies: the throttle body and the fuel metering assembly.

The throttle body contains a single throttle valve, controlled by the accelerator pedal, again, similar to a carburetor. Attached to the exterior of the body are the throttle position sensor (TPS) which sends

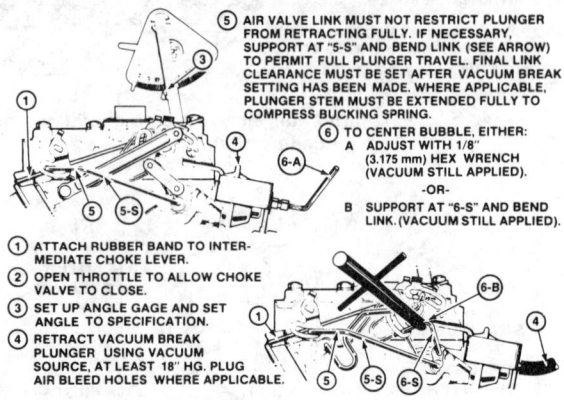

Fig. 4.20 E2SE carburetor primary vacuum break adjustment producure (for models with dual vacuum break units) (Sec 8)

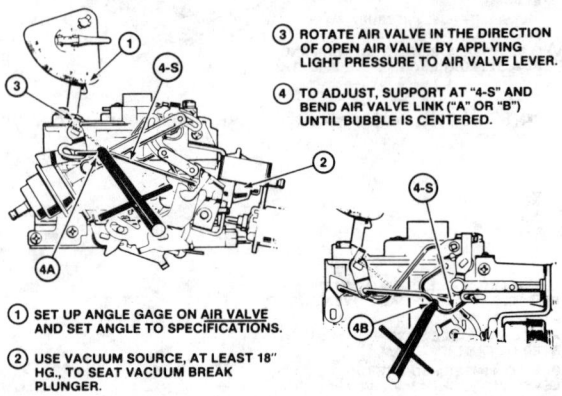

Fig. 4.21 E2SE carburetor air valve link adjustment procedure (Sec 8)

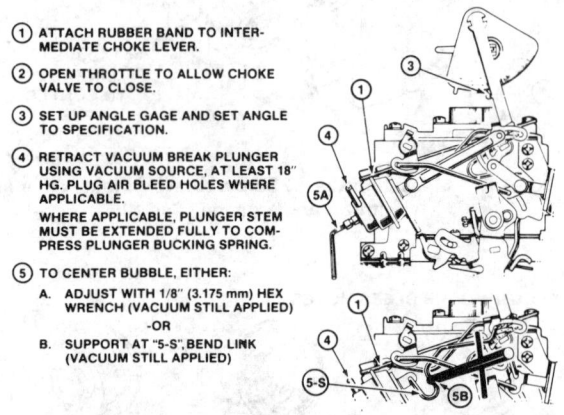

Fig. 4.22 E2SE carburetor secondary vacuum break adjustment procedure (Sec 8)

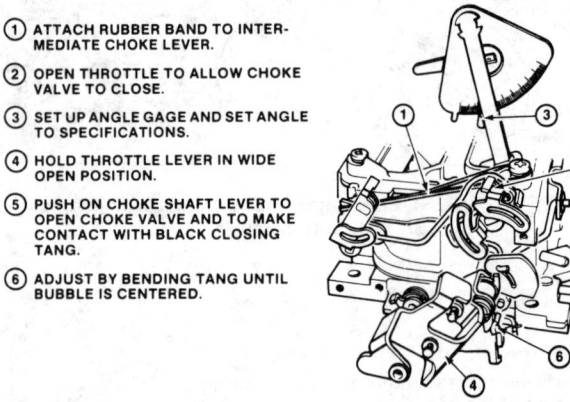

Fig. 4.23 E2SE carburetor unloader adjustment procedure (Sec 8)

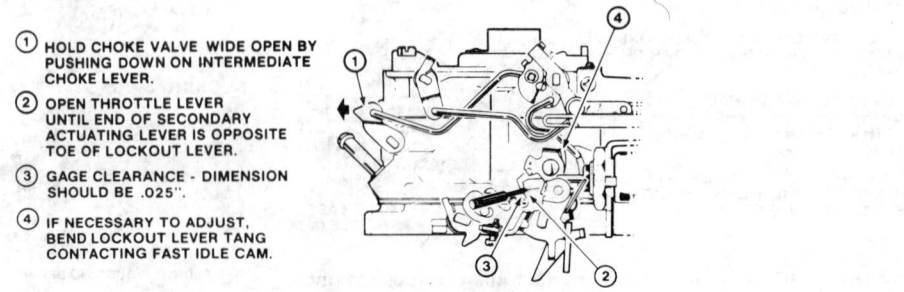

Fig. 4.24 E2SE carburetor secondary lockout adjustment procedure (Sec 8)

throttle position information to the ECM and the idle air control assembly (IAC), which is used by the ECM to maintain a constant idle speed during normal engine operation.

The fuel metering assembly contains the fuel pressure regulator and the single fuel injector. The regulator dampens the pulsations of the fuel pump and maintains a steady pressure at the injector. The fuel injector is controlled by the ECM through an electrically operated solenoid. The amount of fuel injected into the intake manfiold is varied by the length of time the injector plunger is held open.

The ECM controlling the EFI system has a learning capability for certain performance conditions. If the battery is disconnected, part of the ECM memory is erased, which makes it necessary to reteach the computer. This is done by thoroughly warming up the engine and operating the vehicle at part throttle, stop and go and idle conditions.

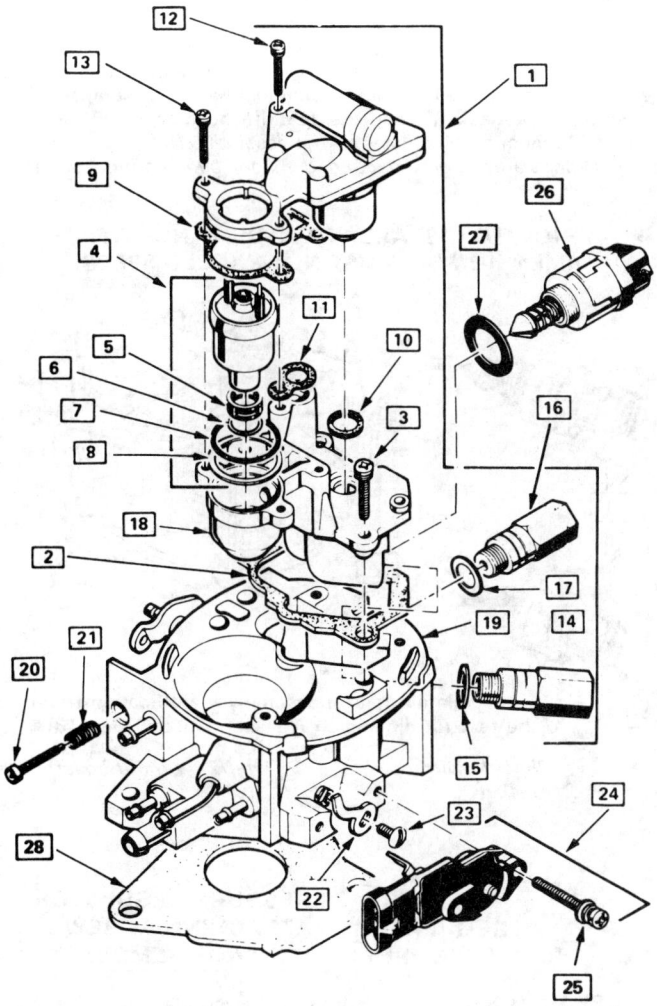

Fig. 4.25 Model 300 TBI assembly — exploded view

1 *Fuel meter assembly*
2 *Fuel meter body gasket*
3 *Fuel meter body attaching screw and washer (3)*
4 *Fuel injector assembly*
5 *Fuel injector filter*
6 *Small O-ring*
7 *Large O-ring*
8 *Fuel injector back-up ring washer*
9 *Fuel meter cover gasket*
10 *Pressure regulator dust seal*
11 *Fuel meter outlet gasket*
12 *Ruel meter cover long screw and washer (3)*
13 *Fuel meter cover short screw*
14 *Fuel inlet nut*
15 *Fuel inlet nut gasket*
16 *Fuel outlet nut*
17 *Fuel outlet nut gasket*
18 *Fuel meter body assembly*
19 *Throttle body assembly*
20 *Idle screw stop*
21 *Idle stop screw spring*
22 *Throttle position sensor (TPS) lever*
23 *TPS lever attaching screw*

No mechanical fuel pump is employed in this system. Rather, an electric pump, located in the fuel tank, is used. When the ignition is turned on, the fuel pump relay immediately supplies current to the fuel pump. If the engine doesn't start after two seconds, the fuel pump will automatically shut off to avoid flooding. If the fuel pump relay fails, the fuel pump will still operate after about four pounds of oil pressure has built up.

The fuel filter is an in-line type and is located at the rear of the engine on the left side of the vehicle.

The throttle stop screw, used to regulate the minimum idle speed, is adjusted at the factory and sealed with a plug to discourage unnecessary readjustment. For this reason, this adjustment, if necessary, should be left to a dealer or other qualified mechanic.

The complexity of the EFI system prevents many problems from being accurately diagnosed by the home mechanic. Therefore, if a problem should develop in the system, take the vehicle to a dealer to locate the fault.

Caution: *Prior to any operation in which a fuel line will be disconnected, the high pressure in the system must first be relieved as described in Section 11. Also, disconnect the negative battery cable to eliminate the possibility of sparks occurring when fuel vapors are present.*

11 Fuel injection pressure relief procedure

Caution: *To reduce the risk of fire and personal injury, relieve the pressure in the fuel system before servicing any fuel injection components.*

1 Remove the fuel pump fuse from the fuse block located in the passenger compartment under the left side of the dash.
2 Start the engine and let it run until lack of fuel causes it to stop.
3 Engage the starter again for about three (3) seconds to ensure that all pressure has been relieved.
4 Turn the ignition off and disconnect the negative cable from the battery.
5 After servicing the fuel system, reinstall the fuel pump fuse and reconnect the battery cable.

12 Fuel pump (fuel-injected models) — removal and installation

Note: *Before starting this procedure, the fuel system pressure must be relieved as outlined in Section 11.*

1 On fuel injected models, the electric fuel pump is an integral part of the fuel gauge tank unit, but it may be replaced separately after the fuel gauge/fuel pump assembly has been removed from the fuel tank.
2 Remove the fuel tank (refer to Section 5).
3 Unbolt and remove the fuel pump/gauge assembly from the fuel tank.
4 Installation is the reverse of the removal procedure.
5 When the installation is complete, carefully check all lines, hoses and fittings for leaks.

13 Idle Air Control (IAC) assembly — removal and installation

1 Relieve the pressure in the fuel system (refer to Section 11).
2 Disconnect the cable from the negative battery terminal.
3 Remove the air cleaner.
4 Disconnect the wire from the IAC assembly.
5 Use a wrench to unscrew the assembly from the TBI.
6 Before installing the IAC, measure the distance between the end of the motor assembly housing and the tip of the conical valve. The valve should be extended from the housing no more than 1.260-inch (32 mm) for 1982 models, or 1.125-inch (28 mm) for 1983 and 1984 models or damage may occur to the motor when it is installed. If the distance is excessive, first determine if the assembly is a Type 1 (with a collar at the end of the motor) or Type 2 (without a collar). The pintle on a Type 1 assembly can be retracted simply by pushing on the end of the cone until it's in position. For a Type 2, push the pintle in and attempt to turn it clockwise. If it turns, continue turning until it is properly set. If it will not turn, exert firm hand pressure to retract it. **Note:**

If the pintle was turned, be sure the spring is in its original position with the straight portion of the spring end aligned with the flat surface under the pintle head.

7 Installation of the IAC assembly is the reverse of removal. Be sure to install the gasket with the assembly.

8 Following installation, start the engine and allow it to reach normal operating temperature. On manual transmission-equipped vehicles, the idle speed will automatically be controlled when operating temperature is reached. On automatic transmission-equipped vehicles, the assembly will begin controlling idle speed when the engine is at operating temperature and the transmission is shifted into Drive.

9 If the idle speed is abnormally high and does not regulate back to normal after a few moments, operate the vehicle at a speed of 40 mph (64 kph). At that speed the ECM will command the IAC pintle to extend fully to the mating seat in the throttle body and allow the ECM to establish an accurate reference with respect to the pintle position. Proper idle regulation will result.

14 Fuel pressure regulator — removal and installation

1 Relieve the pressure in the fuel system (refer to Section 11).

2 Remove the cable from the negative battery terminal.

3 Remove the air cleaner assembly.

4 Disconnect the wire from the injector by squeezing the two tabs on the connector and pulling straight up.

5 The fuel meter cover assembly, including the fuel pressure regulator, is serviced only as a complete unit; if either the regulator or cover requires replacement, the entire assembly must be replaced. **Caution:** *The fuel pressure regulator is enclosed in the fuel meter cover. It is under heavy spring tension and the four screws securing the pressure regulator to the fuel meter cover should not be removed, as personal injury could result.*

6 Remove the five fuel meter cover mounting screws and separate the cover from the body. Note the positions of the two shorter screws because they must be replaced in the same positions.

7 Do not immerse the fuel meter cover in solvent or cleaner, as it will damage the pressure regulator diaphragm and gaskets.

8 Installation is the reverse of the removal procedure. Be sure to use new gaskets and a new regulator dust seal.

15 Injector — removal and installation

1 Relieve the pressure in the fuel system (refer to Section 11).

2 Remove the cable from the negative battery terminal.

3 Remove the fuel meter cover as described in Section 14.

4 With the fuel meter cover gasket still in place to prevent damage to the casting, place a screwdriver shank atop the gasket to serve as a fulcrum and carefully pry the injector from the fuel meter body with another screwdriver as shown in the accompanying illustration.

5 Do not push the injector out from underneath as this could damage the injector tip.

6 Do not immerse the injector in solvent or carburetor cleaner. If it is defective, the injector must be replaced as a unit.

7 Check and clean the fuel filter on the base of the injector. To remove the filter, carefully rotate it back-and-forth to pull it off. The filter is installed by pushing it into the injector until it's seated.

8 Whenever the injector is removed, new O-rings should be installed. Remove the large O-ring and steel backup washer from the injector cavity, then remove the small O-ring from the bottom of the cavity.

9 Lubricate a new small O-ring with lithium grease and push it over the nozzle end of the injector so it is seated against the fuel filter.

10 Install the steel backup ring washer in the injector cavity. Lubricate the large O-ring with lithium grease and install it in the cavity, directly above the washer. When properly installed, the O-ring is flush with the fuel meter body surface. **Note:** *Do not attempt to reverse this procedure by installing the backup washer and O-ring after the injector is located in the cavity, as this would prevent proper seating of the O-ring.*

11 Use a pushing and twisting motion to install the injector into the cavity. Be sure it is fully seated and that the raised lug in the injector base is aligned with the notch in the fuel meter body.

12 The remainder of the installation is the reverse of removal.

16 Throttle Body Injection (TBI) 300 assembly — removal and installation

1 Relieve the pressure in the fuel system (refer to Section 11).

2 Remove the cable from the negative battery terminal.

3 Remove the air cleaner.

4 Disconnect the wires leading to the IAC, throttle position sensor and injector included in the TBI assembly.

5 Disconnect the throttle linkage and return spring from the TBI.

6 If so equipped, disconnect the cruise control linkage from the TBI.

7 Use tape to label the installed locations of all vacuum hoses leading to the TBI, then disconnect them.

8 Disconnect the fuel lines from the TBI.

9 Remove the three (3) TBI mounting bolts and lift it off, along with the gasket.

10 Prior to installing the TBI, be sure the intake manifold mating surface is clean and that all old gasket material has been scraped off.

11 Using a new gasket, place the TBI into position on the intake manifold. Connect the fuel lines to the assembly. **Note:** *Be sure the fuel line O-rings are not nicked, cut or damaged before connecting the lines.*

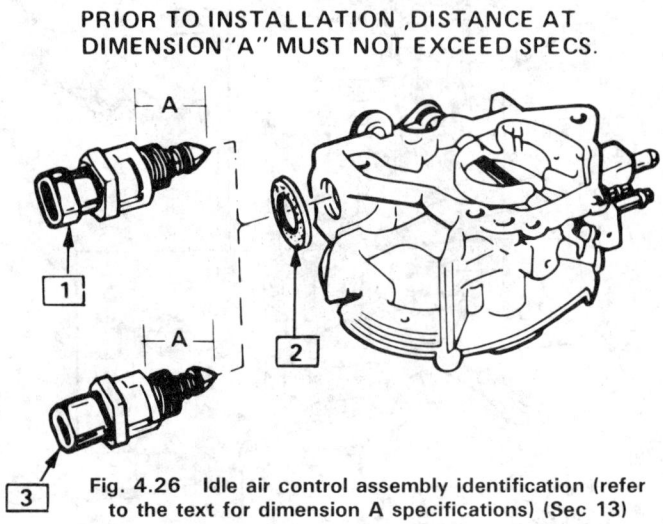

Fig. 4.26 Idle air control assembly identification (refer to the text for dimension A specifications) (Sec 13)

1 Type 1 (with collar)
2 Gasket
3 Type 2 (without collar)

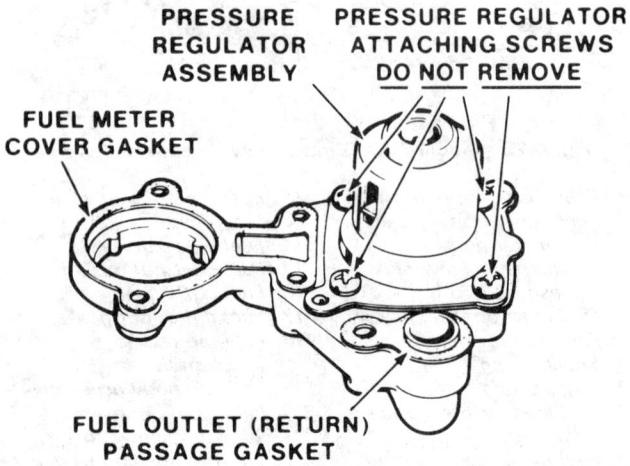

Fig. 4.27 Bottom view of the fuel meter body, showing the fuel pressure regulator (Sec 14)

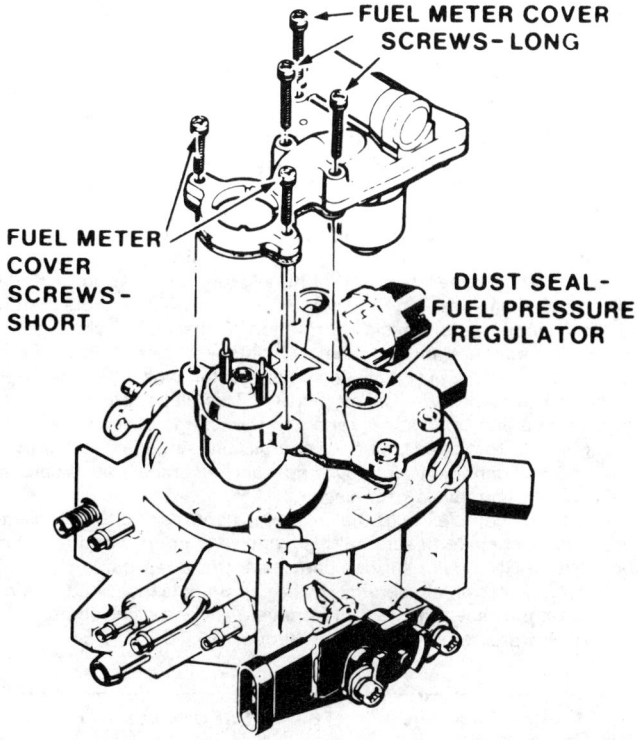

Fig. 4.28 Fuel meter cover-to-fuel meter body screw locations (Sec 14)

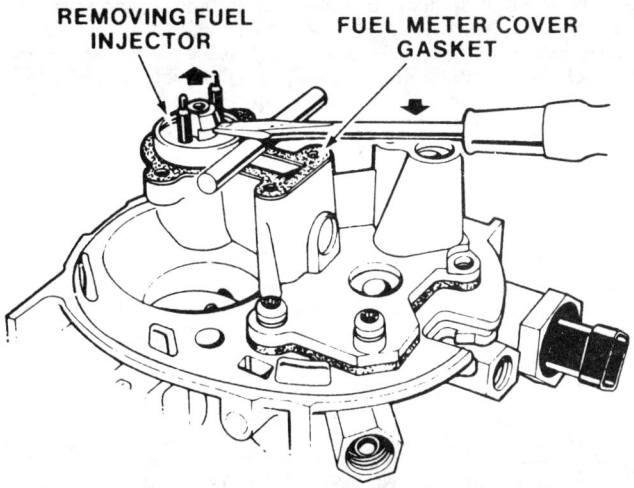

Fig. 4.29 Correct procedure for removing the fuel injector assembly from the fuel meter body (Sec 15)

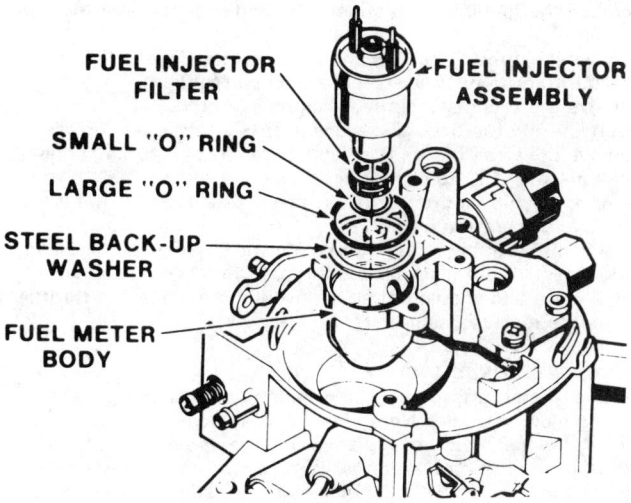

Fig. 4.30 Fuel injector assembly mounting details (Sec 15)

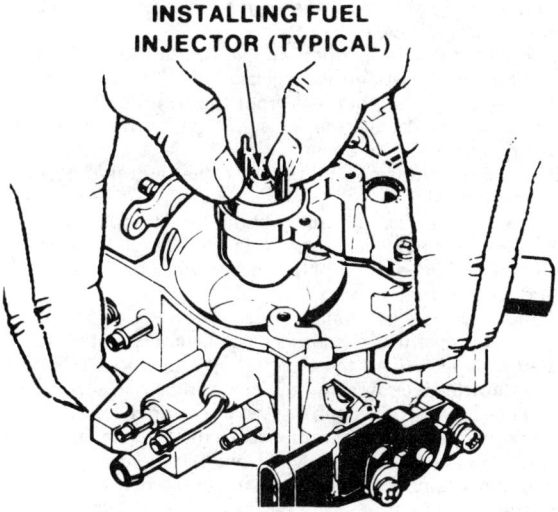

Fig. 4.31 Correct procedure for attaching the fuel injector assembly to the fuel meter body (typical) (Sec 15)

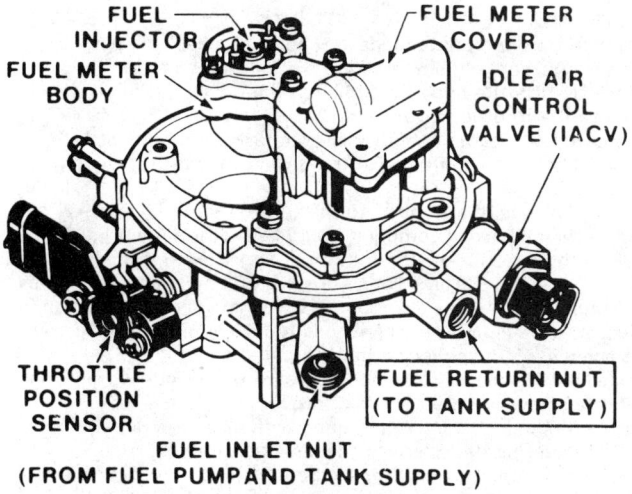

Fig. 4.32 Model 300 TBI assembly (Sec 16)

12 Install the TBI mounting bolts and tighten them to the specified torque.
13 The remainder of the installation procedure is the reverse of removal.
14 Following installation, start the engine and check for fuel leaks.

17 Throttle Body Injection (TBI) 300 assembly — overhaul

Note: *The following procedures cover the complete disassembly, cleaning, inspection and reassembly of the TBI 300 assembly after it has been removed from the engine. Refer to the illustrations for location and identification of parts. In many cases, service and repair of individual TBI systems may be completed without removing the TBI unit from the engine. If such service is indicated, refer to the appropriate Section in this Chapter.*

1 The TBI assembly must be supported on a fixture to prevent damage to the throttle valve.

Disassembly

2 Refer to the appropriate Steps in Section 14 and remove the fuel meter cover.
3 Refer to the appropriate Steps in Section 15 and remove the fuel injector assembly.
4 If not previously done, remove the air cleaner stud.
5 Remove the fuel inlet and outlet nuts and gaskets from the fuel meter body.
6 Remove the three (3) screws and washers retaining the fuel meter body to the throttle body assembly and separate them.
7 Remove the fuel meter body insulator gasket.
8 Invert the throttle body and place it on a clean, level surface.
9 Remove the two Throttle Position Sensor (TPS) retaining screws and separate the TPS from the throttle body.
10 Remove the Idle Air Control (IAC) valve and gasket from the throttle body and discard the gasket.
11 Further disassembly of the throttle body is not required for cleaning and inspection purposes. The throttle valve retaining screws are welded in place and should not be removed. If defective, the throttle body must be replaced as a unit.

Cleaning and inspection

12 Clean all metal parts in a cold immersion-type cleaner, such as Carbon X, and blow them dry. **Note:** *The TPS, IAC assembly, fuel meter cover (with the fuel pressure regulator), fuel injector, fuel injector filter and all diaphragms and other rubber parts should not be immersed in cleaner, as it will cause swelling, hardening and distortion. Clean these parts by hand with clean shop rags.* Make sure all air and fuel passages are clear; however, any thread locking compound on the IAC mounting threads should be allowed to remain.

Reassembly

13 Refer to the appropriate Steps in Section 13, check the pintle position in the IAC and install the IAC in the throttle body.
14 Place the throttle body assembly on the holding fixture to avoid damage to the throttle valve.
15 Install a new fuel meter body insulator gasket on the throttle body. Make sure the cutout portions of the gasket match the openings in the throttle body.
16 Attach the fuel meter body to the throttle body.
17 Apply thread locking compound, supplied in the service kit, to the threads on the three fuel meter body mounting screws. If locking compound is not provided, Threadlock Adhesive 262 or its equivalent may be substituted. Do not use a compound of higher strength than recommended, as it may promote screwhead breakage or prevent removal of the screws when service is again required.
18 Install the fuel meter body mounting screws and lock washers and tighten the screws to the specified torque.
19 Install the fuel feed and return nuts (with new gaskets) in the fuel meter body and tighten them to the specified torque.
20 Refer to the appropriate Steps in Section 15 and reassemble and install the injector in the injector cavity in the fuel meter body.
21 Refer to the appropriate Steps in Section 14 and install the fuel meter cover.
22 With the throttle valve in the normal closed idle position, install the TPS on the throttle body assembly, making sure the TPS pick-up lever is located above the tang on the throttle actuator lever.
23 Install two new screws, retainers and lock washers to secure the TPS to the throttle body. Use thread locking compound on the screw threads.

18 Crossfire Injection (CFI) system — general information

Some 1982 and 1983 V8-equipped vehicles employ CFI in place of the conventional carburetor-fed intake system. The principal component of the CFI system is the Model 400 throttle body injector (TBI).

The CFI consists of two throttle body injection (TBI) units (one feeding the right bank of cylinders and one feeding the left), an Electronic Command Module (ECM) and a number of sensors.

The ECM (which is a miniature computer) and the sensors control the amount of fuel sprayed into the intake manifold through the injectors by monitoring such things as throttle position, vehicle speed and load, coolant temperature and altitude. During operation of the vehicle, the oxygen sensor in the exhaust and the other sensors continually send electrical feedback information to the ECM, which refers to stored program memory and makes high-speed calculations of fuel requirements. The ECM signals each fuel injector, which then cycles open or closed to adjust the fuel/air mixture. This all happens in a fraction of a second and the cycle is repeated thousands of times per minute. The result is to achieve, as near as possible, an ideal fuel/air mixture regardless of conditions, improved driveability compared to a carbureted vehicle and low levels of emissions.

Due to the complexity of the electronic controls and the special tools required to check and service this system, diagnosis and repair must be left to a GM dealer service department or a reputable repair shop. However, if off-vehicle service to the TBI units is indicated, the home mechanic may save the cost of removing and installing them by referring to the procedure outlined in Section 19.

19 Throttle Body Injection (TBI) 400 assemblies — removal and installation

Note: *The following procedures should be employed only if the original TBI units are to be reinstalled on the original manifold. If replacement TBI(s) are to be installed, throttle valve synchronization will be required. This operation is beyond the scope of the home mechanic and should be performed by a dealer.*

1 Relieve the pressure in the fuel system (refer to Section 11).
2 Remove the cable from the negative battery terminal.
3 Remove the air cleaner.
4 Disconnect the wires from the Idle Air Control (IAC), the Throttle Position Sensor (TPS) and the injector.
5 Remove the wiring harness from the throttle body.
6 Disconnect and label the vacuum lines from the TBI.
7 Remove the throttle cable.
8 Remove the transmission detent cable (automatic transmission-equipped models).
9 Remove the cruise control cable (if so equipped).
10 With a shop towel placed to catch spilled fuel, disconnect the fuel inlet and return lines, then plug the openings to prevent contamination and further fuel loss.
11 Remove the fuel line running between the two TBI units. Use two wrenches to prevent the fuel nuts from turning in the TBI units.
12 Remove and discard the throttle rod retaining clip from the front TBI unit throttle lever stud. A new clip, supplied in service repair kits, must be used when reassembling the front and rear TBIs. One throttle rod bearing is permanently attached to the throttle lever stud on the rear unit and must not be removed from it. If the throttle rod and bearing assembly requires replacement, the entire rear throttle body must be replaced.
13 Remove the three TBI-to-cover mounting bolts from each of the TBIs.
14 Lift the rear TBI unit off the manifold cover locator pins and raise it above the front TBI unit. Slide the throttle rod end bearing off the front unit throttle lever stud and remove the rear TBI unit and gasket.
15 Remove the front TBI unit and gasket.
16 Begin installation by installing new front and rear throttle body mounting gaskets on the manifold cover.

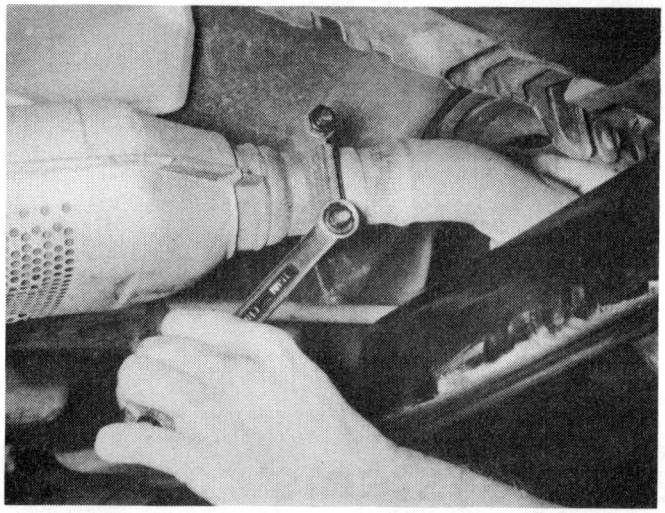

20.6 Removing the catalytic converter-to-front exhaust pipe clamp

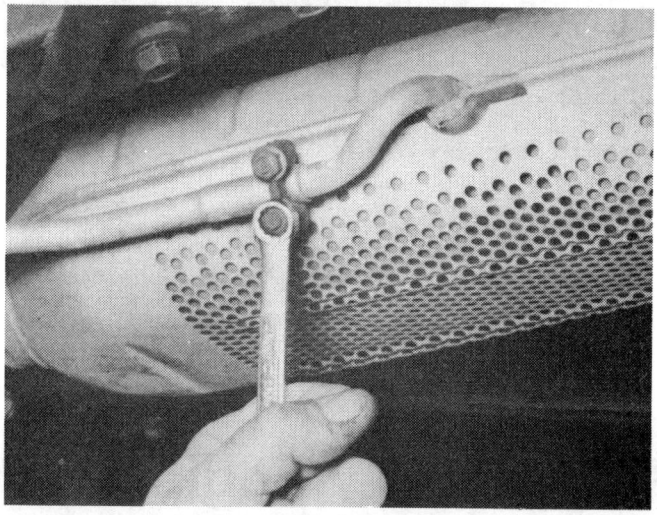

20.7 Removing the air management pipe-to-converter clamp

17 Install the front TBI unit on the manifold cover locator pins and tighten the mounting bolts to the specified torque.
18 While holding the rear unit above the front unit, install the throttle rod end bearing on the front unit throttle lever stud, then install the rear unit onto the manifold cover locator pins and tighten the mounting bolts to the specified torque.
19 The remaining installation steps are the reverse of those for removal.

20 Exhaust system components — removal and installation

Caution: *The vehicle's exhaust system generates very high temperatures and should be allowed to cool down completely before any of the components are touched. Be especially careful around the catalytic converter, where the highest temperatures are generated.*

Due to the high temperatures and exposed locations of the exhaust system components, rust and corrosion can 'freeze' parts together. Liquid penetrating oils are available to help loosen frozen fasteners; however, in some cases it may be necessary to cut the pieces apart with a hacksaw or cutting torch. The latter method should be employed only by persons experienced in this work.

1 Raise the vehicle and place it securely on jackstands.

Front exhaust pipe

2 Remove the bolts securing the exhaust pipe to the exhaust manifold(s).
3 Remove the clamp securing the exhaust pipe to the catalytic converter.
4 Separate the exhaust pipe from the exhaust manifold(s) and from the catalytic converter.
5 Installation is the reverse of the removal procedure. Be sure to install new bolts, gaskets and nuts.

Catalytic converter

Caution: *Make sure the catalytic converter has been allowed sufficient time to cool before attempting removal.*

6 Remove the converter-to-front exhaust pipe clamp (photo).
7 Remove the air management pipe-to-converter clamp (photo).
8 Remove the bolts securing the rear of the converter to the converter/intermediate pipe bracket and remove the converter (photo).
9 Installation is the reverse of the removal procedure. Be sure to use new gaskets, bolts and nuts.

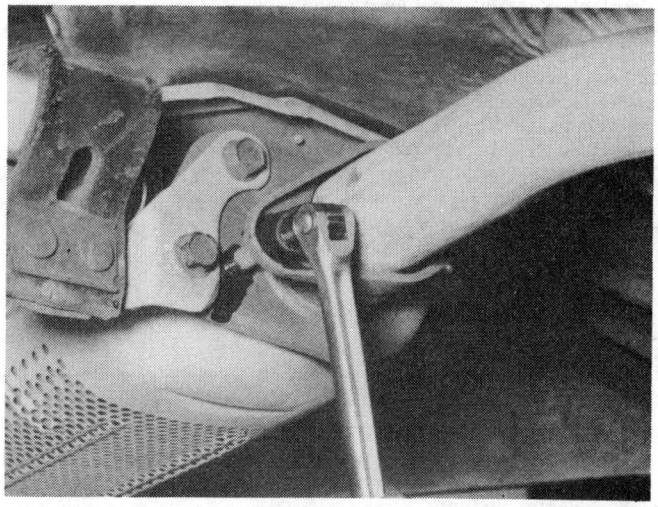

20.8 Removing the bolts from the converter/intermediate pipe bracket

Intermediate exhaust pipe

10 Remove the clamp securing the intermediate pipe to the muffler/tailpipe assembly.
11 Remove the bolts securing the intermediate pipe to the intermediate pipe/converter bracket and remove the intermediate pipe.
12 Installation is the reverse of the removal procedure. Be sure to use new gaskets, bolts and nuts.

Muffler and tailpipe

13 The tailpipe is actually part of the muffler assembly.
14 Remove the bolts and nuts which connect the muffler to the intermediate exhaust pipe.
15 Remove the muffler/tailpipe assembly rear hanger bolts and separate the muffler/tailpipe assembly from the vehicle.
16 Installation is the reverse of the removal procedure. Be sure to use new nuts and bolts.

Chapter 5 Engine electrical systems

Contents

Alternator — removal and installation 12
Alternator brushes — replacement 13
Battery — emergency jump starting 3
Battery — removal and installation 2
Battery cables — check and replacement 4
Charging system — check 11
Charging system — general information and precautions 10
'Check engine' light . . . . . Chapter 6
Distributor — removal and installation 6
Ignition coil — removal and installation 9
Ignition key lock cylinder — removal and installation . . . . . Chapter 10
Ignition module — replacement 8
Ignition pick-up coil — replacement 7
Ignition switch — replacement . . . . . Chapter 10
Ignition system — check 5
Ignition system — general information and precautions 1
Ignition timing check and adjustment . . . . . Chapter 1
Spark plug replacement . . . . . Chapter 1
Spark plug wires, distributor cap and rotor — check and replacement . . . . . Chapter 1
Starter motor — removal and installation 16
Starter motor — testing in vehicle 15
Starter motor brushes — replacement 18
Starter solenoid — removal, repair and installation 17
Starting system — general information 14

1 Ignition system — general information and precautions

The ignition system is composed of the battery, distributor, ignition switch, spark plugs and the primary (low tension) and secondary (high tension) wiring circuits.

A high energy ignition (HEI) distributor is used on all vehicles. Some vehicles use a coil mounted integral with the distributor, while others employ a separately mounted coil.

The HEI works in conjunction with the electronic spark timing (EST) system and uses a magnetic pick-up assembly (located inside the distributor) which contains a permanent magnet, a pole piece with internal teeth and a pick-up coil in place of the traditional ignition point assembly.

All ignition timing changes in the HEI-EST distributor are accomplished by an electronic control module (ECM), which monitors data from various engine sensors, computes the desired spark timing and signals the distributor to change the timing accordingly. The electronic spark control (ESC) used on some engines retards the spark advance when detonation occurs. This retard mode is held for approximately 20 seconds, after which the spark control will revert to the EST.

There are three basic components involved in the ESC system: the sensor, the controller and the distributor. The ESC sensor detects the presence (or absence) and intensity of detonation through vibrations in the engine and sends the data to the controller. The ESC controller changes the sensor signal into a command signal to the distributor, which then adjusts the spark timing accordingly.

The distributor in the ECS system is modified to respond to the ESC controller signal. This command is delayed when detonation occurs, thus providing the level of retard required. The degree of retard is based on the detonation.

The secondary (spark plug) wire used with the HEI system is a carbon-impregnated cord conductor encased in an 8 mm (5/16-inch) diameter rubber jacket. This type of wire will withstand very high temperatures and still provide insulation for the HEI's high voltage. For more information on spark plug wiring, refer to Chapter 1.

Resistor type, tapered seat spark plugs without gaskets are used in all engines. Refer to the underhood Vehicle Emissions Control Information label for the proper plug for your particular engine.

Note: *Because of the very high voltage generated by the HEI system, extreme care should be taken whenever an operation involving ignition components is performed. This not only includes the distributor, coil, control module and spark plug wires, but related items that are connected to the system as well (such as the plug connections, tachometer and testing equipment). Consequently, before any work is performed, the ignition should be turned off or the battery ground cable disconnected.*

2 Battery — removal and installation

1 The battery is located at the front of the engine compartment. It is held in place by a hold-down clamp near the bottom of the battery case.

2 Hydrogen gas is produced by the battery, so keep open flames and lighted cigarettes away from it at all times.

3 Always keep the battery in an upright position. Spilled electrolyte should be rinsed off immediately with large quantities of water. Always wear eye protection when working around a battery.

4 Always disconnect the negative (-) battery cable first, followed by the positive (+) cable.

5 After the cables are disconnected from the battery, remove the hold-down clamp.

6 Carefully lift the battery out of the engine compartment.

7 Installation is the reverse of removal. The cable clamps should be tight, but do not overtighten them as damage to the battery case could occur. The battery posts and cable ends should be cleaned prior to connection (see Chapter 1).

3 Battery — emergency jump starting

Refer to the *Booster battery (jump) starting* procedure at the front of this manual.

4 Battery cables — check and replacement

1 Periodically inspect the entire length of each battery cable for damage, cracked or burned insulation and corrosion. Poor battery cable connections can cause starting problems and decreased engine performance.
2 Check the cable-to-terminal connections at the ends of the cables for cracks, loose wire strands and corrosion. The presence of white, fluffy deposits under the insulation at the cable terminal connection is a sign the cable is corroded and should be replaced. Check the terminals for distortion, missing mounting bolts or nuts and corrosion.
3 If only the positive cable is to be replaced, be sure to disconnect the negative cable from the battery first.
4 Disconnect and remove the cable(s) from the vehicle. Make sure the replacement cable(s) is the same length and diameter.
5 Clean the threads of the starter or ground connection with a wire brush to remove rust and corrosion. Apply a light coat of petroleum jelly to the threads to ease installation and prevent future corrosion. Inspect the connections frequently to make sure they are clean and tight.
6 Attach the cable(s) to the starter or ground connection and tighten the mounting nut(s) securely.
7 Before connecting the new cable(s) to the battery, make sure they reach the terminals without having to be stretched.
8 Connect the positive cable first, followed by the negative cable. Tighten the nuts and apply a thin coat of petroleum jelly to the terminal and cable connection.

5 Ignition system — check

Caution: *Because of the very high voltage generated by the HEI system, extreme care should be taken whenever an operation is performed involving ignition components. This not only includes the distributor, coil, control module and spark plug wires, but related items that are connected to the system as well, such as the plug connections, tachometer and any test equipment. Consequently, before any work is performed, the ignition should be turned off or the battery ground cable disconnected.*

1 If the engine turns over but will not start, remove the spark plug wire from a spark plug and, using an insulated tool, hold the wire about 1/4-inch from a good ground and have an assistant crank the engine.
Note: *A special tool, ST-125, is available for making the above test. It is available from your dealer and auto parts stores.*
2 If there is no spark, check another wire in the same manner. A few sparks, then no spark, should be considered as no spark.
3 If there is good spark, check the spark plugs (refer to Chapter 1) and/or the fuel system (refer to Chapter 4).
4 If there is a weak spark or no spark in a system with an integral coil, the system must be checked further by a dealer or repair shop.
5 If there is a weak spark or no spark in a system with a remote coil, unplug the coil lead from the distributor, hold it about 1/4-inch from a good ground and check for spark as described above.
6 If there is no spark, have the system checked by a dealer or repair shop.
7 If there is a spark, check the distributor cap and/or rotor (refer to Chapter 1).
8 Further checks of the HEI ignition system must be done by a dealer or repair shop.

6 Distributor — removal and installation

Removal

1 Disconnect the ignition switch/battery feed wire (BAT) and the tachometer lead (TACH), if so equipped, from the distributor cap.
2 Release the coil connectors from the cap. Depress the locking tabs by hand; do not use a screwdriver or other tool.
3 Turn the four distributor cap locking latches counterclockwise, remove the cap and position it out of the way.
4 Disconnect the four-terminal ECM wiring harness connector from the distributor.
5 If necessary for clearance (on distributor caps with the secondary wiring harness attached to the cap), release the wiring harness latches and remove the wiring harness retainer. Note that the spark plug wire numbers are indicated on the retainer.
6 Remove the distributor hold-down clamp bolt and the hold-down clamp (photo). Mark the distributor body and engine to make sure they are aligned properly during installation.
7 Mark the position of the rotor on the outside of the distributor body, then slowly lift the distributor from the engine and note again the position of the rotor when it stops turning. Mark this position also. To ensure correct timing of the distributor upon reinstallation, the rotor should be in the second mark position prior to reinserting the distributor in the engine.
8 Avoid turning the crankshaft with the distributor removed, as the ignition timing will be changed.

Installation if the crankshaft was not turned after distributor removal

9 Position the rotor in the exact location it was in (second mark, Step 7) when the distributor was removed.
10 Lower the distributor into the engine. To mesh the gears at the bottom of the distributor, it may be necessary to turn the rotor slightly.
11 With the base of the distributor all the way down against the engine block, the rotor should point to the first alignment mark (Step 7). If not, refer to Step 17.
12 Place the hold-down clamp in position and loosely install the hold-down bolt.
13 Connect the ignition feed wire and tachometer wire, if so equipped, to the distributor.
14 Install the distributor cap. If the secondary wiring harness was removed from the cap, install it.
15 Reconnect the coil connector.
16 With the distributor in its original position, tighten the hold-down bolt and check the ignition timing (Chapter 1).

Installation if the crankshaft was turned after distributor removal

17 Remove the number 1 spark plug.
18 Place your finger over the spark plug hole while turning the crankshaft with a wrench on the pulley bolt at the front of the engine.
19 When you feel compression, continue turning the crankshaft slowly until the timing mark on the crankshaft pulley is aligned with the 0 on the engine timing indicator.
20 Position the rotor to point between the number 1 and 8 spark plug terminals in the distributor cap on V8 engines; between the number

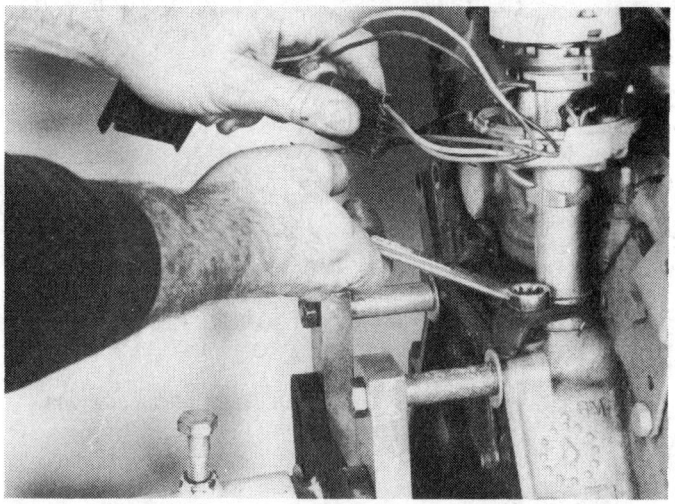

6.6 Removing the distributor hold-down clamp bolt (V6 engine shown out of vehicle)

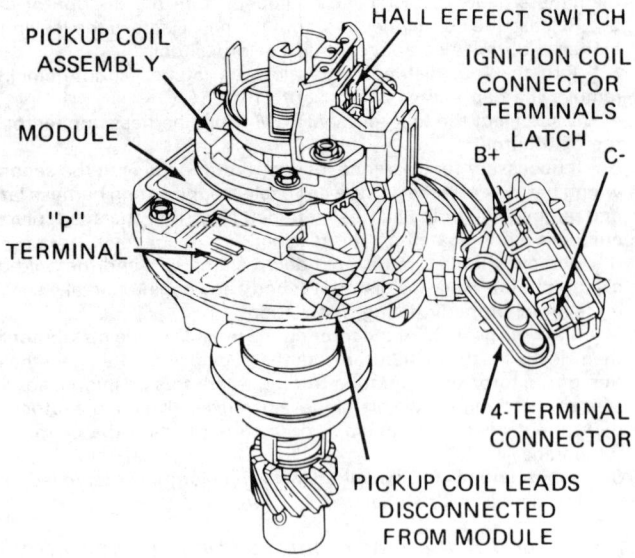

Fig. 5.1 Separately mounted coil distributor with Hall effect switch installed (Sec 7)

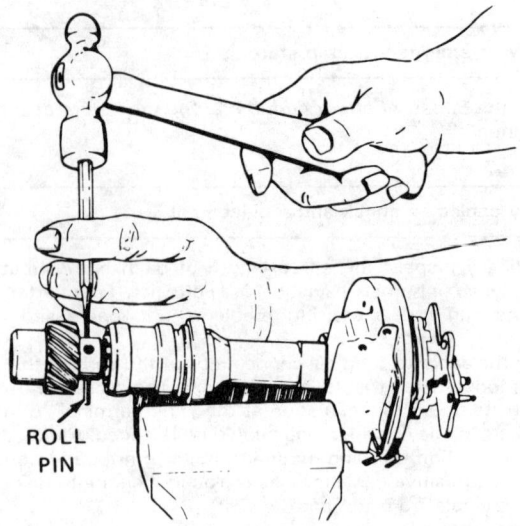

Fig. 5.2 Removing the roll pin from the distributor shaft and gear (Sec 7)

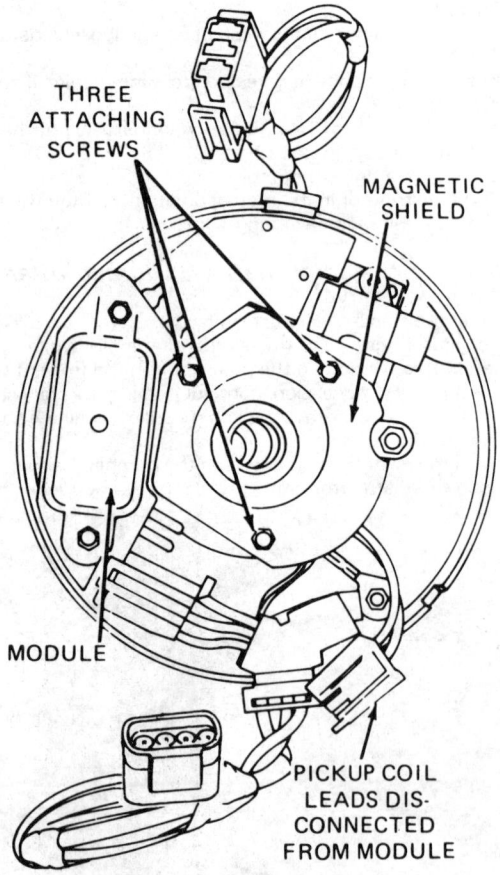

Fig. 5.3 Magnetic shield installation details (distributor with coil-in-cap construction) (Sec 7)

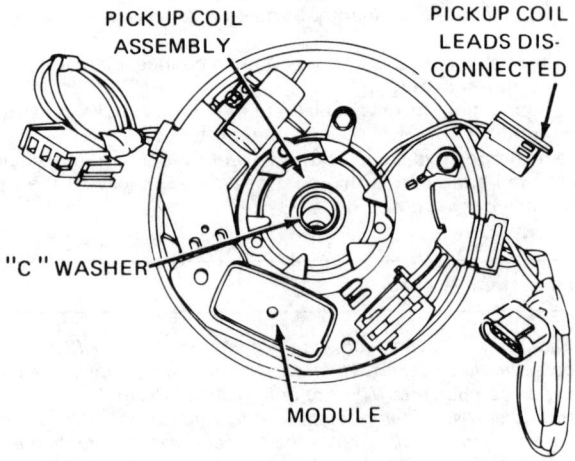

Fig. 5.4 Coil-in-cap distributor with the magnetic shield removed (Sec 7)

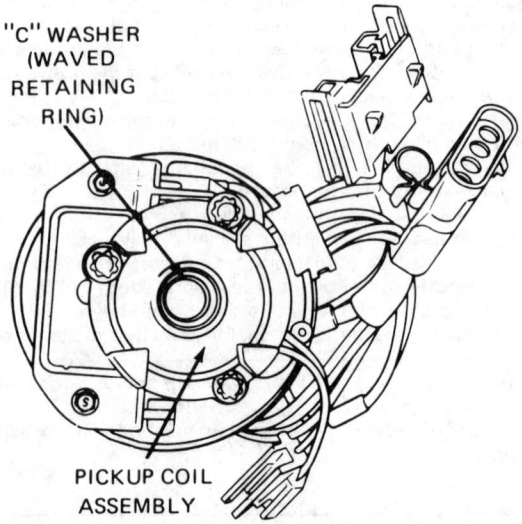

Fig. 5.5 Pick-up coil installation details (distributors with separately mounted coil) (Sec 7)

1 and 6 terminals on the V6 engine; or between the number 1 and 3 terminals on L4 engines.
21 Complete the installation by referring to Steps 10 through 16.

7 Ignition pick-up coil — replacement

1 Remove the distributor from the engine as described previously.
2 Remove the two rotor mounting screws and remove the rotor.
3 Disconnect the pick-up coil leads from the module.
4 On models with a separately mounted coil and a Hall-effect switch, remove the switch retaining screws and the switch.
5 Mark the distributor gear and shaft so they can be reassembled in the same position.
6 Carefully mount the distributor in a soft-jawed vise and, using a hammer and punch, remove the roll pin from the distributor shaft and gear.
7 Remove the gear and washers from the shaft.
8 Carefully pull the shaft out through the top of the distributor.
9 On distributors with coil-in-cap construction, remove the three mounting screws and the magnetic shield.
10 Remove the C-washer retaining ring at the center of the distributor and remove the pick-up coil.
11 Installation is the reverse of the removal procedure.

8 Ignition module — replacement

1 Remove the distributor.
2 Remove the pick-up coil assembly.
3 On distributors with an internal capacitor, remove the module mounting screws and the capacitor mounting screw, then separate the module, capacitor and harness assembly from the distributor base and disconnect the wiring harness from the module.
4 On distributors without an internal capacitor, disconnect the wiring harness from the module, remove the two mounting screws and remove the module.
5 Installation is the reverse of the removal procedure. Before installing the module, apply silicone lubricant to the housing on which the module mounts.

9 Ignition coil — removal and installation

Coil-in-cap models

1 Disconnect the ignition switch/battery feed wire (BAT) from the distributor cap.
2 Disconnect the coil connector from the distributor cap.
3 Disconnect the four-terminal ECM wiring harness connector from

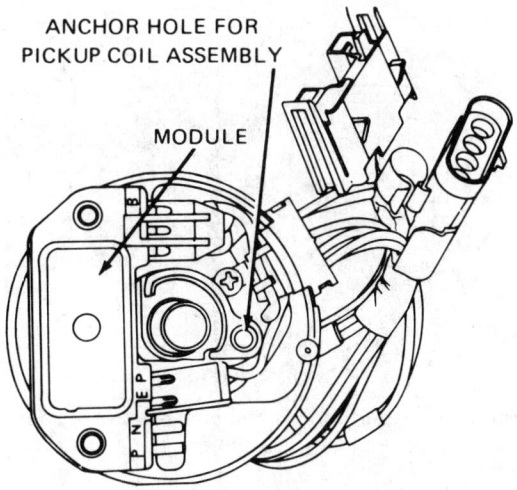

Fig. 5.6 Separately mounted coil distributor with the pick-up coil removed (Sec 7)

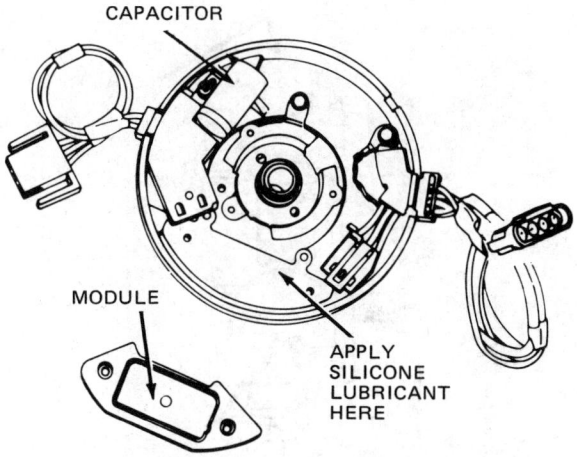

Fig. 5.7 Coil-in-cap distributor with an internal capacitor (Sec 8)

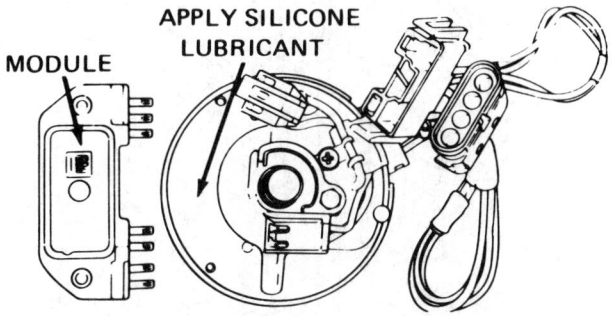

Fig. 5.8 Module installation on a distributor without an internal capacitor (Sec 8)

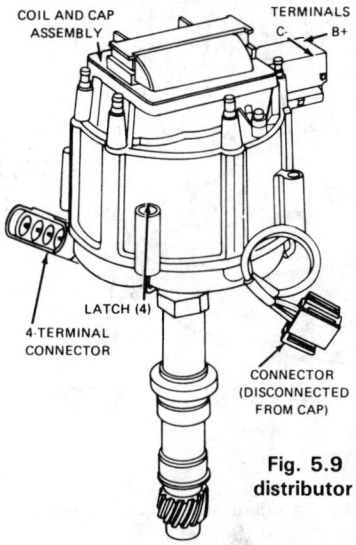

Fig. 5.9 Typical coil-in-cap distributor (V6 shown) (Sec 9)

the distributor.
4 Release the spark plug wiring harness retainers from the distributor cap and remove the wiring retainer. To simplify installation, note that the spark plug wire numbers are indicated on the retainer.
5 If necessary for working clearance, turn the four distributor cap locking latches counterclockwise, remove the cap from the distributor and position it so that the coil cover and coil may be removed.
6 Remove the coil cover mounting screws and lift the cover off.
7 Remove the coil mounting screws and separate the coil and leads from the cap.
8 Before installing the coil, remove the arc seal and clean it with a soft cloth.
9 Installation is the reverse of the removal procedure.

Models with separately mounted coil

10 Disconnect the secondary wire from the coil.
11 Loosen the coil bracket mounting nuts and remove the coil.
12 Installation is the reverse of the removal procedure.

10 Charging system — general information and precautions

The charging system is made up of the alternator, voltage regulator and battery. These components work together to supply electrical power for the engine ignition, lights, radio, etc.

The alternator is turned by a drivebelt at the front of the engine. When the engine is operating, voltage is generated by the internal components of the alternator to be sent to the battery for storage.

The purpose of the voltage regulator is to limit the alternator voltage to a pre-set value. This prevents power surges, circuit overloads, etc., during peak voltage output. On all models with which this manual is concerned, the voltage regulator is contained within the alternator housing.

The charging system does not ordinarily require periodic maintenance. The drivebelts, electrical wiring and connections should, however, be inspected at the intervals suggested in Chapter 1.

Take extreme care when making circuit connections to a vehicle equipped with an alternator and note the following. When making connections to the alternator from a battery, always match correct polarity. Before using electric-arc welding equipment to repair any part of the vehicle, disconnect the wires from the alternator and the battery terminal. Never start the engine with a battery charger connected. Always disconnect both battery leads before using a battery charger.

11 Charging system — check

1 If a malfunction occurs in the charging circuit, do not immediately assume that the alternator is causing the problem. First check the following items:

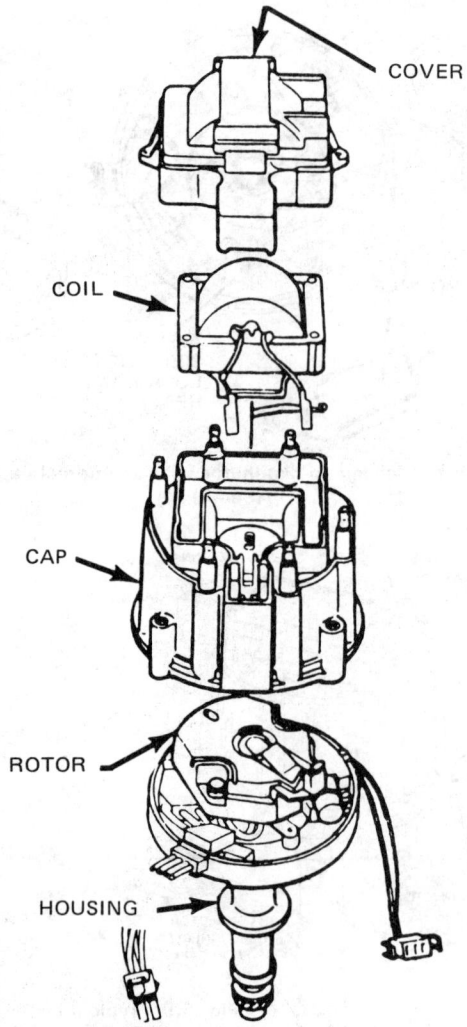

Fig. 5.10 Typical coil-in-cap distributor — exploded view (V6 shown) (Sec 9)

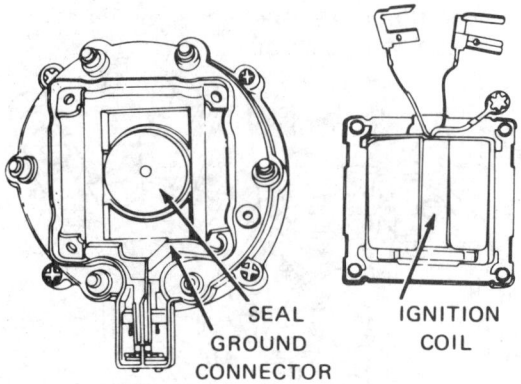

Fig. 5.11 Coil-in-cap distributor cap with the coil removed (Sec 9)

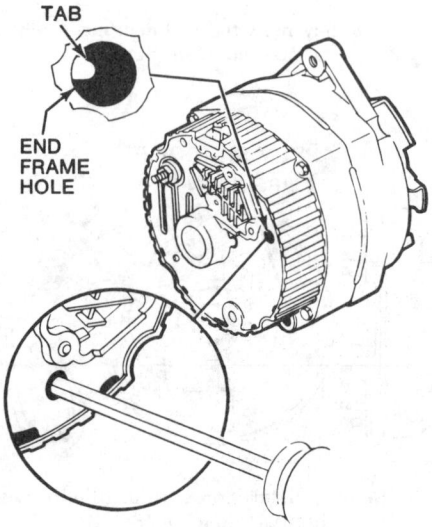

Fig. 5.12 Location of the test hole in the rear of the alternator (inset shows grounding procedure) (Sec 11)

The battery cables where they connect to the battery (make sure the connections are clean and tight)
The battery electrolyte specific gravity (if it is low, charge the battery)
Check the external alternator wiring and connections (they must be in good condition)
Check the drivebelt condition and tension (see Chapter 1)
Check the alternator mount bolts for tightness
Run the engine and check the alternator for abnormal noise

2 Using a voltmeter, check the battery voltage with the engine off. It should be approximately 12-volts.
3 Start the engine and check the battery voltage again. It should now be approximately 14 to 15-volts. If it does not rise when the engine is started or if it exceeds 15-volts, proceed to Step 4.
4 Locate the test hole in the back of the alternator and ground the tab that is located inside the hole by inserting a screwdriver blade into the hole and touching the tab and the case at the same time. **Note:** *Do not run the engine with the tab grounded any longer than necessary to obtain a voltmeter reading. The alternator, if it is charging, is running unregulated at this point. This condition may overload the electrical system and cause damage to the components.*
5 The reading on the voltmeter should be 15-volts or higher with the tab grounded in the test hole.
6 If the voltmeter indicates battery voltage, the alternator is faulty and should be replaced with a new one (refer to Section 12).
7 If the voltage reading is 15-volts or higher and a no-charge condition is present, the regulator or field circuit is the problem. Remove the alternator and have it checked further by an auto electric shop.

12 Alternator — removal and installation

1 Remove the bolt retaining the heater hose clamp to the alternator adjusting bracket (photo).
2 Remove the bolt retaining the negative battery cable to the alternator adjusting bracket, if so equipped. If not, disconnect the cable from the negative battery terminal.
3 Disconnect the cable from the positive battery terminal.
4 Disconnect the two electrical connectors from the rear of the alternator (photos).
5 Remove the bolt retaining the alternator to the adjusting bracket (photo).
6 Remove the alternator pivot bolt and remove the alternator from the vehicle (photo).
7 Installation is the reverse of the removal procedure, noting the following.
8 Reconnect the positive battery cable *before* reattaching the negative cable at the alternator bracket.
9 Adjust the alternator drivebelt (refer to Chapter 1).

12.1 Arrow (A) indicates the bolt retaining the heater hose clamp to the alternator adjusting bracket; arrow (B) indicates the bolt retaining the negative battery cable (ground) (V6 engine shown)

12.4A Disconnect the wires from the bottom rear of the alternator

12.4B Disconnect the BAT terminal wire from the rear of the alternator

12.5 Removing the bolt retaining the alternator to the adjusting bracket

12.6 Removing the alternator from the vehicle (arrow denotes pivot bolt)

13.3A Separating the alternator front and rear end frames

13.3B Details of the alternator rear end frame

A	*Brush holder*	*D*	*Resistor (not all models)*
B	*Paper clip retaining brushes*	*E*	*Diode trio*
C	*Regulator*	*F*	*Rectifier bridge*

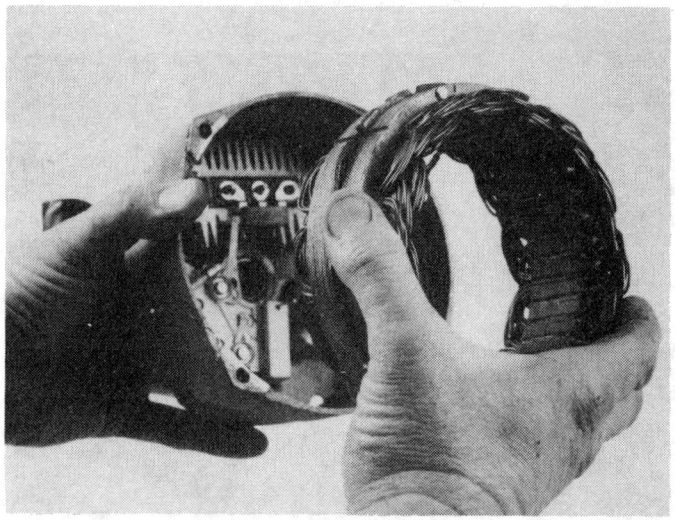

13.4 Separating the stator from the rear end frame

13.5 Removing the diode trio from the rectifier bridge

13.6 Removing the brush holder from the rear end frame

13.7 Removing a brush assembly from the brush holder

13.10 Paper clip inserted from rear of the alternator to hold the brushes in place during disassembly and reassembly

13 Alternator brushes – replacement

1 Remove the alternator from the vehicle (refer to Section 12).
2 Scribe a match mark on the front and rear end frame housings of the alternator to facilitate reassembly.
3 Remove the four (4) through–bolts holding the front and rear end frames together, then separate the end frames (photos).
4 Remove the bolts holding the stator to the rear end frame and separate the stator from the end frame (photo).
5 Remove the nuts attaching the diode trio to the rectifier bridge and remove the trio (photo).
6 Remove the screws retaining the brush holder and resistor (if equipped) to the end frame and remove the brush holder (photo).
7 Remove the brushes from the brush holder by slipping the brush retainer off the brush holder (photo).
8 Remove the springs from the brush holder.
9 Installation is the reverse of the removal procedure, noting the following.
10 When installing the brushes in the brush holder, install the brush closest to the end frame first. Slip a paper clip through the rear of the end frame to hold the brush, then insert the second brush and push the paper clip in to hold both brushes while reassembly is completed (photo). The paper clip should not be removed until the front and rear end frames have been bolted together.

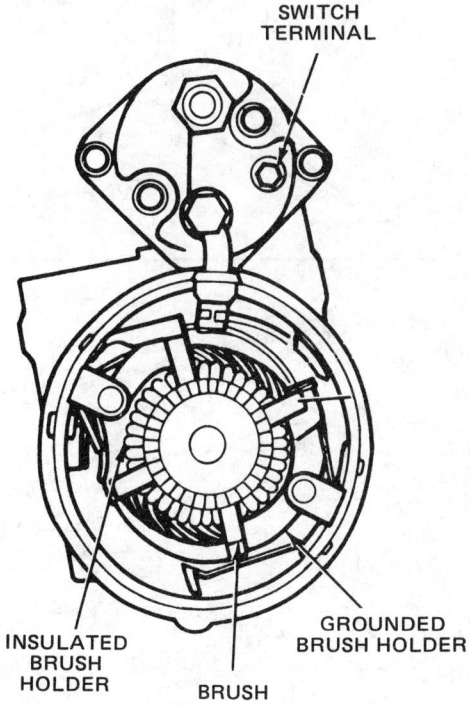

Fig. 5.13 View of the starter and solenoid showing the S (switch) terminal on the solenoid and details of the brush installation in the starter field frame housing (Sec 15)

14 Starting system — general information

The function of the starting system is to crank the engine. This system is composed of a starting motor, solenoid and battery. The battery supplies the electrical energy to the solenoid, which then completes the circuit to the starting motor which does the actual work of cranking the engine.

The solenoid and starting motor are mounted together at the lower right side of the engine. No periodic lubrication or maintenance is required.

The electrical circuitry of the vehicle is arranged so that the starter motor can only be operated when the clutch pedal is depressed (manual transmission) or the transmission selector lever is in Park or Neutral (automatic transmission).

Never operate the starter motor for more than 30 seconds at a time without pausing to allow it to cool for at least two minutes. Excessive cranking can cause overheating, which can seriously damage the starter.

15 Starter motor — testing in vehicle

1 If the starter motor does not turn at all when the switch is operated, make sure that the shift lever is in Neutral or Park (automatic transmission) or that the clutch pedal is depressed (manual transmission).
2 Make sure that the battery is charged and that all cables, both at the battery and starter solenoid terminals, are secure.
3 If the motor spins but the engine is not being cranked, then the overrunning clutch in the starter motor is slipping and the motor must be removed from the engine and disassembled.
4 If, when the switch is actuated, the starter motor does not operate at all but the solenoid clicks, then the problem is in the main solenoid contacts or the starter motor itself.
5 If the solenoid plunger cannot be heard when the switch is ac-

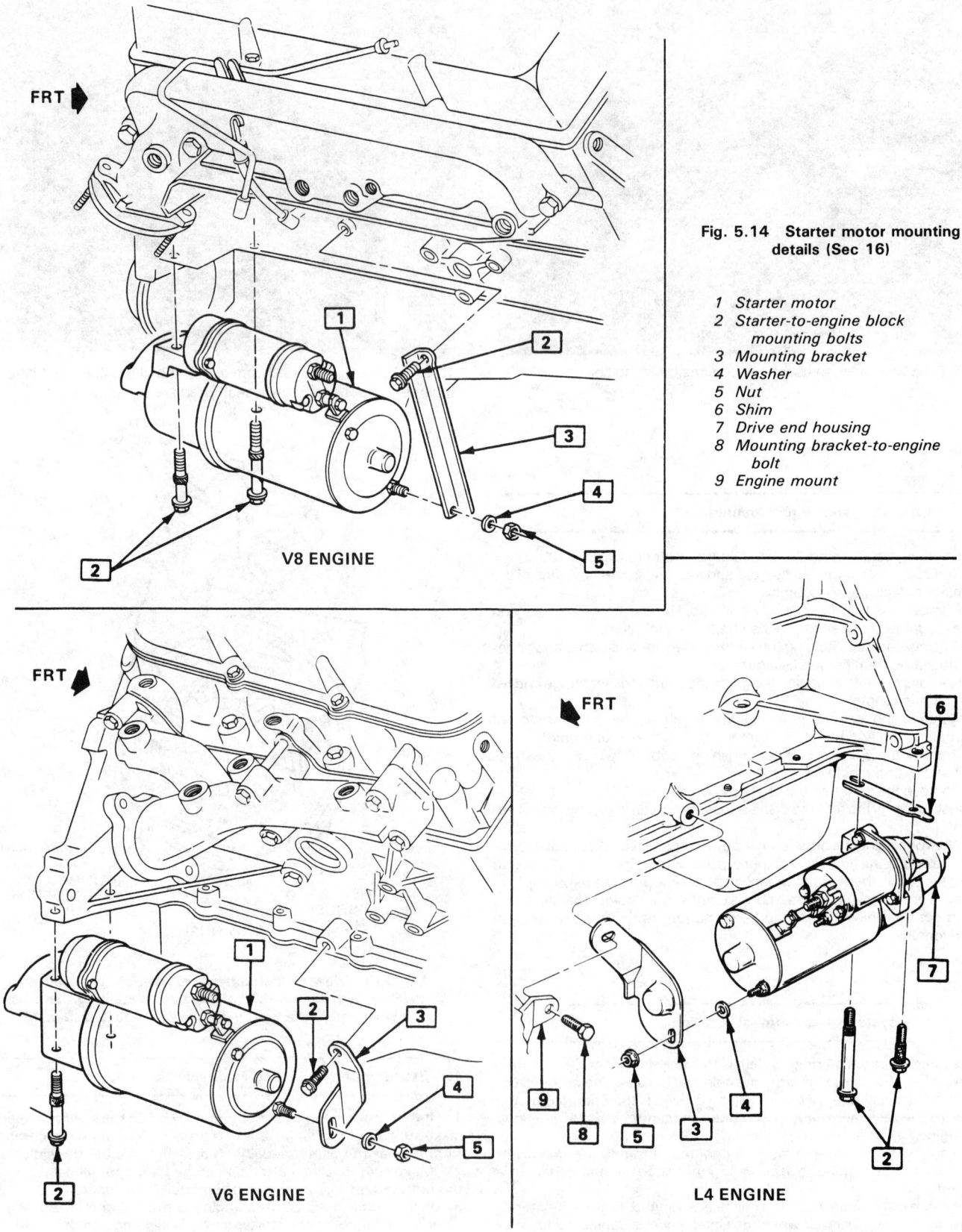

Fig. 5.14 Starter motor mounting details (Sec 16)

1 Starter motor
2 Starter-to-engine block mounting bolts
3 Mounting bracket
4 Washer
5 Nut
6 Shim
7 Drive end housing
8 Mounting bracket-to-engine bolt
9 Engine mount

tuated, the solenoid itself if defective or the solenoid circuit is open.
6 To check out the solenoid, connect a jumper lead between the battery (+) and the S terminal on the solenoid. If the starter motor now operates, the solenoid is OK and the problem is in the ignition or neutral start switches or in the wiring.
7 If the starter motor still does not operate, remove the starter/solenoid assembly for disassembly, testing and repair.
8 If the starter motor cranks the engine at an abnormally slow speed, first make sure that the battery is charged and that all terminal connections are tight. If the engine is partially seized, or has the wrong viscosity oil in it, it will crank slowly also.
9 Run the engine until normal operating temperature is reached, then disconnect the coil wire from the distributor cap and ground it on the engine (remote coil only). On models with the coil in the distributor cap, disconnect the pink wire from the distributor.
10 Connect a voltmeter positive lead to the starter motor terminal of the solenoid and then connect the negative lead to ground.
11 Actuate the ignition switch and take the voltmeter readings as soon as a steady figure is indicated. Do not allow the starter motor to turn for more than 30 seconds at a time. A reading of 9-volts or more, with the starter motor turning at normal cranking speed, is normal. If the reading is 9-volts or more but the cranking speed is slow, the motor is faulty. If the reading is less than 9-volts and the cranking speed is slow, the solenoid contacts are probably burned (refer to Section 17).

16 Starter motor — removal and installation

1 Disconnect the cable from the negative battery terminal.
2 Raise the vehicle and place it securely on jackstands.
3 Refer to the accompanying illustration and remove the brackets supporting the starter.
4 Remove the two starter motor-to-engine bolts and let the starter drop down far enough to allow removal of the nuts attaching the wires to the starter solenoid and battery cable. Support the starter while the wires are removed.
5 Separate the starter from the engine.
6 Installation is the reverse of the removal procedure. Make sure that any removed shims are reinstalled.

17 Starter solenoid — removal, repair and installation

1 After removing the starter and solenoid as described in Section 16, disconnect the strap from the solenoid MOTOR terminal.
2 Remove the two screws which secure the solenoid housing to the end frame assembly.
3 Twist the solenoid in a clockwise direction to disengage the flange key and then withdraw the solenoid (photo).
4 Remove the nuts and washers from the solenoid terminals and then unscrew the two solenoid end cover retaining screws and washers and pull off the end cover.
5 Unscrew the nut from the battery terminal on the end cover and remove the terminal.
6 Remove the resistor bypass terminal and contact.
7 Remove the motor connector strap terminal and solder a new terminal in position.
8 Install a new battery terminal. Install the bypass terminal and contact.
9 Install the end cover and the remaining terminal nuts.
10 Attach the solenoid to the starter motor. First make sure that the return spring is in position on the plunger and then insert the solenoid body into the drive housing and turn the body counterclockwise to engage the flange key.
11 Install the two solenoid screws and connect the MOTOR connector strap.

18 Starter motor brushes — replacement

1 Remove the starter and solenoid assembly from the vehicle (refer to Section 16).

17.3 Withdrawing the solenoid assembly from the starter housing

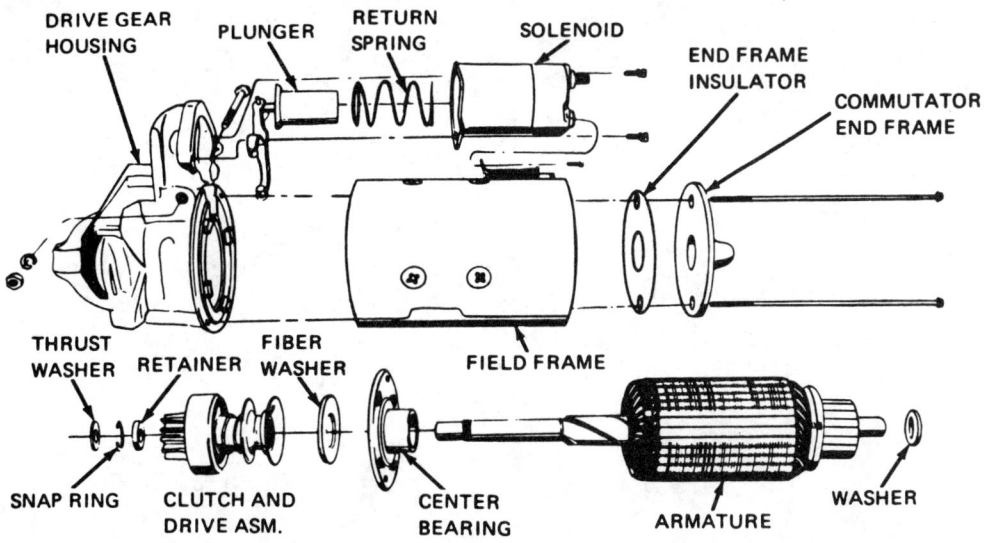

Fig. 5.15 Typical starter motor components — exploded view (Sec 18)

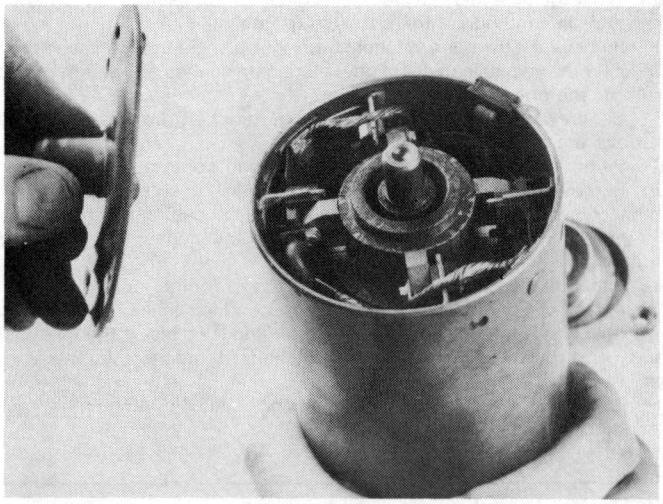

18.4 Removing the end frame from the field frame housing

18.6 Unbolting a brush and brush support from the brush holder

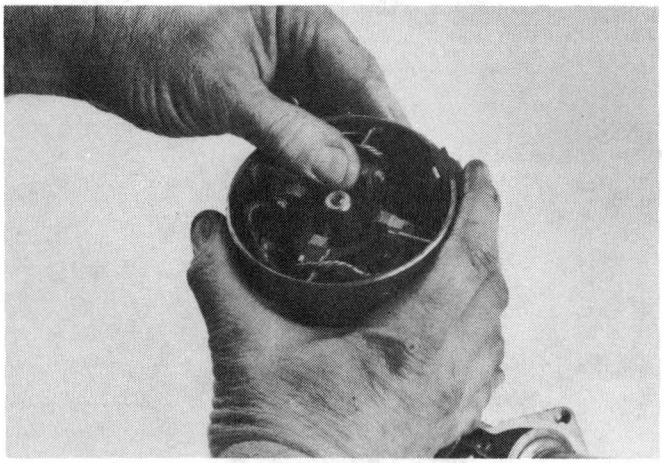

18.8 Installing the field frame over the armature (note the position of the brushes on the armature collar)

2 Remove the solenoid from the starter housing (refer to Section 17).
3 Remove the starter motor through-bolts after marking the relationship of the commutator end frame to the field frame housing to simplify reassembly.
4 Remove the end frame from the field frame housing (photo).
5 Mark the relationship of the field frame housing to the drive end housing and pull the field frame housing away from the drive end housing and over the armature.
6 Unbolt the brushes and brush supports from the brush holders in the field frame housing (photos).
7 To install new brushes, attach the brushes to the brush supports, making sure they are flush with the bottom of the supports, and bolt the brushes/supports to the brush holders.
8 Install the field frame over the armature, with the brushes resting on the first step of the armature collar at this point (photo).
9 Make sure the field frame is properly aligned with the drive end housing, then push the brushes off the collar into place on the armature.
10 The remaining installation steps are the reverse of those for removal.

Chapter 6 Emissions control systems

Refer to Chapter 13 for information related to 1985 and later models

Contents

Air Injection Reaction (AIR)/PULSAIR reactor systems 5
Computer Command Control System (CCCS) 2
Early Fuel Evaporation (EFE) system 6
Electronic Spark Timing (EST) 4
Evaporative Emissions Control System (EECS) 8
Exhaust Gas Recirculation (EGR) system 7
Fuel control system 3
General information 1
Positive Crankcase Ventilation (PCV) system 9
Thermostatic Air Cleaner (THERMAC) 10
Transmission Converter Clutch (TCC) 11

Specifications

Torque specifications

	Ft-lb
EGR valve	25
AIR pump pulley	24
Air Management Valve	20 to 35
Air pump	13
Oxygen sensor	30

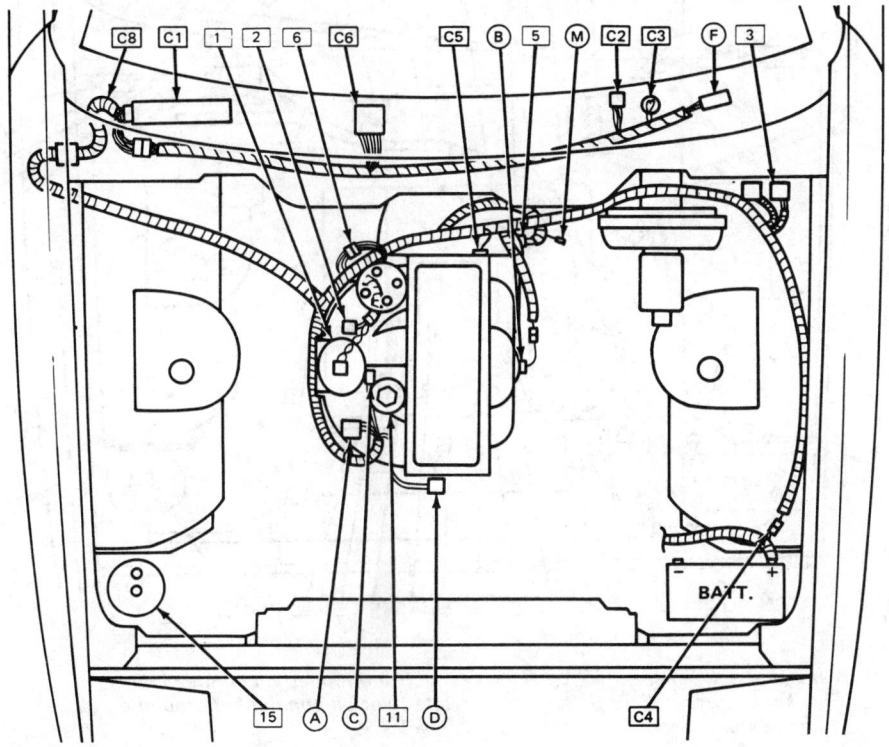

Fig. 6.1 Locations of emission system and related components (L4) (1984 engine shown, others similar)

C1 Electronic control module (ECM)
C2 ALCL connector
C3 'Check engine' light
C4 Code memory power
C5 Computer system ground
C6 Fuse panel
C7 Lamp driver
C8 Computer control harness
1 Throttle body injection
2 Idle air control
3 Fuel pump relay
5 Transmission converter clutch connector
6 Electronic spark timing connector
11 Exhaust gas recirculation valve
15 Vapor canister
A Manifold pressure sensor
B Exhaust oxygen sensor
C Throttle position sensor
D Coolant sensor
F Vehicle speed sensor
M Fuel pump test connector

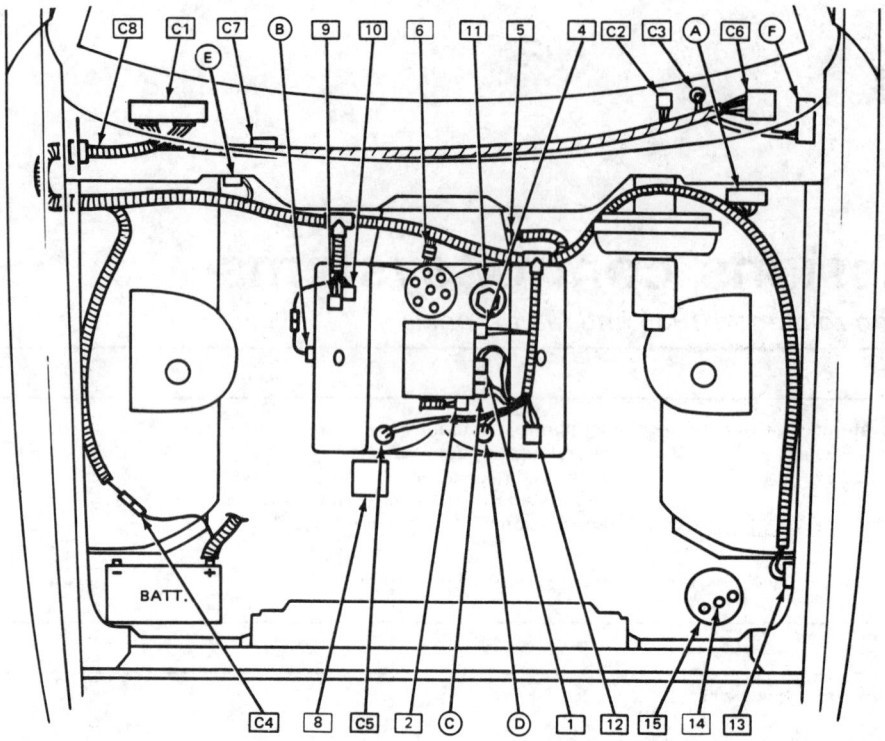

Fig. 6.2 Locations of emissions system and related components (V6) (1984 engine shown, others similar)

C1 Electronic control module (ECM)
C2 ALCL connector
C3 'Check engine' light
C4 Code memory power
C5 Computer system ground
C6 Fuse panel
C7 Lamp driver
C8 Computer control harness
1 Mixture control
2 Idle speed solenoid
4 Heated grid (EFE)
5 Transmission converter clutch connector
6 Electronic spark timing connector
8 Air injection pump
9 Air control solenoid (divert) valve
10 Air switching solenoid valve
11 Exhaust gas recirculation valve
12 Exhaust gas recirculation solenoid valve
13 Canister purge solenoid valve
14 Fuel vapor control line from tank
15 Vapor canister
A Differential pressure sensor
B Exhaust oxygen sensor
C Throttle position sensor
D Coolant sensor
E Barometric pressure sensor
F Vehicle speed sensor
Note: *Not all components found on all vehicles*

Fig. 6.3 Locations of emissions system and related components (carbureted V8) (1984 engine shown, others similar)

C1 Electronic control module (ECM)
C2 ALCL connector
C3 'Check engine' light
C4 Code memory power
C5 Computer system ground
C6 Fuse panel
C8 Computer control harness
C9 Engine harness
1 Mixture control
2 Idle speed solenoid
2A Throttle kicker control
3 Early fuel evaporation valve
4 Air conditioner clutch control
4A Engine fan relay
5 Transmission converter clutch connector
6 Electronic spark timing connector
7 Electronic spark control (ESC) (L69 [HO] engines only)
8 Air injection pump
9 Air control solenoid valve
10 Air switching solenoid valve
11 Exhaust gas recirculation valve
12 Exhaust gas recirculation solenoid valve
13 Canister purge valve
15 Vapor canister
A Differential pressure sensor
B Exhaust oxygen sensor
C Throttle position sensor
D Coolant sensor
E Barometeric pressure sensor

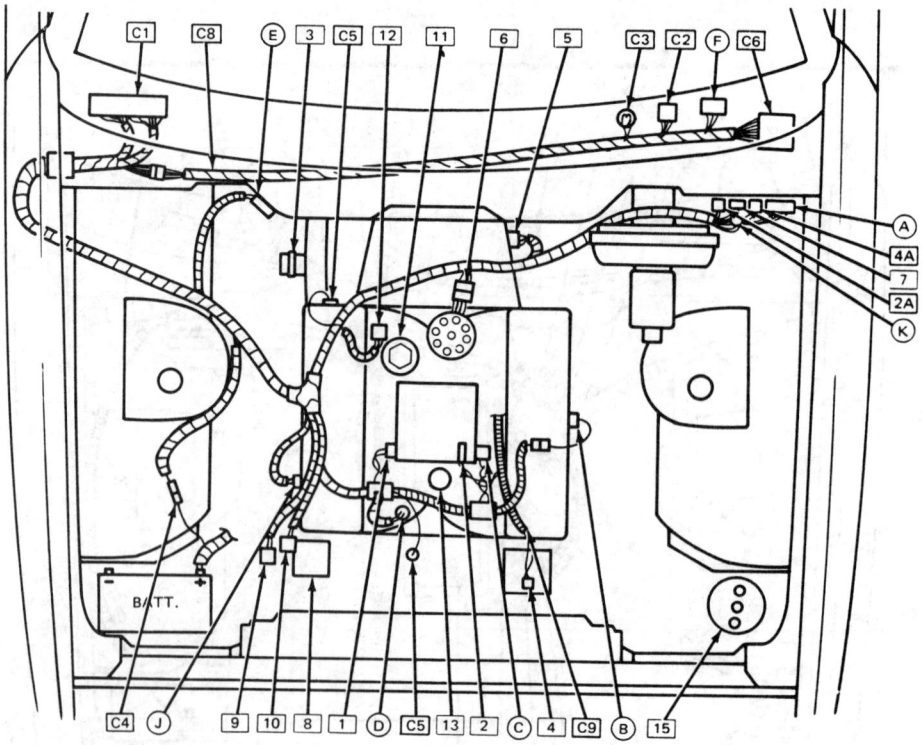

F Vehicle speed sensor
J ESC knock sensor (L69 [HO] engines only)
K Hood louver vacuum switch
Note: *Not all components found on all vehicles*

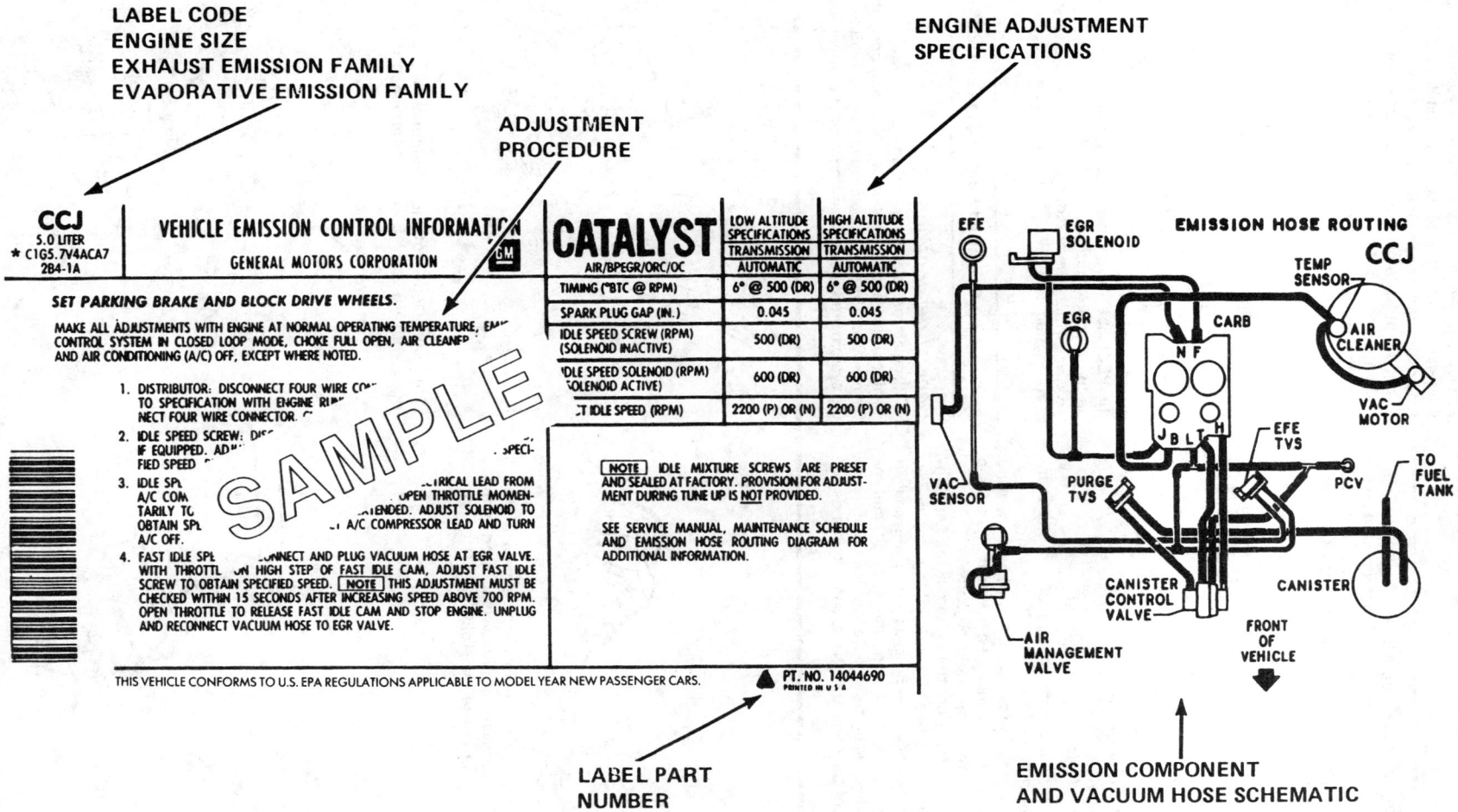

Fig. 6.4 A Vehicle Emissions Control Information label will be found in the engine compartment of all vehicles (Sec 1)

1 General information

To prevent pollution of the atmosphere from burned and evaporating gases, a number of emissions control systems are incorporated on the vehicles covered by this manual. The combination of systems used depends on the year in which the vehicle was manufactured, the locality to which it was originally delivered and the engine type. The major systems incorporated on the vehicles with which this manual is concerned include the:

Fuel Control System
Electronic Spark Timing (EST)
Air Injection Reaction (AIR)
Early Fuel Evaporation (EFE)
Exhaust Gas Recirculation (EGR)
Evaporative Emissions Control (EECS)
Transmission Converter Clutch (TCC)
Positive Crankcase Ventilation (PCV)
Thermostatic Air Cleaner (THERMAC)

All of these systems are linked, directly or indirectly, to the Computer Command Control System (CCCS).

The Sections in this Chapter include general descriptions, checking procedures (where possible) and component replacement procedures (where applicable) for each of the systems listed above.

Before assuming that an emissions control system is malfunctioning, check the fuel and ignitions systems carefully. In some cases, special tools and equipment, as well as specialized training, are required to accurately diagnose the causes of a rough running or difficult-to-start engine. If checking and servicing become too difficult or if a procedure is beyond the scope of the home mechanic, consult your dealer service department. This does not necessarily mean, however, that the emissions control systems are particularly difficult to maintain and repair. You can quickly and easily perform many checks and do most (if not all) of the regular maintenance at home with common tune-up and hand tools. **Note:** *The most frequent cause of emissions system problems is simply a loose or broken vacuum hose or wiring connection. Therefore, always check the hose and wiring connections first.*

Pay close attention to any special precautions outlined in this Chapter. It should be noted that the illustrations of the various systems may not exactly match the system installed on your particular vehicle, due to changes made by the manufacturer during production or from year to year.

A Vehicle Emissions Control Information label is located in the engine compartment of all vehicles with which this manual is concerned. This label contains important emissions specifications and setting procedures, as well as a vacuum hose schematic with emissions components identified. When servicing the engine or emissions systems, the VECI label in your particular vehicle should always be checked for up-to-date information.

Unless otherwise noted, procedures in this Chapter referring to carbureted models also apply to fuel injected (TBI) models.

2 Computer Command Control System (CCCS)

General description

This electronically controlled emissions system is linked with as many as nine other related emissions systems. It consists mainly of sensors (as many as 15) and an Electronic Control Module (ECM). Completing the system are various engine components which respond to commands

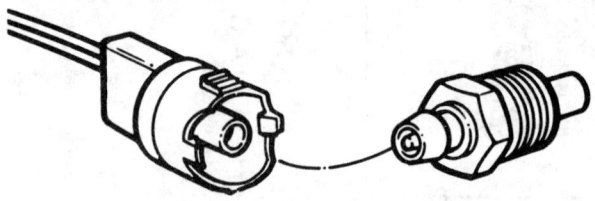

Fig. 6.5 The engine coolant temperature sensor (right), shown with its ECM harness connector is located in the cylinder head or intake manifold (Sec 2)

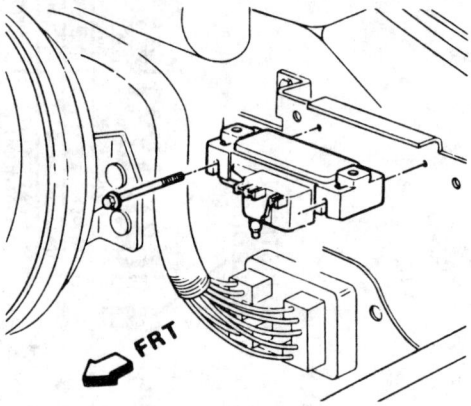

Fig. 6.6 Details of the differential pressure (vacuum) sensor used with V6 and V8 engines, attached to the firewall (Sec 2)

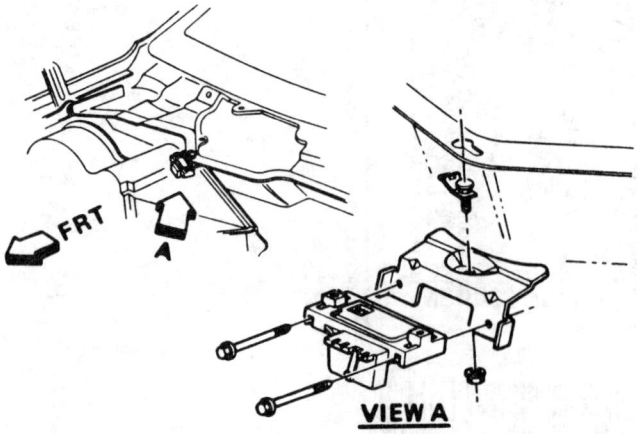

Fig. 6.7 Details of a typical firewall-mounted Baro sensor installation (Sec 2)

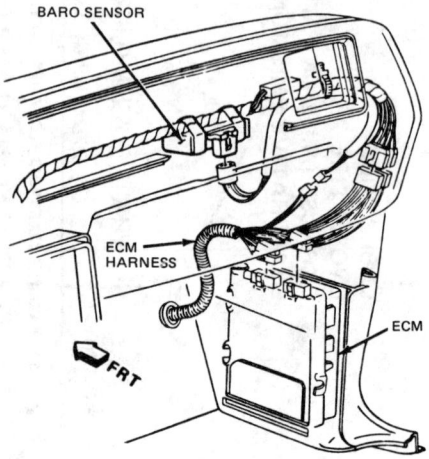

Fig. 6.8 Position of the Baro sensor relative to the ECM (Sec 2)

from the ECM.

In many ways, this system can be compared to the central nervous system in the human body. The sensors (nerves) constantly gather information and send this data to the ECM (brain) which processes the data and, if necessary, sends out a command for some type of vehicle (body) change.

Here's a specific example of how one portion of this system operates. An oxygen sensor, mounted in the exhaust manifold and protruding into the exhaust gas stream, constantly monitors the oxygen content of the exhaust gas as it travels through the exhaust pipe. If the percentage of oxygen in the exhaust gas is incorrect, an electrical signal is sent to the ECM. The ECM takes this information, processes it and then sends an electrical command to the carburetor Mixture Control (M/C) solenoid telling it to change the fuel/air mixture. To be effective, all this happens in a fraction of a second, and it goes on continuously while the engine is running. The end result is a fuel/air ratio which is constantly kept at a predetermined "exact' proportion, regardless of driving conditions.

Testing

One might think that a system which uses exotic electrical sensors and is controlled by an on-board computer would be difficult to diagnose. This is not necessarily the case.

The Computer Command Control System has a built-in diagnostic system which indicates a problem by flashing a "Check Engine' light on the instrument panel. When this light comes on during normal vehicle operation, a fault has been detected.

Perhaps more importantly, the ECM will recognize this fault in a particular system monitored by one of the various information sensors and store it in its memory in the form of a trouble code. Although the trouble code cannot reveal the exact cause of the malfunction, it greatly facilitates diagnosis as you or a dealer mechanic can "tap into" the ECM's memory and be directed to the problem area.

To extract this information from the ECM memory, you must use a short jumper wire to ground a 'Test' terminal. This terminal is part of a wiring connector located just behind the dashboard, next to the steering column. A small, rectangular plate is used to cover the connector and must be pried out of place to provide access to the terminals.

With the connector exposed to view, push one end of the jumper wire into the "Test' terminal and the other end into the "Ground' terminal. **Note:** *Do not start the engine with the "Test' terminal grounded.*

Turn the ignition to the On position – *not* the Start position. The "Check Engine' light should flash Trouble Code 12, indicating that the diagnostic system is working. Code 12 will consist of one flash, followed by a short pause, and then two flashes in quick succession. After a longer pause, the code will repeat itself two more times.

If no other codes have been stored, Code 12 will continue to repeat itself until the jumper wire is disconnected. If additional Trouble Codes have been stored, they will follow Code 12. Again, each Trouble Code will flash three times before moving on.

The ECM can also be checked for stored codes on carbureted models with the engine running. Completely remove the jumper wire from the "Test' and "Ground' terminals, start the engine and then plug the jumper wire back in. With the engine running, all stored Trouble Codes will flash. However, Code 12 will flash only if there is a fault in the distributor reference circuit.

Once the code(s) have been noted, use the Trouble Code Identification information which follows to locate the source of the fault. **Note:** *Whenever the positive battery cable is disconnected, all stored Trouble Codes in the EMC are erased. Be aware of this before you disconnect the battery for servicing or replacement of electrical components, engine removal, etc.*

It should be noted that the self-diagnosis feature built into this system does not detect all possible faults. If you suspect a problem with the Computer Command Control System, but a "Check Engine' light has not come on, have your local dealer perform a "system performance check.'

Furthermore, when diagnosing an engine performance, fuel economy or exhaust emissions problem (which is not accompanied by a "Check Engine' light) do not automatically assume the fault lies in this system. Perform all standard troubleshooting procedures, as indicated elsewhere in this manual before turning to the Computer Command Control System.

Finally, since this is an electronic system, you should have a basic knowledge of automotive electrics before attempting any diagnosis. Damage to the ECM, PROM or related components can easily occur is care is not exercised.

Trouble Code Identification

Following is a list of the Trouble Codes which may be encountered while diagnosing the Computer Command Control System. Also included are simplified troubleshooting procedures. If the problem persists after these checks have been made, the vehicle must be diagnosed by a professional mechanic who can use specialized diagnostic tools and advanced troubleshooting methods to check the system. Procedures marked with an asterisk (*) indicate component replacements which may not cure the problem in all cases. For this reason, you may want to seek professional advice before purchasing replacement parts.

To clear the Trouble Code(s) from the ECM memory, unplug the ECM electrical pigtail at the positive (+) battery cable.

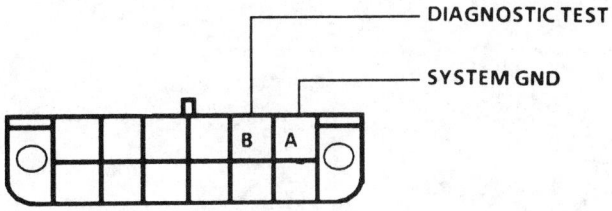

Fig. 6.9 Locations of the test terminals in the diagnostic connector (Sec 2)

Trouble Code	Circuit or system	Probable cause
12 (1 flash, pause, 2 flashes)	No reference pulses to ECM	This code should flash whenever the "Test' terminal is grounded with the ignition On and the engine not running. If additional Trouble Codes are stored (indicating a problem), they will appear after this code has flashed three times With the engine running, the appearance of this code indicates that no references from the distributor are reaching the ECM Carefully check the four-terminal EST connector at the distributor
13 (1 flash, pause, 3 flashes)	Oxygen sensor circuit	Check for a sticking or misadjusted throttle position sensor Check the wiring and connectors from the oxygen sensor Replace oxygen sensor (see Chapter 1)
14 (1 flash, pause, 4 flashes)	Coolant sensor circuit	**Note:** *If the engine is experiencing overheating problems (as indicated by high temperature gauge readings or the "hot' light coming on), rectify by referring to Chapters 1 and 3 before continuing.*

Trouble Code	Circuit or system	Probable cause
14 (1 flash, pause, 4 flashes)	Coolant sensor circuit	Check all wiring and connectors from the coolant sensor *Replace coolant sensor (located at front of left-hand cylinder head — V6 and V8; front of engine block — four-cylinder)
15 (1 flash, pause, 5 flashes)	Coolant sensor circuit	See above, plus: Check the wiring connections at the ECM
21 (two flashes, pause, one flash)	TPS circuit	Check for sticking or misadjusted TPS plunger Check all wiring and connections at the TPS and at the ECM *Adjust or replace TPS (see Chapter 4)
22 (two flashes, pause, two flashes)	TPS circuit	Check TPS adjustment (Chapter 4) Check ECM connector *Replace TPS (Chapter 4)
23 (two flashes, pause, three flashes)	M/C solenoid circuit	Check the electrical connections at the M/C solenoid (see Chapter 4). If OK, clear the ECM memory and recheck for code(s) after driving the vehicle Check wiring connections at the ECM Check wiring from M/C solenoid (Chapter 4)
24 (two flashes, pause, four flashes)	Vehicle speed sensor (VSS) circuit	**Note:** *A fault in this circuit should be indicated only while the vehicle is in motion. Disregard code 24 if set when drive wheels are not turning.* Check connections at the ECM Check the TPS setting (Chapter 4)
32 (three flashes, pause, two flashes)	Baro sensor circuit	Check for a short between sensor terminals B and C or the wires leading to these terminals Check the wire leading to ECM terminal 1 Check the ECM connections Check the wires leading to ECM terminals 21 and 22 *Replace Baro sensor (located in the engine compartment, attached to the firewall)
33 (three flashes, pause, three flashes)	MAP sensor	Check vacuum hose(s) from MAP sensor Check electrical connections at ECM *Replace MAP sensor
34 (three flashes, pause, four flashes)	Vacuum sensor circuit	Check the wiring leading to terminals 20, 21 and 22 of the ECM Check the connections at the ECM Check the vacuum sensor wiring and connections *Replace vacuum sensor (located in the engine compartment)
41 (four flashes, pause, one flash)	No distributor signals	Check all wires and connections at the distributor Check distributor pick-up coil connections (Chapter 5) Check vacuum sensor circuit (see above)

Trouble Code	Circuit or system	Probable cause
42 (four flashes, pause, two flashes)	Bypass or EST problem	**Note:** *If the vehicle will not start and run, check the wire leading to ECM terminal 12.* **Note:** *An improper HEI module can cause this trouble code.* Check the EST wire leading to the HEI module E terminal Check all distributor wires Check the wire leading from EST terminal A to ECM terminal 12 and the wire from EST terminal C to ECM terminal 11 Check all ECM connections *Replace HEI module
43 (four flashes, pause, three flashes)	Electronic Spark Control (ESC) system	Check wire leading to ECM terminal L Check the wiring connector at the ESC controller and at the ECM Check wire from knock sensor to ESC controller; if necessary, reroute it away from other wires such as spark plug, etc.
44 (four flashes, pause, four flashes)	Lean exhaust	*On carburetor-equipped vehicles:* Check for a sticking M/C solenoid (Chapter 4) Check ECM wiring connections, particularly terminals 14 and 9 Check for vacuum leakage at carburetor base gasket, vacuum hoses or intake manifold gasket Check for air leakage at air management system-to-exhaust ports and at decel valve *Replace oxygen sensor *On fuel-injected vehicles:* Check ECM wiring connections Check for incorrect fuel pressure Check for water in the fuel *Replace oxygen sensor
45 (four flashes, pause, five flashes)	Rich exhaust	*On carburetor-equipped vehicles:* Check for a sticking M/C solenoid (Chapter 4) Check wiring at M/C solenoid connector Check the evaporative charcoal canister and its components for the presence of fuel (Chapters 1, 6) *Replace oxygen sensor *On fuel-injected models:* Check ECM wiring connections Check fuel pressure (Chapter 4) Check injectors for leakage (Chapter 4) Check for intermittent bursts of fuel from the injectors at idle, indicating a faulty TPS
51 (five flashes, pause, one flash)	PROM problem	The PROM is located inside the ECM and is very delicate and easily broken. All diagnostic procedures should be done by a dealer mechanic.
54 (five flashes, pause, four flashes)	M/C solenoid	Check all M/C solenoid (carburetor-equipped vehicles) and ECM wires and connections *Replace the M/C solenoid (see Chapter 4)

Trouble Code	Circuit or system	Probable cause
55 (five flashes, pause, five flashes)	Oxygen sensor circuit	Check for corrosion on the ECM connectors and terminals Make sure that the four-terminal EST wiring harness is not too close to electrical signals such as spark plug wires, distributor housing, alternator, etc. Check the wiring of the various sensors *Replace the oxygen sensor

3 Fuel Control System

General description

1 The function of this system is to control the flow of fuel through the carburetor idle main metering circuits. The major components of the system are the mixture control (M/C) solenoid and the oxygen sensor.

2 The M/C solenoid changes the fuel/air mixture by allowing more or less fuel to flow through the carburetor. The M/C solenoid, located in the carburetor air horn, is in turn controlled by the ECM, which provides a ground for the solenoid. When the solenoid is energized, the fuel flow through the carburetor is reduced, providing a leaner mixture, and when the ECM removes the ground path, the solenoid de-energizes and allows more fuel flow.

3 The ECM determines the proper fuel mixture required by monitoring a signal sent by the oxygen sensor located in the exhaust stream. When the mixture is lean, the oxygen sensor voltage is low and the ECM commands a richer mixture. Conversely, when the mixture is rich, the oxygen sensor voltage is higher and the ECM commands a leaner mixture.

Checking

Oxygen sensor

4 Make sure that the oxygen sensor has been replaced at the proper maintenance interval (refer to Chapter 1).

5 The proper operation of the sensor depends on the four conditions which follow:

6 *Electrical conditions:* The low voltages and low currents generated by the sensor depend upon good, clean connections which should be checked whenever a malfunction of the sensor is suspected or indicated.

7 *Outside air supply:* The sensor is designed to allow air circulation to the internal portion of the sensor. Whenever the sensor is removed and installed or replaced, make sure the air passages are not restricted.

8 *Proper operating temperature:* The ECM will not react to the sensor signal until the sensor reaches approximately 600°F (360°C). This factor must be taken into consideration when evaluating the performance of the sensor.

9 *Non-leaded fuel:* The use of non-leaded fuel is essential for proper operation of the sensor. Make sure the fuel you are using is of this type.

10 In addition to observing the above conditions, special care must be taken whenever the sensor is handled, as directed in Chapter 1. Violation of any of these cautionary procedures may lead to sensor failure. **Note:** *Do not attempt to measure the voltage output of the oxygen sensor, because the current drain from a conventional voltmeter would be enough to permanently damage the sensor. For the same reason, never hook up test leads, jumpers or other electrical connections.*

Mixture control solenoid

11 Check the wiring connectors and wires leading to the mixture control solenoid for looseness, fraying and other damage. Repair or replace any damaged wiring as necessary.

12 Check the mixture control solenoid for apparent physical damage. Replace it if damage is found.

Component replacement

Oxygen sensor

13 To replace the sensor, refer to Chapter 1.

Mixture control solenoid

14 For E2SE carburetors, refer to Chapter 4.

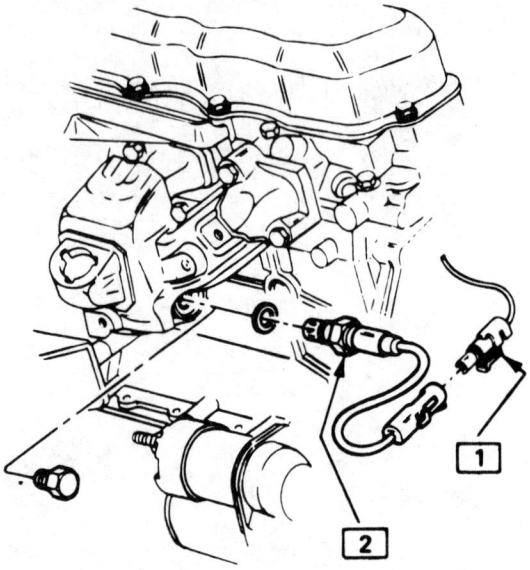

Fig. 6.10 Details of a typical oxygen sensor installation (L4 engine shown) (Sec 3)

1 TBI harness connector *2 Oxygen sensor*

15 For E4ME carburetors, due to the need for special tools, this procedure must be performed by a dealer or other suitably equipped repair shop.

4 Electronic Spark Timing (EST)

1 Electronic Spark Timing is used on all engines with which this manual is concerned. The EST distributor contains no vacuum or centrifugal advance, depending on commands from the ECM instead. The ECM receives a reference pulse form the distributor, indicating both engine RPM and crankshaft position, determines the proper spark advance for the engine operating conditions and sends an EST pulse to the distributor.

2 Under normal operating conditions, the ECM will always control the spark advance; however, under certain conditions such as cranking or setting base timing, the distributor can operate independent of ECM control. This condition is called 'bypass' and is determined by the bypass lead from the ECM to the distributor. When the bypass lead is over two volts, the ECM will control the spark; however, disconnecting the four-terminal EST connector, or grounding the bypass lead, will cause the engine to operate in the bypass mode.

3 The V8 engines equipped with crossfire fuel injection (Model 400 TBI) or the high output (HO) option use Electronic Spark Control (ESC) in conjunction with EST to reduce spark advance under detonation conditions. A knock sensor signals a separate ESC controller to retard the timing if it senses knock. The ESC controller, in turn, signals the ECM, which reduces spark advance until no more signals are received from the knock sensor.

4 For further information (and checking and component replacement procedures) regarding the EST distributor, refer to Chapter 5.

5 Air Injection Reaction (AIR)/PULSAIR reactor systems

General description

Note: *If your engine is equipped with an air pump, your concern in this Section will be with the AIR system; if no air pump is present, refer to the procedures involving the PULSAIR system.*

AIR system

1 The AIR system helps reduce hydrocarbons and carbon monoxide levels in the exhaust by injecting air into the exhaust ports of each cylinder during cold engine operation or directly into the catalytic converter during normal operation. It also helps the catalytic converter reach proper operating temperature quickly during warm-up.

2 The AIR system uses an air pump to force the air into the exhaust stream. An Air Management Valve, controlled by the vehicle's Electronic Control Module (ECM), directs the air to the correct location depending on engine temperature and driving conditions. During certain situations such as deceleration, the air is diverted to the air cleaner to prevent backfiring from too much oxygen in the exhaust stream. On V6 engines only, a deceleration (gulp) valve is used to help prevent backfiring during high vacuum conditions. The deceleration valve allows air to flow into the intake manifold and enter the air/fuel mixture to lean the rich condition created at high vacuum when the throttle closes upon deceleration. One-way check valves are also used in the AIR system's air lines to prevent exhaust gases from being forced back through the system.

3 The following components are utilized in the AIR system: an engine-driven air pump; air control, air switching and divert management valves; a deceleration valve (V6 only); air flow and control hoses; check valves; and a dual bed catalytic converter.

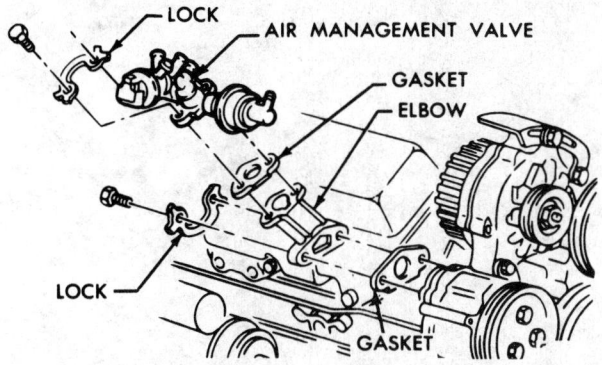

Fig. 6.11a Details of a typical AIR management valve installation (Sec 5)

PULSAIR system

4 This system performs some of the same functions as the AIR system, but utilizes exhaust pressure pulses to draw air into the exhaust system. Fresh air that is filtered by the air cleaner, to avoid the build-up of dirt on the check valve seat, is supplied to the system on a command from the ECM. The air cleaner also serves as a muffler to reduce noise in the system.

5 Components utilized in the system include the PULSAIR valve and external tubes and hoses.

6 The PULSAIR system's operation begins with the engine's firing, creating a pulsating flow of exhaust gases which are of positive or negative pressure. The pressure or vacuum is transmitted through the external tubes to the PULSAIR valve, which reacts as follows:

7 If the pressure is positive, the disc in the valve is forced to the closed position and no exhaust gas is allowed to flow past the valve and into the air supply.

8 If there is negative pressure (vacuum) present in the exhaust system at the valve, the disc will open, allowing fresh air to mix with the exhaust gases.

9 The disc, due to the inertia of the system, ceases to follow the pressure pulsations at high engine rpm. At this point, the disc remains closed, preventing any further flow of fresh air.

Checking

AIR system

10 Because of the ECM's influence on this system, it is difficult for the home mechanic to make a proper diagnosis. If the system is suspected of not operating properly, individual components can be checked.

11 Begin any inspection by carefully checking all hoses, vacuum lines and wires. Be sure they are in good condition and that all connections are tight and clean. Also make sure that the pump drivebelt is in good condition and properly adjusted.

12 To check the pump, allow the engine to reach normal operating temperature and run it at about 1500 rpm. Locate the air hose running to the catalytic converter and squeeze it to feel for pulsations. Have an assistant increase the engine speed and check for a parallel increase in air flow. If this is observed as described, the pump is functioning properly. If it is not operating in this manner, a faulty pump is indicated.

13 The check valve can be inspected by first removing it from the air line. Attempt to blow through it from both directions. Air should only pass through in the direction of normal air flow. If it is either stuck open or stuck closed, the valve should be replaced.

14 To check the Air Management Valve, disconnect the vacuum signal line at the valve. With the engine running, see if vacuum is present in the line. If not, the line is clogged. If vacuum is present, have the valve checked out by a dealer or other qualified mechanic.

15 To check the deceleration valve, remove the air cleaner and plug the air cleaner vacuum source. With the engine running at the specified idle speed, remove the small deceleration valve signal hose from the

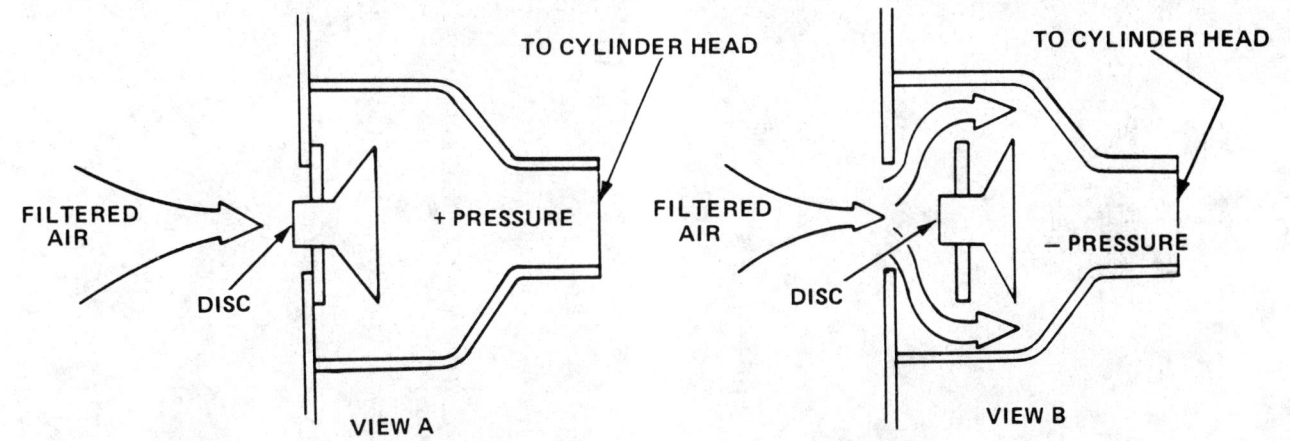

Fig. 6.11b With positive pressure present in the PULSAIR valve (left), the valve disc remains closed and no air enters the valve; when a vacuum is present in the valve (right), the disc opens and air mixes with the exhaust gases (Sec 5)

5.22 Removing the air pump pulley (note that the drivebelt has previously been removed)

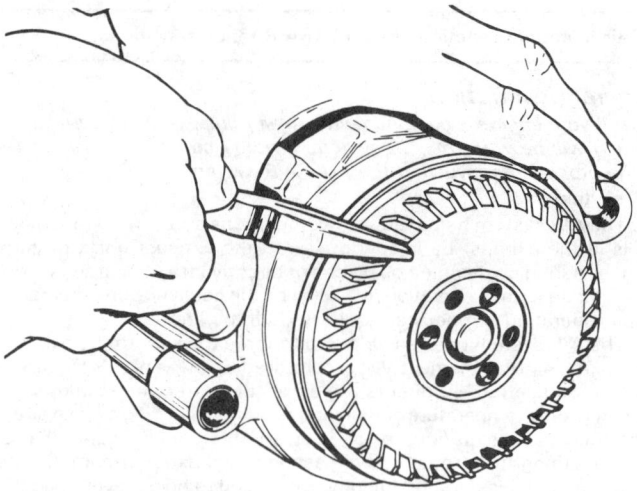

Fig. 6.12 Removing the AIR pump filter (remove as shown — do not insert any tool behind the filter, as damage to the pump may occur) (Sec 5)

5.29 Removing the hose from the AIR system check valve

5.33 Removing a hose from the AIR management valve

5.39 Removing the adjusting bolt from the AIR system pump (the mounting bolt is above the pump)

5.42 AIR system deceleration valve (arrow) — V6 engines only manifold vacuum source, then reconnect the signal hose and listen for air flow through the ventilation pipe and into the deceleration valve. There should also be a noticeable engine speed drop when the signal hose is reconnected. If the air flow does not continue for at least one second, or the engine speed does not drop noticeably, check the deceleration valve hoses for restrictions and leaks. If no restrictions or leaks are found, replace the deceleration valve with a new one.

PULSAIR system

16 A simple, functional test of this system can be performed with the engine running. Disconnect the rubber hose from the air valve and hold your hand over the valve's inlet hole. With the engine idling, there should be a steady stream of air being sucked into it. Have an assistant apply throttle and as the engine gains speed, see if the suction increases. If this does not occur, the lines are leaking or restricted or the check valves are sticking. Also make sure that air is not being blown out of the air valve as this also is an indication that the check valves are sticking open. Service or replace the components as necessary. If other PULSAIR problems are suspected, have a dealer or repair shop diagnose the problems, as they might relate to the ECM/CCCS.

Component replacement (AIR system)

Drivebelt

17 Loosen the pump mounting bolt and the pump adjustment bracket bolt.

18 Move the pump inboard until the belt can be removed.

19 Install the new belt and adjust it (refer to Chapter 1).

AIR pump pulley and filter

20 Compress the drivebelt to keep the pulley from turning and loosen the pulley bolts.

21 Remove the drivebelt as described above.

22 Remove the mounting bolts and lift off the pulley (photo).

23 If the fan-like filter must be removed, grasp it firmly with needle-nose pliers, as shown in the accompanying illustration, and pull it from the pump. **Note:** *Do not insert a screwdriver between the filter and pump housing as the edge of the housing could be damaged. The filter will usually be distorted when pulled off. Be sure no fragments fall into the air intake hole.*

24 The new filter is installed by placing it in position on the pump, placing the pulley over it and tightening the pulley bolts evenly to draw the filter into the pump. Do not attempt to install a filter by pressing or hammering it into place. **Note:** *It is normal for the new filter to have an interference fit with the pump housing and, upon initial operation, it may squeal until worn in.*

25 Install the drivebelt and, while compressing the belt, tighten the pulley bolts to the specified torque.

26 Adjust the drivebelt tension.

Hoses and tubes

27 To replace any tube or hose, always note how it is routed first, either with a sketch or with numbered pieces of tape.

28 Remove the defective hose or tube and replace it with a new one of the same material and size and tighten all connections.

Check valve

29 Disconnect the pump outlet hose at the check valve (photo).

30 Remove the check valve from the pipe assembly, making sure not to bend or twist the assembly.

31 Install a new valve, after making sure that it is a duplicate of the part removed, and tighten all connections.

Air management valve

32 Remove the air cleaner.

33 Disconnect the vacuum signal line from the valve. Also disconnect the air hoses and wiring connectors (photo).

34 If the mounting bolts are retained by tabbed lock washers, bend the tabs back, then remove the mounting bolts and lift the valve off of the adapter or bracket.

35 Installation is the reverse of the removal procedure. Be sure to use a new gasket when installing the valve.

Air pump

36 Remove the air management valve and adapter, if equipped.

37 If the pulley must be removed from the pump, it should be done prior to removing the drivebelt as described elsewhere in this Section.

38 If the pulley is not being removed, remove the drivebelt.

39 Remove the pump mounting bolts and separate the pump from the engine (photo).

40 Installation is the reverse of the removal procedure. **Note:** *Do not tighten the pump mounting bolts until all components are installed.*

41 Following installation, adjust the drivebelt tension as described in Chapter 1.

Deceleration valve

42 Disconnect the vacuum hoses from the valve (photo).

43 Remove the screws retaining the valve to the engine bracket (if present) and remove the valve.

44 Install a new valve and reconnect all hoses.

Component replacement (PULSAIR system)

45 Remove the air cleaner and disconnect the negative battery cable from the battery.

46 Disconnect the rubber hose from the pulse air valve.

47 If equipped, disconnect the support bracket.

48 If equipped, remove the PULSAIR solenoid and bracket from the PULSAIR unit.

49 Loosen the nuts that secure the air tubes to the cylinder head and remove the assembly. Due to the high temperature at this area, these connections may be difficult to loosen. Penetrating oil applied to the threads of the nuts may help.

50 Before installing, apply a light coat of oil to the ends of the air tubes and anti-seize compound to the threads of the attaching nuts.

51 Installation is the reverse of the removal procedure.

Note: *For information relating to the replacement of catalytic converter components, refer to Chapter 4.*

6 Early Fuel Evaporation (EFE) system

General description

Servo type

If your vehicle is equipped with a vacuum servo type EFE system, located between the exhaust manifold and the exhaust pipe, refer to Chapter 1 for a general description of the system and system checking procedures.

Electrically heated type

1 This unit provides rapid heat to the intake air supply on carbureted engines by means of a ceramic heater grid which is integral with the carburetor base gasket and located under the primary bore.

2 The components involved in the EFE's operation include the heater grid, a relay, electrical wires and connectors and the ECM.

3 The EFE heater unit is controlled by the vehicle's Electronic Control Module (ECM) through a relay. The ECM senses the coolant temperature level and applies voltage to the heater unit only when the engine is below a pre-determined level. At normal operating temperatures, the heater unit is off.

4 If the EFE heater is not coming on, poor cold engine performance will be experienced. If the heater unit is not shutting off when the engine is warmed up, the engine will run as if it is out of tune (due to the constant flow of hot air through the carburetor).

Checking

Servo type

5 To check the operation of the EFE/TVS, allow the engine temperature to fall below 80°F (26°C).

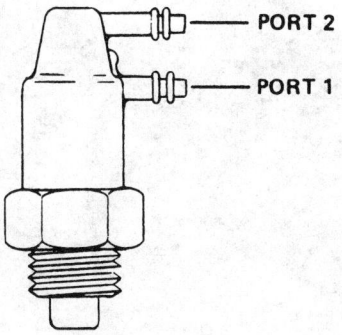

Fig. 6.13 A typical EFE/TVS switch employed in servo-type EFE systems, typically located in the intake manifold (Sec 6)

6 Drain the coolant from the engine until the level is below the level of the switch.
7 Disconnect and label the vacuum lines, then remove the switch.
8 Blow into either of the TVS ports. Air should flow through the valve.
9 Heat the TVS in hot water until the temperature of the valve is above 90°F (32°C).
10 Blow into either port of the TVS. No air should flow through the valve.
11 If the condition in either Step 9 or 10 is not met, replace the valve with a new one.
12 For other checking procedures for the servo type EFE system, refer to Chapter 1.

Electrically heated type

13 If the EFE system is suspected of malfunctioning while the engine is cold, first visually check all electrical wires and connectors to be sure they are clean, tight and in good condition.
14 With the ignition switch On, use a circuit tester or voltmeter to be sure current is reaching the relay. If not, there is a problem in the wiring leading to the relay, in the ECM's thermo switch, or in the ECM itself.
15 Next, with the engine cold but the ignition switch On, disconnect the heater unit wiring connector and use a circuit tester or voltmeter to see if current is reaching the heater unit. If so, use a continuity tester to check for continuity in the wiring connector attached to the heater unit. If continuity exists, the system is operating correctly in the cold engine mode.
16 If current was not reaching the heater unit, but was reaching the relay, replace the relay.
17 To check that the system turns off at normal engine operating temperature, first allow the engine to warm up thoroughly. With the engine idling, disconnect the heater unit wiring connector and use a circuit tester or voltmeter to check for current at the heater unit.
18 If current is reaching the heater unit, a faulty ECM is indicated.
19 For confirmation of the ECM's condition, refer to Section 2 or have the system checked by a dealer or automotive repair shop.

6.37 Removing the EFE system under-carburetor heater element

Component replacement

Servo type EFE switch

20 **Note:** *The oxygen sensor may be located near the EFE valve on some vehicles. If so, care should be taken not to damage the sensor or its connections during this procedure.*
21 Disconnect the vacuum hose from the actuator fitting.
22 Remove the exhaust pipe-to-manifold nuts, washers (if used) and tension springs (if used).
23 Lower the exhaust (crossover) pipe. In some cases, complete removal is not necessary.
24 Remove the nuts retaining the valve assembly to the exhaust manifold.
25 Remove the valve assembly. If only the actuator assembly is to be replaced, separate it from the valve.
26 Installation is the reverse of the removal procedure.

TVS switch (V8 models equipped with carburetor only)

27 Drain the engine coolant until the level is below the level of the switch.
28 Disconnect the hoses from the TVS switch, making note of their positions for reassembly.
29 Using a wrench, remove the TVS switch.
30 Apply a soft-setting sealant uniformly to the threads of the new TVS switch. Be careful that none of the sealant gets on the sensor end of the switch.
31 Install the switch.
32 Connect the vacuum hoses to the switch in their original positions and add coolant as necessary.

Heater element

33 Remove the air cleaner.
34 Disconnect all wires, vacuum hoses and fuel lines from the carburetor.
35 Disconnect the wiring connector leading to the EFE unit.
36 Remove the carburetor, referring to Chapter 4 if necessary.
37 Lift off the heater unit (photo).
38 Installation is the reverse of the removal procedure.
39 Following installation, start the engine and check for air and fuel leaks around the carburetor.

7.8 Removing the EGR valve from the manifold

7.15 Removing the EGR TVS from the manifold (V6 engines only)

Heater relay

40 Disconnect the cable from the negative battery terminal.
41 Disconnect and label the relay electrical connections.
42 Remove the bolts retaining the relay and remove the relay.
43 Installation is the reverse of the removal procedure.

Solenoid

44 Disconnect the cable from the negative battery terminal.
45 If necessary to gain access to the solenoid, remove the air cleaner.
46 Disconnect and label the electrical connections and vacuum hoses at the solenoid.
47 Remove the solenoid.
48 Installation is the reverse of the removal procedure.

7 Exhaust Gas Recirculation (EGR) system

General description

1 An EGR system is used on all engines with which this manual is concerned. The system meters exhaust gases into the engine induction system through passages cast into the intake manifold and into the fuel/air mixture for the purpose of lowering combustion temperatures, thereby reducing the amount of oxides of nitrogen (NOX) formed.
2 The amount of exhaust gas admitted is regulated by a vacuum or backpressure controlled (EGR) valve in response to engine operating conditions. The EGR valve, in turn, is under control of the CCCS/ECM.
3 Common engine problems associated with the EGR system are rough idling or stalling when at idle, rough engine performance during light throttle application and stalling during deceleration.

Checking

4 Refer to Chapter 1 for EGR valve checking procedures.
5 If the EGR valve appears to be in proper operating condition, carefully check all hoses connected to the valve for breaks, leaks and kinks. Replace or repair the valve/hoses as necessary.
6 Due to the interrelationship of the EGR system and the ECM, further checks of the system should be made by referring to Section 2 or having the system checked by a dealer or repair shop.

Component replacement

EGR valve

7 Disconnect the vacuum hose at the EGR valve.
8 Remove the nuts or bolts which secure the valve to the intake manifold or adapter (photo).
9 Lift the EGR valve from the engine.
10 Clean the mounting surfaces of the EGR valve. Remove all traces of gasket material (photo).
11 Place the new EGR valve, with a new gasket, on the intake manifold or adapter and tighten the attaching nuts or bolts.
12 Connect the vacuum signal hose.

TVS (V6 engines only)

13 Drain sufficient coolant from the radiator to bring the level below the bottom of the TVS (the TVS is located at the front of the intake manifold).
14 Remove the hoses from the valve, labeling them to ensure proper installation.
15 Remove the TVS with a wrench and replace it with a new one (photo).
16 Installation is the reverse of the removal procedure.

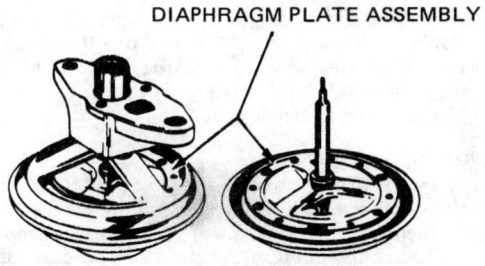

Fig. 6.14 Different types of EGR valves are identified by their diaphragm plate assembly designs (Be sure to obtain a valve of identical type when replacing it) (Sec 7)

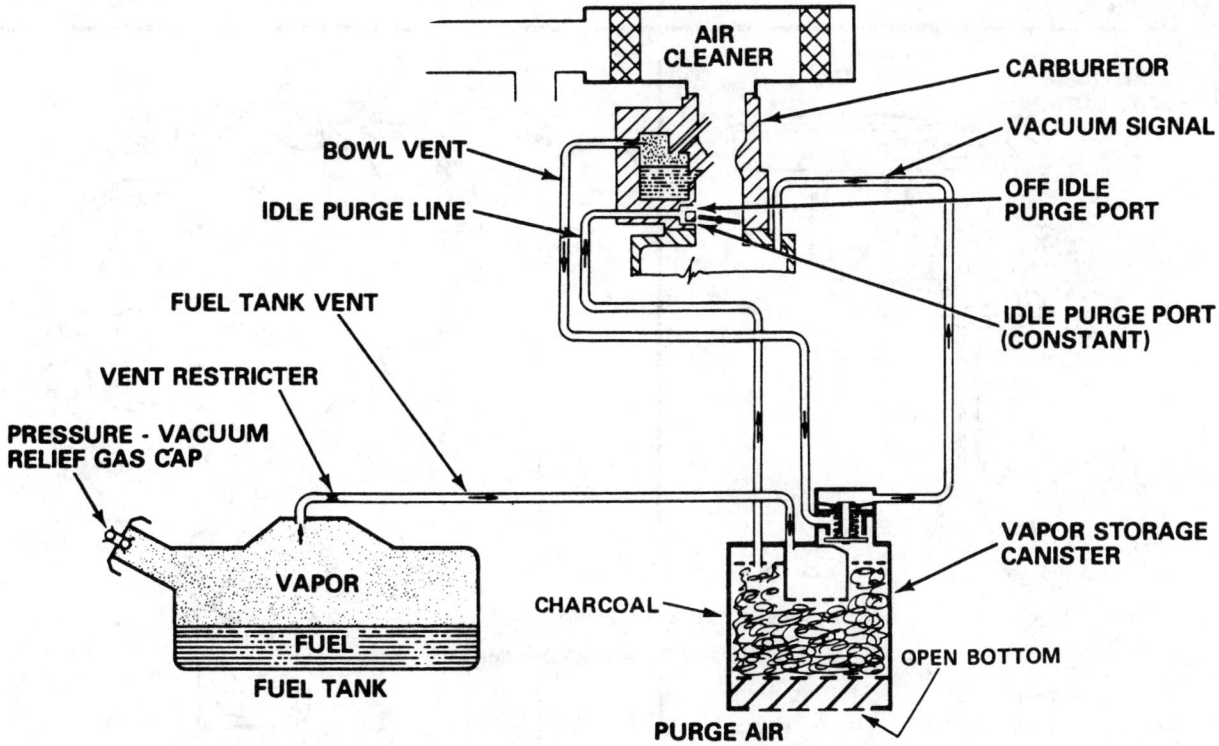

Fig. 6.15 Details of a typical EECS system (Sec 8)

8 Evaporative Emissions Control System (EECS)

General description

1 This system is designed to trap and store fuel that evaporates from the carburetor and fuel tank which would normally enter the atmosphere and contribute to hydrocarbon (HC) emissions.

2 The system consists of a charcoal-filled canister and lines running to and from the canister. These lines include a vent line from the gas tank, a vent line from the carburetor float bowl or injection unit, an idle purge line into the vehicle's induction system and a vacuum line to the manifold. In addition, there is a purge valve in the canister. The CCCS/ECM controls the vacuum to the purge valve with an electrically operated solenoid. The fuel tank cap is also an integral part of the system.

3 A tip-off that the system is not operating properly is a strong fuel odor.

Checking

4 Maintenance and replacement of the charcoal canister filter is covered in Chapter 1.

5 Check all lines in and out of the canister for kinks, leaks and breaks along their entire lengths. Repair or replace as necessary.

6 Check the gasket in the gas cap for signs of drying, cracking or breaks. Replace the gas cap with a new one if defects are found.

7 Due to its interrelationship with the CCCS/ECM, other system checks should be made by referring to Section 2 or having the system checked by a dealer or repair shop.

Component replacement

8 Replacement of the canister filter is covered in Chapter 1.

9 When replacing any line running to or from the canister, make sure the replacement line is a duplicate of the one you are replacing. These lines are often color-coded to denote their particular usage.

9 Positive Crankcase Ventilation (PCV) system

General description

1 The positive crankcase ventilation system, or PCV as it is more commonly called, reduces hydrocarbon emissions by circulating fresh air through the crankcase to pick up blow-by gases which are then rerouted through the carburetor to be burned in the engine.

2 The main components of this simple system are vacuum hoses and a PCV valve which regulates the flow of gases according to engine speed and manifold vacuum.

Checking

3 The PCV system can be checked quickly and easily for proper operation. This system should be checked regularly as carbon and gunk deposited by the blow-by gases will eventually clog the PCV valve and/or system hoses. When the flow of the PCV system is reduced or stopped, common symptoms are rough idling or reduced engine speed at idle.

4 To check for proper vacuum in the system, remove the top plate of the air cleaner and locate the small PCV filter on the inside of the air cleaner housing.

5 Disconnect the hose leading to this filter. Be careful not to break the molded fitting on the filter.

6 With the engine idling, place your thumb lightly over the end of the hose. You should feel a slight pull or vacuum. The suction may be heard as your thumb is released. This will indicate that air is being drawn all the way through the system. If a vacuum is felt, the system is functioning properly. Check that the filter inside the air cleaner housing is not clogged or dirty. If in doubt, replace the filter with a new one, an inexpensive safeguard (refer to Chapter 1).

7 If there is very little vacuum or none at all at the end of the hose, the system is clogged and must be inspected further.

8 Shut off the engine and locate the PCV valve. Carefully pull it from its rubber grommet. Shake it and listen for a clicking sound. That is the rattle of the valve's check needle. If the valve does not click freely, replace it with a new one.

9 Now start the engine and run it at idle speed with the PCV valve removed. Place your thumb over the end of the valve and feel for suction. This should be a relatively strong vacuum which will be felt immediately.

10 If little or no vacuum is felt at the PCV valve, turn off the engine and disconnect the vacuum hose from the other end of the valve. Run the engine at idle speed and check for vacuum at the end of the hose just disconnected. No vacuum at this point indicates that the vacuum hose or inlet fitting at the engine is plugged. If it is the hose which is blocked, replace it with a new one or remove it from the engine and

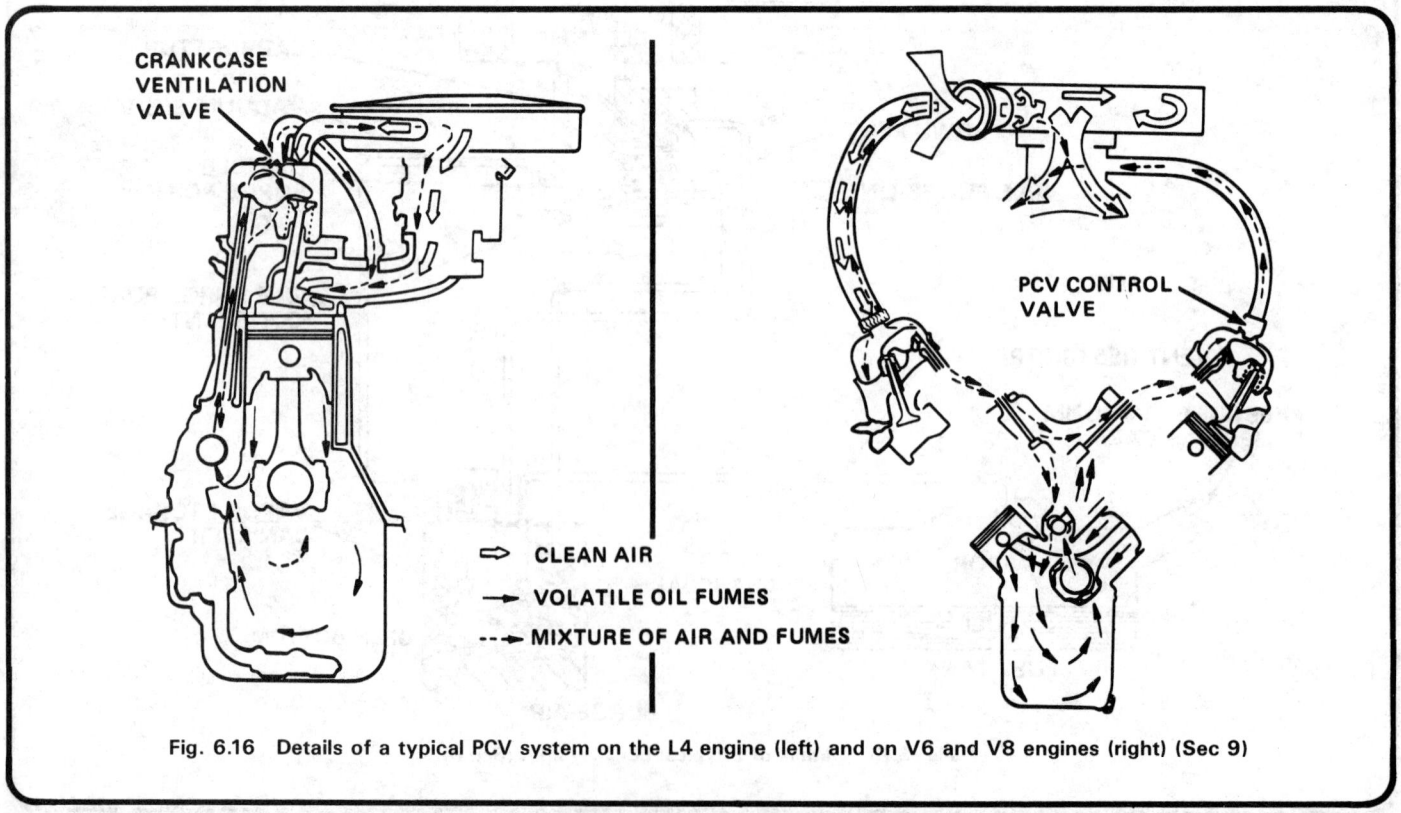

Fig. 6.16 Details of a typical PCV system on the L4 engine (left) and on V6 and V8 engines (right) (Sec 9)

blow it out sufficiently with compressed air. A clogged passage at the carburetor or manifold requires that the component be removed and thoroughly cleaned to remove carbon build-up. A strong vacuum felt going into the PCV valve, but little or no vacuum coming out of the valve, indicates a failure of the PCV valve requiring replacement with a new one.

11 When purchasing a new PCV valve, make sure it is the correct one for your engine. An incorrect PCV valve may pull too little or too much vacuum, possibly leading to engine damage.

Component replacement

12 The replacement procedures for both the PCV valve and filter are covered in Chapter 1.

10 Thermostatic Air Cleaner (THERMAC)

General description

1 The thermostatic air cleaner (THERMAC) system is provided to improve engine efficiency and reduce hydrocarbon emissions during the initial warm-up period by maintaining a controlled air temperature at the carburetor. This temperature control of the incoming air allows leaner carburetor and choke calibrations.

2 The system uses a damper assembly located in the snorkel of the air cleaner housing to control the ratio of cold and warm air directed into the carburetor. This damper is controlled by a vacuum motor which is, in turn, modulated by a temperature sensor in the air cleaner. On some engines, a check valve is used in the sensor, which delays the opening of the damper flap when the engine is cold and the vacuum signal is low.

3 It is during the first few miles of driving (depending on outside temperature) that this system has its greatest effect on engine performance and emissions output. When the engine is cold, the damper flap blocks off the air cleaner inlet snorkel, allowing only warm air from the exhaust manifold to enter the carburetor. Gradually, as the engine warms up, the flap opens the snorkel passage, increasing the amount of cold air allowed in. Once the engine reaches normal operating temperature, the flap opens completely, allowing only cold, fresh air to enter.

4 Because of this cold-engine-only function, it is important to periodically check this system to prevent poor engine performance when cold or overheating of the fuel mixture once the engine has reached operating temperatures. If the air cleaner valve sticks in the 'no heat' position, the engine will run poorly, stall and waste gas until it has warmed up on its own. A valve sticking in the 'heat' position causes the engine to run as if it is out of tune (due to the constant flow of hot air to the carburetor).

Checking

5 Refer to Chapter 1 for maintenance and checking procedures for this system. If problems were encountered in the system's performance while performing the routine maintenance checks, refer to the procedures which follow.

6 If the damper door did not close off snorkel air when the cold engine was first started, disconnect the vacuum hose at the snorkel vacuum motor and place your thumb over the hose end, checking for vacuum. If there is vacuum going to the motor, check that the damper door and link are not frozen or binding within the air cleaner snorkel. Replace the vacuum motor if the hose routing is correct and the damper door moves freely.

7 If there was no vacuum going to the motor in the above test, check the hoses for cracks, crimps and proper connection. If the hoses are clear and in good condition, replace the temperature sensor inside the air cleaner housing.

Component replacement

Air cleaner vacuum motor

8 Remove the air cleaner assembly from the engine and disconnect the vacuum hose from the motor.

9 Drill out the two spot welds which secure the vacuum motor retaining strap to the snorkel tube.

10 Remove the motor attaching strap.

11 Lift up the motor, cocking it to one side to unhook the motor linkage at the control damper assembly.

12 To install, drill a 7/64-inch hole in the snorkel tube at the center of the retaining strap.

13 Insert the vacuum motor linkage into the control damper assembly.

14 Using the sheet metal screw supplied with the motor service kit, attach the motor and retaining strap to the snorkel. Make sure the sheet metal screw does not interfere with the operation of the damper door. Shorten the screw if necessary.

15 Connect the vacuum hose to the motor and install the air cleaner assembly.

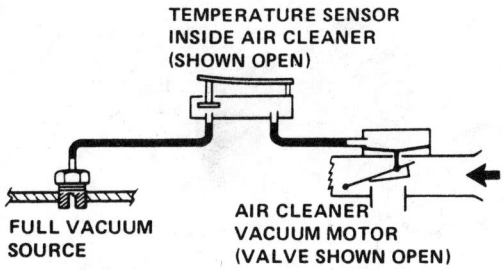

Fig. 6.17 Details of a typical THERMAC system (Sec 10)

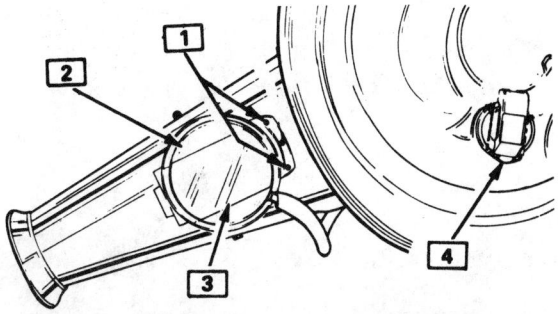

Fig. 6.18 Components involved in removing the THERMAC vacuum motor on a typical carbureted vehicle (Sec 10)

1 Spot welds *3 Retaining strap*
2 Motor assembly *4 Sensor*

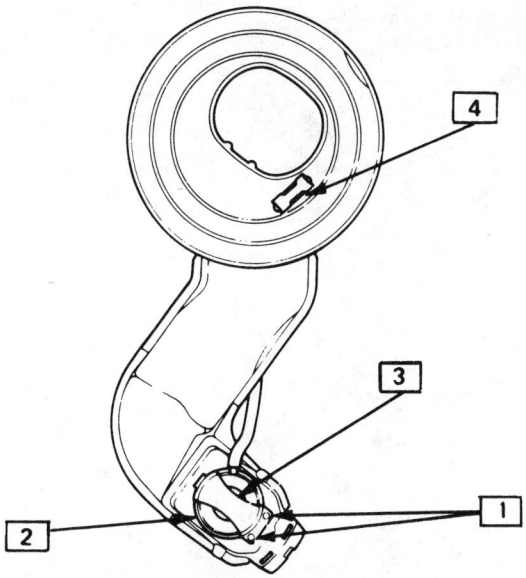

Fig. 6.19 Components involved in removing the THERMAC vacuum motor on a typical fuel-injected vehicle (Sec 10)

1 Spot welds *3 Retaining strap*
2 Motor assembly *4 Sensor*

Air cleaner temperature sensor

16 Remove the air cleaner from the engine and disconnect the vacuum hoses at the sensor.

17 Carefully note the position of the sensor. The new sensor must be installed in exactly the same position.

18 Pry up the tabs on the sensor retaining clip and remove the sensor and clip from the air cleaner.

19 Install the new sensor with a new gasket in the same position as the old one.

20 Press the retaining clip onto the sensor. Do not damage the control mechanism in the center of the sensor.

21 Connect the vacuum hoses and attach the air cleaner to the engine.

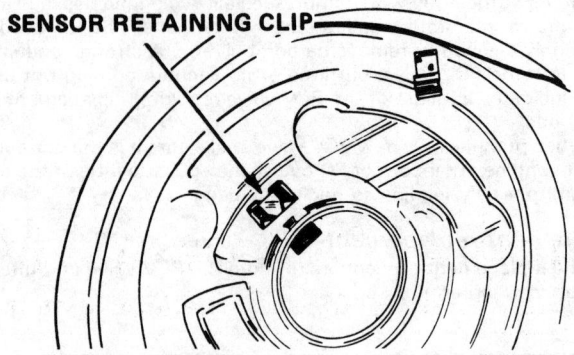

Fig. 6.20 Carefully note the position of the sensor before removing the sensor retaining clip (Sec 10)

11 Transmission Converter Clutch (TCC)

1 Toward optimizing the efficiency of the emissions control network, the ECM controls an electrical solenoid mounted in the automatic transmission of vehicles so equipped. When the vehicle reaches a specified speed, the ECM energizes the solenoid and allows the torque converter to mechanically couple the engine to the transmission, under which conditions emissions are at their minimum. However, because of other operating condition demands (deceleratiion, passing, idle, etc.), the transmission must also function in its normal, fluid-coupled mode. When such latter conditions exist, the solenoid de-energizes, returning the transmission to fluid coupling. The transmission also returns to fluid-coupling operation whenever the brake pedal is depressed.

2 Due to the requirement of special diagnostic equipment for the testing of this system, and the possible requirement for dismantling of the automatic transmission to replace components of this system, checking and replacing of the components should be handled by a dealer or automotive repair shop.

Chapter 7 Part A Manual transmission

Refer to Chapter 13 for specifications related to 1985 and later models

Contents

Extension housing oil seal – replacement 7
General information 1
Neutral start switch – replacement and adjustment 9
Shift control lever (5-speed) – removal and installation 4
Speedometer gear seal – replacement 8
Transmission fluid level check ..... Chapter 1
Transmission (4-speed) – removal and installation 5
Transmission (5-speed) – removal and installation 6
Transmission linkage (4-speed) – removal, installation and adjustment 3
Transmission mounts – check and replacement (with transmission in vehicle) 2

Specifications

Torque specifications

76 mm 4-speed	Ft-lbs
Oil filler plug	15
Transmission-to-clutch housing bolts	55
Crossmember-to-frame bolts	35
Mount-to-crossmember bolts	35
Mount-to-transmission bolts	35
Shift lever-to-shifter shaft bolts	25

83 mm 4-speed	
Oil filler plug	15
Oil drain plug	18
Transmission-to-clutch housing bolts	55
Crossmember-to-frame bolts	35
Mount-to-crossmember bolts	35
Mount-to-transmission bolts	35
Shift lever-to-shifter shaft bolts	18

77 mm 5-speed	
Oil filler plug	20
Dust cover bolts	8
Transmission-to-engine bolts	55
Shift lever cover-to-case bolts	10
Crossmember-to-frame bolts	35
Mount-to-crossmember bolts	35
Mount-to-transmission bolts	35
Dampener-to-transmission bolts	30

1 General information

Camaros, for the years covered in this manual, are equipped with either a 4 or 5-speed manual transmission as standard equipment, depending on year, model and engine. The 5-speed is optionally available in some years for models designated to be equipped with a 4-speed as the standard transmission.

The 76-mm and 83-mm model transmissions are 4-speeds operated through adjustable floor-mounted linkage attached to the left side plate of the transmission.

The 77-mm model transmission is a 5-speed operated through a floor mounted gearshift lever assembly mounted atop the extension housing.

2 Transmission mounts – check and replacement (with transmission in vehicle)

1 Raise the vehicle and support it securely with stands. Make sure the vehicle is stable as you may jostle it somewhat to check the mounts.

2 Push up and pull down on the transmission extension housing and watch the transmission mount.

3 If the extension can be pushed up but cannot be pulled down, it is an indication that the rubber is worn and the mount has bottomed out.

4 If the rubber portion of the mount separates from the metal plate, it means that the mount should be replaced.

5 Check that all of the mounting bolts or nuts are tight on the crossmember and the transmission.

3 Transmission linkage (4)speed) — removal, installation and adjustment

1 Raise the vehicle and support it securely on jackstands.
2 Disconnect the shift rods from the control levers, labeling each one with tape to facilitate reinstallation.
3 Remove the knob from the shift lever.
4 Remove the grip from the parking brake lever.
5 Remove the retaining screws from the console cover.
6 Raise the console cover, disconnect the wire connectors and remove the console cover.
7 Remove the shift boot retaining bolts and the boot.
8 Remove the upper and lower shift control mounting bolts and remove the shifter.
9 Install the shifter by reversing the removal steps, then refer to the accompanying illustration and adjust the linkage as follows.
10 Set levers **L, M** and **N** in the Neutral position.
11 Move lever **F** to the Neutral position.
12 Align the holes in levers **E, H** and **J** with the notch in the shifter assembly and insert the gauge (refer to the accompanying illustration) through the notch and holes to secure the levers in Neutral.
13 Attach rod **C** to lever **N** with washer **A** and retainer **B**.
14 Loosely assemble nuts **R** and **T** and swivel **S** on rod **C** (refer to view A).

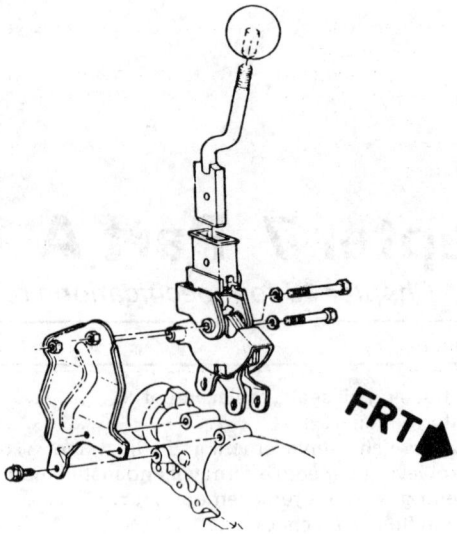

Fig. 7A.1 Details of the shift control mounts on 4-speed manual transmissions (76 mm transmission shown, others similar) (Sec 3)

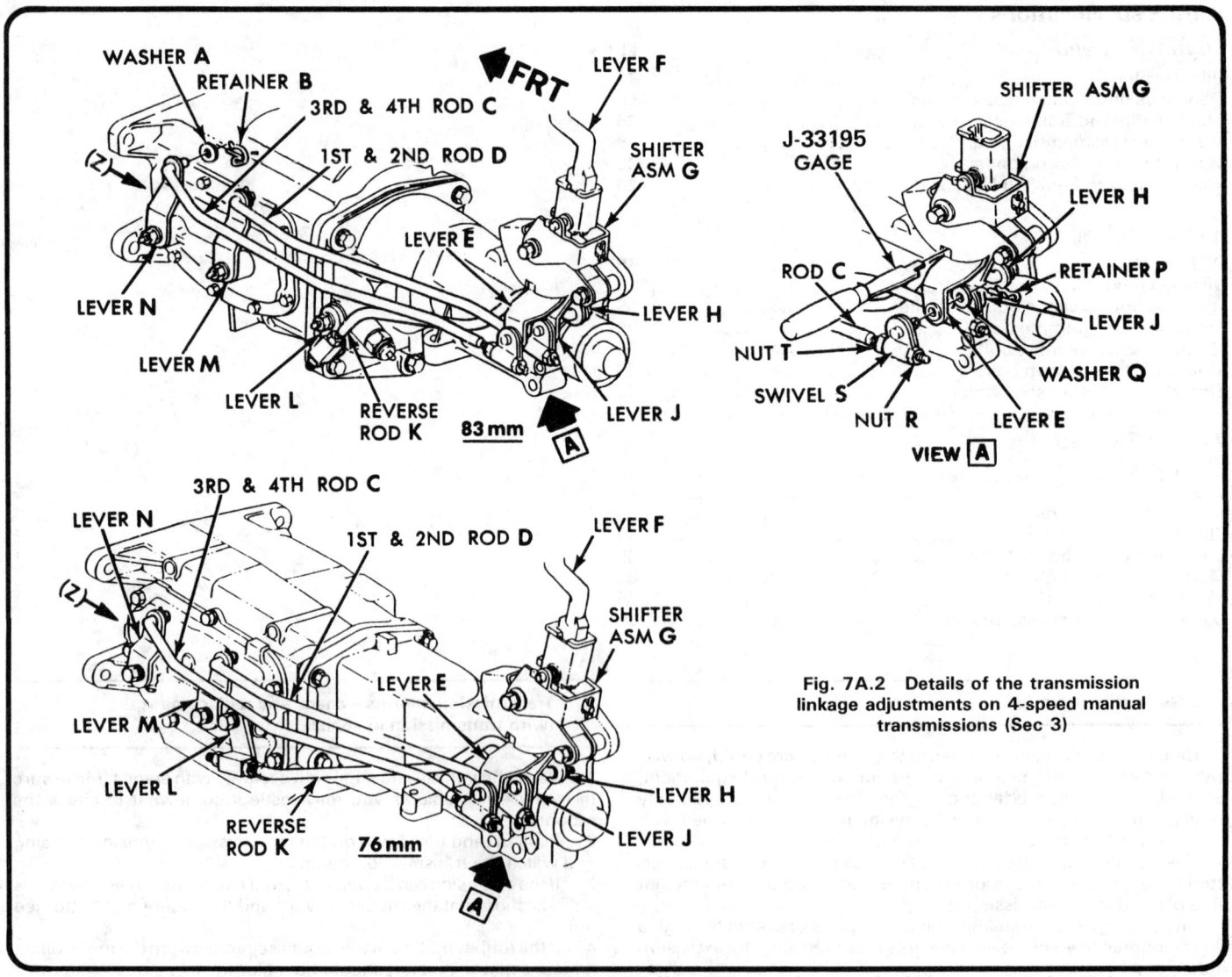

Fig. 7A.2 Details of the transmission linkage adjustments on 4-speed manual transmissions (Sec 3)

15 Insert swivel **S** into lever **E**, attach washer **Q** and secure with retainer **P**.
16 Apply pressure on lever **N** in the direction of arrow **Z** while tightening nuts **T** and **R** against swivel **S**.
17 Tighten nuts **T** and **R** to 25 ft-lbs with a torque wrench.
18 Repeat Steps 14 through 18 for rod **D** and levers **J** and **M** and their respective nuts.
19 Repeat Steps 14 through 18 for rod **K** and levers **H** and **L** and their respective nuts.
20 Remove the gauge.
21 After the adjustments have been made, make sure the centerlines of the shift levers are aligned with each other, providing free crossover motion.

4 Shift control lever (5-speed) — removal and installation

1 With the transmission in Neutral, remove the retaining screws from the shift lever boot retainer and slide the boot up the lever.
2 Remove the shift lever retaining bolts at the transmission and remove the lever assembly by pulling it straight up and out of the transmission.
3 Installation is the reverse of the removal procedure.

5 Transmission (4-speed) — removal and installation

1 Raise the vehicle and support it securely on jackstands.
2 Drain the lubricant from the transmission (refer to Chapter 1, if necessary).
3 Remove the torque arm from the vehicle (refer to Chapter 11).
4 Remove the driveshaft (refer to Chapter 8).
5 Disconnect the speedometer cable from the transmission.
6 Disconnect and label the wire connectors at the transmission.
7 Remove the exhaust brace.
8 Remove the shift linkage and shifter assembly (refer to Section 3).
9 Support the engine/transmission by placing a jack and block of wood under the engine oil pan.
10 Remove the crossmember retaining bolts.
11 Remove the transmission mount retaining bolts, then remove the crossmember and mount from the vehicle.
12 Remove the transmission-to-clutch housing upper bolts, install guide pins in the holes, then remove the lower bolts.
13 Lower the supporting jack until the transmission can be withdrawn to the rear and removed.
14 Installation is the reverse of the removal procedure. Apply a light coat of high-temperature grease to the input shaft bearing retainer and to the splined portion of the shaft to assure free movement of the clutch and transmission components during installation. Be sure to tighten all nuts and bolts to the specified torque.
15 Adjust the shift linkage (refer to Section 3).
16 Refill the transmission with the recommended lubricant.

6 Transmission (5-speed) — removal and installation

1 Remove the shift control lever from the transmission (refer to Section 4).
2 Raise the vehicle and support it securely on jackstands.
3 Remove the torque arm from the vehicle (refer to Chapter 11).
4 Remove the driveshaft (refer to Chapter 8).
5 Disconnect the speedometer cable and wire connector at the transmission.
6 Disconnect the clutch cable at the transmission.
7 Support the engine/transmission by placing a jack and block of wood under the engine oil pan.
8 Remove the transmission mount bolts.
9 Remove the hanger supporting the catalytic converter. Take care not to damage the converter.
10 Remove the crossmember bolts, then remove the crossmember and transmission mounts.
11 Remove the dust cover bolts.
12 Remove the transmission-to-clutch housing upper bolts, install guide pins in the holes, then remove the lower bolts.
13 Lower the supporting jack until the transmission can be withdrawn to the rear and removed.
14 Installation is the reverse of the removal prodecure. Apply a light coat of high-temperature grease to the input shaft bearing retainer and to the splined portion of the shaft to assure free movement of the clutch and transmission components during installation. Be sure to tighten all nuts and bolts to the specified torque.
15 Refill the transmission with the recommended lubricant.

7 Extension housing oil seal — replacement

1 The extension housing oil seal can be replaced without removing the transmission from the vehicle.
2 Raise the vehicle and support it securely on jackstands.
3 Drain the lubricant from the transmission (refer to Chapter 1, if necessary).
4 Remove the driveshaft, as described in Chapter 8.

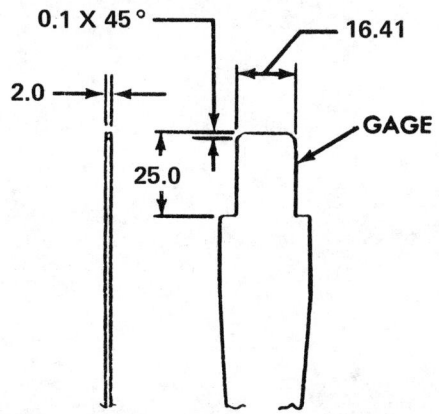

Fig. 7A.3 Specifications of the gauge used to secure levers in Neutral during linkage adjustment (all measurements shown are in millimeters); the gauge may be purchased from a dealer or parts store, or it may be fabricated (Sec 3)

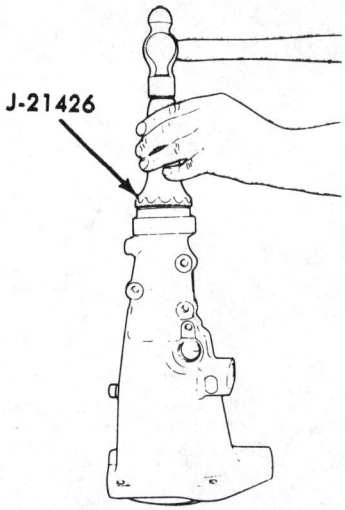

Fig. 7A.4 Installing the extension housing oil seal using special tool J-21426 (a piece of pipe of the proper diameter may be substituted) (Sec 7)

5 Carefuly pry out the old seal.
6 Clean the counterbore and examine it for damage.
7 Lubricate the area between the lips of the new seal with transmission oil and coat the outer diameter with sealant.
8 Carefully install the seal with the lip in, until the flange seats. A special tool, J-21426, is available for this purpose; however, a piece of pipe of the proper diameter used with a hammer may be substituted.
9 Reinstall the driveshaft (refer to Chapter 8).
10 Refill the transmission with the recommended lubricant.

8 Speedometer gear seal — replacement

1 The speedometer gear seal can be replaced without removing the transmission from the vehicle.
2 Raise the vehicle and support it securely on jackstands.
3 Disconnect the speedometer cable, remove the lock plate-to-extension bolt and lock washer, then remove the lock plate.
4 Insert a screwdriver in the lock plate fitting and pry the fitting, gear and shaft from the extension.
5 Pry out the O-ring.
6 Installation is the reverse of removal, but lubricate the new seal with transmission lubricant and hold the assembly so that the slot in the fitting is toward the lock plate boss on the extension.

9 Neutral start switch — replacement and adjustment

1 This switch is a safety device intended to prevent the vehicle from being started without the clutch pedal fully depressed. It is attached to the clutch pedal mounting bracket and is activated by a plastic shaft which is part of the switch.
2 Remove the screws retaining the hush panel and remove the panel.
3 Disconnect the wiring harness connector at the switch.
4 Pull the switch from the tubular retainer in the mounting bracket.

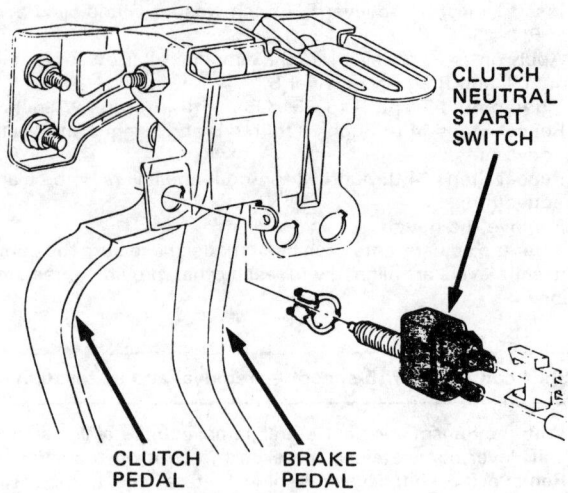

Fig. 7A.5 Details of the Neutral start switch installation (Sec 9)

5 With the clutch pedal depressed, insert the new switch into the tubular retainer until the switch is completely seated. Audible clicks will be heard as the threaded portion of the switch is pushed toward the clutch pedal.
6 Allow the clutch pedal to return to its normal, fully engaged position. The switch will automatically be moved in the tubular retainer providing adjustment.
7 Depress and release the clutch pedal once more to ensure correct adjustment.
8 Reconnect the wiring harness connector and reinstall the hush panel.

Chapter 7 Part B Automatic transmission

Contents

Automatic transmission fluid change . . . . . Chapter 1
Diagnosis — general 2
Extension housing oil seal — replacement . . . . . Chapter 7A
Fluid level check . . . . . Chapter 1
General information 1
Neutral safety and back-up light switch — adjustment 5
Shift linkage — check and adjustment 3
Speedometer gear seal — replacement . . . . . Chapter 7A
Throttle valve (TV) cable assembly — description and adjustment 4
Transmission — removal and installation 6
Transmission mounts — check and replacement . . . Chapter 7A

Specifications

Transmission type

1982	200C
1983	200C; 700-R4
1984	700-R4

Torque specifications

	Ft-lbs
Transmission-to-engine bolts	30
Driveplate-to-converter bolts	35
Inspection cover bolts	8
Dipstick tube brace-to-transmission bolt	8
Transmission crossmember-to-frame bolts	40
Transmission mount-to-crossmember bolts	25
Catalytic converter/exhaust pipe bracket bolts	15
TV cable-to-transmission bolts/nuts	8
Torque arm bracket-to-transmission bolts	31
Shift cable pin-to-shift lever nut	8
Torque arm-to-rear axle bracket bolts	100
Driveshaft strap bolts	16

1 General information

Due to the complexity of the clutches and the hydraulic control system, and because of the special tools and expertise required to perform an automatic transmission overhaul, it should not be undertaken by the home mechanic. Therefore, the procedures in this Chapter are limited to general diagnosis, routine maintenance and adjustment and transmission removal and installation.

If the transmission requires major repair work, it should be left to a dealer service department or a reputable automotive or transmission repair shop. You can, however, remove and install the trnasmission yourself and save the expense, even if the repair work is done by a transmission specialist.

Adjustments that the home mechanic may perform include those involving the throttle valve cable, the shift linkage and the neutral safety switch.

Caution: *Never tow a disabled vehicle at speeds greater than 30 mph or distances over 50 miles unless the driveshaft has been removed. Failure to observe this precaution may result in severe transmission damage caused by lack of lubrication.*

2 Diagnosis — general

Automatic transmission malfunctions may be caused by four general conditions: poor engine performance, improper adjustments, hydraulic malfunctions and mechanical malfunctions. Diagnosis of these problems should always begin with a check of the easily repaired items: fluid level and condition, shift linkage adjustment and throttle linkage adjustment. Next, perform a road test to determine if the problem has been corrected or if more diagnosis is necessary. If the problem persists after the preliminary tests and corrections are completed, additional diagnosis should be done by a dealer service department or a reputable automotive or transmission repair shop.

3 Shift linkage — check and adjustment

Note: *Apply the parking brake and block the wheels to prevent the vehicle from rolling.*

1 Working under the vehicle, loosen the nut attaching the shift lever

to the pin on the shift cable assembly.
2 Place the console shifter (inside the vehicle) in Neutral.
3 Place the manual shifting shaft on the transmission in Neutral.
4 Move the pin to give a 'free pin' fit, making sure that the console shifter is still in the Neutral position, and tighten the pin-to-lever retaining nut to 8 ft-lbs.
5 Make sure the engine will start in the Park and Neutral positions only.
6 If the engine can be started in any of the drive positions (as indicated by the shifter inside the vehicle), repeat the steps above or have the vehicle examined by a dealer, because improper linkage adjustment can lead to band or clutch failure and possible personal injury.

4 Throttle valve (TV) cable assembly — description and adjustment

1 The throttle cable should not be thought of as just an automatic downshift cable, but rather as a cable that controls line pressure, shift feel and shift timing, as well as part throttle and detent downshifts. The function of the cable combines the functions of the vacuum modulator and the downshift (detent) cable found on other model transmissions.
2 The TV cable is attached to the link at the throttle lever and bracket assembly at the transmission and to either the throttle idler lever (L4 engines) or directly to the carburetor/TBI throttle lever on V6 and V8 engines.

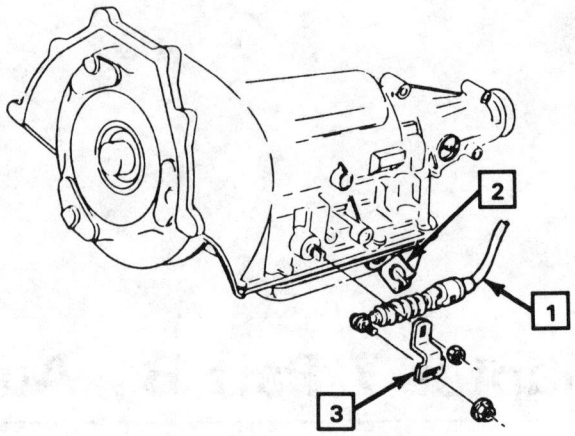

Fig. 7B.1 Exploded view of components involved in shift linkage adjustment (Sec 12)

1 Shift cable
2 Shift cable bracket
3 Shift lever

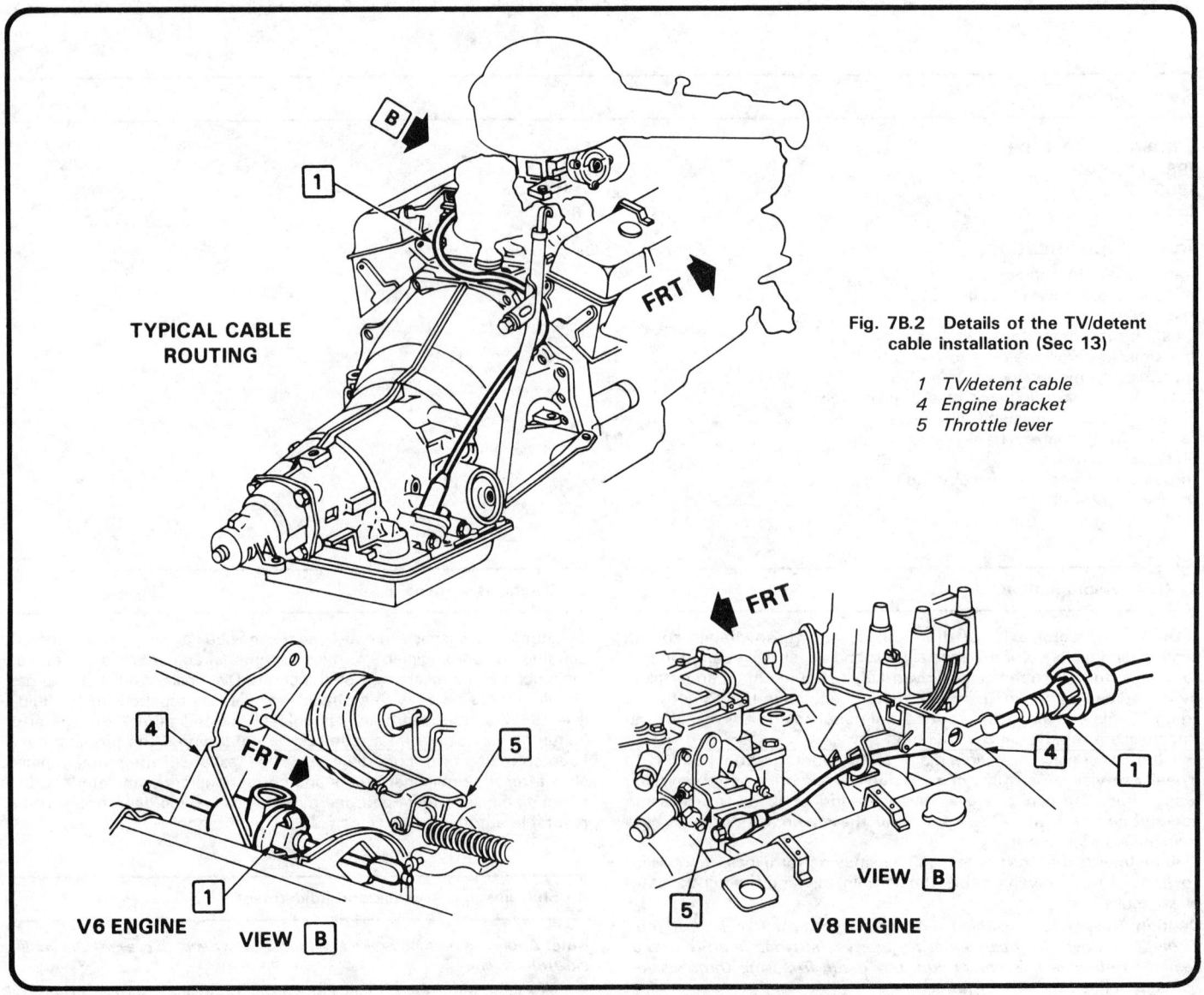

Fig. 7B.2 Details of the TV/detent cable installation (Sec 13)

1 TV/detent cable
4 Engine bracket
5 Throttle lever

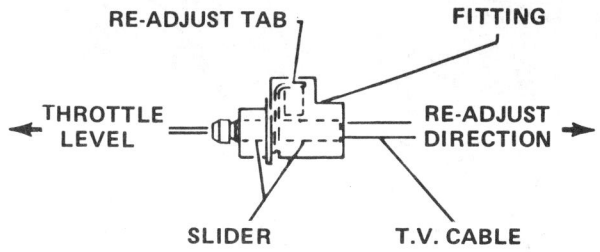

Fig. 7B.3 Details of the components involved in adjusting the TV cable (Sec 13)

3 Whenever the TV cable has been disconnected from the carburetor/TBI unit, it must be adjusted after installation.
4 The engine should be off.
5 If not previously done, remove the air cleaner, labeling all hoses as they are removed to simplify installation.
6 Depress and hold down the metal readjusting tab at the engine end of the TV cable.
7 While holding the tab down, move the slider until it stops against the fitting.
8 Release the readjustment tab.

V6 and V8 engine

9 Manually open the carburetor/TBI lever to the full throttle stop position. The cable will ratchet through the slider and automatically readjust itself.
10 Release the carburetor lever.

L4 engine

11 Rotate the idler lever to the maximum travel stop position. The cable will ratchet through the slider and automatically readjust itself.
12 Release the idler lever. **Note:** *Do not attempt to adjust the L4 engine's TV cable using the TBI lever. Be sure to use the idler lever.*

All engines

13 Road test the vehicle. If delayed or only full-throttle shifts still occur, have the vehicle checked by a dealer.

5 Neutral safety and back-up light switch — adjustment

1 Remove the screws retaining the shift indicator plate to the console.
2 Slide the plate up the shifter handle and secure it to the shifter knob with a rubber band.
3 Position the shifter assembly in the Neutral notch in the detent plate (refer to the accompanying illustration).
4 Loosen but do not remove the screws retaining the switch assembly to the shifter assembly.
5 Rotate the switch assembly so that the service adjustment hole on the switch is aligned with the carrier tang hole.
6 Insert a 3/32-inch gauge pin (a drill bit will work fine) through the service adjustment hole to a depth of 19/32-inch.
7 Tighten the switch assembly retaining screws.
8 Remove the gauge pin.
9 Reinstall the shift indicator plate.

6 Transmission — removal and installation

1 Disconnect the cable from the negative battery terminal.
2 Remove the air cleaner, labeling all hoses as they are disconnected to simplify installation.
3 Disconnect the TV cable at the carburetor/TBI unit.
4 Remove the transmission fluid dipstick from the tube.
5 Remove the upper bolt retaining the dipstick tube and separate the tube from the transmission.
6 Raise the vehicle and support it securely on jackstands.
7 Remove the driveshaft (refer to Chapter 8).

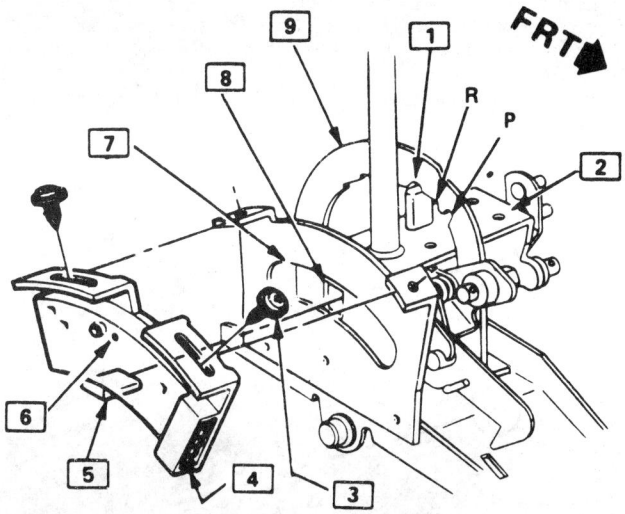

Fig. 7B.4 Details of the components involved in adjusting the Neutral safety and back-up light switch (Sec 14)

1 *Neutral notch*
2 *Transmission control shifter*
3 *Bolt screw*
4 *Switch assembly*
5 *Carrier tang*
6 *Service adjustment hole*
7 *Tang slot*
8 *Tang hole*
9 *Detent plate*

8 Disconnect the speedometer cable at the transmission.
9 Disconnect the TCC wire connector at the transmission.
10 Remove the torque arm-to-transmission bracket bolts. **Caution:** *Rear spring force will cause the torque arm to move toward the floorpan when the arm is disconnected from the bracket. Carefully place a wooden block between the torque arm and floorpan when disconnecting the arm to avoid possible injury and damage to the floorpan.*
11 Loosen the rear torque arm bolts and pull the front of the arm free of the bracket, heeding the caution in Step 10.
12 Remove the catalytic converter/exhaust pipe bracket at the rear of the transmission.
13 Remove the support bracket bolts at the inspection cover.
14 Remove the bolts retaining the inspection cover and remove the cover.
15 Remove the now-exposed converter-to-driveplate bolts. It will be necessary to turn the crankshaft to bring each of the bolts into view (use a wrench on the large bolt at the front of the crankshaft). Mark the relative position of the converter and driveplate with a scribe so they can be reinstalled in the same position. Engage a large screwdriver in the teeth of the driveplate to prevent movement as the bolts are loosened.
16 Using a floor jack and a piece of wood placed between the jack and the transmission, support the transmission and remove the rear mount bolt.
17 Remove the crossmember bolts and the crossmember.
18 Lower the transmission as far as possible without causing the engine or transmission to contact the firewall, then disconnect the TV cable assembly and the oil cooler lines at the transmission.
19 Support the engine at the oil pan rail with a jack and remove the transmission-to-engine bolts. The upper bolts should be removed first.
20 Move the transmission to the rear and down. If necessary, carefully pry it free from the driveplate. Keep the rear of the transmission down at all times to prevent the converter from falling out. The converter can be held in place with a strap.
21 Installation is the reverse of the removal procedure, with the following additional instructions.
22 Before installing the driveplate-to-converter bolts, make sure that the weld nuts on the converter are flush with the driveplate and that the converter can be turned freely by hand in this position. Start the mounting bolts and tighten them finger tight, then tighten them to the specified torque. This will ensure proper alignment of the converter.
23 Tighten all nuts and bolts.
24 Adjust the shift linkage (refer to Section 3).
25 Adjust the TV cable (refer to Section 4).

Fig. 7B.5 Locations of the automatic transmission and related components (200-C transmission shown) (Sec 15)

1 Engine oil pan
2 Inspection cover
3 Support rods
4 Shift lever
5 Shift cable
6 Shift cable bracket
7 Torque arm clamp
8 Driveshaft
9 Catalytic converter/exhaust pipe bracket
10 Rear transmission mount bolt
11 Transmission mount-to-crossmember bolt
12 Crossmember-to-frame bolts
13 Transmission oil pan
14 Transmission oil cooler lines

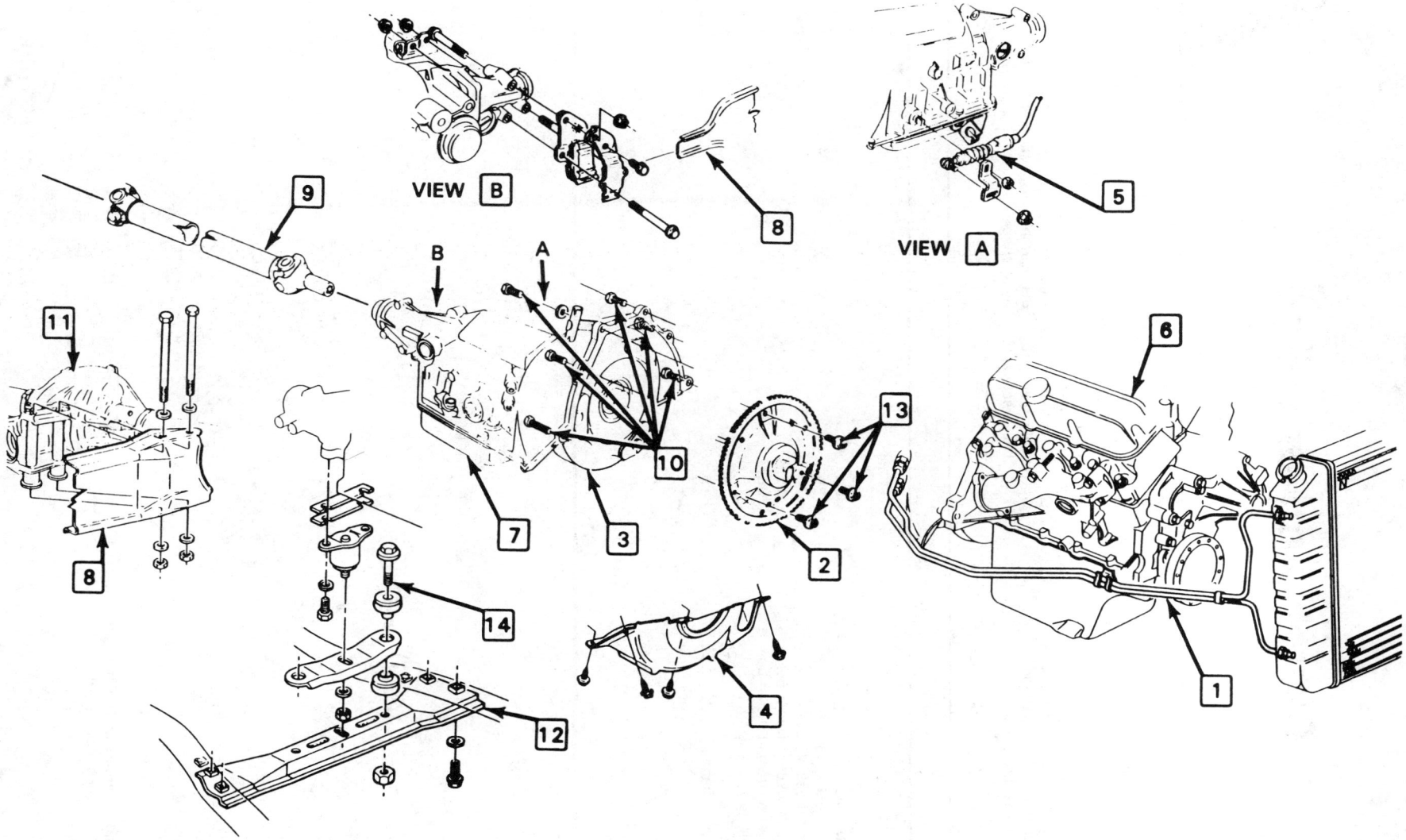

Fig. 7B.6 Details of automatic transmission installation (typical) (Sec 15)

1 *Oil cooler lines*
2 *Flywheel*
3 *Converter*
4 *Inspection cover*
5 *Shift linkage*
6 *Engine*
7 *Transmission*
8 *Torque arm*
9 *Driveshaft*
10 *Transmission-to-engine bolts*
11 *Rear axle*
12 *Crossmember*
13 *Converter-to-flywheel bolts*
14 *Transmission mount bolts*

Chapter 8 Clutch and driveline

Refer to Chapter 13 for specifications related to 1985 and later models

Contents

Axleshaft — removal and installation 13
Axleshaft bearing — replacement 15
Axleshaft oil seal — replacement 14
Clutch — general information 1
Clutch — removal, inspection and installation 4
Clutch cross-shaft — removal and installation 3
Clutch pedal — removal and installation 2
Clutch pilot bearing — removal and installation 5
Driveshaft — general information 7
Driveshaft — removal and installation 10
Driveshaft out-of-balance correction 9
Hydraulic clutch assembly — removal and installation 6
Rear axle — general information and identification 12
Rear axle assembly — removal and installation 16
Rear axle oil change . . . . . Chapter 1
Rear axle oil level check . . . . . Chapter 1
Universal joints — disassembly, inspection and reassembly . . . 11
Universal joints — wear check 8

Specifications

Torque specifications

	Ft-lbs
Clutch pedal-to-mounting bracket	25
Clutch master cylinder-to-cowl	10
Clutch cross-shaft inboard ball stud	25
Clutch cross-shaft outboard ball stud nut	20
Clutch cross-shaft frame bracket-to-frame bolt	15
Clutch cross-shaft bracket-to-frame bracket bolt	35
Slave cylinder heat shield-to-bellhousing	15
Slave cylinder-to-bellhousing	15
Universal joint-to-rear pinion flange	15
Brake assembly-to-rear axle housing	35
Differential pinion shaft lock screw	20
Differential cover bolts	20
Filler plug	20

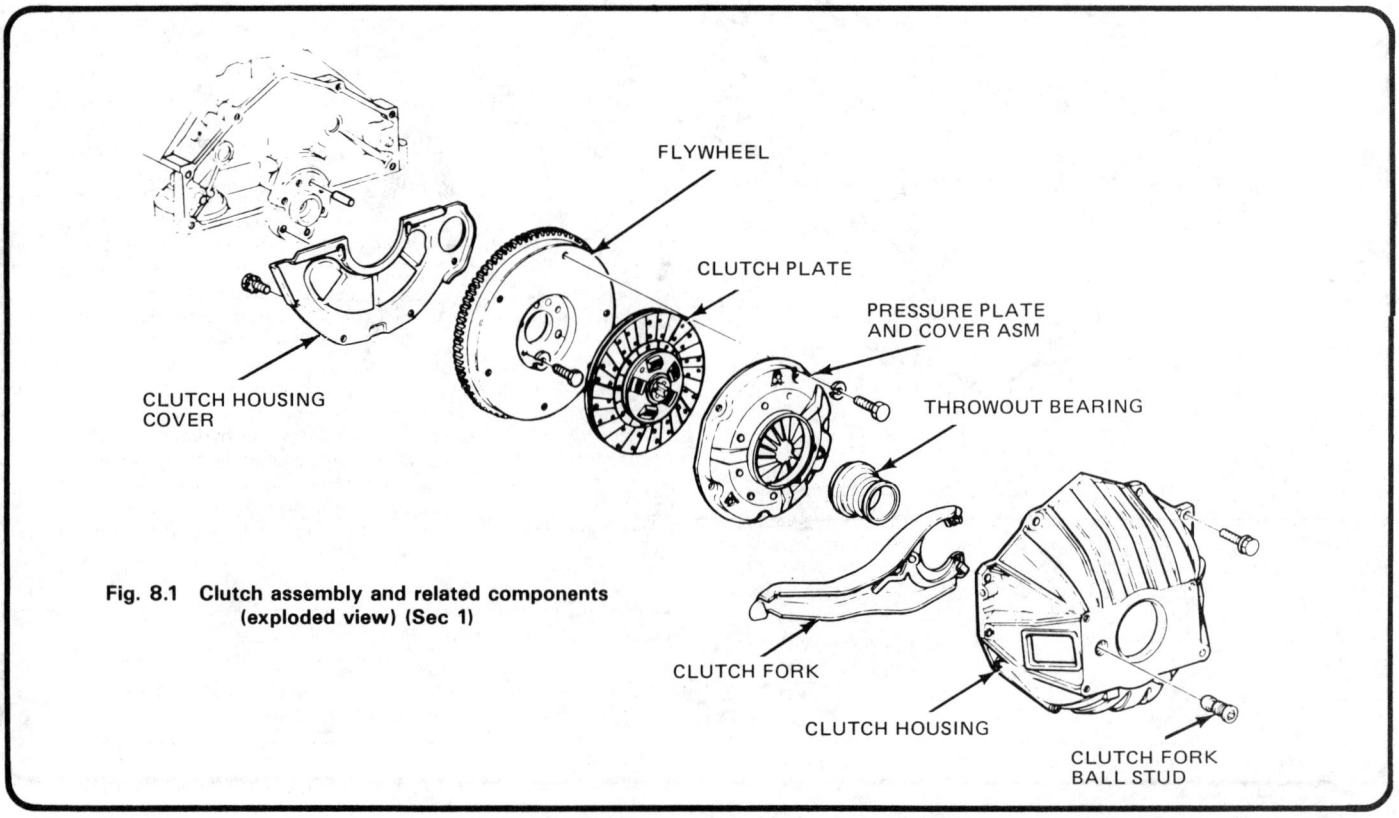

Fig. 8.1 Clutch assembly and related components (exploded view) (Sec 1)

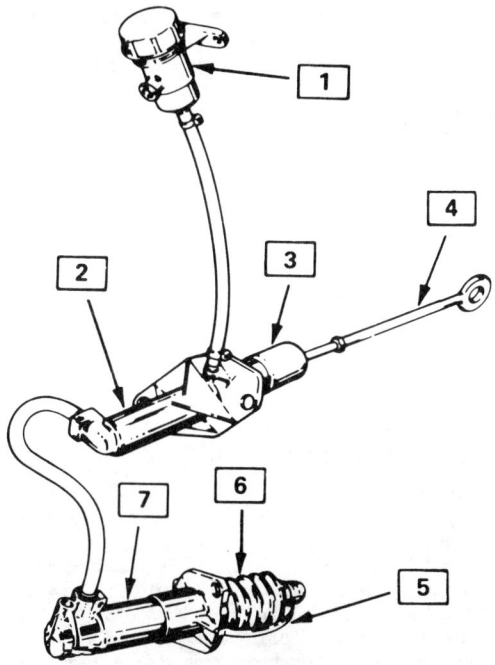

Fig. 8.2 Hydraulic clutch system components (employed on 1984 vehicles) (Sec 1 and 6)

1 *Fluid reservoir*
2 *Clutch master cylinder*
3 *Rubber boot*
4 *Clutch pedal-to-master cylinder pushrod*
5 *Shipping strap*
6 *Rubber boot*
7 *Clutch slave cylinder*

1 Clutch — general information

All manual transmission-equipped vehicles utilize a single dry plate, diaphragm spring-type clutch. Operation is through a foot pedal and rod linkage. The unit consists of a pressure plate assembly which contains the pressure plate, diaphragm spring and fulcrum rings. The assembly is bolted to the rear face of the flywheel.

The driven plate (friction or clutch plate) is free to slide along the transmission input shaft and is held in place between the flywheel and pressure plate by the pressure exerted by the diaphragm spring. The friction lining material is riveted to the clutch plate, which incorporates a spring-cushioned hub designed to absorb driveline shocks and to assist in ensuring smooth starts.

The circular diaphragm spring assembly is mounted on shouldered pins and held in place in the cover by fulcrum rings. The spring itself is held in place by spring steel clips. On high performance vehicles, or where a heavy duty clutch is used, the diaphragm spring has bent fingers, where the standard version has flat fingers. The bent finger design is used to gain a centrifugal boost to aid rapid re-engagement of the clutch at high rotational speeds.

1982 and 1983 models

Depressing the clutch pedal pushes the throwout bearing forward to bear against the fingers of the diaphragm spring. This action causes the diaphragm spring outer edge to deflect and move the pressure plate to the rear to disengage the pressure plate from the clutch plate.

When the clutch pedal is released, the diaphragm spring forces the pressure plate into contact with the friction linings of the clutch plate and at the same time pushes the clutch plate forward on its splines to ensure full engagement with the flywheel. The clutch plate is now firmly sandwiched between the pressure plate and the flywheel and the drive is taken up.

1984 models

A hydraulic clutch assembly is employed on all 1984 models. The assembly consists of a remote reservoir, a clutch master cylinder and a slave cylinder.

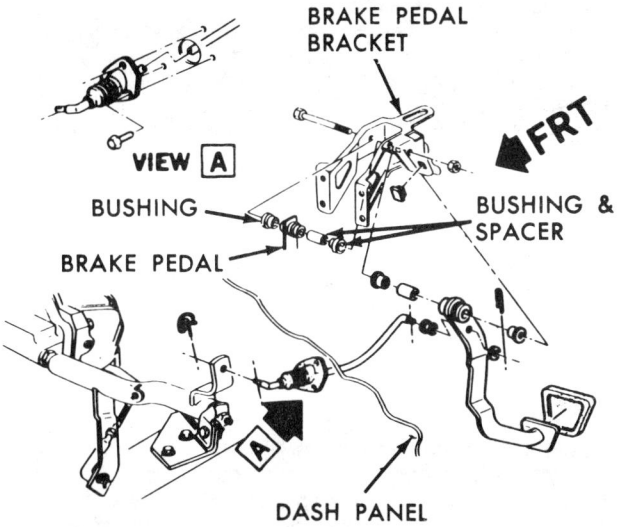

Fig. 8.3 Clutch pedal assembly (exploded view) (Sec 2)

The clutch master cylinder is mounted on the cowl panel in the engine compartment and the slave cylinder is mounted on the engine bellhousing. The master cylinder operates directly off the clutch pedal.

When the clutch pedal is pushed in, hyradulic fluid (under pressure from the clutch master cylinder) flows into the slave cylinder. Because the slave cylinder is also connected to the clutch fork, the fork moves the throwout bearing into contact with the pressure plate release fingers, disengaging the clutch plate.

The hydraulic clutch system locates the clutch pedal and provides clutch adjustment automatically, so no adjustment of the clutch linkage or pedal position is required. **Caution:** *Prior to servicing the clutch or vehicle components that require the removal of the slave cylinder (i.e. transmission and clutch housing removal), the master cylinder pushrod must be disconnected from the clutch pedal. If this is not done, permanent damage to the slave cylinder will occur if the clutch pedal is depressed while the slave cylinder is disconnected.*

If a malfunction of the hydraulic clutch assembly is suspected, verify it by removing the clutch housing dust shield and measuring the clutch slave cylinder pushrod travel as follows.

Raise the vehicle and place it securely on jackstands. Remove the bolts retaining the clutch housing dust shield to the clutch housing and remove the shield. Note the position of the slave cylinder pushrod. Have an assistant push the clutch pedal all the way to the floor. Measure the distance the pushrod travels while engaging the clutch fork. If the distance is 0.57-inch or more, the clutch hydraulic system is operating correctly. Do not remove it. If the slave cylinder pushrod does not travel at least 0.57-inch, check the reservoir fluid level (the slave cylinder must be in place when the fluid level is checked). **Note:** *Carefully clean the top of the reservoir before removing it to prevent contamination of the system.* Fill the reservoir to the proper level, indicated by a step in the reservoir. Use GM Delco Supreme no. 11 or its equivalent. Do not overfill the reservoir, as the upper portion must accept fluid that is displaced from the slave cylinder as the clutch wears.

If the reservoir required fluid, examine the hydraulic system components for leakage by removing the rubber boots from the master and slave cylinders and checking for leakage past the pistons. A slight wetting of the surfaces is acceptable, but if excessive leakage is evident, the clutch hydraulic system must be replaced, as a whole, with a new one.

2 Clutch pedal — removal and installation

1 Disconnect the cable from the negative battery terminal.
2 Disconnect the clutch pedal return spring.
3 Remove the screws retaining the underdash hush panel, then remove the hush panel.
4 If equipped with cruise control, remove the cruise control switch

from the mount at the clutch pedal.
5 Disconnect and remove the neutral start switch and the clutch pedal (refer to Chapter 7A).
6 Remove the turn signal and hazard warning flasher mounting bracket screws.
7 Disconnect the clutch actuator rod (clutch master cylinder pushrod on models with hydraulic clutch) from the clutch pedal.
8 Remove the nut from the clutch pedal pivot bolt.
9 Pull the bolt out only far enough to allow removal of the clutch pedal, leaving it engaged through the brake pedal pivot and bracket.
10 Before installing the clutch pedal assembly, inspect and clean all components. Replace all components showing wear with new ones. **Note:** *Do not clean the bushings with solvent; simply wipe them off with a clean rag.*
11 Installation is the reverse of the removal procedure. After installation is complete, adjust the clutch pedal free play (refer to Chapter 1).

3 Clutch cross-shaft — removal and installation

1 Remove the lever and shaft return spring.
2 Disconnect the clutch pedal and fork pushrods from their respective cross-shaft levers.
3 Loosen the outboard ball stud nut and slide the stud out of the slot in the bracket.
4 Move the cross-shaft outboard far enough to clear the inboard ball stud, then lift it out and remove it from the vehicle.
5 Check the ball stud seats on the cross-shaft, the engine bracket ball stud assembly and the anti-rattle O-ring for damage and wear; replace parts as necessary.
6 When installing, reverse the removal procedure. Lubricate the ball studs and seats with graphite grease.
7 When installation is complete, adjust the clutch pedal free play (refer to Chapter 1).

4 Clutch — removal, inspection and installation

1 Access to the clutch is normally accomplished by removing the transmission, leaving the engine in the vehicle. If, of course, the engine is being removed for major overhaul, then the opportunity should always be taken to check the clutch assembly for wear at the same time.
2 Disconnect the clutch fork pushrod and spring, then remove the clutch housing from the engine block. On models equipped with a hydraulic clutch assembly, remove the slave cylinder heat shield and slave cylinder from the clutch housing before removing the clutch housing.
3 Slide the clutch fork from the ball stud and remove the fork from the dust boot.
4 If necessary, the ball stud can be removed from the clutch housing by unscrewing it.
5 If there are no alignment marks on the clutch cover (an X-mark or white-painted letter) scribe or center-punch marks for indexing purposes during installation.
6 Unscrew the bolts securing the pressure plate and cover assembly one turn at a time in a diagonal sequence to prevent distortion of the clutch cover.
7 With all the bolts and lock washers removed, carefully lift the clutch away from the flywheel. Be careful not to drop the clutch plate.
8 It is not practical to dismantle the pressure plate assembly.
9 If a new clutch plate is being installed, replace the throwout bearing at the same time. This will preclude having to replace it at a later date when wear on the clutch plate is still very little.
10 If the pressure plate assembly must be replaced, a rebuilt unit may be available.
11 Examine the clutch plate friction lining for wear and loose rivets, and the disc for rim distortion, cracks, broken hub springs and worn splines. The surface of the friction linings may be highly glazed, but as long as the clutch material pattern can be seen clearly, this is satisfactory. Compare the amount of lining remaining with a new clutch plate, if possible. If in doubt, replace the clutch plate with a new one.
12 Check the machined faces of the flywheel and the pressure plate. If either are grooved, they should be machined until smooth or replaced.
13 If the pressure plate is cracked or split, replace it with a new one; also, if the condition of the diaphragm spring is suspect, the spring should be checked and replaced if necessary.
14 Check the throwout bearing for smoothness of operation. There should be no harsh or loose spots in it. It should spin reasonably freely, bearing in mind that it has been prepacked with grease. If in doubt, replace the bearing with a new one.
15 It is important that no oil or grease gets on the clutch plate friction linings or the pressure plate and flywheel faces. Always handle the clutch with clean hands and wipe down the pressure plate and flywheel faces with a clean, dry rag before assembly begins.
16 Place the clutch plate against the flywheel, making sure that the longer splined boss faces toward the flywheel (thicker torsional spring assembly projection toward the transmission).
17 Install the pressure plate and clutch cover assembly so that the marks are in alignment. Tighten the bolts only finger tight so that the driven plate is gripped, but can still be moved sideways.
18 The clutch plate must now be aligned so that when the engine and

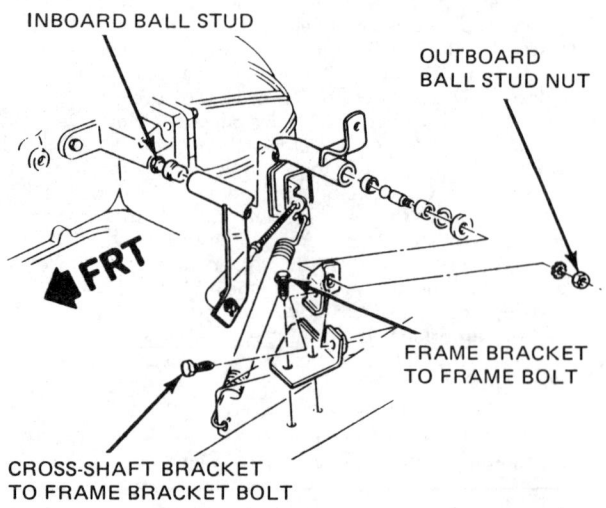

Fig. 8.4 Clutch cross-shaft installation details (Sec 3)

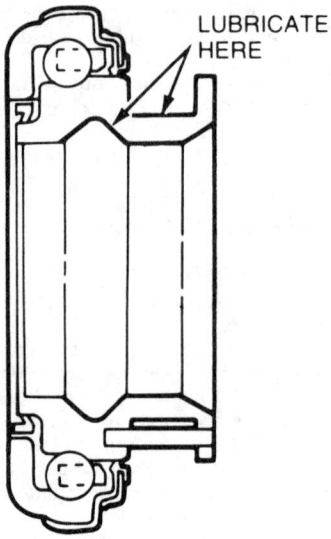

Fig. 8.5 Arrows indicate areas of the throwout bearing to be lubricated prior to installation (Sec 4)

transmission are mated, the input shaft splines will pass through the splines in the center of the clutch hub plate.
19 Alignment can usually be carried out by inserting a round bar or long screwdriver through the hole in the center of the clutch so that the end of the bar rests in the small hole in the end of the crankshaft containing the input shaft pilot bushing. Ideally, an old transmission input shaft or alignment tool should be used.
20 Using the bearing as a fulcrum, move the bar sideways or up and down to move the clutch plate in whichever direction is necessary to align (center) it.
21 Alignment can be verified by removing the bar and viewing the clutch plate hub in relation to the hole in the center of the clutch cover plate diaphragm spring. When the hub appears exactly in the center of the hole, all is correct. On pressure plate covers which have cutaway edges, the clutch plate can be centered by using the fingers to line up its edges with the edge of the flywheel.
22 Tighten the clutch cover bolts in a diagonal sequence to ensure that the cover plate is pulled down evenly and without distortion of the flange. Torque the bolts to Specifications.
23 Lubricate the clutch fork fingers at the throwout bearing end, and the ball and socket, with a high melting point grease. Also lubricate the throwout bearing collar and groove (see the accompanying illustration).
24 Install the clutch fork and dust boot in the clutch housing and attach the throwout bearing to the fork.
25 Install the clutch housing.
26 Install the transmission (refer to Chapter 7A).
27 Connect the fork pushrod and spring, lubricating the spring and pushrod ends. On models with hydraulic clutch assemblies, attach the slave cylinder and slave cylinder heat shield to the clutch housing.
28 Adjust the shift linkage if necessary (4-speed only). Refer to Chapter 7A.
29 Adjust the clutch linkage (refer to Chapter 1) on models not equipped with a hydraulic clutch assembly.

5 Clutch pilot bearing — removal and installation

Note: *If the engine has been removed from the vehicle, disregard the following steps that do not apply.*

1 Remove the transmission (refer to Chapter 7).
2 Remove the clutch (refer to Section 4).
3 The clutch pilot bearing (bushing), which is pressed into the end of the crankshaft, supports the forward end of the transmission input shaft. It requires attention whenever the clutch is removed from the vehicle.
4 Clean the bearing thoroughly and inspect it for excessive wear and damage. If wear is noted, the bearing must be replaced with a new one.
5 Removal can be accomplished with a special puller, but an alternative method, and one that works very well, is to remove the bearing hydraulically. First, locate a solid steel bar with a diameter that is slightly less than the inside diameter of the bearing (19/32-inch should be very close). The bar should just slip into the bearing with very little clearance. Next, pack the bearing and the area behind it (in the crankshaft recess) with heavy grease. Try to eliminate as much air as possible from the recess behind the bearing. Insert the bar into the bearing bore and rap the end of the bar with a hammer. The pressure exerted on the grease will be transferred to the back side of the bearing, forcing it out of the recess. Be sure to clean the grease out of the crankshaft after the bearing has been removed.
6 To install the new bearing, lubricate its outside surface with oil, then drive it into the recess with a socket and a soft-faced hammer (the radius in the bore of the bearing must face *out*). Select a socket that is slightly smaller than the outside diameter of the bearing.
7 The remaining installation steps are the reverse of those for removal.

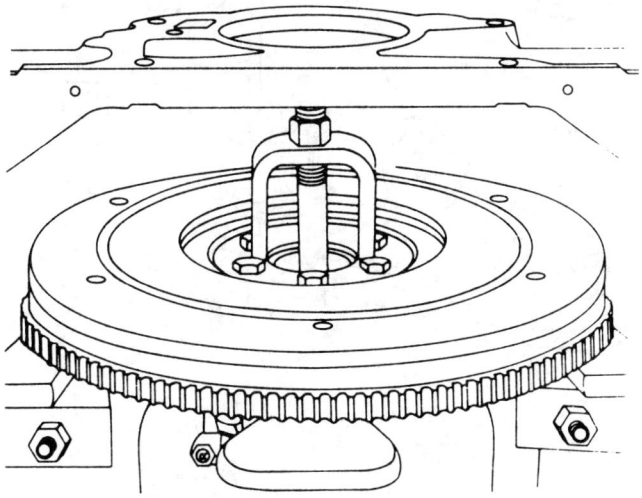

Fig. 8.6 A special bearing puller can be used to remove the pilot bearing (see text for an alternate method that can be used if the puller is not available) (Sec 5)

6 Hydraulic clutch assembly — removal and installation

Note: *The hydraulic clutch assembly is serviced as a complete unit — individual components of the system cannot be serviced separately.*

1 Remove the cable from the negative battery terminal.
2 Remove the steering column trim cover.
3 Remove the underdash hush panel and disconnect the brake switch.
4 Disconnect the master cylinder pushrod from the clutch pedal.
5 Remove the clutch master cylinder-to-cowl nuts.
6 Remove the brake booster-to-cowl nuts.
7 Remove the clutch fluid reservoir from the bracket.
8 Pull the brake master cylinder forward to gain access to the clutch master cylinder. Be careful not to kink the hydraulic lines.
9 Disconnect the clutch master cylinder from the cowl.
10 Raise the vehicle and place it securely on jackstands.
11 Remove the slave cylinder heat shield.
12 Remove the slave cylinder from the bellhousing.
13 Remove the jackstands and lower the vehicle.
14 Remove the hydraulic clutch assembly from the engine compartment.
15 To begin installation, position the clutch master cylinder at the cowl and feed the slave cylinder down to the bellhousing mounting area. Do not install the master cylinder-to-cowl nuts at this time.
16 Feed the clutch master cylinder pushrod through the cowl and loosely attach the clutch master cylinder to the cowl with the U-bolt. Do not attach the U-bolt nuts at this time.
17 Attach the brake booster to the cowl.
18 Install the clutch master cylinder U-bolt through the clutch braces and torque tighten the U-bolt nuts.
19 Install and torque tighten the master cylinder-to-cowl nuts.
20 Attach the clutch master cylinder pushrod to the clutch pedal.
21 Install and adjust the brake switch (refer to Chapter 9).
22 Install the hush panel.
23 Install the steering column trim cover.
24 Attach the clutch reservoir to its bracket.
25 Raise the vehicle and place it securely on jackstands.
26 Guide the slave cylinder pushrod into the clutch fork. **Note:** *When installing a new hydraulic clutch assembly, the slave cylinder contains a shipping strap that pre-positions the pushrod for proper installation and also provides a bearing insert. Following installation of the slave cylinder, the first actuation of the clutch pedal will break the shipping strap and provide normal hydraulic system operation.*
27 Attach the slave cylinder and heat shield to the bellhousing.
28 Remove the jackstands and lower the vehicle.
29 Hook up the negative battery cable.

7 Driveshaft — general information

All vehicles with which this manual is concerned are equipped with one-piece driveshafts. A splined yoke and universal joint are located at the transmission end of the shaft and a second universal joint is used where the driveshaft mates with the pinion flange at the differential.

The universal joints are lubricated for life and are not serviceable on the vehicle. If a universal joint becomes worn or noisy, a service kit

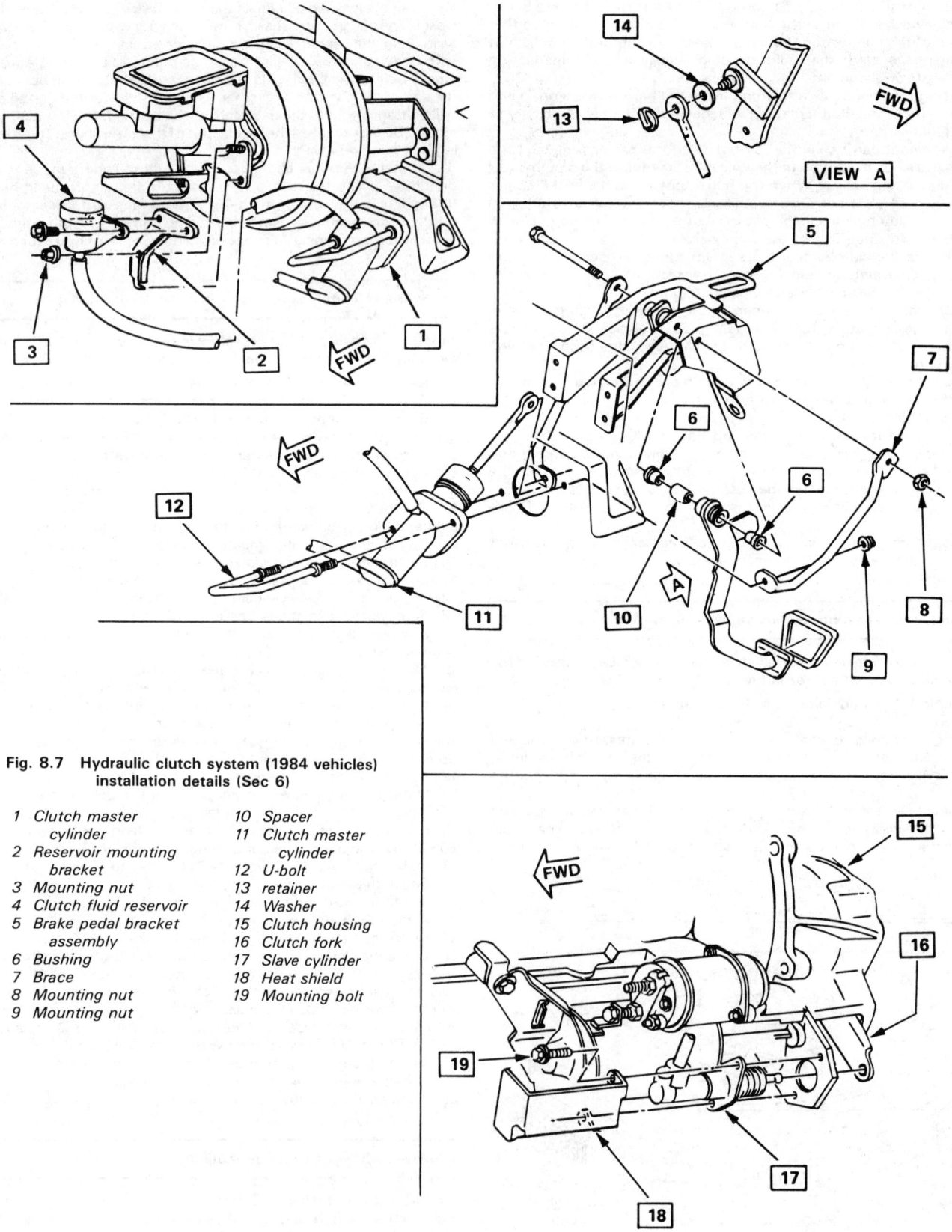

Fig. 8.7 Hydraulic clutch system (1984 vehicles) installation details (Sec 6)

1 *Clutch master cylinder*
2 *Reservoir mounting bracket*
3 *Mounting nut*
4 *Clutch fluid reservoir*
5 *Brake pedal bracket assembly*
6 *Bushing*
7 *Brace*
8 *Mounting nut*
9 *Mounting nut*
10 *Spacer*
11 *Clutch master cylinder*
12 *U-bolt*
13 *retainer*
14 *Washer*
15 *Clutch housing*
16 *Clutch fork*
17 *Slave cylinder*
18 *Heat shield*
19 *Mounting bolt*

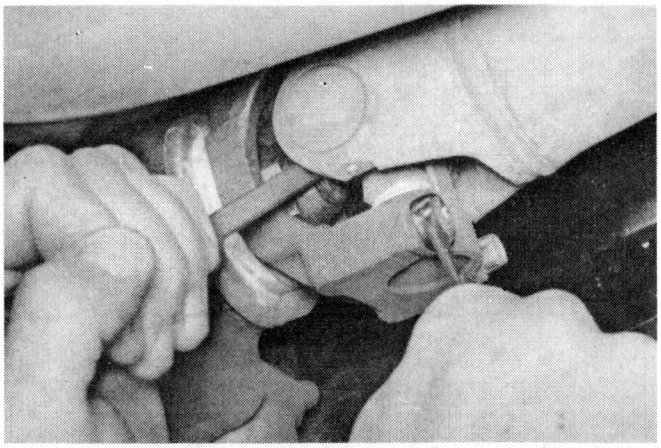

10.2A Insert a screwdriver through the U-joint to keep the driveshaft from turning and remove the retaining bolts with a wrench

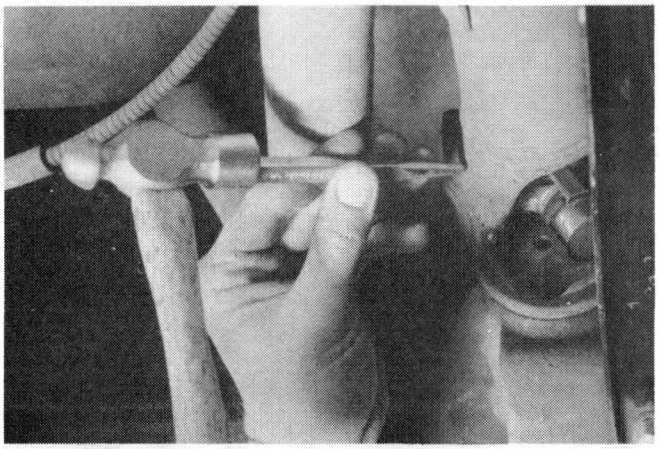

10.2B Use a punch and hammer to mark the relationship of the driveshaft to the pinion flange

containing cross and bearing assemblies is available. The kit also contains snap-rings which must be installed to substitute for the nylon injection rings installed at the factory during shaft assembly. The entire driveshaft must be removed from the vehicle whenever servicing is necessary and care should be taken when handling it to preserve the fine balance produced at the factory. Care should also be taken if the vehicle is undercoated. Never allow undercoating or any other foreign material to adhere to the driveshaft as it will disturb the factory balance.

8 Universal joints — wear check

1 Universal joint problems are usually caused by worn or damaged needle bearings. These problems are revealed as vibration in the driveline or clunking noises when the transmission is put in Drive or the clutch is released. In extreme cases they are caused by lack of lubrication. If this happens, you will hear metallic squeaks and ultimately, grinding and shrieking sounds as the bearings are destroyed.
2 It is easy to check the needle bearings for wear and damage with the driveshaft in place on the vehicle. To check the rear universal joint, turn the driveshaft with one hand and hold the differential yoke with the other. Any movement between the two is an indication of wear. The front universal joint can be checked by holding the driveshaft with one hand and the sleeve yoke in the transmission with the other. Any movement here indicates the need for universal joint repair.
3 If they are worn or damaged, the universal joints will have to be disassembled and the bearings replaced with new ones. Read over the procedure carefully before beginning.

9 Driveshaft out-of-balance correction

1 Vibration of the driveshaft at certain speeds may be caused by any of the following:

Undercoating or mud on the shaft
Loose rear strap mounting bolts
Worn universal joints
Bent or dented driveshaft

2 Vibration which is thought to be coming from the driveshaft is sometimes caused by improper tire balance. This should be one of your first checks.
3 If the shaft is in a good, clean, undamaged condition, it is worth disconnecting the rear end mounting bolts and turning the shaft 180 degrees to see if an improvement is noticed. Be sure to mark the original position of each component before disassembly so the shaft can be returned to the same location.
4 If the vibration persists after checking for obvious causes and changing the position of the shaft, the entire assembly should be checked out by a professional shop or replaced.

10 Driveshaft — removal and installation

1 Raise the vehicle and place it securely on jackstands.
2 Refer to the accompanying photos for the driveshaft removal procedure.
3 Make sure that the driveshaft is handled carefully during the removal and installation procedures, as it must maintain the factory balance to operate smoothly and quietly.
4 Make sure that the outer diameter of the transmission sleeve yoke is not burred, as it could damage the transmission seal when the driveshaft is installed.
5 Lubricate the sleeve yoke splines with engine oil, then slide the yoke onto the transmission output shaft. Do not force it or use a hammer. If resistance is met, check the splines for burrs.
6 Attach the shaft to the pinion flange. Be sure to align the marks on the shaft and pinion flange that were made before removal.
7 Install the rear bolts and tighten them to the specified torque.
8 Remove the jackstands and lower the vehicle.

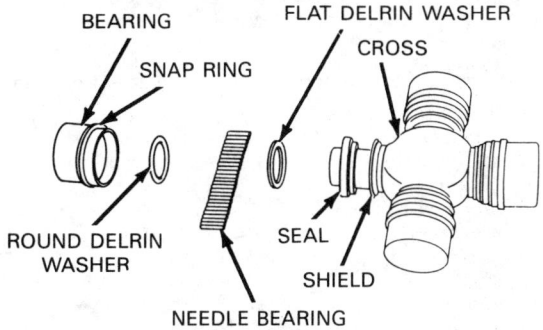

Fig. 8.8 Internal snap-ring type universal joint (exploded view) (Sec 11)

11 Universal joints — disassembly, inspection and reassembly

Note: *Always purchase a universal joint service kit(s) for your model vehicle before starting the procedure which follows. Also, read through the entire procedure before beginning work.*

1 Remove the driveshaft (refer to Section 10).
2 Place the shaft on a workbench equipped with a vise.
3 Place the universal joint in the vise with a 1-1/8 inch socket against one ear of the shaft yoke and a cross press tool placed on the open horizontal bearing cups (photo). A cross press tool, No. J-9522-3, is available from your dealer, or a substitute may be fashioned from 2-1/2

11.3A Pressing a bearing cup out of the front universal yoke

11.3B A cross press tool can be fabricated from a 3/4-inch long section of 2-1/2 inch square steel stock with a wall thickness of 3/16-inch

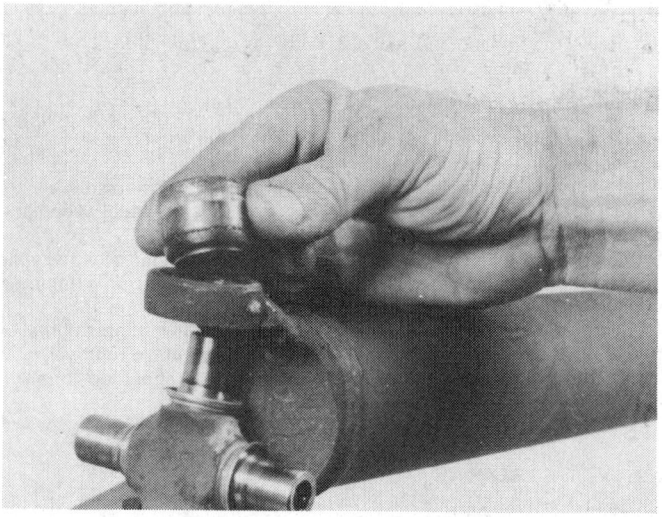

11.4 Removing a bearing cup from the yoke ear after it has been pressed out

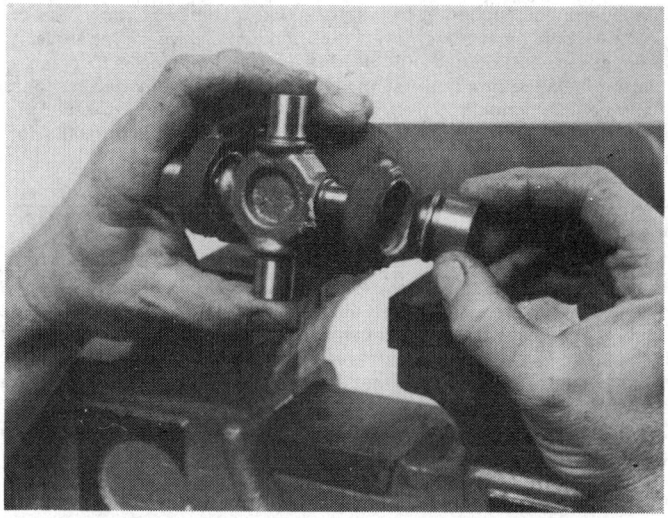

11.11 Assembling a new cross and bearing cups to the driveshaft

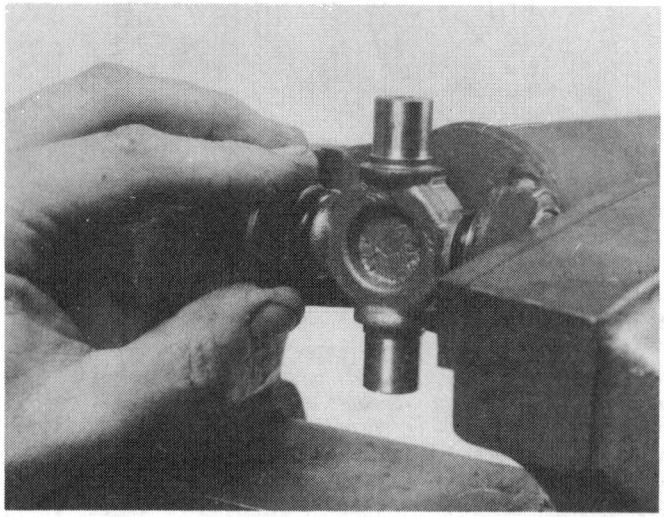

11.13 Installing a snap-ring in the snap-ring groove

11.15 Installing the grease fitting in a U-joint

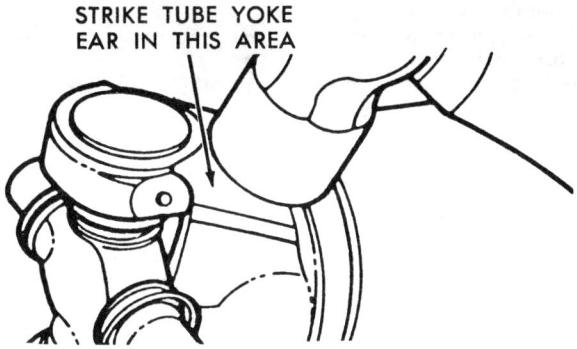

Fig. 8.9 Seating a U-joint snap-ring (Sec 11)

inch by 3/16-inch wall square tube or channel steel (photo). **Note:** *Never clamp the driveshaft tubing itself in a vise, as the tube may be bent.*

4 Press the bearing cup out of the yoke ear, shearing the plastic retaining ring on the bearing. **Note:** *If the cup does not come all the way out of the yoke, it may be pulled free with channel lock pliers, then removed (photo).*

5 Turn the driveshaft 180 degrees and press the opposing bearing cup out of the yoke, again shearing the plastic retainer.

6 Disengage the cross from the yoke and remove the cross. **Note:** *Production universal joints cannot be reassembled because there are no bearing retaining grooves in the production bearing cups. All U-joint service kits contain snap-rings, which will be used when the replacement U-joints are assembled. If your driveshaft has previously had the original U-joints replaced with snap-ring types, remove the snap-rings before pressing the bearing cups out.*

7 If the remaining universal joint is being replaced, press the bearing cups from the slip yoke as detailed above.

8 When reassembling the driveshaft, always install all parts included in the U-joint service kit.

9 Remove all remnants of the plastic bearing retainers from the grooves in the yokes. Failure to do so may keep the bearing cups from being pressed into place and prevent the bearing retainers from seating properly.

10 Using multi-purpose grease to retain the needle bearings, assemble the bearings, cups and washers. Make sure the bearings do not become dislodged during the assembly and installation procedures.

11 In the vise, assemble the cross and cups in the yoke, installing the cups as far as possible by hand (photo).

12 Move the cross back and forth horizontally to assure alignment, then press the cups into place a little at a time, continuing to center the cross to keep the proper alignment.

13 As soon as one snap-ring groove clears the inside of the yoke, stop pressing and install the snap-ring (photo).

14 Continue to press on the bearing cup until the opposite snap-ring can be installed. If difficulty is encountered, strike the yoke sharply with a hammer. This will spring the yoke ears slightly and allow the snap-ring groove to move into position.

15 Install the grease fitting (photo).

16 Install the driveshaft (refer to Section 10).

12 Rear axle — general information and identification

1 The rear axle is a hypoid, semi-floating type. The differential carrier is a casting with a pressed steel cover and the axle tubes are made of steel.

2 Due to the need for special gauges and equipment and the fact that individual differential components are no longer supplied (only a complete rear axle unit) the servicing and repair operations should be confined to those described in this Chapter.

3 An optional 'Positraction' (limited slip) differential unit is also available. Basically, the inclusion of clutch cones or plates and springs slow the rotation of the differential case when one wheel has traction and the other does not. This slowing action applies additional force to the pinion gears and through the medium of the cone which is splined to the axleshafts, exerts equalizing rotational power to the axleshaft which is driving the wheel under traction.

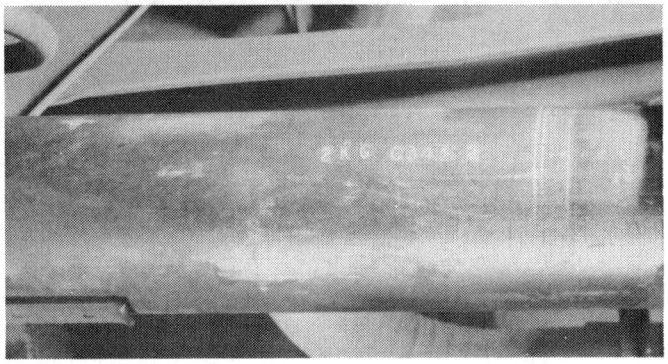

12.4 The axle identification number is stamped on the front of the right axle housing

4 In order to be able to undertake certain operations, particularly removal of the axleshafts, it is important to know the axle identification number. This is located on the front face of the right-hand axle tube about three inches from the differential cover. After the dirt and road grime have been brushed away, the code can be read (photo). A typical General Motors axle code for 1982 (others similar) will read: 2 AB 0 103 2.

5 The initial number and following two letters (2 AB) indicate the axle code; the next number or letter (0) identifies the manufacturer; the next three-digit number (103) indicates the date of manufacture and the final digit (2) denotes the shift during which the axle was built.

6 By using the axle code and the accompanying chart, you can determine the gear ratio of the ring and pinion gears installed in your vehicle.

7 Any inquiries about your axle assembly to the dealer or to the factory must include the full code letters and build date numbers.

Axle code	*Axle ratio*	*Axle type*
1982		
2JE	2.73	Std
2JG	3.08	Std
2JH	3.23	Std
2JJ	3.42	Std
2JR	3.42	Posi
2JW	2.73	Posi
2JX	2.93	Posi
2JY	3.08	Posi
2JZ	3.23	Posi
2KE	2.73	Std
2KF	2.93	Std
2KG	3.08	Std
2KH	3.23	Std
2KH	3.42	Std
2PE	2.73	Std
2PW	2.73	Posi
2PX	2.93	Posi
2PY	3.08	Posi
2PZ	3.23	Posi
1983		
4JF	2.93	Std
4JG/5JG	3.08	Std
4JH/5JH	3.23	Std
4JJ/5/JJ	3.42	Std
4JK/5JK	3.73	Std
4JQ/5JQ	3.73	Posi
4JR/5JR	3.42	Posi
4JX/5JX	2.93	Posi
4JY/5JY	3.08	Posi
4JZ/5JZ	3.23	Posi
5KM	3.73	Std
5KN	3.73	Posi
5PN	3.73	Posi
4PQ/5PQ	3.73	Posi
4PX/5PX	2.93	Posi
4PY/5PY	3.08	Posi
4PZ/5PZ	3.23	Posi

Axle code	*Axle ratio*	*Axle type*
1984		
6EA	3.42	Std
6EF	3.42	Posi
6EP	3.42	Posi
6JG	3.08	Std
6JH	3.23	Std
6JJ	3.42	Std
6JK	3.73	Std
6JQ	3.73	Posi
6JR	3.42	Posi
6JY	3.08	Posi
6JZ	3.23	Posi
6KM	3.73	Posi
6PM	3.73	Std
6PQ	3.73	Posi
6PY	3.08	Posi
6PZ	3.23	Posi

13 Axleshaft — removal and installation

1 Raise the rear of the vehicle, support it securely and remove the wheel and brake drum (refer to Chapter 9).
2 Unscrew and remove the pressed steel cover from the differential carrier and allow the oil to drain into a container.
3 Unscrew and remove the lock screw from the differential pinion shaft. Remove the pinion shaft.
4 Push the outer (flanged) end of the axleshaft in and remove the C-ring from the inner end of the shaft.
5 Withdraw the axleshaft, taking care not to damage the oil seal in the end of the axle housing as the splined end of the axleshaft passes through it.
6 Installation is the reverse of removal. Tighten the lock screw to the specified torque.
7 Always use a new cover gasket and tighten the cover bolts to the specified torque.
8 Refill the axle with the correct quantity and grade of lubricant (Chapter 1).

14 Axleshaft oil seal — replacement

1 Remove the axleshaft as described in the preceding Section.
2 Pry out the old oil seal from the end of the axle housing, using a large screwdriver or the inner end of the axleshaft itself as a lever.
3 Apply high melting point grease to the oil seal recess and tap the seal into position so that the lips are facing in and the metal face is visible from the end of the axle housing. When correctly installed, the face of the oil seal should be flush with the end of the axle housing.
4 Installation of the axleshaft is described in the preceding Section.

15 Axleshaft bearing — replacement

1 Remove the axleshaft (refer to Section 13) and the oil seal (refer to Section 14).
2 A bearing puller will be required or a tool which will engage behind the bearing will have to be fabricated.
3 Attach a slide hammer and pull the bearing from the axle housing.
4 Clean out the bearing recess and drive in the new bearing using a piece of pipe *applied against the outer bearing race.* Lubricate the new bearing with gear lubricant. Make sure that the bearing is tapped into the full depth of its recess and that the numbers on the bearing are visible from the outer end of the housing.
5 Discard the old oil seal and install a new one, then install the axleshaft.

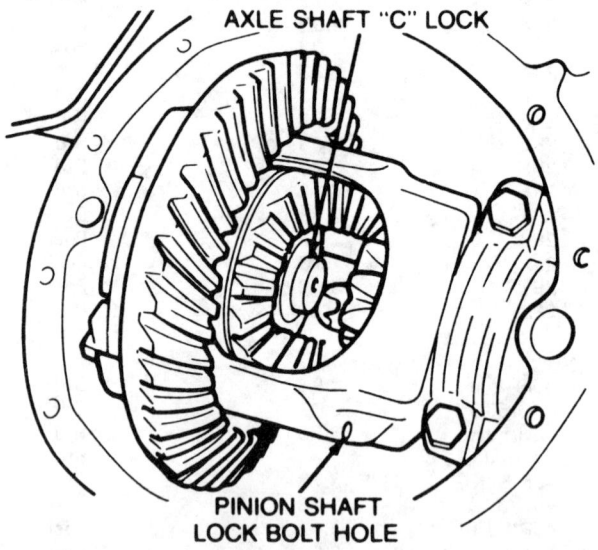

Fig. 8.10 Locations of the pinion shaft lock bolt and axleshaft C-ring lock (Sec 13)

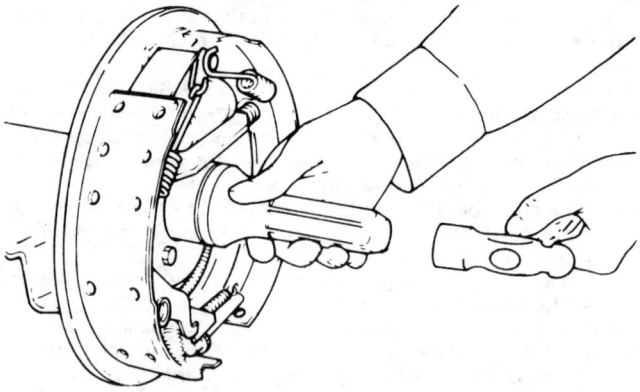

Fig. 8.11 Installing a rear axleshaft oil seal

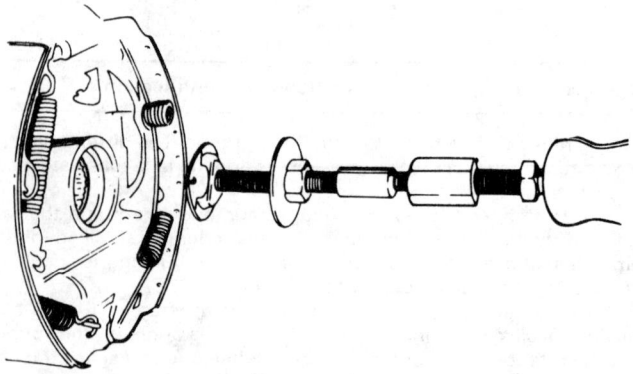

Fig. 8.12 Using a bearing puller to remove a rear axleshaft bearing (Sec 15)

16 Rear axle assembly — removal and installation

1 Raise the rear of the vehicle and support it securely on jackstands placed under the frame rails.
2 Position an adjustable floor jack under the differential housing. Raise the jack just enough to take up the weight, but not far enough to take the weight of the vehicle off the jackstands.
3 Disconnect both shock absorbers.
4 Remove the bolt that attaches the left side of the track bar to the axle.
5 Remove the brake line junction block bolt at the axle housing, then disconnect the brake lines at the junction block.
6 Lower the jack under the differential housing enough to remove the springs.
7 Remove the rear wheels and brake drums (refer to Chapter 9).
8 Remove the cover from the rear axle assembly and drain the lubricant into a container.
9 Remove the axleshafts (refer to Section 13).
10 Disconnect the brake lines from the axle housing clips.
11 Remove the brake backing plates.
12 Disconnect the lower control arms from the axle housing.
13 Disconnect the torque arm at the axle housing.
14 Mark the driveshaft and companion flange, unbolt the driveshaft and support it out of the way on a wire hanger.
15 Remove the rear axle assembly from under the vehicle.
16 Installation is the reverse of the removal procedure. When installation is complete, fill the differential with the recommended oil and bleed the brake system.

Chapter 9 Brakes

Refer to Chapter 13 for specifications and information related to 1985 and later models

Contents

Brake check . . . . . Chapter 1
Brake pedal — removal and installation 17
Combination valve — check and replacement 13
Disc brake rotor — inspection, removal and installation 8
Fluid level check . . . . . Chapter 1
Front disc brake caliper — overhaul 4
Front disc brake caliper — removal and installation 3
Front disc brake pads — replacement 2
General information 1
Hydraulic brake hoses and lines — inspection and replacement 12
Hydraulic system — bleeding 16
Master cylinder — removal, overhaul and installation 11
Parking brake — adjustment 14
Power brake booster — inspection, removal and installation . . 15
Rear disc brake caliper — overhaul 7
Rear disc brake caliper — removal and installation 5
Rear disc brake pads — replacement 6
Rear drum brake shoes — replacement 9
Stop light switch — removal, installation and adjustment 18
Wheel cylinder (drum brakes) — removal, overhaul and installation 10

Specifications

Disc brakes

Rotor thickness after resurfacing (minimum)	0.980 in
Discard thickness	0.965 in
Disc runout (maximum)	0.005 in
Disc thickness variation (maximum)	0.0005 in

Rear drum brakes

Drum diameter	
Standard	9.50 in
Service limit	9.59 in
Drum taper (maximum)	0.003 in
Out-of-round (maximum)	0.002 in
Wheel cylinder bore diameter	0.748 in

Master cylinder

Disc/drum piston diameter	0.945 in
4-wheel disc piston diameter	1.000 in

Torque specifications

	Ft-lbs
Brake pedal-to-bracket nut	22
Brake pedal bracket-to-dash screw	18
Parking brake control assembly-to-console screw	9
Combination valve-to-master cylinder screw	15
Brake line-to-master cylinder nut	18
Brake line-to-combination valve nut	17
Brake line-to-brake hose nut	18
Front flexible hose bracket-to-frame	8
Rear flexible hose bracket-to-frame	8
Junction block-to-axle housing screw	20
Lug nuts	
Standard wheel	80
Aluminum wheel	105
Power cylinder housing-to-master cylinder nut	20
Power cylinder studs-to-dash attaching nuts	15
Caliper mount-to-steering knuckle bolt	35
Splash shield-to-steering knuckle bolt	10
Caliper bleeder valve	11
Front brake hose-to-caliper bolt	32
Parking brake lever retaining nut	30 to 40
Parking brake lever mounting bracket bolt	24 to 38
Banjo housing retaining bolt	18 to 30

1 General information

All vehicles covered by this manual are equipped with hydraulically operated front and rear brake systems. All front brake systems are disc type, while rear brake systems are drum brakes on some models and disc type on others. A quick visual inspection will tell you what type of brakes you have at the rear of your vehicle.

The hydraulic system consists of two separate front and rear circuits. The master cylinder has separate reservoirs for the two circuits and in the event of a leak or failure in one hydraulic circuit, the other circuit will remain operative. A visual warning of circuit failure or air in the system is given by a warning light activated by displacement of the piston in the brake distribution (pressure differential warning) switch from its normal 'in balance' position.

The parking brake mechanically operates the rear brakes only. It is activated by a pull-handle in the center console between the front seats.

A combination valve, located in the engine compartment, consists of three sections providing the following functions: The metering section limits pressure to the front brakes until a predetermined front input pressure is reached and until the rear brakes are activated. There is no restriction at inlet pressures below 3 psi, allowing pressure equalization during non-braking periods. The proportioning section proportions outlet pressure to the rear brakes after a predetermined rear input pressure has been reached, preventing early rear wheel lock-up under heavy brake loads. The valve is also designed to assure full pressure to the rear brakes should the front brakes fail, and vice versa.

The pressure differential warning switch is designed to continuously compare the front and rear brake pressure from the master cylinder and energize the dash warning light in the event of either front or rear brake system failure. The design of the switch and valve are such that the switch will stay in the 'warning' position once a failure has occurred. The only way to turn the light off is to repair the cause of the failure and apply a brake pedal force of 450 psi.

The power brake booster, utilizing engine manifold vacuum and atmospheric pressure to provide assistance to the hydraulically operated brakes, is located in the engine compartment, adjacent to the cowl. All brakes are self-adjusting.

After completing any operation involving the disassembly of any part of the brake system, always test drive the vehicle to check for proper braking performance before resuming normal driving. When testing the brakes, perform the tests on a clean, dry, flat surface. Conditions other than these can lead to inaccurate test results. Test the brakes at various speeds with both light and heavy pedal pressure. The vehicle should stop evenly without pulling to one side or the other. Avoid locking the brakes because this slides the tires and diminishes braking efficiency and control.

Tires, vehicle load and front-end alignment are factors which also affect braking performance.

Torque values given in the Specifications section are for dry, unlubricated fasteners.

2 Front disc brake pads — replacement

Note: *Disc brake pads should be replaced on both wheels at the same time.*

1 Whenever you are working on the brake system, be aware that asbestos dust is present and be careful not to inhale any of it, because it has been proven to be harmful to your health.

2 Remove the cover from the brake fluid reservoir and siphon off about two ounces of the fluid into a container and discard it.

3 Raise the front of the vehicle and place it securely on jackstands.

4 Remove the front wheel, then reinstall two wheel lugs (flat side toward the rotor) to hold the rotor in place. Work on one brake assembly at a time, using the assembled brake for reference if necessary.

5 Push the piston back into its bore. If necessary, a C-clamp can be used, but a flat bar will usually do the job. As the piston is depressed to the bottom of the caliper bore, the fluid in the master cylinder will rise. Make sure that it does not overflow. If necessary, siphon off more of the fluid as directed in Step 2.

6 Now refer to the accompanying photographs and perform the procedure illustrated. Start with photograph 2.6/1.

2.6/1 Use an Allen wrench to remove the two caliper mounting bolts (check the bolt threads for damage)

2.6/2 Remove the brake line retaining clip

2.6/3 Release the hose from the retainer (not from the caliper), pull away the caliper and suspend it with wire. Remove inside pad and plate.

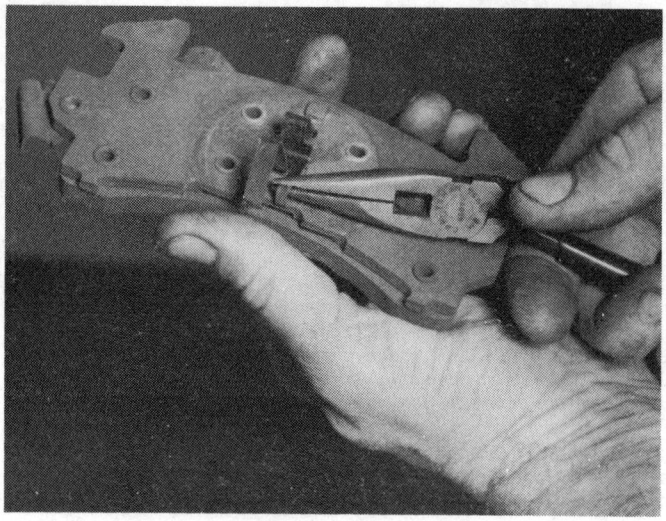

2.6/4 Remove the anti-rattle clip with pliers

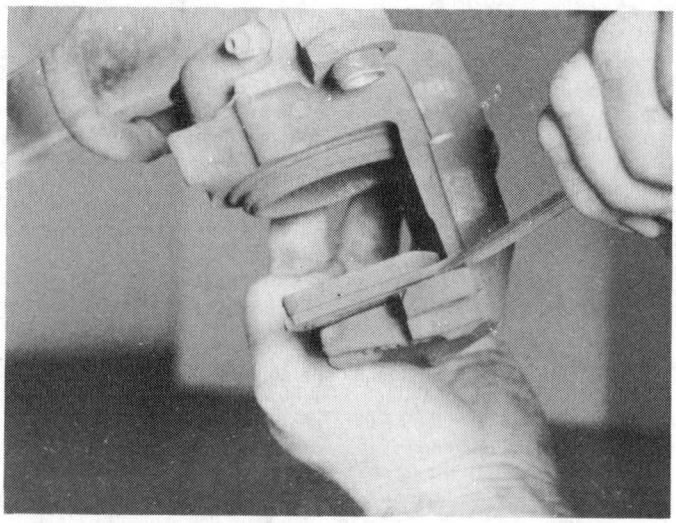

2.6/5 Break the outer pad and plate assembly loose with a screwdriver

2.6/6 Separate the outer pad and plate assembly from the caliper

2.6/7 After cleaning the caliper, attach the inner pad to the caliper after making sure the new anti-rattle clip is in place (caliper removed for clarity)

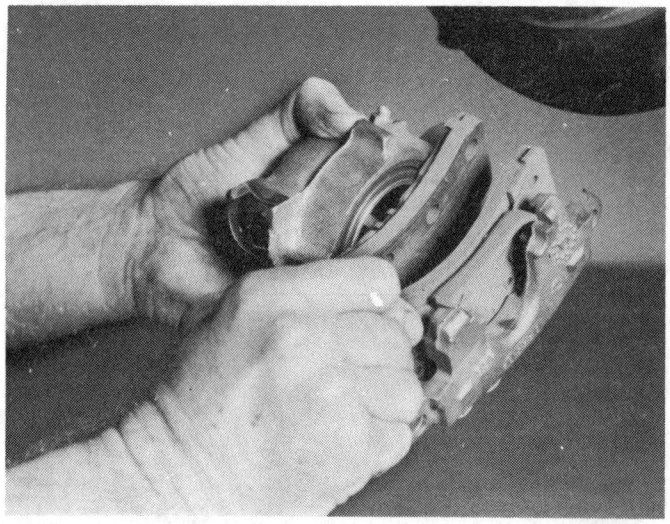

2.6/8 Attach the outer pad to the caliper (caliper removed for clarity)

2.6/9 Apply multi-purpose grease to the ends of the caliper mounting bracket surfaces

2.6/10 Install the caliper on the mounting bracket (do not get any grease on the brake pad linings)

2.6/11 Install the mounting bolts and tighten them securely

2.6/12 Attach the brake hose to the bracket, then install the clip with a hammer, but be careful not to damage the hose

2.6/13 After refilling the master cylinder with fluid and pumping the brake pedal to seat the pad and plate assemblies, use channel lock pliers to bend the upper ears of the outboard plate assembly until the ears are flush with the caliper housing, with no radial clearance

3.6A Removing the banjo fitting bolt

3.6B Be sure to use new copper seal rings when installing the brake hose

3 Front disc brake caliper — removal and installation

1 Whenever you are working on the brake system, be aware that asbestos dust is present and be careful not to inhale any of it because it has been proven to be harmful to your health.
2 Remove the cover from the brake fluid reservoir and siphon off two-thirds of the fluid into a container and discard it.
3 Raise the front of the vehicle and place it securely on jackstands.
4 Remove the front wheel. Remove the caliper from one brake assembly at a time, using the assembled brake for reference, if necessary.
5 Push the caliper piston back into its bore. If necessary, a C-clamp can be used, but a flat bar will usually do the job. As the piston is depressed to the bottom of the caliper bore, the fluid in the master cylinder will rise. Make sure that it does not overflow.
6 Remove the banjo fitting bolt holding the brake hose (photo), then remove and discard the copper seal rings found on either side of the banjo fitting. **Note:** *Always use new copper seals when reinstalling the brake hose (photo).*
7 Remove the Allen head bolts holding the caliper to the mounting bracket and separate the caliper from the rotor.
8 Installation is the reverse of removal. Lubricate the ends of the caliper mounting bracket surfaces and tighten all bolts to the specified torque. Bleed the brake system as described in Section 16.

4 Front disc brake caliper — overhaul

Note: *Purchase a brake caliper overhaul kit for your particular vehicle before beginning this procedure.*
1 Refer to Section 3 and remove the caliper.
2 Refer to Section 2 and remove the brake pad and plate assemblies.
3 Clean the exterior of the brake caliper with brake fluid (*never use gasoline, kerosene or cleaning solvents*), then place the caliper on a clean workbench.
4 Place a wooden block or shop rag in the caliper as a cushion, then use compressed air to remove the piston from the caliper (photo). Use only enough air pressure to ease the piston out of the bore. If the piston is blown out, even with the cushion in place, it may be damaged. **Caution:** *Never place your fingers in front of the piston in an attempt to catch or protect it when applying compressed air — serious injury could occur.*
5 Carefully pry the dust boot out of the caliper bore (photo).
6 Using a wood or plastic tool, remove the piston seal from the groove in the caliper bore. Metal tools may cause bore damage.
7 Remove the caliper bleeder valve, then remove and discard the sleeves and bushings from the caliper ears. Also discard all rubber parts.
8 Clean the remaining parts with brake fluid. Allow them to drain and then shake them vigorously to remove as much fluid as possible.
9 Carefully examine the piston for nicks and burrs and loss of plating.

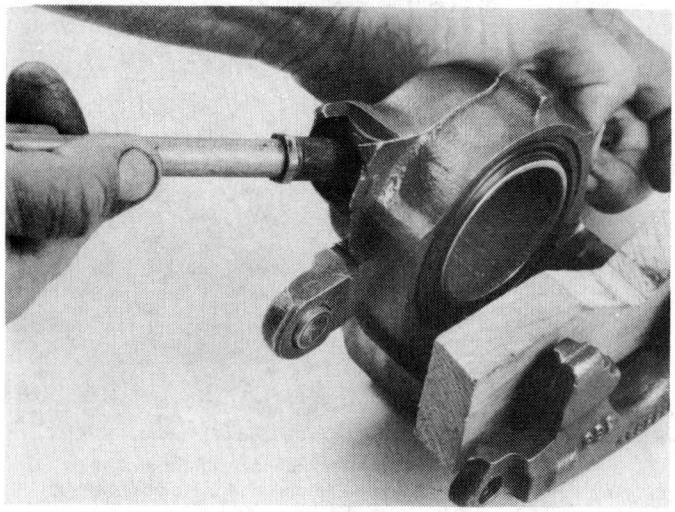

4.4 Compressed air is used to move the piston out of the caliper bore (note use of wooden block as a cushion)

4.5 Prying the dust boot out of the caliper bore

4.6 Removing the piston seal from the caliper groove

4.10 Positioning the seal in the caliper bore groove

If surface defects are present, parts must be replaced. Check the caliper bore in a similar way, but light polishing with crocus cloth is permissible to remove light corrosion and stains. Discard the mounting bolts if they are corroded or damaged.

10 When assembling, lubricate the piston bores and seal with clean brake fluid; position the seal in the caliper bore groove (photo).

11 Lubricate the piston with clean brake fluid, then install a new boot in the piston groove with the fold toward the open end of the piston (photo).

12 Insert the piston squarely into the caliper bore, then apply force to bottom the piston in the bore (photo).

13 Position the dust boot in the caliper counterbore, then use a drift to drive it into position (photo). Make sure that the boot is evenly installed below the caliper face.

14 Install the bleeder valve.

15 The remainder of the installation procedure is the reverse of the removal procedure. Always use new copper gaskets when connecting the brake hose and bleed the system as described in Section 16.

5 Rear disc brake caliper — removal and installation

1 Whenever you are working on the brake system, be aware that asbestos dust is present and be careful not to inhale any of it, as it has been proven to be harmful to your health.

2 Remove the cover from the brake fluid reservoir and siphon off two-thirds of the fluid into a container and discard it.

3 Raise the rear of the vehicle and place it securely on jackstands, then release the parking brake handle.

4 Remove the wheel, then reinstall one wheel lug (flat side toward the rotor) to hold the rotor in place. Work on one brake assembly at a time, using the assembled brake for reference, if necessary.

5 Loosen the tension on the parking brake cable by loosening the nuts at the equalizer.

6 Remove the cable, damper and spring from the cable lever.

7 Hold the lever and remove the nut, then remove the lever, lever seal and anti-friction washer from the caliper actuator screw.

8 Refer to the accompanying illustration and position a C-clamp on the caliper housing. **Note:** *Make sure the C-clamp does not contact the actuator screw.*

9 Tighten the C-clamp until the piston bottoms in the cylinder bore, then remove the clamp.

4.11 Installing a new dust boot in the piston groove (note that folds are toward the open end of the piston)

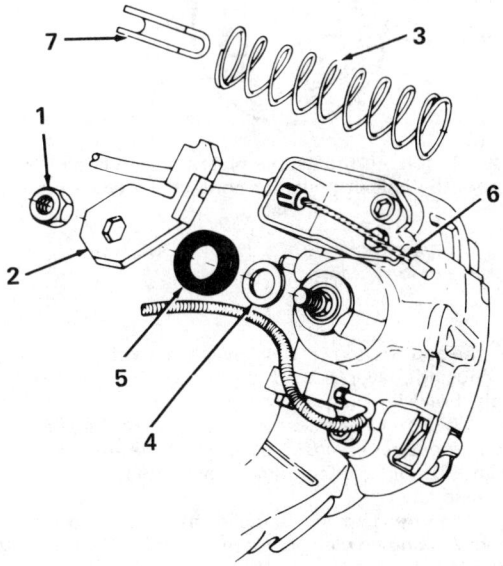

Fig. 9.1 Rear disc brake caliper parking brake assembly (Sec 5)

1 *Parking brake lever retaining nut*
2 *Parking brake lever*
3 *Return spring*
4 *Anti-friction washer*
5 *Parking brake lever seal*
6 *Parking brake cable*
7 *Damper*

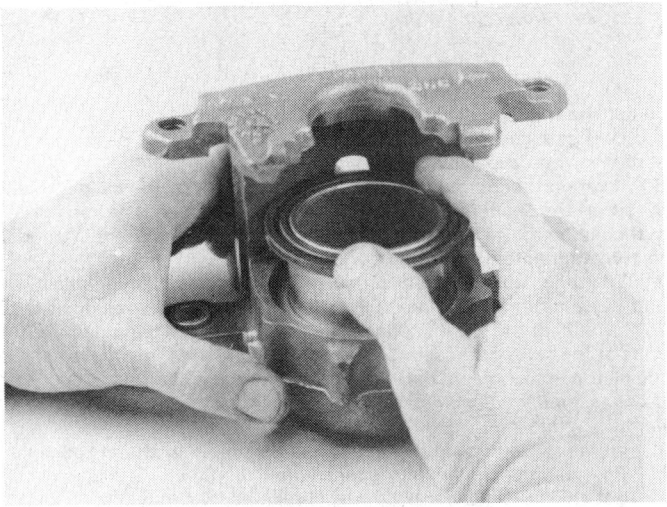

4.12 Squaring the piston in the bore

4.13 Using a hammer and driver to seat the boot in the caliper housing counterbore

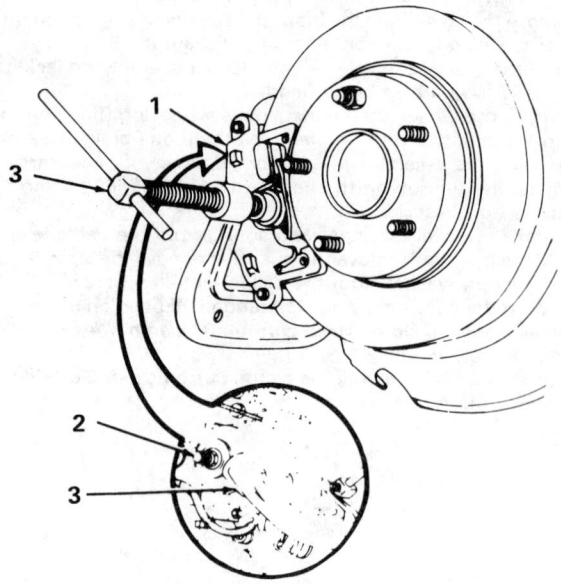

Fig. 9.2 Bottoming the caliper piston with a large C-clamp (make sure the caliper does not contact the actuator screw) (Sec 5)

1 Caliper housing *3 C-clamp*
2 Actuator screw

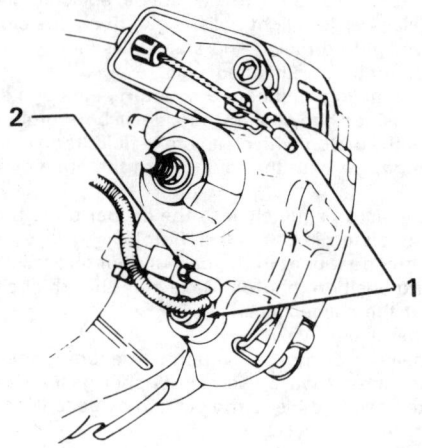

Fig. 9.3 Brake line and caliper mounting details (Sec 5)

1 Caliper mounting bolts *2 Brake line nut*

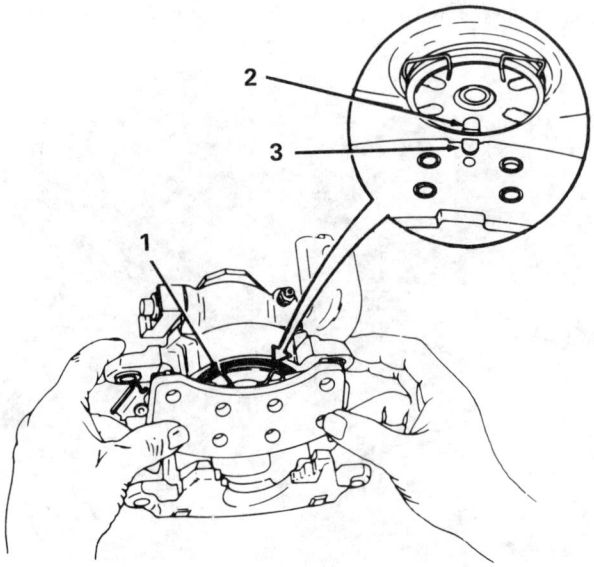

Fig. 9.4 Pad and plate assembly installation details (make sure the D-notch and D-tab engage) (Sec 6)

1 Shoe dampening spring *3 D-tab on shoe*
2 D-notch in piston

10 Reinstall the anti-friction washer, lever seal (with the sealing bead against the housing), lever and nut on the actuating screw.
11 Loosen the brake line nut, then disconnect the brake line from the caliper. **Note:** *If the brake line is seized, remove the banjo fitting bolt, banjo fitting and copper washers in order to free the line.*
12 To prevent fluid loss and contamination, plug the openings in the caliper and brake line.
13 Using an Allen head socket, remove the caliper mounting bolts. **Note:** *If you are working on the right brake assembly, it will be necessary to remove the bolt from the rear of the lower control arm and move the lower control arm to gain access to the lower caliper mounting bolt (refer to Chapter 11 if necessary).*
14 Remove the caliper from the vehicle.
15 Installation is the reverse of removal. Always use a new lever seal and lubricate it with silicone brake lube before installing it. Also use a new anti-friction washer. If the banjo fitting was removed, use new copper washers when reinstalling it.
16 When installing the parking brake lever on the actuator screw hex, install the lever pointing down, then rotate the lever toward the front of the vehicle and hold it while installing the retaining nut. Tighten the nut to the specified torque, then rotate the lever back against the stop on the caliper.
17 After connecting the parking brake cable, tighten it at the equalizer until the lever starts to move off the caliper stop, then loosen the adjustment until the lever just moves back against the stop.
18 Fill the master cylinder and bleed the hydraulic system (Sec 16).

6 Rear disc brake pads — replacement

Note: *Disc brake pads should be replaced on both wheels at the same time.*

1 Remove the brake caliper (refer to Section 5).
2 Remove the brake pad and plate assemblies from the caliper, using a screwdriver if necessary.
3 Remove the sleeves and bushings from the caliper ears.
4 Using a small screwdriver, remove the flexible two-way check valve from the end of the caliper piston.
5 If leakage is noted at the caliper assembly, refer to Section 7 and overhaul the caliper.
6 Lubricate the new sleeves and bushings with silicone grease and install them in the caliper ears.
7 Lubricate a new two-way check valve with silicone grease and install it in the caliper piston end.
8 Position the inboard pad and plate assembly in the caliper. Make sure that the D-shaped tab on the shoe engages the D-shaped notch in the piston. If the tab and notch do not line up, use a spanner wrench to turn the piston until they do.
9 With the tab and notch properly aligned, install the inboard pad and plate assembly in the caliper, with the wear sensor positioned so that it is at the leading edge. Slide the edge of the plate under the ends of the dampening spring and snap the assembly into place against the piston, making sure that the plate lies flat against the piston. **Note:** *If the assembly fails to lie flat, recheck the alignment of the D-shaped notch and tab..*
10 Install the outboard pad and plate assembly in the caliper.
11 Reinstall the caliper (refer to Section 5).
12 Set the caliper by applying the brake pedal firmly.
13 Using channel lock pliers, bend the ears on the upper edge of the plate assembly until they are flush with the caliper housing, with no radial clearance.

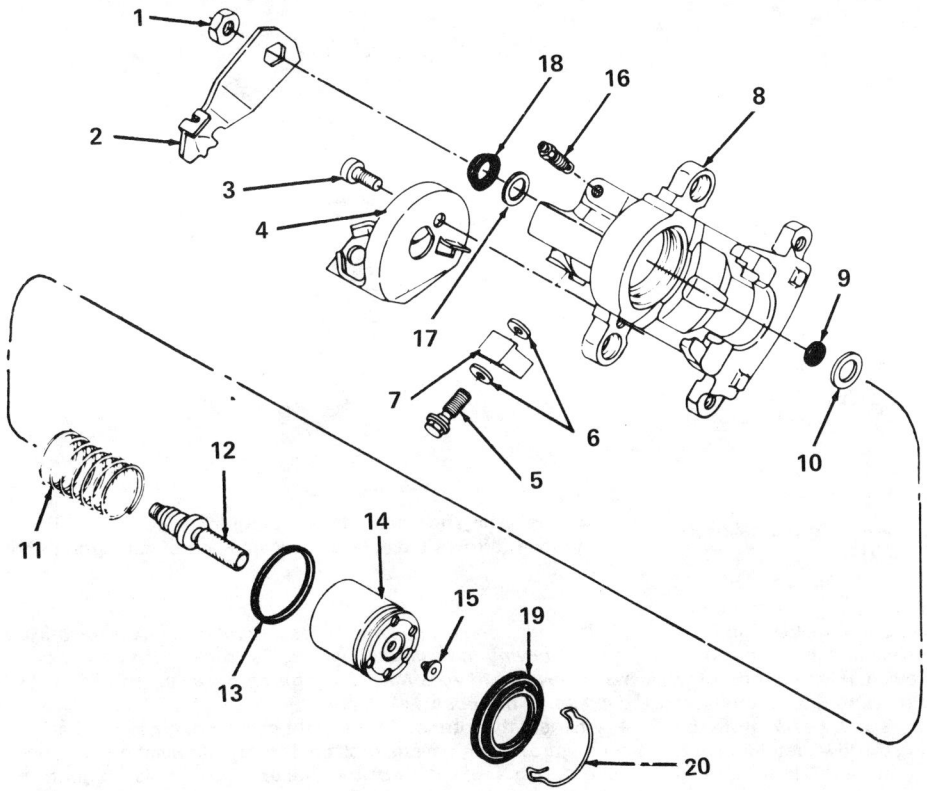

Fig. 9.5 Rear brake caliper components — exploded view (Sec 7)

1 *Parking brake lever retaining nut*
2 *Parking brake lever*
3 *Parking brake lever mounting bracket bolt*
4 *Parking brake lever mounting bracket*
5 *Banjo housing retaining bolt*
6 *Washers*
7 *Banjo housing*
8 *Caliper housing*
9 *Shaft seal*
10 *Thrust washer*
11 *Balance spring*
12 *Actuator screw*
13 *Piston seal*
14 *Piston assembly*
15 *Two-way check valve*
16 *Bleeder valve*
17 *Anti-friction washer*
18 *Lever seal*
19 *Boot*
20 *Shoe dampening spring*

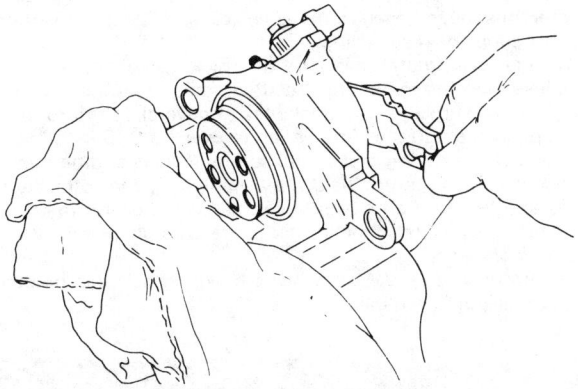

Fig. 9.6 Working the piston out of the caliper bore with the parking brake lever (Sec 7)

7 Rear disc brake caliper — overhaul

Note: *Purchase a brake caliper overhaul kit for your particular vehicle before beginning this procedure.*

1 Remove the caliper (refer to Section 5) and place it on a workbench.
2 Remove the pad and plate assemblies from the caliper (refer to Section 6).
3 Remove the shoe dampening spring from the end of the caliper piston.
4 Place the caliper in a vise equipped with soft jaws and cushion the interior with a piece of wood or shop towels.
5 Move the parking brake lever back and forth to work the piston out of the caliper bore. If the piston will not come out, remove the lever retaining nut and lever and use a wrench to rotate the actuator screw in the same direction as the parking brake is applied until the piston comes out of the cylinder.
6 Remove the balance spring.
7 If not previously removed, remove the retaining nut and lever, then remove the lever seal and anti-friction washer.
8 Press on the threaded end of the actuator screw to remove it from the housing.
9 Remove the dust boot from the caliper, using a screwdriver.
10 Use a wood or plastic tool to remove the piston seal from the bore.
11 If not previously removed, remove the banjo housing retaining bolt, banjo housing and copper washers.
12 Remove the parking brake lever mounting bracket only if it is damaged.
13 Carefully check the caliper bore for score marks, nicks, corrosion and excessive wear. Light corrosion may be removed with crocus cloth; otherwise, replace the caliper housing with a new one.
14 Clean all parts not included in the caliper repair kit with denatured alcohol, clean brake fluid or brake system cleaner. *Do not, under any circumstances, use petroleum-based solvents.*
15 Use compressed air to dry the parts and blow out all the passages in the caliper housing and bleeder valve.
16 Install the bleeder valve and tighten it to the specified torque.
17 If removed for replacement, install the parking brake lever mounting bracket and tighten the mounting bolt to the specified torque.
18 Install the banjo housing assembly, making sure to use new copper washers. Tighten the retaining bolt to the specified torque.
19 Lubricate the new piston seal with clean brake fluid and install it in the caliper bore groove. Make sure that the seal is not twisted.
20 Attach the boot to the piston with the inside lip of the boot in the piston groove and the boot fold toward the end of the piston that contacts the inboard brake shoe.
21 Install the thrust washer on the actuator screw with the bearing surface toward the caliper housing.
22 Lubricate the shaft seal with brake fluid and install it on the actuator screw.
23 Lubricate the actuator screw with brake fluid and install it in the piston.
24 Install the balance spring in the piston recess.
25 Lubricate the piston and caliper bore with brake fluid and start the piston assembly into the caliper bore.
26 Using a piston compressor or other suitable tool, push the piston

Fig. 9.7 Using a piston compressor to bottom the caliper piston in the bore (Sec 7)

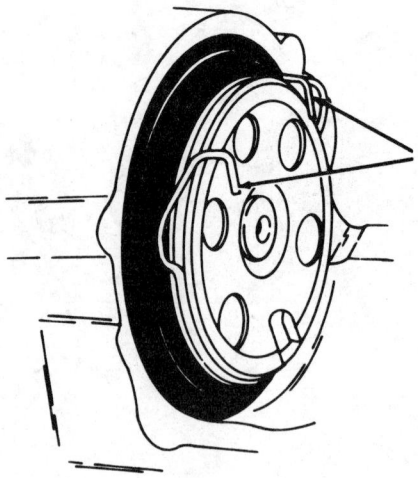

Fig. 9.8 The shoe dampening spring installed in the piston end groove (arrows indicate correct position of the spring ends) (Sec 7)

in until it bottoms in the caliper bore. Before removing the tool, lubricate the anti-friction washer and lever seal with silicone lubricant and install them over the end of the actuating screw, making sure that the sealing bead on the lever seal is against the housing. Install the lever over the actuating screw, then rotate the lever slightly away from the stop on the housing and hold it while installing the lever retaining nut. Tighten the nut to the specified torque, then rotate the lever back to the stop.

27 Position the outside of the boot in the caliper recess. Using a driver and hammer, seat the boot in the recess.

28 Install the dampening spring in the groove in the piston end. **Note:** *It may be necessary to move the parking brake lever off its stop so that the piston may be extended slightly, making the spring groove accessible. If this is necessary, push the piston back into the bottom of the caliper bore before installing the caliper.*

29 Install the caliper on the vehicle (refer to Section 5).

8 Disc brake rotor — inspection, removal and installation

1 Raise the vehicle and place it securely on jackstands.

2 Remove the appropriate wheel and tire.

3 Remove the brake caliper assembly (refer to Section 3 or 5). **Note:** *On front disc brakes, it is not necessary to disconnect the brake hose. After removing the caliper mounting bolts, hang the caliper out of the way on a piece of wire. Never hang the caliper by the brake hose because damage to the hose will occur.*

4 Inspect the rotor surfaces. Light scoring or grooving is normal, but deep grooves or severe erosion is not. If pulsating has been noticed during application of the brakes, suspect disc runout.

5 Attach a dial indicator to the caliper mounting bracket, turn the rotor and note the amount of runout. Check both inboard and outboard surfaces (photo). If the runout is more than the maximum allowable, the rotor must be removed from the vehicle and taken to an automotive machine shop for resurfacing.

6 Using a micrometer, measure the thickness of the rotor (photo). If it is less than the minimum specified, replace the rotor with a new one. Also measure the disc thickness at several points to determine variations in the surface. Any variation over 0.0005-inch may cause pedal pulsations during brake application. If this condition exists and the disc thickness is not below the minimum, the rotor can be removed and taken to an automotive machine shop for resurfacing.

7 To remove and install the front rotor, refer to Chapter 1, *Wheel bearing check and repack.*

8 The rotor on the rear wheel disc brake models can be pulled off after the caliper is removed.

8.5 Checking the disc brake rotor pad surface runout with a dial indicator

8.6 Using a micrometer to check rotor thickness

Fig. 9.9 Components of a typical drum brake assembly (left side shown, right side is the reverse)

1 *Primary brake shoe return spring*
2 *Wheel cylinder*
3 *Shoe guide*
4 *Anchor pin*
5 *Secondary brake shoe return spring*
6 *Actuator link*
7 *Brake hose*
8 *Secondary brake shoe*
9 *Actuator lever pivot*
10 *Actuator lever*
11 *Lever return spring*
12 *Pawl*
13 *Adjusting screw wheel*
14 *Adjusting screw*
15 *Adjusting screw spring*
16 *Primary brake shoe*
17 *Hold-down pin*
18 *Hold-down spring*

9 Rear drum brake shoes — replacement

1 Whenever you are working on the brake system, be aware that asbestos dust is present. It has been proven to be harmful to your health, so be careful not to inhale any of it.
2 Raise the vehicle and place it securely on jackstands.
3 Release the parking brake handle.
4 Remove the wheel and tire assembly. **Note:** *All four rear shoes should be replaced at the same time, but to avoid mixing up parts, work on only one brake assembly at a time.*
5 Now refer to the accompanying photographs and perform the brake shoe replacement procedure. Start with Photo 9.5/1. **Note:** *If the brake drum cannot be easily pulled off the axle and shoe assembly, make sure that the parking brake is completely released, then squirt some penetrating oil around the center hub area. Allow the oil to soak in and try to pull the drum off again. If the drum still cannot be pulled off, the brake shoes will have to be retracted. This is accomplished by first removing the lanced cutout in the backing plate with a hammer and chisel. With this lanced area punched in, pull the lever off the adjusting screw wheel with one small screwdriver while turning the adjusting wheel with another small screwdriver, moving the shoes away from the drum. The drum may now be pulled off.*
6 Before reinstalling the drum, it should be checked for cracks, score marks, deep scratches and 'hard spots,' which will appear as small discolored areas. If the hard spots cannot be removed with fine emery cloth and/or if any of the other conditions listed above exist, the drum must be taken to an automotive machine shop to have it turned. If the drum will not 'clean up' before the maximum drum diameter is reached in the machining operation, the drum will have to be replaced with a new one. **Note:** *The maximum diameter is cast into each brake drum (photo).*
7 Install the brake drum, lining up the marks made before removal if the old brake drum is used, and install the wheel stud lock washers.

9.5/1 Remove the wheel mounting stud lock washers

9.5/2 Mark the relationship of the axle and brake drum

9.5/3 Remove the brake drum (if it cannot be pulled off, refer to copy)

9.5/4 Use a brake spring tool to remove the primary shoe return spring from the anchor pin pivot

9.5/5 Removing the spring after unhooking it from the primary brake shoe

9.5/6 Push on the bottom of the automatic adjuster actuator lever (bottom hand) and remove the actuator link from the top of the actuator lever

9.5/7 Remove the actuator link and secondary shoe return spring from the anchor pin pivot with a brake spring tool

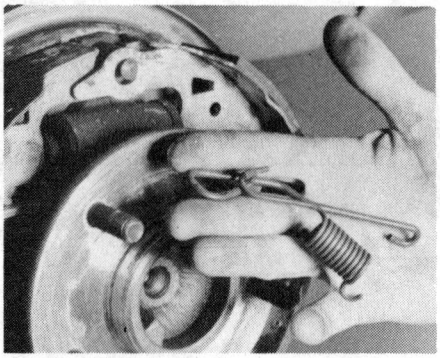
9.5/8 The actuator link and secondary shoe return spring after removal

9.5/9 Remove the shoe guide from the anchor pin (the pin is removed from the rear of the backing plate)

9.5/10 Remove the primary shoe hold-down spring and pin

9.5/11 The primary shoe hold-down spring and pin after removal

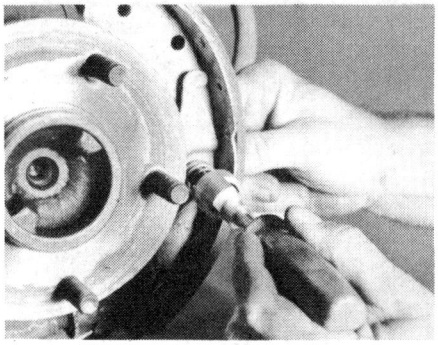

9.5/12 Remove the secondary shoe hold-down spring and pin (note that this spring is shorter than the primary shoe hold-down spring)

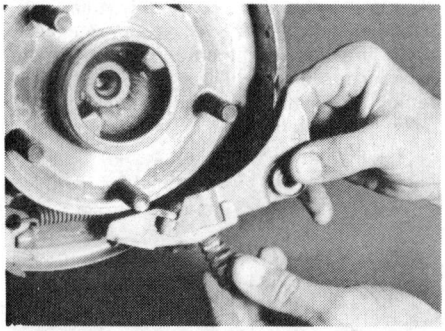

9.5/13 Remove the actuator lever, pawl and lever return spring (note L mark on the actuator lever denoting left side brake assembly)

9.5/14 Turn the adjuster wheel all the way in (toward the right in this case)

9.5/15 Pivot the secondary shoe to the rear and down

9.5/16 Remove the secondary shoe from the adjusting screw spring

9.5/17 Remove the adjusting screw and spring

9.5/18 Note that the longer hook goes toward the rear

9.5/19 Pull the parking brake strut and spring assembly from behind the axle flange

9.5/20 Pull out on the primary shoe and pivot it down until it is free of the backing plate

9.5/21 Remove the spring from the parking brake lever

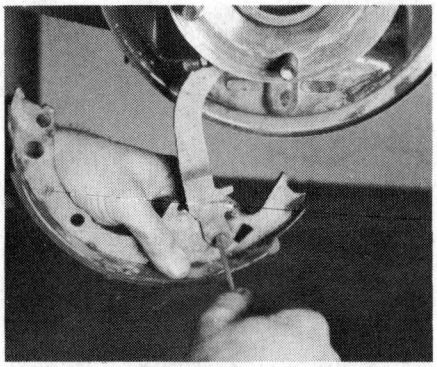

9.5/22 Remove the E-clip retaining the parking brake lever to the primary shoe

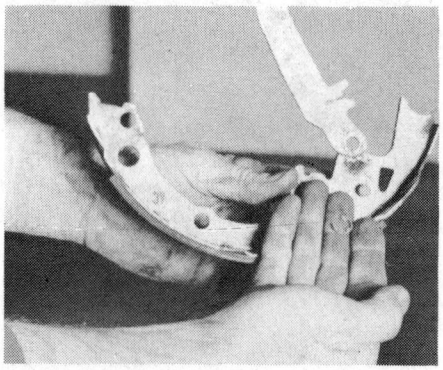

9.5/23 With the E-clip removed, the shoe may be separated from the parking brake lever

9.5/24 Check the shoe guide pads on the backing plate for wear, making sure a ridge has not formed which could hang up the brake shoe (if a sharp ridge is present, it may be removed with 150-grit emery paper). **This step completes the shoe removal procedure. At this time, check all springs for tension and replace them as necessary, then check the wheel cylinder for leakage. If leakage is noted, remove the wheel cylinder for rebuilding (refer to Section 10)**

9.5/25 Lubricate the shoe guide pads and wheel cylinder ends with multi-purpose grease (application on the wheel cylinder ends should be light)

9.5/26 Lightly lubricate the pivot pin with multi-purpose grease

9.5/27 Attach the parking brake lever to the primary brake shoe and the parking brake lever spring to the lever (note that the primary shoe has the shorter lining)

9.5/28 Install the primary shoe in position on the backing plate and install the hold-down spring

9.5/29 Place the parking brake strut into position in the primary shoe slot

9.5/30 Lubricate both ends of the adjuster

9.5/31 Engage the short hook end of the adjusting spring in the hole in the primary shoe, with the hook inserted from the front

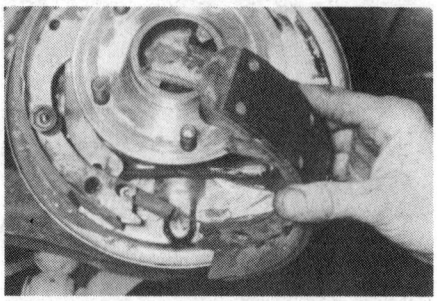

9.5/32 Engage the long hook end of the spring in the hole in the secondary shoe, with the hook inserted from the rear

9.5/33 With the adjuster positioned between the shoes, pivot the secondary shoe into place on the backing plate

9.5/34 Attach the pawl to the actuator lever and install the actuator assembly in place inside the secondary shoe

9.5/35 Attach the bushing to the actuator and shoe

9.5/36 Install the shoe hold-down spring

9.5/37 Install the shoe guide, with the rounded side in

9.5/38 Install the rear hook of the secondary shoe return spring in the shoe

9.5/39 Assemble the actuating link to the spring

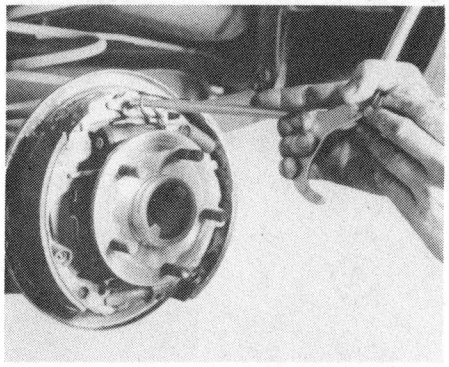

9.5/40 Attach the spring and actuating link to the anchor pin

9.5/41 Attach the actuating link to the parking brake actuating lever

9.5/42 Attach the primary shoe return spring to the shoe . . .

9.5/43 . . . and to the anchor pin

9.5/44 With all parts installed, rock the assembly back and forth with your hands to ensure that all parts are seated (to complete the job, refer to Steps 6 through 9 in the accompanying copy)

9.6 Location of the maximum diameter mark inside the brake drum

8 Mount the wheel and tire, install the wheel lugs and tighten to the Specifications, then lower the vehicle.
9 Make a number of forward and reverse stops to adjust the brakes until a satisfactory pedal action is obtained.

10 Wheel cylinder (drum brakes) — removal, overhaul and installation

Note: *Obtain the wheel cylinder rebuild kits before beginning this procedure.*

1 Remove the brake shoes (refer to Section 9).
2 Remove the brake line fitting from the rear of the wheel cylinder (photo). Cap the brake line to prevent contamination.
3 Using curved needle-nose pliers, remove the wheel cylinder retaining clip (photo), from the rear of the wheel cylinder. Alternatively, insert two awls into the access slots between the cylinder pilot and the retainer locking tabs and bend both tabs away at the same time (photo).
4 Remove the cylinder and place it on a clean workbench.
5 Remove the bleeder valve, seals, pistons, boots and spring assembly from the cylinder body (photo).
6 Clean the wheel cylinder with brake fluid, denatured alcohol or brake system cleaner. *Do not, under any circumstances, use petroleum-based solvents to clean brake parts.*
7 Use compressed air to remove excess fluid from the wheel cylinder

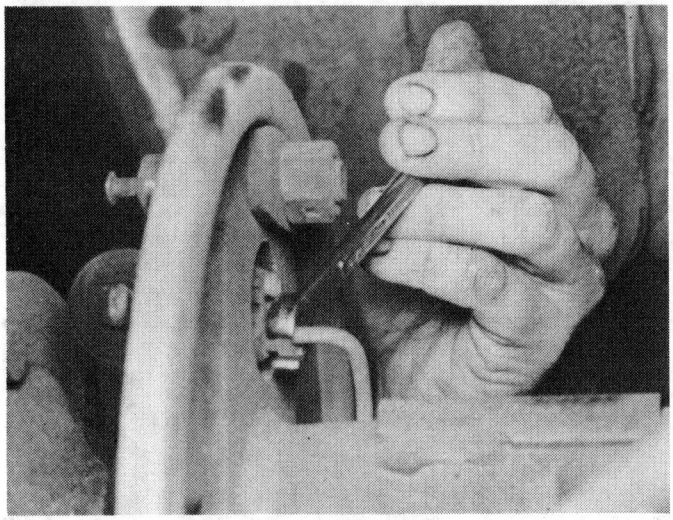

10.2 Removing the brake line fitting from the rear of the wheel cylinder

10.3A Removing the cylinder retaining clip with needle-nose pliers

10.3B Method of removing the retaining clip using two awls (cylinder has been removed from vehicle for illustration purposes)

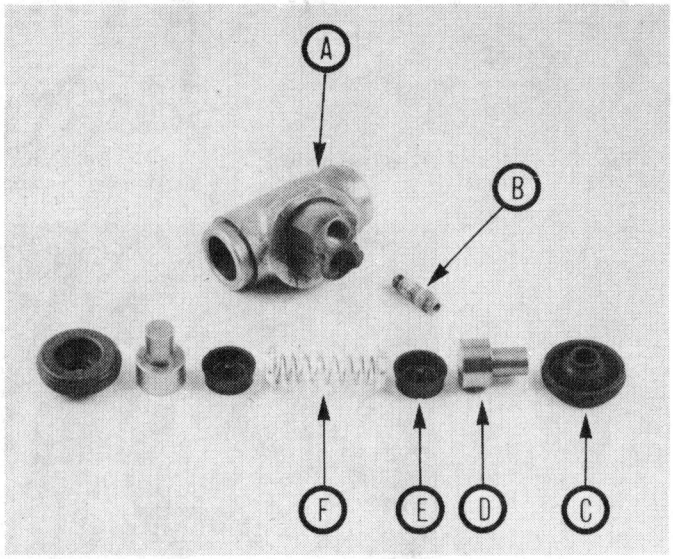

10.5 Wheel cylinder components — exploded view

A Wheel cylinder body *C Boot* *E Seal*
B Bleeder screw *D Piston* *F Spring assembly*

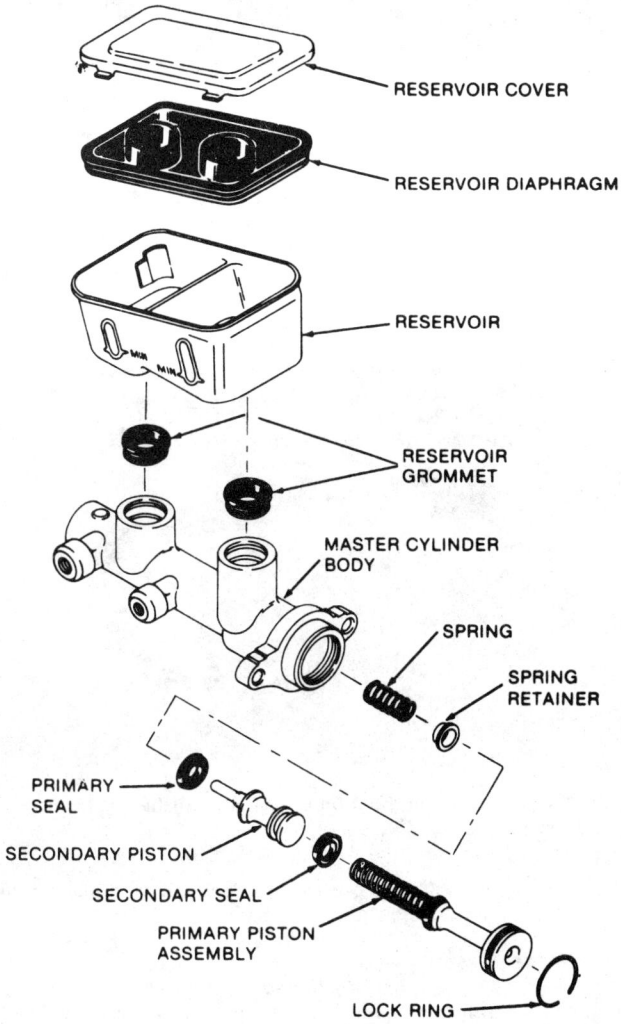

Fig. 9.10 Master cylinder components — exploded view (Sec 11)

and to blow out the passages.

8 Check the cylinder bore for corrosion and scoring. Crocus cloth may be used to remove light corrosion and stains, but the cylinder must be replaced with a new one if the defects cannot be removed easily, or if the bore is scored.

9 Lubricate the new seals with clean brake fluid.

10 Assemble the brake cylinder, making sure the boots are properly seated.

11 Place the wheel cylinder in position in the backing plate.

12 Secure the cylinder in place with a new retaining clip by first starting the brake line fitting into the threads of the wheel cylinder, then placing a screwdriver handle between the wheel cylinder and the axle flange to hold the cylinder in place while the retaining clip is pushed into place with curved needle-nose pliers (photo). Make sure the clip and cylinder are securely in place, then remove the screwdriver and tighten the brake line fitting.

13 Bleed the brake system (refer to Section 16).

11 Master cylinder — removal, overhaul and installation

1 A master cylinder overhaul kit should be purchased before beginning this procedure. The kit will include all the replacement parts necessary for the overhaul procedure. The rubber replacement parts, particularly the seals, are the key to fluid control within the master cylinder. As such, it's very important that they be installed securely and facing in the proper direction. Be careful during the rebuild procedure that no grease or mineral-based solvents come in contact with the rubber parts.

2 Completely cover the front fender and cowling area of the vehicle, as brake fluid can ruin painted surfaces if it is spilled.

3 Disconnect the brake line connections. Rags or newspapers should be placed under the master cylinder to soak up the fluid that will drain out.

4 Remove the two master cylinder mounting nuts, move the bracket retaining the combination valve forward slightly, taking care not to bend the hydraulic lines running to the combination valve, and remove the master cylinder from the vehicle.

5 Remove the reservoir cover and reservoir diaphragm, then discard any remaining fluid in the reservoir.

6 Remove the primary piston lock ring by depressing the piston and prying the ring out with a screwdriver.

7 Remove the primary piston assembly with a wire hook, being careful not to scratch the bore surface.

8 Remove the secondary piston assembly in the same manner.

9 Place the master cylinder in a vise and pry the reservoir from the cylinder body with a pry bar (photo).

10 Do not attempt to remove the quick take-up valve from the cylinder body, as this valve is not serviceable.

10.12 Placement of screwdriver handle between wheel cylinder and axle flange while retaining clip is installed from rear

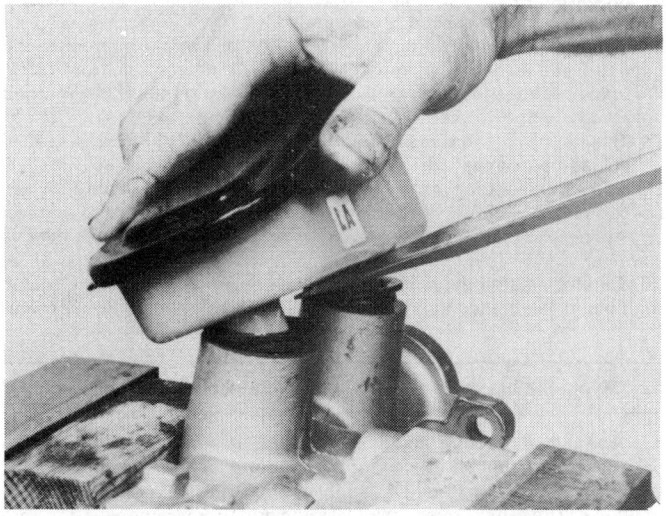

11.9 Prying the plastic reservoir from the master cylinder body

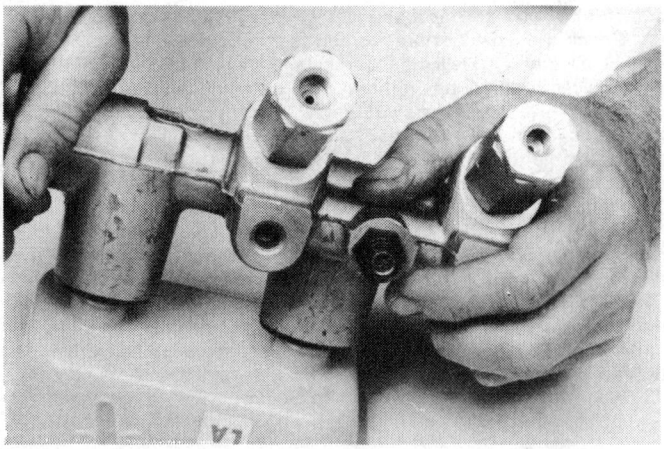

11.14 Pressing the master cylinder body onto the reservoir

11 Remove the reservoir grommets.
12 Inspect the cylinder bore for corrosion and damage. If any corrosion or damage is found, replace the master cylinder body with a new one, as abrasives cannot be used on the bore.
13 Lubricate the new reservoir grommets with silicone brake lube and press the grommets into the master cylinder body, making sure they are properly seated.
14 Lay the reservoir on a hard surface and press the master cylinder body onto the reservoir, using a rocking motion (photo).
15 Remove the old seals from the secondary piston assembly and install the new seals so that the cups face *out* (photo).
16 Attach the spring retainer to the secondary piston assembly.
17 Lubricate the cylinder bore with clean brake fluid and install the spring and secondary piston assembly in the cylinder.
18 Disassemble the primary piston assembly, noting the position of the parts, then lubricate the new seals with clean brake fluid and install them on the piston (photo).
19 Install the primary piston assembly in the cylinder bore, depress it and install the lock ring.
20 Inspect the reservoir cover and disphragm for cracks and deformation. Replace any damaged parts with new ones and attach the diaphragm to the cover.
21 **Note:** *Whenever the master cylinder is removed, the complete hydraulic system must be bled. The time required to bleed the system can be reduced if the master cylinder is filled with fluid and 'bench bled' (refer to Steps 22 through 25) before the master cylinder is installed on the vehicle.*
22 Insert threaded plugs of the correct size into the cylinder outlet holes and fill the reservoirs with brake fluid (the master cylinder should be supported in such a manner that brake fluid will not spill out of it during the bench bleeding procedure).
23 Loosen one plug at a time and push the piston assembly into the bore to force air from the master cylinder. To prevent air from being drawn back into the cylinder, the appropriate plug must be tightened before allowing the piston to return to its original position.
24 Stroke the piston three or four times for each outlet to assure that all air has been expelled.
25 Refill the master cylinder reservoirs and install the diaphragm and cover assembly. **Note:** *The reservoirs should only be filled to the top of the reservoir divider to prevent overflowing when the cover is installed.*
26 Carefully install the master cylinder by reversing the removal steps, then bleed the brakes at the wheel bleed valves (refer to Section 16).

12 Hydraulic brake hoses and lines — inspection and replacement

1 About every six months, with the vehicle raised and placed securely on jackstands, the flexible hoses which connect the steel brake lines with the front and rear brake assemblies should be inspected for cracks, chafing of the outer cover, leaks, blisters and other damage. These are important and vulnerable parts of the brake system and inspection should be complete. A light and mirror will prove helpful for a thorough check. If a hose exhibits any of the above conditions, replace it with a new one as follows:

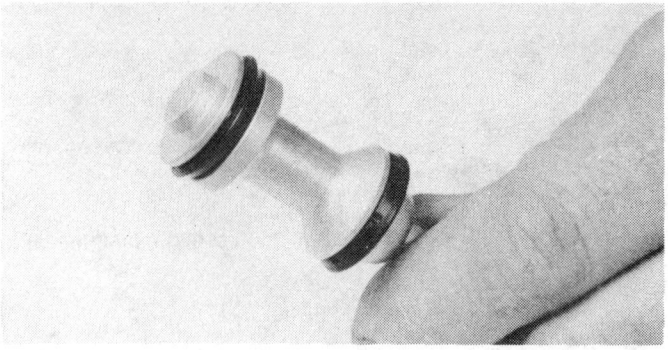

11.15 The secondary seals must be installed with the lips facing *out*

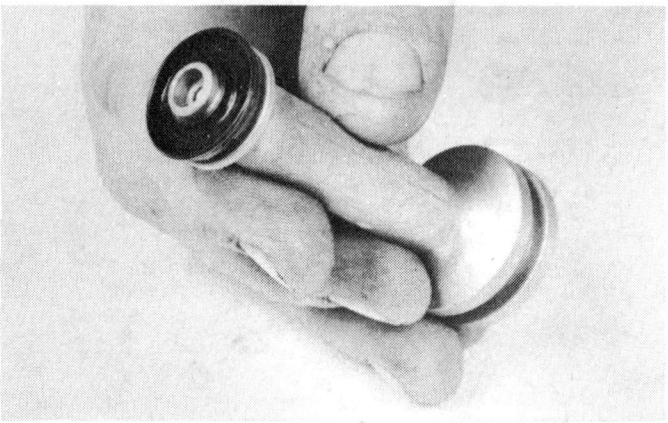

11.18 The primary seal must be installed with the lip facing *away from the piston*

Front brake hose

2 Using a back-up wrench, disconnect the brake line from the hose fitting, being careful not to bend the frame bracket or brake line.
3 Use pliers to remove the U-clip from the female fitting at the bracket, then remove the hose from the bracket.
4 At the caliper end of the hose, remove the bolt from the fitting block, then remove the hose and the copper gaskets on either side of the fitting block.
5 When installing the hose, always use new copper gaskets on either side of the fitting block and lubricate all bolt threads with clean brake fluid before installing them.
6 With the fitting flange engaged with the caliper locating ledge, attach the hose to the caliper and tighten it to the specified torque.
7 Without twisting the hose, install the female fitting in the hose bracket (it will fit the bracket in only one position).
8 Install the U-clip retaining the female fitting to the frame bracket.
9 Using a back-up wrench, attach the brake line to the hose fitting and tighten it to the specified torque.
10 When the brake hose installation is complete, there should be no kinks in the hose. Also, make sure the hose does not contact any part of the suspension. Check this by turning the wheels to the extreme left and right positions. If the hose makes contact, remove the hose and correct the installation as necessary.

Rear brake hose

11 Locate the junction block at the rear axle and disconnect the two steel brake lines from the block.
12 Using a back-up wrench, remove the hose at the female fitting, being careful not to bend the bracket or steel lines.
13 Remove the U-clip with pliers and separate the female fitting from the bracket.
14 Note the position of the junction block carefully so that it may be reinstalled in precisely the same position.

15 Remove the bolt attaching the junction block to the axle and remove the hose from the block.
16 When installing, thread both steel line fittings into the junction block at the rear axle.
17 Bolt the junction block to the axle, then tighten the block bolt and steel lines to the specified torque.
18 Without twisting the hose, install the female end of the hose in the frame bracket (it will fit the bracket in only one position).
19 Install the U-clip retaining the female end to the bracket.
20 Using a back-up wrench, attach the steel line fitting to the female fitting, tightening it to the specified torque. Again, be careful not to bend the bracket or steel line.
21 Check that the hose installation did not loosen the frame bracket. Retorque the bracket if necessary.
22 Fill the master cylinder reservoirs and bleed the system (refer to Section 16).

Steel brake lines

23 When it becomes necessary to replace steel lines, use only double-walled steel tubing. Never substitute copper tubing because copper is subject to fatigue cracking and corrosion. The outside diameter of the tubing is used for sizing.
24 Auto parts stores and brake supply houses carry various lengths of prefabricated brake line. Depending on the type of tubing used, these sections can either be bent by hand into the desired shape or must be bent in a tubing bender.
25 If prefabricated lengths are not available, obtain the recommended steel tubing and fittings to match the line to be replaced. Determine the correct length by measuring the old brake line, and cut the new tubing to length, leaving about 1/2-inch extra for flaring the ends.
26 Install the fittings onto the cut tubing and flare the ends using an ISO flaring tool.
27 Using a tubing bender, bend the tubing to match the shape of the old brake line.
28 Tube flaring and bending can usually be performed by a local auto parts store if the proper equipment mentioned in Steps 26 and 27 is not available.
29 When installing the brake line, leave at least 3/4-inch clearance between the line and any moving parts.

13 Combination valve — check and replacement

Check

1 Disconnect the electrical connector from the pressure differential switch (photo). **Note:** *When disconnecting the connector, squeeze the connector side lock releases, moving the inside tabs away from the switch, then pull up.* Pliers may be used as an aid if necessary.
2 Using a jumper wire, connect the switch wire to a good ground, such as the engine block.
3 Turn the ignition key to the On position. The warning light in the instrument panel should light up.
4 If the warning light does not light, either the bulb is burned out or the electrical circuit is defective. Replace the bulb (refer to Chapter 10) or repair the electrical circuit as necessary.
5 When the warning light functions correctly, turn the ignition switch off, disconnect the jumper wire and reconnect the wire to the switch terminal.
6 Make sure the master cylinder reservoirs are full, then attach a bleeder hose to one of the rear wheel bleeder valves and immerse the other end of the hose in a container partially filled with clean brake fluid.
7 Turn the ignition switch on.
8 Open the bleeder valve while a helper applies moderate pressure to the brake pedal. The brake warning light on the instrument panel should light.
9 Close the bleeder valve before the helper releases the brake pedal.
10 Reapply the brake pedal with moderate to heavy pressure. The brake warning light should go out.
11 Attach the bleeder hose to one of the front brake bleeder valves

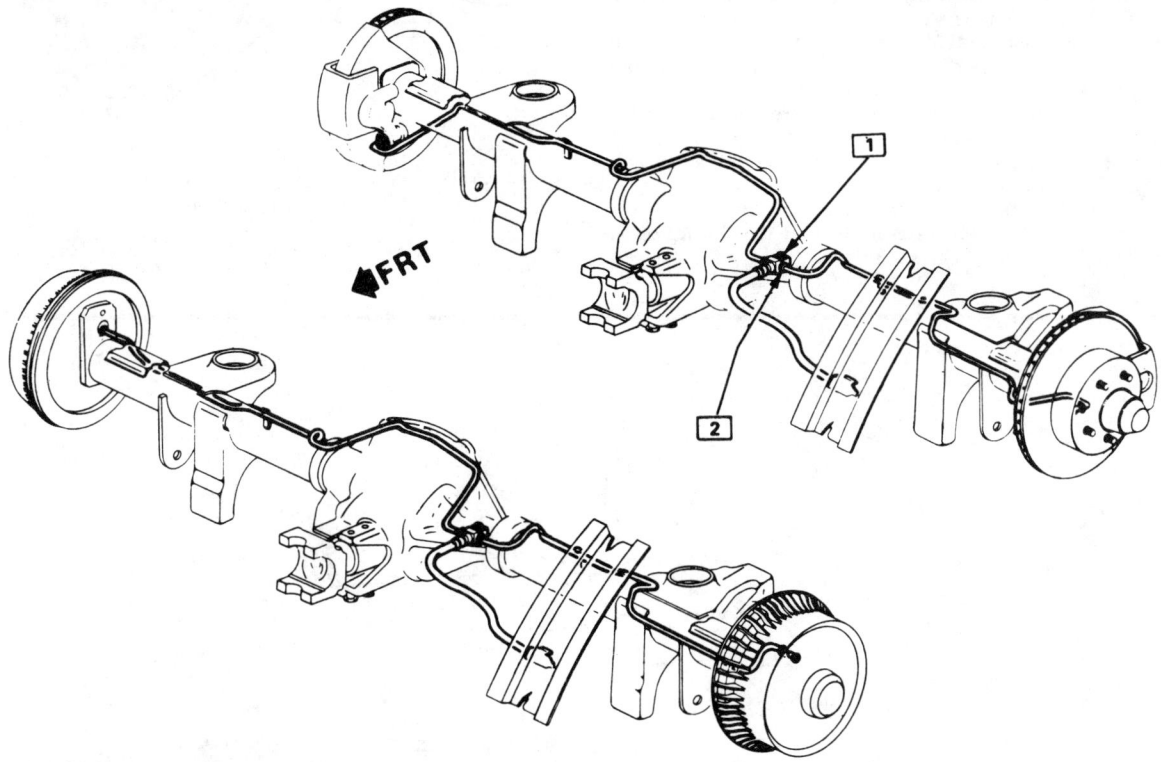

Fig.11 Rear brake line installation details on vehicles with disc brakes (top) and drum brakes (bottom) (Sec 12)

1 Junction block *2 Brake line fitting*

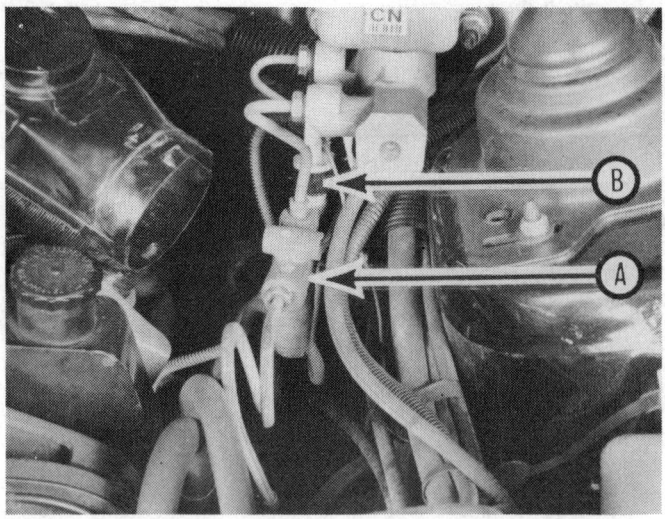

13.1 The combination valve (A) is located just under the master cylinder, with the pressure differential switch (B) located at the center of the valve

14.4 Hold the brake cable stud (front) with one wrench while turning the adjusting nut with another

and repeat Steps 8 through 10. The warning light should react in the same manner as in Steps 8 and 10.

12 Turn the ignition switch off.

13 If the warning light did not come on in Steps 8 and 11, but does light when a jumper is connected to ground, the warning light switch portion of the combination valve is defective and the combination valve must be replaced with a new one since the components of the combination valve are not individually serviceable.

Replacement

14 Place a container under the combination valve and protect all painted surfaces with newspapers or rags.

15 Disconnect the hydraulic lines at the combination valve, then plug the lines to prevent further loss of fluid and to protect the lines from contamination.

16 Disconnect the electrical connector from the pressure differential switch (refer to Step 1 if necessary).

17 Remove the bolt holding the valve to the mounting bracket and remove the valve from the vehicle.

18 Installation is the reverse of the removal procedure.

19 Bleed the entire brake system (refer to Section 16). *Do not move the vehicle until a firm brake pedal is attained.*

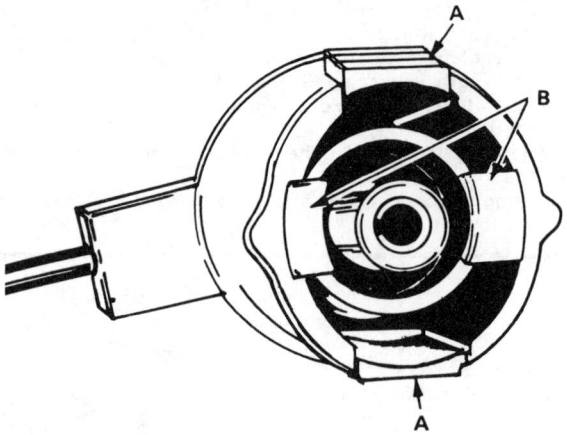

Fig. 9.12 Pressure differential switch electrical connector details (Sec 13)

A Side lock releases *B Inside tabs*

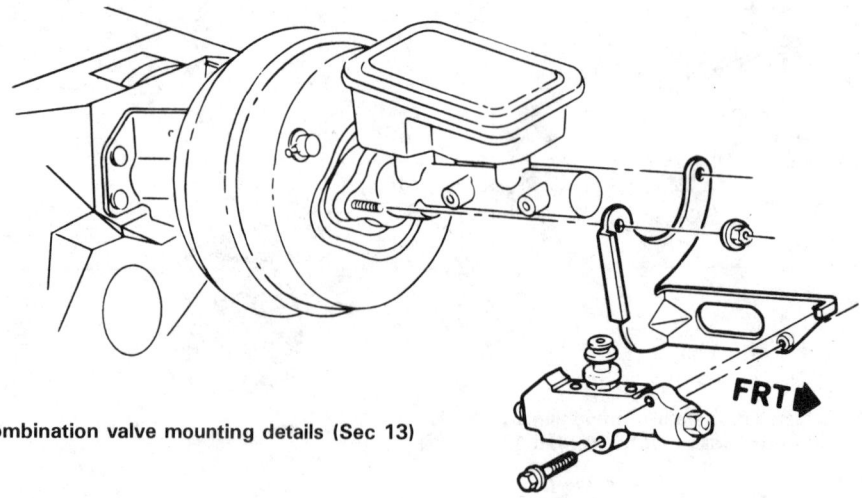

Fig. 9.13 Combination valve mounting details (Sec 13)

14 Parking brake — adjustment

1 The parking brake cables may stretch over a period of time, necessitating adjustment. Also, the parking brake should be checked for proper adjustment whenever the rear brake cables have been disconnected. If the parking brake handle travel is less than 13 or more than 17 ratchet clicks, the parking brake needs adjustment.

Drum rear brakes

2 Pull the parking brake handle exactly two ratchet clicks.
3 Raise the vehicle and support it securely on jackstands.
4 Locate the adjusting screw on the rear side of the equalizer bracket. To keep the brake cable stud from turning, hold it with one wrench and turn the adjusting nut with another wrench until the left rear wheel can just be turned in reverse (using two hands) but is locked when you attempt to turn it forward (photo).
5 Release the parking brake and make sure that both rear wheels turn freely and that there is no brake drag in either direction, then remove the jackstands and lower the vehicle.

Disc rear brakes

6 Make sure the parking brake handle is completely released.
7 Raise the vehicle and place it securely on jackstands.
8 Lubricate the parking brake cables at the rubbing points on the vehicle underbody and at the equalizer hooks, then check for free movement of all cables.
9 Hold the brake cable stud with one wrench to prevent it from turnning and turn the equalizer nut with another wrench until all cable slack is removed.
10 Check that the caliper levers are against the stops on the caliper housings after tightening the equalizer nut. If the levers are off the stops, loosen the cable until the levers return to the stops.
11 Operate the parking brake lever several times to check the adjustment. The parking brake handle should travel approximately 14 clicks with normal handle application effort.
12 Remove the jackstands, lower the vehicle to the ground and again check to see if the levers are on the caliper stops. If not, repeat the above steps as necessary.

15 Power brake booster — inspection, removal and installation

1 The power brake unit requires no special maintenance apart from periodic inspection of the hoses and inspection of the air filter beneath the boot at the pedal pushrod end.
2 Dismantling of the power brake unit requires special tools. If a problem develops, it is recommended that a new or factory-exchange unit be installed rather than trying to overhaul the original booster.
3 Remove the mounting nuts which hold the master cylinder to the power brake unit. Position the master cylinder out of the way, being careful not to strain the lines leading to the master cylinder. If there is any doubt as to the flexibility of the lines, disconnect them at the cylinder and plug the ends.
4 Disconnect the vacuum hose leading to the front of the power brake booster. Cover the end of the hose.
5 Loosen the four nuts that secure the booster to the firewall. Do not remove these nuts at this time.
6 Inside the vehicle, disconnect the power brake pushrod from the brake pedal. Do not force the pushrod to the side when disconnecting it.
7 Now remove the four booster mounting nuts and carefully lift the unit out of the engine compartment.
8 When installing, loosely install the four mounting nuts and then connect the pushrod to the brake pedal. Tighten the nuts to the specified torque and reconnect the vacuum hose and master cylinder. If the hydraulic brake lines were disconnected, the entire brake system should be bled to eliminate any air which has entered the system (refer to Section 16).

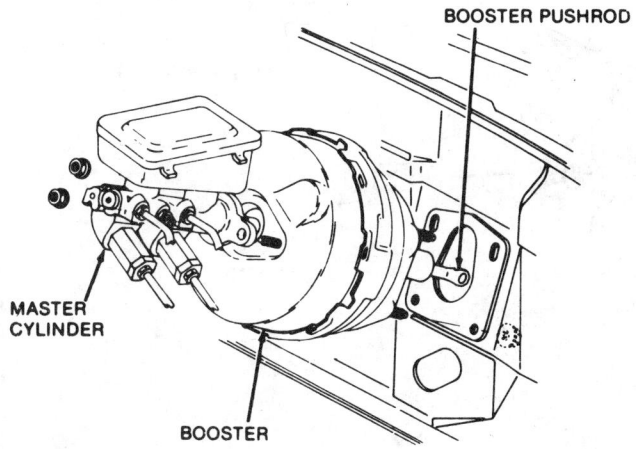

Fig. 9.14 Power brake booster mounting details (Sec 15)

16 Hydraulic system — bleeding

1 Bleeding of the hydraulic system is necessary to remove air whenever it is introduced into the brake system.
2 It may be necessary to bleed the system at all four brakes if air has entered the system due to low fluid level, or if the brake lines have been disconnected at the master cylinder.
3 If a brake line was disconnected only at a wheel, then only that wheel cylinder (or caliper) must be bled.
4 If a brake line is disconnected at a fitting located between the master cylinder and any of the brakes, that part of the system served by the disconnected line must be bled.
5 If the master cylinder has been removed from the vehicle, refer to Section 11, Step 21 before proceeding with the procedure which follows.
6 If the master cylinder is installed on the vehicle but is known to have, or is suspected of having air in the bore, the master cylinder must be bled before any wheel cylinder (or caliper) is bled. Follow Steps 7 through 16 to bleed the master cylinder while it is installed on the vehicle.
7 Remove the vacuum reserve from the brake power booster by applying the brake several times with the engine off.
8 Remove the master cylinder reservoir cover and fill the reservoirs with brake fluid, then keep checking the fluid level often during the bleeding operation, adding fluid as necessary to keep the reservoirs full. Reinstall the cover.
9 Disconnect the forward brake line connection at the master cylinder.
10 Allow brake fluid to fill the master cylinder bore until it begins to flow from the forward line connector port (have a container and shop rags handy to catch and clean up spilled fluid).
11 Reconnect the forward brake line to the master cylinder.
12 Have an assistant depress the brake pedal very slowly (one time only) and hold it down.
13 Loosen the forward brake line at the master cylinder to purge the air from the bore, retighten the connection, then have the brake pedal released slowly.
14 Wait 15 seconds (important).
15 Repeat the sequence, including the 15 second wait, until all air is removed from the bore.
16 After the forward port has been completely purged of air, bleed the rear port in the same manner.
17 To bleed the individual wheel cylinders or calipers, first refer to Steps 7 and 8.
18 Have an assistant on hand, as well as a supply of new brake fluid, an empty clear plastic container, a length of 3/16-inch plastic, rubber or vinyl tubing to fit over the bleeder valve and a wrench to open and close the bleeder valve. The vehicle may have to be raised and placed on jackstands for clearance.
19 Beginning at the right rear wheel, loosen the bleeder valve slightly, then tighten it to a point where it is snug but can still be loosened quickly and easily.
20 Place one end of the tubing over the bleeder valve and submerge the other end in brake fluid in the container.
21 Have the assistant pump the brakes a few times to get pressure in the system, then hold the pedal firmly depressed.
22 While the pedal is held depressed, open the bleeder valve just enough to allow a flow of fluid to leave the valve. Watch for air bub-

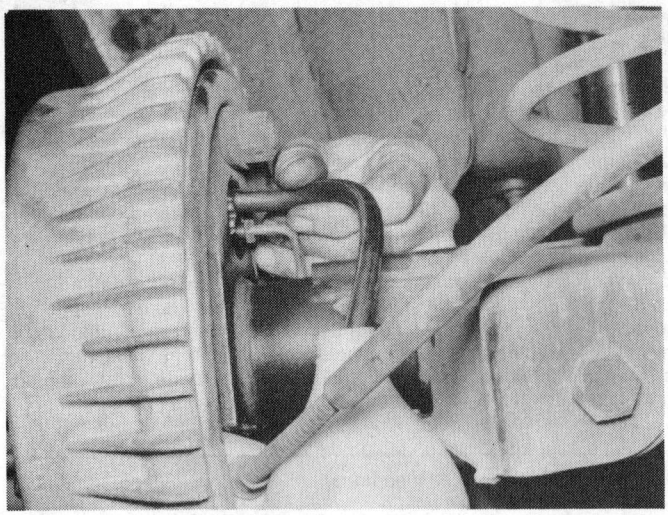

16.23A Attaching a hose to the bleeder valve on a rear drum brake assembly

16.23B Bleeding a front disc brake caliper

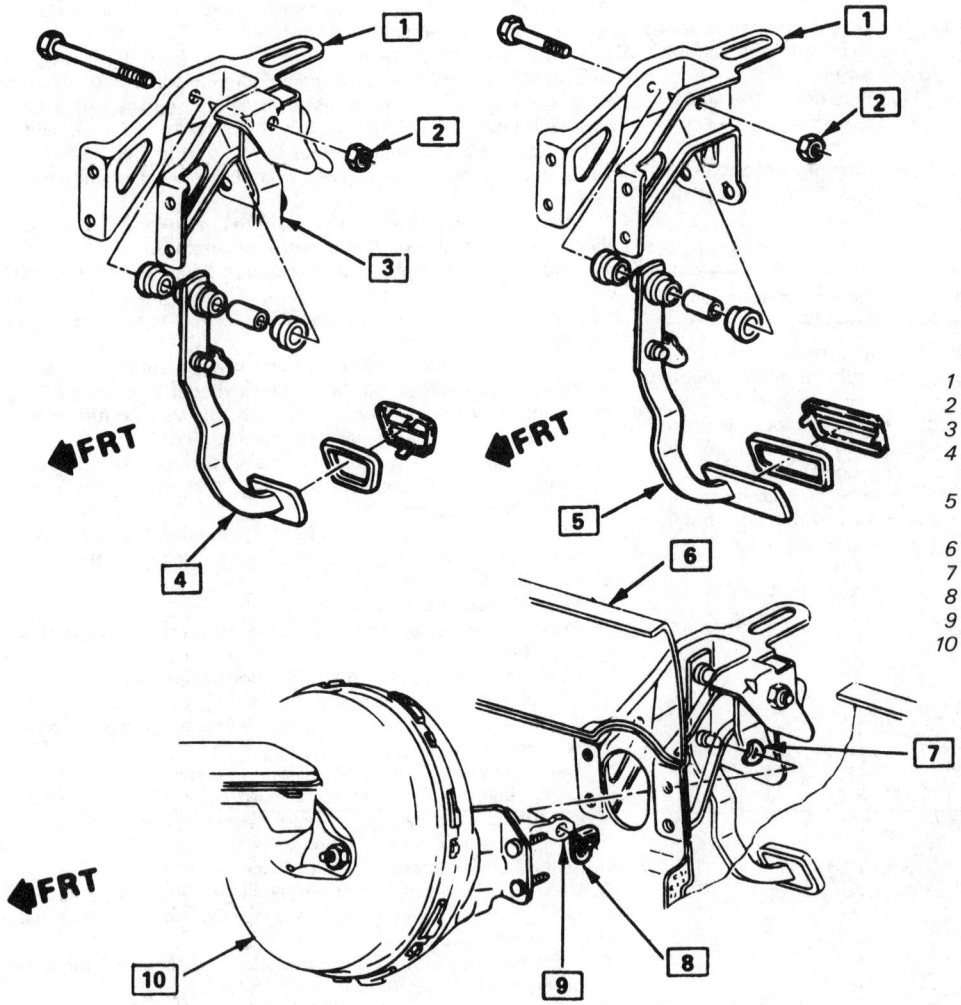

Fig. 9.15 Brake pedal assembly mounting details (Sec 17)

1 *Brake pedal bracket*
2 *Nut*
3 *Clutch pedal*
4 *Brake pedal (with standard transmission)*
5 *Brake pedal (with automatic transmission)*
6 *Dash panel*
7 *Washer*
8 *Retainer*
9 *Pushrod*
10 *Power booster assembly*

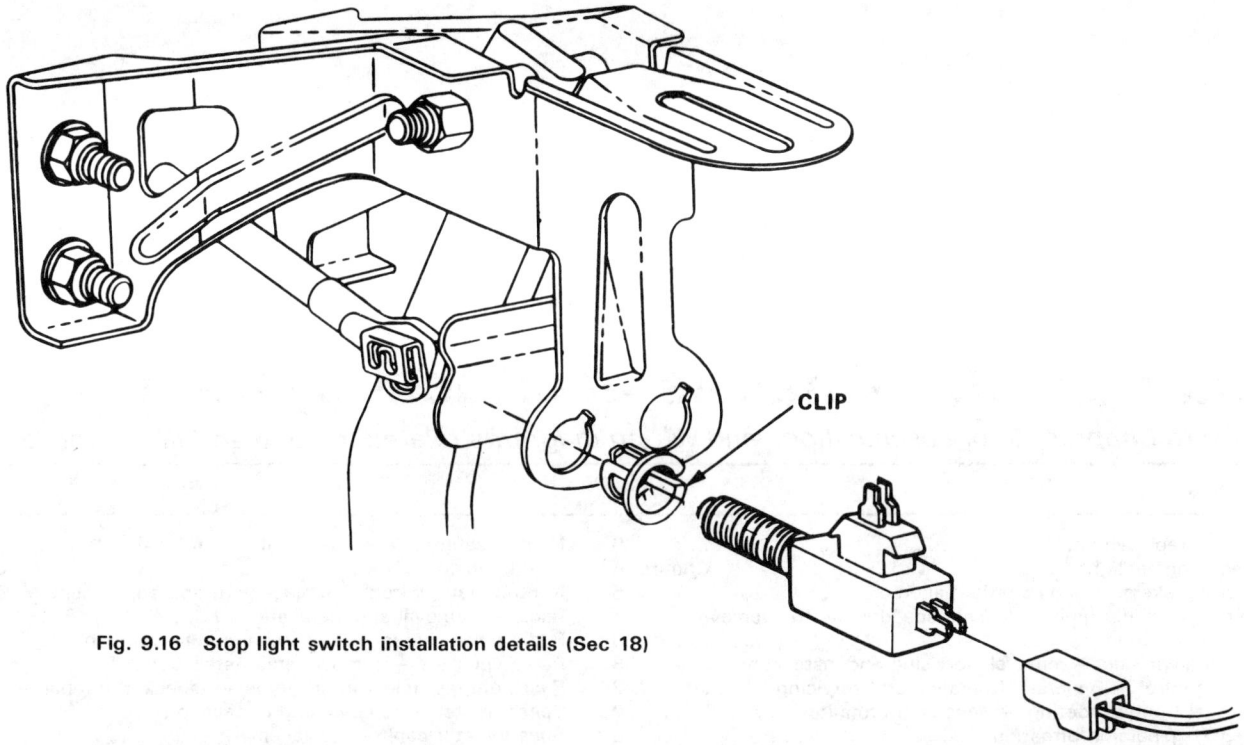

Fig. 9.16 Stop light switch installation details (Sec 18)

bles to exit the submerged end of the tube. When the fluid flow slows after a couple of seconds, close the valve again and have your assistant release the pedal. If the pedal is released before the valve is closed again, air can be drawn back into the system.

23 Repeat Steps 21 and 22 until no more air is seen leaving the tube, then tighten the bleeder valve and proceed to the left rear wheel, the right front wheel and the left front wheel, in that order, and perform the same procedure (photos). Be sure to check the fluid in the master cylinder reservoir frequently.

24 Never use old brake fluid because it attracts moisture which will deteriorate the brake system components.

25 Refill the master cylinder with fluid at the end of the operation.

26 If any difficulty is experienced in bleeding the hydraulic system, or if an assistant is not available, a pressure bleeding kit is a worthwhile investment. If connected in accordance with the instructions, each bleeder valve can be opened in turn to allow the fluid to be pressure ejected until it is clear of air bubbles without the need to replenish the master cylinder reservoir during the process.

17 Brake pedal — removal and installation

1 Disconnect the cable from the negative battery terminal.

2 Disconnect the clutch pedal return spring (if equipped with a manual transmission).

3 Remove the clip retainer from the pushrod pin which travels through the pedal arm.

4 Remove the nut from the pedal shaft bolt. Slide the shaft out far enough to clear the brake pedal arm.

5 The brake pedal can now be removed, along with the spacer and bushing. The clutch pedal (if equipped) will remain in place.

6 When installing, lubricate the spacer and bushings with lightweight grease and tighten the pivot nut to the specified torque.

18 Stop light switch — removal, installation and adjustment

1 The switch is located on a flange or bracket protruding from the brake pedal support.

2 With the brake pedal in the fully released position, the plunger on the body of the switch should be completely pressed in. When the pedal is pushed in, the plunger releases and sends electrical current to the stop lights at the rear of the vehicle.

3 If the stop lights are inoperative and it has been determined that the bulbs are not burned out, push the stop light switch into the tubular clip, noting that audible clicks can be heard as the threaded portion of the switch is pushed through the clip toward the brake pedal.

4 Pull the brake pedal all the way to the rear against the pedal stop until no further clicks can be heard. This will seat the switch in the tubular clip and provide the correct adjustment.

5 Release the brake pedal and repeat Step 4 to ensure that no further clicks can be heard.

6 Make sure that the stop lights are working.

7 If the lights are not working, disconnect the electrical connectors at the stop light switch and remove the switch from the clip.

8 Install a new switch and adjust it by performing Steps 3 through 6, making sure the electrical connectors are hooked up.

Chapter 10 Chassis electrical system

Refer to Chapter 13 for specifications and wiring diagrams related to 1985 and later models

Contents

Bulbs — replacement 9
'Check engine' light . . . . . Chapter 6
Circuit breakers — general information 5
Cluster panel instruments (except speedometer) — removal and installation 13
Console switches — removal, servicing and installation 18
Cruise control — general information and servicing 22
Electrical troubleshooting — general information 2
Fuses — general information 3
Fusible links — general information 4
General information 1
Headlight sealed beam unit — removal and installation 7
Headlight switch — removal and installation 17
Headlights — adjustment 8
Ignition switch — replacement . . . . . Chapter 11
Instrument cluster panel — removal, servicing and installation 16
Neutral safety and back-up light switch (AT) adjustment . . . . . Chapter 7B
Neutral start switch — replacement and adjustment Chapter 7A
Radio — removal and installation 10
Radio power antenna — removal and installation 11
Radio speakers — removal and installation 12
Rear defogger (electric grid-type) — check and repair 23
Speedometer — removal and installation 14
Speedometer cable — replacement 15
Steering column switches — removal and installation . . . . . Chapter 11
Stoplight switch — adjustment and replacement . . . Chapter 9
Turn signals and hazard flashers — check and replacement . . 6
Windshield wiper arm — removal, installation and adjustment 19
Windshield wiper motor — removal and installation 20
Windshield wiper transmission — removal and installation . . . 21

Specifications

Bulb application	Type
Headlight	
Inner, standard	4651
Inner, Halogen	H4651
Outer	4652
Park and directional signal	
Standard Firebird	1157
Formula and Trans Am	1157NA
Side marker	194
Tail and stop	1157
Luggage compartment	561
License plate	194
Back-up	1156
Instrument cluster	161/194
Heater and A/C control panel	194
Indicators	
Headlight high beam	161
Directional signal	194
Temperature	194
Battery	194
Brake	194
Fasten seat belts	194
Check engine	194
Oil pressure	194
Choke	194
Radio (all)	194
Clock	194
Underhood	93
Automatic transmission indicator	194
Dome light	561
Courtesy light	631

1 General information

The electrical system is a 12-volt, negative ground type. Power for the lights and all electrical accessories is supplied by a lead/acid-type battery which is charged by the alternator.

This chapter covers repair and service procedures for the various electrical components not associated with the engine. Information on the battery, alternator, distributor and starter motor can be found in Chapter 5.

It should be noted that whenever portions of the electrical system are worked on, the negative battery cable should be disconnected to prevent electrical shorts and/or fires.

Note: *Information concerning the 1984 Camaro Berlinetta digital instrumentation and dash-related accessories is not included in this manual. Problems involving these components should be referred to your dealer.*

2 Electrical troubleshooting — general information

A typical electrical circuit consists of an electrical component, any switches, relays, motors, etc. related to that component and the wiring and connectors that connect the component to both the battery and the chassis. To aid in locating a problem in any electrical circuit, wiring diagrams for each model are included at the end of this Chapter.

Before tackling any troublesome electrical circuit, first study the appropriate diagrams to get a complete understanding of what makes up that individual circuit. Trouble spots, for instance, can often be narrowed down by noting if other components related to that circuit are operating properly or not. If several components or circuits fail at one time, chances are the problem lies in the fuse or ground connection, as several circuits often are routed through the same fuse and ground connections.

Electrical problems often stem from simple causes, such as loose or corroded connections, a blown fuse or melted fusible link. Prior to any electrical troubleshooting, always visually check the condition of the fuse, wires and connections in the problem circuit.

If testing instruments are going to be utilized, use the diagrams to plan ahead of time where you will make the necessary connections in order to accurately pinpoint the trouble spot.

The basic tools needed for electrical troubleshooting include a circuit tester or voltmeter (a 12-volt bulb with a set of test leads can also be used), a continuity tester (which includes a bulb, battery and set of test leads) and a jumper wire, preferably with a circuit breaker incorporated, which can be used to bypass electrical components.

Voltage checks should be performed if a circuit is not functioning properly. Connect one lead of a circuit tester to either the negative battery terminal or a known good ground. Connect the other lead to a connector in the circuit being tested, preferably nearest to the battery or fuse. If the bulb of the tester goes on, voltage is reaching that point (which means the part of the circuit between that connector and the battery is problem free). Continue checking along the entire circuit in the same fashion. When you reach a point where no voltage is present, the problem lies between there and the last good test point. Most of the time the problem is due to a loose connection. *Keep in mind that some circuits receive voltage only when the ignition key is in the Accessory or Run position.*

A method of finding shorts in a circuit is to remove the fuse and connect a test light or voltmeter in its place to the fuse terminals. There should be no load in the circuit. Move the wiring harness from side-to-side while watching the test light. If the bulb goes on, there is a short to ground somewhere in that area, probably where insulation has rubbed off of a wire. The same test can be performed on other components of the circuit, including the switch.

A ground check should be done to see if a component is grounded properly. Disconnect the battery and connect one lead of a self-powered test light such as a continuity tester to a known good ground. Connect the other lead to the wire or ground connection being tested. If the bulb goes on, the ground is good. If the bulb does not go on, the ground is not good.

A continuity check is performed to see if a circuit, section of circuit or individual component is passing electricity through it properly. Disconnect the battery and connect one lead of a self-powered test light such as a continuity tester to one end of the circuit. If the bulb goes on, there is continuity, which means the circuit is passing electricity through it properly. Switches can be checked in the same way.

Remember that all electrical circuits are composed basically of electricity running from the battery, through the wires, switches, relays, etc. to the electrical component (light bulb, motor, etc.). From there it is run to the body (ground) where it is passed back to the battery. Any electrical problem is basically an interruption in the flow of electricity to and from the battery.

3 Fuses — general information

The electrical circuits of the vehicle are protected by a combination of fuses, circuit breakers and fusible links. The fuse block is located on the underside of the instrument panel on the driver's side (photo).

Each of the fuses is designed to protect a specific circuit and the various circuits are identified on the fuse panel itself.

Miniaturized fuses are employed in the fuse block. These compact fuses with blade terminal design, allow fingertip removal and replacement.

If an electrical component has failed, your first check should be the fuse. A fuse which has 'blown' is easily identified by inspecting the element inside the clear plastic body. Also, the blade terminal tips are exposed in the fuse body, allowing for continuity checks.

It is important that the correct fuse be installed. The different elec-

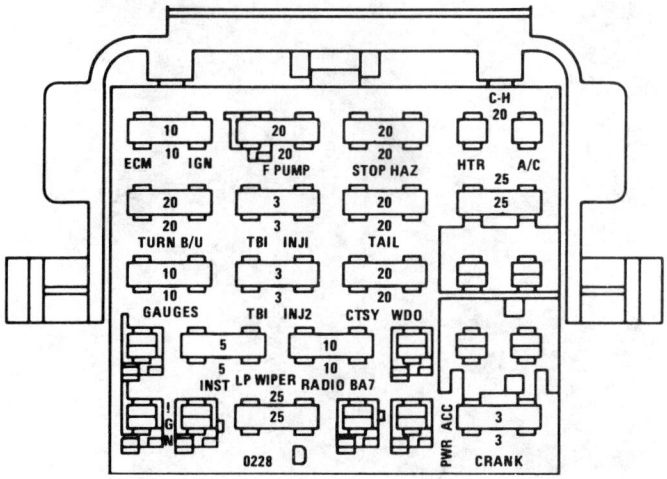

Fig. 10.1 Fuse block layout (typical) (Sec 3)

3.1 The fuse block is located under the left side of the dash (cover has been removed)

trical circuits need varying amounts of protection, indicated by the amperage rating molded in bold, color-coded numbers on the fuse body.

At no time should the fuse be bypassed with pieces of metal or foil. Serious damage to the electrical system could result.

If the replacement fuse immediately fails, do not replace it again until the cause of the problem is isolated and corrected. In most cases, this will be a short circuit in the wiring caused by a broken or deteriorated wire.

4 Fusible links — general information

In addition to fuses, the wiring is protected by fusible links. These links are used in circuits which are not ordinarily fused, such as the ignition circuit.

Although the fusible links appear to be a heavier gauge than the wire they are protecting, the appearance is due to the thick insulation. All fusible links are four wire gauges smaller than the wire they are designed to protect.

The location of the fusible links on your particular vehicle may be determined by referring to the wiring diagram(s) at the end of this Chapter.

The fusible links cannot be repaired, but a new link of the same size wire can be put in its place. The procedure is as follows:

a) Disconnect the battery ground cable.
b) Disconnect the fusible link from the starter solenoid.
c) Cut the damaged fusible link out of the wiring just behind the connector.
d) Strip the insulation approximately 1/2-inch.
e) Position the connector on the new fusible link and crimp it into place.
f) Use rosin core solder at each end of the new link to obtain a good solder joint.
g) Use plenty of electrical tape around the soldered joint. No wires should be exposed.
h) Connect the fusible link at the starter solenoid. Connect the battery ground cable. Test the circuit for proper operation.

5 Circuit breakers — general information

A circuit breaker is used to protect the headlight wiring and is located in the light switch. An electrical overload in the system will cause the lights to go on and off, or in some cases to remain off. If this happens, check the entire headlight circuit immediately. Once the overload condition is corrected, the circuit breaker will function normally.

Circuit breakers are also used with accessories such as power windows, power door locks and rear window defogger.

The circuit breakers in your particular vehicle may be found by referring to the wiring diagram(s) at the end of this Chapter.

6 Turn signals and hazard flashers — check and replacement

1 Small canister-shaped flasher units are incorporated into the electrical circuits for the directional signals and hazard warning lights.
2 When the units are functioning properly, an audible click can be heard with the circuit in operation. If the turn signals fail on one side only and the flasher unit cannot be heard, a faulty bulb is indicated. If the flasher unit can be heard, a short in the wiring is indicated.
3 If the turn signal fails on both sides, the problem may be due to a blown fuse, faulty flasher unit or switch, or a broken or loose connection. If the fuse has blown, check the wiring for a short before installing a new fuse.
4 The hazard warning lights are checked as described in paragraph 3 above.
5 The hazard warning flasher and turn signal flasher are mounted either at the rear of the fuse box or at the convenience center located under the center of the dash.
6 When replacing either of the flasher units, be sure to buy a replacement of the same capacity. Compare the new flasher to the old one before installing it.

7 Headlight sealed beam unit — removal and installation

1 When replacing the headlight, do not turn the spring-loaded adjusting screws or the headlight aim will be changed.
2 Remove the headlight bezel screws and the decorative bezel.
3 Use a cotter pin removal tool to unhook the spring from the retaining ring.
4 Remove the two screws which secure the retaining ring and withdraw the ring. Support the light as this is done.
5 Pull the sealed beam unit out slightly and disconnect the wires from the rear of the light. Remove the light from the vehicle.
6 Position the new unit close enough to connect the wires. Make sure that the numbers molded into the lens are at the top.
7 Install the retaining ring with the mounting screws and spring.
8 Install the bezel and check for proper operation. If the adjusting screws were not turned, the new headlight should not require adjustment.

8 Headlights — adjustment

1 Any adjustments made by the home mechanic that affect the aim of the headlights should be considered temporary only. After adjustment, always have the beams readjusted by a facility with the proper aligning equipment as soon as possible. In some states, these facilities must be state-authorized. Check with your local motor vehicle department concerning the laws in your area.

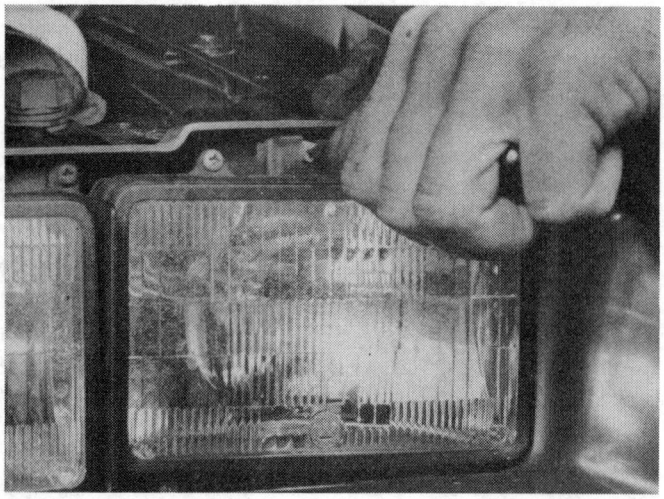

8.2A The screw in the top adjusts the headlight beam up and down

8.2B The side screw adjusts the headlight beam left and right

2 Adjustment screws are provided at the front of each headlight to alter the beam horizontally (side screw) and vertically (top screw) (photos). When making adjustments, be careful not to scratch the paint on the body.

9 Bulbs — replacement

Front end

1 The bulbs for the parking lights and side marker lights are accessible from the rear of each unit.
2 Locate the affected bulb housing and twist it out of the socket.
3 Remove the defective bulb from the housing, install a new bulb and twist/lock the housing back into the socket.

Rear end

4 The bulbs for the taillights, brake lights, directional signal lights and back-up lights are all contained in one assembly on each side of the rear of the vehicle.
5 From inside the vehicle, remove the plastic screws retaining the upholstery at the rear quarter panel.
6 Remove the plastic screws and metal screw retaining the upholstery panel at the rear cover panel (the metal screw is at the top of the panel), revealing the plastic wing nuts retaining the taillight cover panel.
7 Disconnect the wires attached to the panel, remove the wing nuts and pull off the taillight cover panel (photo).
8 Remove the affected bulb socket by squeezing the lock on the bulb socket assembly and turning it counterclockwise, or by simply turning the socket out of the panel (side marker light only) (photos).
9 Replace the burned out bulb with a new one.
10 Installation is the reverse of the removal procedure.

License plate bulb

11 Remove the screw retaining the lens/socket assembly and pull the assembly away from the taillight lens housing.
12 Twist the socket out of the lens, replace the burned out bulb with a new one and reinstall the lens/socket assembly (photo).

Interior light

13 To replace the courtesy light, remove the translucent light panel by squeezing it on both sides and pulling it away from the fixture.
14 Replace the burned out bulb with a new one and reinstall the translucent panel by snapping it into place.

Radio light

15 Remove the radio (refer to Section 10).
16 Remove the bulb access panel and replace the bulb with a new one.
17 Installation is the reverse of the removal procedure.

Console light

18 Refer to Section 18.

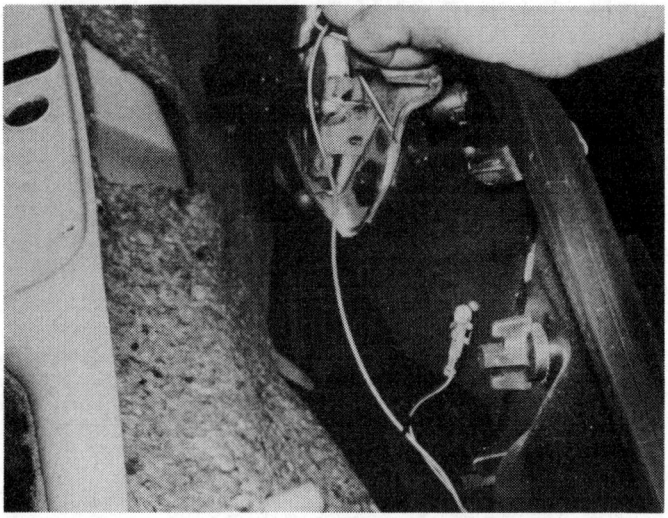

9.7 Disconnecting the wire connectors at the taillight cover panel

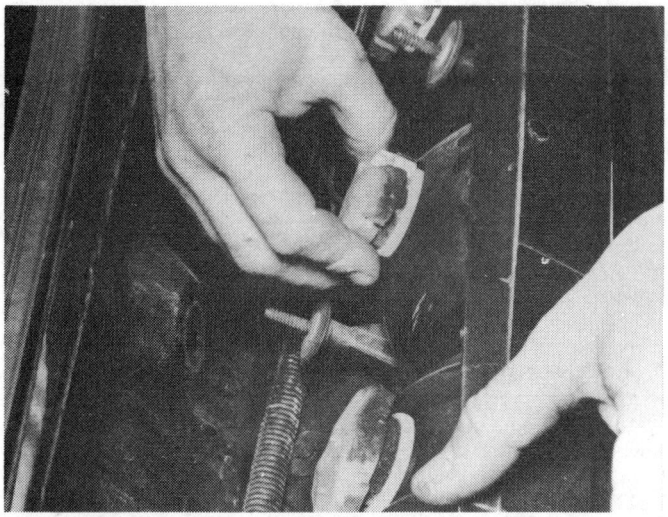

9.8A Removing a lock-type bulb socket from the taillight panel

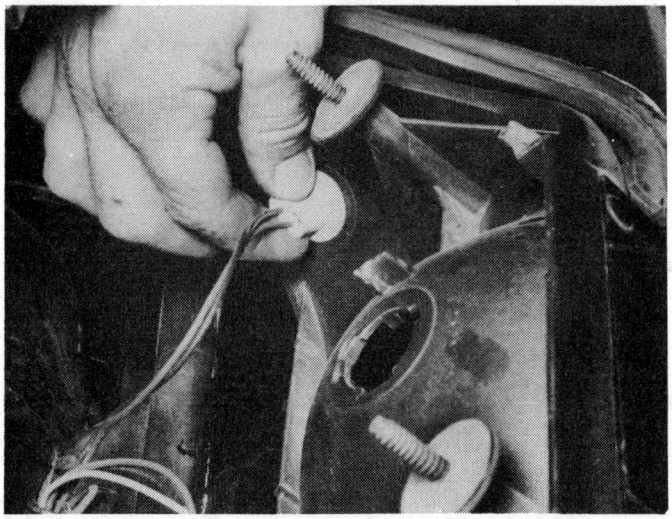

9.8B Removing the bulb socket from the license plate light lens

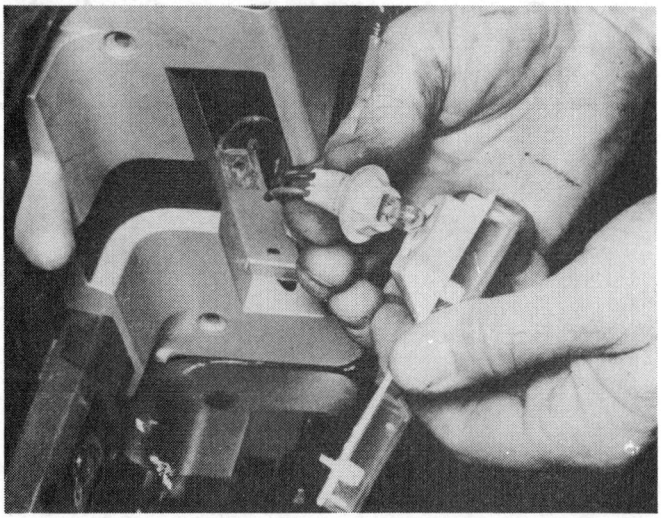

9.12 Removing the bulb socket from the license plate light lens

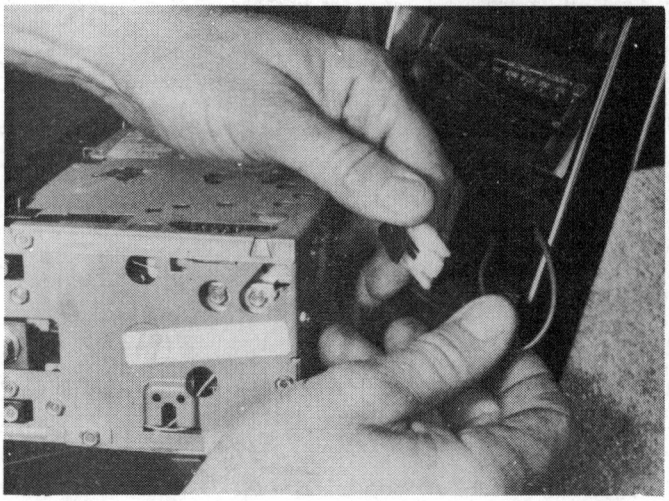

10.4 Removing the wire connector from the rear of the radio

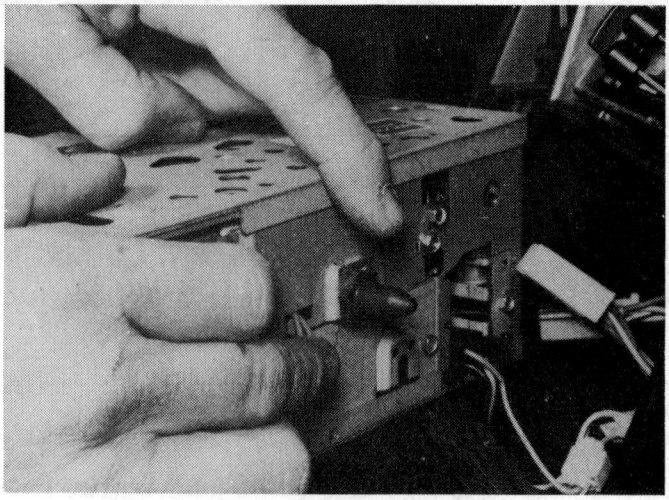

10.5 Location of the radio locating lug

10 Radio — removal and installation

1 Disconnect the cable from the negative battery terminal.
2 Remove the screws retaining the console bezel and remove the bezel.
3 Remove the screws retaining the radio to the console and pull the radio out.
4 Remove all electrical connectors from the rear of the radio (photo). To simplify reinstallation, label the connectors with pieces of numbered tape.
5 Installation is the reverse of removal. The locating lug at the rear of the radio will position it correctly when the lug contacts the receptacle in the console frame (photo). **Caution:** *Always connect the speaker wiring harness* **before** *turning the radio on (to prevent possible radio damage).*

11 Radio power antenna — removal and installation

1 Lower the antenna by turning off the radio or the ignition switch. **Note:** *If the mast has failed in the up position and the mast or the entire antenna assembly is being replaced, the mast can be cut off at fender level to facilitate removal.*
2 Disconnect the cable from the negative battery terminal.
3 Remove the screws attaching the inner fender skirt to the fender.
4 To gain access to the antenna assembly, pull down on the rear of the fender skirt and hold it away from the fender with a block of wood.
5 Remove the escutcheon nut retaining the antenna cable assembly to the fender.

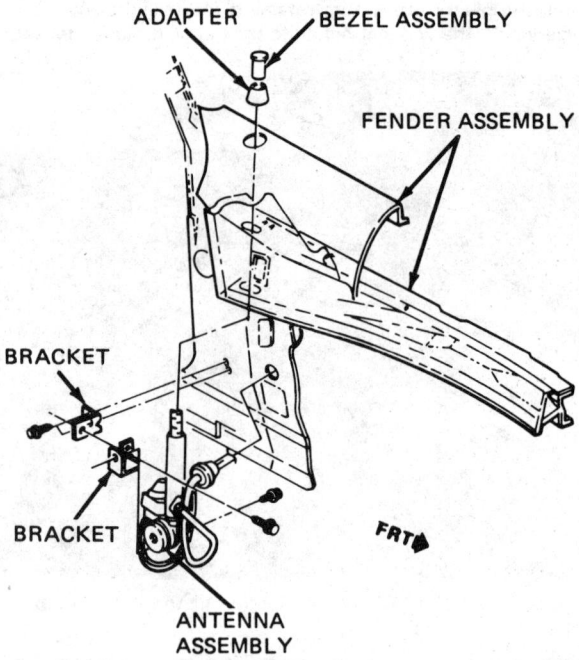

Fig. 10.2 Power antenna installation details (Sec 11)

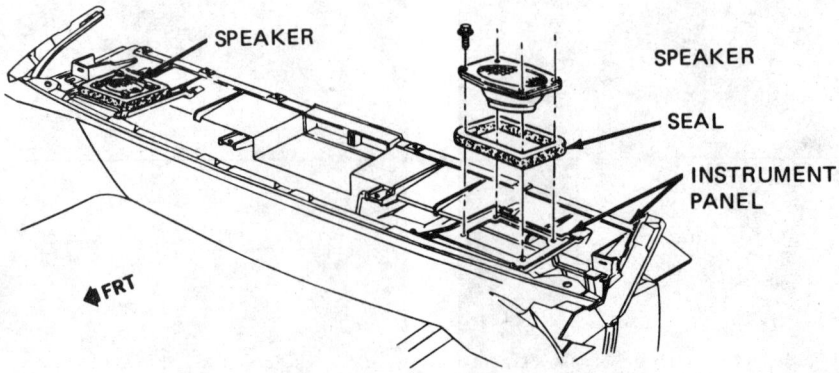

Fig. 10.3 Radio speaker installation details (Sec 12)

6 Disconnect the electrical leads, then remove the bolts retaining the antenna assembly to the bracket and remove the assembly.
7 Installation is the reverse of the removal procedure. Make sure that the mast is in the fully retracted position before installation and that the lower bracket is secured by the fender skirt retaining screw.

12 Radio speakers — removal and installation

1 Disconnect the cable from the negative battery terminal.
2 Remove the dash pad by removing the four retaining screws underneath the lip of the pad.
3 Remove the four bolts retaining the speaker to the speaker braces.
4 Lift the speaker out and disconnect the wire connector.
5 Installation is the reverse of the removal procedure.

13 Cluster panel instruments (except speedometer) — removal and installation

1 Disconnect the cable from the negative battery terminal.
2 Remove the screws retaining the instrument cluster bezel and remove the bezel.
3 Remove the trip odometer reset knob by unscrewing it.
4 Remove the screws retaining the instrument cluster plastic lens.
5 The individual instruments may now be lifted out of the panel after removing the retaining screws.
6 Installation is the reverse of the removal procedure.

14 Speedometer — removal and installation

1 Disconnect the cable from the negative battery terminal.
2 Remove the dash pad by removing the four retaining screws in the defroster ducts and six retaining screws underneath the lip of the dash pad.
3 Remove the left-hand radio speaker, then disconnect the electrical connector.
4 Using a Torx screwdriver of the appropriate size, remove the screws retaining the instrument cluster outer bezel and remove the bezel.
5 Remove the cluster panel retaining screws.
6 Pull the cluster panel out just far enough to reach behind it. With one hand, reach behind the panel and locate the speedometer cable where it joins the speedometer case. Hold the cable and, with the other hand, pinch the cable release and separate the cable from the speedometer.
7 Remove the screws retaining the instrument cluster panel plastic lens.
8 From the rear of the cluster panel, remove the speedometer retaining screws and copper retaining clip, then pull the speedometer out of the panel, disconnect the VSS electrical connector with a screwdriver and remove the speedometer.
9 Installation is the reverse of the removal procedure. When reattaching the speedometer cable connector to the speedometer case, the cluster panel should be positioned so that the speedometer cable connector is visible. Hold the connector in position with one hand and 'feed' the cluster panel into the connector with the other hand until the connector snaps into place.

15 Speedometer cable — replacement

1 Disconnect the cable from the negative battery terminal.
2 On non-Cruise Control-equipped vehicles, disconnect the speedometer cable strap at the power brake booster.
3 On Cruise Control-equipped vehicles, disconnect the speedometer cable at the Cruise Control transducer.
4 Remove the speedometer as described in Section 14 (paragraphs 2 through 6).
5 Slide the old cable out from the upper end of the casing, or, if broken, from both ends of the casing.
6 If speedometer operation has been noisy, but the speedometer cable appears to be in good condition, take a short piece of speedometer cable with a tip to fit the speedometer and insert it in the speedometer socket. Spin the piece of cable between your fingers. If binding is noted, the speedometer is faulty and should be repaired.
7 Inspect the speedometer cable casing for sharp bends and breaks, especially at the transmission end. If breaks are noted, replace the casing with a new one.
8 When installing the cable, perform the following operations to ensure quiet operation.
9 Wipe the cable clean with a lint-free cloth.
10 Flush the bore of the casing with solvent and blow it dry with compressed air.
11 Place some speedometer cable lubricant in the palm of one hand.
12 Feed the cable through the lubricant and into the casing until lubricant has been applied to the lower two-thirds of the cable. Do not overlubricate it.
13 Seat the upper cable tip in the speedometer and snap the retainer onto the casing.
14 The remaining installation steps are the reverse of those for removal.

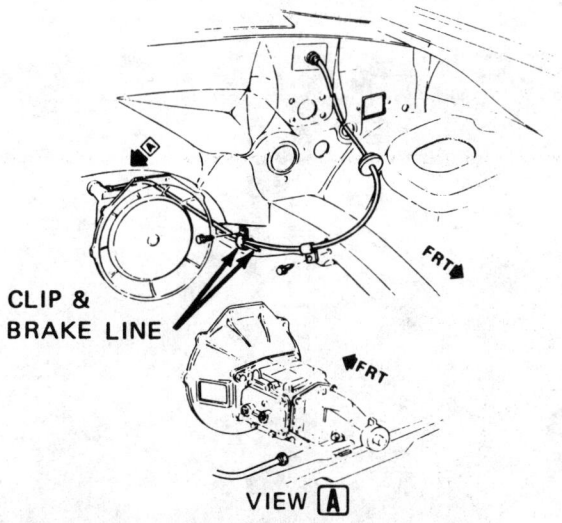

Fig. 10.4 Speedometer cable routing on vehicles without Cruise Control (Sec 15)

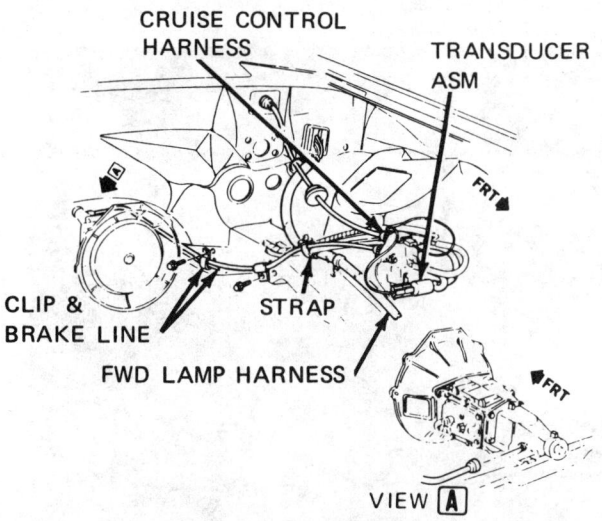

Fig. 10.5 Speedometer cable routing on vehicles with Cruise Control (Sec 15)

16.8A Removing the top retaining nuts from the instrument cluster panel

16.8B Removing the bottom retaining nuts from the instrument cluster panel

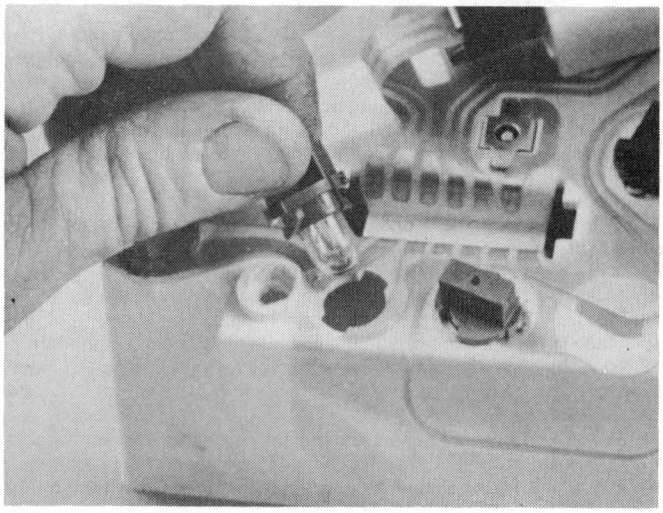

16.12 Removing a bulb socket from the cluster panel

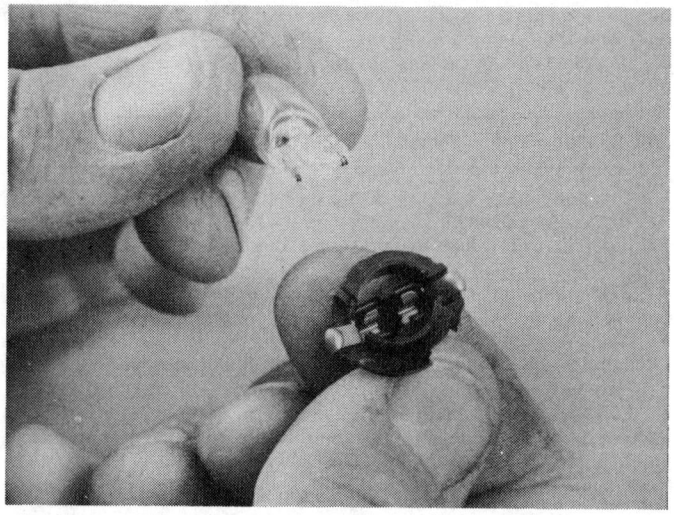

16.13 Removing a bulb from a cluster panel bulb socket

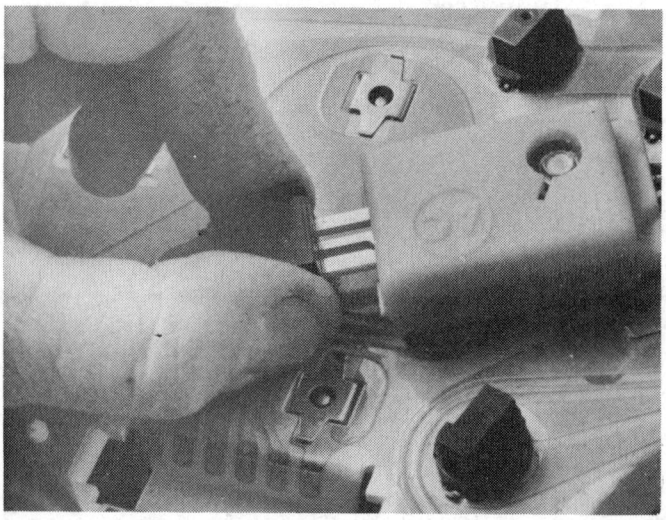

16.14A Removing the VSS printed circuit connector from the male VSS cable lead

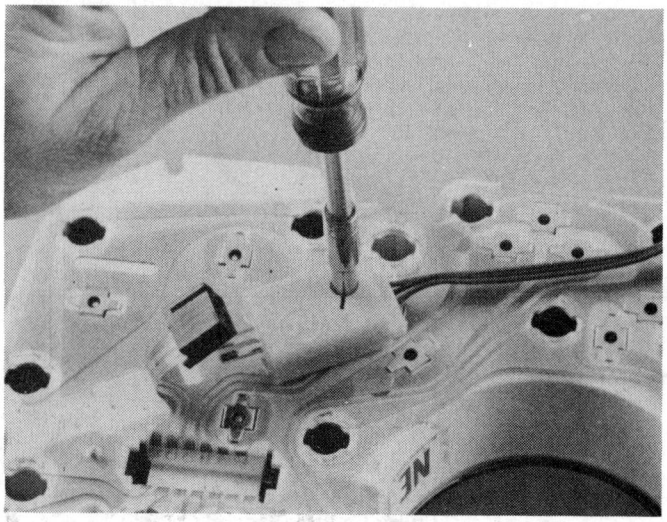

16.14B Removing the male VSS cable lead from the cluster panel

16 Instrument cluster panel — removal, servicing and installation

1 Remove the cable from the negative battery terminal.
2 Although it is not absolutely necessary, access to the instrument cluster panel will be easier if the steering wheel is removed at this point (refer to Chapter 11).
3 Remove the dash pad by removing the four screws in the defroster ducts and the six screws under the lip of the dash pad.
4 Remove the left-hand radio speaker and braces by removing the two retaining screws at the top front braces and the two screws from the rear braces, then disconnect the wire connector.
5 **Note:** *Steps 5 and 6 are not absolutely necessary, but lowering the steering column makes access to the cluster panel much easier.* Remove the screws retaining the plastic trim cover below the steering column and remove the cover.
6 Support the steering column from below with a brace, so as not to damage the undercowl panel, remove the screws retaining the outer instrument cluster bezel and remove the bezel.
7 Using a Torx screwdriver of the appropriate size, remove the screws retaining the outer instrument cluster bezel and remove the bezel.
8 Remove the two nuts retaining the top of the instrument cluster panel, then remove the two nuts retaining the bottom of the cluster panel, located in the warning light housings on either side of the panel (photos).
9 Pull the instrument cluster panel out just far enough to reach behind it. With one hand, reach behind the panel and locate the speedometer cable where it joins the speedometer case. With the other hand, pinch the cable release and separate the cable from the speedometer.
10 Remove the cluster panel by pulling it straight out to prevent damage to the electrical contacts behind it.
11 Place the cluster panel assembly in a clean work area.
12 From the rear of the cluster panel, remove the bulb assemblies by turning the plastic retainers counterclockwise (photo). **Note:** *To keep the bulb assemblies in order, they should be placed in the work area in a pattern simulating the back of the cluster panel. As an alternative, tag each bulb assembly and the recess from which it was removed with numbered pieces of tape.*
13 Replace burned-out bulbs with new ones by pulling them out of the plastic housings and inserting the new ones (photo). Keep the bulb assemblies in order.
14 Disconnect the vehicle speed sensor (VSS) by pinching the printed circuit connector and pulling it away from the male side of the connector, then remove the bolt retaining the male connector (photos).
15 Remove the bolts retaining the speedometer case and the speedometer end of the VSS cable, then remove the copper retaining clip (photos).
16 From the front of the cluster assembly, remove the trip odometer reset knob by unscrewing it.
17 Remove the screws retaining the plastic face plate and remove the face plate from the cluster panel assembly (photo).
18 Remove the screws retaining the metal bezel and separate the bezel

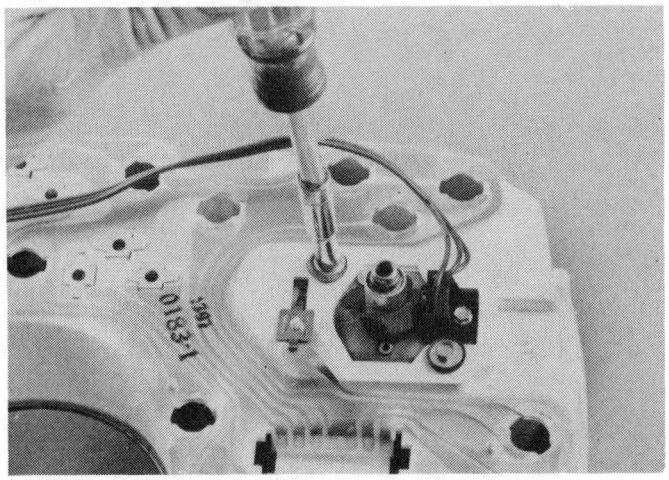

16.15A Removing the speedometer housing bolts (VSS connector is visible at right of speedometer drive housing)

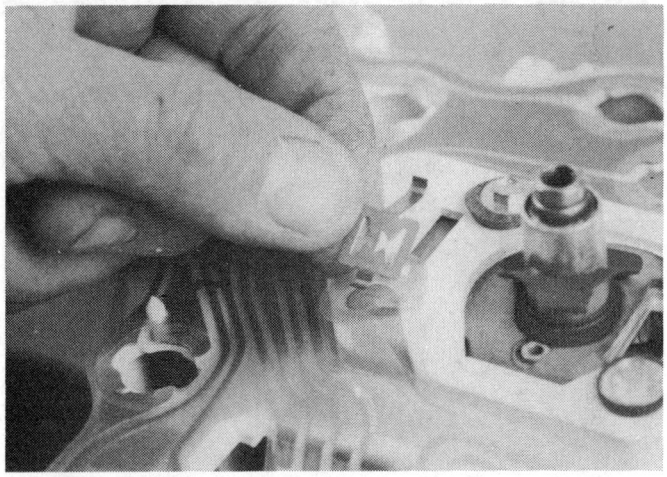

16.15B Removing the copper speedometer housing connector

16.17 Removing the plastic face plate from the cluster panel

16.19 Removing a gauge retaining bolt from the cluster panel

from the panel.
19 Remove the screws retaining the individual gauges and separate the gauges from the panel (photo).
20 Remove the instrument contact clips by pinching them at the base with needle-nose pliers and pulling them out from the rear (photo). Replace any bent or damaged clips with new ones.
21 Remove the printed circuit (photo) and replace it with a new one, if necessary.
22 Installation is the reverse of the removal procedure. When installing the speedometer, after the instrument cluster has been placed in position, pull it out just far enough so the speedometer cable connector is visible, hold the connector in position with one hand and 'feed' the instrument cluster into the connector with the other hand until the connector snaps into place.

17 Headlight switch — removal and installation

1 Disconnect the negative battery cable and remove the instrument cluster panel (refer to Section 16).
2 Pull the headlight switch knob to the full On position.
3 Reach behind the switch with one hand and push the release button, then pull the switch knob out of the dash with the other hand (photos).
4 Remove the screws retaining the knob panel and remove the panel (photo).
5 Working behind the switch, hold it in position with one hand and, using the index finger of the other hand, unscrew the retainer from the front and remove it (photos).
6 Remove the switch from the dash (photo). Note the angle at which it was removed to simplify installation.
7 Disconnect the wire connector (photo).
8 Installation is the reverse of the removal procedure.

18 Console switches — removal, servicing and installation

1 Remove the cable from the negative battery terminal.
2 If the vehicle is equipped with a manual transmission, unscrew the shifter ball from the shifter handle.
3 If the vehicle is equipped with an automatic transmission, pull up on the shift detent button assembly and remove it from the shifter (photo).
4 Using a screwdriver and an awl, spread the ends of the snap-ring at the bottom of the shifter detent button hole with the screwdriver and remove the snap-ring by passing the tip of the awl through one of the snap-ring eyes and pulling the snap-ring out of the hole (this method is required because the hole is too small to accept normal snap-

16.20 Removing a gauge contact clip from the rear of the cluster panel

16.21 Removing the printed circuit from the cluster panel

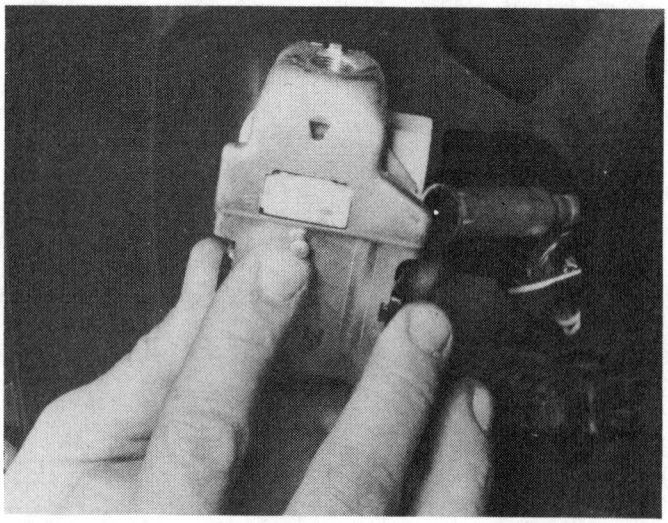

17.3A Location of the switch knob release button on the bottom of the headlight switch (shown with switch removed)

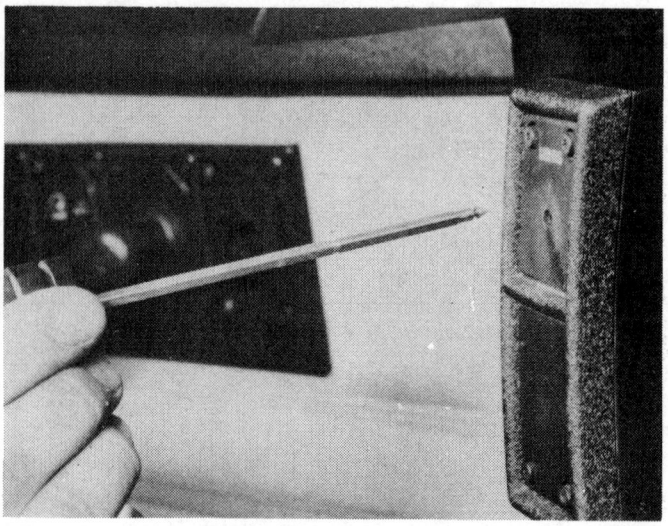

17.3B Removing the headlight switch knob from the dash

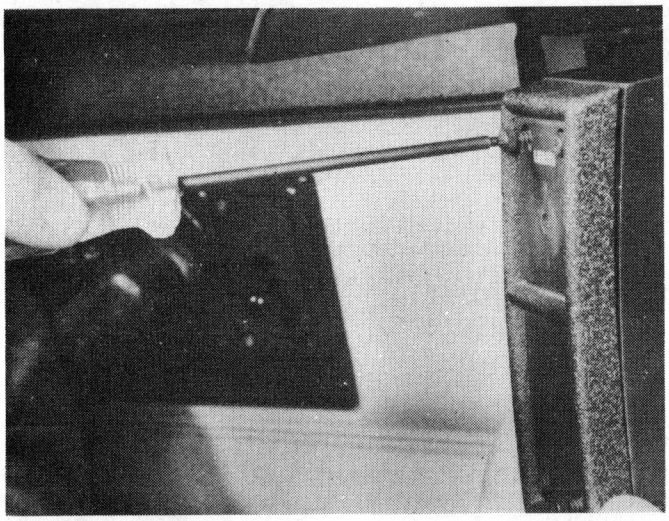

17.4 Removing the headlight switch knob panel

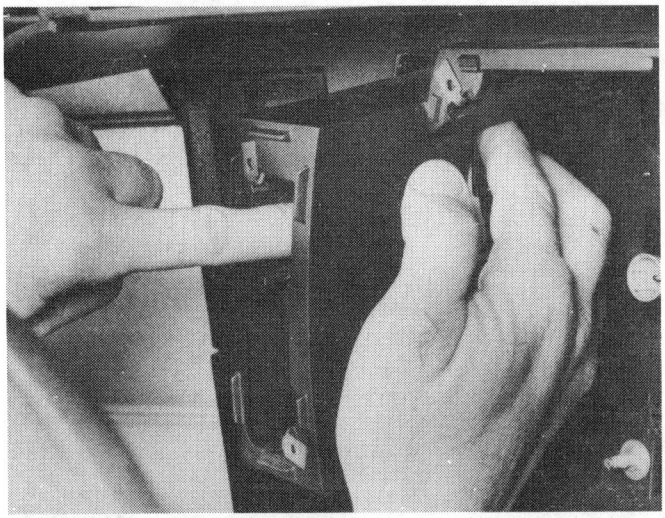

17.5A Unscrewing the headlight switch retainer

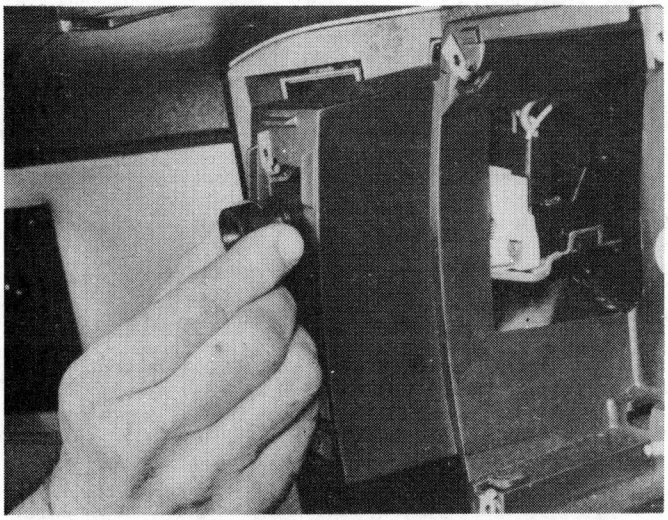

17.5B Removing the headlight switch retainer

17.6 Removing the headlight switch from the dash

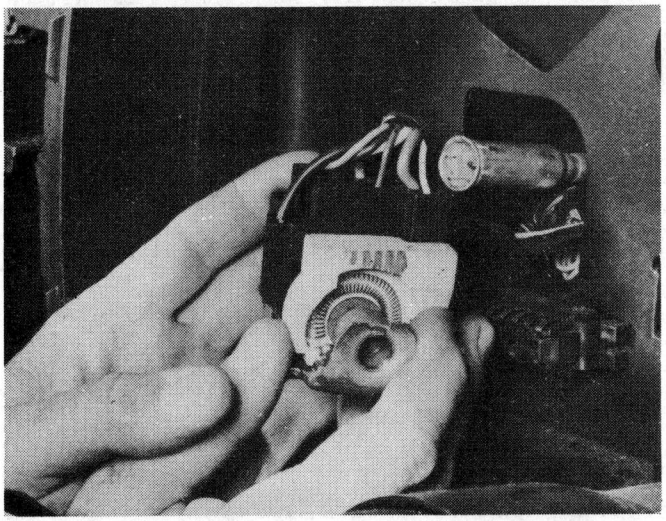

17.7 Disconnecting the headlight switch electrical connector

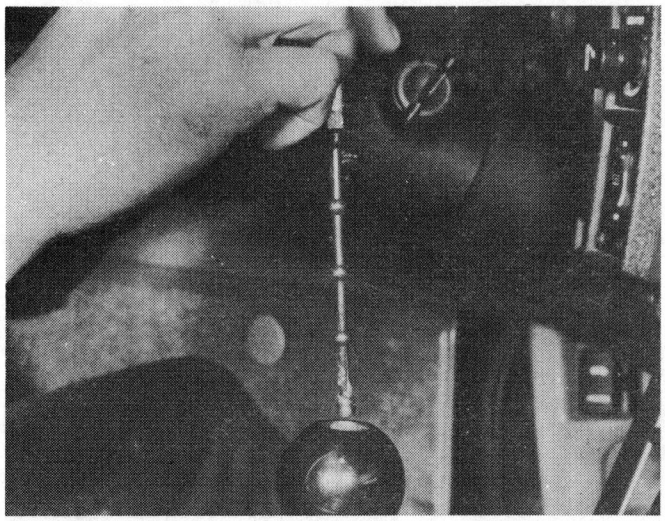

18.3 Removing the shift detent button assembly from the shift handle

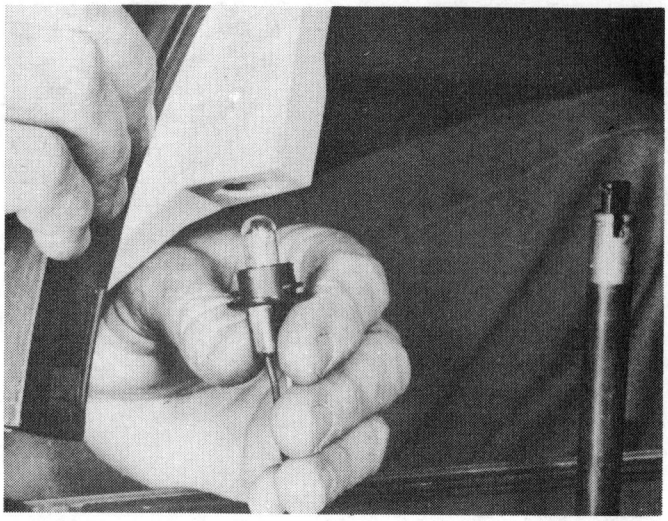

18.6 Removing the console light bulb socket from the console cover

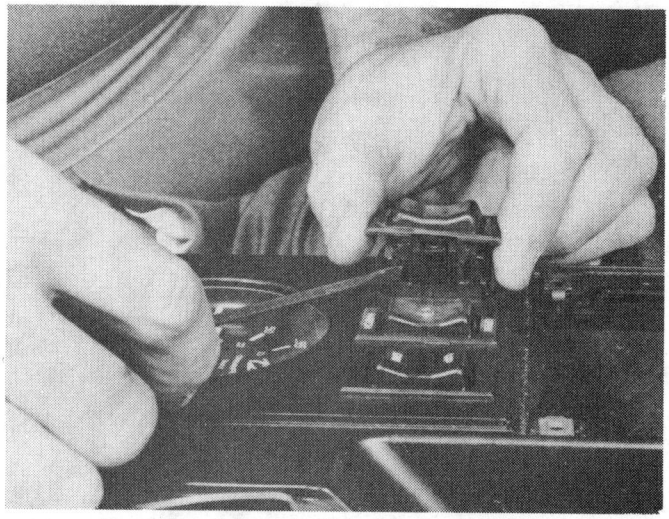

18.9 Prying the top half of the up/down switch from the bollom half

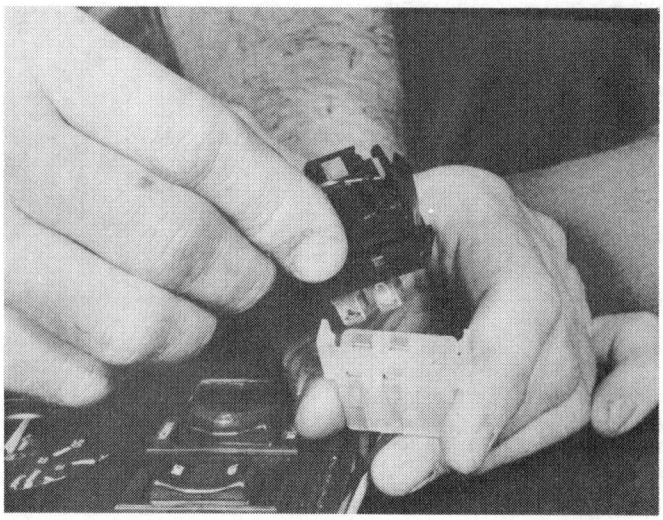

18.10 Removing the electrical connector from the bottom half of the up/down switch

ring pliers).
5 Pull the knob off the shifter handle.
6 Lift the console cover, twist the console light bulb socket out of the cover and remove the cover from the console (photo).
7 To replace the left-or-right up-or-down rocker switch, it must be removed in two pieces.
8 Raise the rear of the switch with a screwdriver while pushing on the front release lever to elevate the entire switch.
9 With the switch elevated, use a screwdriver to pry the top half of the switch from the bottom half (photo).
10 Remove the bottom half of the switch from the underside of the console cover and separate it from the electrical connector (photo).
11 The center open/close switch may be removed in one piece through the top of the console cover while pressing the release lever from the bottom.
12 Installation is the reverse of the removal procedure.

19 Windshield wiper arm — removal, installation and adjustment

1 Raise the hood.
2 Lift the wiper blade off the glass and disconnect the wiper arm locking retainer from the wiper transmission shaft.
3 Using tool J-8966, or its equivalent, lift the arm off the transmission shaft.
4 When installing the arm, the windshield wiper blade assembly release lever must be facing the base of the arm assembly for correct matching of the blade and glass contours.
5 Align the right wiper arm assembly in the parked position and attach the wiper arm to the transmission shaft (the same tool used for removal may be used for installation).
6 If the wiper arms and blades were in the correct positions prior to wiper arm removal, no adjustment is required. However, if adjustment is required, proceed as follows.
7 To adjust the left wiper arm, loosen and raise the left side of the cowl vent screen.
8 Loosen but do not remove the transmission rear wiper link-to-motor crank arm attaching nuts.
9 Rotate the right arm assembly to a position 3/4-inch below the ramp stops.
10 Tighten the attaching nuts on the transmission rear wiper link(s)-to-motor crank arm to 4 to 6 ft-lbs.
11 Place the left arm and blade assembly next to the spindle and align the slot in the arm with the keyway in the spindle. Push the arm down and place the blade on the ramp stop on the vent screen.
12 Lift the right arm and blade assemblies over the stop.
13 Refer to the accompanying illustration and check the wiper pattern and parked position. Dimension A should be 1-19/32 inch, plus or minus 19/32-inch. The overlap of the blade tip (Dimension B) should be 1-inch.
14 To adjust the right arm, remove it from the transmission driveshaft. The correct Park position and outwipe dimensions are determined with the wipers operating at low speed on a wet windshield.
15 Reinstall the cowl vent assembly.

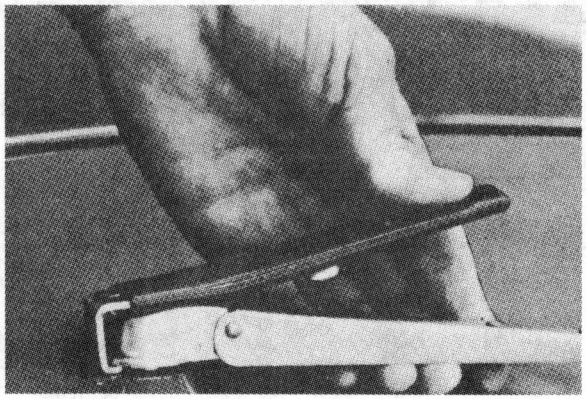

Fig. 10.6 Using a windshield wiper arm removal tool to remove the arm (Sec 19)

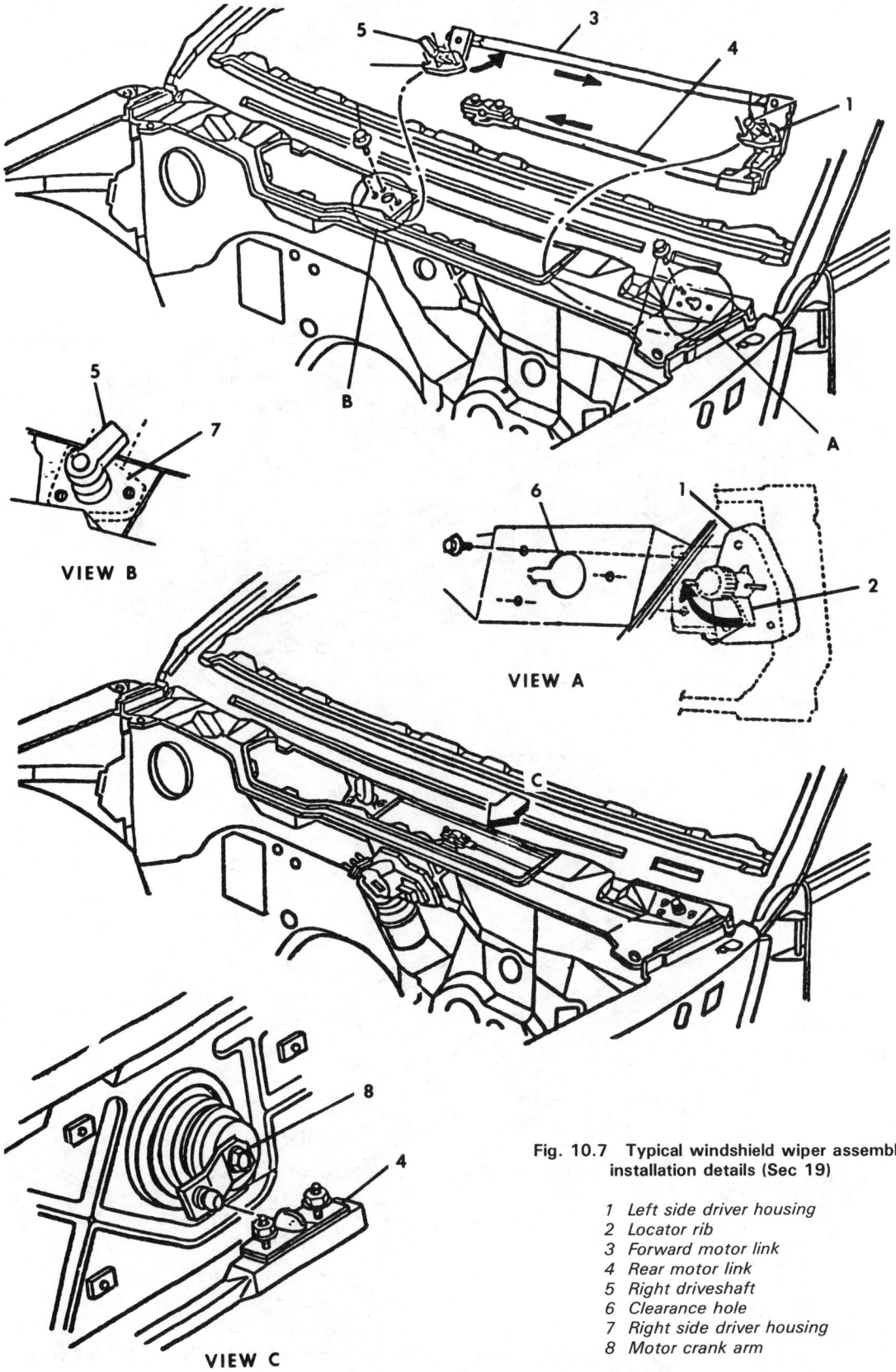

Fig. 10.7 Typical windshield wiper assembly installation details (Sec 19)

1 Left side driver housing
2 Locator rib
3 Forward motor link
4 Rear motor link
5 Right driveshaft
6 Clearance hole
7 Right side driver housing
8 Motor crank arm

20 Windshield wiper motor — removal and installation

1 Lift the hood.
2 Remove the cowl vent screen.
3 Loosen the wiper transmission rear wiper link-to-motor crank arm retaining nuts.
4 Remove the transmission rear wiper link from the motor crank arm.
5 Disconnect the wire connectors.
6 Disconnect the washer hoses.
7 Remove the three motor mounting bolts.
8 While guiding the crank arm through the hole in the cowl, remove the wiper motor.
9 Installation is the reverse of the removal procedure. The motor must be in the parked position before attaching the crank arm to the transmission rear wiper link.

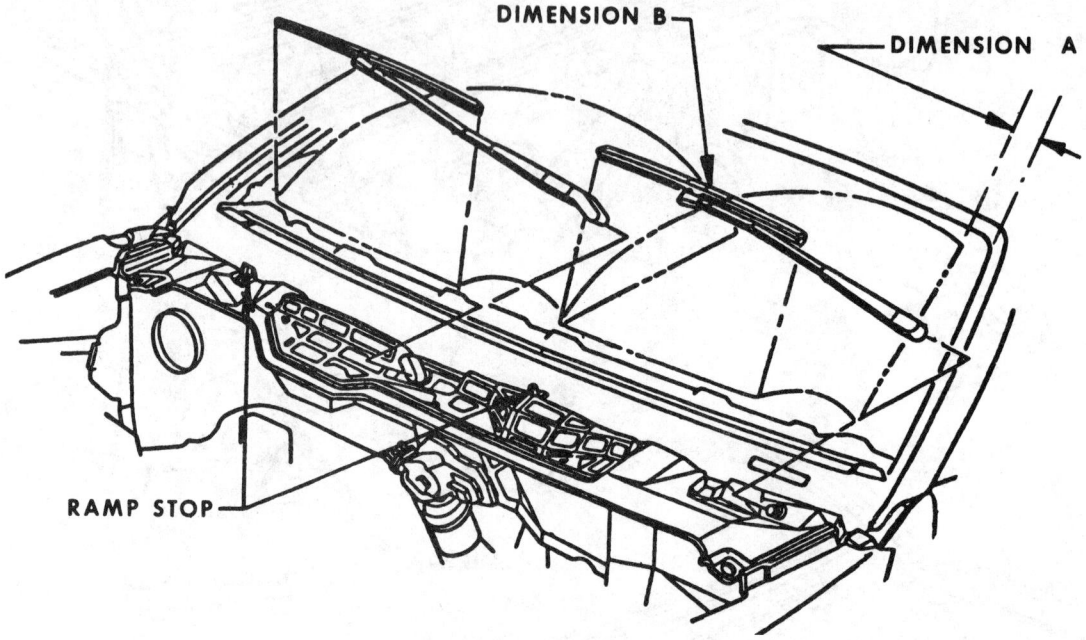

Fig. 10.8 Wipe pattern dimensions (Sec 19)

A 1-19/32 inch (plus or minus 19/32 inch) *B 1-inch*

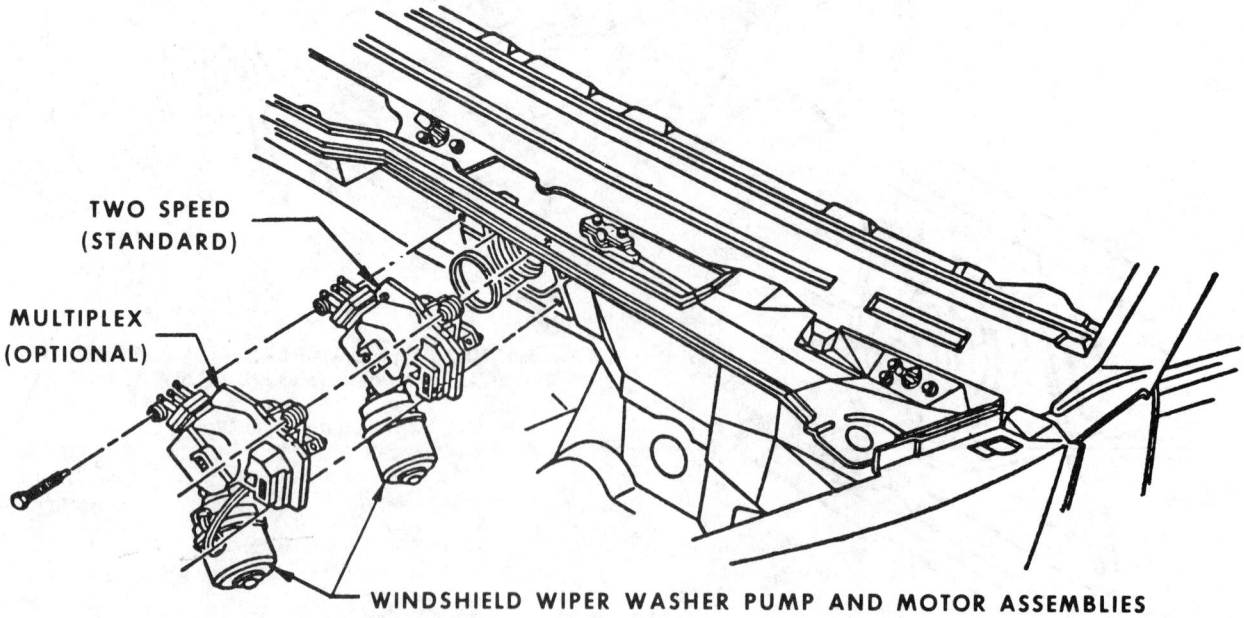

Fig. 10.9 Identification of two-speed standard and optional multiplex windshield wiper motors (Sec 20)

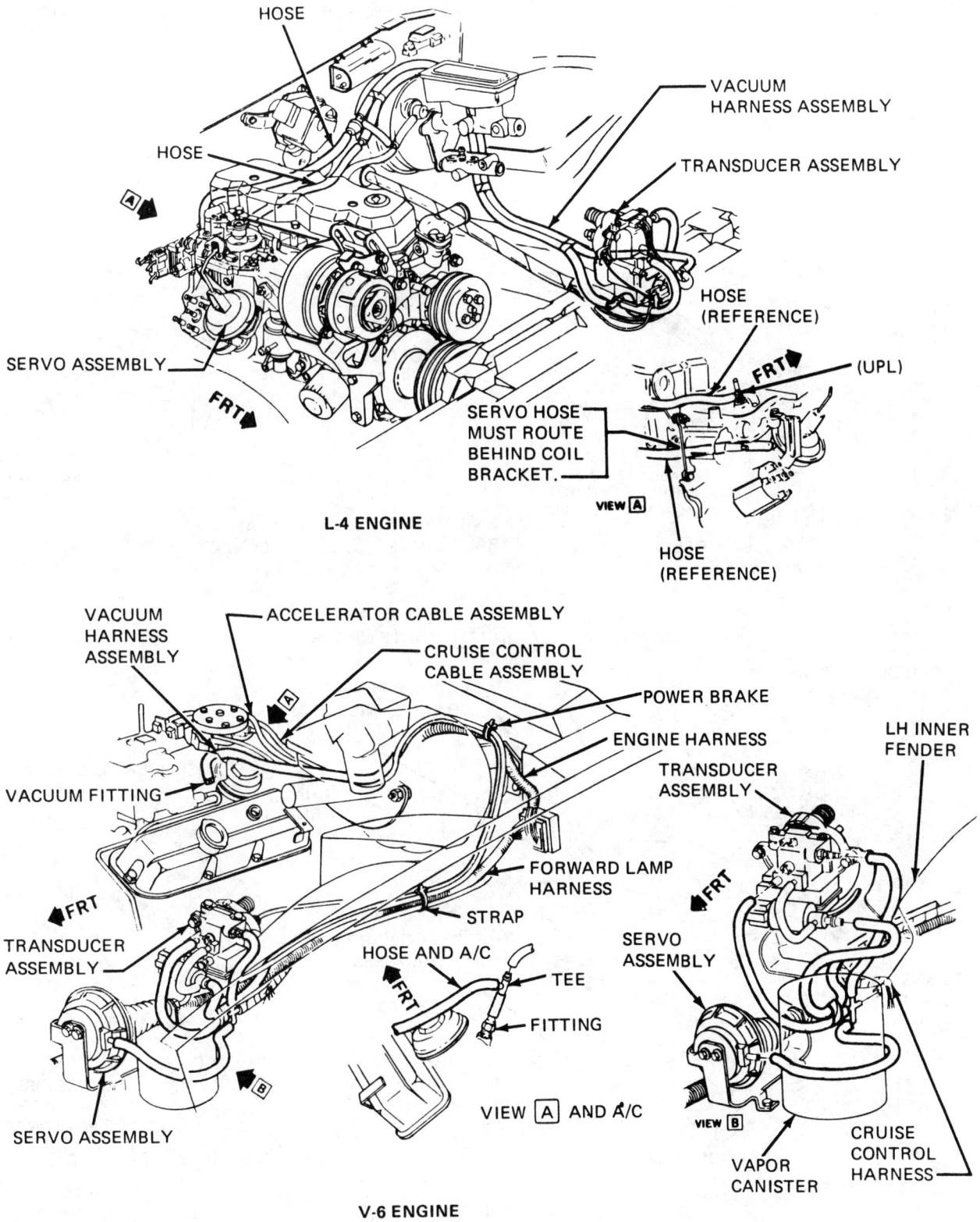

Fig. 10.10 Cruise Control vacuum hose routing on L4 and V6-equipped vehicles (Sec 22)

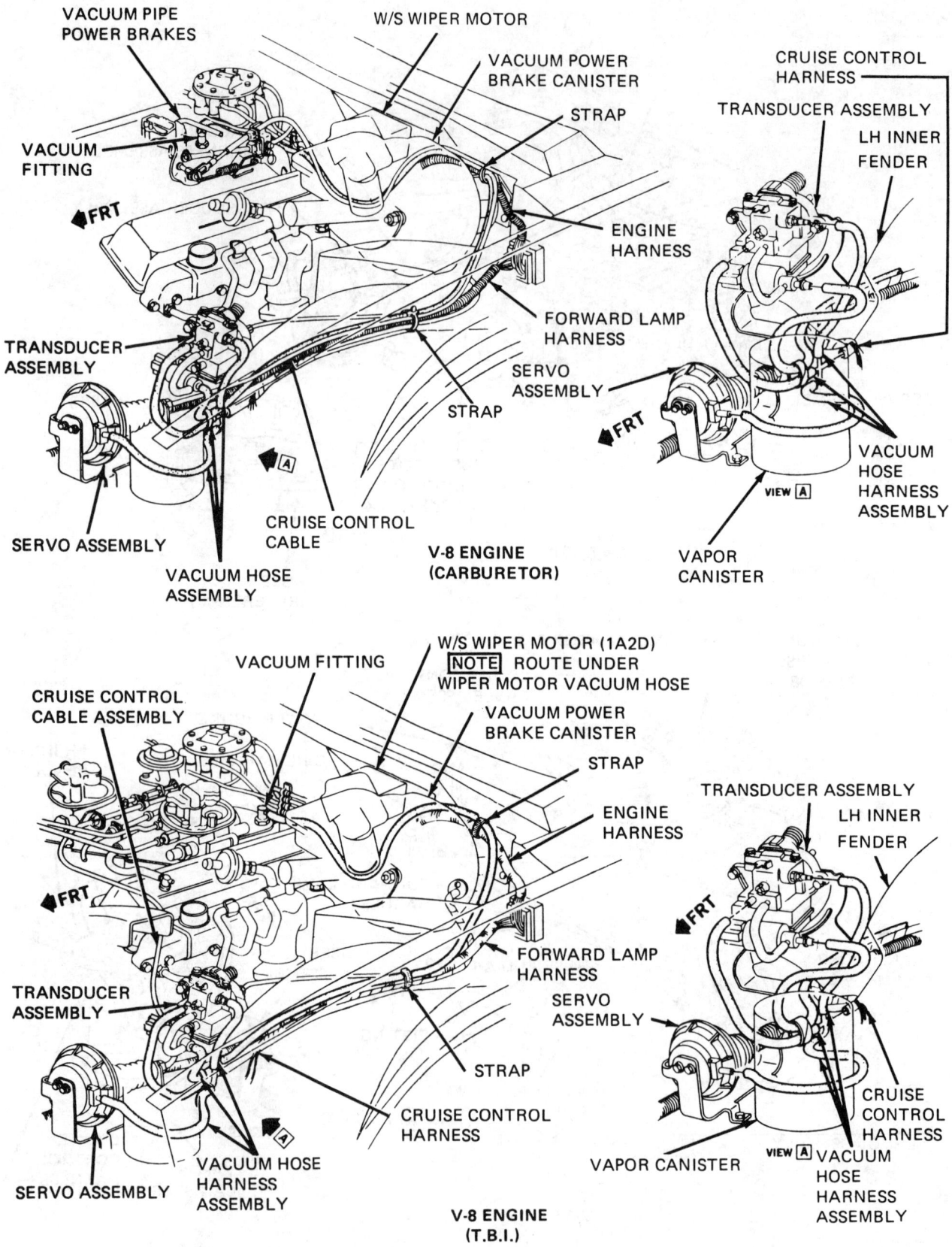

Fig. 10.11 Cruise Control vacuum hose routing on V8-equipped vehicles (Sec 22)

21 Windshield wiper transmission — removal and installation

1 Lift the hood.
2 Remove the right wiper arm and the cowl vent screen.
3 Loosen but do not remove the nuts retaining the transmission rear wiper link from the motor crank arm.
4 Pull the transmission rear wiper link from the motor crank arm.
5 Remove the transmission-to-cowl bolts.
6 Remove the transmission and linkage assembly by guiding it through the plenum chamber opening.
7 To install, extend the forward motor link and rear wiper link and rotate the left-side driver housing, aligning the locator rib with the cutout in the roof panel.
8 Collapse the assembly by turning the forward motor link clockwise, the rear wiper link counterclockwise and the right driveshaft in — in effect making the assembly as short as possible.
9 Feed the assembly through the access hole in the plenum chamber from the left side.
10 Turn the right driveshaft clockwise and insert the driver through the hole so that the driver rests on the top of the roof panel.
11 Install the bolts on both the right and left driver housing driving screws and tighten them to 4 to 6 ft-lbs.
12 Install the transmission rear wiper link on the motor crank arm and tighten the bolts to 4 to 6 ft-lbs.
13 Align the transmission and attach it to the cowl.
14 Install and adjust the wiper arm and blade assemblies (refer to Section 19).
15 Check the wiper operation and wiper pattern (refer to Section 19).
16 Install the cowl vent screen.

22 Cruise Control — general information and servicing

The Cruise Control with Resume system is an option which maintains a desired vehicle speed under normal driving conditions. The system also has the capability of resuming a preset speed upon driver demand after the system has been disengaged. This is accomplished by moving a slide on the Cruise Control lever handle to the Resume position. Steep grades, up or down, may cause variations in the selected speed, which is considered normal.

The main components of the Cruise Control system are a transducer assembly, a resume solenoid valve, a vacuum servo with linkage, an engagement switch button and an On/Off/Resume switch on the turn signal lever, and vacuum and electric release switches attached to the brake pedal (automatic) or clutch pedal (manual).

Because of the variations in installations, it is not possible to include all the service procedures in this manual. However, those elements of the system most often requiring service and/or adjustment are covered.

If the servo unit requires replacement and/or adjustment, refer to the appropriate diagram and instructions accompanying this Section for your vehicle.

The transducer is calibrated in such a manner during production that overhaul operations are impractical. A defective transducer must be replaced with a new one; however, one adjustment is possible. If there is a difference betweeen the engagement speed selected and the actual cruising speed, proceed as follows:

a) Check all hoses for kinks and cracks. If there is still a difference between the engagement and cruising speeds, proceed to b.
b) If the cruising speed is lower than the engagement speed, loosen the orifice tube locknut and turn the tube *out*.
c) If the cruising speed is higher than the engagement speed, loosen the orifice tube locknut and turn the tube *in*.
d) Each 90-degree (1/4-turn) rotation will alter the engagement/cruising speed one (1) mph.
e) Tighten the locknut after adjustment has been made and check the system operation at 55 mph.

To remove, install and adjust the brake and clutch release switches and valves, refer to the accompanying illustration.

To remove the Cruise Control switch, vacuum valve assembly/TCC switch (automatic transmission) or vacuum release valve (manual transmission), pull the component from the bracket under the dash, then disconnect the wiring and/or vacuum connector and discard the faulty component and the retainer.

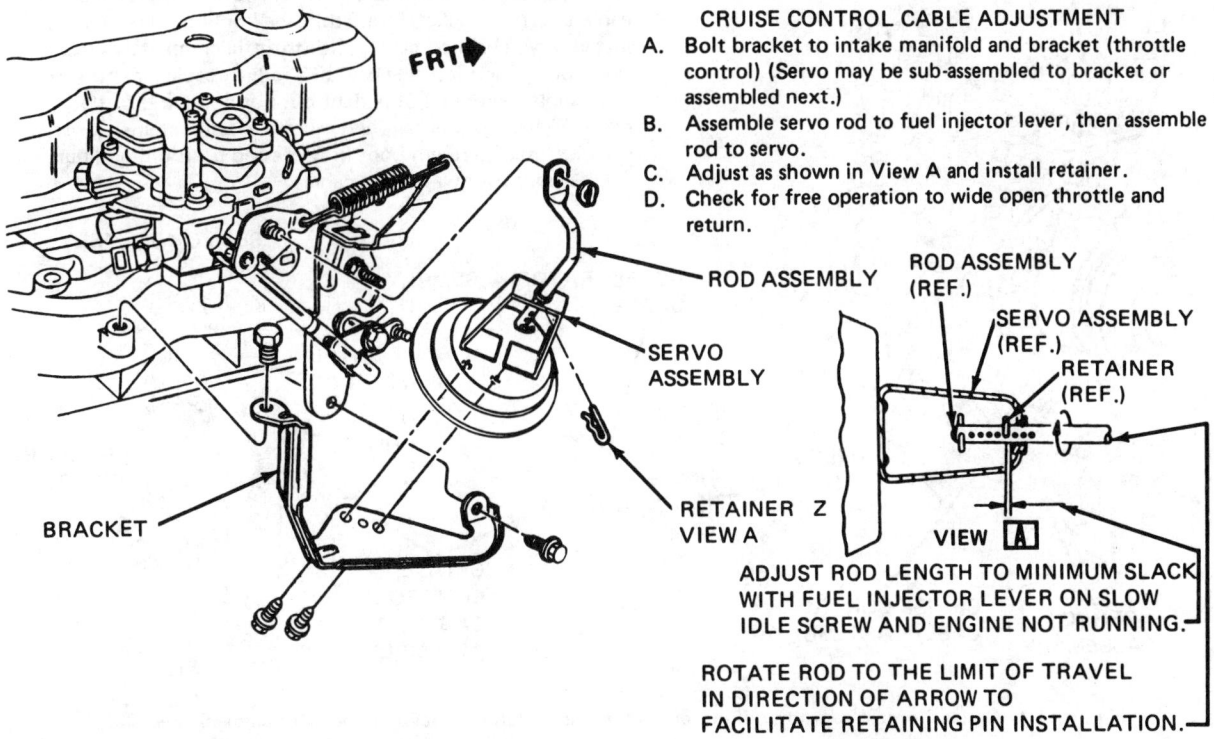

Fig. 10.12 Cruise Control servo mounting and adjustment details (L4-equipped vehicles) (Sec 22)

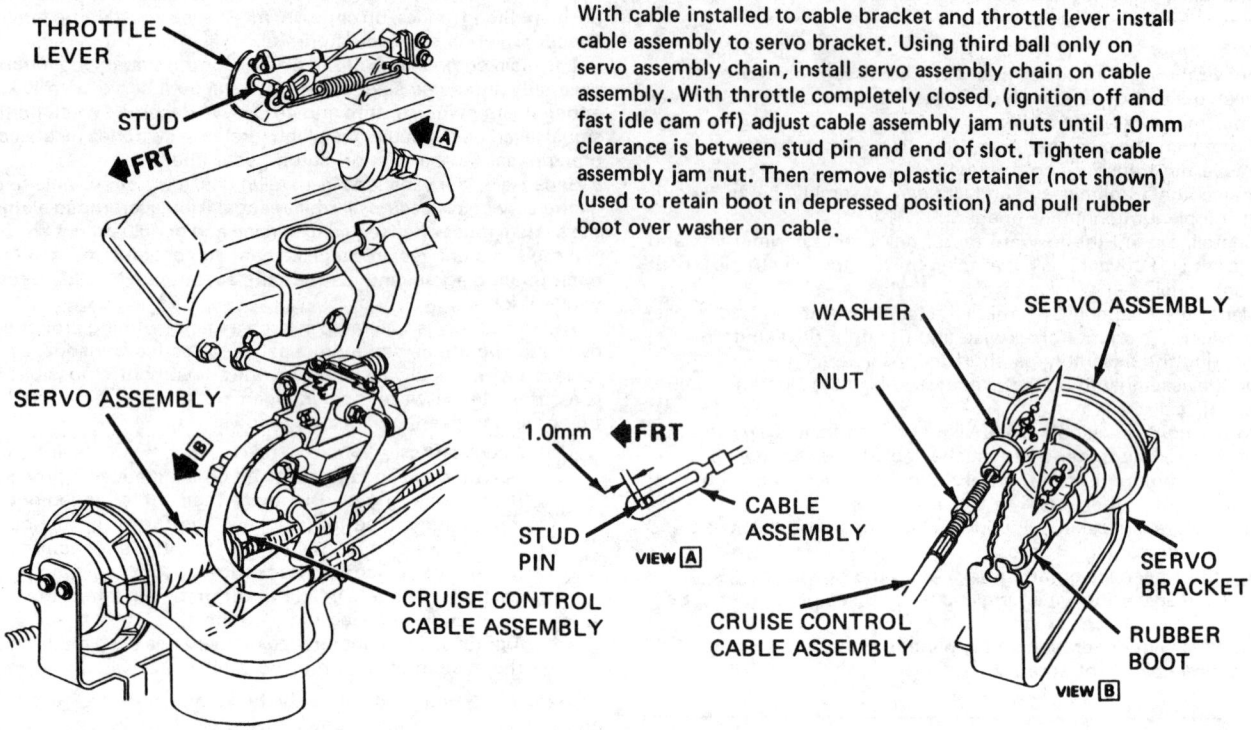

CRUISE CONTROL CABLE ADJUSTMENT

With cable installed to cable bracket and throttle lever install cable assembly to servo bracket. Using third ball only on servo assembly chain, install servo assembly chain on cable assembly. With throttle completely closed, (ignition off and fast idle cam off) adjust cable assembly jam nuts until 1.0mm clearance is between stud pin and end of slot. Tighten cable assembly jam nut. Then remove plastic retainer (not shown) (used to retain boot in depressed position) and pull rubber boot over washer on cable.

Fig. 10.13 Cruise Control servo mounting and adjustment details (carbureted V6 and V8-equipped vehicles) (Sec 22)

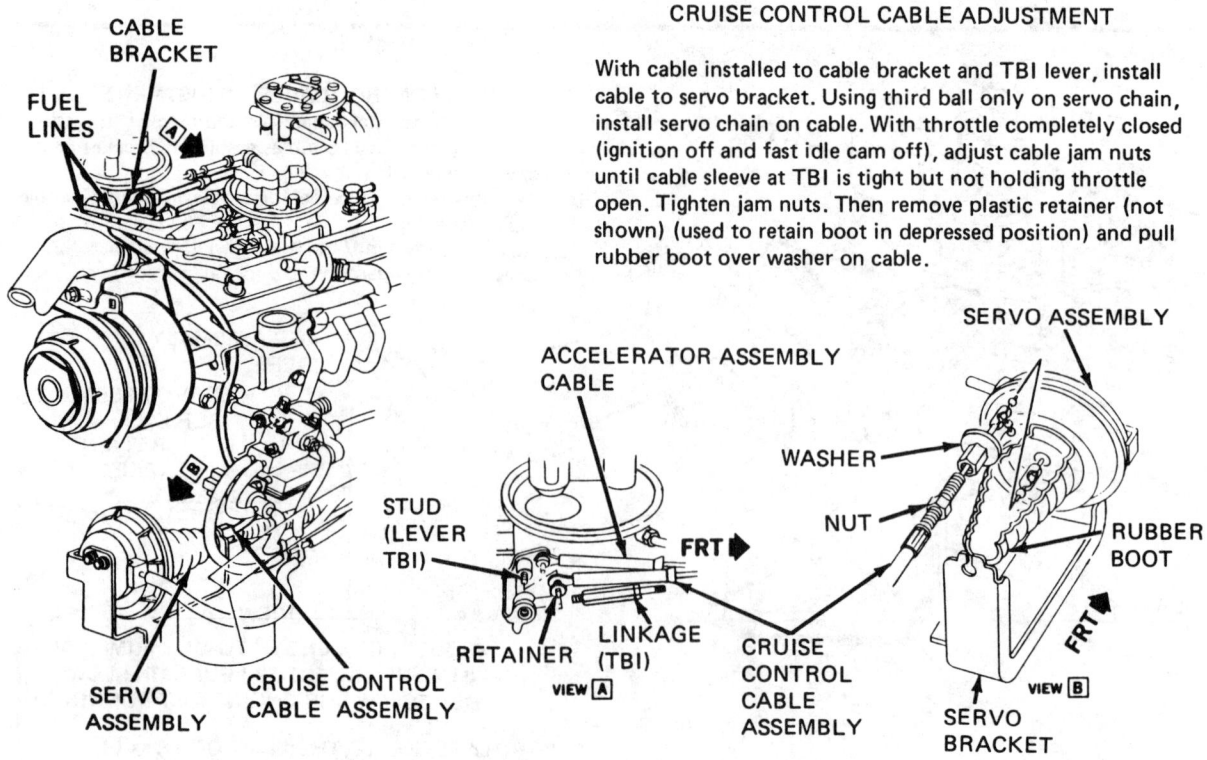

CRUISE CONTROL CABLE ADJUSTMENT

With cable installed to cable bracket and TBI lever, install cable to servo bracket. Using third ball only on servo chain, install servo chain on cable. With throttle completely closed (ignition off and fast idle cam off), adjust cable jam nuts until cable sleeve at TBI is tight but not holding throttle open. Tighten jam nuts. Then remove plastic retainer (not shown) (used to retain boot in depressed position) and pull rubber boot over washer on cable.

Fig. 10.14 Cruise Control servo mounting and adjustment details (fuel-injected V8 vehicles) (Sec 22)

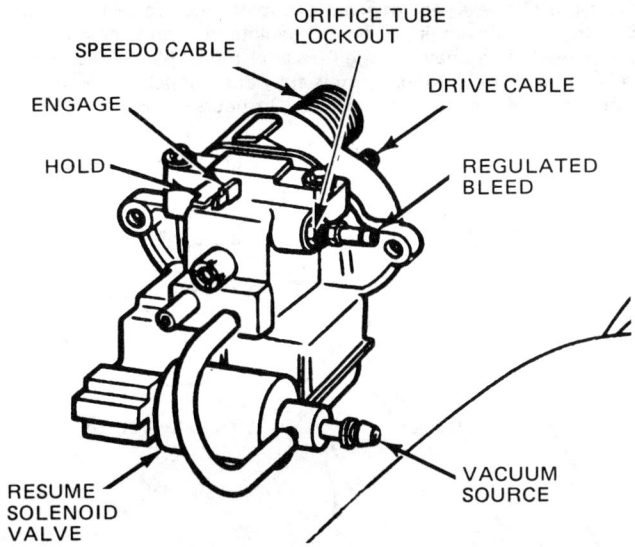

Fig. 10.15 Cruise Control transducer assembly details (Sec 22)

Install a new retainer in the bracket.

With the brake or clutch pedal depressed, install the new component in the retainer and make sure it is seated. Note that audible clicks can be heard as the component is pressed into the retainer.

Pull the brake or clutch pedal all the way up against the stop until the clicks can no longer be heard.

Release the clutch or brake pedal, then repeat the procedure to ensure that no more clicks can be heard. The component is now properly seated and adjusted.

Reconnect the wiring and/or vacuum connectors.

Other servicing of the Cruise Control system components should be done by your dealer.

23 Rear defogger (electric grid type) — check and repair

1 This option consists of a rear window with a number of horizontal elements that are baked into the glass surface during the glass forming operation.

2 Small breaks in the element can be successfully repaired without removing the rear window.

3 To test the grids for proper operation, start the engine and turn on the system.

4 Ground one lead of a test light and carefuly touch the other lead

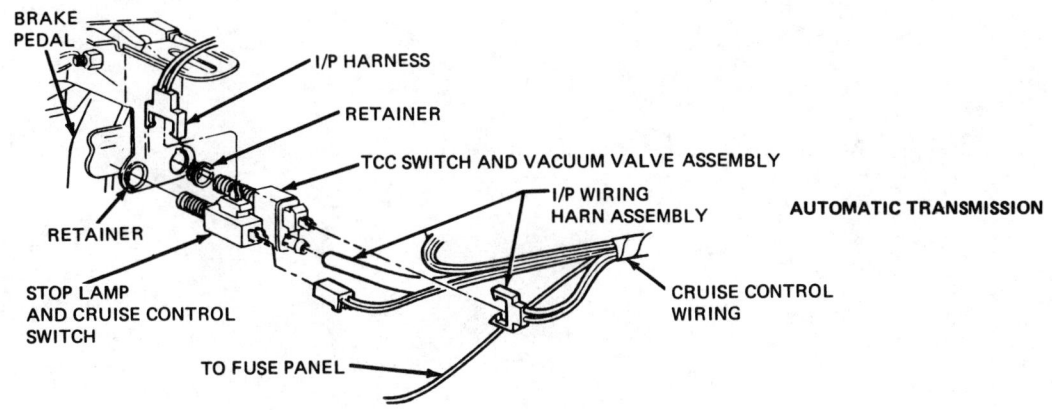

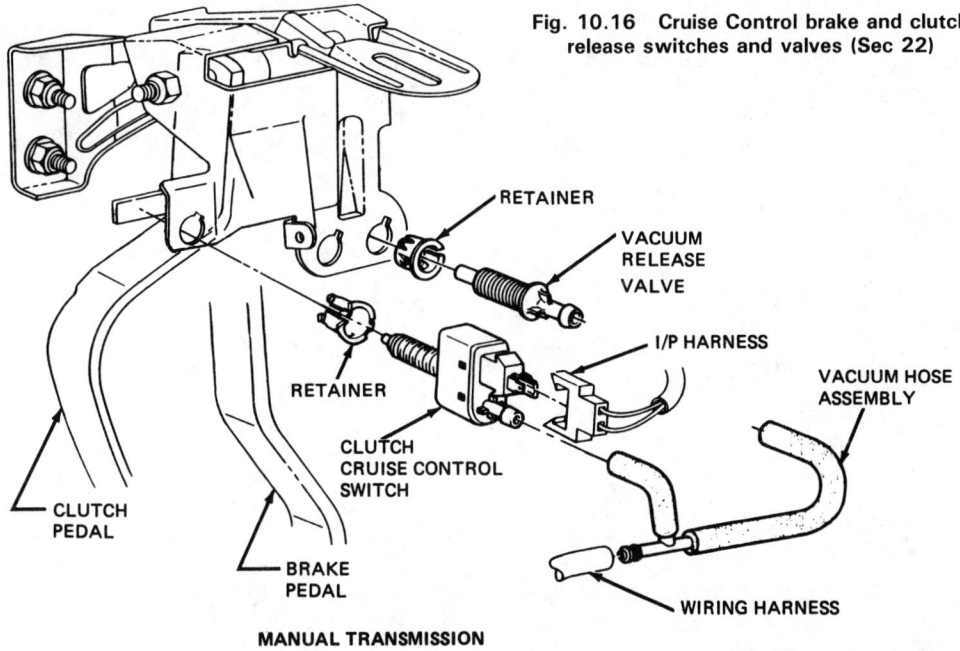

Fig. 10.16 Cruise Control brake and clutch release switches and valves (Sec 22)

to each element line.

5 The brilliance of the test light should increase as the lead is moved across the element from right to left. If the test light glows brightly at both ends of the lines, check for a loose ground wire. All of the lines should be checked in at least two places.

6 To repair a break in a line, it is recommended that a repair kit specifically for this purpose be purchased from a GM dealer. Included in the repair kit will be a decal, a container of silver plastic and hardener, a mixing stick and instructions.

7 To repair a break, first turn off the system and allow it to de-energize for a few minutes.

8 Lightly buff the element area with fine steel wool and then clean it thoroughly with alcohol.

9 Use the decal supplied in the repair kit, or apply strips of electrician's tape above and below the area to be repaired. The space between the pieces of tape should be the same width as the existing lines. This can be checked from outside the vehicle. Press the tape tightly against the glass to prevent seepage.

10 Mix the hardener and silver plastic thoroughly.

11 Using the wood spatula, apply the silver plastic mixture between the pieces of tape, overlapping the undamaged area slightly on either end.

12 Carefully remove the decal or tape and apply a constant stream of hot air directly to the repaired area. A heat gun set at 500 to 700 degrees Fahrenheit is recommended. Hold the gun about one (1) inch from the glass for one to two minutes.

13 If the new element appears off color, tincture of iodine can be used to clean the repair and bring it back to the proper color. This mixture should not remain on the repair for more than 30 seconds.

14 Although the defogger is now fully operational, the repaired area should not be disturbed for at least 24 hours.

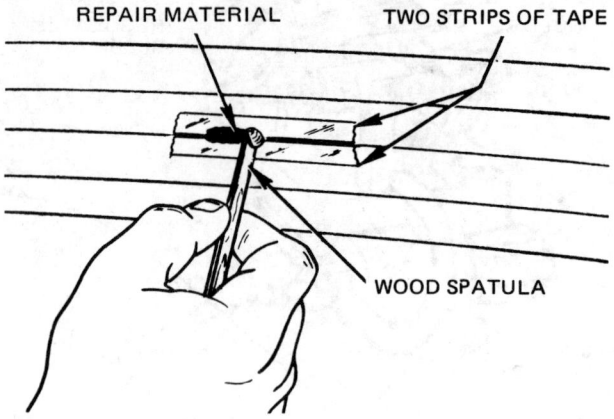

Fig. 10.17 Applying repair material to a broken rear window defogger grid (Sec 23)

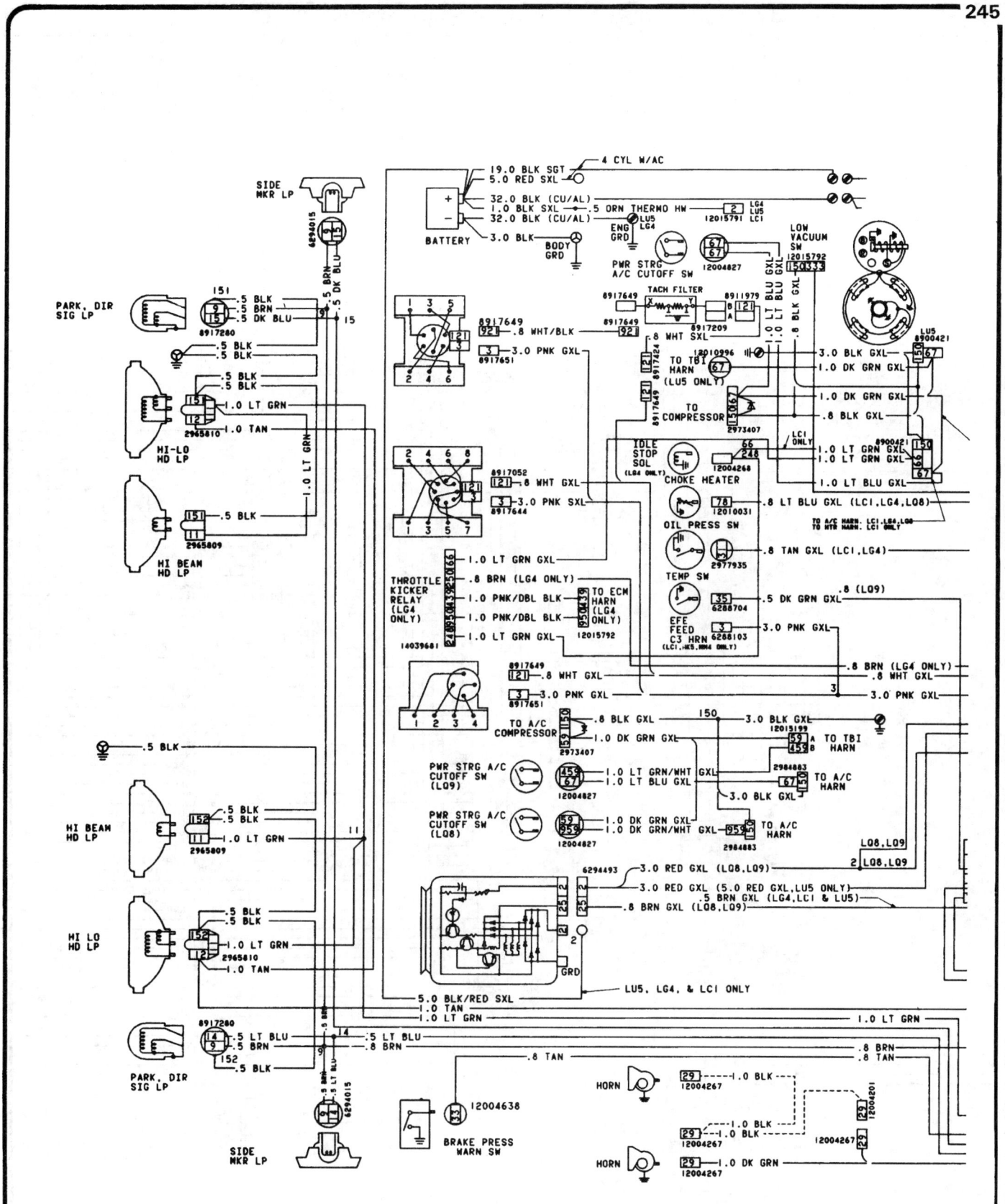

Fig. 10.18 Wiring diagram for 1982 and 1983 models — front end and engine compartment

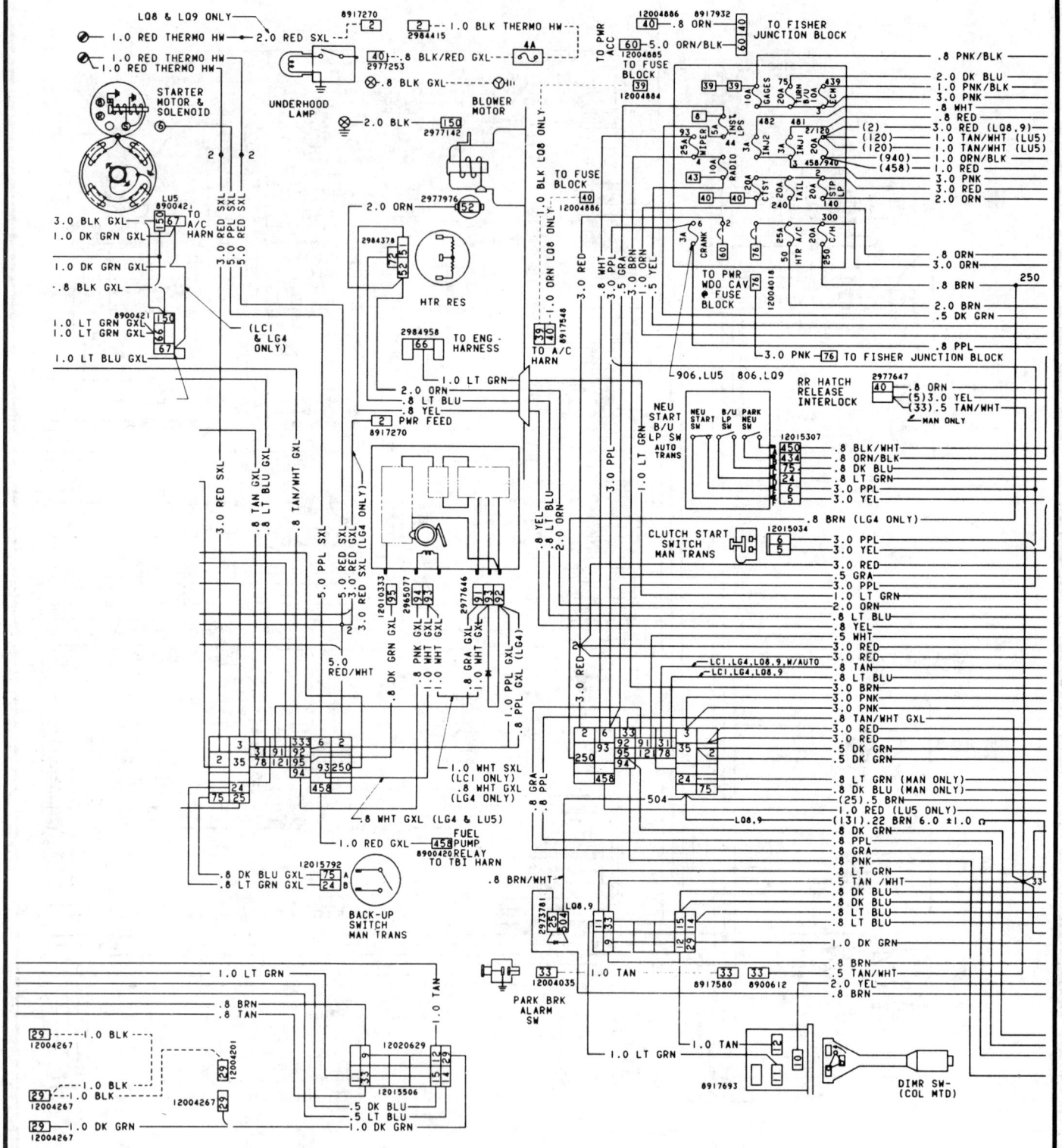

Fig. 10.19 Wiring diagram for 1982 and 1983 models — engine compartment and instrument panel

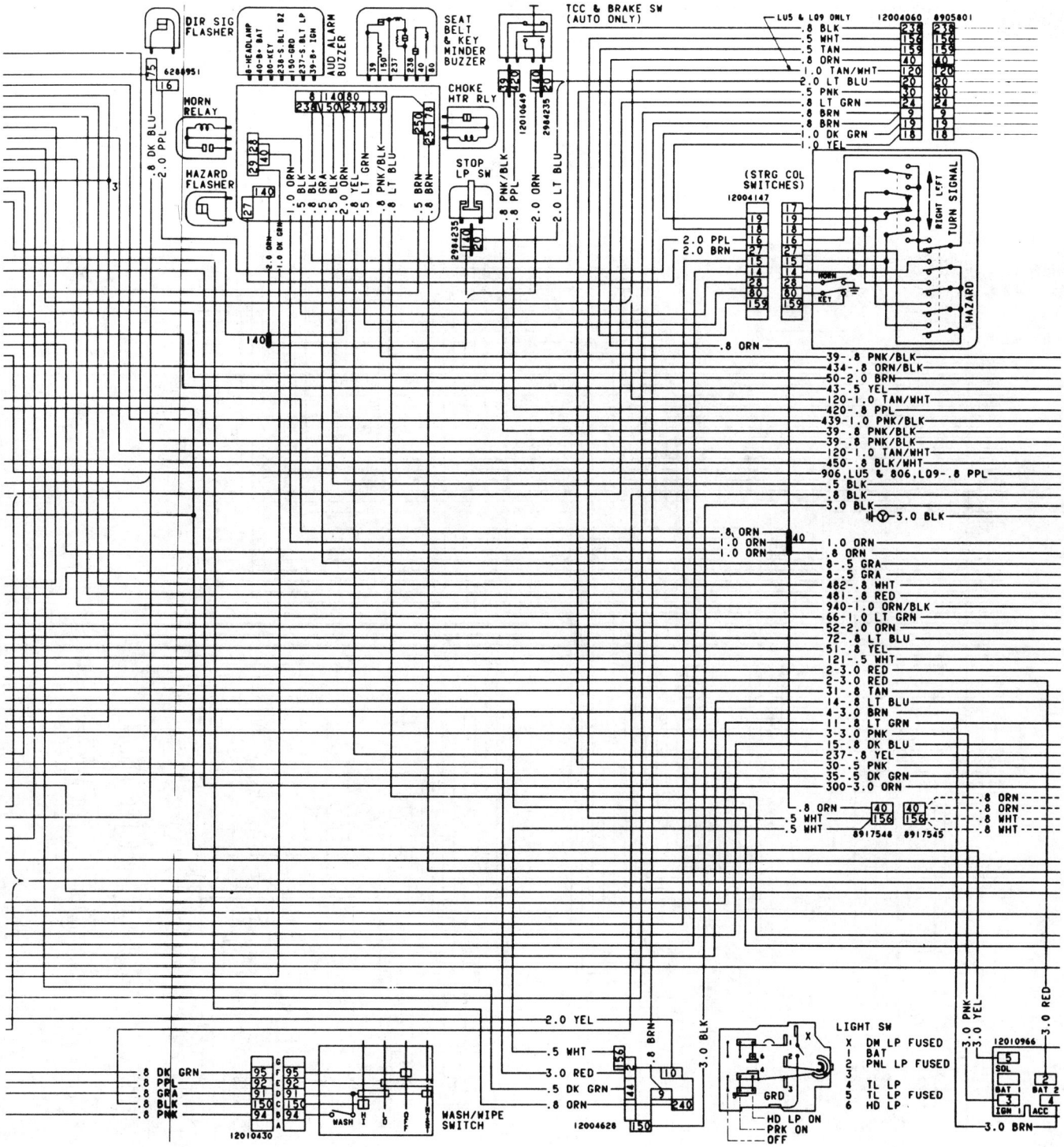

Fig. 10.20 Wiring diagram for 1982 and 1983 models — instrument panel and interior

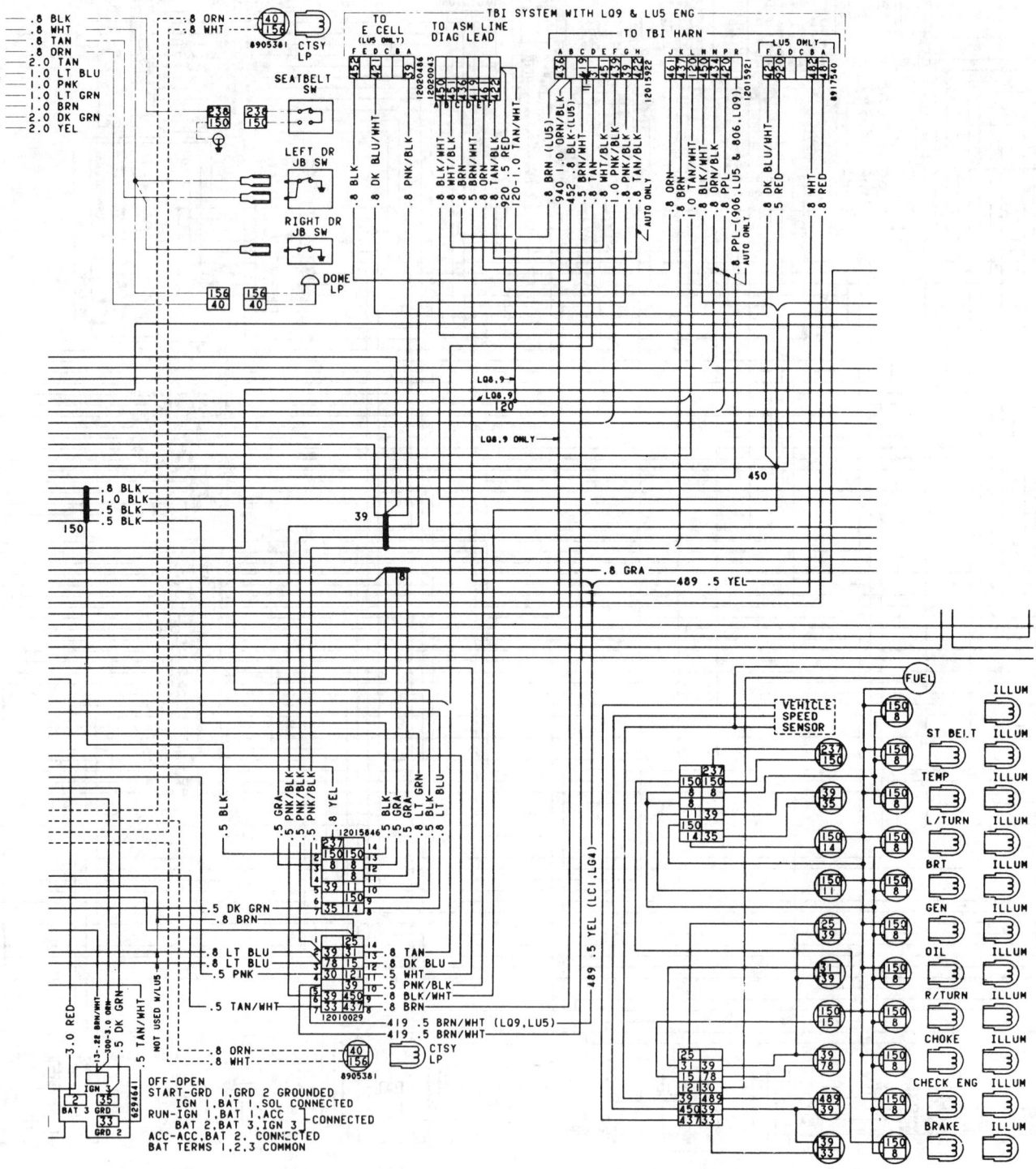

Fig. 10.21 Wiring diagram for 1982 and 1983 models — instrument panel and interior (continued)

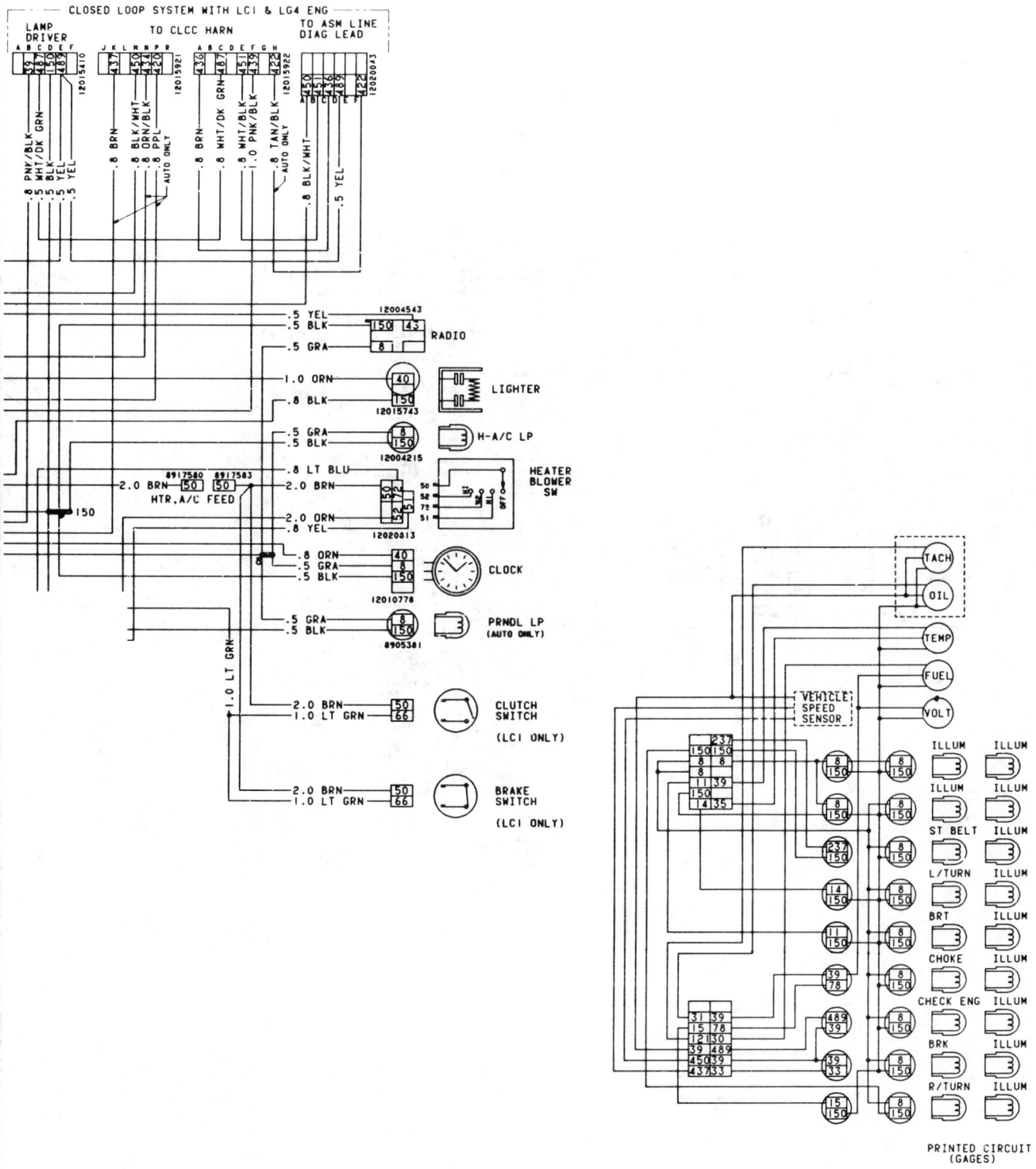

Fig. 10.22 Wiring diagram for 1982 and 1983 models — gauges

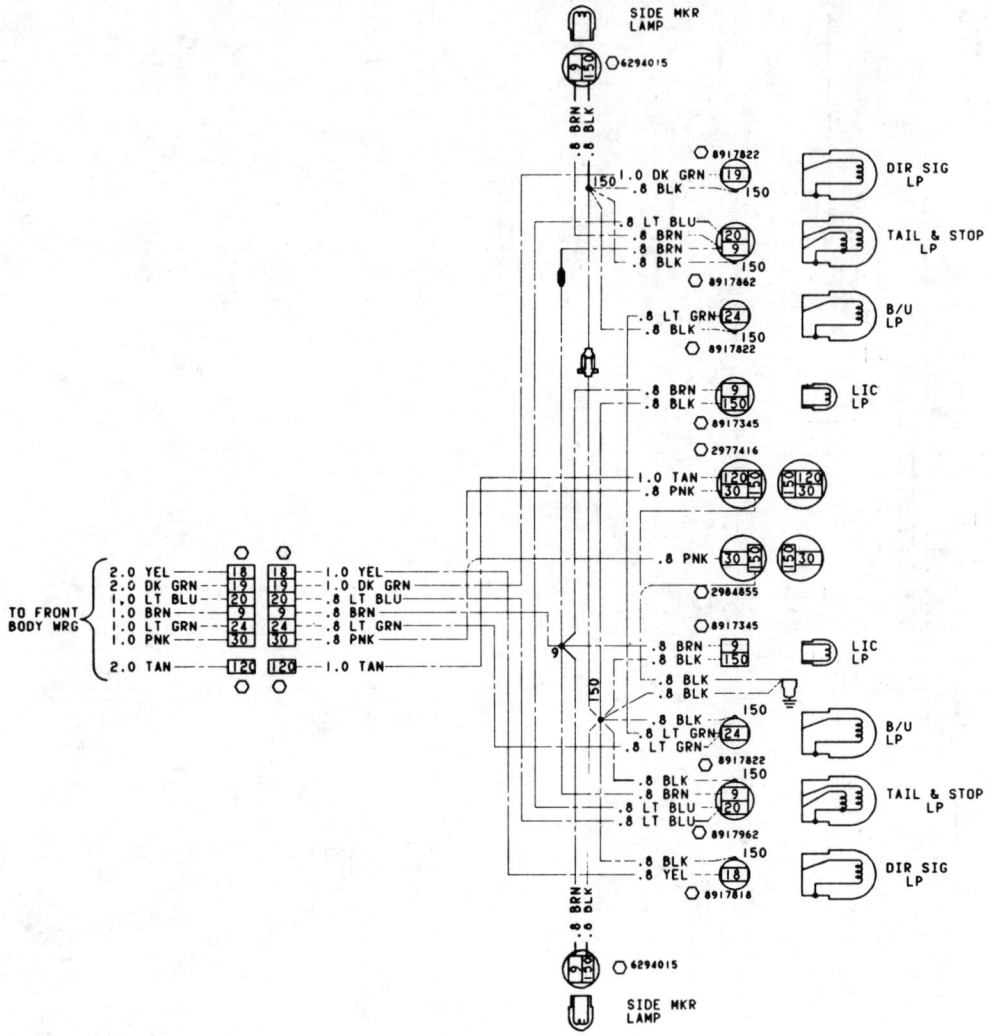

Fig. 10.23 Wiring diagram for 1982 and 1983 models — rear end

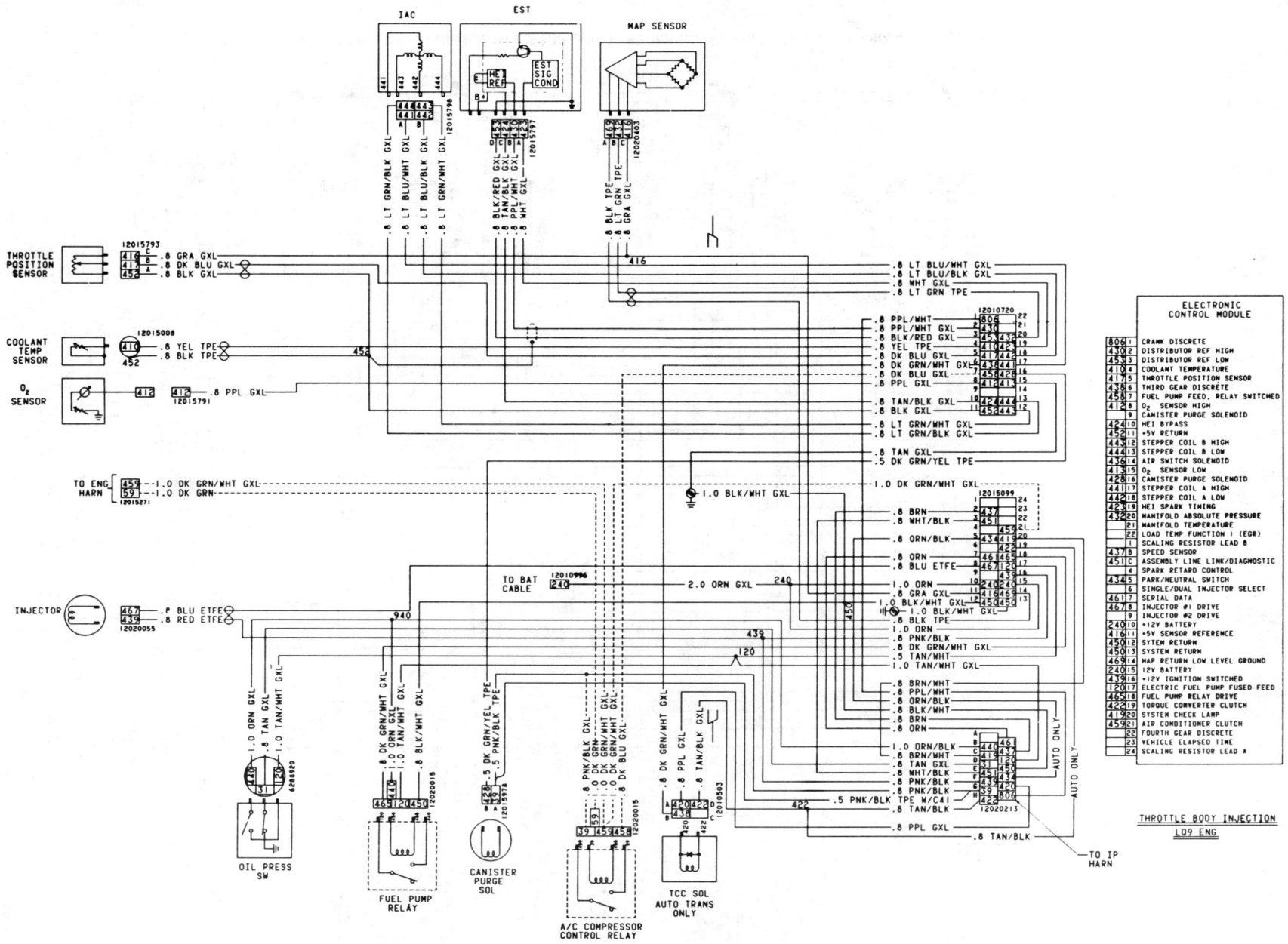

Fig. 10.24 Wiring diagram for 1982 and 1983 models ECM — (L4 engine)

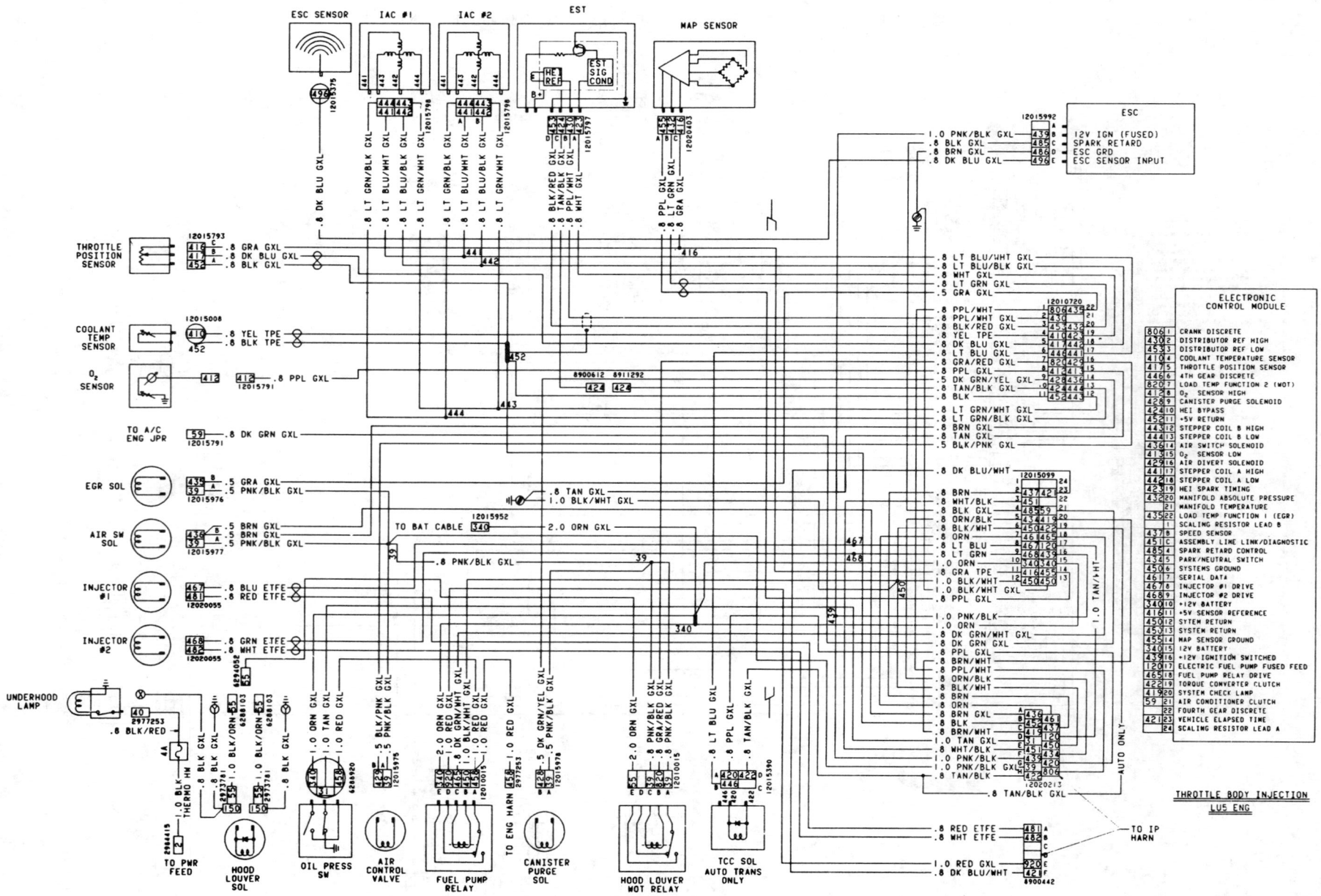

Fig. 10.25 Wiring diagram for 1982 and 1983 models — ECM (V8 engine with fuel injection)

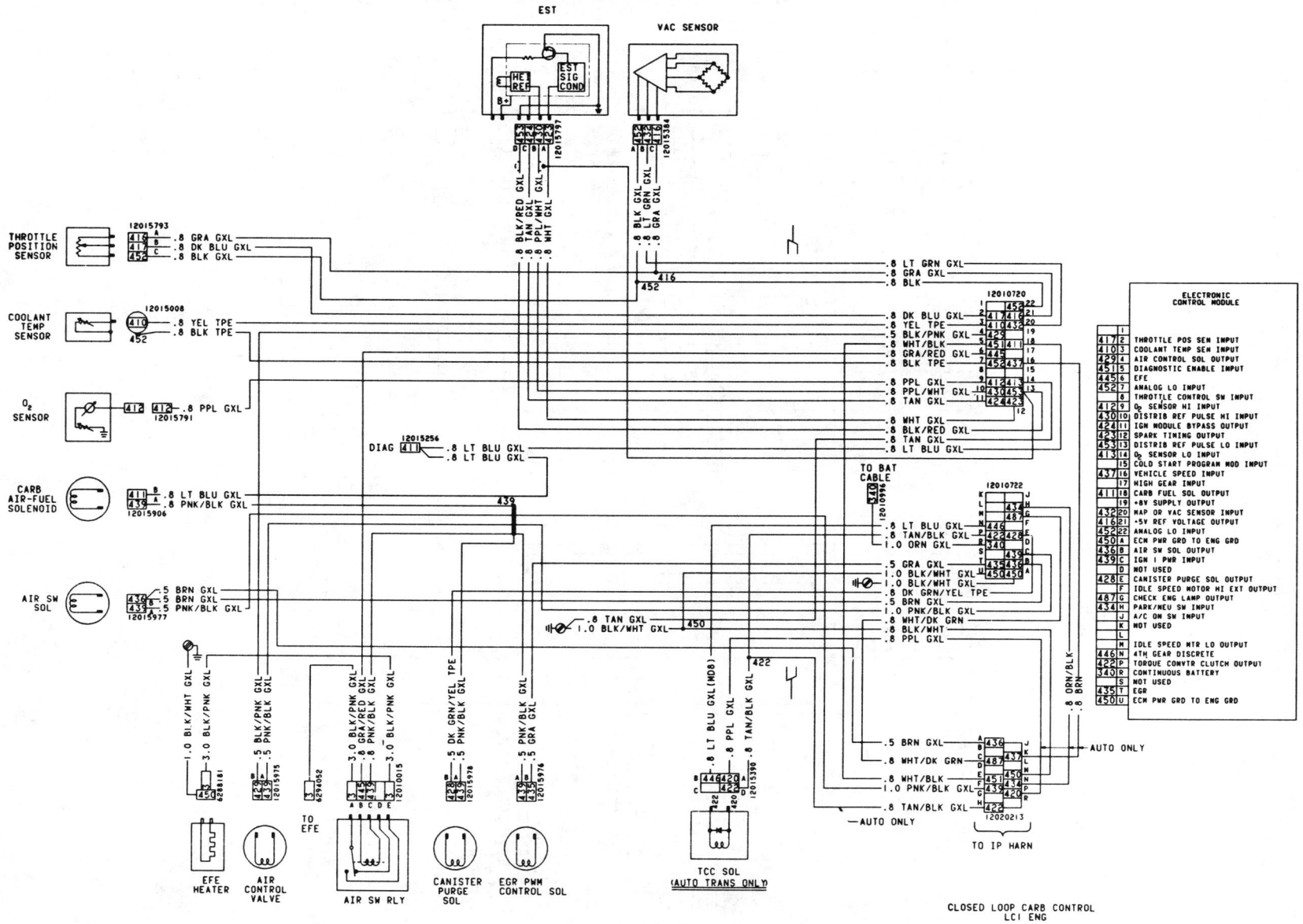

Fig. 10.26 Wiring diagram for 1982 and 1983 models — ECM (V6 engine)

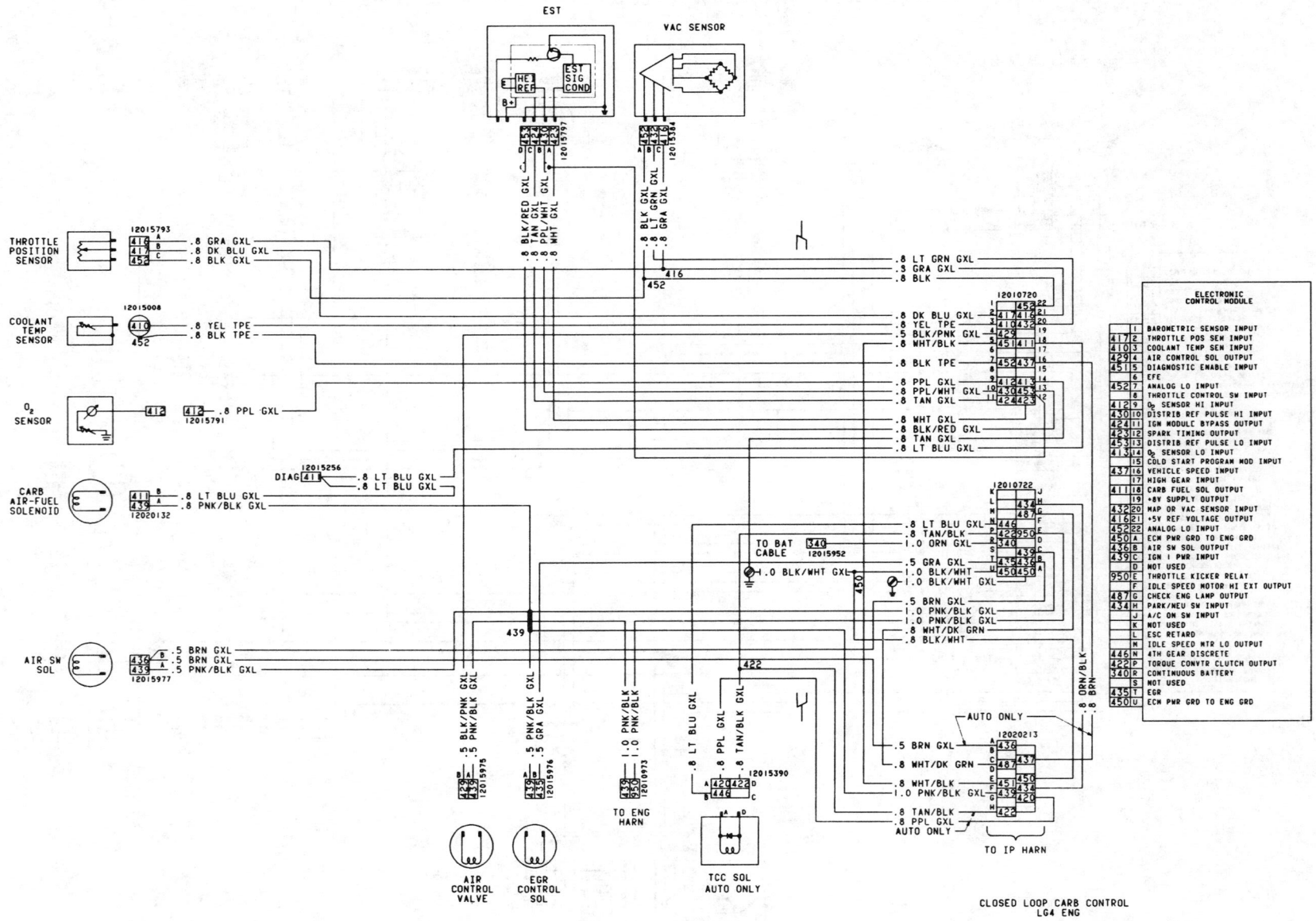

Fig. 10.27 Wiring diagram for 1982 and 1983 models — ECM (V8 carbureted engine)

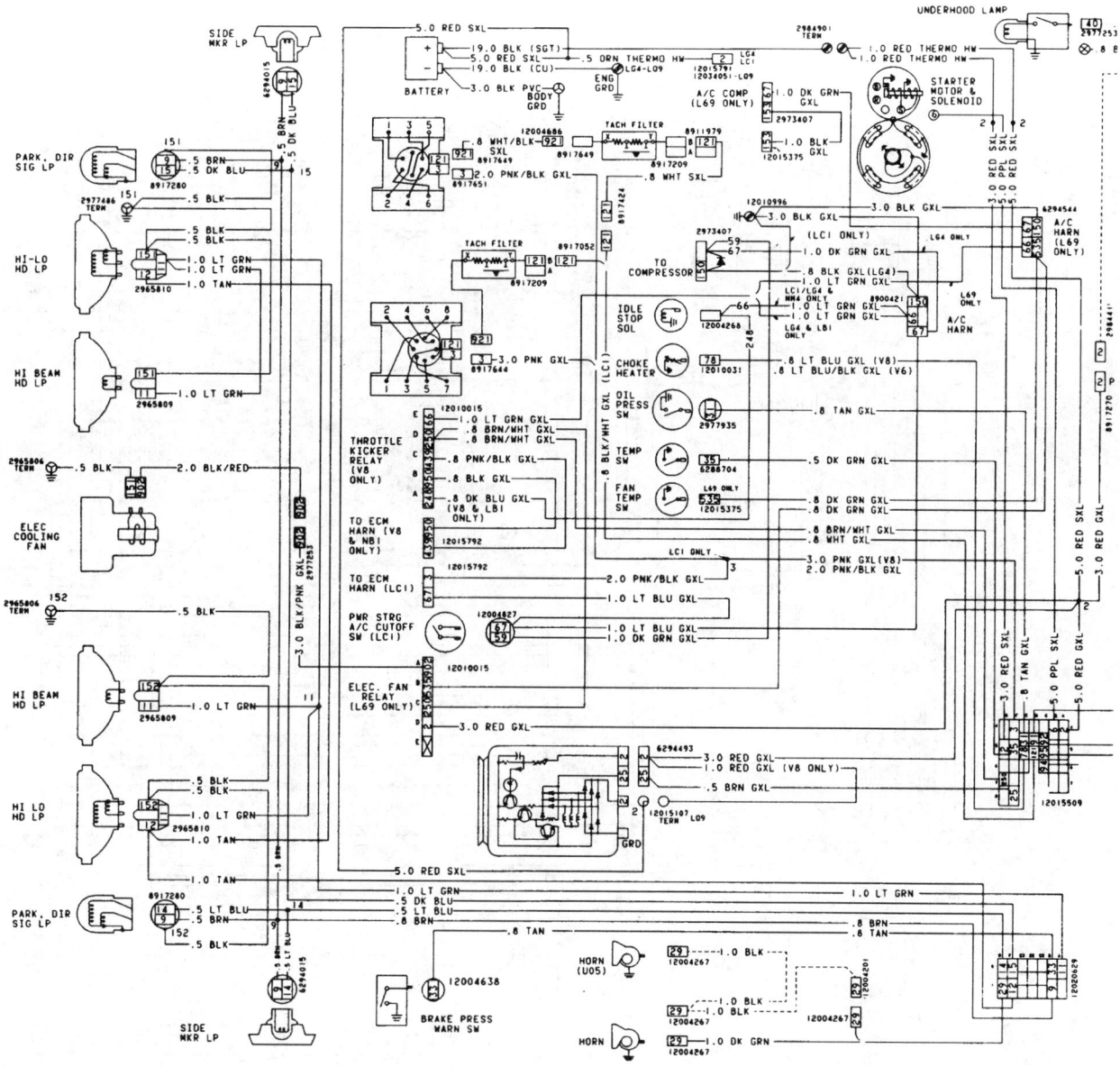

Fig. 10.28 Wiring diagram for 1984 models — front end and V6 and V8 engine compartment

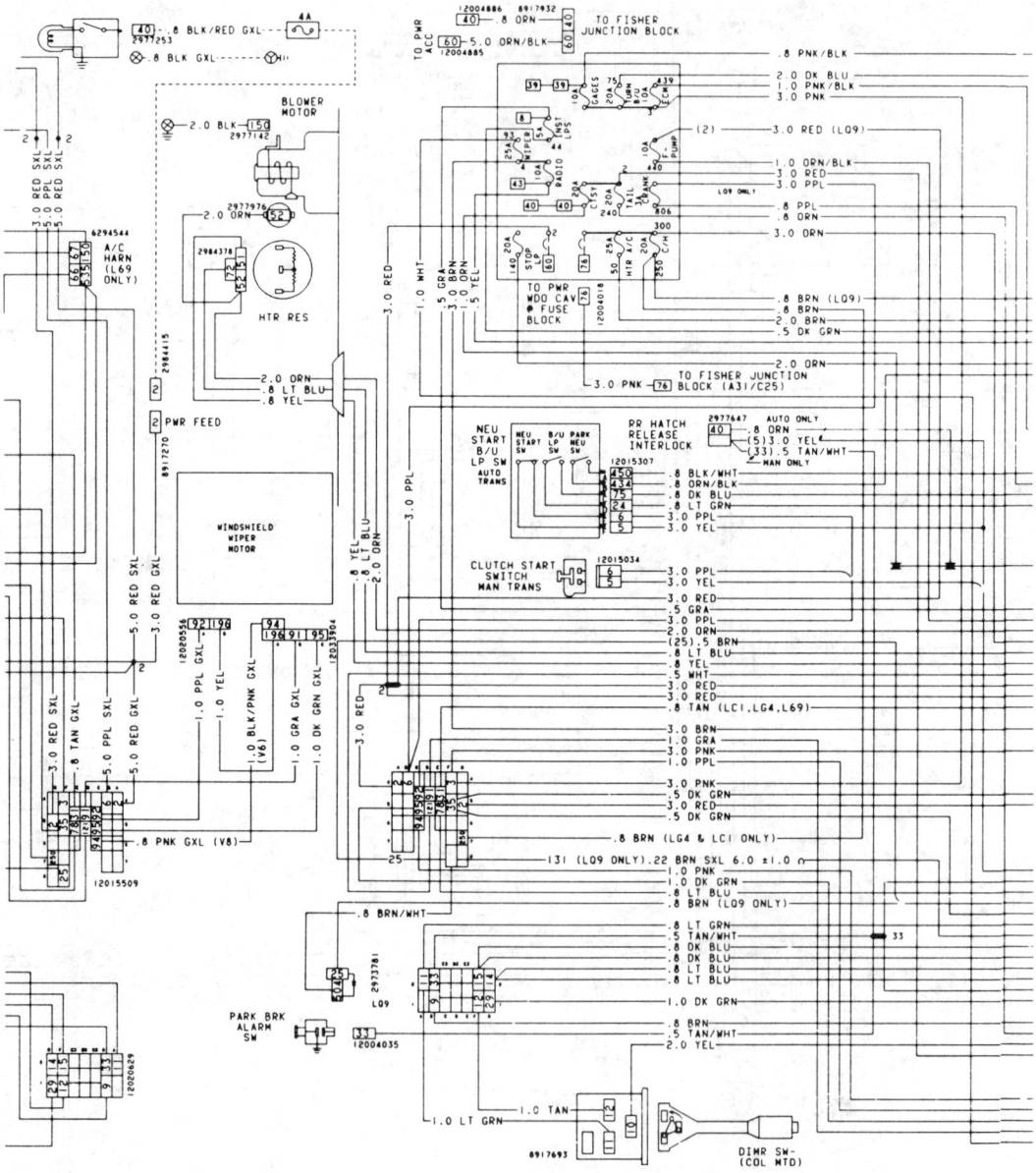

Fig. 10.29 Wiring diagram for 1984 models — V6 and V8 engine compartment and instrument panel

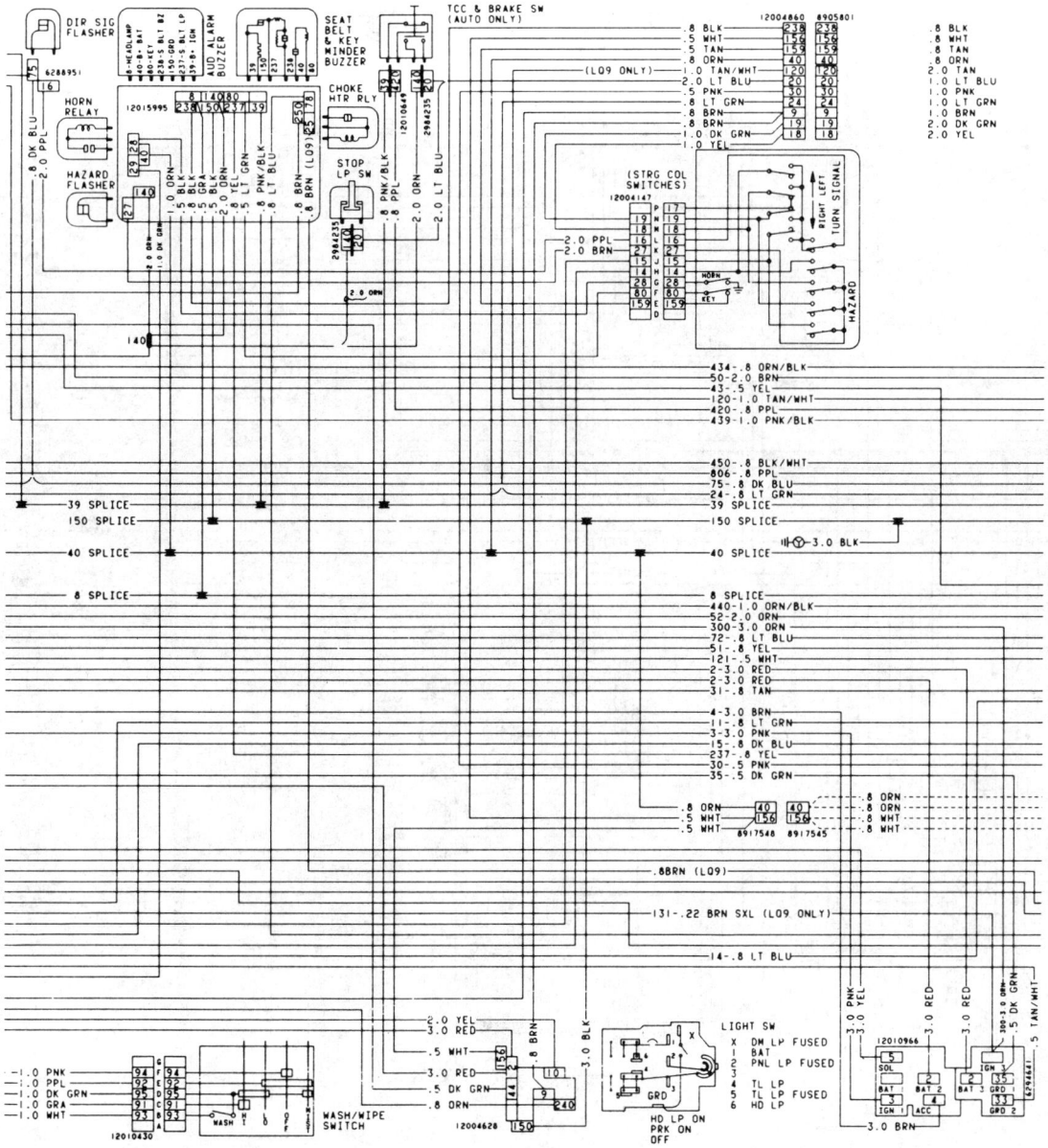

Fig. 10.30 Wiring diagram for 1984 models — instrument panel and interior

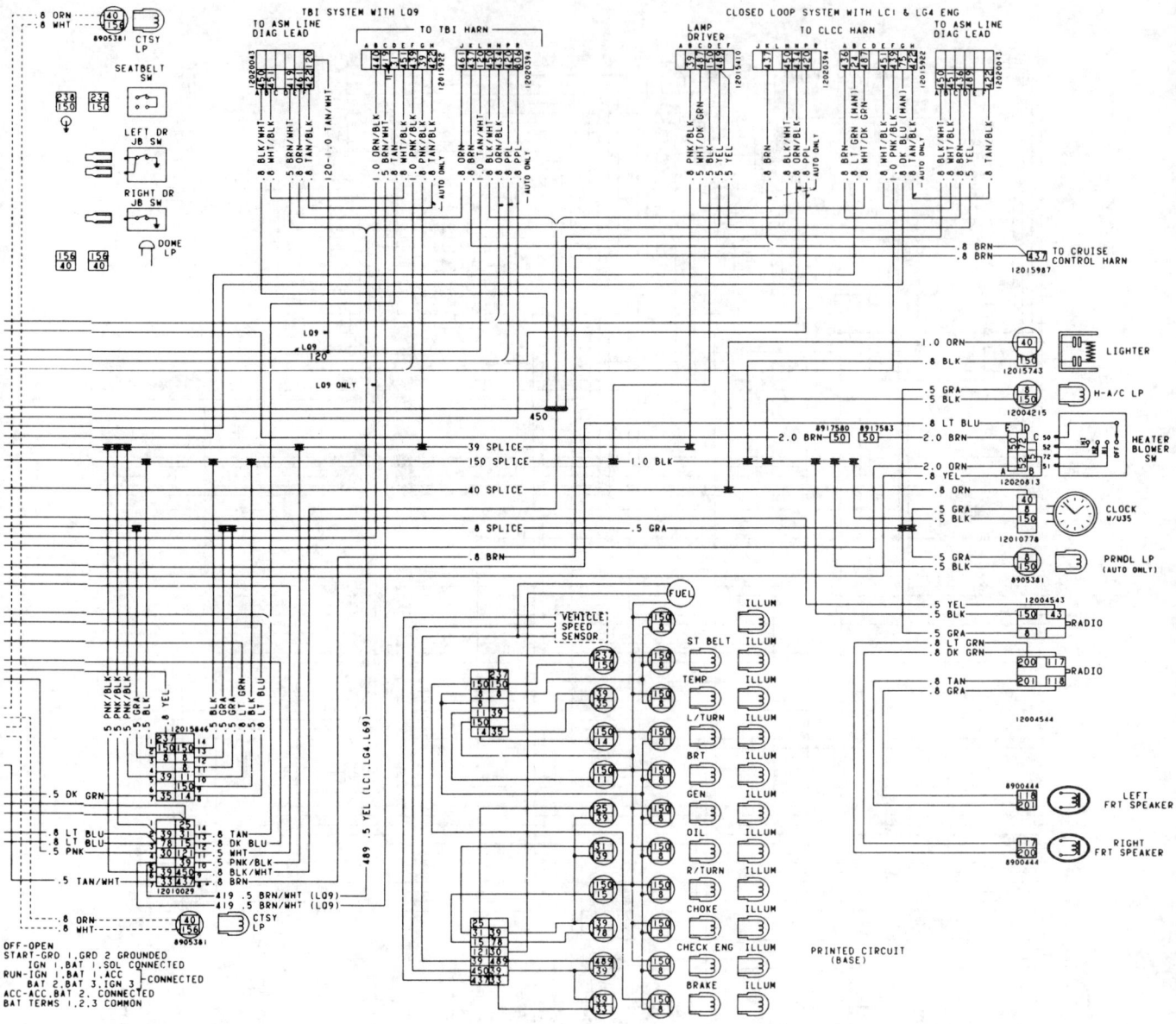

Fig. 10.31 Wiring diagram for 1984 models — instrument panel and interior (continued)

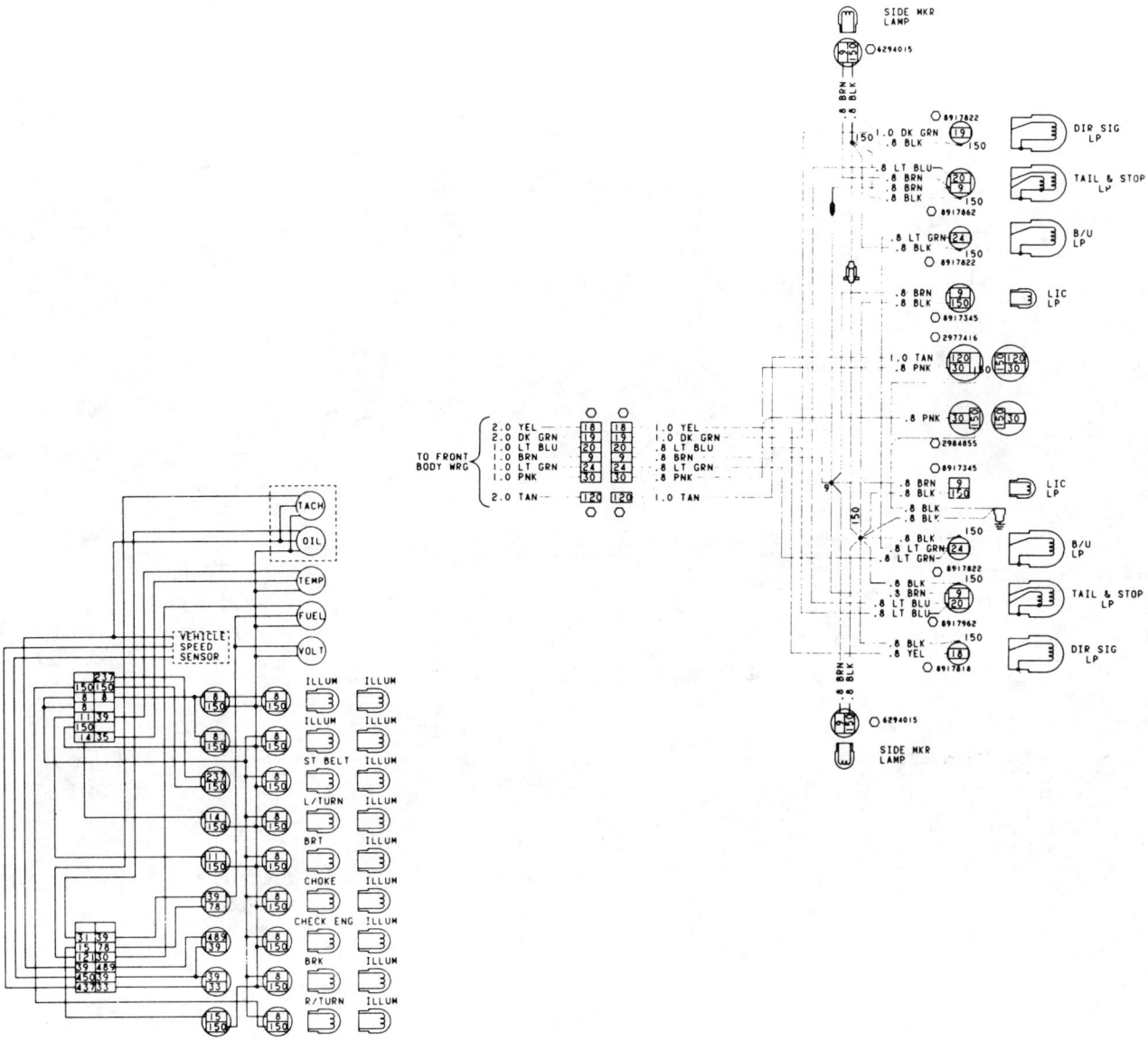

Fig. 10.32 Wiring diagram for 1984 models — gauges and rear end

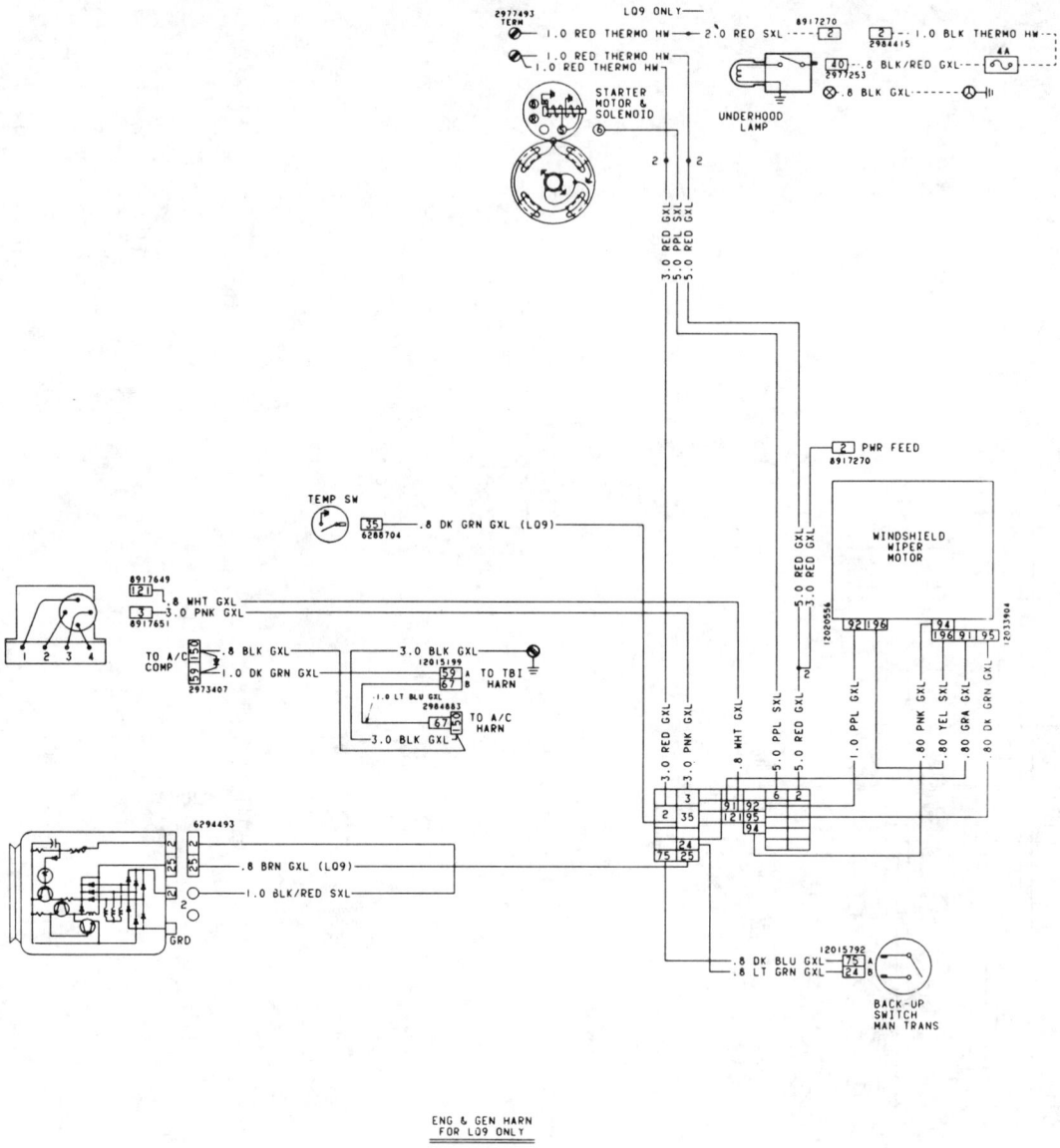

Fig. 10.33 Wiring diagram for 1984 models — engine and alternator wiring for L4 engine only

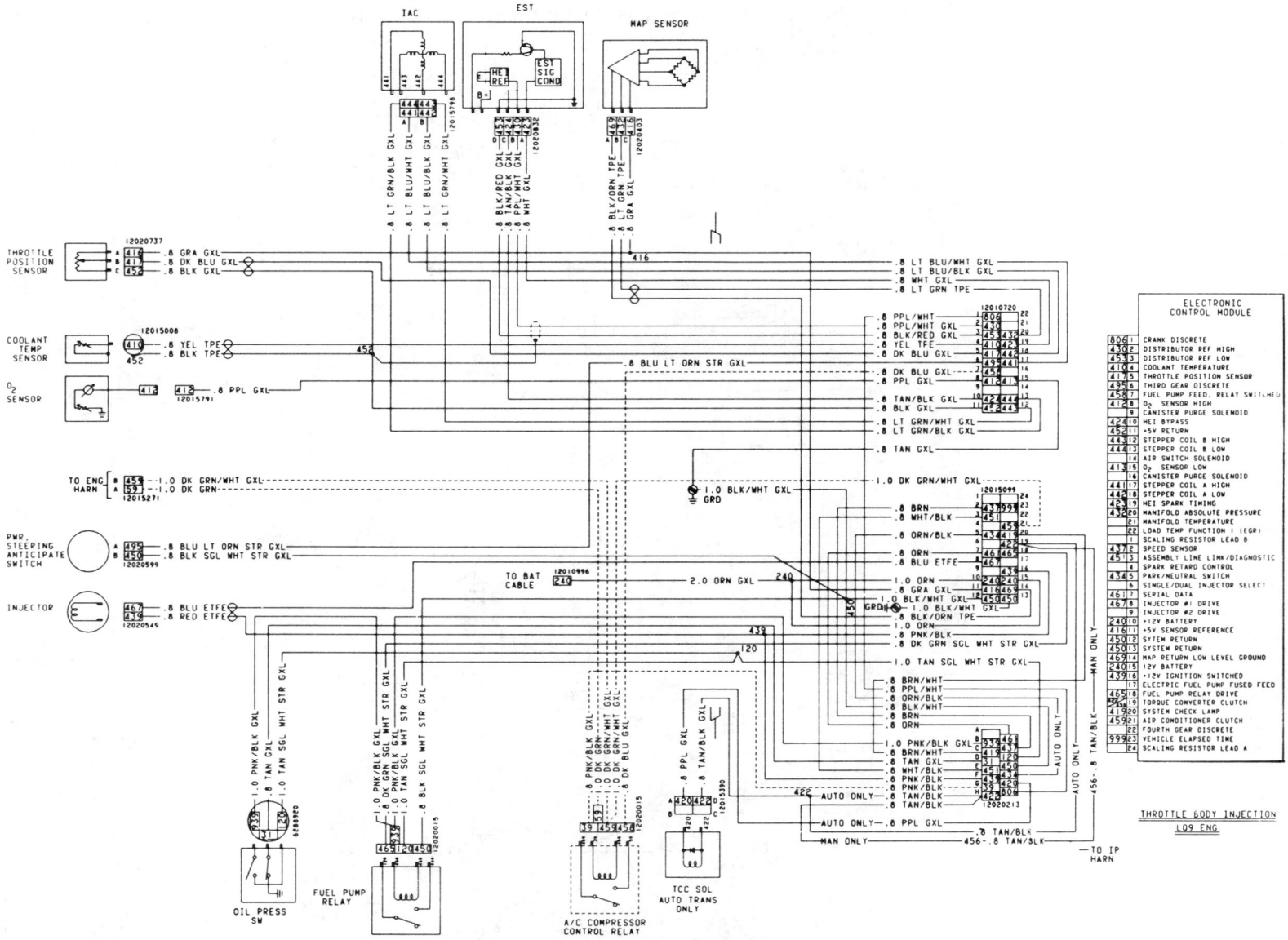

Fig. 10.34 Wiring diagram for 1984 models — ECM (L4 engine)

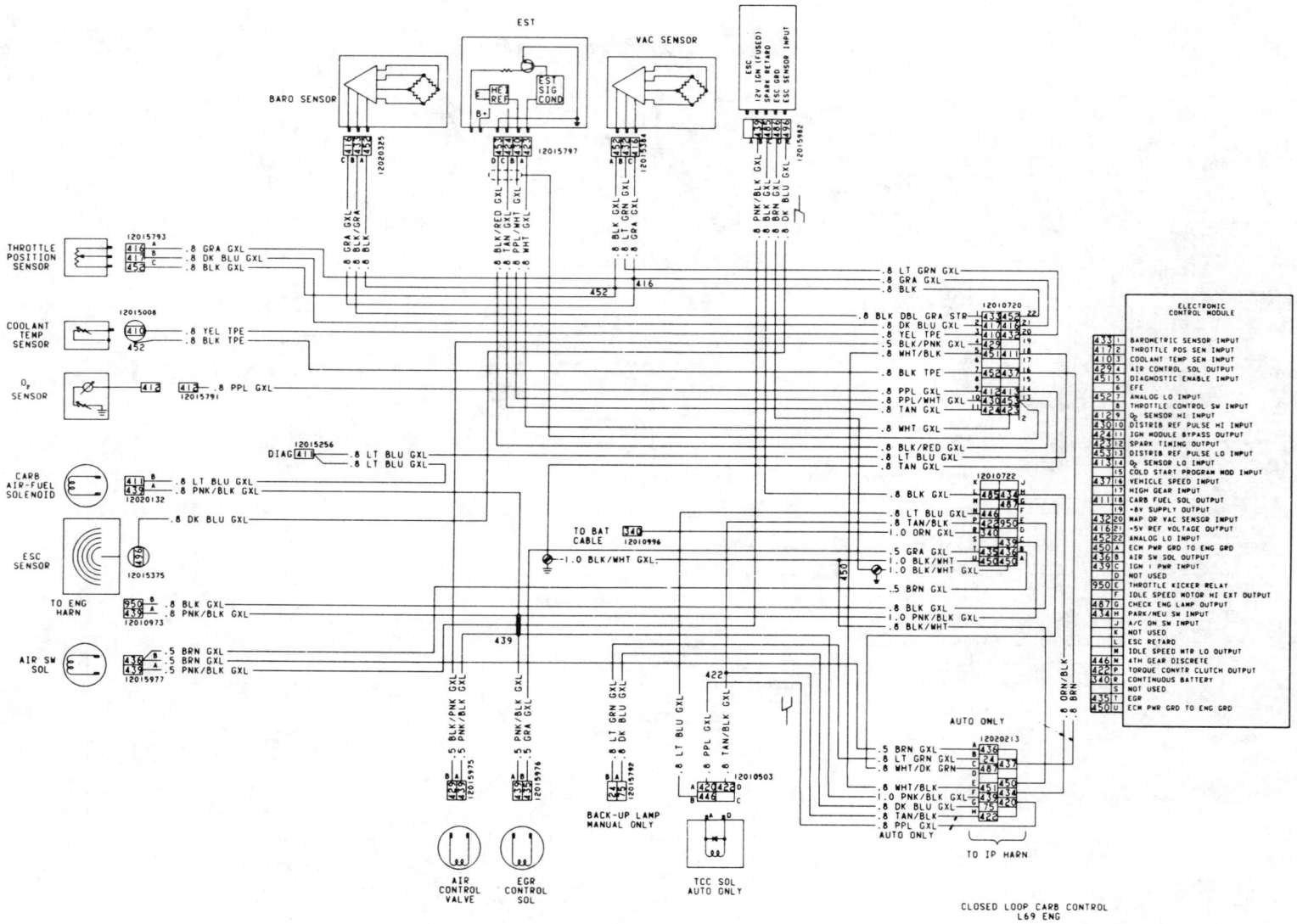

Fig. 10.35 Wiring diagram for 1984 models — ECM (HO V8 engine)

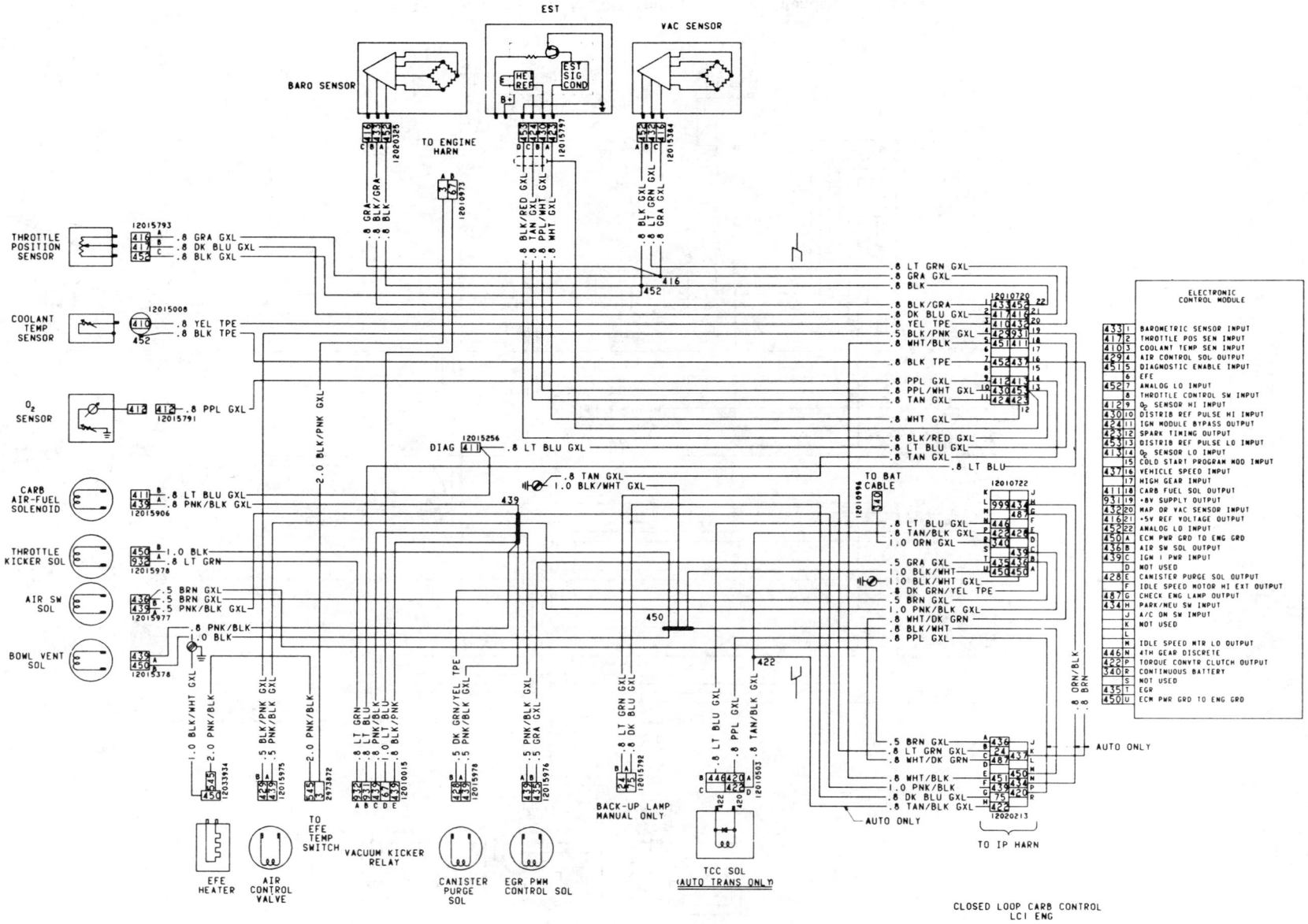

Fig. 10.36 Wiring diagram for 1984 models — ECM (V6 engine)

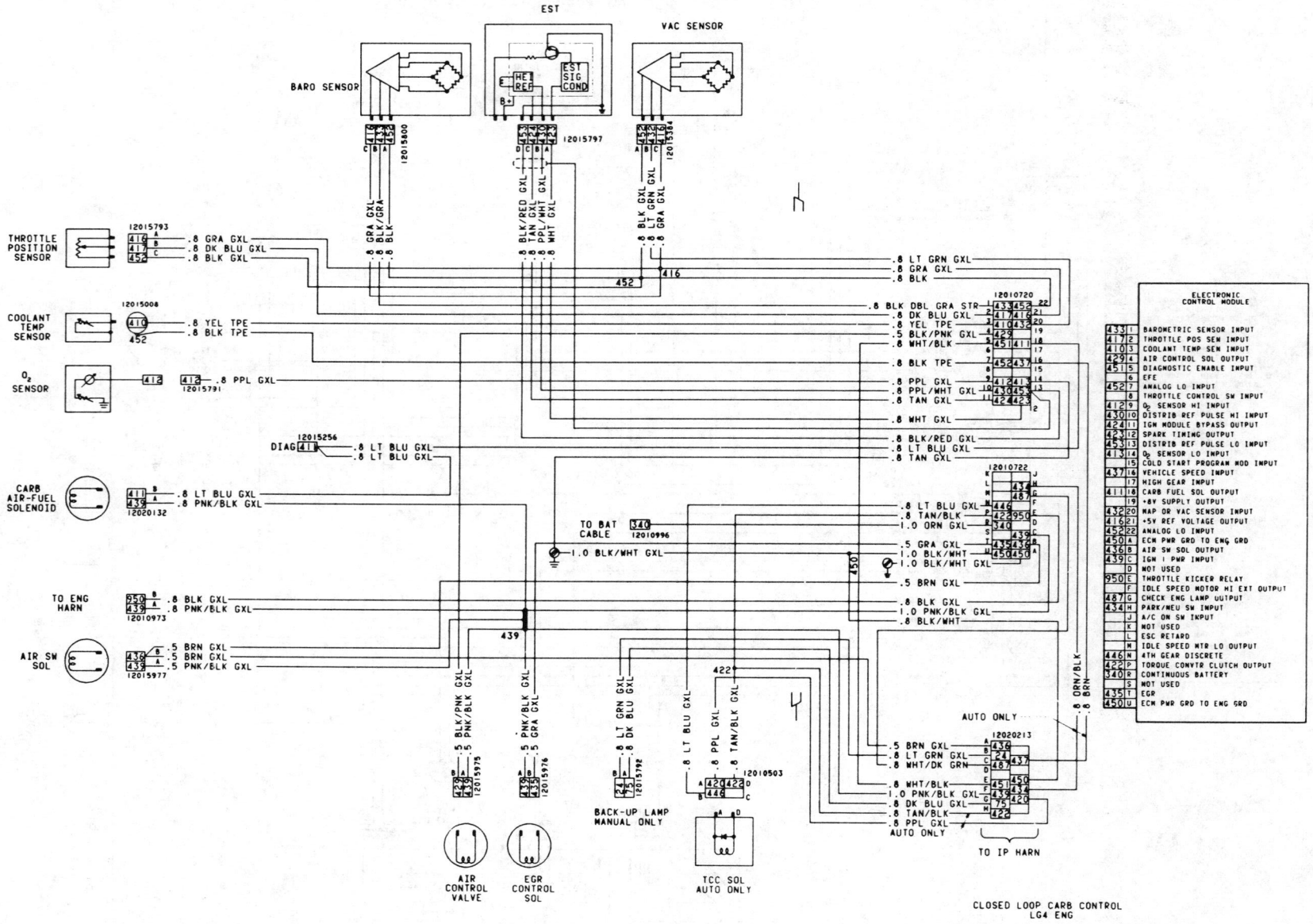

Fig. 10.37 Wiring diagram for 1984 models — ECM (standard V8 engine)

Chapter 11 Suspension and steering systems

Refer to Chapter 13 for specifications and information related to 1985 and later models

Contents

Balljoints — replacement 4
Front coil springs — removal and installation 5
Front lower control arms — removal and installation 6
Front stabilizer — removal and installation 2
Front suspension — general information 1
Power steering gear — removal and installation 22
Power steering pump — removal and installation 23
Power steering system — bleeding 21
Power steering system — general information 20
Rear coil springs and insulators — removal and installation . . . 9
Rear lower control arm — removal and installation 12
Rear lower control arm bushing — replacement 13
Rear shock absorbers — removal and installation 8
Rear stabilizer bar — removal and installation 15
Rear suspension — general information 7
Steering column switches — removal and installation 24
Steering knuckle — removal and installation 18
Steering system — general information 16
Steering wheel — removal and installation 19
Strut assembly — removal and installation 3
Tie-rods — removal and installation 17
Torque arm — removal and installation 14
Track bar — removal and installation 10
Track bar brace — removal and installation 11

Specifications

Tires

Standard	P195/75R14
Optional	P205/70R14, P215/65R15

Front wheel alignment

Caster	+3° ± 0.5°
Camber	+1° ± 0.5°
Toe-in (per wheel)	
Trans Am	+0.15° ± 0.05°
All others	+0.2° ± 0.05°

Torque specifications

	Ft-lbs
Steering linkage	
Steering knuckle-to-tie-rod end nut	40
Tie-rod clamp nuts	14
Tie-rod-to-intermediate rod nut	40
Pitman arm-to-intermediate rod nut	40
Pitman arm-to-steering gear nut	185
Idler arm-to-intermediate rod nut	40
Idler arm-to-frame nut	50
Power steering pump	
Reservoir bolt	35
Flow control fitting	35
Pressure hose fitting	20
Power steering gear	
Gear-to-frame bolts	80
High pressure line fitting at gear	20
Oil return line fitting at gear	20
Coupling flange bolt (if equipped)	30
Coupling flange nut (if equipped)	20
Steering wheel and column	
Steering wheel-to-shaft nut	30
Bracket-to-steering column support nuts	25
Toe-pan-to-dash screws	45 in-lbs
Toe-pan clamp screws	60 in-lbs
Bracket-to-steering column bolt	30
Clamp-to-steering shaft nut	55
Support-to-lock plate screws	60 in-lbs

Fig. 11.1 Suspension and steering components

1 Idler arm
2 Stabilizer bar
3 Pitman arm
4 Tie-rod
5 Balljoint
6 Lower control arm
7 Crossmember
8 Relay rod
9 Spring
10 Strut
11 Steering knuckle (arm)

Front suspension

Strut assembly	
Strut-to-knuckle bolts	195
Strut-to-upper mount nut	50
Upper mount-to-wheelhouse tower	20
Lower control arm	
Pivot bolt nuts	65
Lower control arm balljoint-to-knuckle	90
Stabilizer bar	
Link bolt/nut	14
Insulator bracket bolt	25

Rear suspension

Shock absorber-to-upper mount nut	13
Shock absorber-to-axle nut	70
Track bar-to-axle nut	93
Track bar-to-body bracket	58
Track bar brace-to-body bracket nut	58
Track bar brace-to-body brace bracket screws	34
Control arm-to-rear axle bolt	68
Control arm-to-underbody bolt	68
Torque arm front bracket-to-transmission bolt	
With 200C automatic transmission	20
All except 200C transmission	31
Torque arm-to-rear axle nut	100
Stabilizer shaft-to-body bracket link bolt	12
Stabilizer bracket-to-body screw	35
Stabilizer shaft clamp-to-U-bolt nut	20
Axle bumper bracket-to-underbody bolt	20

1 Front suspension — general information

The front suspension is independent, allowing each wheel to compensate for road surface changes without appreciably affecting the opposite wheel.

Each wheel is connected independently to the frame by a steering knuckle, strut assembly, balljoint and lower control arm. The front wheels are held in proper relationship to each other by tie-rods, which are connected to the steering knuckles, and to a relay rod assembly.

Coil springs are mounted between spring housings located on the front frame crossmember and the lower control arms. Shock absorbtion is provided by double, direct-acting strut assemblies, with the upper portion of each strut extending through the fender well and attaching to the upper mount assembly with a nut.

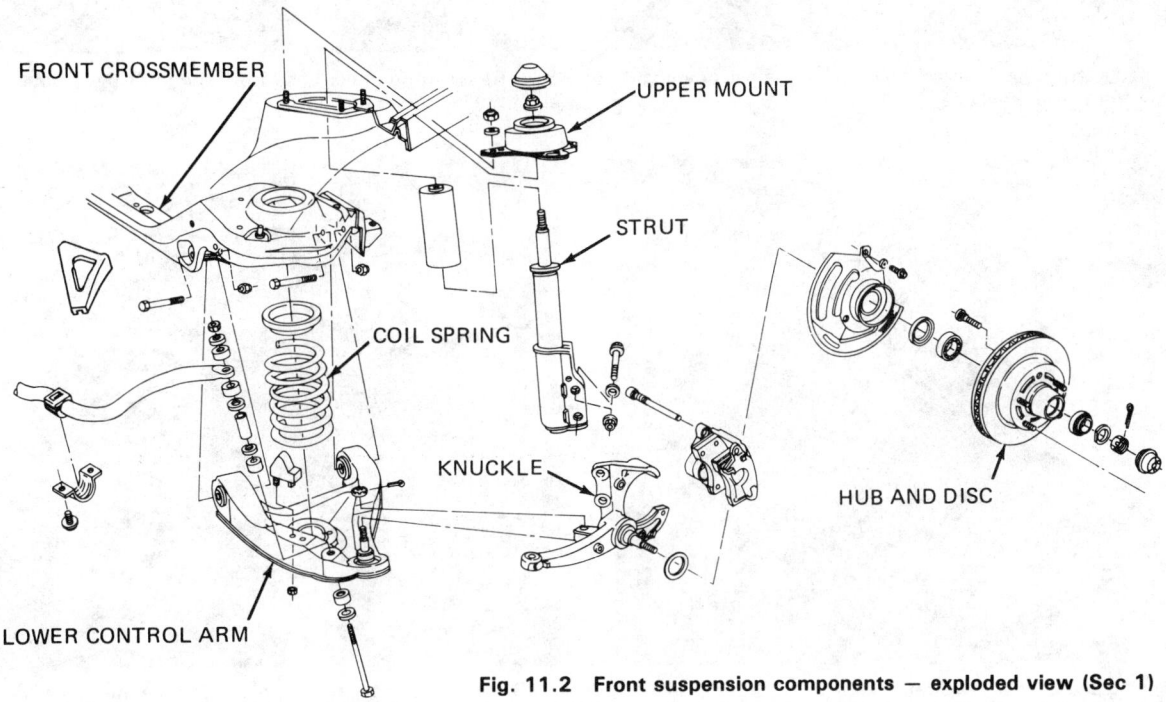

Fig. 11.2 Front suspension components — exploded view (Sec 1)

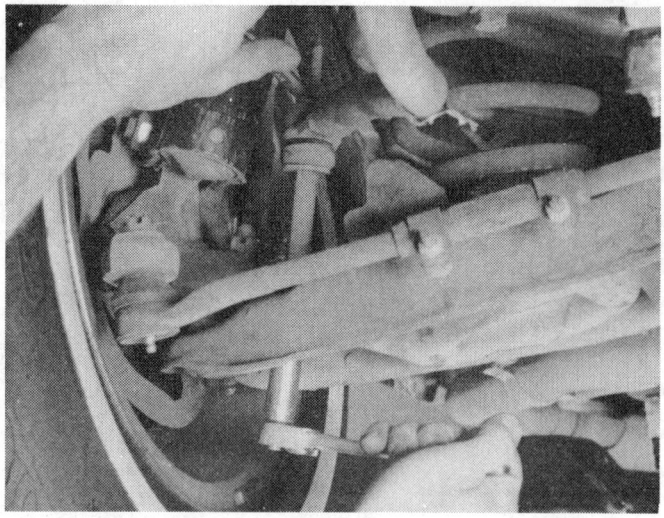

2.2 Removing the link bolt from the front stabilizer bar

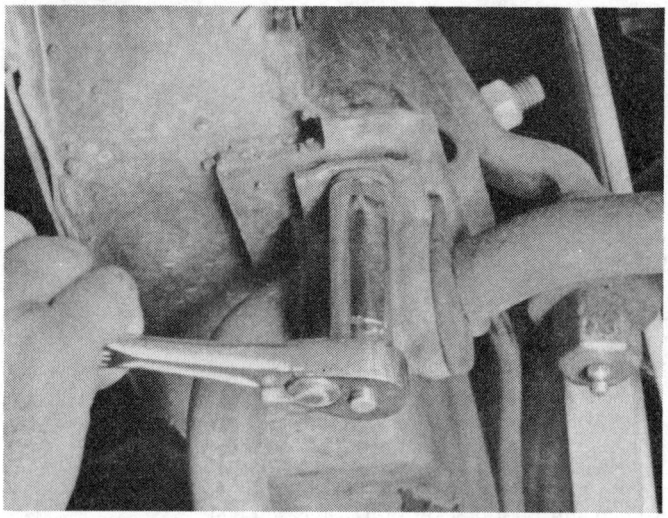

2.3 Removing the bracket and insulator bolts from the stabilizer bar

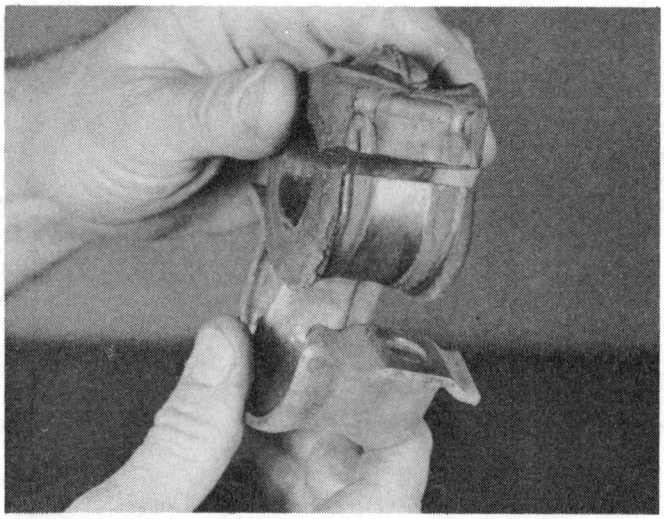

2.7 When reinstalling the insulator, make sure the slit opening is toward the front of the vehicle

3.3 Remove the brake hose from the strut by removing the clip attaching it to the bracket

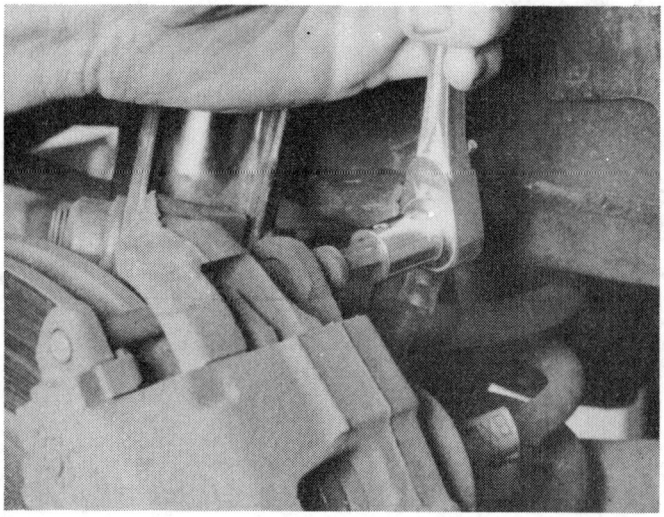

3.4A Removing the caliper mounting bolts

3.4B Hang the caliper by a piece of wire, never by the brake hose

Front suspension side roll is controlled by a steel stabilizer shaft mounted in rubber bushings which are held to the frame side rails by brackets. The stabilizer ends are connected to the lower control arms by link bolts isolated by rubber grommets.

Pressed-in bushings are located in the inner ends of the lower control arms. Bolts passing through the bushings attach the arm to the suspension crossmember. The lower balljoint assembly is press-fit in the arm and attaches to the steering knuckle with a torque-prevailing nut.

To keep dirt and moisture from entering the joints and damaging bearing surfaces, rubber grease seals are provided at the ball socket assemblies.

2 Front stabilizer — removal and installation

1 Raise the vehicle and support it securely on jackstands.
2 Remove the link bolt, nut, grommets, spacer and retainers from each side of the stabilizer bar, noting their positions relative to each other to simplify installation (photo).
3 Remove the insulator and bracket from each side of the stabilizer bar, noting their positions to simplify installation (photo).
4 Remove the stabilizer bar.
5 Installation is the reverse of the removal procedure, noting the following.
6 Apply multi-purpose grease to the insulator mounting points on the stabilizer bar before installing the insulators and brackets.
7 Make sure the slits in the insulators are toward the front of the vehicle (photo). Also make sure the locater on top of the insulator seats properly on the frame pad.
8 Make sure the stabilizer is centered in the brackets before tightening the mounting bolts.
9 Tighten all bolts to the specified torque.

3 Strut assembly — removal and installation

1 Raise the front of the vehicle and support it securely on jackstands, then remove the wheel and tire.
2 Place a hydraulic jack under the lower control arm to support the control arm.
3 Remove the brake hose retaining clip from the strut mount and remove the hose from the mount (photo).
4 Remove the brake caliper mounting bolts and hang the caliper out of the way with wire (photos). **Caution:** *Do not allow the caliper to hang by the brake hose.*
5 Mark the relationship of the strut to the steering knuckle with white paint.
6 Loosen the top strut-to-steering knuckle bolt.
7 Loosen and remove the bottom strut-to-knuckle bolt, then remove

3.8 Removing the cover from the upper end of the strut

3.9 Removing the strut upper retaining nut

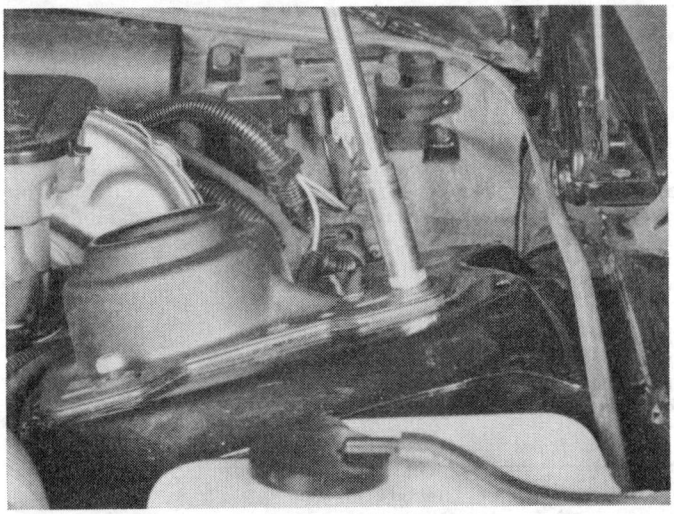

3.13A Removing the upper strut mount bolts

3.13B Removing the upper strut mount and retaining bracket

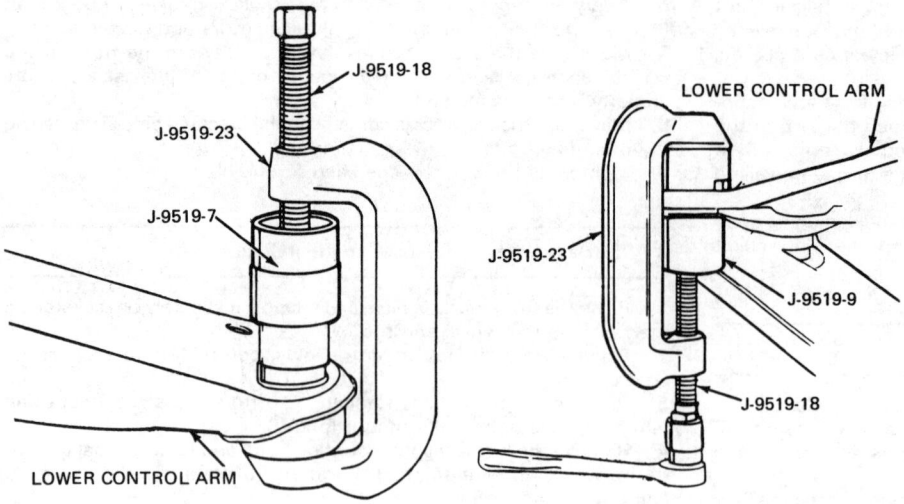

Fig. 11.3 Balljoint removal (left) and installation (right) (Sec 4)

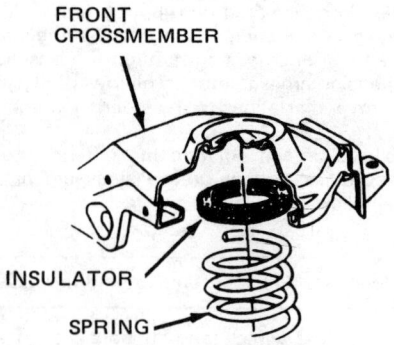

Fig. 11.4 Relationship of the spring, insulator and front crossmember (Sec 5)

the top bolt.
8 From under the hood, remove the cover from the upper end of the strut (photo).
9 Hold the top of the strut piston with one wrench while removing the upper mount retaining nut with another (photo).
10 Remove the strut.
11 Check the bearing in the upper mount by turning it with your finger.
12 If the upper mount must be removed, carefully scribe around it before removing the retaining bolts.
13 Remove the mount and retaining bracket (photos).
14 If the bearing is defective, the entire top strut mount assembly must be replaced because the bearing alone is not replaceable.
15 Installation is the reverse of the removal procedure. To assist in installing the upper end of the strut, carefully lower the jack supporting the lower control arm enough to get a hand on the piston, then feed it through the upper mount and attach the retaining nut.

4 Balljoints — replacement

1 Raise the front of the vehicle and support it securely on jackstands.
2 Remove the wheel and tire.
3 Place a floor jack under the lower control arm spring seat. **Note:** *The jack must remain under the spring seat during removal and installation of the balljoint to hold the spring and control arm in position.*
4 Remove the cotter pin from the top of the balljoint and loosen, but do not remove, the castellated nut.
5 Using tool number J-24292A or equivalent, according to the accompanying instructions, break the balljoint loose from the steering knuckle, then remove the tool and the castellated nut and separate the joint from the knuckle.
6 Using a putty knife or similar tool, guide the lower control arm out of the opening in the splash shield.
7 Remove the grease fittings.
8 Refer to the accompanying illustration and use the tools indicated to press the balljoint out of the lower control arm.
9 Inspect the tapered hole in the steering knuckle, removing any accumulated dirt first. If out-of-roundness, deformation or other damage is noted, the knuckle must be replaced with a new one (refer to Section 18).
10 To install, refer to the illustration and, using the tools illustrated, position the balljoint in the lower control arm and press it in until it bottoms. **Note:** *The grease purge seal must face in.*
11 Position the balljoint stud in the steering knuckle, install the ball stud nut and tighten it to the specified torque.
12 If the cotter pin hole in the stud does not line up with an opening in the castellated nut, tighten the nut enough to allow installation of the cotter pin.
13 Install the balljoint grease fitting and lubricate it until grease appears at the seal.
14 Install the wheel and tire and lower the vehicle.
15 The front end alignment should be checked by a GM dealer or alignment shop.

5 Front coil springs — removal and installation

1 Raise the vehicle and support it on jackstands placed under the outside frame rails.
2 Remove the wheel and tire.
3 Disconnect the stabilizer bar at the lower control arm, keeping the bolt, nut, grommet spacer and retainers in order to simplify installation.
4 Remove the nuts from the lower control arm-to-front crossmember pivot bolts, *but do not remove the pivot bolts at this time.*
5 Place a hydraulic jack in position to support the lower control arm/crossmember junction.
6 Raise the jack enough to remove both pivot bolts.
7 Carefully lower the jack and remove the spring and insulator. The spring's position should be carefully noted as it is removed.
8 Installation is the reverse of the removal procedure.
9 Note that after assembly, the end of the bottom coil of the spring must cover all or part of one inspection drain hole. The other hole must be partly or completely uncovered.

6 Front lower control arms — removal and installation

Note: *Before starting the following procedure, measure the distance from the center of the lower control arm front bushing bolt head to the ground and record the figure.*
1 Remove the coil spring (refer to Section 5).
2 Remove the balljoint securing the steering arm to the lower control arm (refer to Section 4).
3 Be sure to replace the bushings in the lower control arm before installation.
4 Installation is the reverse of the removal procedure, noting the following.
5 When attaching the lower control arm to the front crossmember, install the front leg first, then the rear leg.
6 When the bolts are tightened, the control arm must be positioned at the height recorded before the procedure was started.
7 Tighten all bolts to the specified torque.

7 Rear suspension — general information

The rear axle assembly is attached to the vehicle through a link-type suspension system. A track bar and two lower control arms attach the

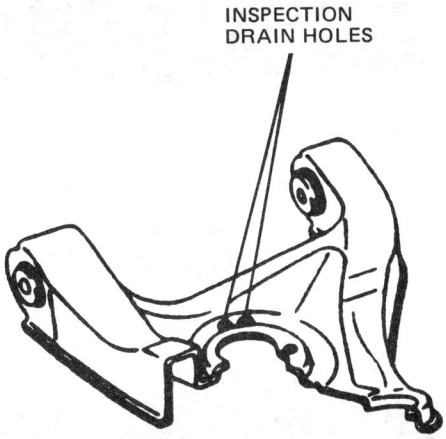

Fig. 11.5 Location of the drain holes in the lower control arm (Sec 5)

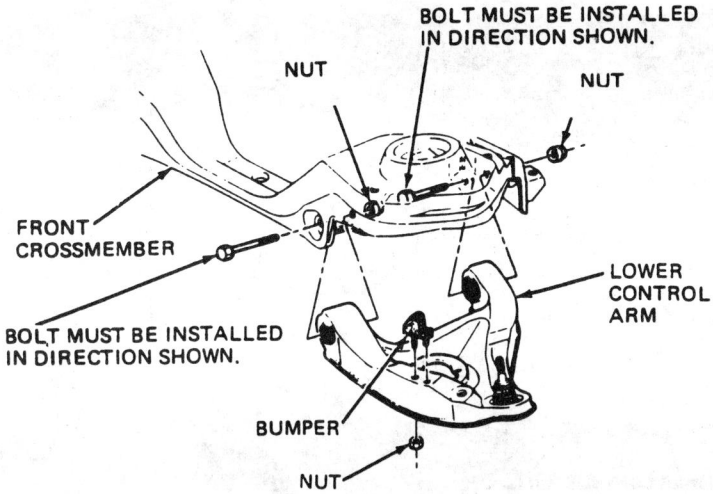

Fig. 11.6 Lower control arm-to-crossmember installation details (Sec 6)

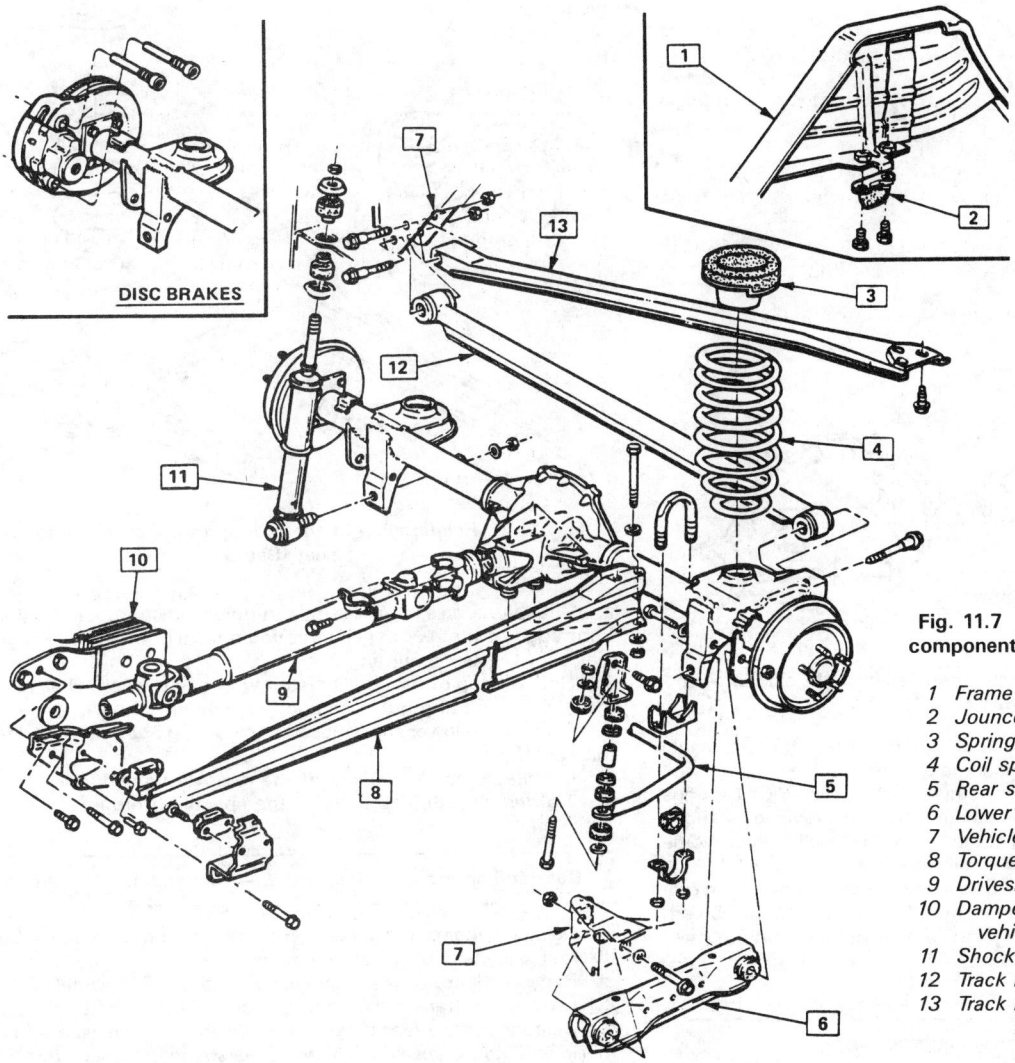

Fig. 11.7 Rear axle and suspension components — exploded view (Sec 7)

1 *Frame rail*
2 *Jounce bumper*
3 *Spring insulator assembly*
4 *Coil spring*
5 *Rear stabilizer bar (if equipped)*
6 *Lower control arm*
7 *Vehicle underbody*
8 *Torque arm*
9 *Driveshaft*
10 *Damper assembly (L4-equipped vehicles)*
11 *Shock absorber*
12 *Track bar*
13 *Track bar brace*

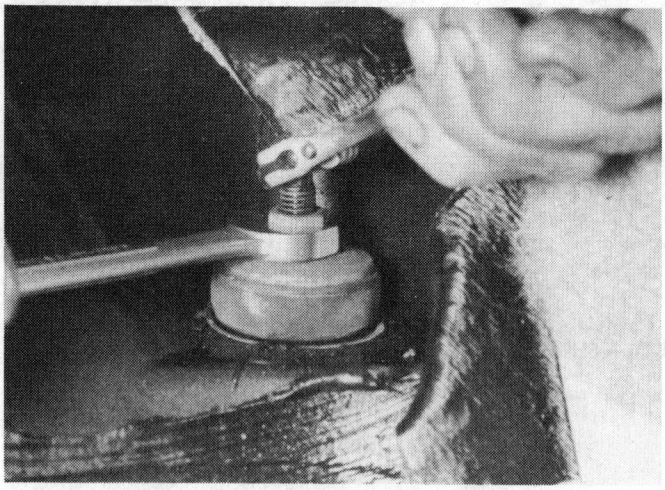

8.3 Removing the rear shock upper mounting nut

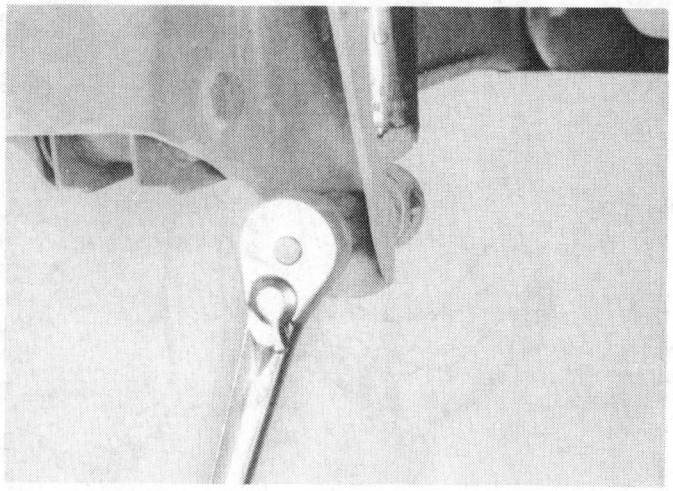

8.4 Removing the rear shock lower mounting nut

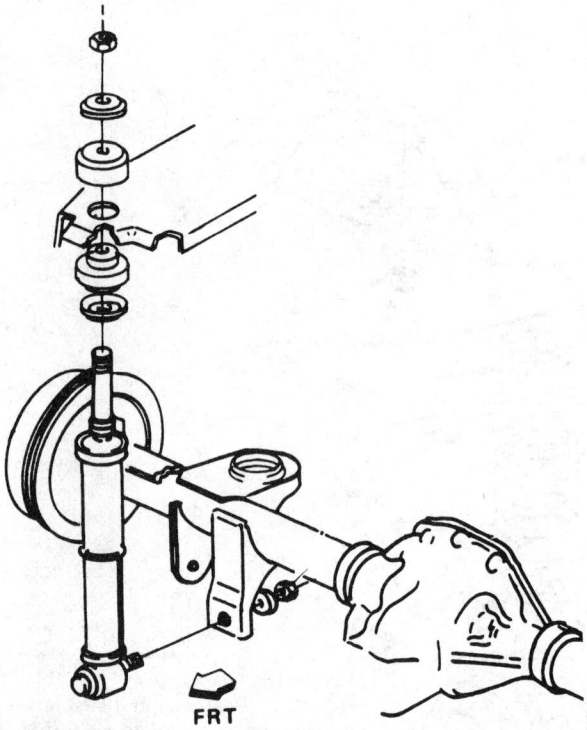

Fig. 11.8 Typical shock absorber installation (right side shown) (Sec 8)

axle housing to the body. In place of upper control arms, a single torque arm is employed, rigidly mounted to the rear axle housing and, through a rubber bushing, to the transmission at the front. Coil springs support the weight of the vehicle and ride control is provided by shock absorbers mounted on both sides of the rear of the axle housing. Some models are equipped with a rear stabilizer bar.

Both standard and heavy-duty shock absorbers are sealed during production. They contain a measured amount of fluid and cannot be adjusted, refilled or disassembled. If they are damaged or worn out, they must be replaced with new units.

8 Rear shock absorbers — removal and installation

1 Raise the rear of the vehicle and support it on jackstands. Be sure to support the rear axle with a second jack to avoid damage to the brake lines, track bar and driveshaft.

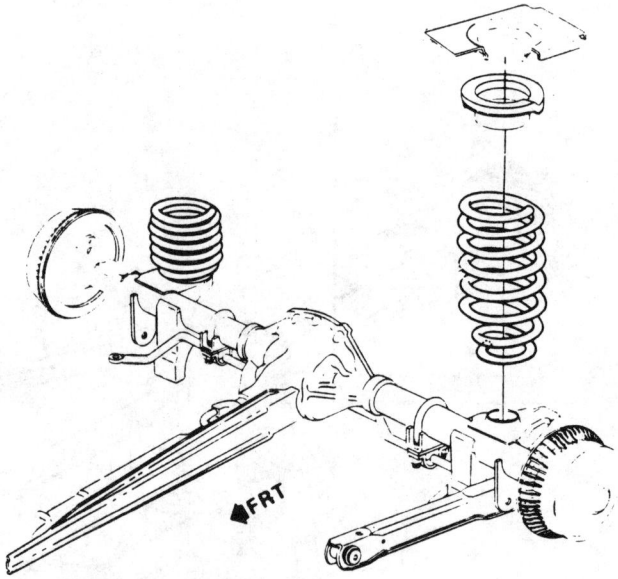

Fig. 11.9 Relationship of the rear spring, insulator and spring seat (Sec 9)

2 With the rear seat in the folded-down position, carefully lift the front edge of the rear panel carpeting and fold it back to expose the rear shock upper mounting nut.

3 Hold the top of the shock piston with one wrench while removing the upper mounting nut with another wrench (photo).

4 Remove the lower shock absorber mounting nut (photo) and remove the shock.

5 Installation is the reverse of the removal procedure.

6 Tighten all nuts and bolts to the specified torque.

9 Rear coil springs and insulators — removal and installation

1 Raise the rear of the vehicle and support it on jackstands. Support the rear axle assembly on a hydraulic jack.

2 Remove the track bar mounting bolt at the axle assembly (left side).

3 Loosen the track bar bolt at the body brace (right side).

4 Disconnect the rear brake hose clip from the vehicle underbody to allow additional axle drop. **Note:** *Never suspend the axle by the brake hose, as damage will occur.*

5 Remove the lower shock absorber mounting nut from both rear shocks.

6 On vehicles equipped with a 4-cylinder engine, remove the

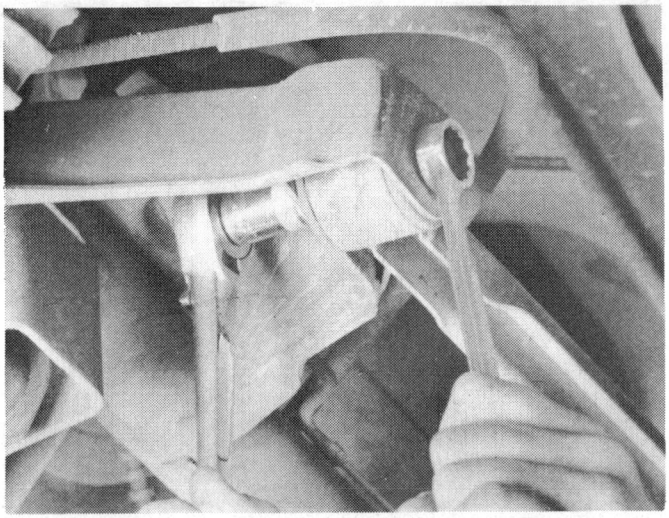

10.2 Removing the track bar at the rear axle mount

10.3 Removing the track bar at the body bracket

11.3 Removing the track bar brace at the left body brace

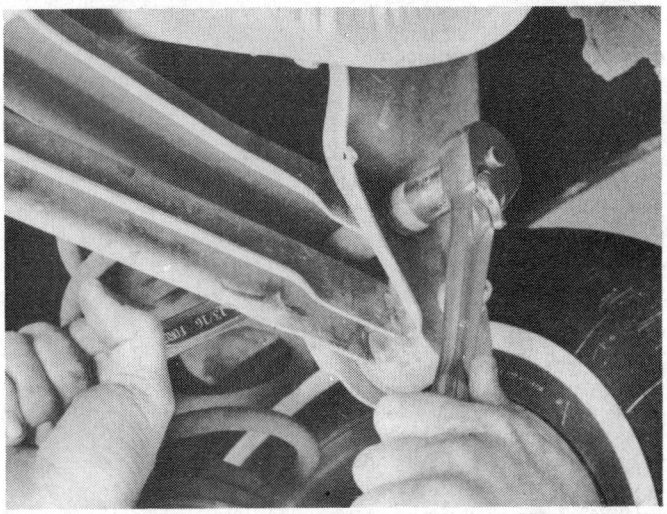

11.4 Removing the track bar brace at the right body bracket

driveshaft (refer to Chapter 8).
7 Carefully lower the jack under the rear axle assembly until the spring and/or insulator can be removed.
8 To install, position the spring and insulator in the spring seat and raise the hydraulic jack until the spring is normally compressed.
9 Attach the shocks to the rear axle and tighten the nuts to the specified torque.
10 Install the track bar mounting bolt at the axle and tighten the nut to the specified torque.
11 Tighten the track bar-to-body bracket nut to the specified torque.
12 Attach the brake hose clip to the underbody.
13 On 4-cylinder vehicles, install the driveshaft.
14 Remove the jackstands and lower the vehicle.

10 Track bar — removal and installation

1 Raise the rear of the vehicle and support it with jackstands under the axle housings.
2 Remove the track bar mounting bolt and nut at the rear axle (photo).
3 Remove the track bar mounting bolt and nut at the body bracket and remove the track bar (photo).
4 Installation is the reverse of the removal procedure. Tighten all fasteners to the specified torque.

11 Track bar brace — removal and installation

1 Raise the rear of the vehicle and support it with jackstands placed under the rear axle housings.
2 If equipped, remove the heat shield screws from the track bar brace.
3 Remove the three track bar brace-to-body brace bolts on the left side of the vehicle (photo).
4 Remove the nut and bolt attaching the track bar brace to the body bracket (right side) and remove the track bar brace (photo).
5 Installation is the reverse of the removal procedure. Tighten all fasteners to the specified torque.

12 Rear lower control arm — removal and installation

Note: *If both lower control arms are to be removed, remove and install them one at a time to prevent the axle assembly from rolling or slipping sideways, complicating the procedure.*
1 Raise the rear of the vehicle and place the rear axle housings securely on jackstands.
2 Remove the lower control arm-to-axle housing retaining bolt and nut (photo).
3 Remove the lower control arm-to-body brace bolt and nut and

12.2 Removing the lower control arm from the axle housing bracket

12.3 Removing the lower control arm from the front body brace

remove the control arm (photo).
4 Installation is the reverse of the removal procedure. Tighten all bolts to the specified torque.

13 Rear lower control arm bushing — replacement

Note: *This procedure requires the use of an arbor press. If you do not have access to one, take the lower control arm to a machine shop to have the bushing replaced.*

1 Refer to Section 12 and remove the control arm.
2 Refer to the accompanying illustration and, using the tools indicated, remove the bushing. The driver used (J-21465-8 in the illustration) must contact the outside diameter of the bushing outer sleeve.
3 To install the bushing, refer to the accompanying illustration and press the bushing into place, using the tools indicated.
4 Refer to Section 12 and install the lower control arm.

14 Torque arm — removal and installation

1 Remove both rear coil springs (refer to Section 9). **Note:** *The coil springs must be removed before the torque arm is removed to avoid rear axle forward twist, which could lead to vehicle damage.*
2 Remove the torque arm rear mounting bolts and nuts (photo).
3 Remove the torque arm front outer bracket bolts and remove the torque arm (photo).
4 To install, position the torque arm and loosely install the rear bolts.
5 Install the front torque arm bracket and tighten the bolts to the specified torque.
6 Tighten the rear torque arm bolts to the specified torque. **Caution:** *Do not overtighten these bolts, as it could lead to distortion of the differential case and subsequent pinion bearing damage.*
7 The remaining installation steps are the reverse of those for removal. Tighten all bolts to the specified torque.

15 Rear stabilizer bar — removal and installation

Note: *The rear stabilizer is an option and is not installed on all vehicles.*

1 Raise the vehicle and support it securely on jackstands.
2 Remove the link bolt, nut, grommets, spacer and retainers from each side of the stabilizer bar, noting their positions relative to each other to simplify installation.
3 Remove the insulator and bracket from each side of the stabilizer bar, noting their positions to simplify installation.

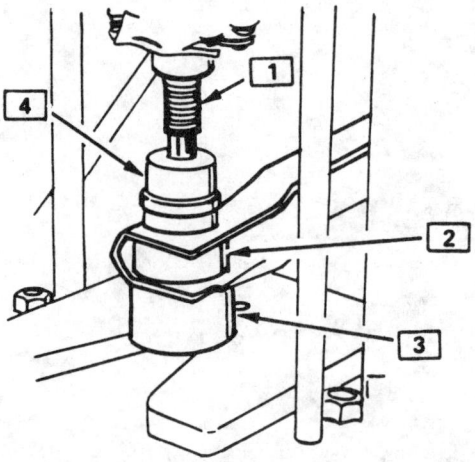

Fig. 11.10 Tools required for removal of the bushing from the rear control arm (Sec 13)

1 Press
2 J-23737
3 J-25317-2
4 J-21465-8

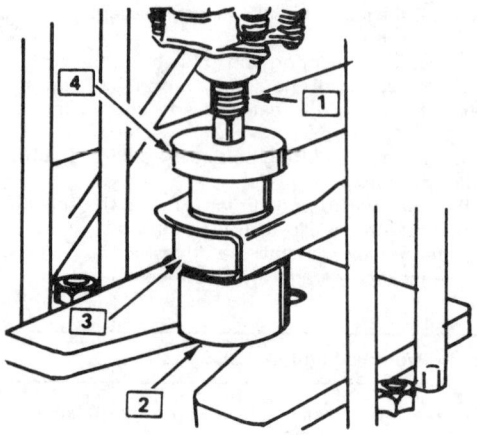

Fig. 11.11 Tools required for installation of the bushing in the rear control arm (Sec 13)

1 Press
2 J-25317-2
3 J-23737
4 J-25317-1

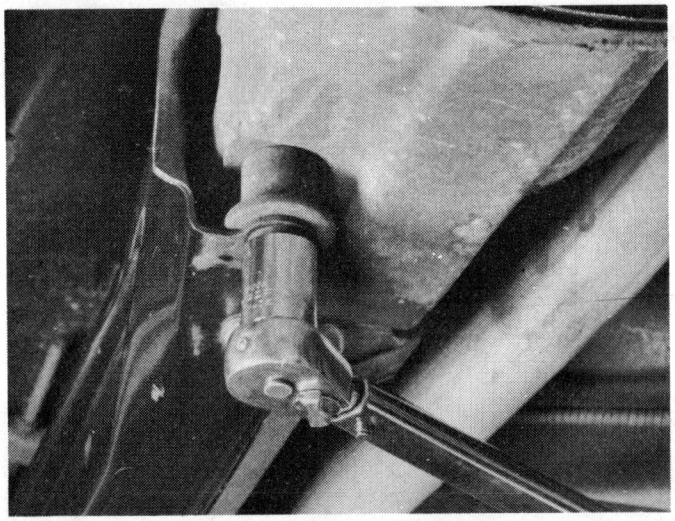

14.2 Removing the torque arm from the axle housing

14.3 Removing the torque arm outer bracket from the transmission

4 Remove the stabilizer bar.
5 Installation is the reverse of the removal procedure, noting the following.
6 Apply multi-purpose grease to the insulator mounting points on the stabilizer bar before installing the insulators and brackets.
7 Make sure the slits in the insulators are toward the front of the vehicle.
8 Be sure the stabilizer is centered in the brackets before tightening the mounting bolts.
9 Tighten all bolts to the specified torque.

16 Steering system — general information

The steering linkage connects both front wheels to the steering gear through a Pitman arm. The right and left tie-rods are attached to the steering knuckles (arms) and to the relay rod by balljoint studs. The left end of the relay rod is supported by the Pitman arm, which is driven by the steering gear. The right end of the relay rod is supported by the idler arm, which pivots on a support bolted to the frame rail. The Pitman arm and idler arm move in symmetrical arcs and remain parallel

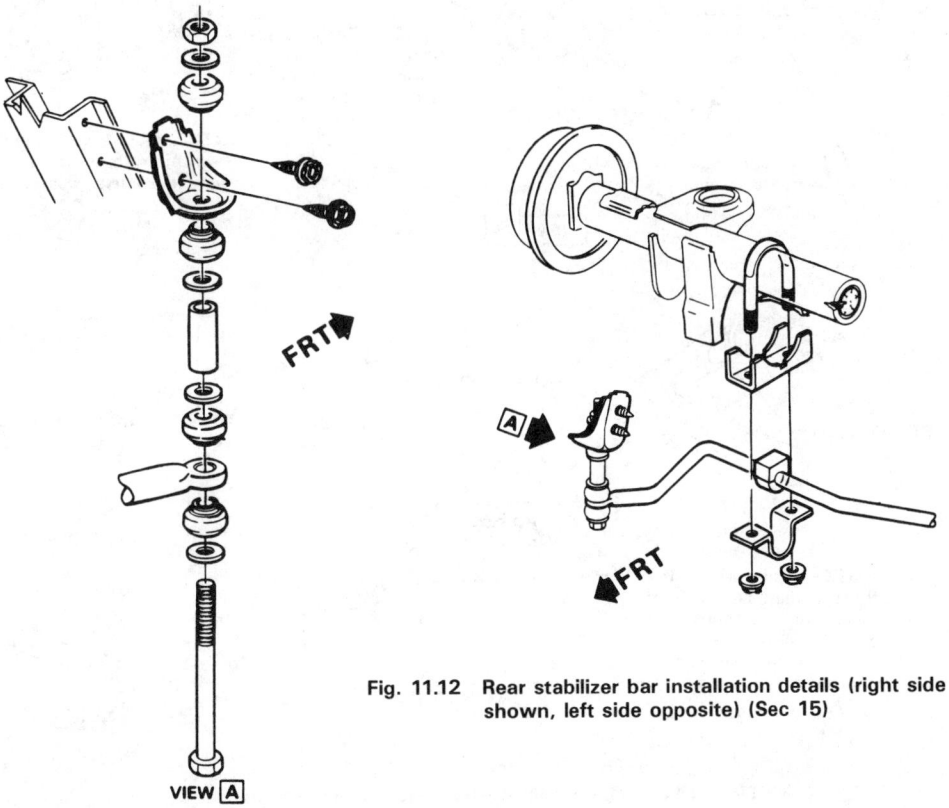

Fig. 11.12 Rear stabilizer bar installation details (right side shown, left side opposite) (Sec 15)

to each other.

The tie-rods and tie-rod ends may be removed and installed by carefully following the procedures set forth in the Section that follows; however, the replacement and alignment of the relay rod, Pitman arm and/or the idler arm requires the use of special tools and alignment equipment not normally in the possession of the home mechanic. These procedures, therefore, should be done by a GM dealer or a front-end alignment shop.

17 Tie-rods — removal and installation

1 Raise the vehicle and place it securely on jackstands.
2 Remove the cotter pins and nuts from the relay rod and steering knuckle ends of the tie-rod (photos).
3 Using the special tool as shown in the accompanying illustration, remove the outer end of the tie-rod from the steering knuckle. **Note:** *Do not use a 'pickle fork' type tool to remove the tie-rod end, as seal damage may result.*
4 Using the same tool referred to in Step 3, remove the inner end of the tie-rod from the relay rod.
5 If the tie-rods are to be disassembled and reassembled using the same tie-rod ends, mark the threads of the tie-rod ends with white paint so they can be reassembled to the original length.
6 If new tie-rod ends are to be installed, carefully measure the overall length of the tie-rod assembly before removing the tie-rod ends so the assemblies can be adjusted to the original length.

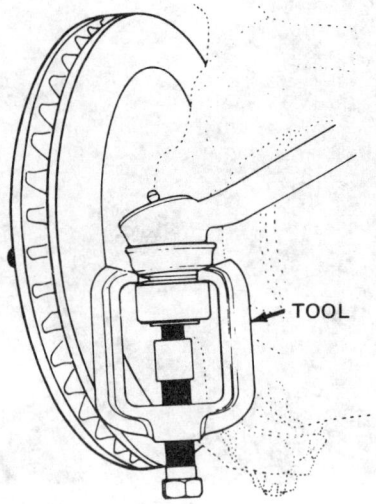

Fig. 11.13 Removing the outer end of the tie-rod from the steering knuckle (use this type of tool only — 'pickle fork' type would damage the seals) (Sec 17)

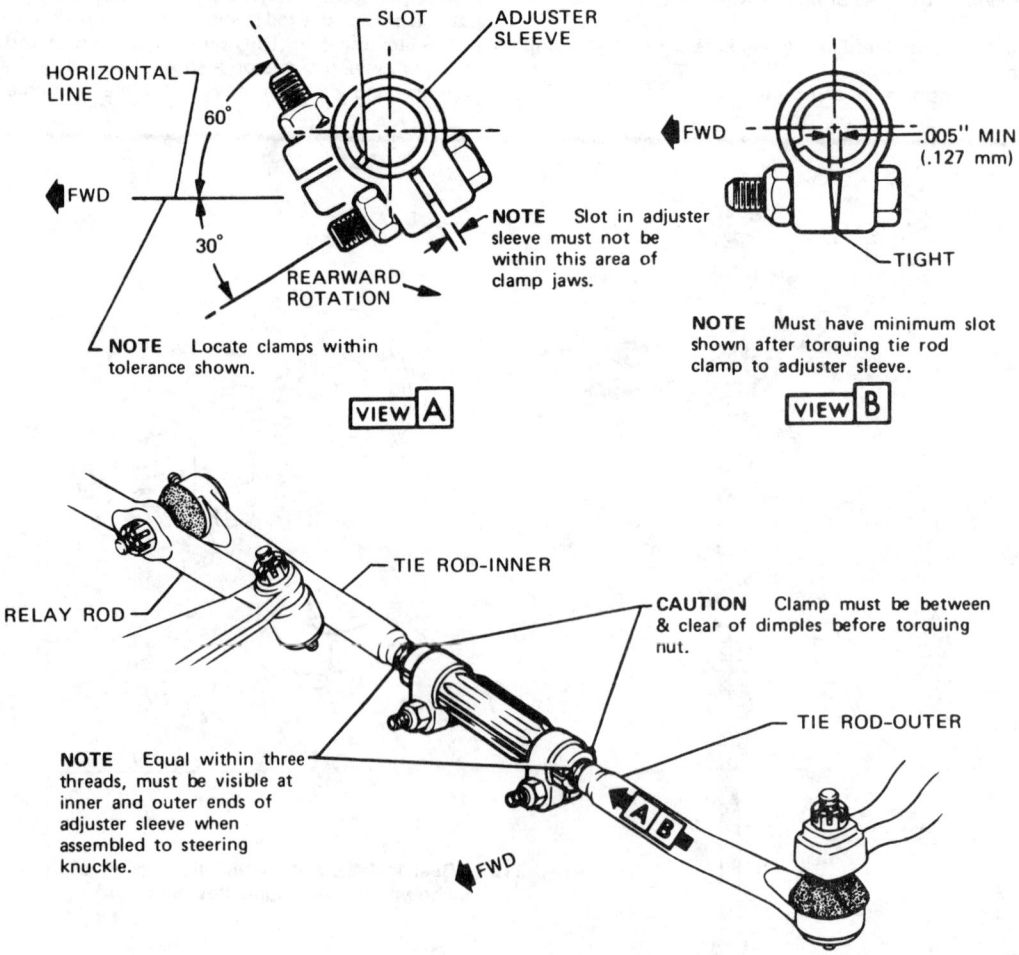

Fig. 11.14 Relationship of the tie-rod clamp and sleeve (Sec 17)

7 To remove the tie-rod ends, loosen the clamp nuts and unscrew the tie-rod ends.
8 If new tie-rod ends are used, lubricate the threads with multipurpose chassis grease and install the ends, making sure both ends are threaded an equal distance into the tie-rod and that the overall length is the same as recorded in Step 6. If the old tie-rod ends were removed and are being reinstalled, apply grease to the threads and reinstall the ends to the depth marked by the white paint (Step 5). Install the clamps but do not tighten them at this time.
9 Carefully remove all dirt from the threads of the balljoint studs and nuts. If the threads are not clean, binding may occur and the studs may turn in the tie-rod ends when attempting to tighten the nuts. Also make sure the tapered surfaces of the balljoints are clean and that the seals are installed on the studs.
10 Attach the assembled tie-rod to the steering arm and relay rod.
11 Tighten the retaining nuts to the specified torque, then, if necessary, tighten the nuts farther to permit insertion of the cotter pins.
12 Refer to the accompanying illustration and adjust the tie-rod clamps as shown, then tighten the clamps to the specified torque.
13 Lubricate the tie-rod ends with the specified grease.
14 Remove the jackstands and lower the vehicle.
15 A front-end alignment must now be performed by a GM dealer or an alignment shop.

18 Steering knuckle — removal and installation

1 Raise the vehicle and place it securely on jackstands under the side frame rails.
2 Remove the wheel and tire.
3 Remove the brake hose bracket from the strut assembly.
4 Remove the caliper retaining bolts and hang the caliper out of the way on a piece of wire. *Do not allow the caliper to hang by the brake hose.*
5 Remove the hub and disc assembly (refer to Chapter 1, if necessary).
6 Unbolt and remove the splash shield.
7 Disconnect the tie-rod from the steering knuckle (refer to Section 17).
8 Place a hydraulic jack under the lower control arm to support the control arm.
9 Disconnect the balljoint from the steering knuckle (refer to Section 4).
10 Remove the two bolts retaining the strut to the steering knuckle and remove the knuckle.
11 Installation is the reverse of the removal procedure. Tighten all bolts to the specified torque.

17.2A Removing the nut retaining the tie-rod to the steering knuckle

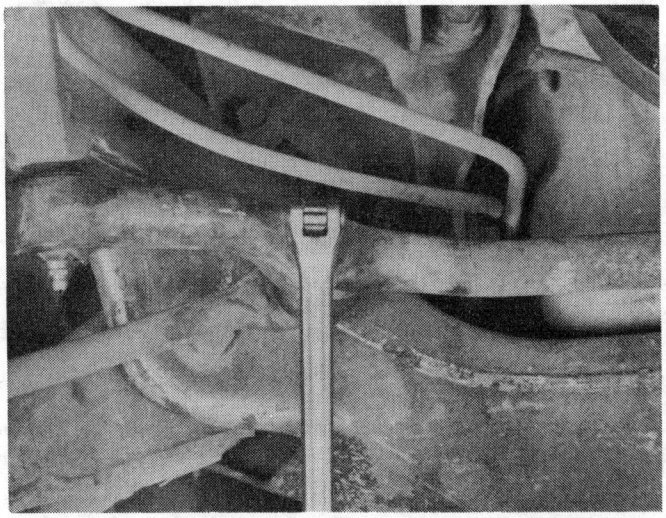

17.2B Removing the nut retaining the tie-rod to the relay rod

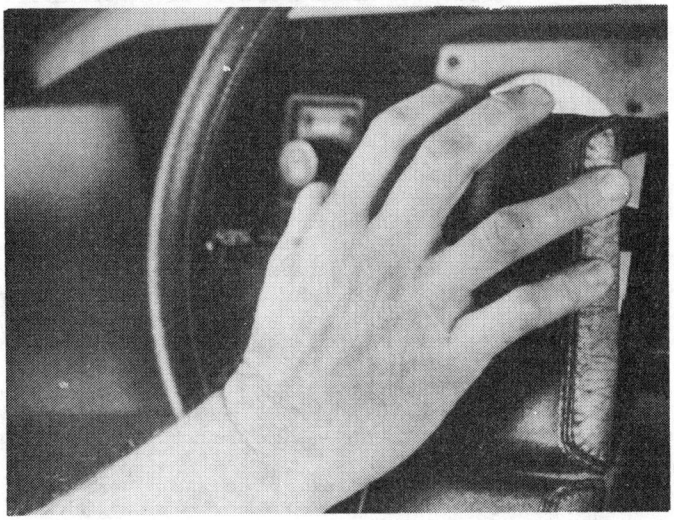

19.1 Removing the horn pad from the steering wheel

19.2 Removing the retaining clip from the steering wheel nut

19.3 Removing the steering wheel nut

19.5 Using a puller to remove the steering wheel

19 Steering wheel – removal and installation

Warning: *Later models (1990 on) are equipped with a driver's side Supplemental Inflatable Restraint (SIR) (airbag) installed in the center of the steering wheel. The SIR must be removed before the steering wheel is removed (see Chapter 13 for additional information).*

1 Pull off the horn pad by pushing down on it from the top and pulling it away from the steering wheel (photo). On models so equipped, remove the two screws securing the steering wheel shroud from the underside of the steering wheel, then lift the steering wheel shroud and horn contact lead assembly from the steering wheel.
2 Remove the retaining clip from the top of the steering wheel nut (photo).
3 Using a socket and breaker bar, remove the steering wheel nut (photo).
4 Thread the puller anchor screws into the threaded holes provided in the steering wheel.
5 Turn the center bolt of the puller clockwise with a wrench to remove the wheel (photo).
6 To install the wheel, make sure the turn signal lever is in the neutral position, then position the steering wheel on the shaft, making sure the flats on the wheel and shaft are aligned.
7 Install the steering wheel nut and tighten it to the specified torque.
8 The remaining installation steps are the reverse of removal.

20 Power steering system — general information

With the optional power steering gear, hydraulic pressure is generated in an engine-driven vane type pump and supplied through hoses to the steering box spool valve. The valve is normally positioned in the neutral mode by a torsion bar but when the steering wheel is turned and force is applied to the steering shaft, the spool moves in relation to the body and allows oil to flow to the appropriate side of the piston nut. The greater the movement of the steering wheel, the greater the hydraulic pressure which is applied and therefore the greater the power assistance given to the drive.

The Sections which follow concern procedures that can be performed by the home mechanic; however, due to the requirements for special tools and diagnostic procedures, major repairs to the power steering system should be performed by a dealer or reputable repair shop.

21 Power steering system — bleeding

1 This is not a routine operation and normally will only be required when the system has been dismantled and reassembled.
2 Fill the reservoir to the correct level with fluid of the recommended type and allow it to remain undisturbed for at least two (2) minutes.
3 Start the engine and run it for two or three seconds only. Check the reservoir and add more fluid as necessary.
4 Repeat the operations described in the preceding paragraph until the fluid level remains constant.
5 Raise the front of the vehicle until the wheels are clear of the ground.
6 Start the engine and increase the speed to about 1500 rpm. Now turn the steering wheel gently from stop-to-stop. Check the reservoir fluid level.
7 Lower the vehicle to the ground and, with the engine still running, move the vehicle forward sufficiently to obtain full right lock followed by full left lock. Recheck the fluid level. If the fluid in the reservoir is extremely foamy, allow the vehicle to stand for a few minutes with the engine switched off and then repeat the previous operations. At the same time, check the belt tightness and check for a bent or loose pulley. Check also to make sure the power steering hoses are not touching any other part of the vehicle, especially sheet metal or the exhaust manifold.
8 The procedures above will normally remedy an extreme foam condition and/or an objectionably noisy pump (low fluid level and/or air in the power steering fluid are the leading causes of this condition). If, however, either or both conditions persist after a few trials, the power steering system will have to be checked by a dealer. Do not drive the vehicle until the condition(s) have been remedied.

22 Power steering gear — removal and installation

1 Disconnect the cable from the negative battery terminal.
2 If equipped, remove the coupling shield.
3 Disconnect the pressure and return hoses attached to the power steering gear assembly.
4 Plug the ends of the disconnected hoses and the holes in the power steering housing to prevent contamination.
5 Remove the nuts, lock washers and bolts at the steering coupling-to-steering shaft flange.
6 Remove the Pitman arm locknut and washer. Mark the position of the Pitman arm in relation to the shaft and disconnect the Pitman arm with a puller.
7 Remove the bolts securing the steering gear to the frame and separate it from the vehicle.
8 When installing the gear, place it into position so that the coupling mounts properly to the flanged end of the steering shaft. Secure the gear to the frame, install the washers and bolts, then tighten the bolts to the specified torque.
9 Secure the steering coupling to the flanged end of the column with the lock washers and nuts. Tighten the nuts.
10 Install the Pitman arm and tighten the nut.
11 Connect the negative battery cable (and the coupling shield, if applicable).

24.3 Removing the horn contact bushing and spring

24.4 Removing the plastic cover from the steering column

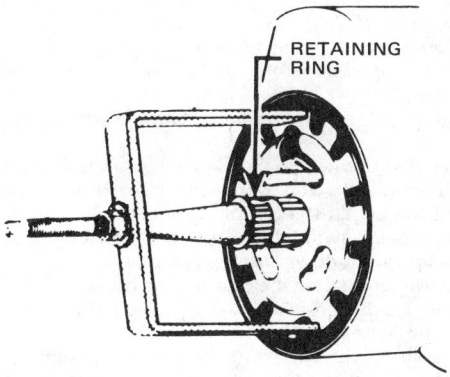

Fig. 11.15 This tool is used to depress the steering shaft lock in order to remove or install the lock retaining snap-ring (Sec 24)

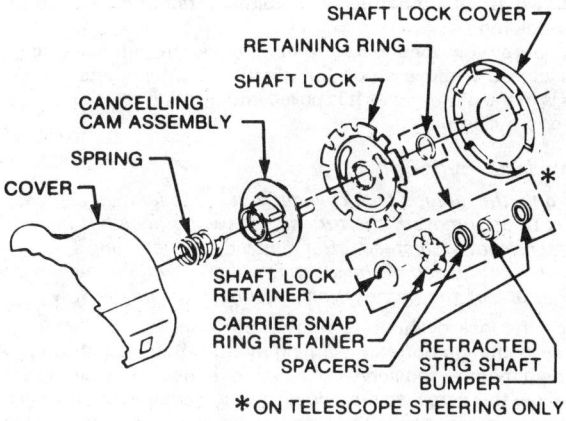

Fig. 11.16 Steering column shaft lock assembly — exploded view (Sec 24)

23 Power steering pump — removal and installation

1 Disconnect the hydraulic hoses from the pump and keep them in the raised position to prevent the fluid from leaking out until they can be plugged.
2 Remove the pump drivebelt by loosening the pump mounts and pushing it in toward the engine.
3 Unscrew and remove the pump mounting bolts and braces and remove the pump.
4 To remove the pump pulley, a puller will be needed.
5 Installation is the reverse of removal.
6 Prime the pump by turning the pulley in the reverse direction to that of normal rotation (counterclockwise as viewed from the front) until air bubbles cease to emerge from the fluid when observed through the reservoir filler cap.
7 Install the drivebelt and adjust it as described in Chapter 1.
8 Bleed the system as described in Section 21.

24 Steering column switches — removal and installation

Note: *Removal and installation of the switches included in this Section are presented in a progressive manner. Those nearest the top of the steering column are generally covered first.*

Canceling cam

1 Disconnect the cable from the negative battery terminal.
2 Remove the steering wheel (refer to Section 19).
3 Remove the horn contact bushing and spring by pulling it straight out of the recess in the steering column cover plate (photo).
4 Using a screwdriver in the slots provided, pry the plastic cover out of the steering column (photo).
5 Remove the shaft lock from the steering column. On vehicles with standard steering columns, it is held in place with a snap-ring which fits into a groove in the steering shaft. The shaft lock must be depressed to relieve pressure on the snap-ring. A special U-shaped tool which fits on the shaft is used to depress the shaft lock as the snap-ring is removed from the groove (refer to the accompanying illustration). On models with telescopic steering wheel assemblies, remove the spacers, retracted steering shaft bumper, carrier snap-ring retainer and the shaft lock retainer.
6 Pull out and remove the canceling cam assembly and upper bearing preload spring.
7 If no other components are to be removed from the steering column, reverse the above steps for installation, making sure the turn signal switch is in the neutral position before installing the upper bearing preload spring and cenceling cam. Always use a new snap-ring for the shaft lock.

Turn signal switch

8 Remove the turn signal lever attaching screw and withdraw the turn signal lever from the side of the column.
9 Push in on the hazard warning knob and unscrew the knob from the threaded shaft.
10 Remove the three turn signal assembly mounting screws.
11 Pull the switch wiring connector out of the bracket on the steering column jacket. Tape the connector terminals to prevent damage.

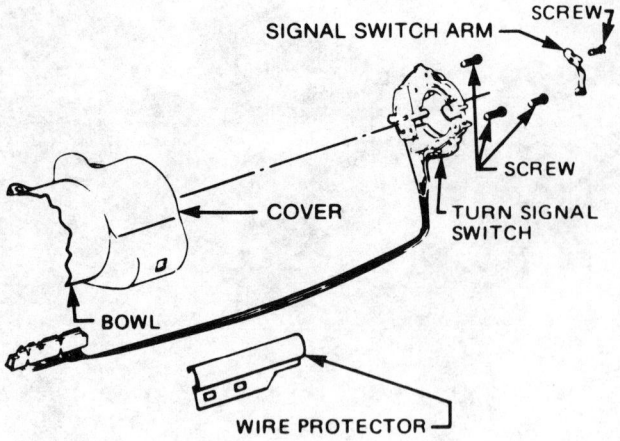

Fig. 11.17 Turn signal switch mounting details (Sec 24)

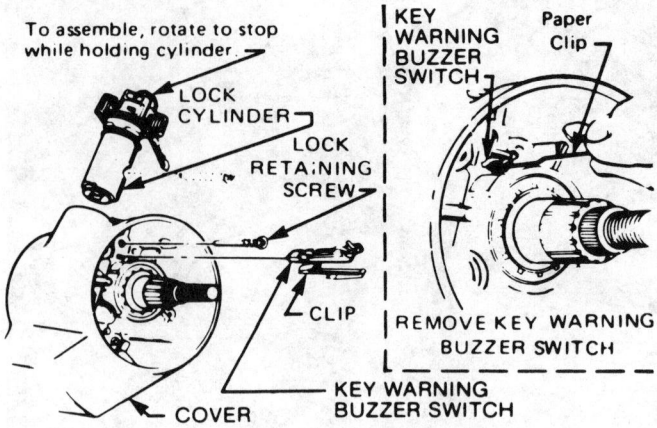

Fig. 11.18 Ignition lock cylinder installation details (Sec 24)

Feed the wiring connector up through the column support bracket and pull the switch, wiring harness and connectors out of the top of the steering column.

12 If no other components are to be removed from the steering column, reverse the above steps for installation. Make sure the wiring harness is in the protector as it is pulled into position and that the warning knob is pulled out.

Ignition lock cylinder

Note: *If only the ignition lock cylinder and/or the ignition key warning switch is to be removed, the turn signal switch does not have to be completely removed. Instead, after removing the retaining screws, pull the switch over the end of the steering column (without having pulled the wiring out of the harness) and secure it out of the way.*

13 Place the lock cylinder in the Run position.

14 Remove the ignition key warning switch (Steps 21 through 23).

15 Using a magnetized screwdriver, remove the lock retaining screw. Do not allow this screw to drop down into the column or you will have to completely disassemble the steering column to retrieve the screw.

16 Pull the lock cylinder out of the side of the steering column.

17 To install it, rotate the lock cylinder set and align the cylinder key with the keyway in the steering column housing.

18 Push the lock all the way in and install the retaining screw.

19 Install the ignition key warning switch (Steps 24 and 25).

20 Reverse the removal steps to complete the installation.

Ignition key warning switch

21 The key warning switch may be removed independently of the ignition key lock, or it may be removed at the same time if the key lock is to be removed.

22 If the ignition lock cylinder is still intact, set the key to the Run position.

23 Use a piece of stiff wire (a paper clip will work fine) to remove the switch. Make a hook at the end of the wire and attach the hooked end into the loop of the clip at the top of the switch. Pull up on the wire and remove the clip and switch together from the steering column. Do not allow the clip to fall down into the column.

24 If the lock cylinder is still in the column, the buzzer switch actuating button (on the lock cylinder) must be depressed before the new warning switch can be installed.

25 Install the switch with the contacts toward the upper end of the steering column and with the formed end of the spring clip at the lower end of the switch. Reinstall the remaining components.

Note: *Further dismantling of the steering column is not advised. If other switches attached to the column require servicing, it should be done by a dealer or repair facility.*

Chapter 12 Body

Contents

Body — maintenance 2
Body repair — major damage 7
Body repair — minor damage 6
Door connecting rods and locking rods — removal and installation 15
Door lock assembly — replacement 17
Door lock cylinder — removal and installation 16
Door lock knob — removal and installation 12
Door outside handle assembly — removal and installation . . . 19
Door remote control spring clips — disengagement and engagement 14
Door trim panel — removal and installation 13
Door window glass replacement 20
Door window regulator — adjustment 21
General information 1
Hinges and locks — maintenance 5
Hood — removal and installation 8
Hood damper — replacement 9
Hood latch release cable — removal and installation 10
Power door lock actuator — removal and installation 18
Rear hatch — removal, installation and adjustment 11
Upholstery and carpets — maintenance 3
Vinyl trim — maintenance 4
Windshield replacement 22

Specifications

Torque specifications

	Ft-lbs
Hinge-to-rear hatch/glass bolts	11
Hinge-to-body bolts	15 to 20
Gas damper-to-rear hatch assembly bolts	4 to 5
Door lock assembly screws	7 to 9
Power door lock actuator mounting bolts	8 to 10

1 General information

The Camaro, for the years covered by this manual, was available in three models: Sport Coupe, Berlinetta and Z28, all based on a two-door body style. Differences between the various models are noted where appropriate in the service procedures within this Chapter.

Camaro bodies are of unitized construction in which the body is designed to provide vehicle rigidity so that a separate frame is not necessary. Front and rear frame side rails integral with the body support the front end sheet metal, front and rear suspension systems and other mechanical components. Due to this type of construction, it is very important that, in the event of collision damage, the underbody be thoroughly checked by a facility with the proper equipment to do so.

Component replacement and repairs possible for the home mechanic are included in this Chaper.

2 Body — maintenance

1 The condition of your vehicle's body is very important, because it is on this that the second-hand value will mainly depend. It is much more difficult to repair a neglected or damaged body than it is to repair mechanical components. The hidden areas of the body, such as the fender wells, the frame, and the engine compartment, are equally important, although obviously do not require as frequent attention as the rest of the body.

2 Once a year, or every 12 000 miles, it is a good idea to have the underside of the body and the frame steam cleaned. All traces of dirt and oil will be removed and the underside can then be inspected carefully for rust, damaged brake lines, frayed electrical wiring, damaged cables, and other problems. The front suspension components should be greased after completion of this job.

3 At the same time, clean the engine and the engine compartment using either a steam cleaner or a water soluble degreaser.

4 The fender wells should be given particular attention, as undercoating can peel away and stones and dirt thrown up by the tires can cause the paint to chip and flake, allowing rust to set in. If rust is found, clean down to the bare metal and apply an anti-rust paint.

5 The body should be washed once a week (or when dirty). Wet the vehicle thoroughly to soften the dirt, then wash it down with a soft sponge and plenty of clean soapy water. If the surplus dirt is not washed off very carefully, it will in time wear down the paint.

6 Spots of tar or asphalt coating thrown up from the road should be removed with a cloth soaked in solvent.

7 Once every six months, give the body and chrome trim a thorough wax job. If a chrome cleaner is used to remove rust from any of the vehicle's plated parts, remember that the cleaner also removes part of the chrome, so use it sparingly.

These photos illustrate a method of repairing simple dents. They are intended to supplement *Body repair - minor damage* in this Chapter and should not be used as the sole instructions for body repair on these vehicles.

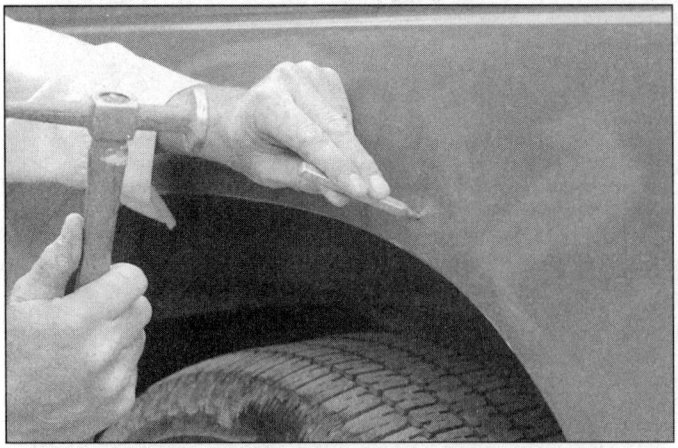

1 If you can't access the backside of the body panel to hammer out the dent, pull it out with a slide-hammer-type dent puller. In the deepest portion of the dent or along the crease line, drill or punch hole(s) at least one inch apart . . .

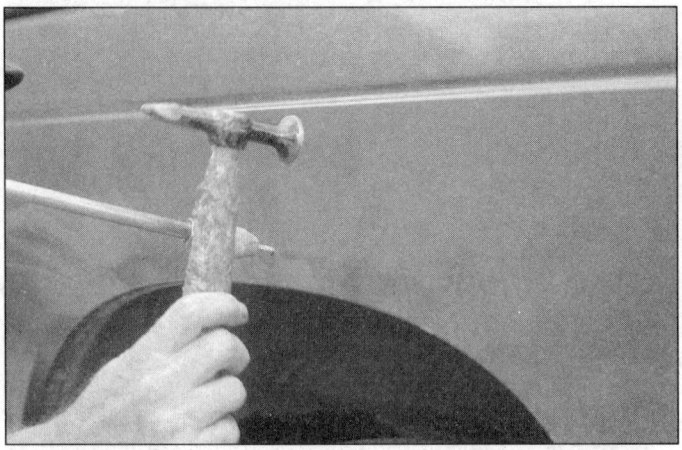

2 . . . then screw the slide-hammer into the hole and operate it. Tap with a hammer near the edge of the dent to help 'pop' the metal back to its original shape. When you're finished, the dent area should be close to its original contour and about 1/8-inch below the surface of the surrounding metal

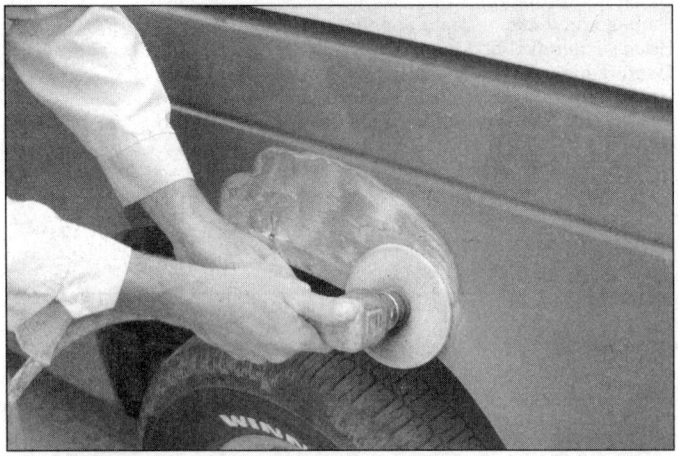

3 Using coarse-grit sandpaper, remove the paint down to the bare metal. Hand sanding works fine, but the disc sander shown here makes the job faster. Use finer (about 320-grit) sandpaper to feather-edge the paint at least one inch around the dent area

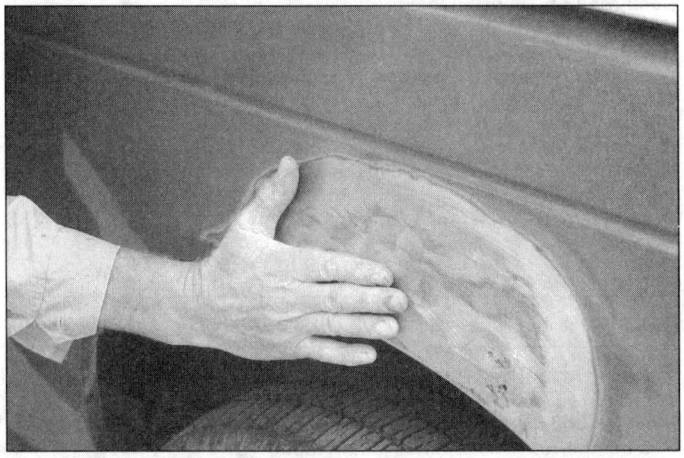

4 When the paint is removed, touch will probably be more helpful than sight for telling if the metal is straight. Hammer down the high spots or raise the low spots as necessary. Clean the repair area with wax/silicone remover

5 Following label instructions, mix up a batch of plastic filler and hardener. The ratio of filler to hardener is critical, and, if you mix it incorrectly, it will either not cure properly or cure too quickly (you won't have time to file and sand it into shape)

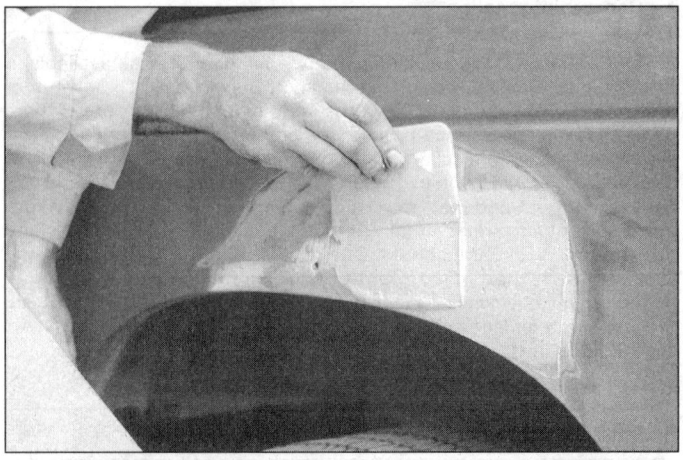

6 Working quickly so the filler doesn't harden, use a plastic applicator to press the body filler firmly into the metal, assuring it bonds completely. Work the filler until it matches the original contour and is slightly above the surrounding metal

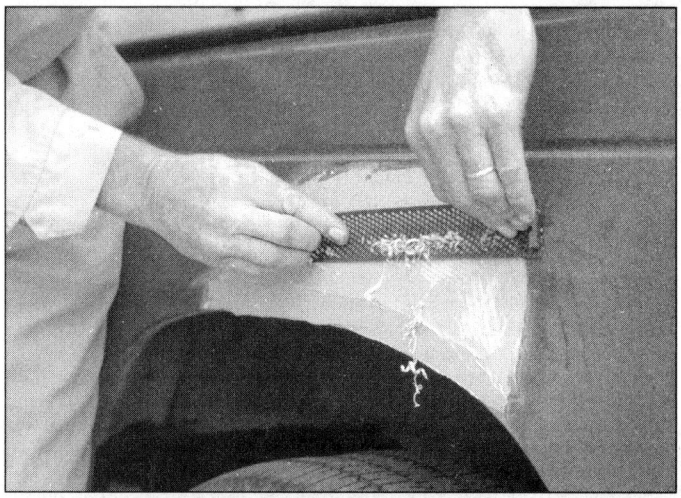

7 Let the filler harden until you can just dent it with your fingernail. Use a body file or Surform tool (shown here) to rough-shape the filler

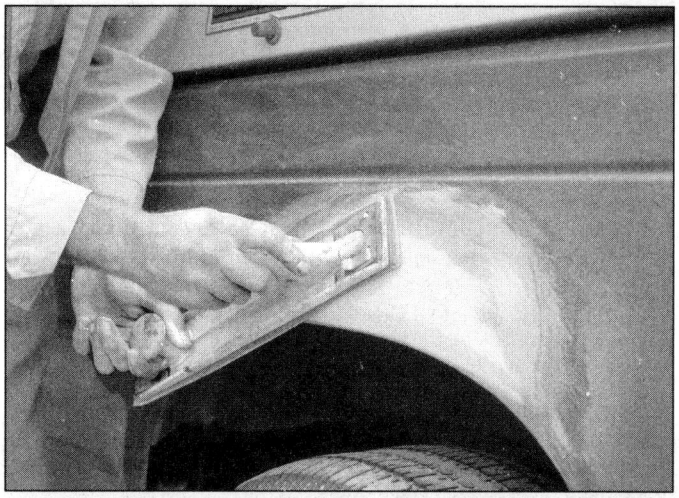

8 Use coarse-grit sandpaper and a sanding board or block to work the filler down until it's smooth and even. Work down to finer grits of sandpaper - always using a board or block - ending up with 360 or 400 grit

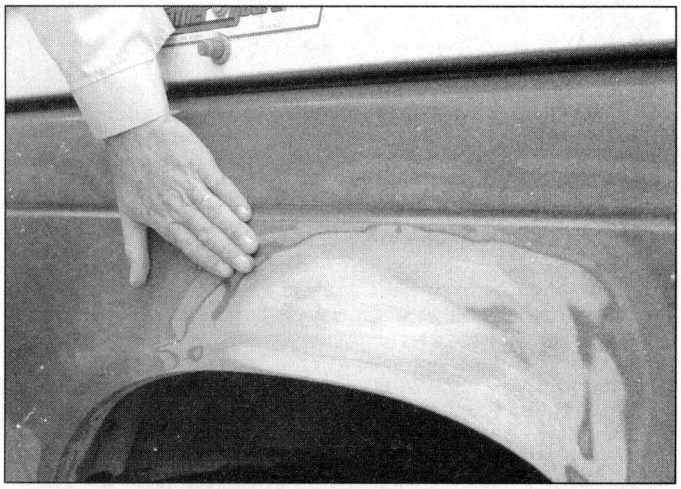

9 You shouldn't be able to feel any ridge at the transition from the filler to the bare metal or from the bare metal to the old paint. As soon as the repair is flat and uniform, remove the dust and mask off the adjacent panels or trim pieces

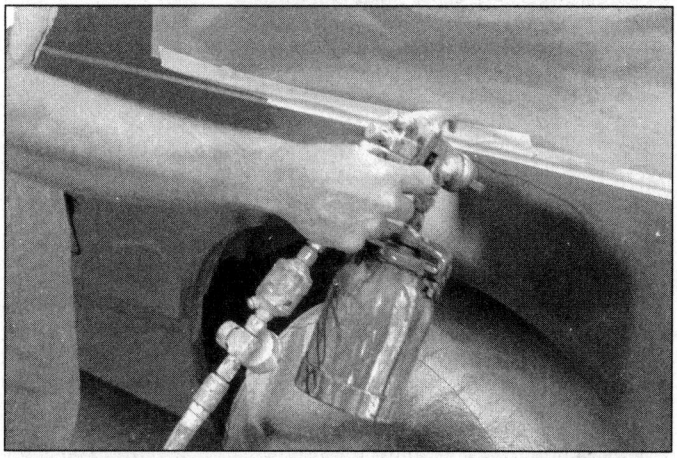

10 Apply several layers of primer to the area. Don't spray the primer on too heavy, so it sags or runs, and make sure each coat is dry before you spray on the next one. A professional-type spray gun is being used here, but aerosol spray primer is available inexpensively from auto parts stores

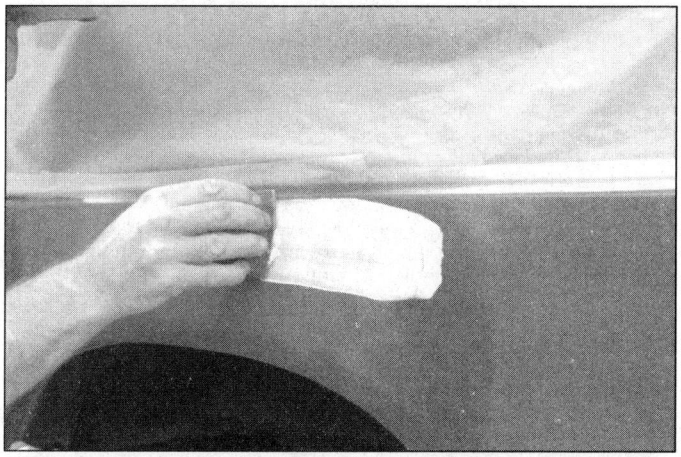

11 The primer will help reveal imperfections or scratches. Fill these with glazing compound. Follow the label instructions and sand it with 360 or 400-grit sandpaper until it's smooth. Repeat the glazing, sanding and respraying until the primer reveals a perfectly smooth surface

12 Finish sand the primer with very fine sandpaper (400 or 600-grit) to remove the primer overspray. Clean the area with water and allow it to dry. Use a tack rag to remove any dust, then apply the finish coat. Don't attempt to rub out or wax the repair area until the paint has dried completely (at least two weeks)

3 Upholstery and carpets — maintenance

1 Every three months, remove the carpets or mats and clean the interior of the vehicle (more frequently if necessary). Vacuum the upholstery and carpets to remove loose dirt and dust.
2 If the upholstery is soiled, apply upholstery cleaner with a damp sponge and wipe it off with a clean, dry cloth.

4 Vinyl trim — maintenance

Vinyl trim should not be cleaned with detergents, caustic soaps or petroleum-based cleaners. Plain soap and water or a mild vinyl cleaner is best for stains. Test a small area for color fastness. Bubbles under the vinyl can be corrected by piercing them with a pin and then working the air out.

5 Hinges and locks — maintenance

Every 3000 miles or three months, the door, hood and rear hatch hinges and locks should be lubricated with a few drops of oil. The door and rear hatch striker plates should also be given a thin coat of grease to reduce wear and ensure free movement.

6 Body repair — minor damage

See photo sequence

Repair of minor scratches

If the scratch is very superficial and does not penetrate to the metal of the body, repair is very simple. Lightly rub the scratched area with a fine rubbing compound to remove loose paint and built-up wax. Rinse the area with clean water.

Apply touch-up paint to the scratch, using a small brush. Continue to apply thin layers of paint until the surface of the paint in the scratch is level with the surrounding paint. Allow the new paint at least two weeks to harden, then blend it into the surrounding paint by rubbing with a very fine rubbing compound. Finally, apply a coat of wax to the scratch area.

If the scratch has penetrated the paint and exposed the metal of the body, causing the metal to rust, a different repair technique is required. Remove all loose rust from the bottom of the scratch with a pocket knife, then apply rust-inhibiting paint to prevent the formation of rust in the future. Using a rubber or nylon applicator, coat the scratched area with glaze-type filler. If required, the filler can be mixed with thinner to provide a very thin paste, which is ideal for filling narrow scratches. Before the glaze filler in the scratch hardens, wrap a piece of smooth cotton cloth around the tip of a finger. Dip the cloth in thinner and then quickly wipe it along the surface of the scratch. This will ensure that the surface of the filler is slightly hollow. The scratch can now be painted over as described earlier in this section.

Repair of dents

When repairing dents, the first job is to pull the dent out until the affected area is as close as possible to its original shape. There is no point in trying to restore the original shape completely as the metal in the damaged area will have stretched on impact and cannot be restored to its original contours. It is better to bring the level of the dent up to a point which is about 1/8-inch below the level of the surrounding metal. In cases where the dent is very shallow, it is not worth trying to pull it out at all.

If the back side of the dent is accessible, it can be hammered out gently from behind using a soft-faced hammer. While doing this, hold a block of wood firmly against the opposite side of the metal to absorb the hammer blows and prevent the metal from being stretched out.

If the dent is in a section of the body which has double layers, or some other factor that makes it inaccessible from behind, a different technique is required. Drill several small holes through the metal inside the damaged area, particularly in the deeper sections. Screw long, self-tapping screws into the holes just enough for them to get a good grip in the metal. Now the dent can be pulled out by pulling on the protruding heads of the screws with locking pliers.

The next stage of repair is the removal of paint from the damaged area and from an inch or so of the surrounding metal. This is easily done with a wire brush or sanding disk in a drill motor, although it can be done just as effectively by hand with sandpaper. To complete the preparation for filling, score the surface of the bare metal with a screwdriver or the tang of a file (or drill small holes in the affected area). This will provide a very good grip for the filler material. To complete the repair, see the Section on filling and painting.

Repair of rust holes or gashes

Remove all paint from the affected area and from an inch or so of the surrounding metal using a sanding disk or wire brush mounted in a drill motor. If these are not available, a few sheets of sandpaper will do the job just as effectively. With the paint removed, you will be able to determine the severity of the corrosion and decide whether to replace the whole panel, if possible, or repair the affected area. New body panels are not as expensive as most people think and it is often quicker to install a new panel than to repair large areas of rust.

Remove all trim pieces from the affected area (except those which will act as a guide to the original shape of the damaged body, i.e. headlight shells, etc.). Then, using metal snips or a hacksaw blade, remove all loose metal and any other metal that is badly affected by rust. Hammer the edges of the hole in to create a slight depression for the filler material.

Wire brush the affected area to remove the powdery rust from the surface of the metal. If the back of the rusted area is accessible, treat it with rust-inhibiting paint.

Before filling is done, block the hole in some way. This can be done with sheet metal riveted or screwed into place, or by stuffing the hole with wire mesh.

Once the hole is blocked off, the affected area can be filled and painted (see the following sub-section on filling and painting).

Filling and painting

Many types of body fillers are available, but generally speaking, body repair kits which contain filler paste and a tube of resin hardener are best for this type of repair work. A wide, flexible plastic or nylon applicator will be necessary for imparting a smooth and contoured finish to the surface of the filler material.

Mix up a small amount of filler on a clean piece of wood or cardboard (use the hardener sparingly). Follow the manufacturer's instructions on the package, otherwise the filler will set incorrectly.

Using the applicator, apply the filler paste to the prepared area. Draw the applicator across the surface of the filler to achieve the desired contour and to level the filler surface. As soon as a contour that approximates the original one is achieved, stop working the paste. If you continue, the paste will begin to stick to the applicator. Continue to add thin layers of filler paste at 20-minute intervals until the level of the filler is just above the surrounding metal.

Once the filler has hardened, the excess can be removed with a body file. From then on, progressively finer grades of sandpaper should be used, starting with a 180-grit paper and finishing with 600-grit wet-or-dry paper. Always wrap the sandpaper around a flat rubber or wooden block, otherwise the surface of the filler will not be completely flat. During the sanding of the filler surface, the wet-or-dry paper should be periodically rinsed in water. This will ensure that a very smooth finish is produced in the final stage.

At this point, the repair area should be surrounded by a ring of bare metal, which in turn should be encircled by the finely feathered edge of good paint. Rinse the repair area with clean water until all of the dust produced by the sanding operation is gone.

Spray the entire area with a light coat of primer. This will reveal any imperfections in the surface of the filler. Repair the imperfections with fresh filler paste or glaze filler and once more smooth the surface with sandpaper. Repeat this spray-and-repair procedure until you are satisfied that the surface of the filler and the feathered edge of the paint are perfect. Rinse the area with clean water and allow it to dry completely.

The repair area is now ready for painting. Spray painting must be carried out in a warm, dry, windless and dust-free atmosphere. These conditions can be created if you have access to a large indoor work area, but if you are forced to work in the open, you will have to pick the day very carefully. If you are working indoors, dousing the floor in the work area with water will help settle the dust which would otherwise be in the air. If the repair area is confined to one body panel, mask off the surrounding panels. This will help minimize the effects of a slight mismatch in paint color. Trim pieces such as chrome strips, door handles, etc., will also need to be masked off or removed. Use masking tape and several thicknesses of newspaper for the masking operations.

Before spraying, shake the paint can thoroughly, then spray a test area until the spray painting technique is mastered. Cover the repair area with a thick coat of primer. The thickness should be built up using several thin layers of primer rather than one thick one. Using 600-grit wet-or-dry sandpaper, rub down the surface of the primer until it is very smooth. While doing this, the work area should be thoroughly rinsed with water and the wet-or-dry sandpaper periodically rinsed as well. Allow the primer to dry before spraying additional coats.

Spray on the top coat, again building up the thickness by using several thin layers of paint. Begin spraying in the center of the repair area and then, using a circular motion, work out until the whole repair area and about two inches of the surrounding original paint is covered. Remove all masking material 10 to 15 minutes after spraying on the final coat of paint. Allow the new paint at least two weeks to harden, then use a very fine rubbing compound to blend the edges of the new paint into the existing paint. Finally, apply a coat of wax.

7 Body repair — major damage

1 Major damage must be repaired by an auto body shop specifically equipped to perform unibody repairs. These shops have available the specialized equipment required to do the job properly.
2 If the damage is extensive, the underbody must be checked for proper alignment or the vehicle's handling characteristics may be adversely affected and other components may wear at an accelerated rate.
3 Due to the fact that all of the major body components (hood, fenders, etc.) are separate and replaceable units, any seriously damaged components should be replaced rather than repaired. Sometimes these components can be found in a wrecking yard that specializes in used vehicle components (often at considerable savings over the cost of new parts).

8 Hood — removal and installation

1 Raise the hood.
2 Place protective pads along the edges of the engine compartment to prevent damage to the painted surfaces.
3 Disconnect the cable from the negative battery terminal.
4 Remove the clip retaining the underhood light wire harness to the hood (photo).

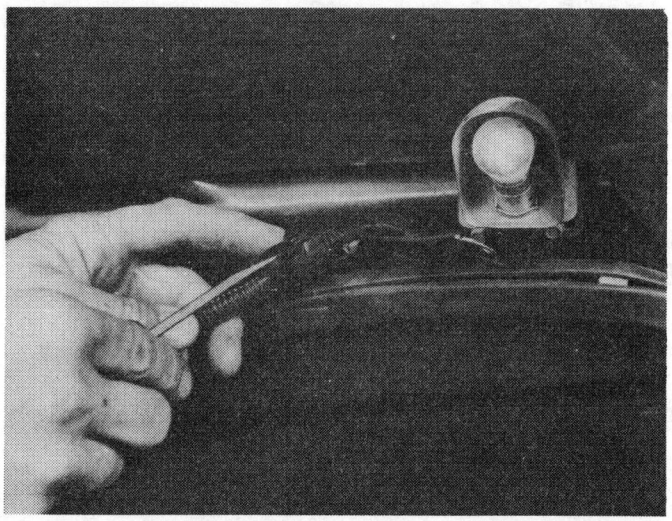

8.4 Removing the clip retaining the light wire harness to the hood

5 Disconnect the underhood light electrical connector and unbolt the bulb assembly from the hood.
6 Scribe lines on the underside of the hood, around the hood mounting bracket, so the hood can be installed in the same position.
7 Apply white paint around the bracket-to-hood bolts so they can be aligned quickly and accurately during installation.
8 Remove the bracket-to-hood bolts and, along with an assistant, separate the hood from the vehicle.
9 Installation is the reverse of the removal procedure.

9 Hood damper — replacement

1 Raise the hood.
2 Position a prop between the radiator support and a reinforced area at the front underside of the hood.
3 Remove the damper pin retaining clips at the top and bottom of the hood damper and remove the pins and damper.
4 Installation is the reverse of the removal procedure.

10 Hood latch release cable — removal and installation

1 The hood latch cable is a one-piece assembly that includes the pull handle, control cable and housing.
2 To remove it, raise the hood and disengage the cable from the primary hood latch assembly. Take precautions to keep the hood from closing and locking while the cable is disconnected.
3 Unclip the cable from the body.
4 Remove the side trim panel to the left of the driver's seat.
5 Disengage the control assembly housing from the cable handle cutout in the trim panel and remove the cable assembly from the panel.
6 Withdraw the cable through the firewall.
7 Installation is the reverse of the removal procedure. When installing the cable, make sure that the sealing grommet attached to the dash panel is in place.

11 Rear hatch — removal, installation and adjustment

1 Open the rear hatch and support it with a prop.
2 Place protective pads along the edges of the hatch opening to prevent damage to the painted surfaces while work is being performed.
3 Using a 13 mm socket, remove the nuts holding the hinges to the glass.
4 While a helper supports the hatch, remove the bolts retaining the top of the dampers to the hatch assembly. **Caution:** *Do not try to remove or loosen the damper attachments with the hatch in any position other than completely open, as personal injury may result.*
5 Disconnect the wiring harness connector for the electric grid defogger, if so equipped.
6 With the aid of a helper, remove the hatch assembly and place it

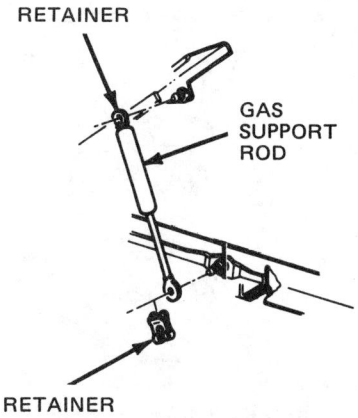

Fig. 12.1 Hood damper installation details (Sec 9)

on a protected surface until it is reinstalled.
7 Installation is the reverse of the removal procedure, noting the following points.
8 When attaching the dampers to the hatch assembly, be sure to tighten the bolts to the specified torque.
9 When installing the nuts retaining the hinges to the glass, tighten the nuts to the specified torque. **Caution:** *Be sure not to overtighten the nuts, as glass breakage and possible personal injury could result.*
10 The hatch assembly height, fore-and-aft and side-to-side adjustments are made at the the hinge-to-body locations.
11 Fore-and-aft and side-to-side adjustments are made by loosening the nut and bolt assemblies in the oversize holes found at these locations, moving the hatch assembly as required, then retightening the nut and bolt assemblies.
12 Hatch height may be adjusted by adding or subtracting the number of spacers at the hinge-to-body locations.
13 Tighten the hinge-to-body bolts to the specified torque when adjustments are complete.

12 Door lock knob — removal and installation

1 Insert a small flat-blade screwdriver behind the leading end of the lock knob and pry it away from the locking rod, located behind the knob.
2 After the end of the rod is free from the knob, slide the knob forward and remove it.
3 To install the knob, insert the small end of the knob through the hole in the escutcheon and slide the knob to the rear until the end of the locking rod engages the depression in the front end of the knob.

13 Door trim panel — removal and installation

1 On models with window crank handles, remove the crank handle. It is secured to the regulator shaft with a clip. The trim panel should be pushed away from the handle to expose the shaft and clip. A special

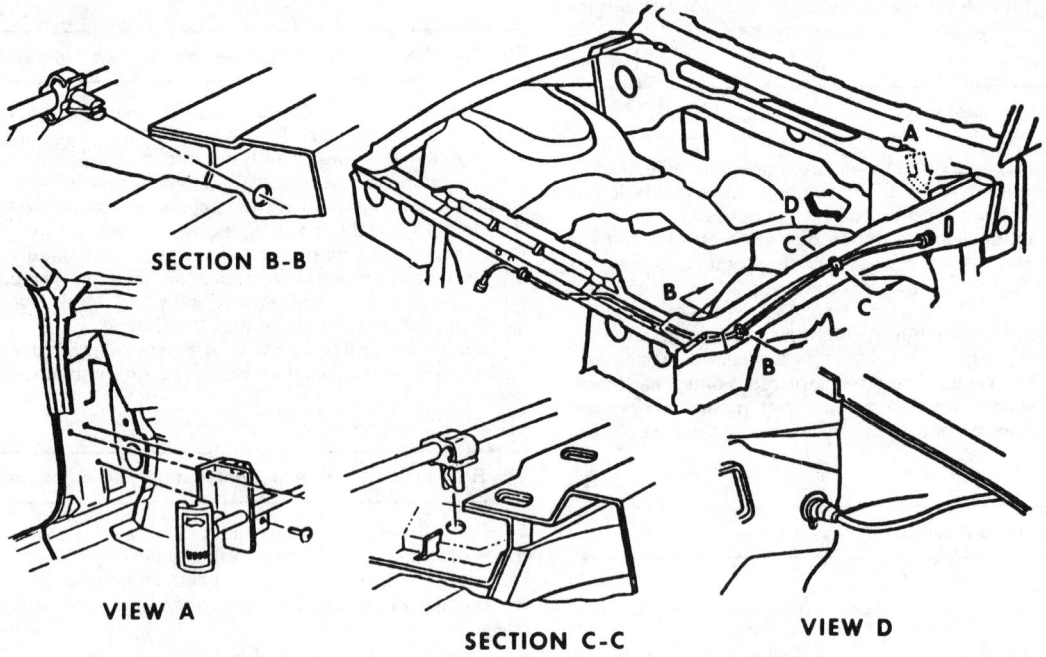

Fig. 12.2 Hood latch release cable routing (Sec 10)

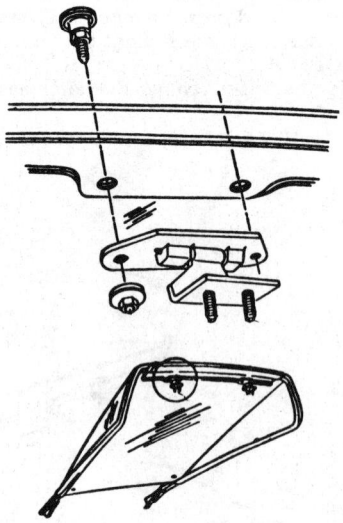

Fig. 12.3 Rear hatch hinge attachment — exploded view (Sec 11)

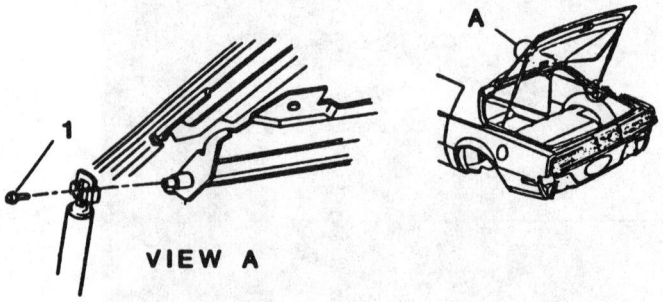

Fig. 12.4 Rear hatch damper installation details (Sec 11)

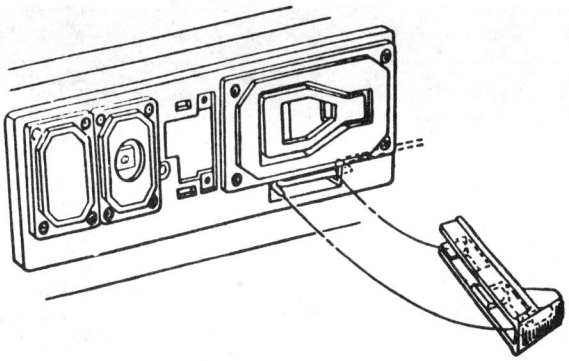

Fig. 12.5 Door lock knob installation details (Sec 12)

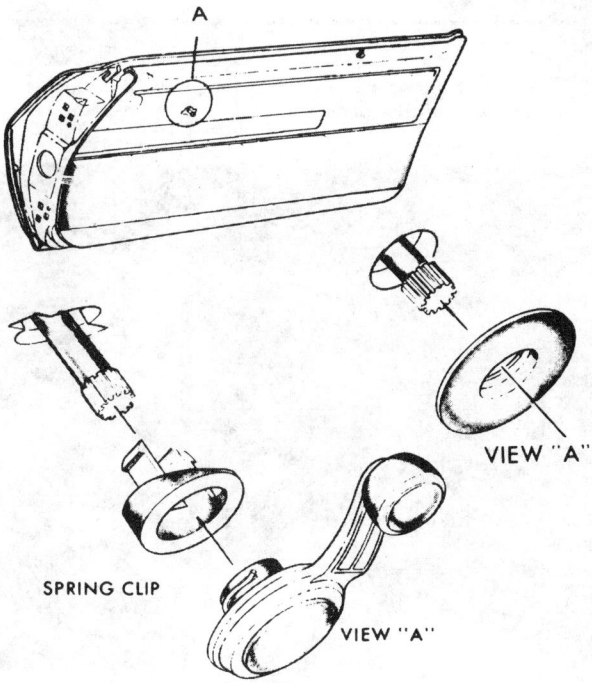

Fig. 12.6 Typical window regulator handle installation (Sec 13)

13.2A Removing the front armrest assembly screw

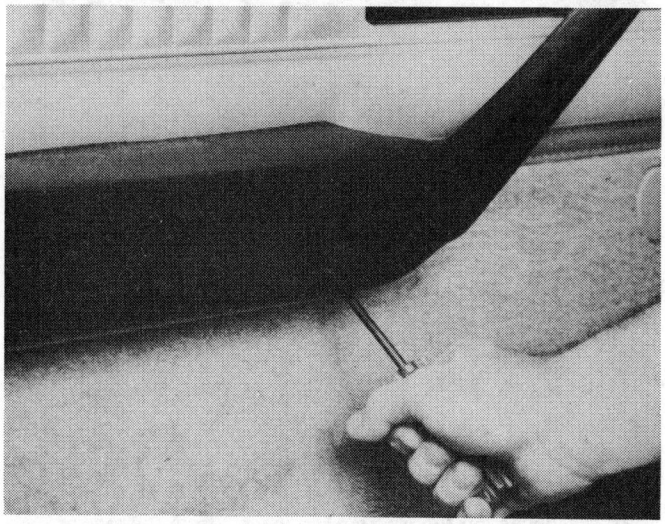

13.2B Removing the bottom armrest assembly screws

13.4 Removing the screws retaining the remote mirror control switch panel

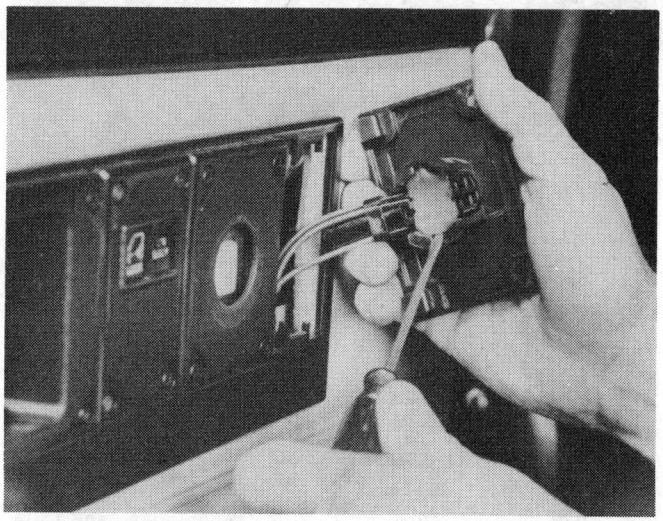

13.6 Removing the snap panel from the rear of the remote mirror control switch

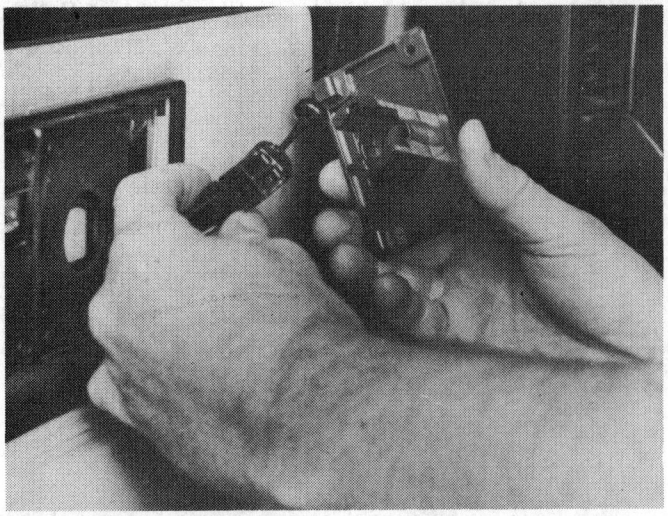

13.7 Disconnecting the switch assembly from the remote mirror switch panel

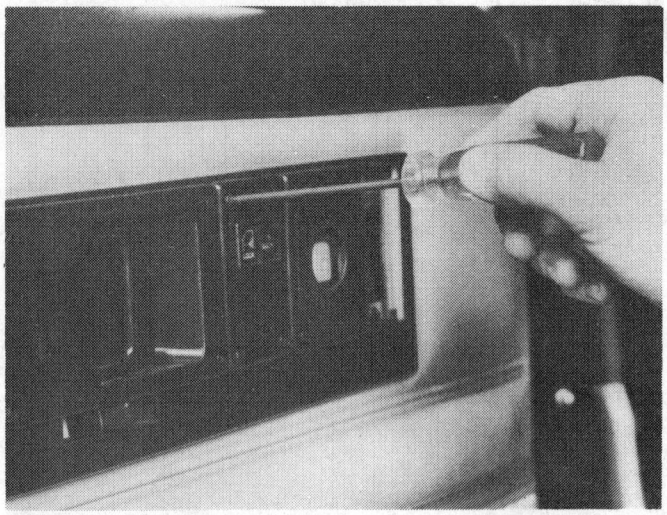

13.8 Removing the screws retaining the power door lock switch panel to the trim panel

13.9 Disconnecting the electrical connector from the rear of the door lock switch

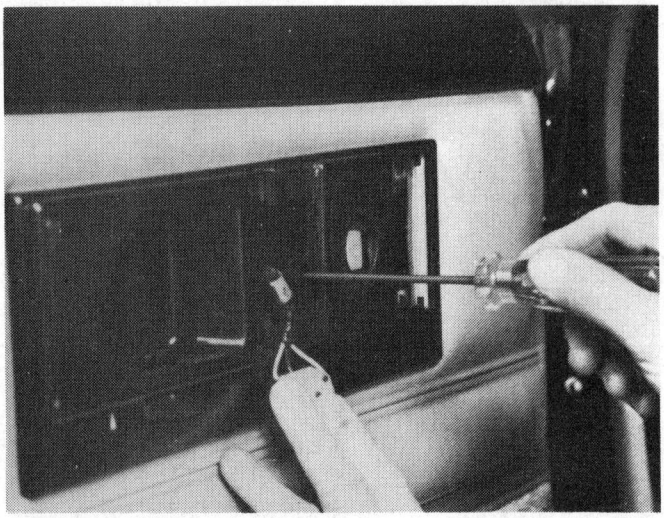

13.10 Removing the screws retaining the trim panel escutcheon to the trim panel

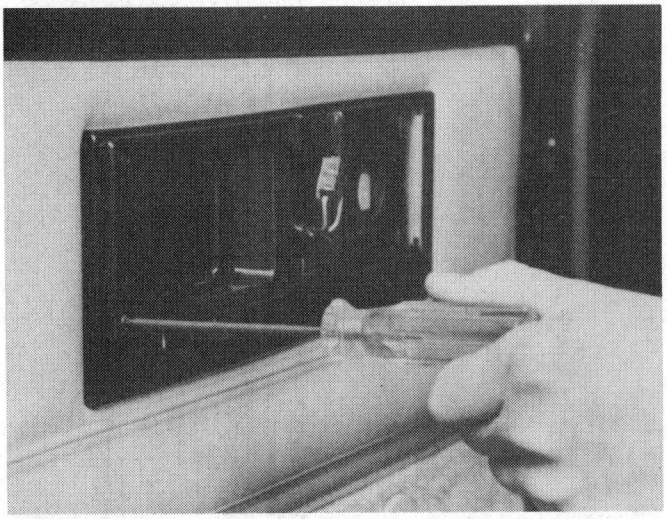

13.11 Removing the screws retaining the remote door handle trim panel

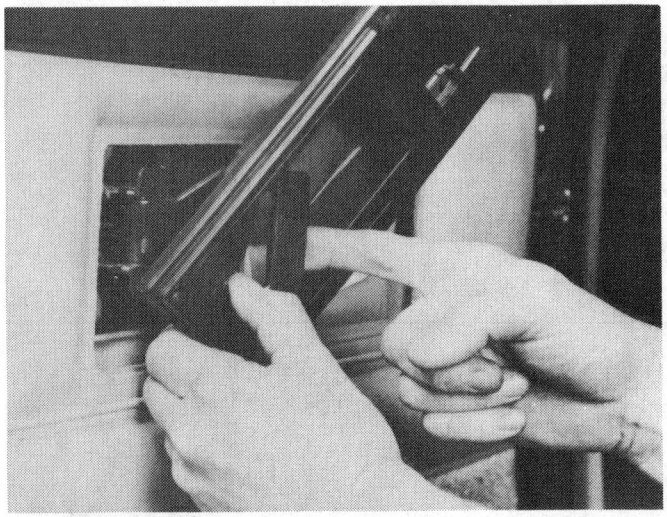

13.13 Removing the trim panel escutcheon from the trim panel

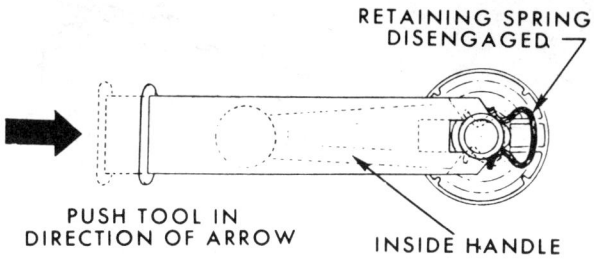

Fig. 12.7 Removing an inside door handle retaining clip with the special tool (Sec 13)

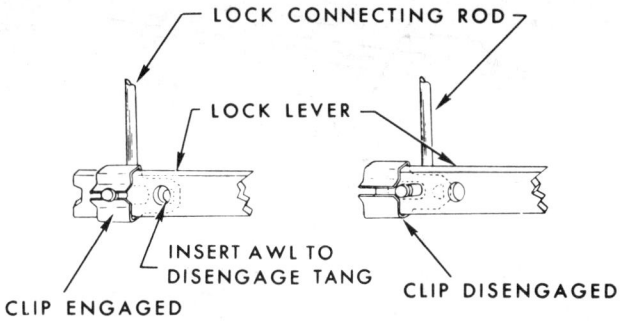

Fig. 12.8 Spring clip removal details (Sec 14)

spring clip removing tool is available to disengage the clip from the groove. A screwdriver can also be used to remove the clip if care is taken not to tear the panel.

2 Remove the three screws retaining the pull handle/armrest assembly (photos).

3 Separate the pull handle/armrest assembly from the trim panel by pulling it away from the retaining clips.

4 If so equipped, remove the two screws retaining the remote mirror control switch panel (photo). **Note:** *On this panel and others, carefully examine the screw heads before removal because some are real and others are 'dummy' heads, cast when the panel was made, which will break if a screwdriver is applied to them.*

5 Pull out the switch panel.

6 Pry off the snap panel at the back of the switch assembly (photo).

7 Disconnect the switch assembly from the switch panel (photo).

8 If so equipped, remove the power door lock switch panel screws and pull out the switch assembly (photo).

9 Disconnect the electrical connector at the rear of the switch panel (photo).

10 Remove the screws retaining the trim panel escutcheon to the trim panel (photo).

11 Remove the screws retaining the remote door handle trim panel (photo).

12 Slide the trim panel escutcheon toward the rear and, using a small screwdriver inserted behind the end of the lock knob, pry the lock knob away from the lock knob rod and remove it.

13 Pull the trim panel escutcheon away from the trim panel, turn it approximately 60 degrees to clear the remote control handle and remove the escutcheon (photo).

14 The trim panel is attached to the door with plastic retaining clips. To disengage these clips, insert a flat, blunt tool (like a screwdriver blade wrapped with tape) between the metal door skin and the trim panel. Carefully pry the door panel away from the door, keeping the tool close to the clips to prevent damage to the panel. Start at the bottom and work around the door toward the top. The top section is secured at the window channel. Once the retaining clips are pried free, lift the trim panel up and away from the door.

15 If work is to be performed on the inner door panel, remove the water deflector.

16 Installation is the reverse of the removal procedure, noting the following.

17 Before attaching the trim panel to the door, make sure that all the trim retainers are undamaged and installed tightly in the panel.

18 If a retainer is to be replaced with a new one, start the retainer flange into the 1/4-inch cutout attachment hole in the trim panel, then rotate the retainer until the flange is fully engaged.

19 When attaching the door trim panel to the door, locate the top of the panel over the upper flange of the inner door panel and press down on the trim panel to engage the upper retaining clips.

20 Position the trim panel on the inner door panel so that the panel retainers are aligned with the holes in the door panel and tap the retainers into the holes with the palm of your hand or a rubber mallet.

14 Door remote control spring clips — disengagement and engagement

1 Spring clips are used to retain remote control connecting rods and inside locking rods to door lock levers and remote handles. They are accessible after the door trim panel and water deflector are removed (refer to Section 13).

2 A slot in the clip provides for simple disengagement, allowing easy detachment of linkage rods.

3 To disengage a spring clip, use an awl or thin-bladed screwdriver to slide the clip out of the slot in the linkage rod end, as shown in the accompanying illustration.

4 To reengage the clip, return it to its original position.

15 Door connecting rods and locking rods — removal and installation

1 Connecting rods are employed to connect the door lock actuating levers with the inside and outside door handles and the inside lock knob. They are accessible after the door trim panel and water deflector have been removed (refer to Section 13).

2 To remove a door connecting or locking rod, disengage the spring clip retaining the rod to the lock lever (refer to Section 13), then disengage the rod.

3 To remove the inside remote handle connecting rod, the spring clip retaining the rod to the remote handle must also be disengaged, using the method just described.

4 To remove the inside lock knob-to-lock rod, first remove the knob (refer to Section 13).

5 To reinstall the rods, first reengage the spring clips on the lock levers, then align the rod in the installed position and push the rod end through the hole in the lock lever until it engages the spring clip.

16 Door lock cylinder — removal and installation

1 Raise the window completely.

2 Remove the door trim panel and detach the water deflector (refer to Section 13) enough to expose the access hole.

3 Disengage the lock cylinder-to-lock rod at the cylinder (refer to the accompanying illustration).

4 Using a screwdriver, slide the lock cylinder retainer forward until it is disengaged. The retainer may also be removed by hand by grasping the anti-theft shield at the top of the retainer and rotating it until the retainer disengages. **Caution:** *If the lock cylinder retainer is being removed by hand, wear gloves to prevent injury from the sharp edges on the retainer.*

5 Remove the lock cylinder by pulling it out of the door from the outside.

6 Installation is the reverse of the removal procedure. Make sure the cylinder gasket is properly installed between the cylinder and the retainer. Lubricate the cylinder with WD-40 or a similar lubricant.

17 Door lock assembly — replacement

Note: *If the door lock assembly fails, it cannot be repaired. Correct door lock failures by replacing the faulty assembly.*

1 Completely raise the window glass.

2 Remove the door trim panel and the water deflector (refer to Section 13).

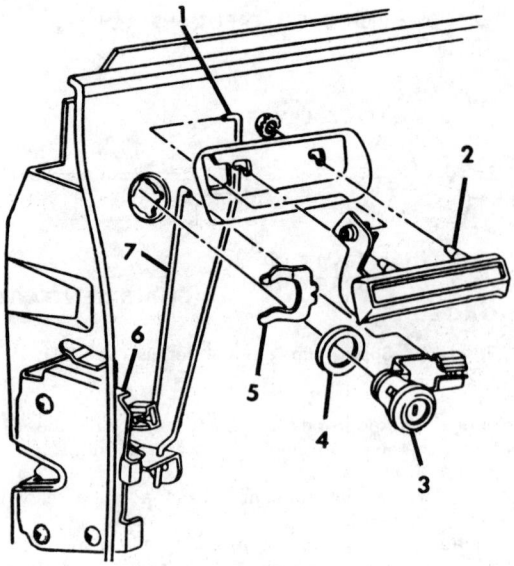

Fig. 12.9 Outside door lock assembly — exploded view (Sec 16)

1 Outside handle-to-lock rod
2 Outside door handle
3 Lock cylinder
4 Lock cylinder gasket
5 Lock cylinder retainer
6 Door lock assembly
7 Lock cylinder-to-lock rod

3 Disengage the inside locking rod at the lock assembly.
4 Disengage the inside handle-to-lock rod at the lock assembly.
5 Disengage the lock cylinder-to-lock rod at the lock assembly.
6 Remove the lock retaining screws, then lower the lock assembly to disengage the outside handle-to-lock rod.
7 Remove the lock assembly from the door.
8 Installation is the reverse of the removal procedure.
9 Tighten the lock screws to the specified torque.

18 Power door lock actuator — removal and installation

1 Completely raise the window glass.
2 Remove the door trim panel and the water deflector (refer to Section 13).
3 Disconnect the electrical connector at the actuator.
4 Locate the rivets retaining the actuator to the inner door panel (refer to the accompanying illustration).
5 Drive the center pins out of the rivets.
6 Using a 1/4-inch drill bit, drill out the rivets.
7 Disengage the bellcrank-to-actuator rod at the actuator.
8 Remove the actuator from the door.
9 To install the actuator, place it in position in the door and reengage the bellcrank-to-actuator rod.
10 If a hand rivet tool is available, attach the actuator to the inner door panel using 1/4 x 0.500-inch aluminum peel-type rivets.
11 If a hand rivet tool is not available, attach the actuator to the inner door panel using 1/4—20 x 1/2-inch nuts and bolts. Tighten the bolts to the specified torque.
12 Reinstall the water deflector and door trim panel.

19 Door outside handle assembly — removal and installation

1 Raise the window completely.
2 Remove the door trim panel as described in Section 13.
3 Pry back the water deflector enough to gain access to the exterior handle mounting hardware. To help prevent tools or parts from falling down into the door cavity, place rags or newspapers in the cavity.
4 Disengage the spring clip from the exterior handle and separate the lock rod from the handle.
5 Remove the two handle mounting nuts.

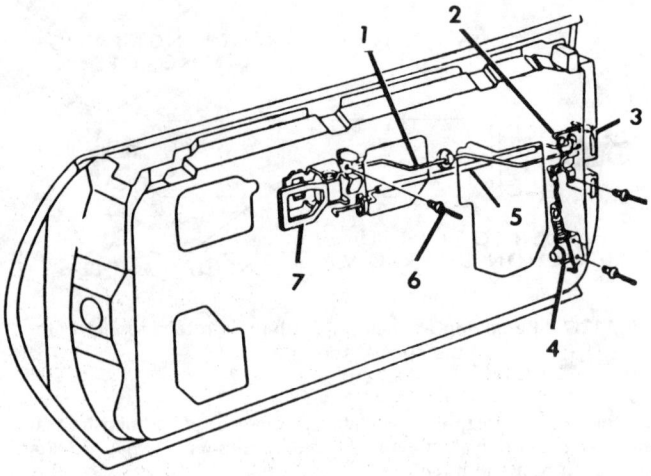

Fig. 12.10 Inside door lock system (Sec 17)

1 Inside handle-to-lock rod
2 Bellcrank (power lock models only)
3 Lock assembly
4 Power lock actuator
5 Inside locking rod
6 Rivet
7 Inside handle

6 Slide the handle to the rear and rotate it up to disengage it from the mounting hole.
7 Installation is the reverse of the removal procedure.

20 Door window glass replacement

Due to the requirements for special handling techniques, window glass should be replaced by a dealer or auto glass shop.

21 Door window regulator — adjustment

Note: *The procedure which follows is for* adjusting *the front door window regulators only. The removal and installation procedure is beyond the scope of the home mechanic and must be left to a professional. One of the reasons for having experienced personnel handle this work is the extreme danger involved in handling the regulator lift arm, which is under tension from the counterbalance spring on motor-driven models. We highly emphasize the need to have all window regulator removal and installation work performed by experienced personnel.*

1 To make any of the adjustments which follow, first remove the door trim panel and the water deflector (refer to Section 13).
2 Depending on the year of the vehicle being worked on, refer to one of the accompanying illustrations.

Rotation (up and down) adjustment

3 If an uneven gap shows at the top of the glass when it is rotated to the top of the channel, loosen the up-stop on-support nut/bolt (no. 2 in the diagram), the rear up-stop nut/bolt (no. 6) and the cam channel nuts/bolts (no. 9) (photos).
4 Adjust the up-stops as necessary and retighten all nuts/bolts.

Channel adjustment

5 If the glass is either too tight or too loose in the channel, rotate the glass until it is half open with the door open.
6 Loosen the glass stabilizer bolts (no. 5) (photo).
7 Loosen the filler assembly bolts (no. 1) (photo).
8 Loosen the adjusting stud and nut-filler assembly bolt (no. 11) (photo).
9 On 1982 vehicles, loosen the front run channel bolts (no. 12) (photos).
10 Loosen the guide channel-to-support bolts (no. 8) (photo).
11 With all fasteners loose, slowly close the door, then align the glass.
12 Open the door and tighten the bolts, reversing the order in which they were loosened.

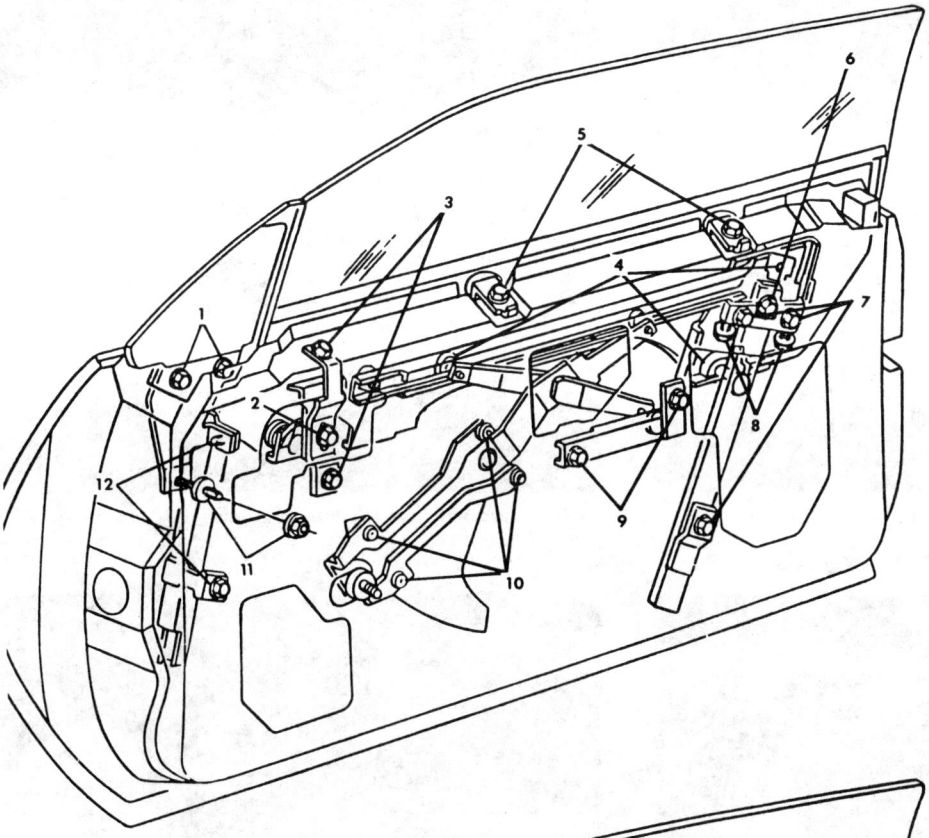

Fig. 12.11 Door hardware mounting details — 1982 models (Sec 21)

1 *Filler at belt bolt*
2 *Up-stop on support bolt*
3 *Support-to-inner panel bolt*
4 *Sash channel-to-glass bolt*
5 *Glass stabilizer bolt*
6 *Rear up-stop bolt*
7 *Rear guide channel bolt*
8 *Guide channel-to-support bolt*
9 *Cam channel bolt*
10 *Regulator rivet*
11 *Adjusting stud and nut-filler at belt*
12 *Front run channel bolt*

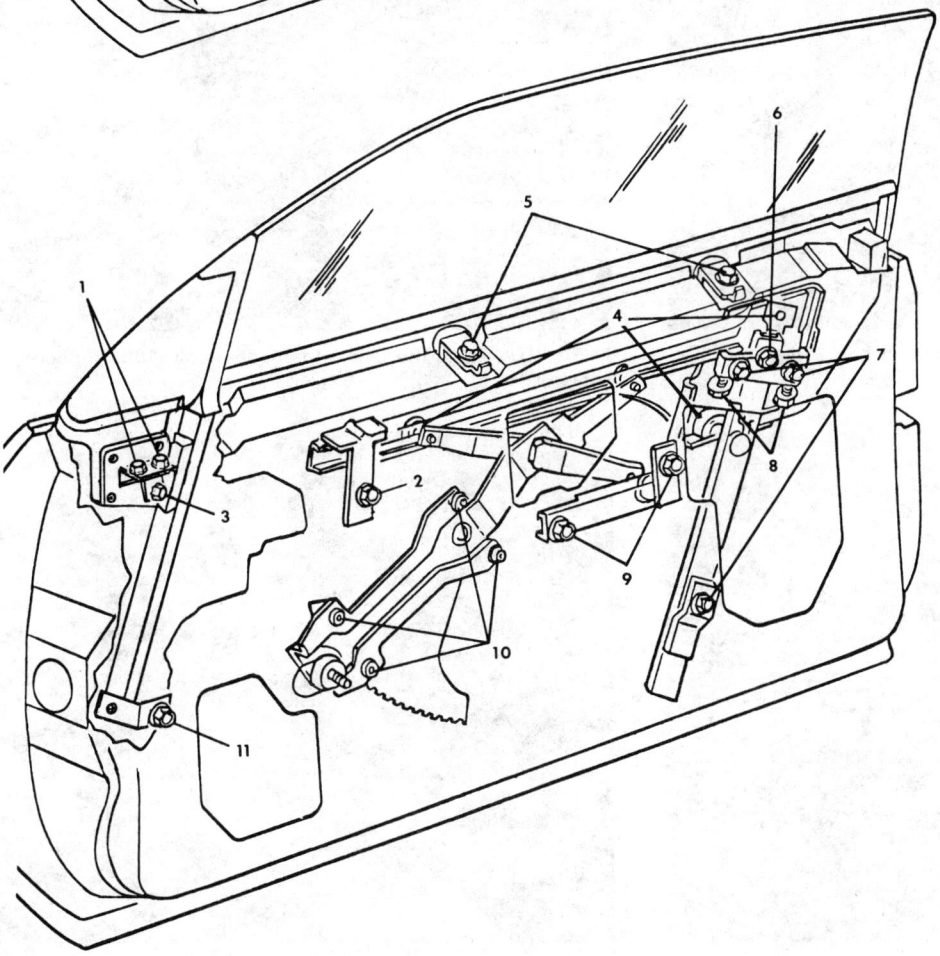

Fig 12.12 Door hardware mounting details — 1983 and 1984 models (Sec 21)

1 *Filler assembly-to-support bolt*
2 *Up-stop on support bolt*
3 *support-to-inner panel bolt*
4 *Sash channel-to-glass rivet*
5 *Glass stabilizer bolt*
6 *Rear up-stop bolt*
7 *Rear guide channel bolt*
8 *Guide channel-to-support bolt*
9 *Cam channel bolt*
10 *Regulator rivet*
11 *Adjusting stud and nut-filler assembly*

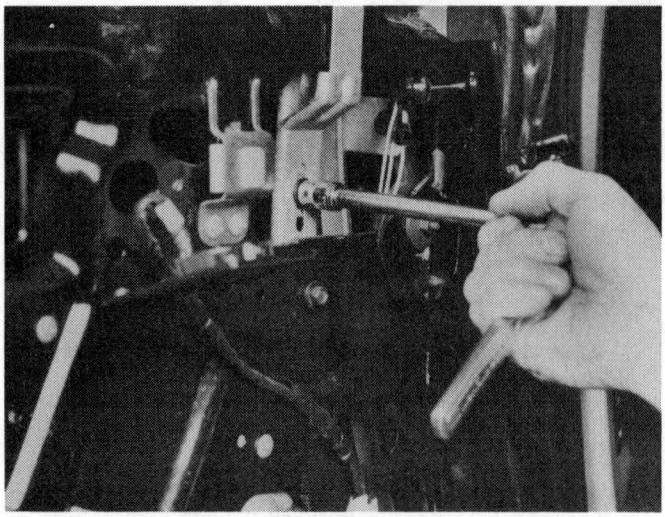

21.3A Loosening the up-stop on-support bolt (1982 model shown)

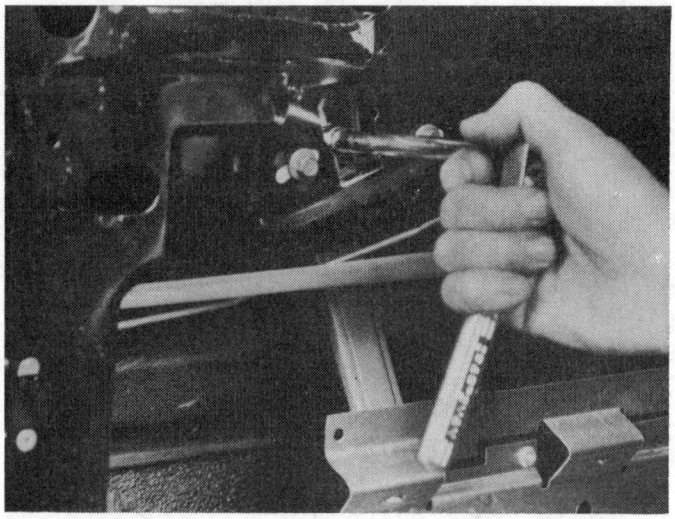

21.3B Loosening the rear up-stop bolt (1982 model shown)

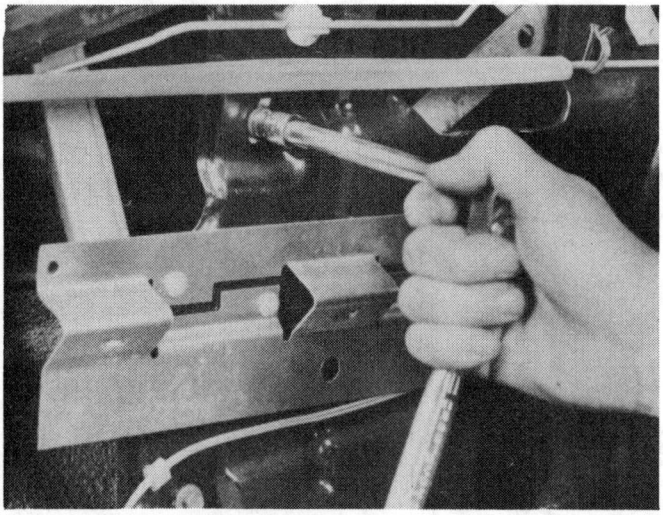

21.3C Loosening the front cam channel bolt (1982 model shown)

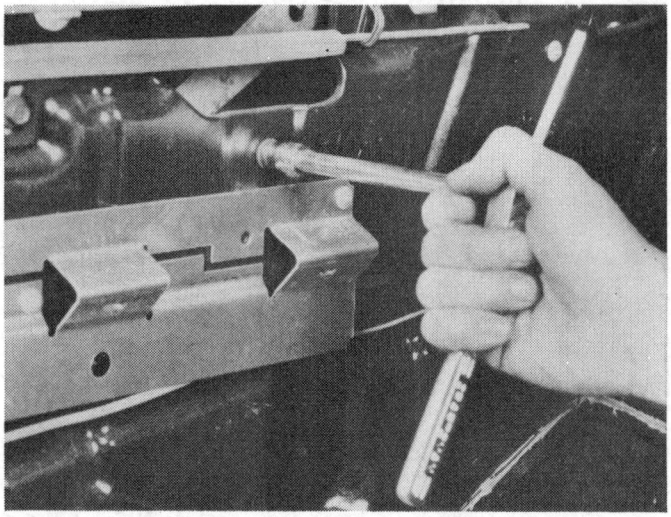

21.3D Loosening the rear cam channel bolt (1982 model shown)

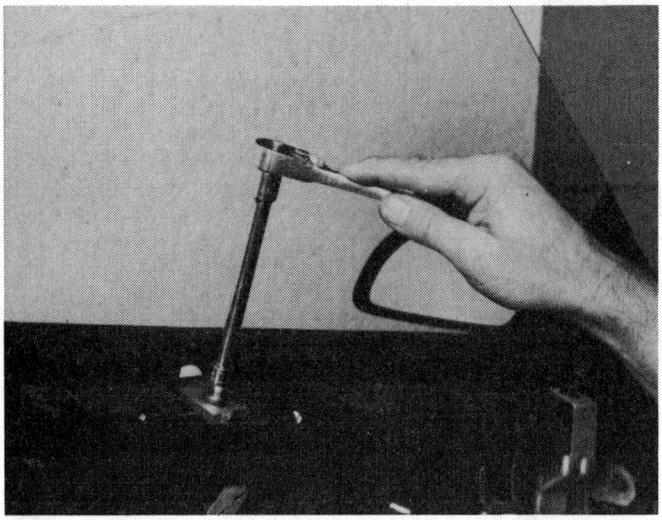

21.6 Loosening the glass stabilizer bolts (1982 model shown)

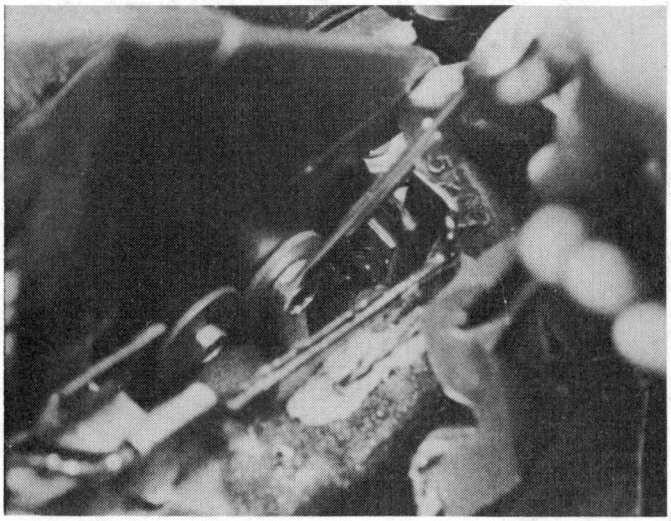

21.7 Loosening the filler assembly bolts (1982 model shown)

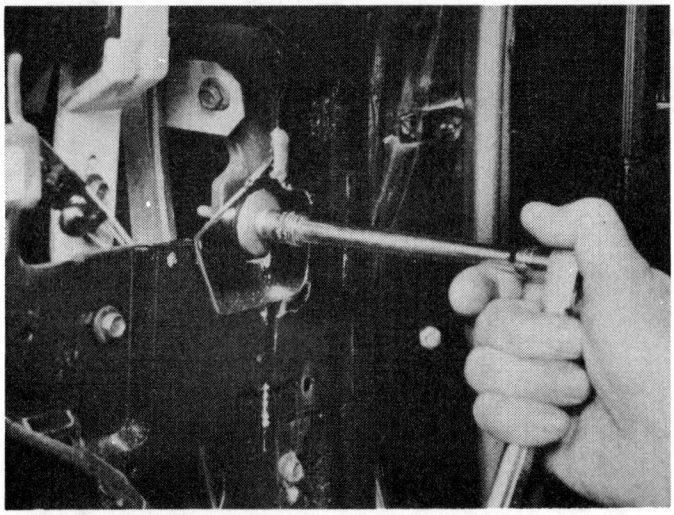

21.8 Loosening the adjusting stud and nut-filler assembly nut (1982 model shown)

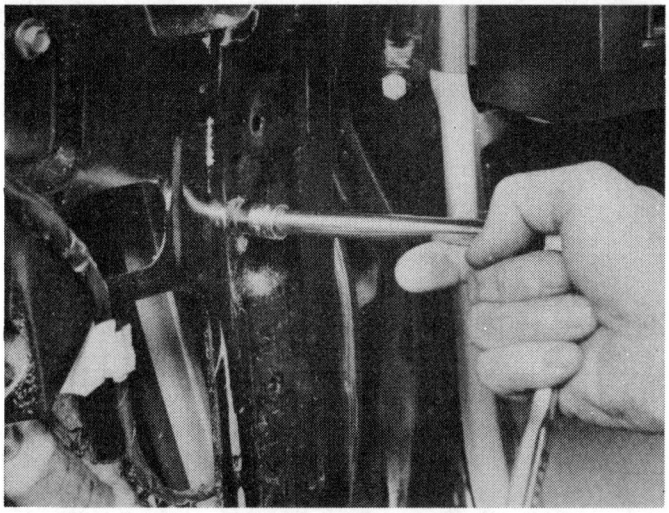

21.9A Loosening the bottom front run channel bolt (1982 model shown)

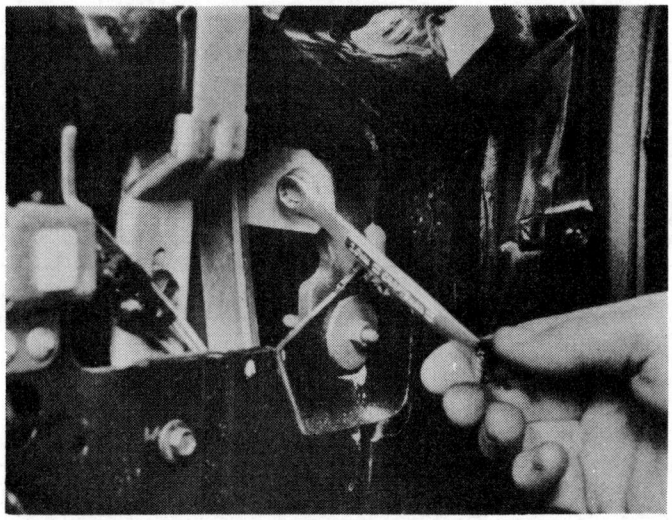

21.9B Loosening the top front run channel bolt (1982 model shown)

21.10 Loosening the guide channel-to-support bolts (1982 model shown)

13 Make sure that the window operates correctly.
14 If misalignment still exists, repeat the above steps.

In/out adjustment

15 If the window is misaligned in (toward the passenger compartment) or out, rotate the window glass until it is half open, with the door open, and loosen the glass stabilizer bolts (no. 5).
16 Loosen the filler assembly bolts (no. 1).
17 Loosen the adjusting stud and nut-filler assembly bolts (no. 11).
18 On 1982 vehicles, loosen the front run channel bolts (no. 12).
19 Loosen the rear guide channel bolts (no. 7).
20 Loosen the guide channel-to-support bolts (no. 8).
21 With all bolts loose, refer to Steps 9 through 12. The glass stabilizer bolts (no. 5) should be tightened last, with the felt bumpers just touching the glass.

22 Windshield replacement

Due to the requirements for special handling techniques, the windshield glass should be replaced by a dealer or auto glass shop.

Chapter 13 Supplement: Revisions and information on 1985 and later models

Contents

Introduction 1
Specifications 2
Tune-up and routine maintenance 3
Special maintenance techniques
Drivebelt check and adjustment
Air filter replacement (TPI and MPFI vehicles only)
Fuel filter replacement
Tire rotation
L4 (four-cylinder) engine 4
General information
Hydraulic lifters — removal, inspection and installation
Oil pan — removal and installation
Timing gear cover — removal and installation
Intake manifold — removal and installation
Exhaust manifold — removal and installation
Cylinder head — removal and installation
V6 engine 5
General information
Rocker arm covers — removal and installation
Intake manifold — removal and installation
Cylinder heads — removal and installation (1987 and later models)
Rear main bearing oil seal — replacement (engine in vehicle)
Crankcase front cover — removal and installation
Timing chain — removal and installation
V8 engine 6
General information
Rocker arm covers — removal and installation
Intake manifold (Tuned-Port Injection) — removal and installation
Hydraulic lifters — removal, inspection and installation
Rear main bearing oil seal — replacement (engine in vehicle)
Crankcase front cover — removal and installation
General engine overhaul procedures 7
Rear main oil seal (V8 engine only) — installation
Cooling, heating and air conditioning systems 8
Electric cooling fan — removal and installation
Electric cooling fan motor and circuit — check
Fuel injection systems 9
General information
Fuel pump — testing
Multi-Port Fuel Injection system — removal, overhaul and installation
Tuned-Port Injection system components — removal
Tuned-Port Injection system components — installation
Minimum idle speed adjustment (TPI and MPFI systems)
Throttle cable — removal and installation
Fuel line — repair and replacement
Emissions control systems 10
General information
Trouble code identification
Air Injection Reaction (AIR) system
Exhaust Gas Recirculation (EGR) system
Early Fuel Evaporation (EFE) system
Clutch and driveline 11
Borg-Warner rear axle — general information and identification
Brakes 12
General information
Front disc brake pads — replacement
Front disc brake caliper — removal and installation
Front disc brake caliper — overhaul
Rear disc brake pads — replacement
Rear disc brake caliper — removal and installation
Rear disc brake caliper — overhaul
Disc brake rotor — inspection, removal and installation
Parking brake free travel — adjustment
Parking brake cables — adjustment
Chassis electrical system 13
SIR (airbag) module – removal and installation
Wiring diagrams

1 Introduction

This Supplement contains specifications and service procedure changes that apply to 1985 and later Chevrolet Camaro vehicles produced for sale in the US. Also included is material related to previous models that was not available at the time of the original publication of this manual.

Where no differences (or very minor differences) exist between the later models and previous models, no information is given. In such instances, the original material included in Chapters 1 through 12 should be used. Therefore, owners of 1985 and later vehicles should refer to this Supplement *before* using the information in the original chapters of the manual.

2 Specifications

Note: *The following specifications are revisions of or supplementary to those listed at the beginning of each Chapter of this manual. The original specifications apply unless alternative figures are included here.*

Tune-up and routine maintenance

Engine oil type	Grade SG, SF/CC or SF/CD
Engine oil viscosity	See accompanying chart
Cooling system capacity (approx. – see owner's manual)	
L4 engine	13.0 qts
V6 engine	
2.8 liter	12.5 qts.
3.1 liter	14.8 qts.
V8 engine	17.2 qts.
Manual transmission oil type	
4-speed	
1985	Dexron II or equivalent ATF
1986 on	GL-5 gear lubricant, SAE 80W or 80W-90
5-speed	
All 1985 and later, except 1987	Dexron II or equivalent ATF
1987	SAE 5W-30 Manual Transaxle Fluid (GM part no. 1052931 or equivalent)

Torque specifications	**Ft-lbs**
Carburetor mounting bolts	
Long	7
Short	11
Wheel lug nuts	
Standard wheels	80
Cast aluminum wheels	100

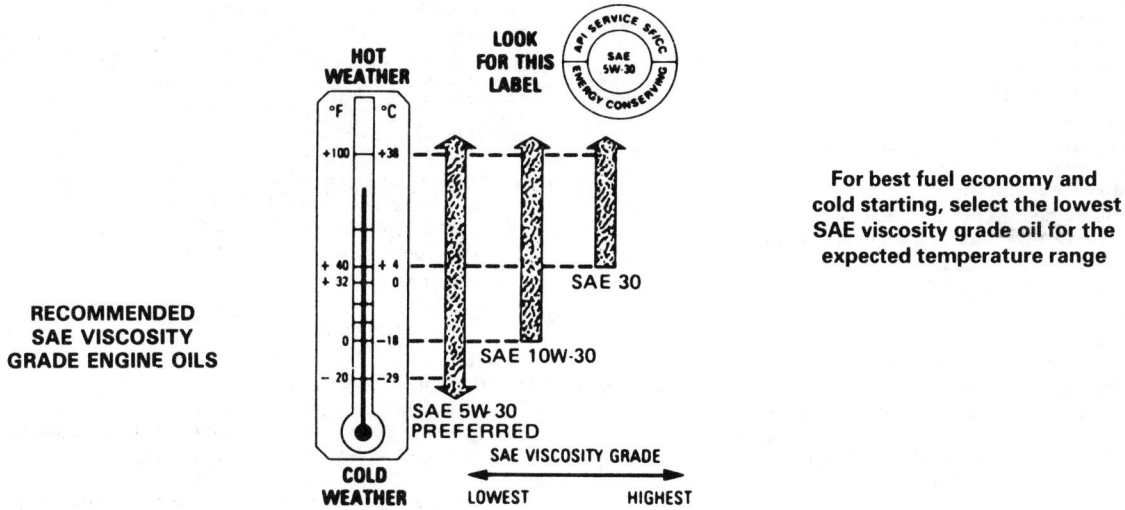

L4 (four-cylinder) engine

Torque specifications	**Ft-lbs**
Flywheel/driveplate-to-crankshaft bolts	55
Intake manifold-to-cylinder head bolts (see Fig. 13.6)	
Bolt 3	25
Bolt 4	37
Bolt 5	28

L4 (four-cylinder) engine (continued)

Torque specifications	Ft-lbs
Exhaust manifold-to-cylinder head bolts	
1985	44
1986 on (see Fig. 13.7)	
Bolts 1, 2, 6 and 7	32
Bolts 3, 4 and 5	37
Cylinder head bolts	
1985	92
1986 on (refer to Chapter 2, Part A for head bolt locations)	
Bolt 9	29 plus 90° additional rotation
All others	22 plus 120° additional rotation
Roller lifter retainer studs	7

2.8 liter V6 engine

Torque specifications	Ft-lbs
Cylinder head bolts (1988 on)	40 plus 90-degrees additional rotation

3.1 liter V6 engine

Torque specifications	Ft-lbs (unless otherwise indicated)
Center intake manifold bolts	15
Lower intake manifold bolts	19
Engine front cover bolts	15
Oil pan bolts/nuts	
Rear two	18
All others	89 in-lbs
Cylinder head bolts	
Step 1	40
Step 2	1/4-turn (90-degrees) additional rotation
Camshaft sprocket bolts	21
Camshaft rear cover bolts	89 in-lbs
Vibration damper bolt	76
Exhaust manifold nuts/bolts	25
Coolant drain plug	115 in-lbs
Thermostat housing bolts	18
Timing chain damper bolt	15
Rocker arm nuts	89 in-lbs
Rocker arm cover nuts	10
Spark plugs	25
Flywheel-to-crankshaft bolts	52

V8 engine

Torque specifications	Ft-lbs (unless otherwise indicated)
Flywheel-to-crankshaft bolts (1986 on)	74
Driveplate-to-torque-converter bolts	35
Flywheel housing bolts	55
Rocker arm cover bolts/nuts	
1985	50 to 65 in-lbs
1986	
L98	80 to 120 in-lbs
All others	50 to 80 in-lbs
1987 and 1988	62 to 115 in-lbs
1989	75 in-lbs
1990 on	89 in-lbs
Cylinder head bolts (1989 on)	68

General engine overhaul procedures

Engine model identification (the eighth character of the Vehicle Identification Number — VIN)

VIN engine code 2	
Type	L4 (four-cylinder)
Displacement	2.5L
Compression ratio	9.0:1
Fuel delivery system	Throttle Body Injection (TBI)
RPO code	LQ9

VIN engine code S	
Type	V6
Displacement	2.8L
Compression ratio	8.9:1
Fuel delivery system	Multi-Port Fuel Injection (MPFI)
RPO code	LB8
VIN engine code T	
Type	V6
Displacement	3.1L
Compression ratio	8.8:1
Fuel delivery system	Multi-Port Fuel Injection (MPFI)
RPO code	LHO
VIN engine code F	
Type	V8
Displacement	5.0L
Compression ratio	9.5:1
Fuel delivery system	Tuned-Port Injection (TPI)
RPO code	LB9
VIN engine code H	
Type	V8
Displacement	5.0L
Compression ratio	8.6:1
Fuel delivery system	4 BBL carburetor
RPO code	LG4
VIN engine code G	
Type	V8
Displacement	5.0L
Compression ratio	9.5:1
Fuel delivery system	4 BBL carburetor
RPO code	L69
VIN engine code 8	
Type	V8
Displacement	5.7L
Compression ratio	9.5:1
Fuel delivery system	Tuned-Port Injection (TPI)
RPO code	L98
VIN engine code E	
Type	V8
Displacement	5.0L
Compression ratio	9.3:1
Fuel delivery system	Throttle body injection (TBI)
RPO code	L03

L4 engine

Valve spring pressure and length (intake and exhaust)	
Valve closed	1.66 in at 76 to 86 lbs.
Valve open	1.254 in at 170 to 180 lbs.
Pushrod length	8.3996 in (213.35 mm)
Cylinder bore taper limit	0.005 in (0.13 mm)
Piston-to-bore clearance	0.0014 to 0.0022 in (measured 1.8-inch down from piston top)
Piston pin-to-piston clearance	0.0002 to 0.0004 in
Pin-to-rod clearance	Press fit
Piston ring end gap	
Top ring	0.010 to 0.020 in
Second ring	0.010 to 0.020 in
Oil ring	0.020 to 0.060 in
Camshaft lobe lift (intake and exhaust)	0.398 in (10.1 mm)

2.8 liter V6 engine

Main bearing journal diameter	2.647 to 2.648 in (67.241 to 67.265 mm)
Main bearing oil clearance	0.0016 to 0.0032 in (0.041 to 0.081 mm)
Camshaft lobe lift	
Intake	0.262 in (6.67 mm)
Exhaust	0.273 in (6.94 mm)

3.1 liter V6 engine

Note: *All specifications for the 3.1 liter V6 engine not listed here are the same as the specifications for the 2.8 liter V6 engine listed in Chapter 2, Part D.*

General

Oil pressure	
At 500 rpm	10 psi minimum
At 2000 rpm	51 to 55 psi
Firing order	1-2-3-4-5-6

Crankshaft and connecting rods

Crankshaft end play	0.0024 to 0.0083 in (0.06 to 0.21 mm)
Connecting rod end play (side clearance)	0.014 to 0.027 in (0.36 to 0.68 mm)
Main bearing	
Journal diameter	2.6473 to 2.6483 in (67.241 to 67.265 mm)
Oil clearance	
Thrust bearing only	0.0016 to 0.0027 in (0.042 to 0.069 mm)
All others	0.0012 to 0.0027 in (0.032 to 0.069 mm)
Connecting rod bearing oil clearance	0.0014 to 0.0036 in (0.038 to 0.093 mm)

Engine block

Cylinder bore diameter	3.5046 to 3.5053 in (89.016 to 89.034 mm)
Out-of-round/taper limit	0.0005 in (0.013 mm)

Pistons and rings

Piston-to-bore clearance	0.0012 to 0.0029 in (0.030 to 0.073 mm)
Compression ring side clearance	0.002 to 0.0035 in (0.05 to 0.09 mm)
Oil ring end gap	0.010 to 0.030 in (0.25 to 0.75 mm)
Piston pin	
Diameter	0.9052 to 0.9054 in (22.9915 to 22.9964 mm)
Clearance in piston	0.0004 to 0.0008 in (0.0096 to 0.0215 mm)

Camshaft lobe lift

Intake	0.2626 in (6.67 mm)
Exhaust	0.2732 in (6.94 mm)

Torque specifications

	Ft-lbs
Cylinder head bolts	
Step 1	40
Step 2	1/4-turn (90-degrees) additional rotation
Main bearing cap bolts	73
Connecting rod cap nuts	39
Oil pump mounting bolts	30
Oil filter adapter bolts	63

V8 engine

Valve spring installed height	
Intake	1-23/32 in
Exhaust	1-19/32 in
Engine block bore diameter	
5.0 Liter	3.7350 to 3.7385 in
5.7 Liter	3.9995 to 4.0025 in
Camshaft lobe lift	
5.0 Liter	
Carbureted	0.234 in
Fuel injected	0.269 in
5.7 Liter	0.273 in

Fuel and exhaust systems

Torque specifications	**Ft-lbs (unless otherwise indicated)**
Carburetor mounting bolts	
Long	7
Short	11
Plenum-to-intake manifold bolts (V6)	18
Intake runner-to-plenum bolts (V8)	15 to 20
Intake runner-to-intake manifold bolts (V8)	15 to 20
Throttle body-to-plenum bolts	18
Fuel rail-to-intake manifold bolts	18
IAC valve	
1985 thru 1989	13
1990 on	30 in-lbs
Fuel pressure connection assembly (Schrader valve)	7
Model 220 TBI mounting bolts	12

Manual transmission

Torque specifications (77 mm only)	**Ft-lbs**
Transmission-to-engine bolts	55
Rear mount-to-crossmember bolts	35

Clutch and driveline

Axle codes	**Axle ratio**	**Axle type**
1985		
8JA	3.08	Std
8JB	3.23	Std
8JC	3.42	Std
8JD	3.73	Std
8JP	3.08	Posi
8JQ	3.23	Posi
8KA	3.42	Std
8KB	3.73	Std
8KR	3.08	Std
8PP	3.08	Posi
8PQ	3.23	Posi
8PR	3.42	Posi
8PS	3.73	Posi
1986		
2HC	3.73	Std
2HD	2.73	Std
2HG	3.23	Std
2HH	3.42	Std
2HL	3.42	Std
2HM	3.73	Std
2HP	2.73	Std
2HS	3.73	Posi
2HT	2.73	Posi
2HV	2.73	Posi
2HY	3.23	Posi
1987		
2HA	3.23	Std
2HL	3.42	Std
2HP	2.73	Std
2HQ	3.23	Posi
2HT	2.73	Posi
4EU*	3.27	Posi
4EW*	3.45	Posi
4HB	3.08	Posi
4HD	2.73	Posi
4HF	3.08	Posi
4HY	3.23	Posi

Clutch and driveline (continued)

Axle codes	Axle ratio	Axle type
1988 and 1989		
6HT	2.73	Posi
6HP	2.73	Std
6HB, 6HF	3.08	Posi
6HK	3.08	Std
6HL	3.42	Std
6HQ, 6HJ	3.23	Posi
GH3*	2.77	Posi
GW6*	3.27	Posi
GM3*	3.45	Posi
1990		
8HP	2.73	Std
8HT, 8RC	2.73	Posi
8HK	3.08	Std
8HF, 8HB	3.08	Posi
2PM, 2PN	3.23	Posi
8HL	3.42	Std
9EQ*	3.27	Posi
9ER*	3.45	Posi
1991 and 1992		
8H3	3.23	Std
8HE	2.73	Posi

**Denotes Borg-Warner axle*

Torque specifications (1989 on)	Ft-lbs
Rear axle housing cover bolts	20

Brakes

Parking brake lever free travel (1989 on)	0.024 to 0.028 in
Rotor discard thickness	
Front	0.965 in
Rear	
1985 thru 1988	0.965 in
1989 on	0.724 in

Torque specifications	Ft-lbs
Brake line-to-master cylinder nut	
1987	12
1988 on	17
Front caliper bracket bolts (twin-piston caliper)	137
Rear caliper guide pin bolt (1989 on)	
Hex head	26
Allen head	16
Rear caliper bracket bolts	70

Chassis electrical system

Bulb application	
Headlight	H6054
Center high-mounted brake light	577
Brake warning light (w/gauges only)	74
Choke telltale light (w/gauges only)	74
Console ashtray light	194
Instrument panel courtesy light	194
Radio illumination	37
Seat belt warning light (w/gauges only)	74

Suspension and steering systems

Tires (optional)	P235/60VR15, P245/50VR16
Front wheel alignment (caster)	
1985	+3° ± 0.5°
1986	+3.5° ± 0.5°
1987 and 1988	+5° ± 0.5°
1989 on	+4.7° ± 0.05°

Torque specifications	**Ft-lbs**
Steering knuckle-to-tie-rod end nut	35
Tie-rod-to-intermediate rod nut	35
Pitman arm-to-intermediate rod nut	35
Pitman arm-to-steering gear nut	180
Idler arm-to-intermediate rod nut	30
Power steering gear-to-frame bolts	70
Strut-to-steering knuckle bolts	203
Track bar-to-axle nut (1988 and later models)	59
Track bar-to-body bracket (1988 and later models)	78
Control arm-to-rear axle bolt	
1988	80
1989 on	85
Control arm-to-underbody bolt	
1988	80
1989 on	85

3 Tune-up and routine maintenance

Special maintenance techniques

Some parts on later model vehicles have an elecrostatic discharge sensitive sticker applied to them. When installing or servicing any parts with this sticker, the following guidelines should be followed:

Avoid touching the electrical terminals of the part

Always ground the package to a known good ground on the vehicle before removing the part from the package

Always touch a known good ground before handling the part

Drivebelt check and replacement

1 As of the 1987 model year, vehicles with V6 engines are equipped with a single "serpentine" drivebelt. No adjustment is necessary. The belt should be replaced when the wear indicator on the tensioner reaches it's maximum travel (Fig 13.1). Inspect the belt for missing ribs, fraying and other signs of abnormal wear.

2 To replace the belt, place a 15 mm socket connected to a breaker bar on the tensioner pulley bolt and rotate the pulley *counterclockwise* to release the belt tension.

3 Remove the drivebelt from the pulleys.

4 Install the new belt starting with the bottom pulleys, then release the tensioner to allow final installation.

Air filter replacement (TPI and MPFI vehicles only)

5 The air filter is located in the air cleaner housing, which is mounted to the right of the radiator. When replacing the air filter, be sure to clean the inside of the housing. Also inspect the ducts between the air filter housing and the mass airflow sensor and between the mass airflow sensor and the throttle body for holes and cracks. Replace the ducts if necessary.

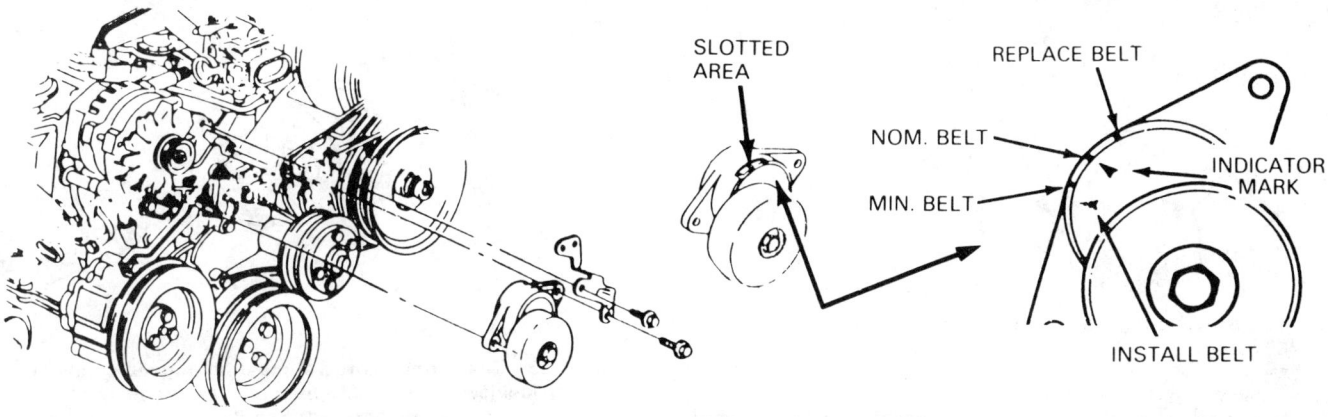

Fig. 13.1 Replace the serpentine drivebelt when the indicator mark on the tensioner reaches the limit of clockwise travel and is aligned with the mark on the tensioner housing (Sec 3)

Fuel filter replacement

6 The fuel filter on V6 and V8 fuel injection equipped vehicles is located forward of the fuel tank on the rear crossmember (Fig 13.2). **Warning:** *Before removing the fuel filter, the fuel system pressure must first be relieved (refer to Chapter 4).*

7 Using a backup wrench on the fuel filter canister, disconnect the fuel lines from each side of the filter. Inspect the small O-rings on each fuel line and replace them if necessary (Fig 13.3).

8 Loosen the fuel filter clamp bolt and remove the filter.

9 Position the new filter in the clamp, tighten the clamp bolt and connect the fuel lines.

10 Pressurize the fuel system by turning the ignition key on for two seconds, then check the fittings for leaks.

Tire rotation

11 If the car is equipped with P245/50VR16 tires, they must be dismounted from the wheels prior to rotation (due to the fact that the front and rear wheels have different offsets and the tires are directional). Directional tires have arrows on both sidewalls that point in the direction the tire must rotate for correct performance. Tire rotation should be done as follows:

12 Dismount the tires from the wheels.

13 Rotate the tires, not the wheels, following the *optional* pattern shown in Chapter 1 (Fig. 1.23). Make sure the tires will rotate in the direction indicated by the arrows as they're remounted on the wheels.

14 Rebalance the tires.

15 Replace the wheels in their original positions.

16 Recheck the tire pressures and lug nut tightness.

4 L4 (four-cylinder) engine

General information

The inline four-cylinder engine used in later models is virtually identical to earlier models, the main difference being the addition of roller type hydraulic lifters and changes to various engine reassembly procedures.

Hydraulic lifters — removal, inspection and installation

1 1985 and later models are equipped with roller type hydraulic valve lifters. Lifter retainers and guides are installed in the block, behind the pushrod cover, to prevent the lifters from rotating on the camshaft lobes (Fig. 13.4). Consequently, the camshaft lobes are ground with no front-to-rear taper. Follow the removal procedures in Chapter 2A, but note that before the lifters are removed, the retainers and guides must first be removed. Operation, disassembly and inspection are similar to a standard hydaulic lifter. However, the roller must be inspected for freedom of movement, excessive looseness of the roller bearings, flat spots and pitting. The camshaft must also be inspected if any abnormal roller wear is indicated.

2 Upon installation, position the lifter guides and retainers and tighten the studs before installing the pushrods. The remaining reassembly sequence is identical to the procedure described in Chapter 2A Section 8.

Oil pan — removal and installation

3 Disconnect the negative battery cable from the battery, then raise the front of the vehicle and support it on jackstands.

4 Drain the engine oil. Disconnect the exhaust pipe at the exhaust manifold-to-exhaust pipe flange and loosen the exhaust pipe hanger bracket.

5 Remove the starter bolts and position the starter out of the way.

6 Remove the flywheel dust cover.

7 Connect an engine hoist or a block-and-tackle set-up to the engine.

8 Remove the engine mount through-bolts.

9 Raise the engine just enough so the oil pan can be removed without any interference from other components.

10 Remove the oil pan bolts and lower the oil pan.

11 Clean the sealing surfaces on the oil pan, engine block and the exposed portion of the front cover.

12 Apply a continuous bead of RTV sealant to the oil pan (Fig 13.5).

13 Install the oil pan and bolts while the sealant is still wet, then tighten the bolts to the specified torque.

14 Lower the engine, align the engine mounts and install and tighten the through-bolts. Remove the engine hoist.

15 Attach the flywheel dust cover.

16 Install the starter.

17 Attach the exhaust pipe to the exhaust manifold, then tighten the hanger bracket.

18 Lower the vehicle and fill the crankcase with oil (see Chapter 1).

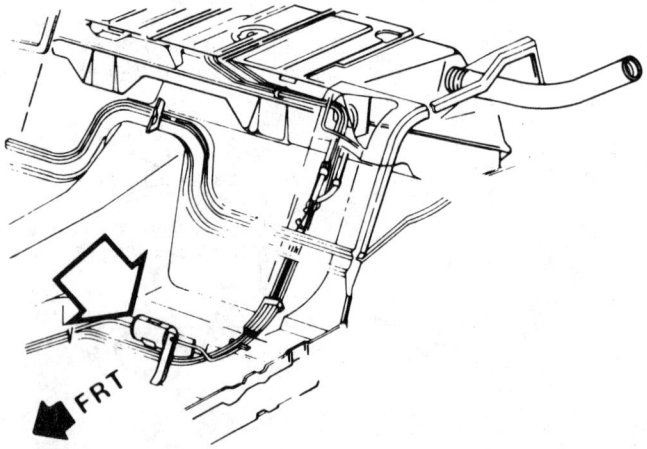

Fig. 13.2 The fuel filter on fuel injected models is located on the rear crossmember (Sec 3)

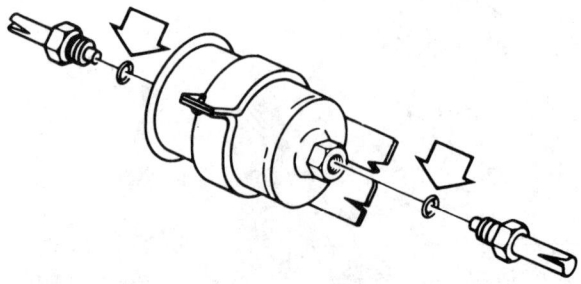

Fig. 13.3 The O-rings (arrows) must be in good condition or fuel leaks will result when the line fittings are retightened (Sec 3)

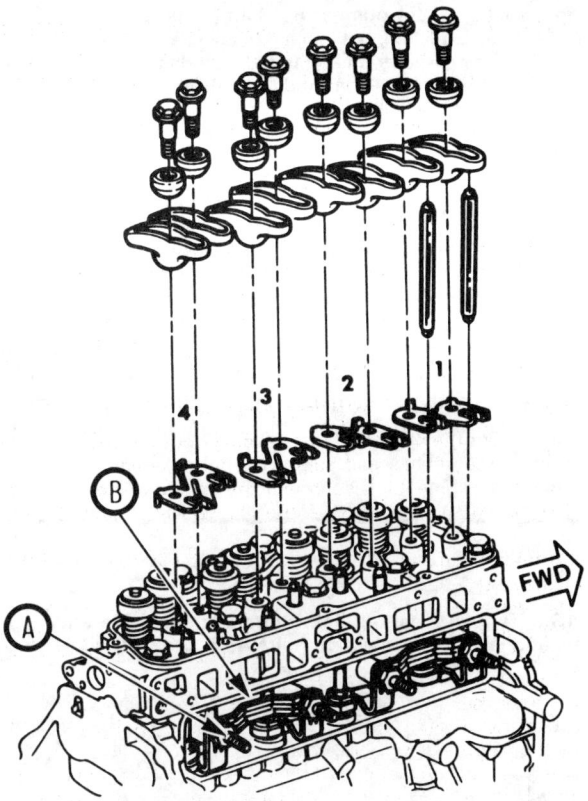

Fig. 13.4 The studs (A) must be unscrewed to remove the lifter retainers (B) (Sec 4)

19 Connect the negative battery cable.
20 Start the engine and check for leaks.

Timing gear cover — removal and installation

21 Disconnect the negative battery cable from the battery.
22 Raise the front of the vehicle and support it on jackstands.
23 Drain the cooling system (Chapter 1).
24 Remove the lower radiator hose and the lower fan shroud.
25 Lower the vehicle.
26 Remove the upper radiator hose and the upper fan shroud.
27 Loosen the accessory drivebelt adjusting screws and remove the drivebelts.
28 Remove the fan assembly.
29 Remove the crank pulley and hub (Chapter 2A).
30 Remove the timing gear cover bolts and detach the cover.
31 Clean the mating surfaces on the cover and the block, removing all traces of old sealing material.
32 If the pulley hub seal is to be replaced, refer to Chapter 2A, Section 12.
33 Apply a 3/8-inch (wide) by 3/16-inch (thick) bead of RTV sealant to the joint at the oil pan and timing gear cover.
34 Apply a 1/4-inch (wide) by 1/8-inch (thick) bead of RTV sealant to the timing gear cover at the block mating surfaces.
35 Lubricate the cover seal with clean engine oil.
36 Insert the hub through the cover seal and place the cover in position on the block as the hub slides onto the crankshaft. This will ensure that the seal is centered evenly around the hub.
37 Install the oil pan-to-cover bolts and partially tighten them.
38 Install the timing cover-to-block bolts, then tighten all of the mounting bolts to the specified torque.
39 The remaining steps are the reverse of removal. Refer to Chapter 2A, Section 12, for the crankshaft pulley hub installation procedure.

Intake manifold — removal and installation

40 The intake manifold removal and installation procedure is identical to the procedure outlined in Chapter 2A, with the exception of the bolt

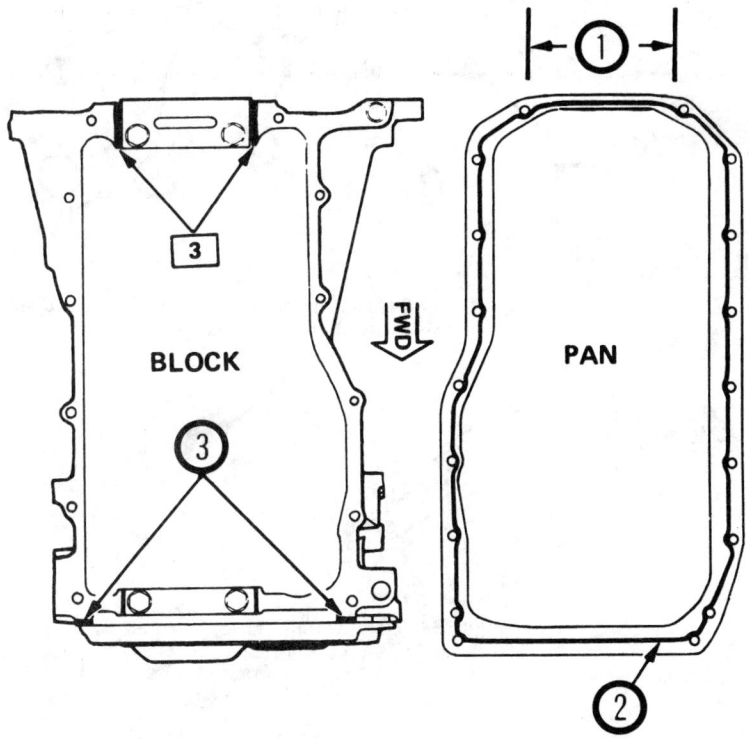

Fig. 13.5 Oil pan sealant application details (Sec 4)

1 3/8-inch wide by 3/16-inch thick
2 3/16-inch wide by 1/8-inch thick
3 1/8-inch bead at points indicated

tightening sequence. Refer to Figure 13.6 for the new tightening sequence and torque specifications.

Exhaust manifold — removal and installation

41 Exhaust manifold removal and installation for 1985 and later vehicles is similar to previous models. However, a different tightening sequence is employed. See Figure 13.7 for the new tightening sequence and torque values.

Cylinder head — removal and installation

42 Cylinder head removal and installation for 1985 and later vehicles is nearly identical to the procedure for earlier models, with the exception of the tightening procedure and torque specifications. Vehicles manufactured in 1985 use the same tightening sequence as shown in Chapter 2A but have a different torque specification. Vehicles manufactured in 1986 also use the same sequence, but a different tightening procedure as follows:

43 Tighten the bolts in sequence to 18 ft-lbs.

44 Repeat the sequence, this time bringing the torque up to 22 ft-lbs. on all bolts except number 9. Tighten number 9 to 29 ft-lbs.

45 Repeat the sequence, but instead of using a final torque value, turn all bolts except number 9 an additional 120-degrees (two flats of the bolt head). Turn number 9 an additional 90-degrees (1/4-turn).

5 V6 engine

General information

The V6 engine used in 1985 and later vehicles is very similar to the earlier model V6 engine, with the main differences being in the external components such as the belt-driven accessories and the Multi-Port fuel injection system. From 1990 on, the displacement is 3.1 liters (up from 2.8 in previous years). The majority of the service procedures are identical to previous years, with the exception of the ones affected by the aforementioned components/systems. They will be covered along with some other minor differences due to various design changes.

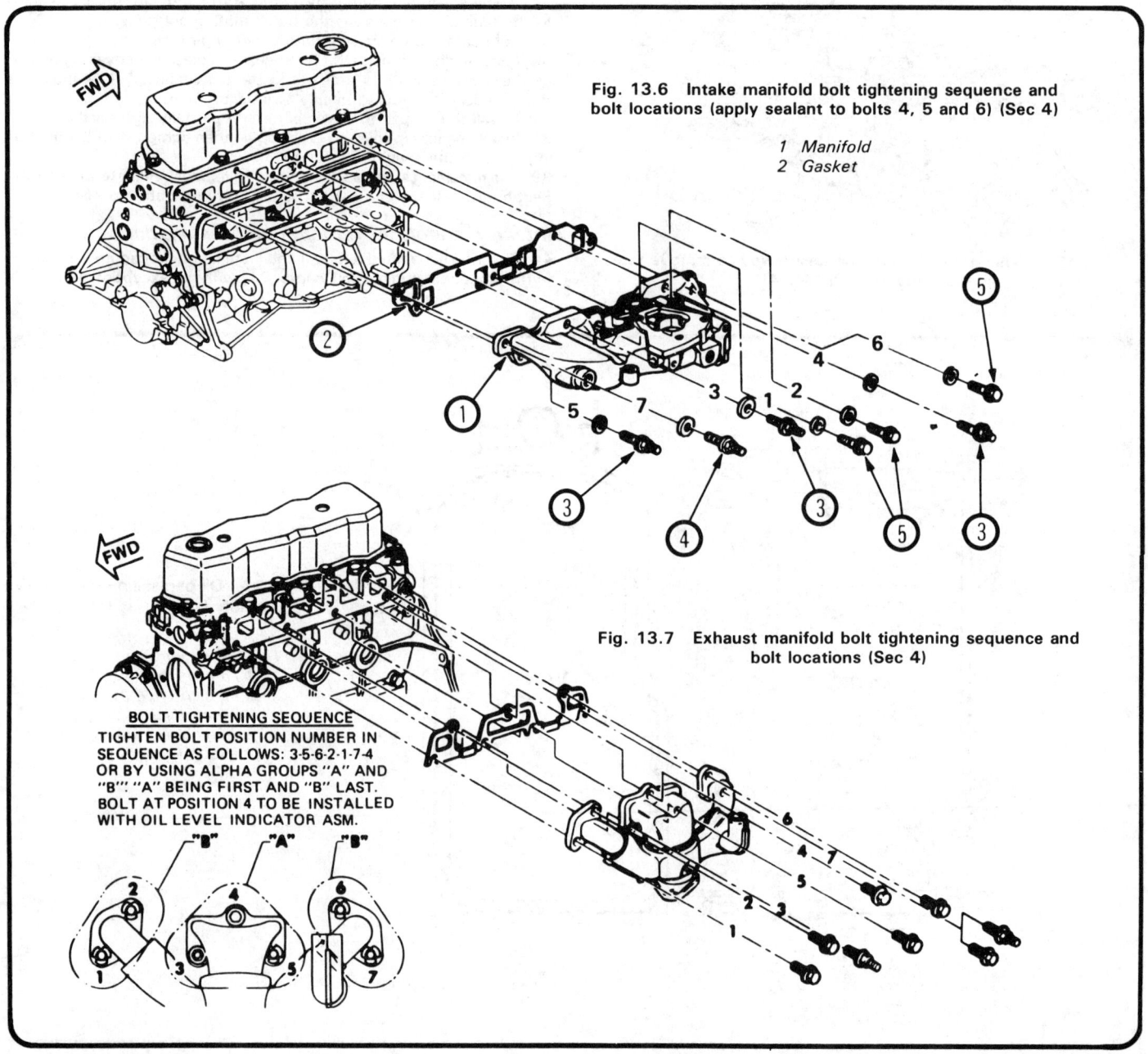

Fig. 13.6 Intake manifold bolt tightening sequence and bolt locations (apply sealant to bolts 4, 5 and 6) (Sec 4)

Fig. 13.7 Exhaust manifold bolt tightening sequence and bolt locations (Sec 4)

Rocker arm covers — removal and installation

Right side

1 Disconnect the negative battery cable from the battery.
2 Remove the EGR valve tube from the plenum.
3 Remove the coil and coil mounting bracket from the right cylinder head, marking the connectors with numbered tape to avoid confusion during reinstallation.
4 Remove the plenum and runners and disconnect the throttle body as outlined in Section 9 of this chapter.
5 Remove the rocker arm cover nuts and retainers.
6 Remove the rocker arm cover. If the cover sticks to the head, use a block of wood and a rubber hammer to dislodge it. If the cover still will not come loose, pry on it carefully, but do not distort the sealing flange surface.
7 Before installing the cover, clean all dirt, oil and old gasket material from the sealing surfaces of the cover and cylinder head with a scraper and degreaser.
8 Apply a continous 3/16-inch (5 mm) diameter bead of RTV sealant to the flange of the cover. Be sure to apply the sealant inboard of the bolt holes.
9 Place the rocker arm cover on the cylinder head while the sealant is still wet and install the mounting bolts. Tighten the bolts a little at a time to the specified torque.
10 Complete the installation by reversing the removal procedure.

Left side

11 Disconnect the negative battery cable from the battery.
12 Disconnect the air management hose (manual transmission only).
13 Remove the plenum and runners and disconnect the throttle body as outlined in Section 9 of this chapter.
14 Remove the air conditioning compressor and bracket and position them out of the way. *Do not, under any circumstances, disconnect the hoses while the system is under pressure!*
15 Remove the rocker arm cover nuts and retainers.
16 Refer to paragraphs 6 through 10 in this Section for the remainder of the procedure.

Intake manifold — removal and installation

17 If the vehicle is equipped with air conditioning, carefully note the routing of the hoses and the mounting of the compressor. You may be able to remove the intake manifold without disconnecting the system. If you are in doubt, take the vehicle to a dealer or air conditioning shop to have the system depressurized. *Do not, under any circumstances, disconnect the hoses while the system is under pressure.*
18 Disconnect the negative battery cable from the battery.
19 Drain the coolant.
20 Remove the port fuel-injection components (plenum, fuel rail and runner — refer to Section 9 in this chapter).
21 Disconnect the spark plug wires at the spark plugs, referring to the procedure described in Chapter 1.
22 Disconnect the wires at the coil, using numbered pieces of tape to label them.
23 Remove the distributor cap and the attached spark plug wires (refer to Chapter 1).
24 Remove the distributor (refer to Chapter 5).
25 Remove the air management hose (manual transmission only).
26 Disconnect the emission canister hoses, marking them with numbered pieces of tape to avoid confusion during installation.
27 Remove the left rocker arm cover.
28 Remove the air management bracket (manual transmission only).
29 Remove the right rocker arm cover.
30 Remove the upper radiator hose from the manifold.
31 Disconnect the heater hose from the manifold.
32 Remove the coolant switches from the front of the manifold.
33 Loosen all of the intake manifold bolts slightly, then remove the bolts.
34 Separate the manifold from the engine. Do not pry between the mating surfaces, because it could damage them. Instead, tap on the manifold using a hammer and a wooden block to break the seal and dislodge the manifold from the engine.
35 Clean the mating surfaces of the manifold and the cylinder heads and remove all traces of old RTV from the engine block ridges.
36 If a new manifold is being installed, transfer the external components from the old manifold to the new one.
37 For the remainder of the procedure, refer to Chapter 2, Part B, Section 4, Paragraphs 28 through 35.

Cylinder heads — removal and installation (1987 and later models)

38 The procedure outlined in Chapter 2, Part B, Section 8 is the correct procedure to follow. However, when removing the left cylinder head, the serpentine belt, air conditioning compressor and bracket and the power steering pump must also be removed. Right side cylinder head removal requires removal of the serpentine belt and belt tensioner, air pump and bracket, and the alternator and bracket.

Rear main bearing oil seal — replacement (engine in vehicle)

Note: *Special tools, as noted in the Steps which follow, are required for this procedure. They are available from your dealer or may in some cases be rented from an auto parts store or tool rental shop.*

39 Beginning in 1985, a 360-degree lip-type seal is utilized, which allows the oil pan to remain in place when performing this procedure.
40 Remove the transmission (refer to Chapter 7).
41 Remove the flywheel or driveplate.
42 Pry out the old seal, taking care not to mar the crankshaft surface (Fig 13.8). Inspect the crankshaft for scratches, burrs and nicks on the sealing surfaces.
43 A special seal installation tool (GM tool J-34686) is required to properly seat the seal in the bore without damaging it (Fig. 13.9).

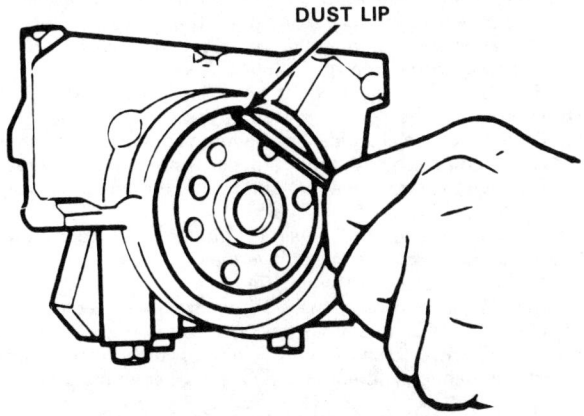

Fig. 13.8 Using a screwdriver, pry the rear main seal from the bore (be careful not to scratch the crankshaft seal surface or the edge of the bore) (Sec 5)

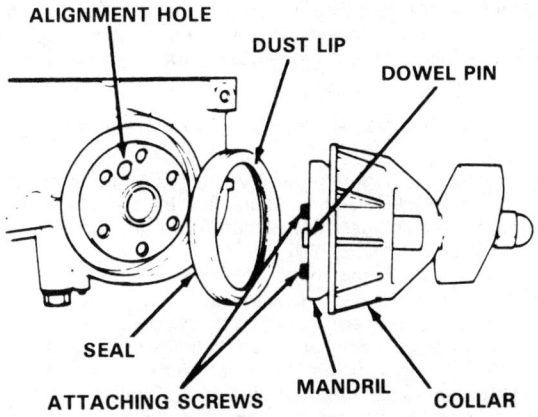

Fig. 13.9 V6 engine rear main seal installation details (Sec 5)

Lubricate the seal bore, seal lip and sealing surface on the crankshaft with engine oil. Slide the seal over the mandril on the tool until the dust lip on the seal bottoms squarely against the collar of the tool.
44 Position the dowel pin on the tool in the dowel pin hole in the crankshaft and secure the tool to the crankshaft.
45 Turn the T-handle of the tool until the collar pushes the seal into the bore completely. Make sure the seal is installed squarely.
46 To complete the operation install the flywheel (or driveplate) and transmission, then start the engine and check for leaks.

Crankcase front cover — removal and installation

47 Disconnect the negative battery cable from the battery.
48 Raise the front of the vehicle and support it on jackstands.
49 Remove the oil pan (1987 and later models — refer to Chapter 2, Part B, Section 9).
50 Drain the engine coolant (refer to Chapter 1).
51 Lower the vehicle.
52 Remove the water pump (refer to Chapter 3).
53 If equipped with air conditioning, remove the air conditioning compressor from the bracket and position the compressor aside, then remove the compressor mounting bracket. **Note:** *Do not disconnect any of the air conditioning lines unless the system has been depressurized by a dealer or repair shop, because personal injury may occur.* Most likely, disconnecting the lines will not be necessary.
54 Remove the vibration damper (refer to Chapter 2B).
55 Disconnect the lower radiator hose from the front cover.
56 Remove the remaining front cover bolts and separate the cover from the engine. If the cover sticks to the engine, use a block of wood and a rubber mallet to dislodge it. Do not pry between the cover and the engine block, as this will damage the sealing surface.
57 Clean the mating surfaces of the front cover and the block with a scraper and degreaser.
58 If the front cover oil seal is to be replaced, refer to Chapter 2, Part B, Section 14.
59 Apply a 3 mm bead of RTV sealant to the oil pan sealing surface of the front cover.
60 Install a new front cover gasket and position the front cover on the engine. Install the front cover bolts and tighten them to the specified torque (Fig. 13.10).
61 Install the water pump.
62 The remaining installation procedures are the reverse of removal.

Timing chain — removal and installation

63 Once the front cover has been removed as in the above described procedure, timing chain removal is identical to the procedure outlined in Chapter 2, Part B, Section 15.

6 V8 engine

General information

Although the V8 engine installed in 1985 and later vehicles is very nearly identical to earlier models, a few changes in component design have been made that consequently affect some service procedures. Where similarities exist in a given procedure, referrals are made to Chapter 2, Part C.

Rocker arm covers — removal and installation

Right side

1 Disconnect the negative battery cable from the battery.
2 Disconnect the EGR solenoid wires and hoses, marking them with numbered pieces of tape to avoid confusion during installation.
3 Remove the EGR solenoid.
4 Remove the air management valve and position it out of the way.
5 Remove the air management tubes.
6 Remove the rocker arm cover bolts and detach the cover. If the cover sticks to the cylinder head, use a block of wood and a rubber hammer to dislodge it. If it still will not come loose, carefully pry it loose, but be very careful not to distort the sealing flange.
7 Clean the rocker arm cover sealing flange and the mating surface on the cylinder head. Remove all traces of sealant and old gasket material.
8 Install a new gasket and position the cover on the head.
9 Install the rocker arm cover retaining bolts (or nuts, depending on application). Tighten them in a criss-cross pattern to the specified torque.
10 To complete the installation, reverse the removal procedure, then run the engine and check for leaks.

Left side

11 Refer to Chapter 2, Part C, Section 2 for removal and installation of the left side rocker arm cover.

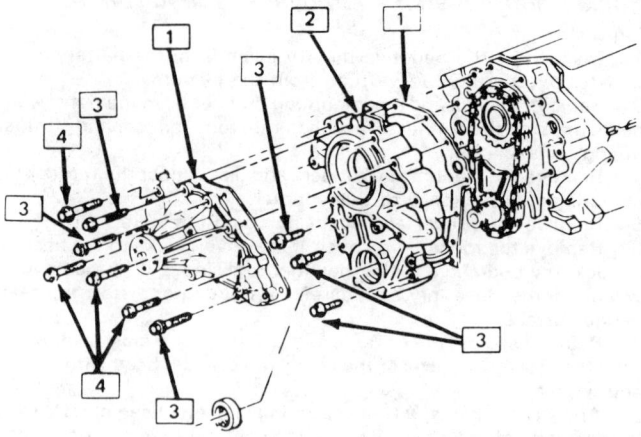

Fig. 13.10 V6 engine front cover and related components (Sec 5)

1 Gasket
2 Front cover
3 Tighten to 13 – 22 ft-lbs
4 Tighten to 20 – 35 ft-lbs

Fig. 13.11 V8 engine intake manifold bolt tightening sequence (Sec 6)

Intake manifold (Tuned-Port Injection) — removal and installation

12 Disconnect the negative battery cable from the battery.
13 Drain the coolant from the radiator (refer to Chapter 1).
14 Disconnect the accelerator cable, throttle valve cable and cruise control cable from the throttle body by removing the retaining clips and the three bolts from the cable bracket.
15 Remove the air intake duct from the front of the throttle body.
16 Disconnect the coolant hoses at the throttle body.
17 Disconnect the electrical and vacuum connectors from the throttle body, marking them with tape to avoid confusion during reassembly.
18 Remove the intake plenum and runners (refer to Section 9 in this Chapter).
19 Mark the distributor housing-to-manifold position and rotor-to-distributor housing position, remove the distributor hold-down bolt and lift the distributor from the engine.
20 Disconnect and mark the EGR electrical connector.
21 Mark and disconnect the vacuum hoses from the fuel pressure regulator, EGR valve and canister vacuum valve (near the left front of the intake manifold), then remove the EGR solenoid and the vacuum hoses.
22 Unplug the electrical connectors from the fuel injectors, the ground wire on the thermostat housing and the electrical connectors from the coolant temperature sensors on the front of the intake manifold. Mark the connectors with pieces of numbered tape to avoid confusion during reassembly. Position the wiring harnesses out of the way.

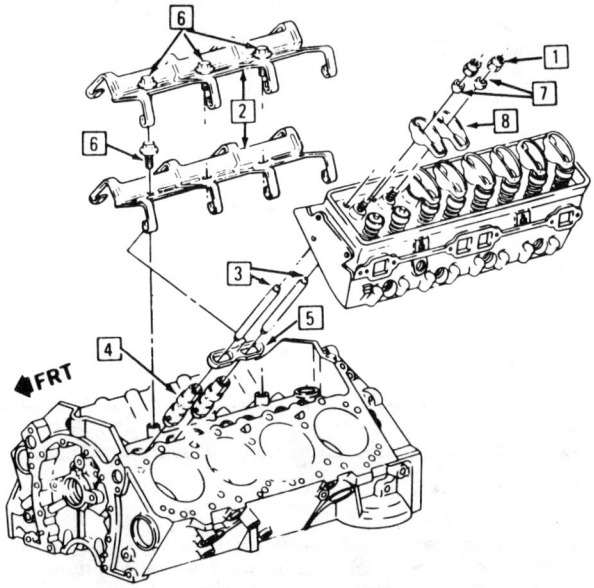

Fig. 13.12 Roller type valve lifters and related components — V8 engine (Sec 6)

1 Rocker arm nuts	*5 Restrictor*
2 Retainer	*6 Retainer bolt*
3 Pushrod	*7 Pivot ball*
4 Roller lifter	*8 Rocker arm*

23 Remove the intake manifold bolts, then separate the manifold from the engine. If the manifold sticks, do not pry between the sealing surfaces of the manifold and head. Instead, break it loose using a block of wood and a rubber hammer.
24 Clean the sealing surfaces on the manifold and heads, and also on the ridges between the cylinder banks. Remove all traces of old gasket and sealing material.
25 Install new intake manifold gaskets on the cylinder heads and apply a 3/16-inch (5 mm) bead of RTV sealant to the front and rear ridges of the block. Extend the bead 1/2-inch up each cylinder head to help hold the gaskets in place.
26 Lower the intake manifold into position and install the bolts.
27 Starting with the center manifold bolts, partially tighten the bolts in a criss-cross pattern to seat the manifold.
28 Using Figure 13.11 as a guide, tighten the intake manifold bolts to the specified torque.
29 The remaining steps are the reverse of removal.

Hydraulic lifters — removal, inspection and installation

30 Beginning in 1987, the V8 engine is equipped with roller type hydraulic valve lifters. Lifter retainers and restrictors are used to prevent the lifters from rotating on the camshaft lobes. Consequently, the cam lobes are ground with no front-to-rear taper.
31 Remove the rocker arm covers as described in this Chapter.
32 Remove the intake manifold as outlined in this Chapter.
33 Loosen the rocker arm nuts and rotate the rocker arms away from the pushrods (Chapter 2, Part C, Section 5).
34 Remove the pushrods and store them in order so they can be reinstalled in the same positions.
35 Remove the valve lifter retainer (Fig. 13.12).
36 Remove the valve lifter restrictors.
37 Remove the valve lifters. A special hydraulic lifter removal tool should be used, or a scribe can be positioned at the top of the lifter and used to force the lifter up. Do not use pliers or other tools on the outside of the lifter body as they will damage the finished surface and render the lifter useless.
38 The lifters should be kept separate for reinstallation in their original positions.
39 Inspection and disassembly are similar to the standard hydraulic lifter (Chapter 2, Part C, Section 5). However, the roller must be inspected for freedom of movement, excessive looseness, flat spots or pitting. The camshaft must also be inspected if any signs of abnormal wear are indicated.
40 When reinstalling the lifters, make sure they are replaced in their original bores and coat them with moly-based grease or engine assembly lube.
41 The remaining installation steps are the reverse of removal.

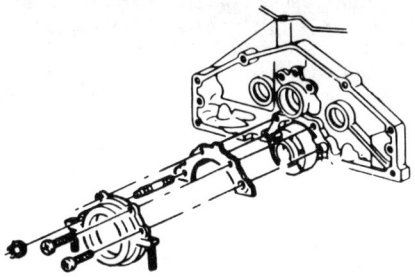

Fig. 13.13 Later model V8's are equipped with a one-piece rear main oil seal that fits into a housing bolted to the block (Sec 7)

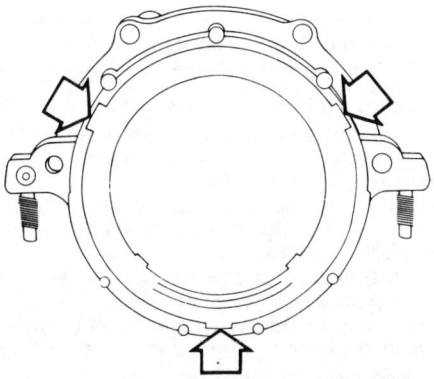

Fig. 13.14 To remove the seal from the housing, insert the tip of the screwdriver into each notch (arrows) and lever the seal out (Sec 7)

Rear main bearing oil seal — replacement (engine in vehicle)

42 Later models (1986 on) are equipped with a one-piece seal that requires an entirely different installation procedure. The transmission (Chapter 7) and oil pan must be removed to gain access to the seal housing. Once the housing is removed, the seal replacement procedure in Section 7 of this Supplement can be followed.

Crankcase front cover — removal and installation

43 On later models (1985 on) it is necessary to first remove the oil pan due to the front cover-to-oil pan seal arrangement. This will facilitate removal and installation and prevent distortion of the front cover. Follow the oil pan and front cover removal procedures in Chapter 2, Part C.

7 General engine overhaul procedures

Rear main oil seal (V8 engine only) — installation

1 Later models are equipped with a one-piece seal that requires an entirely different installation procedure. The crankshaft must be installed first and the main bearing caps bolted in place, then the new seal should be installed in the housing and the housing bolted to the block (Fig. 13.13).
2 Before installing the crankshaft, check the seal contact surface very carefully for scratches and nicks that could damage the new seal lip and cause oil leaks. If the crankshaft is damaged, the only alternative is a new or different crankshaft.
3 The old seal can be removed from the housing by inserting a large screwdriver into the notches provided and prying it out (Fig. 13.14). Be sure to note how far it is recessed into the housing bore before

removing it; the new seal will have to be recessed an equal amount. Be very careful not to scratch or otherwise damage the bore in the housing or oil leaks could develop.

4 Make sure the housing is clean, then apply a thin coat of engine oil to the outer edge of the new seal. The seal must be pressed squarely into the housing bore, so hammering it into place is not recommended. If you do not have access to a press, sandwich the housing and seal between two smooth pieces of wood and press the seal into place with the jaws of a large vise. The pieces of wood must be thick enough to distribute the force evenly around the entire circumference of the seal. Work slowly and make sure the seal enters the bore squarely.

5 The seal lips must be lubricated with clean engine oil or moly-based grease before the seal/housing is slipped over the crankshaft and bolted to the block. Use a new gasket — no sealant is required — and make sure the dowel pins are in place before installing the housing.

6 Tighten the nuts/screws a little at a time until they are all snug.

8 Cooling, heating and air conditioning systems

Electric cooling fan — removal and installation

1 Disconnect the negative battery cable from the battery.

2 Unplug the electrical connector from the fan motor and position the wiring harness out of the way.

3 Remove the fan frame-to-radiator support bolts and lift the fan assembly out of the engine compartment, taking care not to damage the radiator fins (Fig. 13.15).

4 If the fan blade is to be replaced, remove the fan retaining nut and remove the fan from the motor shaft.

5 If the motor is to be replaced, remove the fan retaining nut and the fan, then remove the four motor-to-frame mounting bolts and detach the motor from the frame.

6 Installation is the reverse of removal.

Electric cooling fan motor and circuit — check

Note: *Refer to Section 12 of this Supplement for fan motor circuit wiring diagrams.*

7 The cooling fan temperature switch is located on the right side cylinder head, above the starter solenoid. Disconnect the electrical connector from the switch.

8 Turn the ignition key to the On position.

9 Ground the cooling fan temperature switch wire. The fan should operate. **Note:** *If the fan comes on, but has not been coming on when the engine is operating at above normal temperature, the temperature switch is faulty.*

10 If the fan does not come on, check the fan motor circuit fuse. Replace it if it is blown and search for the cause of the blown fuse.

11 If the fuse is not faulty, unplug the cooling fan relay located at the right side of the radiator support (engine codes F and 8), or on the relay bracket at the left front of the dash (engine codes G, H, E and S). To find your particular engine code, refer to the Specifications at the beginning of this Supplement. Jump terminal A (black/red wire) to terminal E (red wire) (terminal D for engine codes G and H). If the fan now comes on, the relay is faulty. Replace it.

12 If the fan does not come on during either test, check for battery voltage to the red wire with a test light.

13 If battery voltage is present, either the fan motor is faulty, the wire between the relay and the fan motor is faulty or the wire between the motor and ground is broken or disconnected.

14 To check the wire between the relay and fan motor, bypass the relay as in Step 11, attach a test light to ground and probe the black/red wire at the fan motor connector. If the light does not come on, there is an open in this wire. Locate and repair the open.

15 To check the fan motor, connect a fused jumper wire from the battery directly to the fan motor hot terminal (black/red wire). If the motor does not come on, check the ground wire. If the ground wire is good, replace the fan motor.

16 To check the fan motor ground wire, connect a test light to a voltage source and probe the black wire at the fan motor connector. If the light does not come on, there is an open in the ground wire. Repair or replace it.

17 Aside from these checks, further diagnosis of the electric cooling fan motor and circuit is beyond the scope of the home mechanic, because the fan is ECM controlled during operation of the air conditioning system.

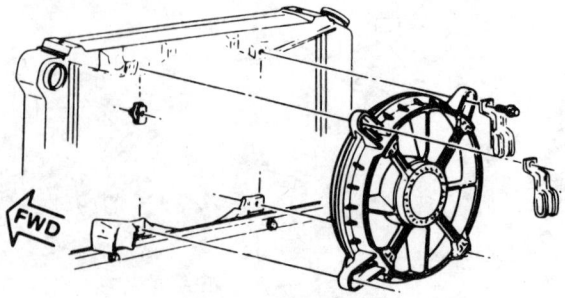

Fig. 13.15 To remove the electric fan, unplug the wire harness and remove the four mounting bolts (Sec 8)

9 Fuel injection systems

General information

Throttle body injection (TBI)

Beginning with the 1988 model year, the 5.0L V8 engine was offered with a throttle body injection system (this engine/fuel injection combination will be denoted by a letter E in the eighth position of the V.I.N.). This setup uses the model 220 TBI unit, which is nearly identical to the system described in Chapter 4 (for the model 300 TBI), except that it contains two fuel injectors instead of one. The service procedures outlined in Chapter 4 are valid for the 220 model TBI, but it is not necessary to relieve the fuel pressure before disconnecting the fuel lines, as the pressure regulator employs a constant bleed feature. However, the negative cable must be disconnected from the battery before any work is performed to avoid the possibility of energizing the fuel pump with the fuel lines disconnected.

Port fuel injection (TPI and MPFI)

The Tuned-Port Injection (V8) and Multi-Port Fuel Injection (V6) systems used on these models incorporate an injector solenoid at each intake port. The Electronic Control Module (ECM) controls the opening time of the injectors, which determines the amount of fuel delivered to the engine.

All of the injectors operate simultaneously. Half of the fuel required by each cylinder is injected with each injector pulse. The pulses occur once every crankshaft revolution. Therefore, the full charge needed is available in the intake port for induction into the cylinder when the intake valve opens every other revolution.

Electronic Control Module

In determining how much fuel is required at any given moment, the Electronic Control Module processes signals from sensors recording engine coolant temperature, exhaust oxygen content, throttle position, intake air mass, engine rpm, vehicle speed and accessory load.

Mass Air Flow (MAF) sensor

The Mass Air Flow (MAF) sensor measures the mass (weight) of the air entering the intake manifold, rather than the volume, which gives better control of the fuel/air mixture. The mass of any given volume of air is dependent upon the temperature of the air. Cold air weighs more than warm or hot air, and therefore the colder the air is, the more fuel is required to maintain the proper fuel/air ratio.

On these models, the air mass sensor system utilizes a heated film to measure the intake air mass. The temperature of the incoming air is constantly monitored and the air mass sensor plate is maintained at exactly 75 degrees above the temperature of the air. Since the incoming air, passing over the plate, acts to cool the plate, the measurement of the amount of electrical energy needed to maintain that 75 degree temperature differential accurately indicates the mass of air entering the engine, and therefore the amount of fuel necessary to mix with that mass to provide the proper fuel/air ratio.

Idle air control

The idle speed on the Tuned-Port and Multi-Port injection systems is also controlled by the electronic control module, through an idle air control bypass channel. When the engine is cold or when accessory loads, such as air conditioning, necessitate a higher idle speed, the ECM, through a stepper motor, opens the bypass valve to provide more air into the system. At the same time, the ECM increases the fuel injector open time to maintain the proper fuel/air ratio.

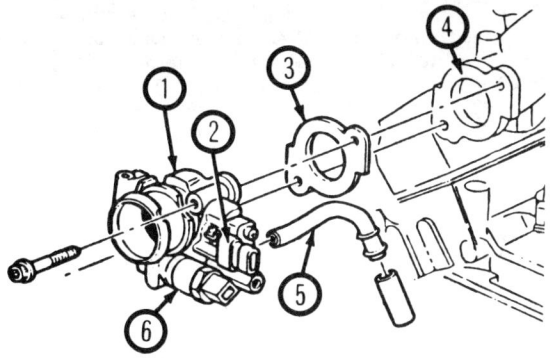

Fig. 13.16 MPFI throttle body and related components (Sec 9)

1 Throttle body
2 TPS
3 Gasket
4 Plenum
5 Cold start valve air tube
6 IAC valve

Fuel pump

While the open time of the injectors is controlled by the electronic control module, the actual fuel delivery is determined by the fuel pressure, which must be maintained at a constant level. This is done by a positive displacement roller vane pump mounted in the fuel tank. The pump intake is mounted in a special sump in the fuel tank so that a low fuel level, combined with hard cornering, acceleration or deceleration, cannot uncover the pump inlet, causing a pressure drop.

Fuel pressure regulator

A fuel pressure regulator maintains the required pressure in the system by returning unneeded fuel to the tank. A fuel accumulator is used to dampen the pulses which would otherwise be set up in the fuel rail and fuel line when the injectors open and close.

Fuel pump — testing

1 Before testing the fuel pump, check the fuel pump fuse (FP in the fuse box on the left underside of the dashboard). If the fuse is blown, replace it before proceeding.

2 Remove the plastic cover from the Assembly Line Communications Link (ALCL) located underneath the instrument panel, next to the steering column. Energize the fuel pump by applying battery voltage to terminal G. You should hear the pump come on inside the fuel tank. If the pump does not come on, there is a loose connection or an open in the fuel pump circuit or the fuel pump is faulty. Check the entire fuel pump electrical circuit (see *Wiring Diagrams*) before diagnosing the fuel pump as faulty. If the circuit checks out, replace the pump.

3 If the pump runs, check the fuel pressure.

TPI and MPFI fuel pressure test

4 Remove the cap from the Schrader valve on the fuel rail and install a pressure gauge. The gauge must be capable of reading between 34 and 39 psi.

5 Start the engine and note the fuel pressure reading. It should be between 34 and 39 psi.

6 If there is no pressure (even though the pump is running), the in-line fuel filter is probably clogged. Replace it and check the pressure again. If there is still zero pressure, the fuel feed line may be clogged. If the fuel feed line is clear, check the fuel pump inlet filter.

7 If the fuel pressure is below 34 psi, the in-line fuel filter may be restricted. Replace it and check the pressure again. If it is still low, check for a loose fuel line fitting or a punctured fuel line. Also inspect the pressure regulator, fuel rail, cold start valve and injectors for a leak. If nothing is leaking, block the fuel return line and apply 12-volts to the fuel pump test connector (terminal G of the ALCL). Note the indicated pressure. If the pressure is now above 39 psi, the fuel pressure regulator is faulty. If the pressure is still less than 34 psi, replace the fuel pump.

8 If the fuel pressure is above 39 psi, disconnect the fuel return line flexible hose and attach a 5/16-inch ID rubber hose to the pressure regulator side of the return line. Insert the other end into an approved gasoline container. Note the fuel pressure within two seconds after the ignition is turned to the On position. If pressure is still above 39 psi, replace the regulator. If the fuel pressure is now between 34 and 39 psi, the return line is restricted. Locate and remove the restriction.

TPI and MPFI fuel flow test

9 Disconnect the fuel feed line from the fuel rail and connect a hose to the feed line. Run the other end of the hose into an unbreakable container.

10 If you have an assistant to monitor the hose and container, the easiest way to activate the fuel pump is at the Assembly Line Communications Link (ALCL) underneath the dashboard, near the steering column. Remove the plastic cover from the ALCL diagnostic connector and apply battery voltage to terminal G of the connector. However, if you do not have an assistant, you can keep an eye on the container while energizing the fuel pump by using the fuel pump relay located in the left rear corner of the engine compartment, on the firewall. Remove both relay mounting nuts, disconnect the relay plug and set the relay aside. Using a jumper wire with alligator clips on both ends, jump terminals A and E of the relay plug to apply battery voltage to the pump.

11 The fuel pump should supply 1/2-pint or more in 15 seconds. If the fuel system flow is below the minimum, inspect the fuel system for a restriction between the pump and the fuel rail. If there is no restriction, check the fuel pump pressure.

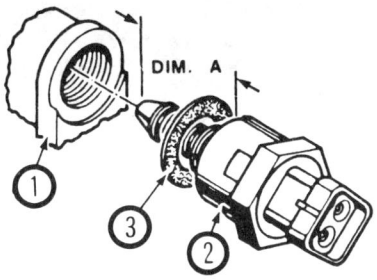

Fig. 13.17 The length of dimension A must be no more than 1-1/8 inch (Sec 9)

1 Idle air/vacuum signal housing
2 Idle Air Control (IAC) valve
3 Gasket

Multi-Port Fuel Injection system — removal, overhaul and installation

Warning: *Before servicing an injector, fuel rail or pressure regulator, the pressure in the fuel system must be relieved to minimize the risk of fire and injury (refer to Chapter 4). After servicing the fuel system, cycle the ignition between On and Off several times (wait 10 seconds between cycles) and check the system for leaks.*

Throttle Body — removal and installation

12 Disconnect the cable from the negative terminal of the battery.
13 Unplug the IAC and TPS connectors.
14 Disconnect the coolant lines.
15 Disconnect the throttle linkage
16 Detach the air inlet duct.
17 Remove the throttle body bolts and detach the throttle body (Fig. 13.16).
18 Install the throttle body and gasket and tighten the bolts to the specified torque.
19 Reconnect the air inlet duct.
20 Reconnect the throttle linkage.
21 Reconnect the coolant lines.
22 Plug in the TPS and IAC electrical connectors.
23 Reconnect the cable to the negative terminal of the battery.

Idle Air Control (IAC) valve — removal and installation

24 Unplug the electrical connector from the Idle Air Control (IAC) valve assembly.

25 Remove the IAC valve assembly from the idle air/vacuum signal housing assembly. **Caution:** *Do not remove any thread locking compound from the threads.*

26 Remove the IAC valve assembly gasket and discard it.

27 Clean the gasket mounting surface of the idle air/vacuum signal housing assembly to ensure a good seal. **Caution:** *The IAC valve assembly itself is an electrical component and must not be soaked in any liquid cleaner or solvent or damage may result.*

28 Before installing the IAC valve assembly, the position of the pintle must be checked. If the pintle is extended too far, damage to the assembly may occur.

29 Measure the distance from the gasket mounting surface of the IAC valve assembly to the tip of the pintle (Fig. 13.17).

30 If the distance is greater than 1-1/8 inch, reduce it as follows:

a) If the IAC valve assembly has a collar around the electrical connector end, apply firm hand pressure on the pintle to retract it (a slight side-to-side motion may help).

b) If the IAC valve assembly has no collar, compress the pintle retaining spring toward the body of the IAC and try to turn the pintle *clockwise*. If the pintle will turn, continue turning it until the 1-1/8 inch dimension is reached. Return the spring to its original position with the straight part of the spring end lined up with the flat surface under the pintle head. If the pintle will not turn, apply firm hand pressure to retract it.

31 Position the new gasket on the IAC valve assembly.
32 Install the IAC valve in the idle air/vacuum signal housing assembly.
33 Tighten the IAC valve to the specified torque.
34 Plug in the electrical connector at the IAC valve assembly. **Note:** *No adjustment is made to the IAC assembly after reinstallation. IAC resetting is controlled by the ECM when the engine is started.*

Plenum — removal and installation

35 Remove the cable from the negative terminal of the battery.
36 Remove the vacuum lines and throttle cable bracket bolts.
37 Remove the EGR pipe-to-EGR valve base bolts.

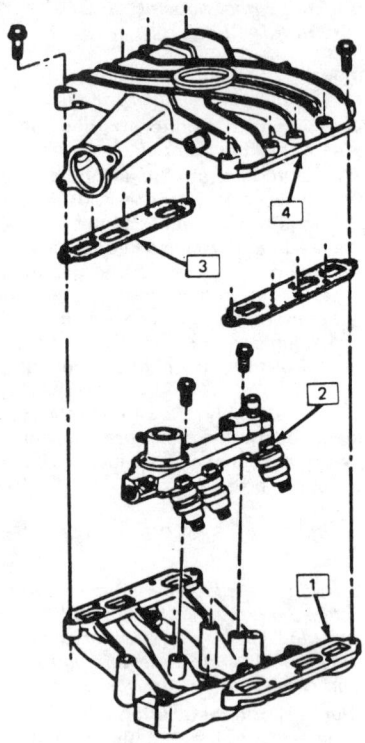

Fig. 13.18 Exploded view of the MPFI plenum, fuel rail and intake manifold components (Sec 9)

1 Intake manifold
2 Fuel rail assembly
3 Gasket
4 Plenum

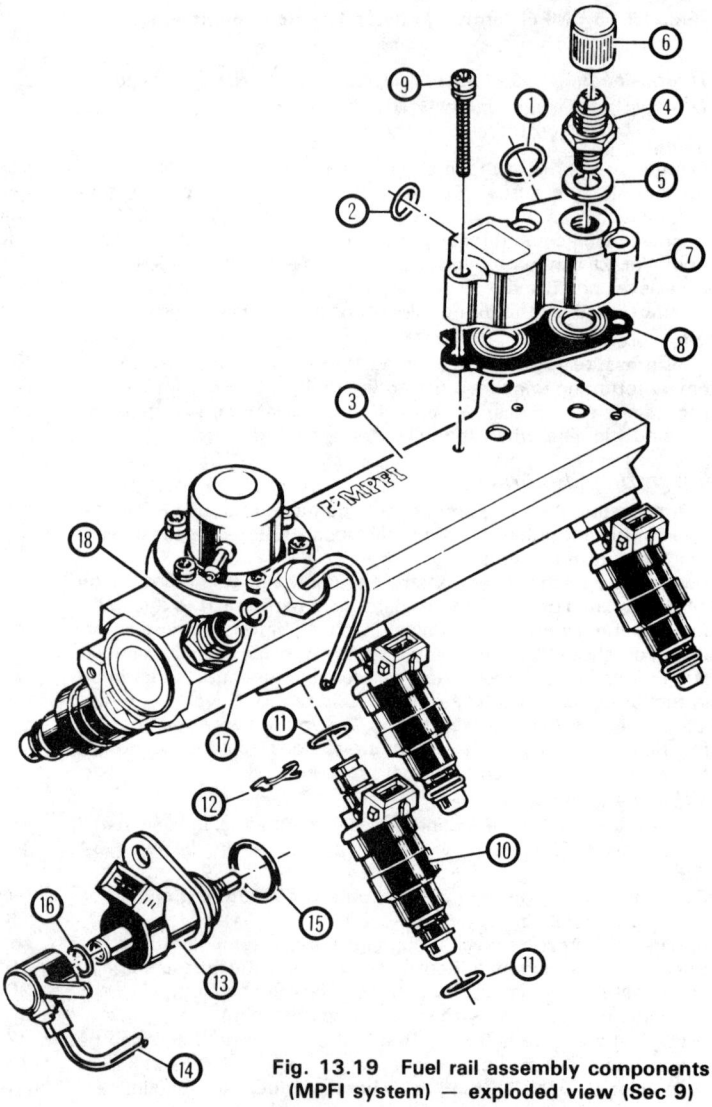

Fig. 13.19 Fuel rail assembly components (MPFI system) — exploded view (Sec 9)

1 Fuel inlet line O-ring
2 Fuel return line O-ring
3 Fuel rail and pressure regulator assembly
4 Fuel pressure connection assembly (Schrader valve)
5 Fuel pressure connection assembly seal
6 Fuel pressure connection cap
7 Fuel block
8 Fuel block seal
9 Fuel block attaching screw
10 Port injector
11 Injector O-ring seal
12 Injector retainer clip
13 Cold start valve
14 Tube and body assembly
15 Valve O-ring seal
16 Body O-ring seal
17 Tube O-ring seal
18 Cold start fitting

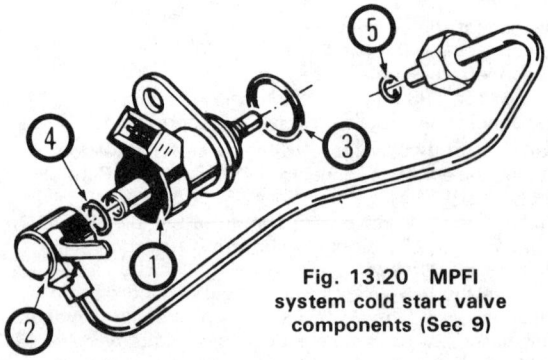

Fig. 13.20 MPFI system cold start valve components (Sec 9)

1 Cold start valve
2 Tube and body assembly
3 Valve O-ring seal
4 Body O-ring seal
5 Tube O-ring seal

38 Remove the throttle body.
39 Remove the plenum bolts.
40 Remove the plenum and gaskets (Fig. 13.18).
41 Install the plenum and gaskets.
42 Install the plenum bolts and tighten them to the specified torque.
43 Install the throttle body. Tighten the throttle body bolts to the specified torque.
44 Install the EGR pipe.
45 Install the throttle cable bracket bolts.
46 Install the vacuum lines.
47 Attach the negative battery cable to the battery.

Fuel rail and related components — removal and installation

Warning: *Before any work is performed on the fuel lines, fuel rail or injectors, the fuel system pressure must be relieved (see Chapter 4).*

48 Detach the negative battery cable from the battery.
49 Remove the plenum.
50 Remove the cold start valve line at the fuel rail.
51 Remove the fuel lines at the fuel rail.
52 Remove the vacuum line at the regulator.
53 Unplug the injector electrical connectors.
54 Remove the fuel rail retaining bolts.
55 Remove the fuel rail and injectors (Fig. 13.19).
56 **Caution:** *Use care when handling the fuel rail assembly to avoid damaging the injectors.*
57 To remove the cold start valve, disconnect the electrical connector.
58 Remove the cold start valve retaining bolt.
59 Disconnect the valve from the tube and body assembly by bending the tab back to permit unscrewing of the valve.
60 Install a new cold start valve O-ring seal and body O-ring seal on the cold start valve (Fig. 13.20). **Note:** *An eight digit identification number is stamped on the side of the fuel rail assembly (Fig. 13.21). Refer to this number if servicing or part replacement is required.*
61 Install a new O-ring seal on the tube and body assembly.
62 Turn the valve completely into the body assembly.
63 Turn the valve back one full turn, until the electrical connector is at the top position.
64 Bend the tang of the body forward to limit rotation of the valve to less than one full turn.
65 Before reinstalling the valve assembly, coat the O-ring seals with engine oil.
66 To remove the fuel injectors, rotate the injector retaining clip(s) to the released position (Fig. 13.22). **Caution:** *To prevent dirt from entering the engine, the area around the injectors should be cleaned before servicing.*
67 Remove the injector(s).
68 Inspect the injector O-ring seal(s). Replace them if they are damaged.
69 Install the new O-ring seal(s), as required, on the injector(s) and lubricate them with engine oil.
70 Install the injectors on the fuel rail.
71 Rotate the injector retainer clips to the locked position.
72 The pressure regulator is factory adjusted and is not serviceable. Do not attempt to remove the regulator from the fuel rail.
73 Before installing the fuel rail, lubricate all injector O-ring seals with engine oil.
74 Install the fuel rail and injector assembly.
75 Install the fuel rail retaining bolts and tighten them to the specified torque.
76 Plug in the injector electrical connectors.
77 Install the vacuum line at the regulator.
78 Install the fuel lines at the fuel rail.
79 Install the cold start valve line at the fuel rail.
80 Install the plenum.
81 Attach the negative battery cable to the battery.

Tuned-Port Injection system components — removal

Note: *Although the following procedure details the complete removal and installation of the principal components of the Tuned-Port Injection (TPI) system, the removal or replacement of some parts is independent from the rest of the system. Therefore, read the following procedures carefully.*

82 Relieve the fuel pressure (Chapter 4).
83 Disconnect the negative cable from the battery.

Fuel pressure connection assembly (Schrader valve)

Note: *Removal and installation of the fuel pressure connection assembly is the only procedure in this Section which requires no preliminary removal of other fuel injection components. Unless you are only replacing the fuel pressure connection assembly seal, or the valve itself, proceed to Step 87.*

84 Thoroughly clean the area around the fuel pressure connection assembly.
85 Remove the fuel pressure connection assembly. Note which end of the valve is threaded into the fuel rail (Fig. 13.23). Discard the old seal.
86 Install a new seal on the valve, install the fuel pressure connection assembly and tighten it to the specified torque.

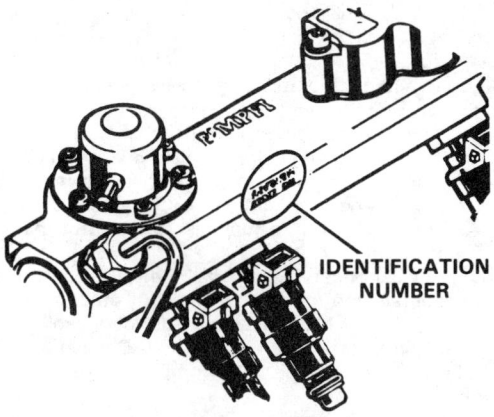

Fig. 13.21 The fuel rail assembly has an eight digit identification number stamped on the side — refer to it every time you buy a part for the fuel rail (Sec 9)

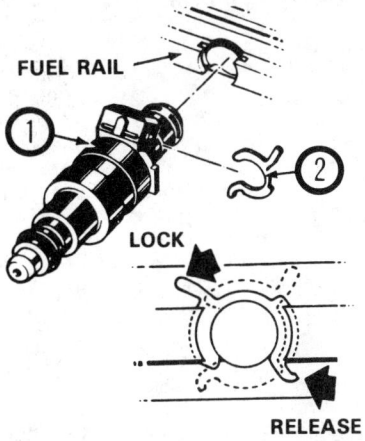

Fig. 13.22 A Port injector with an injector retaining clip is removed by rotating the clip clockwise to the release position (Sec 9)

1 Injector 2 Retaining clip

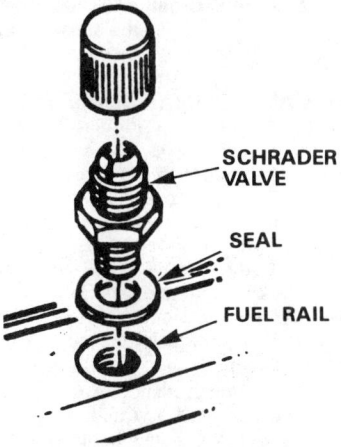

Fig. 13.23 Exploded view of the fuel pressure connection assembly — the ends of the valve are not the same so it is important to note which end threads into the fuel rail (Sec 9)

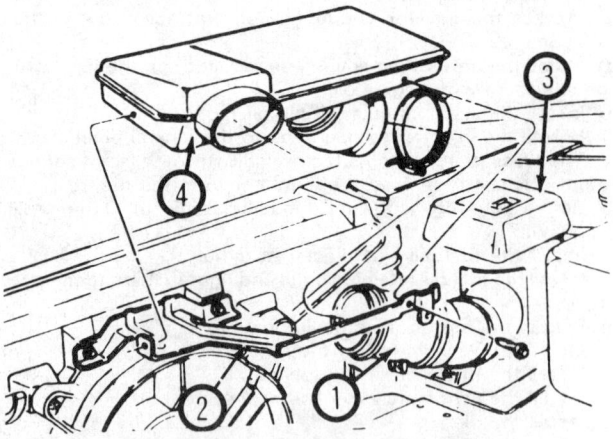

Fig. 13.24 Mass air flow sensor (1), bracket (2), air cleaner housing (3) and intake duct (4) locations (5.0 and 5.7L engines) (Sec 9)

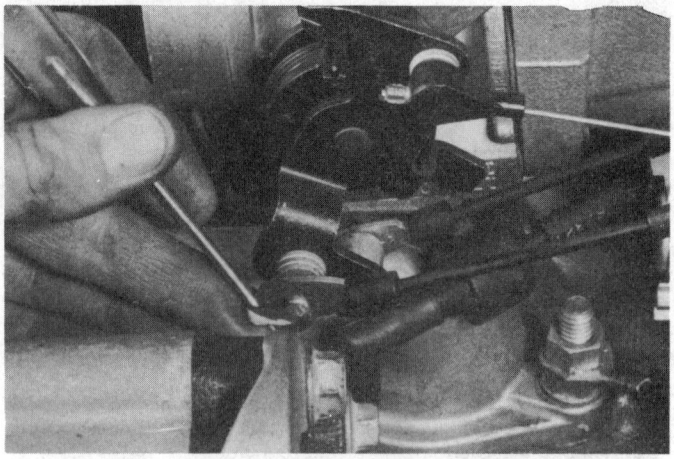

Fig. 13.25 Two retainer clips must be removed — one locks the cruise control cable and the other locks the throttle cable to the throttle lever arm (Sec 9)

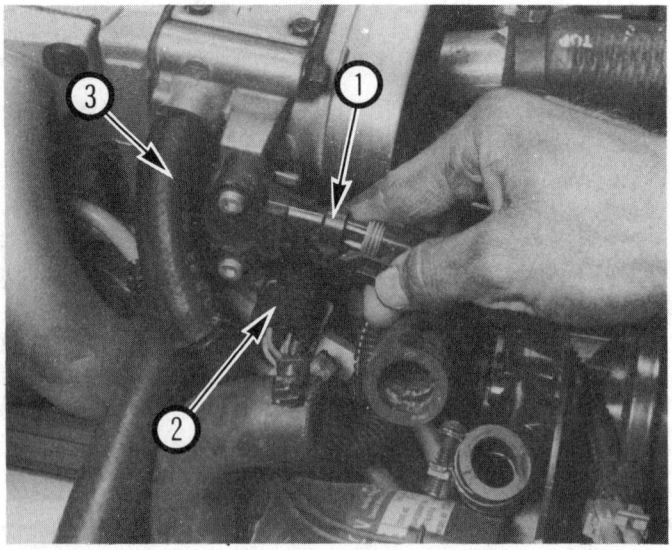

Fig. 13.26 Disconnect the wire harness from the TPS (1) and the IAC valve (2), then detach the PCV hose (3) from the throttle body (Sec 9)

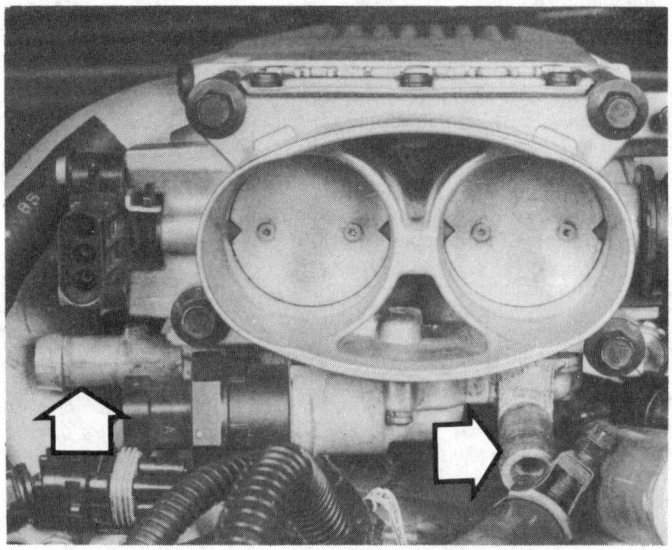

Fig. 13.27 Loosen the hose clamps and detach the coolant hoses (arrows) from the throttle body coolant cover (Sec 9)

Mass Air Flow (MAF) sensor

87 If you are only removing the MAF sensor and flexible air intake duct to gain access to other components, loosen the hose clamp which attaches the duct to the throttle body and the clamp which attaches the MAF sensor to the air cleaner housing. Remove the mounting screws and disconnect the electrical connector (Fig. 13.24). If you are replacing the MAF sensor itself, do not disconnect the duct from the throttle body — loosen the hose clamp attaching the duct to the rear of the MAF sensor and separate the duct from the MAF sensor. **Caution:** *the MAF sensor is delicate — if you are going to reinstall the existing unit, handle it carefully.*

Throttle body

88 Detach the cruise control, accelerator and transmission TV cables from the throttle body (Fig. 13.25).

89 Remove the three cable bracket bolts and separate the cable bracket from the plenum.

90 Unplug the wiring harness connectors from the Idle Air Control (IAC) and Throttle Position Sensor (TPS). Disconnect the PCV hose, which is attached to the throttle body right behind the TPS (Fig. 13.26).

91 Drain the cooling system (Chapter 1), then loosen the hose clamps and detach the two coolant hoses from the throttle body fittings (Fig. 13.27).

92 There is a vacuum hose connected between the left rear underside of the throttle body and a vacuum damper valve situated at the left front corner of the plenum. Disconnect this hose from the damper valve.

93 If you are removing or replacing the throttle body or the plenum, remove the throttle body mounting bolts, then remove the throttle body and gasket from the front of the plenum. If you are removing/replacing an intake runner gasket, the fuel rail or any component attached to the fuel rail (pressure regulator or a fuel injector), do not detach the throttle body from the plenum.

94 The throttle body is not serviceable. If the throttle shaft or either throttle plate is damaged or worn, the throttle body must be replaced.

95 The four gaskets for the IAC valve, the IAC valve coolant cover, the clean air cover, the throttle body itself, the IAC valve or the TPS may be replaced if necessary . The following procedure outlines disassembly of the throttle body.

96 If you intend to reinstall the same Throttle Position Sensor (TPS) switch, scribe alignment marks between the top and bottom edges of the TPS and the throttle body to assure proper reinstallation. **Caution:** *Failure to scribe alignment marks for the proper reinstallation of the TPS will necessitate adjustment with a digital voltmeter — if you don't have this tool, you must drive the vehicle to a dealer immediately after completing reassembly of the fuel injection system.*

97 Remove the TPS mounting screws, lock washers, retainers and the TPS.

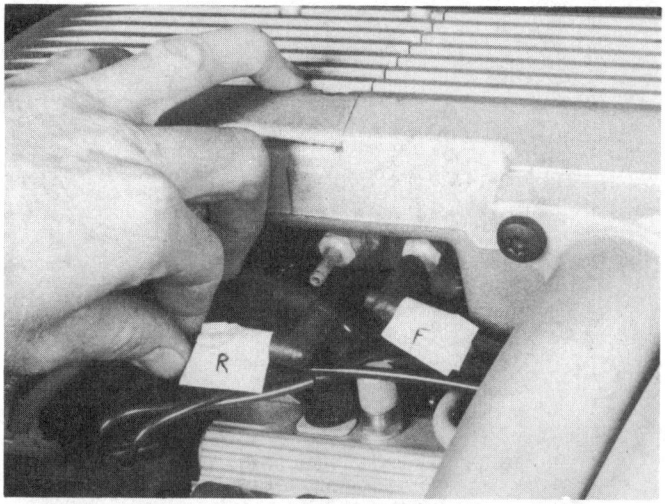

Fig. 13.28 The vacuum lines at the right rear corner of the plenum should be labelled and disconnected (Sec 9)

Fig. 13.29 Unscrew the power brake booster vacuum line fitting (arrow) and pull it back slightly from the plenum (Sec 9)

Fig. 13.30 There are four Torx bolts attaching the upper end of each intake runner assembly to the plenum (only three are visible in this photo of the right side runner) (Sec 9)

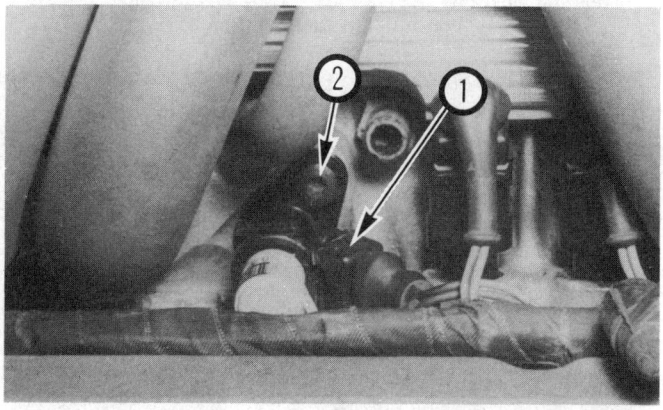

Fig. 13.31 To remove the cold start valve from the intake manifold, disconnect the fuel injector wiring harness connector (1) from the cold start valve electrical terminal, then remove the cold start valve flange bolt (2) . . .

98 Unscrew and remove the IAC valve and gasket.
99 Remove the idle air/vacuum signal housing and coolant cover assembly mounting screws, the cover and the gasket.
100 Remove the clean air cover screws, the clean air cover and the clean air cover gasket.
101 The throttle body may be cleaned in a cold immersion-type solvent.
Caution: *The TPS, IAC valve and all four gaskets must not be soaked in solvent or they will be damaged. If the TPS or the IAC valve are mounted in the throttle body, do not immerse the throttle body.*
102 Clean all metal parts thoroughly and blow them dry. Be sure that all air passages are free of burrs and dirt.
103 Inspect the casting mating surfaces for damage that could affect gasket sealing.
104 Reassembly is the reverse of disassembly. Use new gaskets and tighten all screws securely. Do not install the IAC and TPS until the throttle body has been bolted to the plenum.

Plenum

105 Label and disconnect the two vacuum lines from the right rear corner of the plenum (Fig. 13.28).
106 Unscrew the power brake booster vacuum line threaded fitting at the left rear corner of the plenum (Fig. 13.29).
107 Unplug the wiring harness electrical connector from the Manifold Air Temperature (MAT) sensor protruding from the underside of the plenum, just forward of the pressure regulator.
108 Disconnect the PCV fresh air intake assembly from the right rocker arm cover, disconnect it from the PCV assembly and remove it. Remove the two flanged screws from the right fuel injector wire harness shroud and remove the shroud.
109 Remove the four Torx bolts which attach the upper end of the right intake runner assembly to the plenum (Fig. 13.30) and the five Torx bolts which attach the lower end of the runners to the intake manifold. One of the bolts holding the runners to the manifold can only be removed by reaching under the plenum from the left side of the engine.
110 Remove the four Torx bolts which attach the upper end of the left intake runner assembly to the plenum.
111 Remove the plenum.

Intake runner assemblies and cold start valve assembly

112 Remove the right rocker arm cover upper nuts to free the PCV line brackets. Remove the PCV line to provide clearance for the intake runner flanges. Remove the right intake runner assembly.
113 Label and disconnect the vacuum hoses at the EGR valve and fuel pressure regulator.
114 The left runner has one Torx bolt that is accessible only from the opposite side of the engine. It's located just below the pressure regulator. Remove it.
115 Disconnect the PCV valve assembly from the left rocker arm cover and the PCV line assembly and remove it. Remove the two flanged screws from the left fuel injector wire harness shroud and remove the shroud.
116 Disconnect the wire harness electrical connector from the cold start valve (Fig. 13.31). Disconnect the connectors from the left bank of fuel injectors.
117 Remove the left rocker arm cover upper nuts to free the PCV line

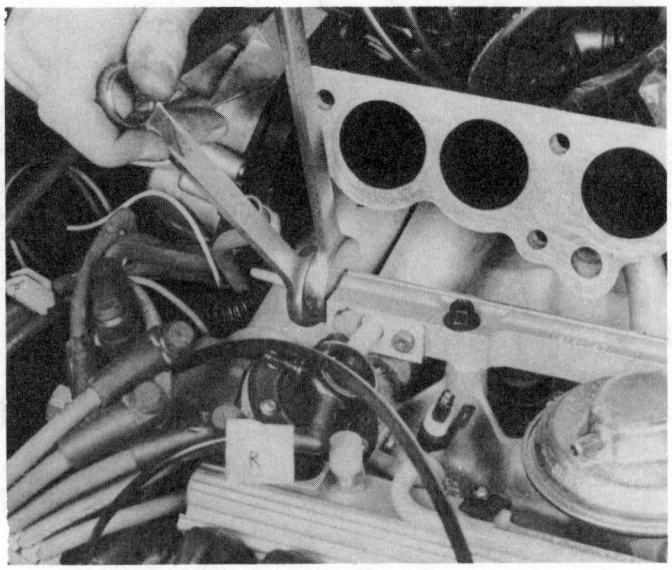

Fig. 13.32 . . . then, using a backup wrench on the fuel rail nut, unscrew the cold start valve line threaded fitting from the left fuel rail (Sec 9)

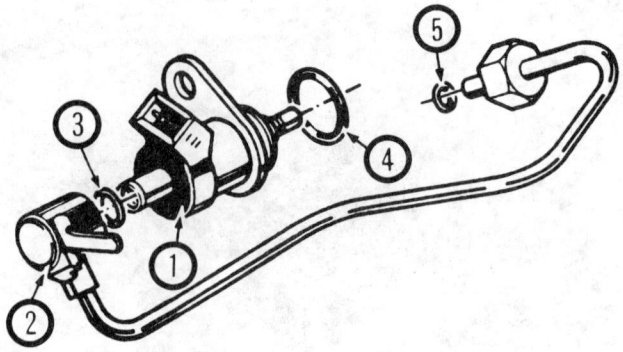

Fig. 13.33 Exploded view of the TPI cold start valve assembly (Sec 9)

1 *Cold start valve*
2 *Tube and body assembly elbow*
3 *Valve-to-elbow O-ring*
4 *Valve-to-intake O-ring*
5 *Tube and body assembly threaded fitting O-ring*

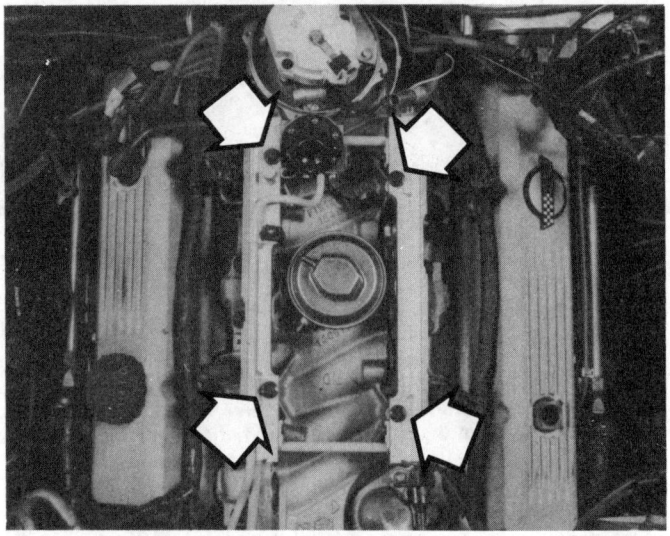

Fig. 13.34 To remove the fuel rail assembly from the intake manifold, remove the four fuel rail mounting bolts (arrows) (Sec 9)

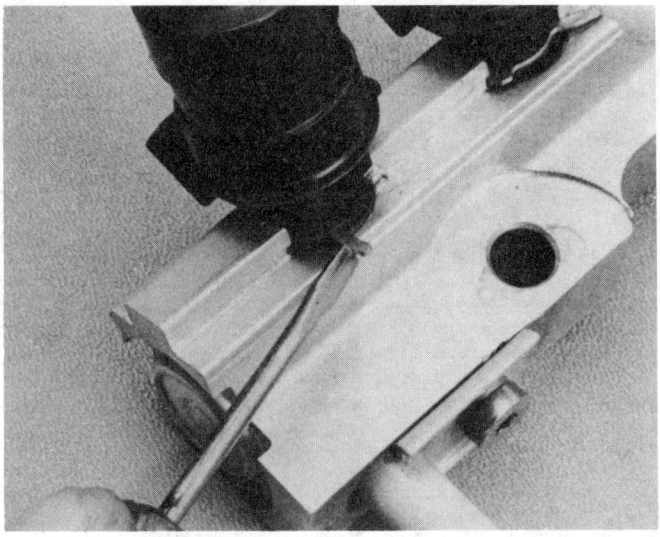

Fig. 13.35 The tangs on the retainer clips are difficult to push with your finger — use a small screwdriver instead (Sec 9)

Fig. 13.36 The O-rings on the ends of the injectors should be inspected and, if damaged, should be replaced (Sec 9)

brackets. Remove the PCV line assembly to provide clearance for removal of the left intake runner assembly.

118 Using a backup wrench on the fuel rail fitting, disconnect the cold start valve fuel line fitting from the rear of the left side fuel rail (Fig. 13.32). Remove the cold start valve mounting bolt. Clean the area around the cold start valve connection, then remove the cold start valve assembly.

119 To replace the cold start valve or check the O-ring between the cold start valve and the elbow fitting of the tube and body assembly, bend back the tab on the elbow fitting of the tube and body assembly and unscrew the valve from the tube (Fig.13.33). Before disassembling the valve from the tube and body assembly, note the relationship of the electrical connector to the tube and body assembly. This is important because the connector must be in the same position once the new valve is adjusted in order for the cold start valve assembly to fit properly.

120 Inspect the O-ring between the cold start valve and the elbow fitting of the tube and body assembly. Replace it if it's damaged.

121 To adjust the new cold start valve, screw it into the elbow fitting of the tube and body assembly until it bottoms. Be sure to lubricate

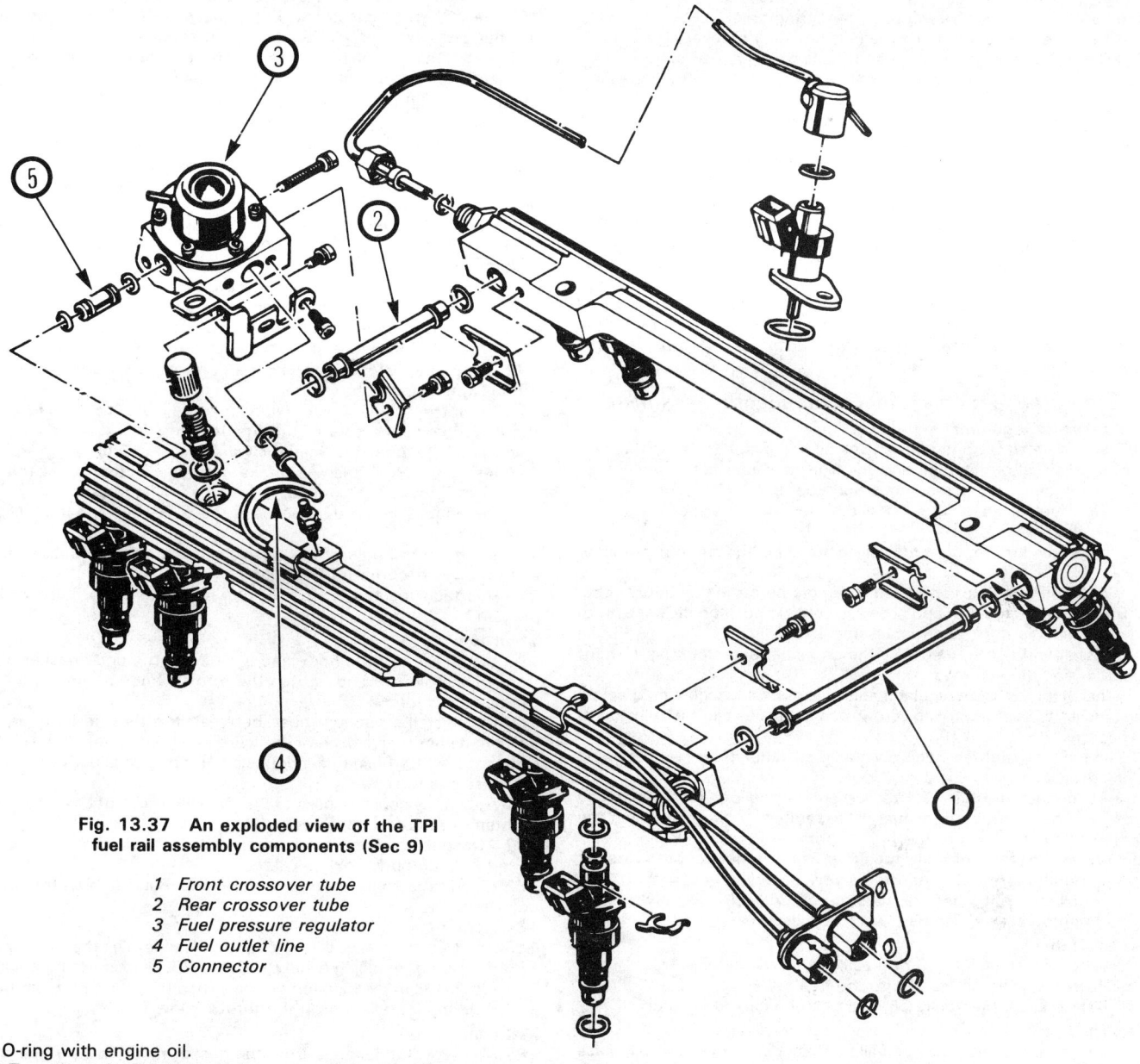

Fig. 13.37 An exploded view of the TPI fuel rail assembly components (Sec 9)

1 Front crossover tube
2 Rear crossover tube
3 Fuel pressure regulator
4 Fuel outlet line
5 Connector

the O-ring with engine oil.
122 Turn the valve back one full turn, until the electrical connector is at the top position you noted before separating the valve from the tube and body assembly.
123 Bend the tab forward to limit rotation of the valve to less than a full turn.
124 Remove the four remaining Torx bolts which attach the lower end of the left intake runner assembly to the intake manifold.
125 Remove the left intake runner assembly.

Fuel rail assembly, fuel injectors and fuel pressure regulator

126 Disconnect the fuel line fittings at the bracket on the right front of the intake manifold, then remove both bracket bolts from the manifold.
127 Disconnect the wiring harness electrical connectors from the right bank of fuel injectors.
128 Remove the four mounting bolts from the fuel rail (Fig. 13.34) and remove the fuel rail assembly.
129 Before disassembly or removal of any component(s), the fuel rail assembly should be thoroughly cleaned with a spray type cleaner. **Caution:** *Do not immerse the fuel rail assembly or its components in solvent.*

Fuel injectors

Note: *Whether you are replacing a malfunctioning fuel injector, a leaking injector O-ring seal or the fuel pressure regulator assembly, it is essential that all O-ring seals at both ends of all injectors be inspected. In view of the relative inaccessibility of these items, it is a good idea to replace all of them, since you have to remove the injectors to inspect the O-rings anyway.*

130 Each port injector is located and held in position by a retainer clip that must be rotated to release and/or lock the injector (Fig. 13.35).
131 Rotate the injector retaining clips to the unlocked position.
132 Pull the injectors from the fuel rail with a slight twisting motion.
133 Inspect all injector O-ring seals for wear. Remove and discard damaged O-rings by prying them off either end of the injector with a small screwdriver (Fig. 13.36).

Fuel pressure regulator

Note: *The fuel pressure regulator is factory adjusted. Do not attempt to remove the regulator cover. If the regulator is malfunctioning, it must be replaced.*

134 Remove either front crossover tube retainer screw and crossover tube retainer (Fig. 13.37).
135 Remove the rear retainer-to-regulator base screw and crossover tube retainer.
136 Separate the left and right fuel rail assemblies.
137 Remove the regulator bracket-to-right fuel rail screw (just below

the fuel pressure connection), the bracket-to-pressure regulator base screws and the pressure regulator mounting bracket.
138 Remove the fuel outlet tube rear retaining bracket stud.
139 Remove the pressure regulator base-to-right fuel rail mounting screw (on the side of the regulator base, just in front of the rear crossover tube).
140 Remove the pressure regulator and base assembly from the right fuel rail assembly.
141 Carefully rotate and pull on the regulator to remove it from the fuel outlet tube.
142 Remove the regulator base-to-fuel rail connector.
143 Remove the remaining front and rear crossover tube retainers and separate the crossover tubes from the fuel rails.
144 Inspect all O-rings on both ends of the connector, the fuel outlet tube and the crossover tubes for any damage that could prevent proper sealing. Replace any damaged O-ring seals. **Caution:** *The fuel rail assembly O-rings are not all the same size. When removing O-rings, be sure to note their location and size to ensure proper replacement.*

Tuned-Port Injection system components — installation

Fuel pressure regulator
145 Lubricate all O-rings with engine oil and install them on the connector, the fuel outlet tube and the front and rear crossover tubes.
146 Insert the front and rear crossover tubes into their respective holes in the left fuel rail assembly. Install the tube retainers and screws and tighten them securely.
147 Install the base-to-right fuel rail connector in the pressure regulator base.
148 Push the regulator and base assembly onto the fuel outlet tube.
149 Rotate the regulator and base assembly to position the base properly against the right fuel rail assembly.
150 Install and tighten securely the pressure regulator base-to-right fuel rail mounting screw.
151 Install the pressure regulator and base assembly mounting bracket, the bracket-to-base mounting screws and the bracket-to-fuel rail mounting screw.
152 Install and tighten securely the fuel outlet tube rear retaining bracket stud.
153 Reconnect the left fuel rail assembly and previously installed crossover tubes to the right fuel rail assembly by gently pushing the crossover tube ends into the right fuel rail.
154 Install the rear crossover tube-to-pressure regulator base retainer and mounting screw. Tighten the screw securely.
155 Install the front crossover tube-to-right fuel rail assembly retainer and mounting screw. Tighten the screw securely.

Fuel injectors
156 Lubricate all injector O-ring seals with engine oil.
157 Push the injectors back into the fuel rail.
158 Rotate each injector retainer clip to the locked position.

Fuel rail
159 Lubricate the nozzle ends of all fuel injectors with engine oil, place the fuel rail assembly into position above the intake manifold and carefully push down on the rails until all injectors are fully seated.
160 Install the fuel rail assembly mounting bolts and tighten them to the specified torque.

Left intake runner
161 Using new intake runner-to-intake manifold gaskets, place the left runner assembly in position and install the four outboard intake runner Torx mounting bolts. From the right side of the engine, install the fifth bolt just below the pressure regulator. Tighten all bolts securely.

Cold start valve assembly
162 Inspect the large O-ring between the cold start valve and the intake manifold and the small O-ring between the threaded fitting of the tube and body assembly and the rear of the right fuel rail. If there is evidence of damage to either O-ring, replace it.
163 Lubricate both O-rings with engine oil and install the cold start valve assembly.
164 Attach the threaded fitting to the rear of the left fuel rail and the cold start valve flange mounting bolt to the intake manifold. Tighten the threaded fitting and the mounting bolt securely.
165 Install the PCV line, retaining brackets and left upper rocker arm cover nuts. Tighten the nuts securely.
166 Place the fuel injector wiring harness in position and connect the electrical connectors to the left bank of fuel injectors. Also connect the electrical connector to the cold start valve.
167 Place the fuel injector wiring harness shroud in position, install the flanged hex bolts and tighten them securely.
168 Install the PCV valve assembly in the left rocker arm cover and connect the hose to the PCV line assembly.

Plenum and right intake runner assemblies
169 Place the right intake runner assembly and runner-to-intake manifold gaskets in position and loosely tighten the four outboard Torx mounting bolts. Don't forget to install the fifth inboard bolt located behind the second runner from the front. It must be installed from the opposite side of the engine.
170 Holding the plenum in position, connect the wiring harness electrical connector to the MAT sensor protruding from the bottom of the plenum.
171 Slide the new runner-to-plenum gaskets into place and bolt the left and right intake runner assemblies to the plenum. Tighten all bolts securely. Don't forget to tighten the bolts that attach the lower end of the right intake runner assembly to the intake manifold.
172 Install the PCV line, brackets and right rocker arm cover upper nuts. Tighten the nuts securely.
173 Lay the right fuel injector wiring harness in position and reconnect the electrical connectors to the fuel injectors.
174 Install the right fuel injector wiring harness shroud.
175 Install the PCV fresh air intake assembly in the top of the right rocker arm cover and connect it to the PCV line assembly.
176 Reconnect the vacuum hoses at the EGR valve and the fuel pressure regulator.
177 Reattach the power brake booster vacuum line threaded fitting at the left rear corner of the plenum.
178 Reattach the two vacuum lines at the right rear corner of the plenum.

Throttle body
179 Install the throttle body and a new throttle body gasket on the front of the plenum and tighten the throttle body mounting bolts to the specified torque.
180 Connect the vacuum hose between the throttle body and the vacuum damper valve beneath the lower left front corner of the plenum. Connect the PCV hose to the right side of the throttle body and tighten the hose clamp securely.
181 Attach the coolant hoses to the IAC valve coolant cover assembly. Tighten the hose clamps securely.
182 Attach the cruise control, accelerator and transmission TV cables to the throttle body linkage.
183 Attach the cable bracket to the plenum and tighten the bracket bolts securely.
184 Add coolant to the cooling system.
185 Hook up the cable to the negative terminal of the battery.
186 If you are installing a new IAC or TPS, they must be adjusted. The new IAC can be adjusted on the bench, but the TPS must be adjusted during installation on the throttle body.

IAC adjustment
187 Measure the distance from the gasket mounting surface of the IAC valve assembly to the tip of the pintle (Fig. 13.38).
188 If this distance is greater than 1-1/8 inch (28 mm), reduce it as follows:
189 If the IAC valve assembly has a collar around the electrical con-

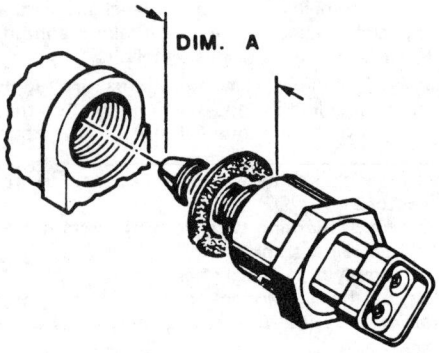

Fig. 13.38 Before installing the IAC valve, make sure that dimension A is 1-1/8 inches (28 mm) (Sec 9)

nector end, retract the pintle by applying firm hand pressure (a slight side-to-side motion may help).
190 If the IAC valve assembly has no collar, compress the pintle retaining spring toward the body of the IAC and try to turn the pintle *clockwise*.
191 If the pintle turns, continue turning it until the specified distance is attained. Return the spring to its original position, with the straight part of the spring end lined up with the flat surface under the pintle head. **Note:** *If the pintle will not turn, apply firm hand pressure to retract it.*
192 Install a new gasket on the IAC valve and screw the IAC valve and gasket into the idle air/vacuum signal housing of the coolant cover assembly. Tighten the IAC valve to the specified torque.
193 Reconnect the wiring harness electrical connector to the IAC valve terminal. **Note:** *No further adjustment of the IAC valve is necessary. It will reset itself after the first time the engine is started and the ignition is turned off.*

TPS adjustment

194 With the throttle valve in the closed idle position, install the TPS on the throttle body assembly, making sure that the TPS lever lines up with the tang on the throttle actuator lever (Fig. 13.39).

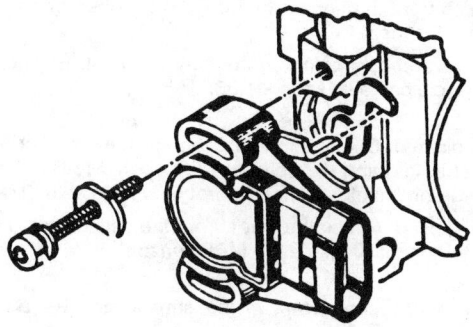

Fig. 13.39 When installing the TPS, make sure that the TPS pickup lever lines up with the tang on the throttle actuator lever (Sec 9)

195 Install the retainers and the TPS mounting screws.
196 Install three jumper wires between the TPS and the harness connector.
197 With the ignition turned on and a digital voltmeter connected to the terminals marked A and B on the connector (the top and middle terminals), adjust the TPS to obtain a reading of 0.54 ± 0.075-volts.
198 Tighten the TPS mounting screws and recheck your reading to ensure that the adjustment has not changed.
199 Turn the ignition off, remove the jumper wires and connect the wiring harness connector to the TPS.
200 Connect the wiring harness connector to the terminal on the MAF sensor and install the MAF sensor/flexible duct assembly between the sensor and the throttle body. Tighten the hose clamps which attach the flexible duct to the throttle body and the clamps that secure the MAF sensor to the air cleaner housing duct.
201 Start the engine and then turn it off. The IAC valve will automatically reset itself.

Minimum idle speed adjustment (TPI and MPFI systems)

202 The idle stop screw, which regulates the minimum idle speed of the engine, is adjusted at the factory and covered with a plug to discourage unnecessary adjustment. However, if it is necessary to gain access to the idle stop screw assembly proceed as follows.
203 Pierce the idle stop screw plug with an awl and apply leverage to remove it.
204 With the IAC valve connected, ground diagnostic terminal B of the ALCL connector under the left side of the dash.
205 Turn the ignition to On but do not start the engine. Wait at least 30 seconds.
206 With the ignition on, disconnect the IAC electrical connector.
207 Remove the ground from the diagnostic lead and start the engine. Run the engine until it reaches normal operating temperature.
208 Adjust the idle screw to obtain either 400 rpm in Drive with an automatic transmission or 450 rpm in Neutral with a manual transmission.
209 Turn the ignition to Off and reconnect the IAC valve connector.
210 Following the Tuned-Port Injection TPS adjustment procedure, adjust the TPS to 0.54 volts ±0.075-volts (V8), or 0.55 volts ± 0.06-volts (V6). Secure the TPS and recheck the voltage.
211 Start the engine and check for proper idle operation.

Throttle cable — removal and installation

212 Remove the clip which secures the throttle cable to the throttle shaft lever arm.
213 Push the throttle shaft lever to the rear and disconnect the cable.
214 Pinch the tabs on the top and bottom of the plastic retainer attaching the cable to the cable bracket. Slide the retainer through the bracket.
215 Unclip the plastic cable bracket attached to the side of the plenum (TPI only). Lift the cable out of the bracket.
216 Peel away the body sealing compound from the retainer at the firewall plate. Pinch the tabs on the top and bottom of the retainer and free the retainer from the plate.
217 Disconnect the throttle cable from the accelerator pedal.
218 Remove the throttle cable.
219 Installation is the reverse of removal.
220 No linkage adjustment is necessary because the replacement cable is a specific length. Therefore, it is imperative that the correct cable be installed. Only the specific replacement cable will work properly on this vehicle.
221 After assembly, the throttle linkage must operate freely without binding between the fully closed and fully open positions. Check for correct opening and closing by operating the accelerator pedal. If any binding is noted, check the routing and make sure there are no kinks in the cable.

Fuel line — repair and replacement

Warning: *If a section of fuel line is damaged, the entire line must be replaced. Do not attempt to repair fuel lines with fuel hose. Fuel injection systems operate under high pressure and replacing fuel lines with anything other than original equipment-quality materials may create a hazardous condition.*

10 Emissions control systems

General information

The emissions control systems on the Tuned-Port and Multi-Port Fuel Injection equipped vehicles are very similar to the systems described in Chapter 6, with a few changes made to accomodate the current fuel system. Different diagnostic procedures, new trouble codes and component descriptions are covered in this Section. Where no differences exist, the information in Chapter 6 can be used.

Trouble Code identification

Following is a list of the Trouble Codes that have been changed or added due to the addition of the above-mentioned fuel delivery systems.

Trouble codes	Circuit or system	Probable cause
Code 21 (2 flashes, pause, 1 flash)	Throttle position sensor	Check for a sticking or misadjusted TPS lever. Check all wiring and connections between the TPS and the ECM. Adjust or replace the TPS (see Section 9 of this Supplement).
Code 22 (2 flashes, pause, 2 flashes)	Throttle position sensor	Check the TPS adjustment. Check the ECM connector. Replace the TPS (see Section 9 of this Supplement).

Trouble codes (con't.)	Circuit or system	Probable cause
Code 23 (2 flashes, pause, 3 flashes)	Manifold Air Temperature sensor	Check the MAT sensor, wiring and connectors for an open circuit. Replace the MAT sensor.*
Code 24 (V8) (2 flashes, pause, 4 flashes)	Vehicle Speed Sensor (VSS) or circuit	Check the sensor and wires for an open circuit.
Code 25 (2 flashes, pause, 5 flashes0	Manifold Air Temperature sensor	Check the voltage signal from the MAT sensor to the ECM. It should be above 4-volts.
Code 32 (3 flashes, pause, 2 flashes)	EGR system	The EGR solenoid should not be energized and vacuum should not pass to the EGR valve. The diagnostic switch should close at about 2-inches of vacuum. With vacuum applied, the switch should close. Replace the EGR valve.*
Code 32 (3.1L V6) (3 flashes, pause, 2 flashes)	Digital EGR circuit	Check for an open or short to ground in EGR valve circuit. Check all wires and connectors. A faulty EGR valve may also set Code 32.
Code 33 **(MPFI V6 and V8 to 1989)** (3 flashes, pause, 3 flashes)	MAF sensor	Excessive airflow indicated. Check terminal C on the MAF sensor: it should be about 0.5-volts at idle and 4.7-volts at wide open throttle. Trace the wire from terminal C and look for an open circuit. Replace the MAF sensor.*
Code 33 (1988 and later TBI V8; 90 and later MPFI V8) (3 flashes, pause, 3 flashes)	MAP sensor	Low vacuum sensed. Check for vacuum leaks and proper idle speed. Check the wires from the sensor to the ECM for open or shorted circuit. Replace the MAP sensor.*
Code 33 (3.1L V6) (3 flashes, pause, 3 flashes)	MAP sensor circuit (high signal voltage low vacuum)	Check the wires and connectors in the sensor circuit. Engine misfire and a low, unstable idle can also set code 33.
Code 34 **(MPFI V6 and V8 to 1989)** (3 flashes, pause, 4 flashes)	MAF sensor	Low airflow indicated. Check terminal C on the MAF sensor; it should be about 0.5-volts at idle and 4.7-volts at wide open throttle. Trace the wire from terminal C and look for a short to ground. Replace the MAF sensor.*-
Code 34 (1988 and later TBI V8; 90 and later MPFI V8) (3 flashes, pause, 4 flashes)	MAP sensor	High vacuum sensed. Check the wires from the sensor to the ECM for an open or shorted circuit. Replace the MAP sensor.*
Code 34 (3.1L V6) (3 flashes, pause, 4 flashes)	MAP sensor circuit (low signal voltage, high vacuum)	Check the wires and connectors in the sensor circuit – an intermittent open will result in a code 34.
Code 35 (3.1L V6) (3 flashes, pause, 5 flashes)	Idle Air Control (IAC) circuit system check	Code 35 will set if the closed throttle engine speed is 125 rpm above or below the desired (commanded) idle speed for 45 seconds. See a dealer service department for trouble diagnosis.
Code 36 (MPFI 1987 to 1989) (3 flashes, pause, 6 flashes)	MAF burn-off circuit	Code 36 can be set due to high resistance in the burn-off relay contacts or connections, the failure of the ECM to properly ground the circuit, or a poor connection at any of the relays, MAF sensor, and ECM.
Code 41 (4 flashes, pause, 1 flash)	Cylinder select error	Terminal D3 of the ECM must be properly connected to engine ground for a V8 engine – disconnect the ECM C-D connector and probe terminal D3 with a test light connected to 12-volts. If it comes on, the terminal connection is faulty.
Code 42 (4 flashes, pause, 2 flashes)	Electronic Spark Timing	If the vehicle will not start and run, check the wires between the HEI module and the EST and by-pass terminal (B4 and D5, respectively). Check for a bad connection at the ignition module. Check and, if necessary, replace the HEI module.*
Code 43 (4 flashes, pause, 3 flashes)	Electronic Spark Control	Check the voltage at the ECM A-B connector terminal B7; it should be over 6-volts unless the system is sensing detonation. Check for a loose connection. Replace the ESC sensor and/or module.*
Code 44 (4 flashes, pause, 4 flashes)	Lean exhaust	Check the ECM wiring connection terminals D6 and D7. Check the fuel pressure (see Section 9 of this Supplement). Check the oxygen sensor wire. Check the throttle body gasket, vacuum hoses and intake manifold gaskets for leaks. Replace the oxygen sensor.*
Code 45 (4 flashes, pause, 5 flashes)	Rich exhaust	Check the evaporative charcoal canister and its components for the presence of fuel. Check the fuel pressure regulator vacuum control hose for the presence of fuel. Replace the oxygen sensor.*
Code 46 (3.1L V6) (4 flashes, pause, 6 flashes)	Vehicle Anti-Theft System (VATS)	Make sure the correct ignition key and starting procedure are being used. If the engine turns over, check for an open or short to ground in the VATS decoder module circuit. If the engine won't turn over, have the VATS checked by a dealer service department.
Code 51 (5 flashes, pause, 1 flash)	PROM	Make sure that the PROM is properly installed in the ECM. If all pins are fully inserted in the sockets, replace the PROM.* Clear the memory and recheck. If Code 51 reappears, replace the ECM.*
Code 52 (5 flashes, pause, 2 flashes)	Fuel CALPAK	Check the CALPAK PROM to ensure proper installation. Replace the PROM.*

Code 53 (all except 5.0 carbureted models) (5 flashes, pause, 3 flashes)	System over-voltage	Code 53 will set if the voltage at ECM terminal B2 is greater than 17.1-volts for two seconds. Check the charging system.
Code 53 (5.0L carbureted) (5 flashes, pause, 3 flashes)	EGR control error	EGR valve operation and vacuum lines to valve and solenoid.
Code 54 (5 flashes, pause, 4 flashes)	Fuel pump circuit	Code 54 will set if the voltage at terminal B2 is less than 2-volts for 1.5 seconds since last reference pulse was received. Check the fuel pump relay, circuit and connections. Check the oil pressure switch. Repair/replace faulty components.*
Code 55 (5 flashes, pause, 5 flashes)	ECM	Be sure that the ECM ground connections are tight. If they are, replace the ECM.*
Code 61 (3.1L V6) (6 flashes, pause, 1 flash)	Degraded oxygen sensor	The oxygen sensor is contaminated and must be replaced. Contamination can be caused by fuel additives or the use of non-GM approved lubricants or sealants.

**Component replacement may not cure the problem in all cases. For this reason, you may want to seek professional advice before purchasing replacement parts.*

Air Injection Reaction (AIR) system

General description

The AIR system helps reduce hydrocarbons, carbon monoxide and oxides of nitrogen exhaust emissions. It also enables the catalytic converter to heat up quickly on initial start-up so conversion of exhaust gases can occur sooner.

The AIR system consists of a belt-driven air pump, a control valve, a check valve, a catalytic converter and related hoses and lines.

The control valve combines two valve functions (divert and air switching) into one housing. The air switching portion directs the air to the exhaust ports during cold engine operation. During warm engine operation, air is directed between the catalytic converter beds.

Air is diverted to the atmosphere during rich operation, when the ECM senses a problem and the "check engine" light comes on, during deceleration and during high RPM operation when air pressure is greater than the setting for the internal relief valve. Both the control valve (directing air to the exhaust ports) and the switching valve (directing air to the converter) are operated by solenoids controlled by the ECM.

Checking

1 To check the air pump, accelerate the engine to approximately 1500 RPM and feel for air flow from the output hose. Air flow should increase with engine speed.
2 Check for proper air pump drivebelt tension.
3 Check for a seized air pump.
4 Check all hoses, tubes, valves and connections for leaks and proper routing. Leaks can sometimes be found by applying a soapy water solution to the suspected area.
5 Inspect the check valves to the exhaust ports by disconnecting the hoses and blowing through the valves with the engine cold. Then attempt to suck (but do not inhale) back through the valve. Flow should only be in one direction, towards the exhaust ports.
6 Check the air pump filter (see Chapter 6).

Component replacement

7 For component replacement, refer to Chapter 6.

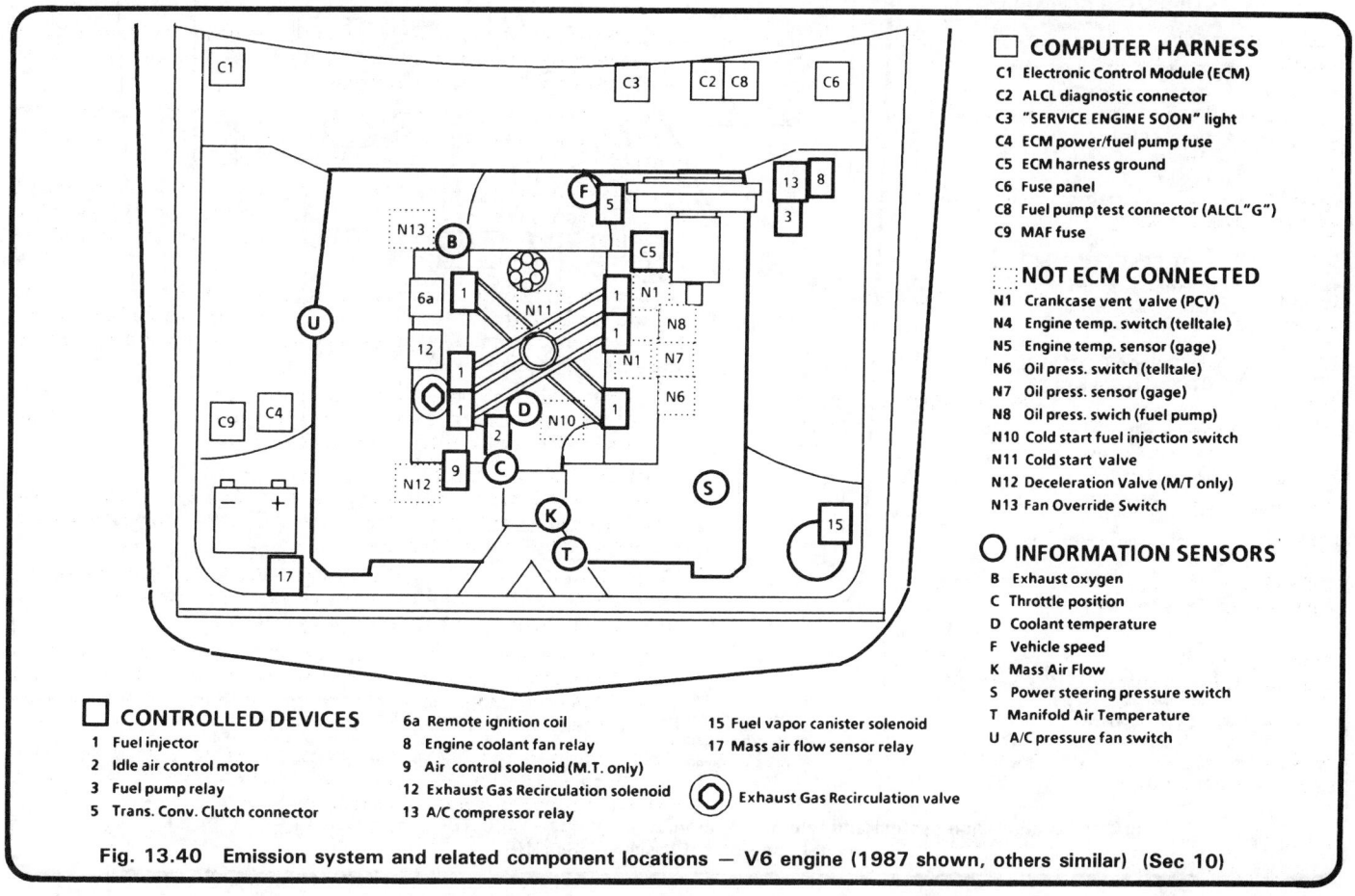

Fig. 13.40 Emission system and related component locations — V6 engine (1987 shown, others similar) (Sec 10)

COMPUTER HARNESS

- C1 Electronic Control Module (ECM)
- C2 ALCL diagnostic connector
- C3 "SERVICE ENGINE SOON" light
- C4 ECM power
- C5 ECM harness ground
- C6 Fuse panel
- C8 Fuel pump test connector (ALCL "G")
- C9 Fuel pump/ECM fuse

NOT ECM CONNECTED

- N1 Crankcase vent valve (PCV)
- N5 Engine temp. sensor (gage.overheat)
- N7 Oil pressure sensor (gage)
- N8 Oil pressure switch (fuel pump)
- N10 Cold start valve
- N11 Cold start fuel injection switch

INFORMATION SENSORS

- B Exhaust oxygen
- C Throttle position
- D Coolant temperature
- F Vehicle speed
- J ESC knock
- K Mass Air Flow
- T Manifold Air Temperature
- U A/C pressure fan switch
- W EGR temp. diagnostic switch

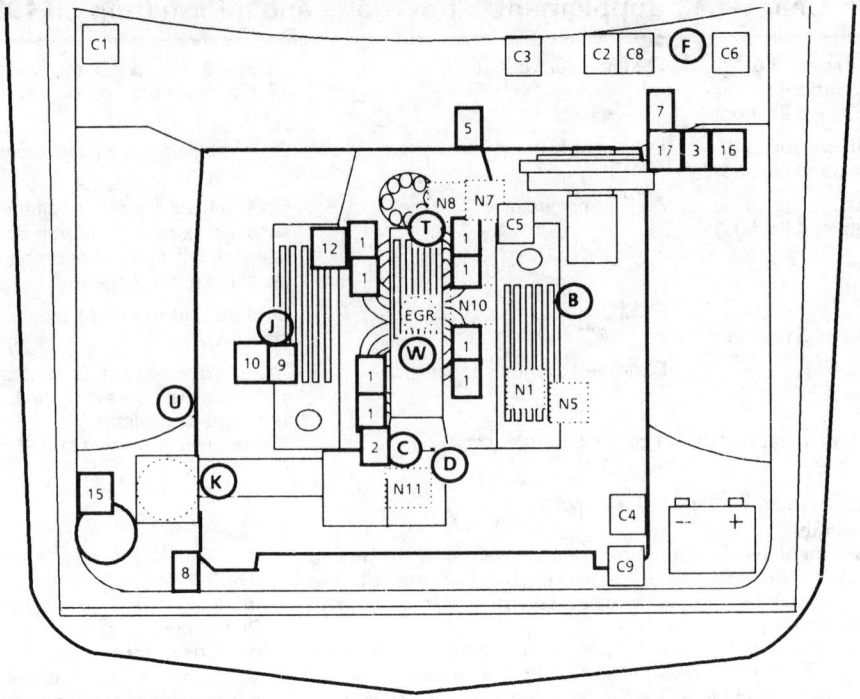

CONTROLLED DEVICES

- 1 Fuel injector
- 2 Idle air control motor
- 3 Fuel pump relay
- 5 Trans. Converter Clutch connector
- 7 Electronic Spark Control module
- 8 Engine cooling fan relay
- 9 Air injection divert valve
- 10 Air injection switching valve
- 12 Exh. Gas Recirc. vacuum solenoid
- 15 Fuel vapor canister solenoid
- 16 MAF sensor power relay
- 17 MAF sensor burn-off relay

Fig. 13.41 Emission system and related component locations — MPFI V8 engine (1987 shown, others similar) (Sec 10)

COMPUTER HARNESS

- C1 Electronic Control Module
- C2 ALDL diagnostic connector
- C3 "SERVICE ENGINE SOON" light
- C5 ECM harness grounds
- C6 Fuse panel
- C8 Fuel pump test connector (Terminal "G" of ALDL Connector)
- C9 Fuel pump/ECM fuse
- C10 Set timing connector

NOT ECM CONNECTED

- N1 Crankcase vent valve (PCV)
- N8 Oil pressure switch

INFORMATION SENSORS

- A Manifold Absolute Pressure
- B Exhaust oxygen
- C Throttle position
- D Coolant temperature
- F Vehicle speed (buffer)
- G MAT (in air cleaner)
- J ESC knock
- S Power Steering Pressure Switch

Exhaust Gas Recirculation valve

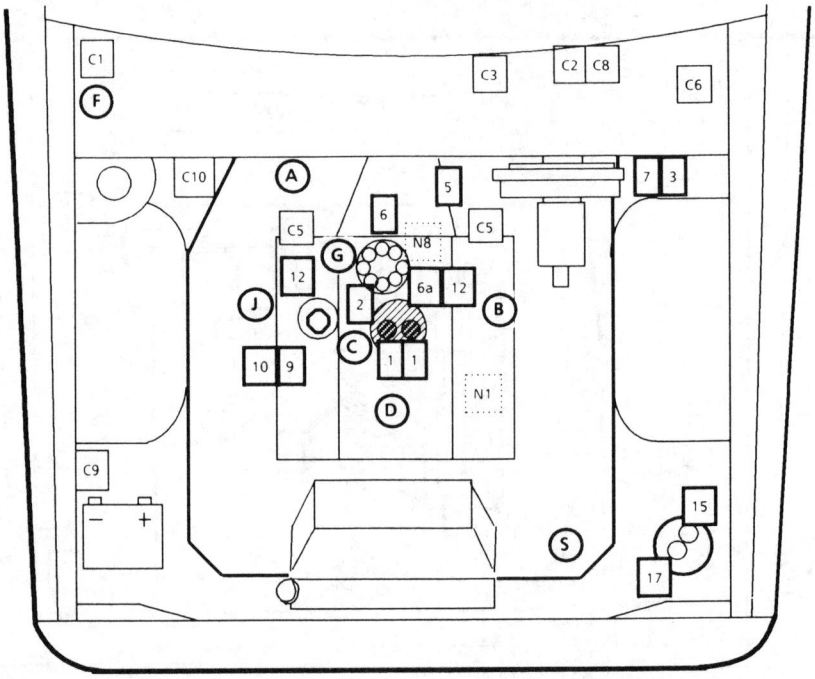

CONTROLLED DEVICES

- 1 Fuel injectors
- 2 Idle air control motor
- 3 Fuel pump relay
- 5 Torque Conv. Clutch connector
- 6 EST distributor
- 6a Remote ignition coil
- 7 Electronic Spark Control module
- 9 AIR port solenoid
- 10 AIR converter solenoid
- 12 Exh. Gas Recirc. vacuum solenoid
- 15 Fuel vapor canister solenoid
- 17 Fuel vapor canister

Fig. 13.42 Emission system and related component locations — 5.0 TBI engine (VIN E) — 1989 model shown, 1988 similar (Sec 10)

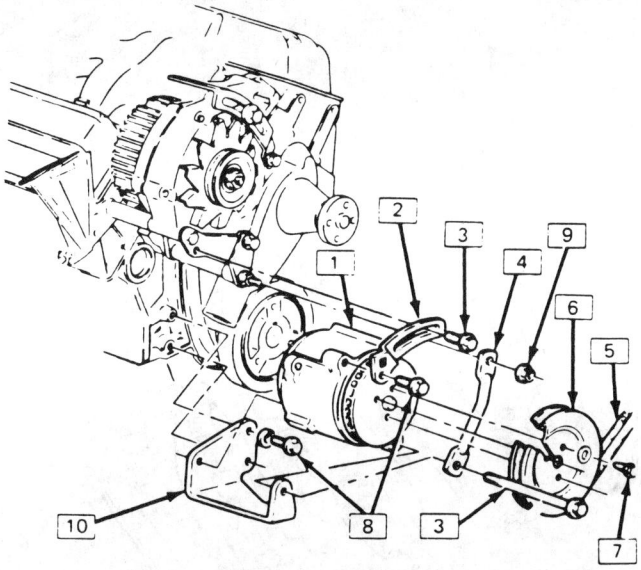

Fig. 13.43 AIR system pump and related components — exploded view (Sec 10)

1 AIR pump	*6 Pulley*
2 Bracket	*7 Bolt*
3 Bolt	*8 Bolt*
4 Brace	*9 Nut*
5 Drivebelt	*10 Support*

Exhaust Gas Recirculation (EGR) system

Refer to Chapter 6 for the EGR system description and checking procedures. Due to the interrelationship of the EGR system and the ECM and because of the inaccessibility of the valve itself (on Tuned-Port Injection equipped V8 engines), diagnosis should be performed by a dealer service department.

All 1990 and later models with a 3.1 liter V6 engine are equipped with a Digital EGR Valve, which controls EGR flow electronically (through the ECM), rather than directly by intake manifold vacuum. The procedures outlined in Chapter 6 do not apply to the Digital EGR Valve. A new trouble code will reveal EGR valve problems.

Early Fuel Evaporation (EFE) system

The EFE system, used to preheat the air/fuel mixture during cold engine operation, is not installed on Tuned-Port and Multi-Port Fuel Injection equipped vehicles.

11 Clutch and driveline

Borg-Warner rear axle — general information and identification

Due to the design of this rear axle assembly, service procedures should be limited to routine maintenance only. Repairs pertaining to the differential assembly, axle shafts, axle bearings and seals require special tools and expertise and should be performed by a professional mechanic. To determine if your vehicle is equipped with a Borg-Warner rear axle, match the axle number (which is stamped on the right side axle housing) with the axle codes in Section 2 of this Supplement.

12 Brakes

General information

The front and rear disc brake calipers used on 1989 and later models are slightly different than the ones installed on earlier models. The pad replacement, caliper removal and installation and overhaul procedures differ slightly as a result, so be sure to note the changes outlined below when working on the brakes. Also, refer to the Specifications at the beginning of this Chapter for information related to later models.

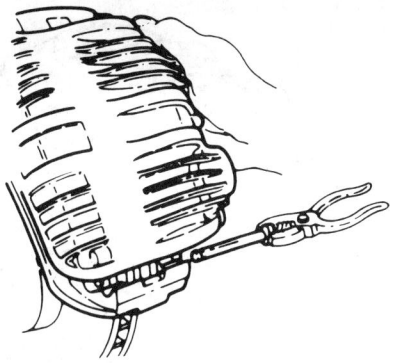

Fig. 13.44 After removing the circlip from the end of the lower retainer pin, pull the pin out with a pair of pliers (Sec 12)

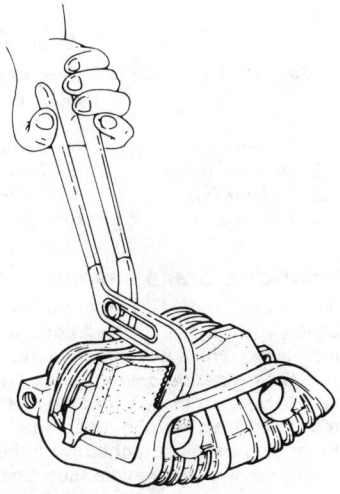

Fig. 13.45 The caliper pistons can be bottomed in their bores by squeezing the old brake pad against the caliper housing (Sec 12)

Front disc brake pads — replacement

1 Refer to Chapter 9, Section 2, and follow the procedure in Steps 1 through 4. Before pulling off the wheel, mark the wheel and hub so the wheel can be reinstalled in the same relative position.

2 Remove the circlip, then pull out the retainer pin with a pair of pliers (Fig. 13.44). Discard the circlip and retainer pin and use new ones when the caliper is reinstalled.

3 Pivot the lower end of the caliper up and disengage the upper end from the bracket. Suspend the caliper from a suspension component with a piece of stiff wire — DO NOT let it hang by the rubber brake hose!

4 Note how they're installed, then remove the old pads from the caliper.

5 Use an old brake pad and a large pair of channel-lock pliers (Fig. 13.45) or a C-clamp to push the pistons into the caliper bores to provide room for the thick linings on the new pads.

6 Use aerosol brake system cleaner to remove all residue from the caliper and bracket, particularly where the pads slide in the caliper. **Warning:** *Dust created by the brake system may contain asbestos, which is harmful to your health. Never blow it out with compressed air and don't inhale it. An approved respirator should be worn when working on the brakes. Do not, under any circumstances, use petroleum based solvents to clean brake parts. Use brake system cleaner or denatured alcohol only!*

7 Install the new pads in the caliper. The outer pad with the bias spring is attached to the caliper housing, while the inner pad with the wear sensor should be pressed into the pistons. Make sure the pads are seated completely and don't get oil or grease on the lining material.

8 Slip the caliper and pad assembly over the rotor and position it in the bracket, then push down on the caliper to compress the bias spring on the outer pad and insert the new retainer pin. Attach the new circlip to the end of the retainer pin.

9 Replace the pads in the remaining front caliper. After the pads on both sides have been replaced, the wheels installed and the vehicle lowered, start the engine and pump the brake pedal slowly and firmly three times to seat the pads against the rotors. Don't forget to add brake fluid to the reservoir.

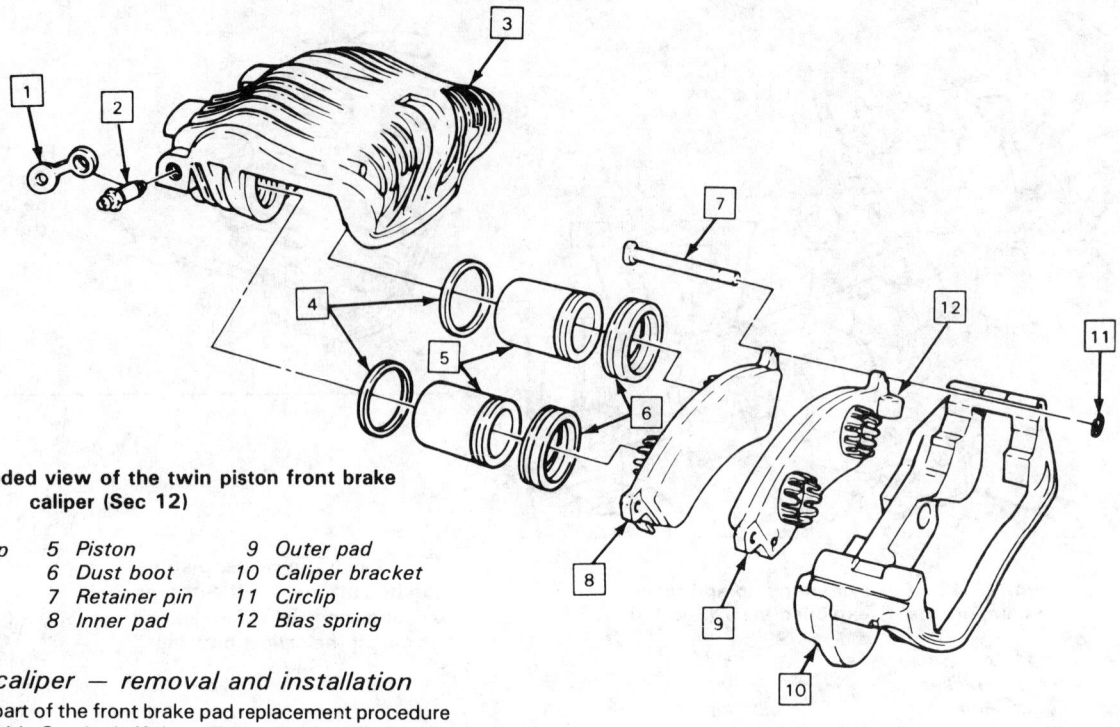

Fig. 13.46 Exploded view of the twin piston front brake caliper (Sec 12)

1 Bleeder valve cap	*5 Piston*	*9 Outer pad*
2 Bleeder valve	*6 Dust boot*	*10 Caliper bracket*
3 Caliper housing	*7 Retainer pin*	*11 Circlip*
4 Piston seal	*8 Inner pad*	*12 Bias spring*

Front disc brake caliper — removal and installation

10 Caliper removal is part of the front brake pad replacement procedure (Steps 1 through 9 of this Section). If the caliper must be overhauled or replaced with a new one, remove the bolt and detach the brake hose from the caliper *before* removing the caliper from the bracket. Plug the hose to prevent the loss of fluid and the entry of dirt. Discard the brake hose fitting washers and use new ones when reinstalling the bolt. If the brake hose is detached from the caliper, the brakes must be bled before driving the vehicle (see Chapter 9).

Front disc brake caliper — overhaul

11 The overhaul procedure in Chapter 9 is also valid for later model calipers, but note that the new style caliper is made of aluminum and can be easily damaged. Also, two pistons and bores are involved (Fig. 13.46), so, when removing the pistons with compressed air, if the pistons don't come out of the caliper at the same time, remove the boot from the empty bore and reinstall the piston far enough to seal it off. Position a spacer between the caliper housing and the piston to hold it in place as compressed air is used to remove the remaining piston.

12 All inspection procedures must be repeated for both pistons and bores and all related components.

13 Do not use a tool to seat the piston boots in the caliper bores — use your fingers only.

Rear disc brake pads — replacement

14 Refer to Chapter 9, Section 5, and follow the procedure in Steps 1 through 4. Before pulling off the wheel, mark the wheel and axle flange so the wheel can be reinstalled in the same relative position.

15 Use a large C-clamp to bottom the piston in the caliper bore. Make sure the frame end of the clamp is resting on the brake hose inlet fitting bolt head and the screw end is pushing on the outer brake pad.

16 Remove the caliper upper guide pin bolt (Fig. 13.47) and discard it. Use a new one when the caliper is reinstalled.

17 Rotate the caliper down so the pads can be removed (be careful not to strain the brake hose or the parking brake cable housing).

18 Lift the pads out of the caliper bracket.

19 Use aerosol brake system cleaner to remove all residue from the caliper and bracket, particularly where the pads slide in the bracket. **Warning:** *Dust created by the brake system contains asbestos, which is harmful to your health. Never blow it out with compressed air and don't inhale it. An approved respirator should be worn when working on the brakes. Do not, under any circumstances, use petroleum-based solvents to clean brake parts. Use brake system cleaner or denatured alcohol only!*

20 Check the guide pins for corrosion and damage — they must be free to slide in the bracket. If they're damaged and/or corroded, replace them with new ones and install new dust boots as well.

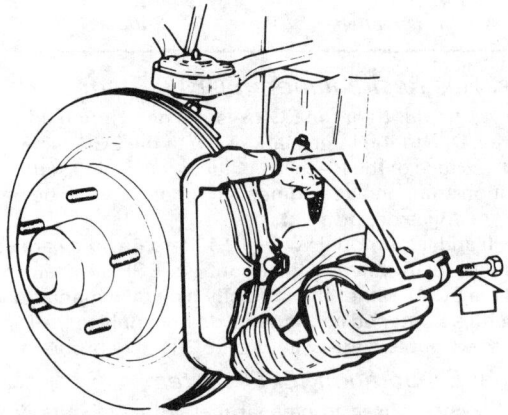

13.47 Remove the upper guide pin bolt (arrow) and rotate the caliper down to expose the brake pads (Sec 12)

21 Install the new pads in the bracket. The outer pad with the insulator must mate with the caliper housing, while the inner pad with the wear sensor should be pressed into the piston. Make sure the pads are seated completely and don't get oil or grease on the lining material. The wear sensor on the inner pad must be at the top of the pad. If it isn't, use the other new inner pad in the set.

22 Rotate the caliper into position and make sure the springs on the pads aren't sticking through the inspection hole in the caliper. If they are, rotate the caliper down and reposition the pads.

23 Align the holes and install a *new* caliper upper guide pin bolt. Tighten it to the specified torque.

24 Repeat the above procedure to replace the pads in the remaining rear caliper.

25 Check the parking brake levers on the calipers to make sure they're against the stops. If not, see if the rear parking brake cables are binding. Free and lubricate the cables, if necessary.

26 If the cables are free, turn the adjusting nut at the cable equalizer until both parking brake levers just touch the stops on the calipers.

27 After the pads on both sides have been replaced, the wheels installed and the vehicle lowered, start the engine and pump the brake pedal slowly and firmly three times to seat the pads against the rotors. Don't forget to add brake fluid to the reservoir.

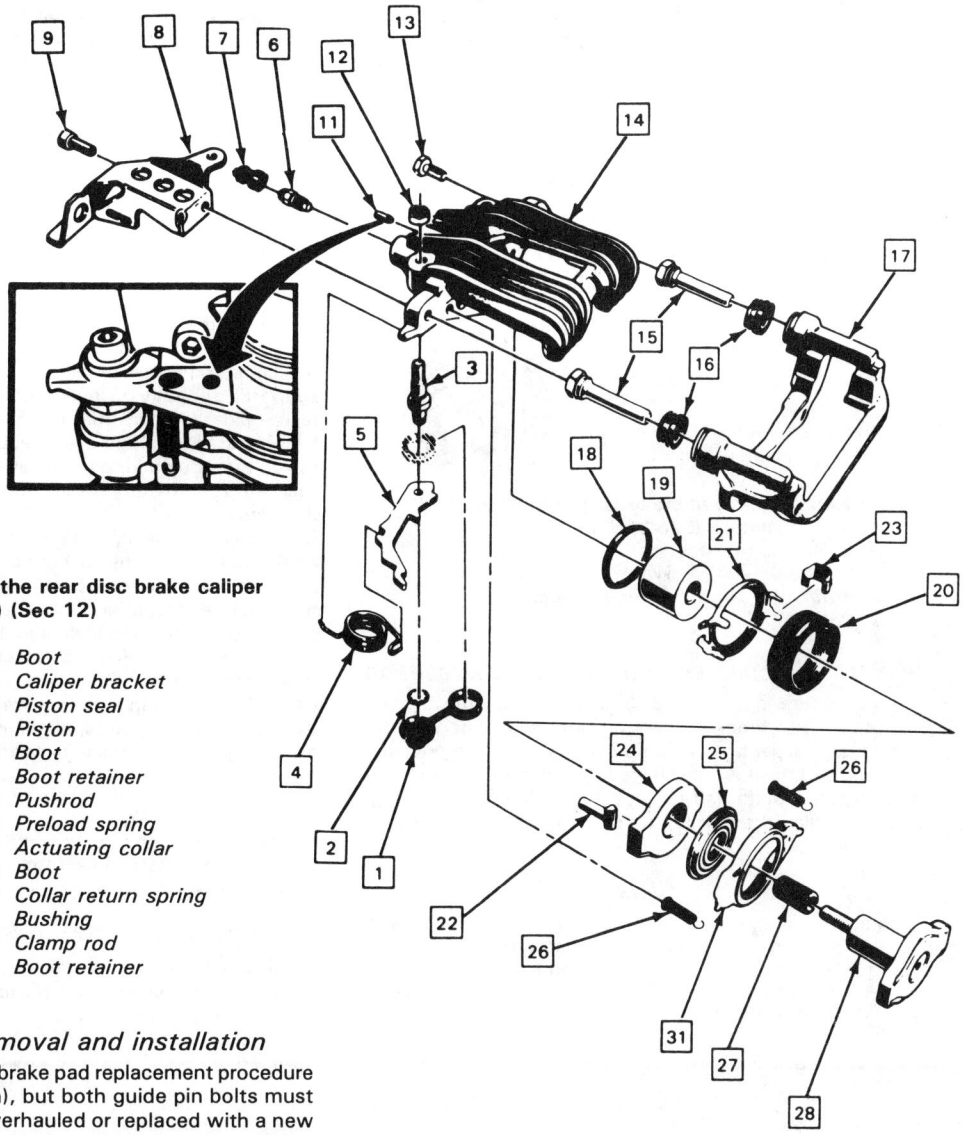

Fig. 13.48 Exploded view of the rear disc brake caliper (1989 models) (Sec 12)

1	*Seal*	16	*Boot*
2	*Sprag clip*	17	*Caliper bracket*
3	*Pivot pin*	18	*Piston seal*
4	*Spring*	19	*Piston*
5	*Parking brake lever*	20	*Boot*
6	*Bleeder valve*	21	*Boot retainer*
7	*Bleeder valve cap*	22	*Pushrod*
8	*Bracket*	23	*Preload spring*
9	*Bolt*	24	*Actuating collar*
11	*Adjuster screw*	25	*Boot*
12	*Nut*	26	*Collar return spring*
13	*Bolt*	27	*Bushing*
14	*Caliper housing*	28	*Clamp rod*
15	*Guide pin*	31	*Boot retainer*

Rear disc brake caliper — removal and installation

28 Caliper removal is part of the rear brake pad replacement procedure (Steps 14 through 16 of this Section), but both guide pin bolts must be removed. If the caliper must be overhauled or replaced with a new one, remove the bolt and detach the brake hose from the caliper *before* removing the caliper from the bracket. Plug the hose to prevent the loss of fluid and the entry of dirt. Discard the brake hose fitting washers and use new ones when reinstalling the bolt. If the brake hose is detached from the caliper, the brakes must be bled before driving the vehicle (see Chapter 9).

29 Also, if the caliper is to be removed from the vehicle, the parking brake cable must be disconnected from the lever at the caliper. To do this, unbolt the cable bracket from the caliper and unhook the cable eye from the hooked end of the lever.

Rear disc brake caliper — overhaul

30 Remove the caliper following the procedure described above, then set the caliper on a clean working surface.

31 Unhook the two collar return springs from the actuating collar (Fig. 13.48).

32 Pull the actuating collar with its attached parts out of the caliper body.

33 Separate the clamp rod and bushing from the actuating collar. Discard the bushing.

34 Bend back the tabs on the boot retainers and remove them from the collar. Discard the retainers and boots.

35 Remove the piston and replace the piston seal as described in the caliper overhaul procedure found in Chapter 9. Be sure to inspect the caliper bore and wash all parts not to be replaced in denatured alcohol, clean brake fluid or brake system cleaner.

36 Inspect the actuating collar and the clamp rod for corrosion and excessive wear, replacing them, if necessary.

37 Lightly lubricate the actuating collar with the liquid lubricant supplied in the rebuild kit. DO NOT use any other type of lubricant. Assemble the pushrod, new boots and retainers on the actuating collar and clamp the retainers firmly against the collar. Bend the tabs on the retainer to hold the assembly together.

38 Attach the preload spring to the boot retainers.

39 Lighlty lubricate the clamp rod with the lubricant supplied in the kit. Insert the clamp rod into the actuator collar and push it in fully, so the boot contacts the reaction plate on the clamp rod.

40 Push a new bushing down over the end of the clamp rod.

41 Lubricate the grooved bead of the inner boot and the corresponding boot groove in the caliper housing with the lubricant furnished in the kit. Also lubricate the actuating collar around the center hole. Push the clamp rod with assembled components into the hole in the piston. Pull on the actuating collar and seat the inner boot into the groove in the caliper housing. While doing this, guide the pushrod into its hole in the caliper.

42 Install the collar return springs to the actuating collar.

43 Refer to the caliper overhaul procedure in Chapter 9 for the remainder of the job. Install the caliper as described above in the brake pad replacement procedure, then bleed the brakes, referring to the appropriate Section in Chapter 9. Adjust the parking brake free travel at the caliper following the procedure beginning at Step 45.

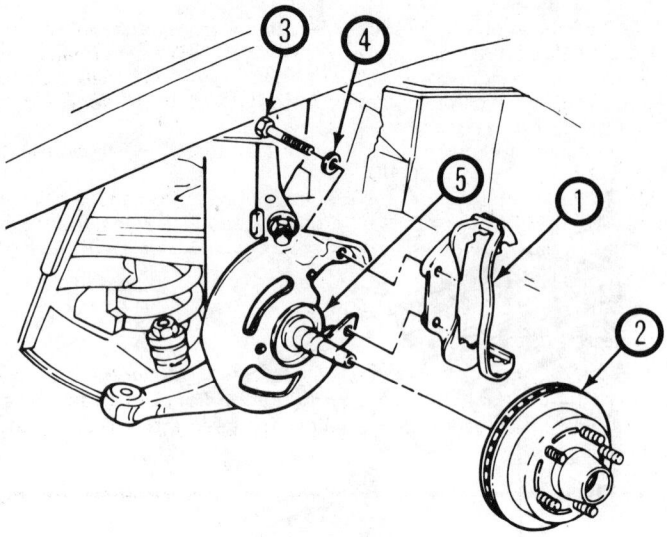

Fig. 13.49 Front disc brake caliper bracket details (Sec 12)

1 Bracket
2 Rotor
3 Bolt
4 Washer
5 Steering knuckle

Disc brake rotor — inspection, removal and installation

44 Follow the procedure in Chapter 9, Section 8, but note that on 1989 models the caliper bracket must be removed prior to rotor removal. To remove a caliper bracket (either front or rear), remove the caliper (don't disconnect the hose) and hang it out of the way. Unscrew the two caliper bracket bolts and remove the caliper bracket (Fig. 13.49). When reinstalling the bracket, be sure to tighten the bolts to the specified torque.

Parking brake free travel — adjustment

45 The only time it is necessary to adjust the parking brake free travel is after the caliper has been overhauled. It is performed to ensure there is no binding between the caliper and the parking brake mechanism. To accurately adjust the free travel, the brake pads must be new or parallel to within 0.006 in (0.15 mm).

46 Have an assistant press on the brake pedal lightly, just enough to take up any clearance between the caliper piston, brake pads and rotor (you should not be able to turn the rotor by hand). Apply light finger pressure to the end of the parking brake lever (Fig. 13.50) and measure the distance the lever stop travels from the caliper housing, comparing the measurement with the figure listed in the specifications section at the beginning of this Chapter.

47 If the free travel is incorrect, remove the adjusting screw (refer to Fig. 13.50), clean the screw threads, apply a thread locking compound and reinstall the screw. Turn it in far enough to obtain the desired clearance between the lever stop and the caliper body.

48 Have the assistant pump the brake pedal a few times, then re-check the free-travel.

Parking brake cables — adjustment

49 Pump the brake pedal firmly three times, then apply and release the parking brake three times.

50 Loosen the rear wheel lug nuts, raise the rear of the vehicle and support it securely on jackstands. Mark the relationship of the wheels to the axle flanges, then remove the wheels.

51 Reinstall two lug nuts on each rotor to hold them in place. Check to see that the parking brake levers on the calipers are seated against their stops — if they aren't, loosen the cables at the equalizer until they are.

52 Turn the adjusting nut on the parking brake equalizer until the levers on the calipers just rise off their stops, then back off the adjusting nut until the levers barely touch their stops.

53 Operate the parking brake lever several times, then re-check the adjustment. The hand lever should travel no more than 16 clicks when properly adjusted. When the lever is released, the rotors should turn freely in both directions.

54 Remove the lug nuts that were installed to retain the rotors. Install the wheels and lug nuts, aligning the matchmarks on the wheels and axle flanges. Lower the vehicle and tighten the lug nuts to the specified torque.

13 Chassis electrical system

SIR (airbag) module – removal and installation

Warning: *The procedure below must be followed exactly as outlined to temporarily disable the Supplemental Inflatable Restraint (SIR) system and prevent false diagnostic codes from being set. Failure to*

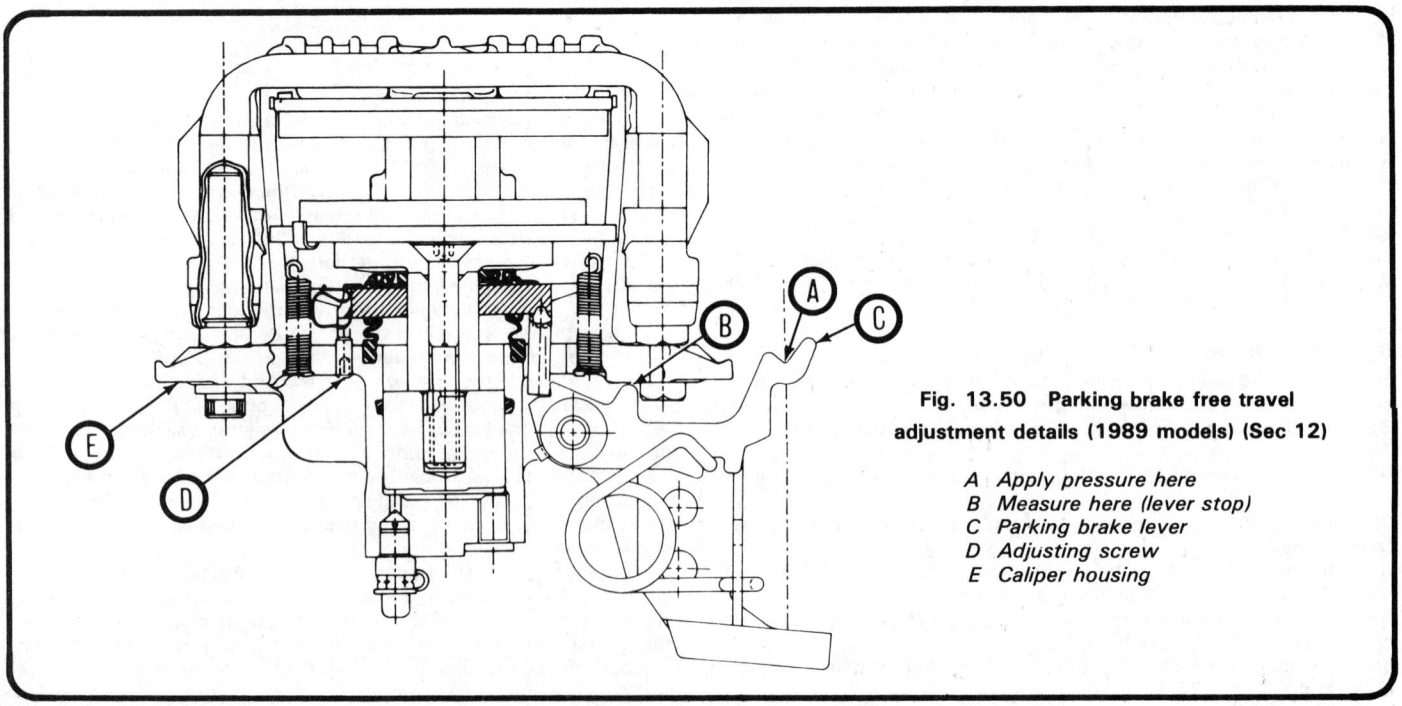

Fig. 13.50 Parking brake free travel adjustment details (1989 models) (Sec 12)

A Apply pressure here
B Measure here (lever stop)
C Parking brake lever
D Adjusting screw
E Caliper housing

follow this procedure could result in air bag deployment, personal injury and/or unneeded SIR system repairs. If air bag deployment has occurred, check the coil assembly wire for scorched or melted insulation and other damage due to excessive heat. If the wire is damaged, the coil assembly must be replaced.

1 Turn the ignition switch off.

2 Detach the negative battery cable from the battery terminal. Wrap tape around the end of the cable to ensure it won't accidentally come into contact with the battery terminal.

3 Remove the SIR fuse from the fuse block.

4 Remove the left sound insulator to gain access to the base of the steering column.

5 Unplug the CPA (connector, positive assurance) and the yellow two-way SIR wire harness connector at the base of the steering column.

6 Remove the screws and nuts from the rear of the steering wheel to release the SIR module. The screws will require a no. 30 Torx driver for removal and installation. **Caution:** *When carrying a live module, make sure the bag and trim cover are pointing away from you. In case of accidental deployment, the chance of injury will be minimal. When placing a live module on a workbench or other surface, make sure the bag and trim cover face up, away from the surface. Never carry the module by the wires or connector.*

7 Installation is the reverse of removal. Tighten the Torx screws to 25 in-lbs.

8 After installation, turn the ignition switch to Run and verify that the "INFLATABLE RESTRAINT" light flashes 7 to 9 times, then turns off. If it doesn't, have the system checked by a dealer service department.

Wiring diagrams

Wiring diagrams for later model vehicles are included at the end of this Supplement. Due to space limitations, not all diagrams for all models are included. However, those included are typical for later models.

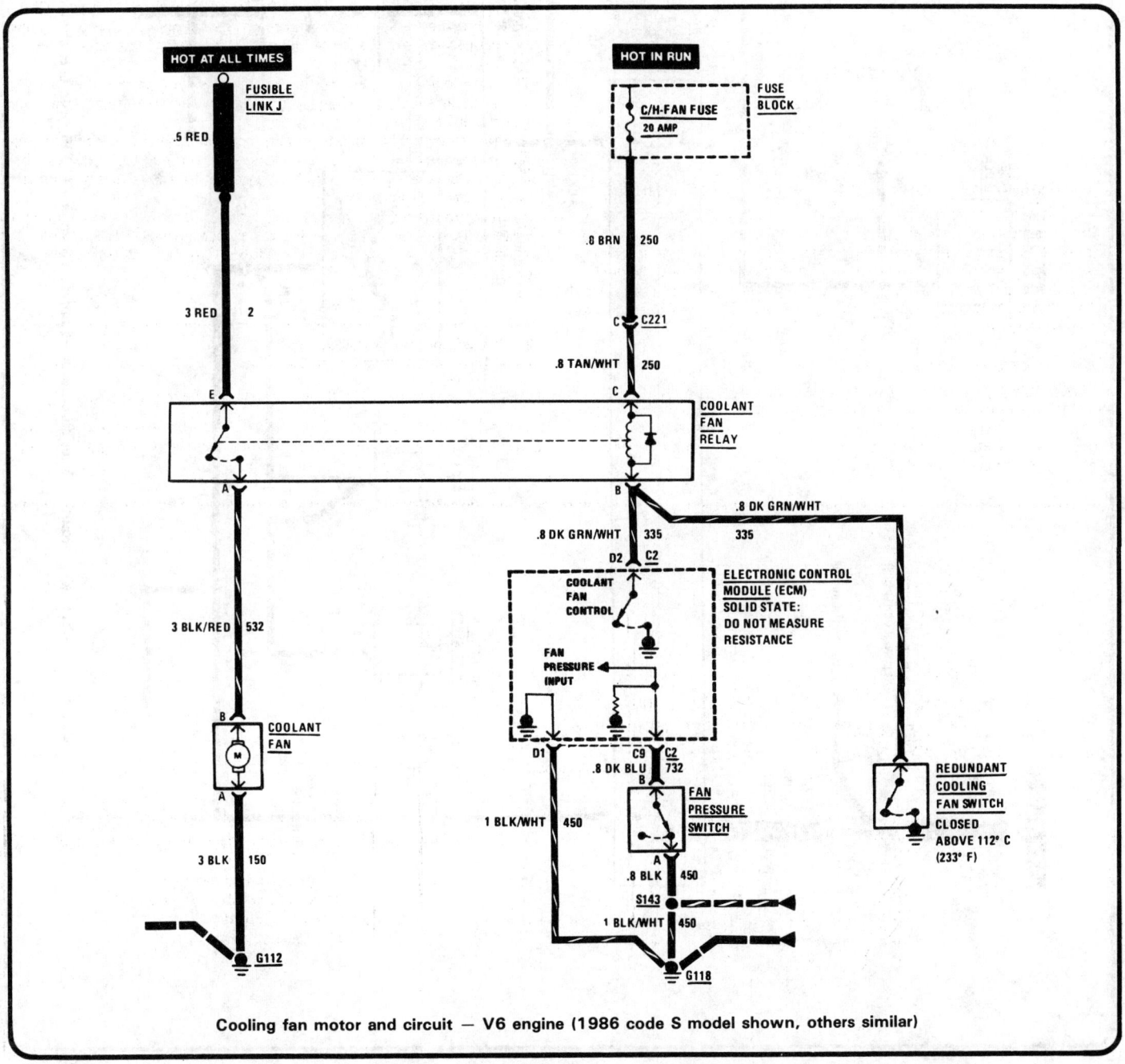

Cooling fan motor and circuit — V6 engine (1986 code S model shown, others similar)

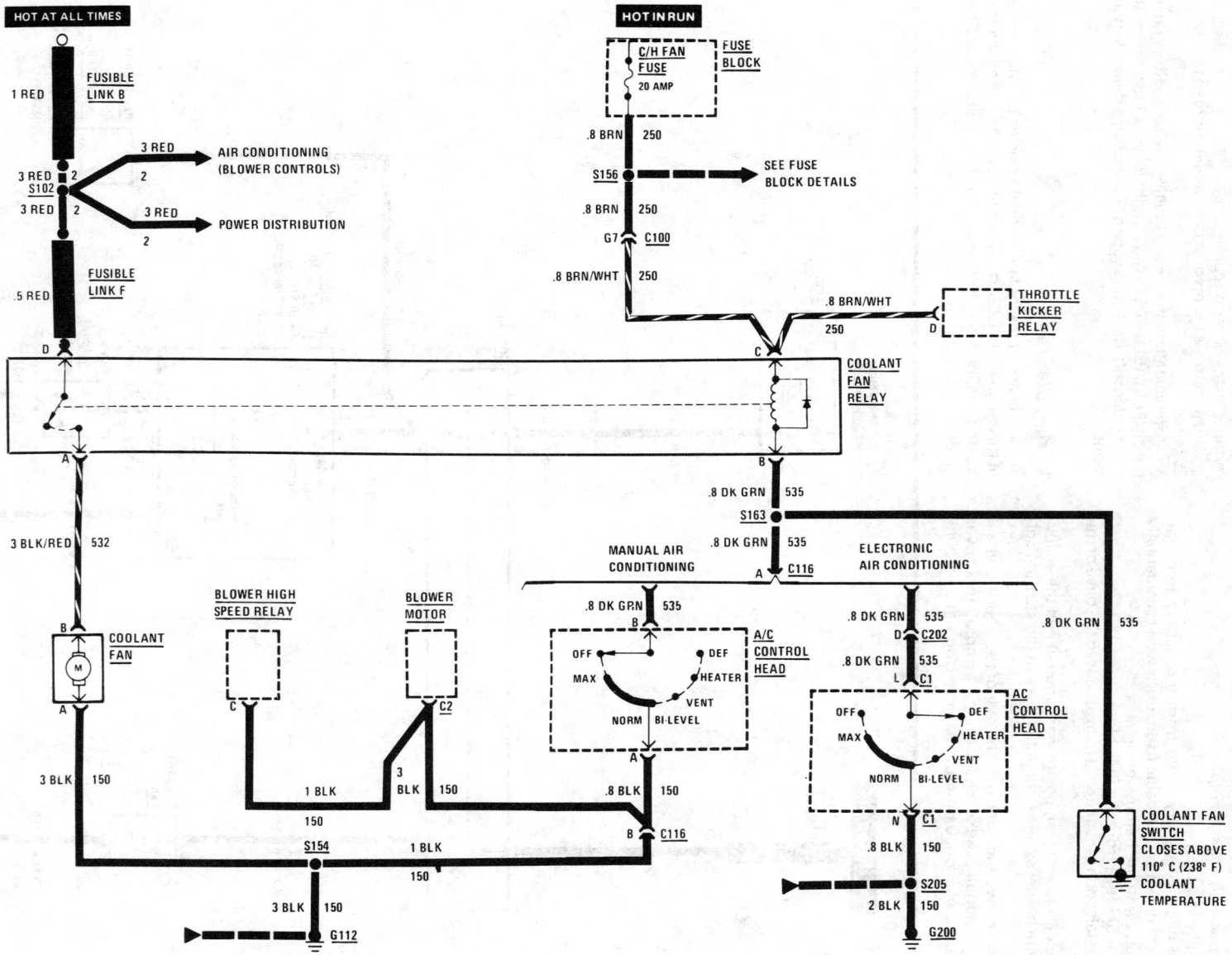

Cooling fan motor and circuit — 5.0L carburetor equipped V8 engine (1986 code G/H shown, others similar)

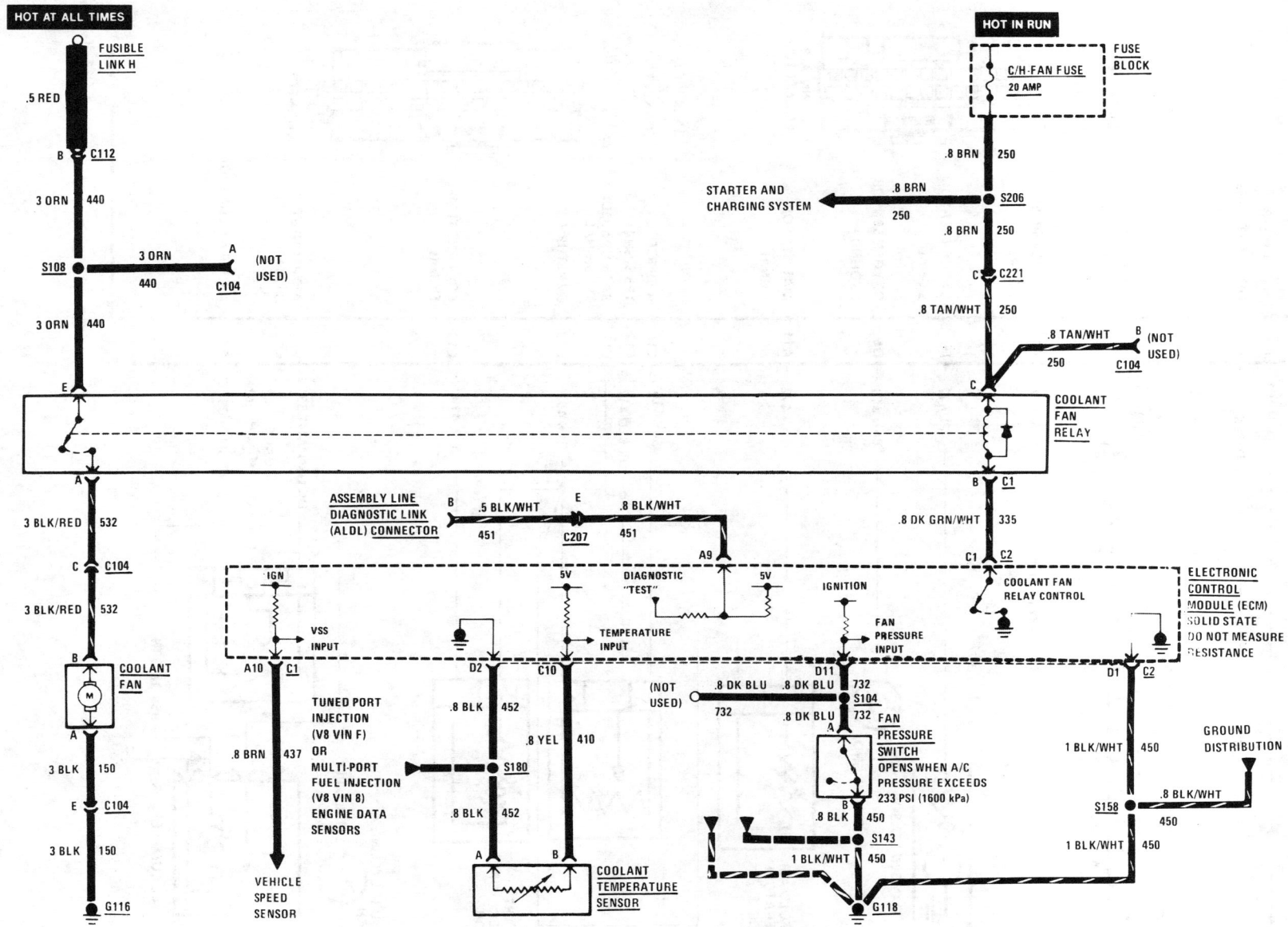

Cooling fan motor and circuit — 5.0/5.7L TPI V8 engine (1987 code F/8 shown, others similar)

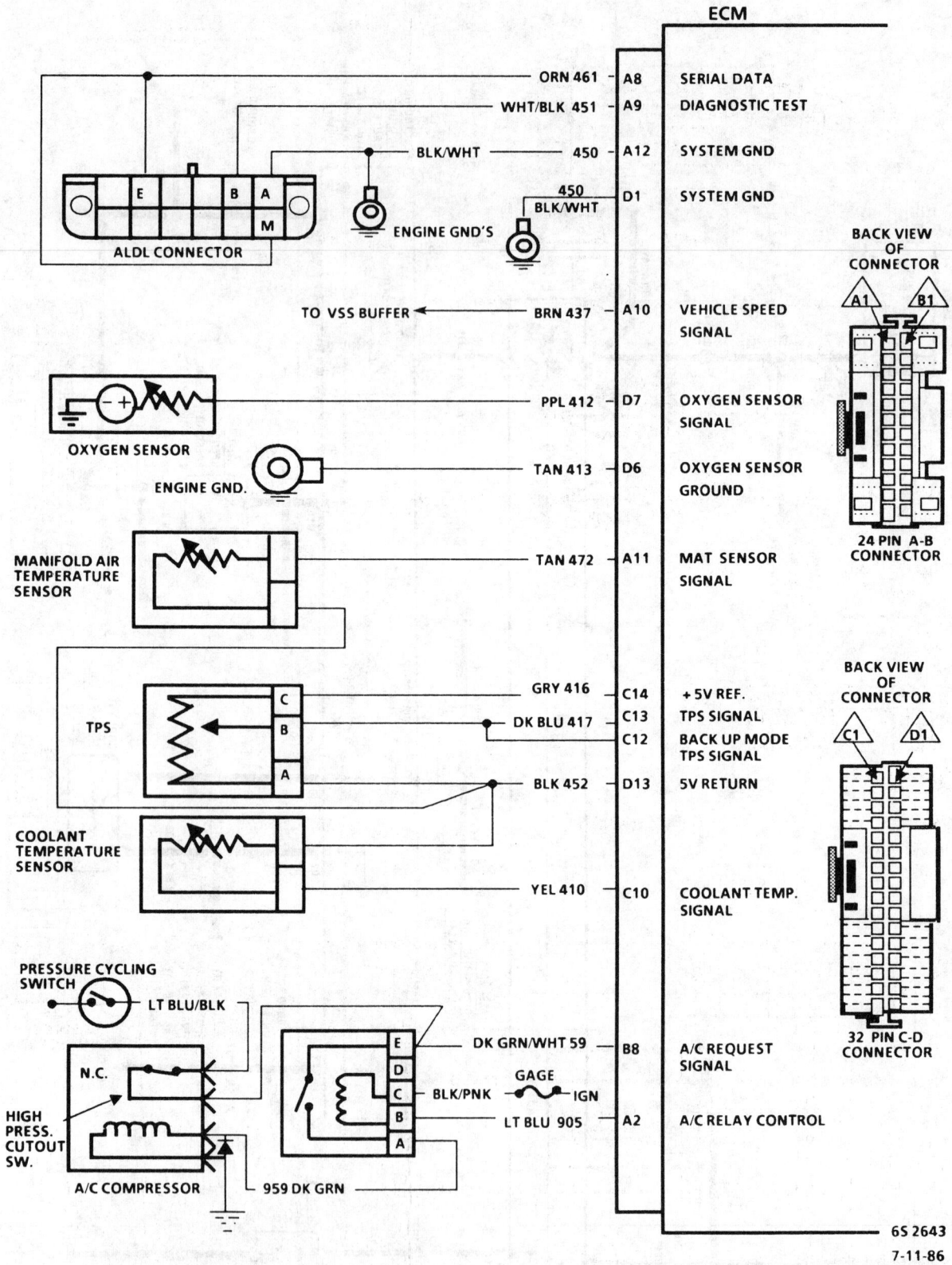

ECM and related components — 2.8L MPFI engine (1 of 3)

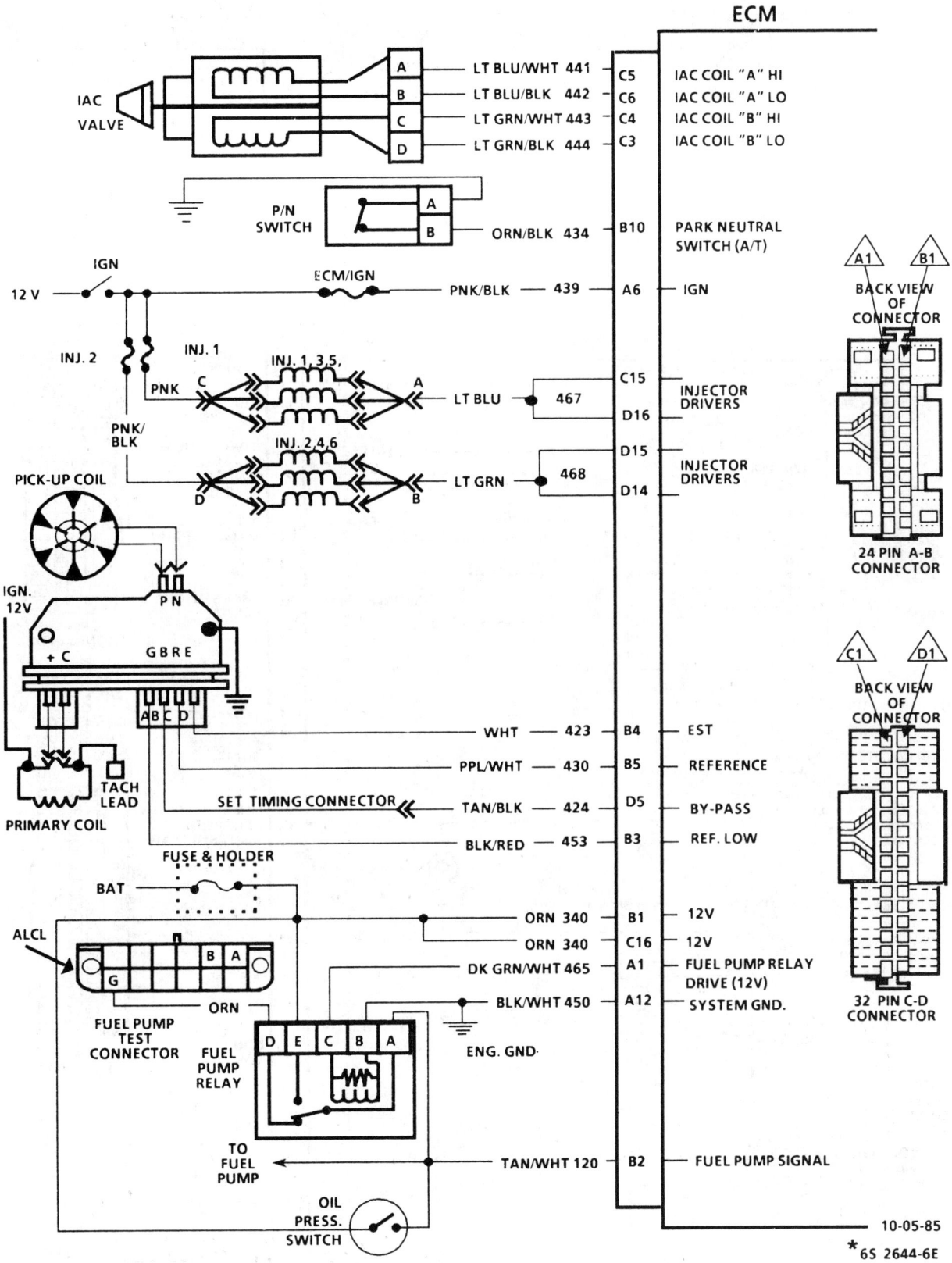

ECM and related components — 2.8L MPFI engine (2 of 3)

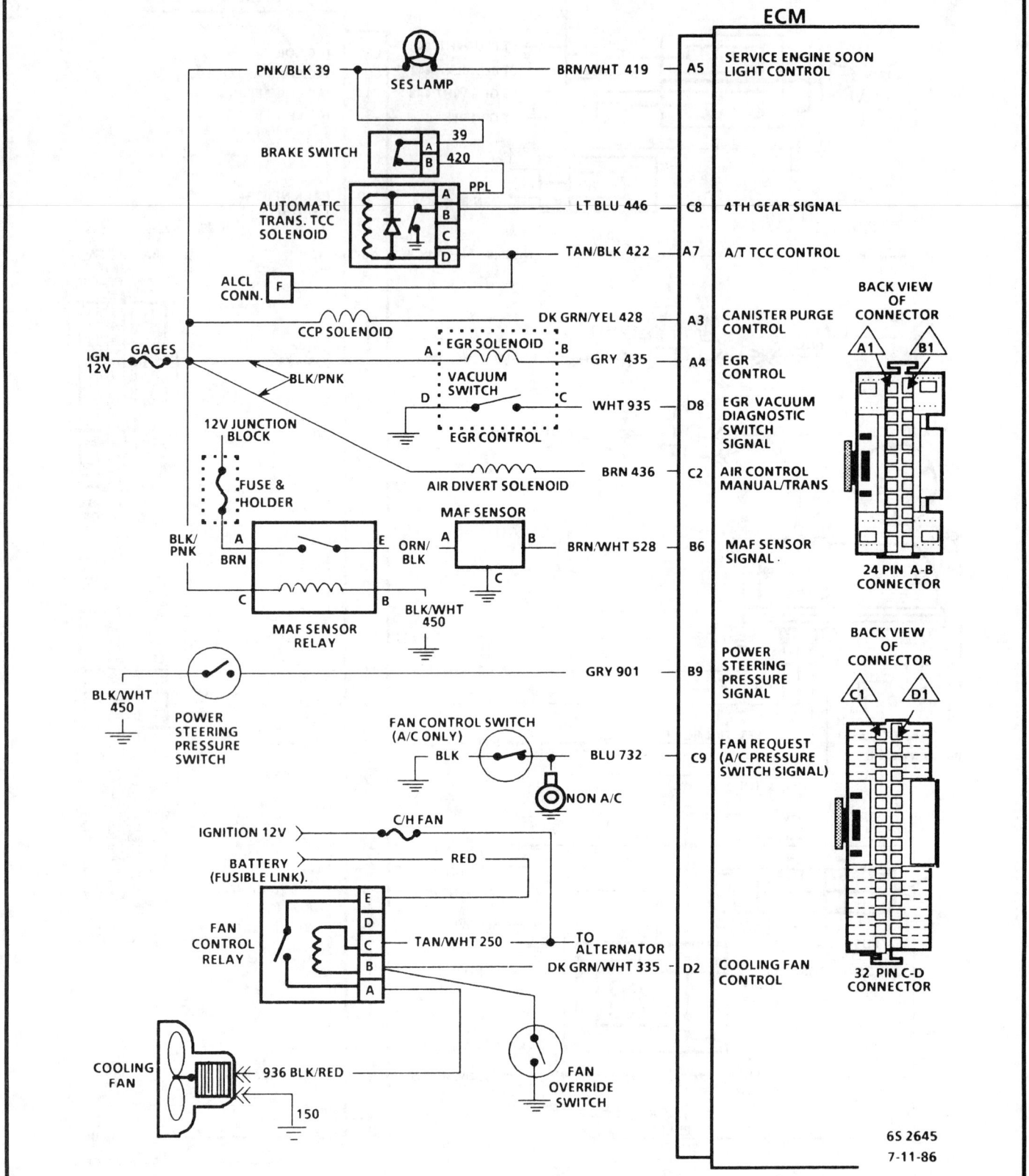

ECM and related components — 2.8L MPFI engine (3 of 3)

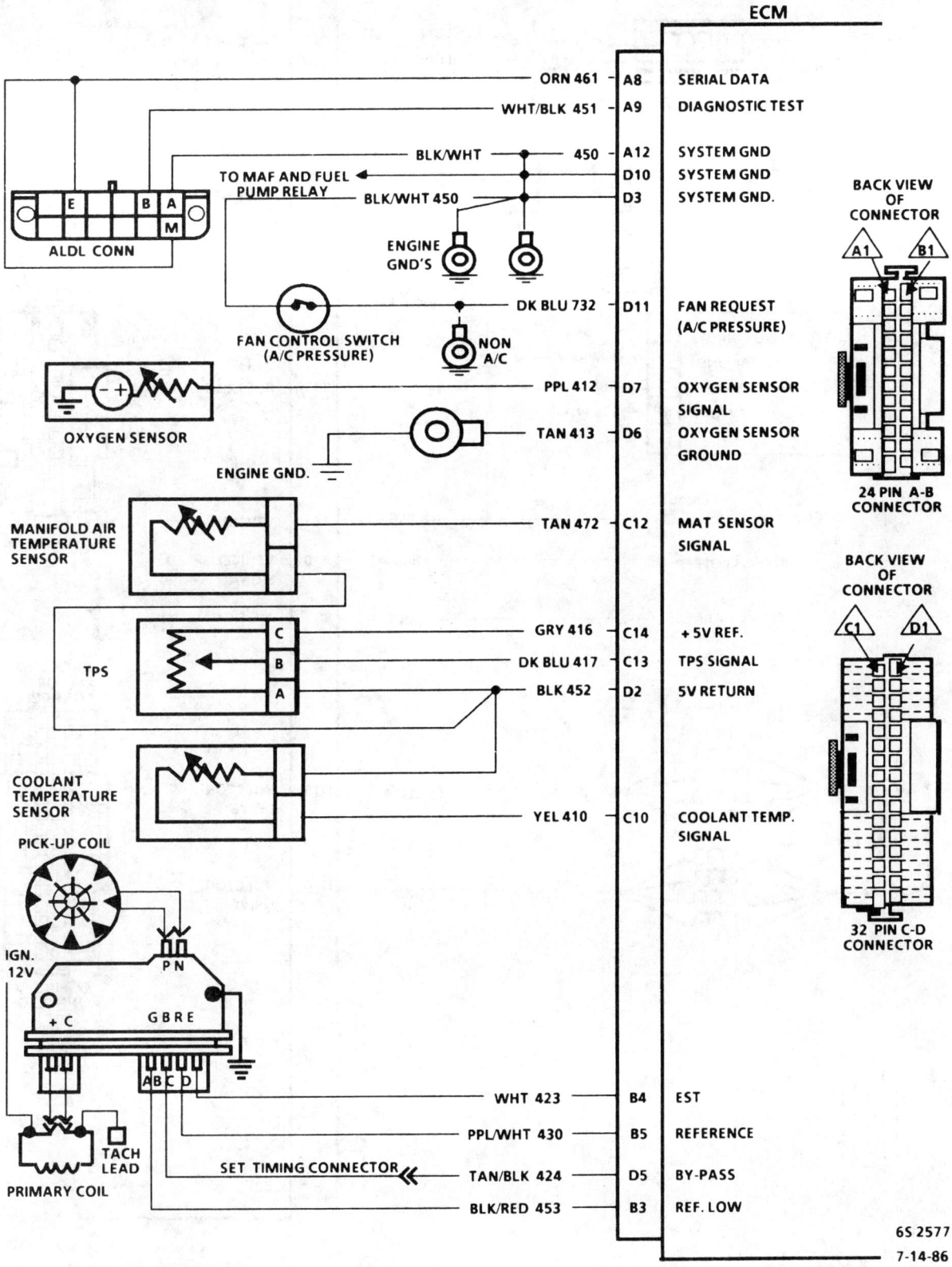

ECM and related components — 5.0/5.7L TPI engine (1 of 4)

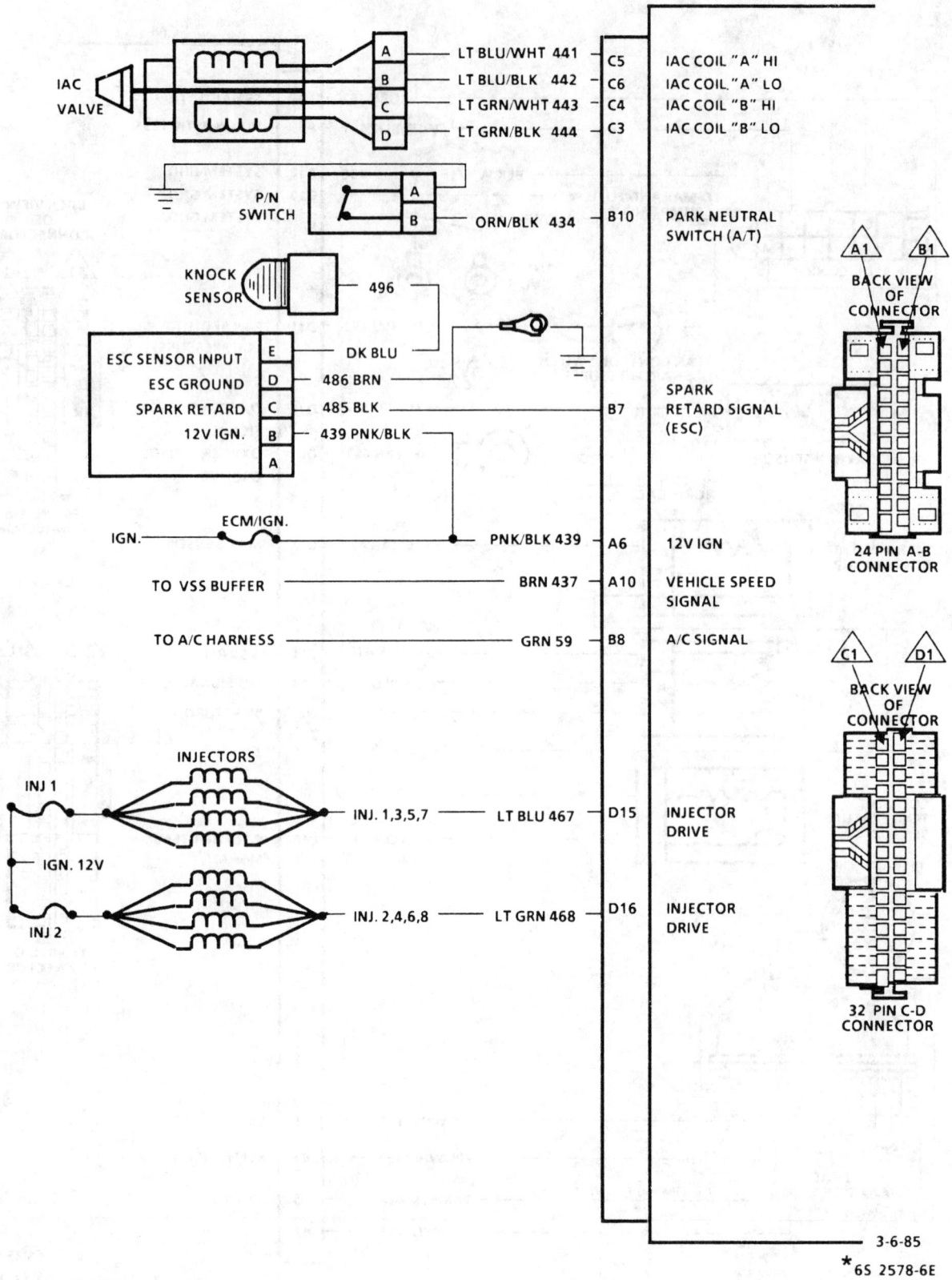

ECM and related components — 5.0/5.7L TPI engine (2 of 4)

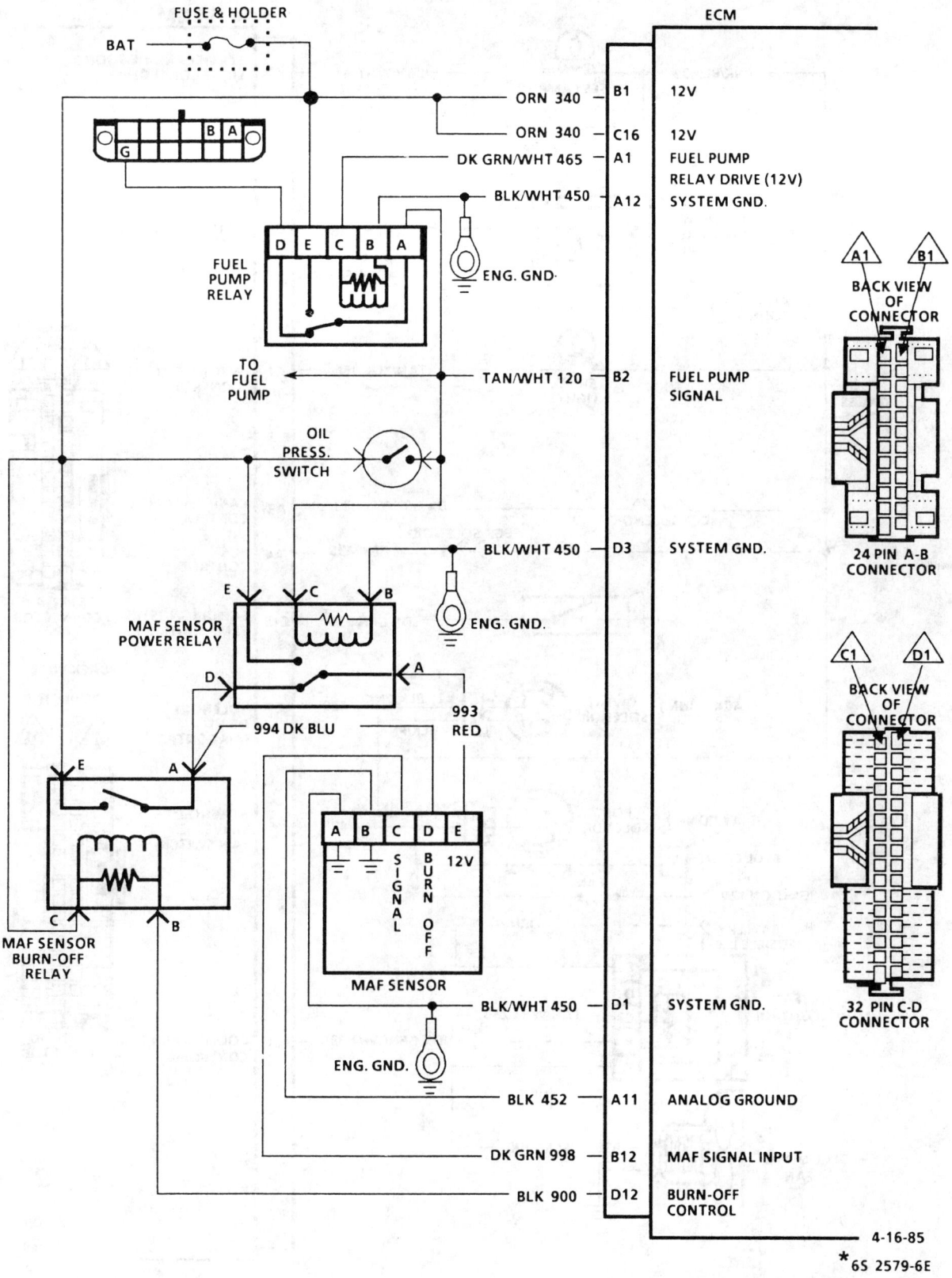

ECM and related components — 5.0/5.7L TPI engine (3 of 4)

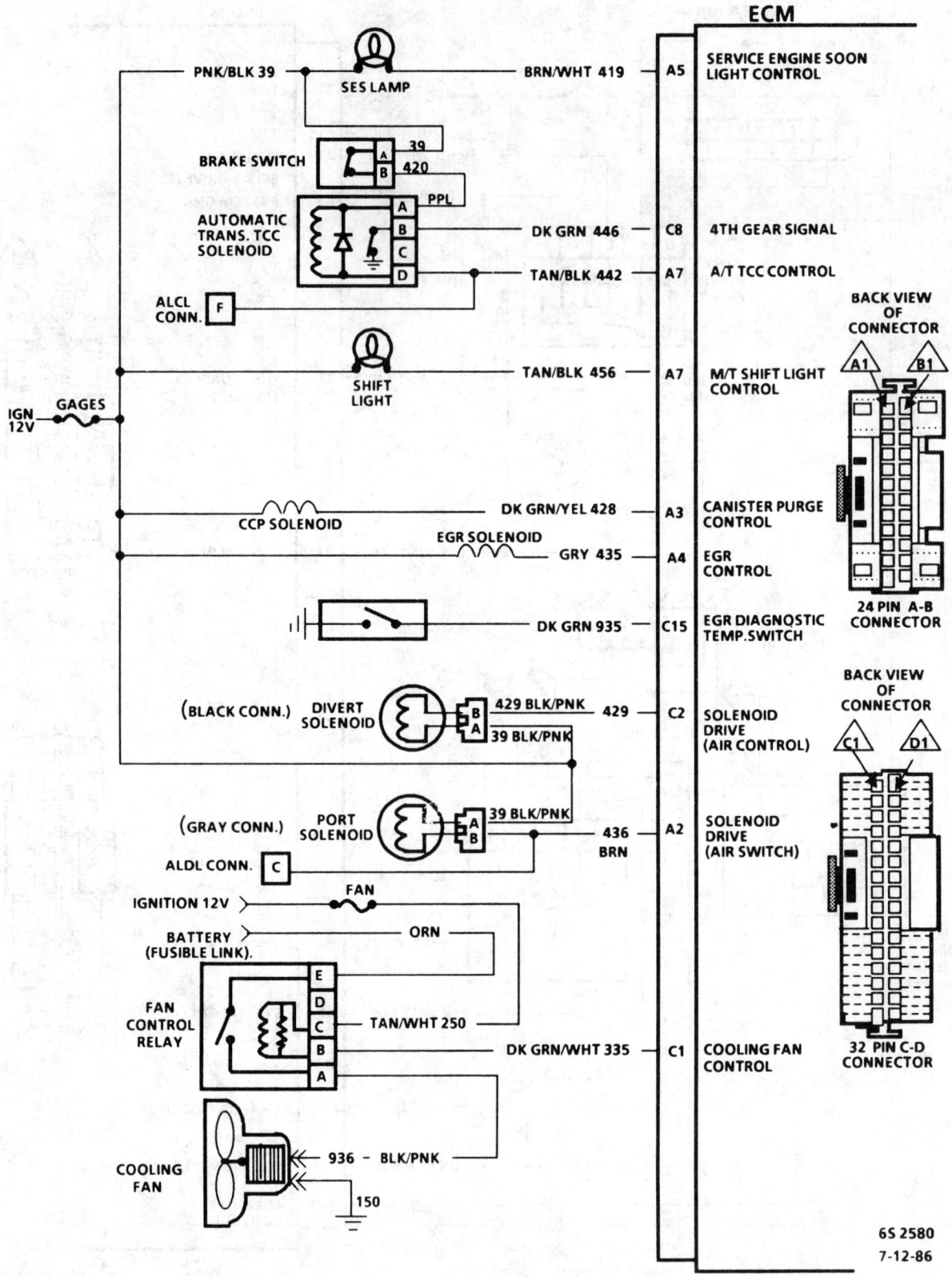

ECM and related components — 5.0/5.7L TPI engine (4 of 4)

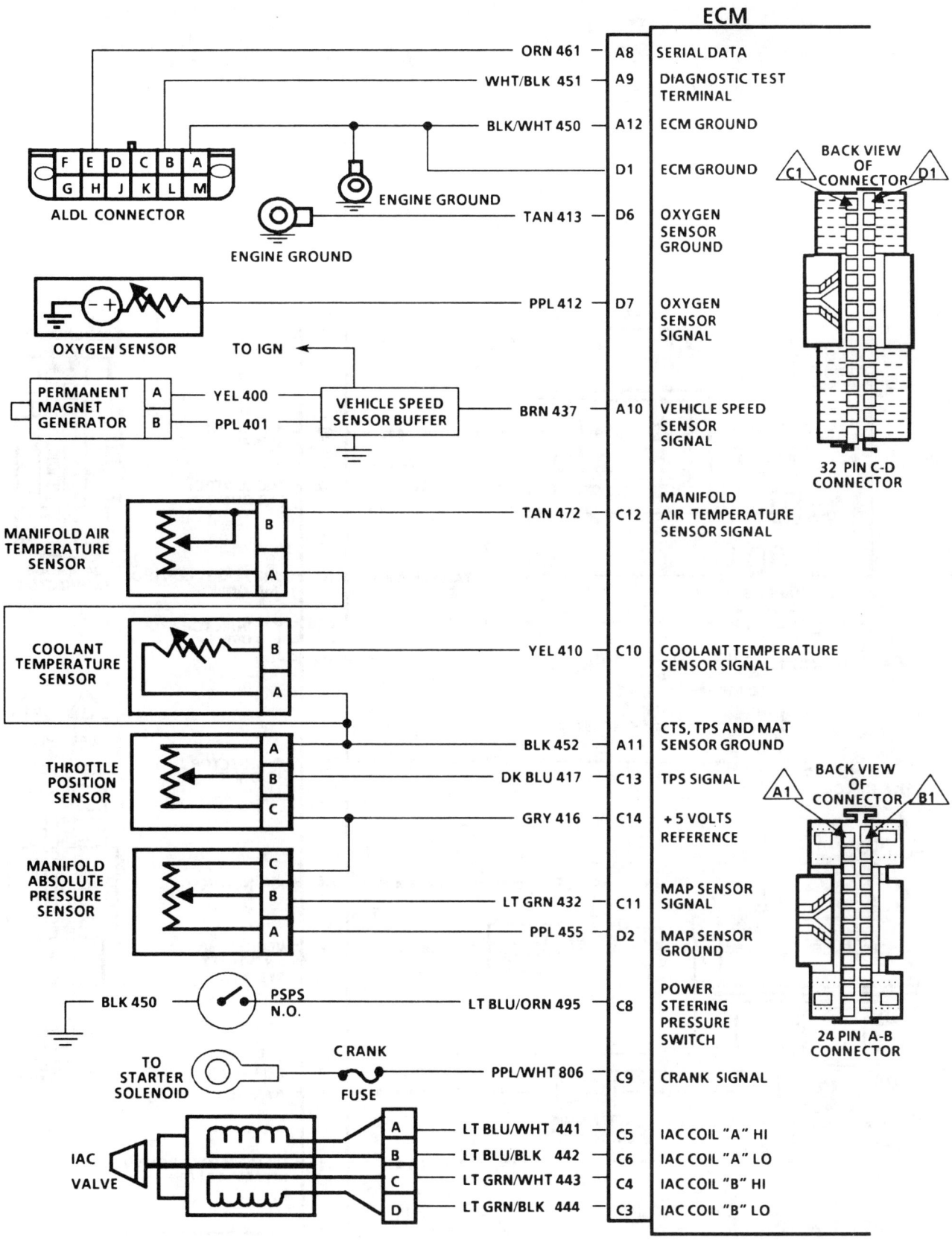

ECM and related components — 5.0L TBI engine (1 of 3)

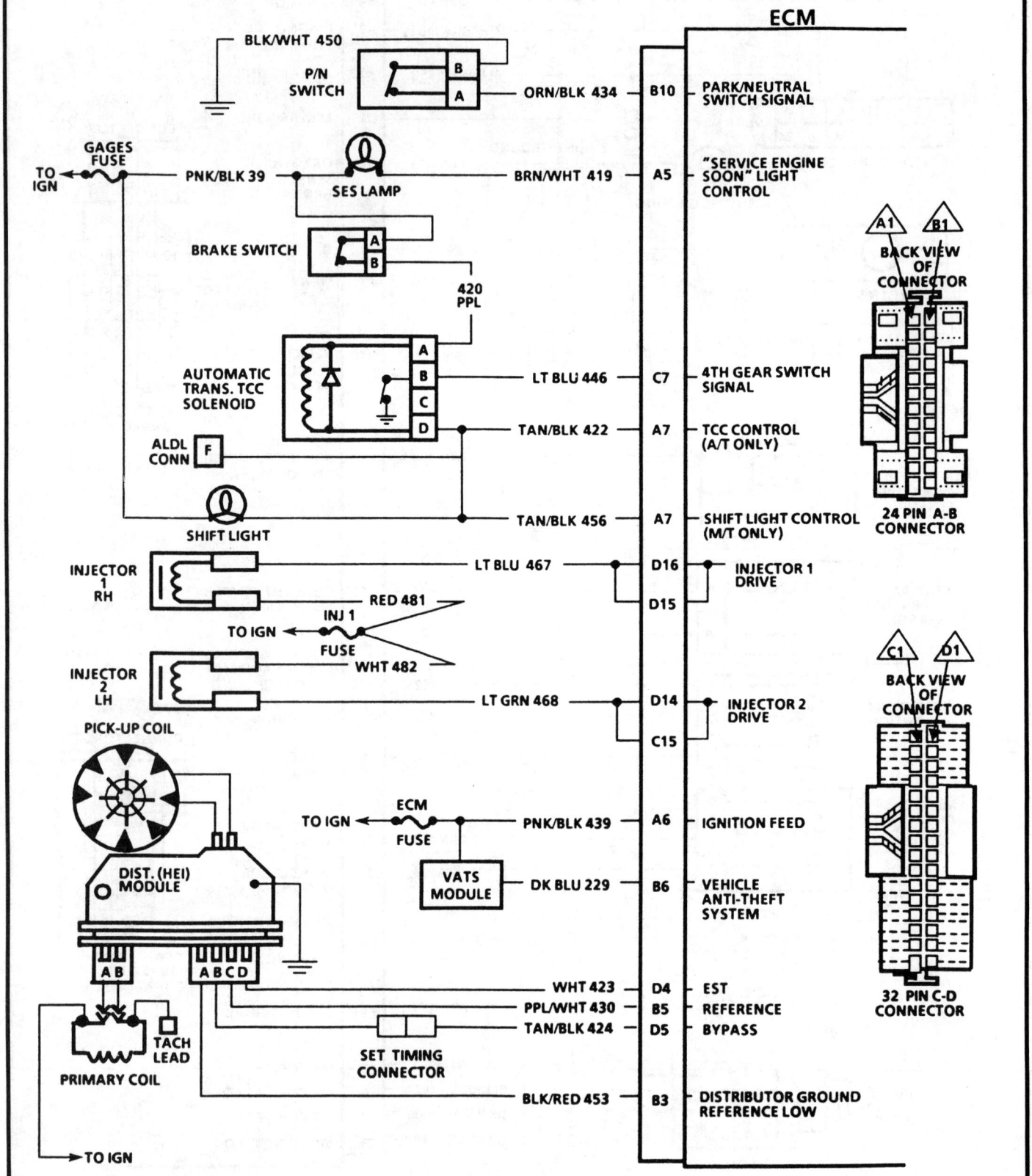

ECM and related components — 5.0L TBI engine (2 of 3)

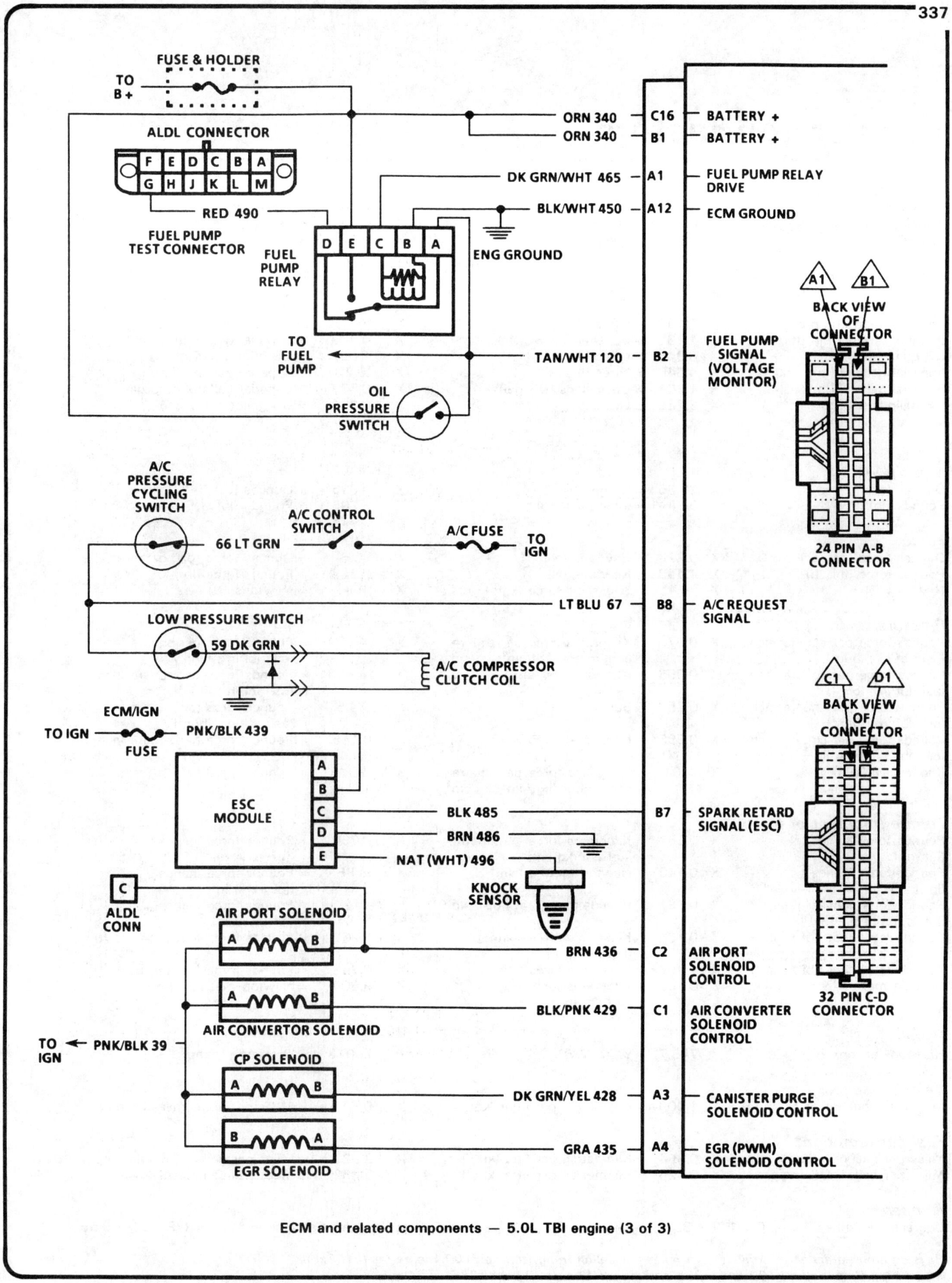

ECM and related components — 5.0L TBI engine (3 of 3)

Conversion factors

Length (distance)

Inches (in)	X 25.4	= Millimetres (mm)	X	0.0394	= Inches (in)
Feet (ft)	X 0.305	= Metres (m)	X	3.281	= Feet (ft)
Miles	X 1.609	= Kilometres (km)	X	0.621	= Miles

Volume (capacity)

Cubic inches (cu in; in^3)	X 16.387	= Cubic centimetres (cc; cm^3)	X	0.061	= Cubic inches (cu in; in^3)
Imperial pints (Imp pt)	X 0.568	= Litres (l)	X	1.76	= Imperial pints (Imp pt)
Imperial quarts (Imp qt)	X 1.137	= Litres (l)	X	0.88	= Imperial quarts (Imp qt)
Imperial quarts (Imp qt)	X 1.201	= US quarts (US qt)	X	0.833	= Imperial quarts (Imp qt)
US quarts (US qt)	X 0.946	= Litres (l)	X	1.057	= US quarts (US qt)
Imperial gallons (Imp gal)	X 4.546	= Litres (l)	X	0.22	= Imperial gallons (Imp gal)
Imperial gallons (Imp gal)	X 1.201	= US gallons (US gal)	X	0.833	= Imperial gallons (Imp gal)
US gallons (US gal)	X 3.785	= Litres (l)	X	0.264	= US gallons (US gal)

Mass (weight)

Ounces (oz)	X 28.35	= Grams (g)	X	0.035	= Ounces (oz)
Pounds (lb)	X 0.454	= Kilograms (kg)	X	2.205	= Pounds (lb)

Force

Ounces-force (ozf; oz)	X 0.278	= Newtons (N)	X	3.6	= Ounces-force (ozf; oz)
Pounds-force (lbf; lb)	X 4.448	= Newtons (N)	X	0.225	= Pounds-force (lbf; lb)
Newtons (N)	X 0.1	= Kilograms-force (kgf; kg)	X	9.81	= Newtons (N)

Pressure

Pounds-force per square inch (psi; lbf/in^2; lb/in^2)	X 0.070	= Kilograms-force per square centimetre (kgf/cm^2; kg/cm^2)	X	14.223	= Pounds-force per square inch (psi; lbf/in^2; lb/in^2)
Pounds-force per square inch (psi; lbf/in^2; lb/in^2)	X 0.068	= Atmospheres (atm)	X	14.696	= Pounds-force per square inch (psi; lbf/in^2; lb/in^2)
Pounds-force per square inch (psi; lbf/in^2; lb/in^2)	X 0.069	= Bars	X	14.5	= Pounds-force per square inch (psi; lbf/in^2; lb/in^2)
Pounds-force per square inch (psi; lbf/in^2; lb/in^2)	X 6.895	= Kilopascals (kPa)	X	0.145	= Pounds-force per square inch (psi; lbf/in^2; lb/in^2)
Kilopascals (kPa)	X 0.01	= Kilograms-force per square centimetre (kgf/cm^2; kg/cm^2)	X	98.1	= Kilopascals (kPa)

Torque (moment of force)

Pounds-force inches (lbf in; lb in)	X 1.152	= Kilograms-force centimetre (kgf cm; kg cm)	X	0.868	= Pounds-force inches (lbf in; lb in)
Pounds-force inches (lbf in; lb in)	X 0.113	= Newton metres (Nm)	X	8.85	= Pounds-force inches (lbf in; lb in)
Pounds-force inches (lbf in; lb in)	X 0.083	= Pounds-force feet (lbf ft; lb ft)	X	12	= Pounds-force inches (lbf in; lb in)
Pounds-force feet (lbf ft; lb ft)	X 0.138	= Kilograms-force metres (kgf m; kg m)	X	7.233	= Pounds-force feet (lbf ft; lb ft)
Pounds-force feet (lbf ft; lb ft)	X 1.356	= Newton metres (Nm)	X	0.738	= Pounds-force feet (lbf ft; lb ft)
Newton metres (Nm)	X 0.102	= Kilograms-force metres (kgf m; kg m)	X	9.804	= Newton metres (Nm)

Power

Horsepower (hp)	X 745.7	= Watts (W)	X	0.0013	= Horsepower (hp)

Velocity (speed)

Miles per hour (miles/hr; mph)	X 1.609	= Kilometres per hour (km/hr; kph)	X	0.621	= Miles per hour (miles/hr; mph)

*Fuel consumption**

Miles per gallon, Imperial (mpg)	X 0.354	= Kilometres per litre (km/l)	X	2.825	= Miles per gallon, Imperial (mpg)
Miles per gallon, US (mpg)	X 0.425	= Kilometres per litre (km/l)	X	2.352	= Miles per gallon, US (mpg)

Temperature

Degrees Fahrenheit = (°C x 1.8) + 32

Degrees Celsius (Degrees Centigrade; °C) = (°F - 32) x 0.56

**It is common practice to convert from miles per gallon (mpg) to litres/100 kilometres (l/100km), where mpg (Imperial) x l/100 km = 282 and mpg (US) x l/100 km = 235*

Index

A

Air cleaner element – 52, 301
Alternator
brush replacement – 165
general information and special precautions – 162
removal and installation – 163
testing in vehicle – 162
Antifreeze solutions – 125
Automatic transmission
checking the fluid level – 31
checking the transmission rear mount – 185
extension oil seal – 187
fault diagnosis – 189
fluid change – 51
general information – 190
neutral start switch – 191
oil change – 51
removal and installation – 191
shift linkage – 189
specifications – 189
speedometer drive gear – 188
throttle valve cable description and adjustment – 190

B

Battery
charging – 162
emergency jump starting – 15
maintenance – 32
removal and installation – 158
state-of-charge indicator – 32
Body
air-conditioner
checks and maintenance – 129
general information – 125
body repairs
major damage – 285
minor damage – 284
door handles – 290
door lock knob – 286
door trim – 286
doors – 286 through 293
front door lock assembly – 289
front door window
assembly – 290
regulator – 290
general information – 281
headlight bezel – 229
heater blower motor – 129
heater general information – 125
hood
damper replacement – 285
latch release cable – 285
removal and installation – 285
maintenance
body – 281
upholstery and carpets – 284
rear hatch removal, installation and adjustment – 285
specifications – 281
spring clips and operating rods – 289
windshield replacement – 293
Body repair photo sequence – 282 and 283
Brakes
brake caliper overhaul
front – 208, 322
rear – 210, 323
combination valve check and replacement – 221
disc brake rotor – 212, 324
disc pads
front – 205, 321
rear – 210, 322
drum brake shoes – 214
fault diagnosis – 205
general information – 205, 321
hydraulic brake hoses and lines – 220
hydraulic system bleeding – 223
master cylinder – 219
parking brake cable – 223, 324
parking brake free travel – adjustment – 324
power brake unit – 223
service brake pedal – 225
specifications – 204, 300
stop light switch – 225
wheel cylinder – 218
Buying spare parts – 9

C

Camshaft
inspection and bearing replacement – 115
removal and installation
L4 – 70
V6 – 88
V8 – 100

Carburetor
choke check – 38
mounting torque check – 44
overhaul
E2SE – 137
E4ME and E4MC – 152
removal and installation – 136
Chapter 13 Supplement: Revisions and information on 1985 and later models – 294
Charging system
check – 162
general information and precautions – 162
'check engine' light – 172
Clutch
adjustment – 46
clutch pedal – 195
cross-shaft – 196
general information – 195
pilot bearing – 197
removal, servicing and installation – 196
specifications – 194
Cold start valve (EFI) – 313, 316
Conversion factors – 338
Cooling system
antifreeze solutions – 125
coolant level – 29
coolant temperature sending unit replacement – 129
cooling system draining, flushing and refilling – 46
fault diagnosis – 34
general information – 125
radiator removal – 127
specifications – 26
thermostat – 126
water pump removal and installation – 128
Crankshaft and main bearings – 113, 117, 121
Cylinder block – 112, 114
Cylinder head
assembly – 111
cleaning and inspection – 109
disassembly – 108
removal and installation
L4 – 64
V6 – 82
V8 – 98

D

Drivebelt check and adjustment – 33, 301
Driveshaft
checking – 199
general information – 197
removal and installation – 199
specifications – 194
universal joints
checking for wear – 199
disassembly, inspection and reassembly – 199

E

Electrical system
alternator
brush replacement – 165
general information and special precautions – 162
removal and installation – 163
testing in vehicle – 162
back-up light warning switch – 191
battery
charging – 146
emergency jump-starting – 15
maintenance – 32
removal and installation – 158
state-of-charge indicator – 32
clutch operated starter switch (manual transmission) – 188
console switches – 234
cooling fan – 308
dome light – 229
electric choke heater
general – 179
testing – 180
fault diagnosis – 227
front lights – 229
fuses – 227
fusible links – 228
general information – 227
headlight replacement and adjustment – 228
instrument panel – 233
light switch – 234
neutral start switch (automatic transmission) – 191
rear lights – 229
specifications – 226, 227
starter motor
general information – 165
removal and installation – 167
solenoid removal, repair and installation – 167
testing in vehicle – 165
stop light switch – 225
turn signal hazard flashers – 228
windshield wiper arm – 236
windshield wiper blades – 35
windshield wiper motor removal and installation – 239
windshield wiper transmission – 239
wiring diagrams – 245 to 264, 325 to 337
Electronic Fuel Injection (EFI) – 152, 308
Emissions control systems
air cleaner element – 52
Air Injection Reactor (AIR) – 177, 319
Computer Command Control System (CCCS) – 172
Early Fuel Evaporation (EFE) system – 179, 321
Electronic Spark Timing (EST) – 176
Evaporative Emissions Control System (EECS) – 182
Exhaust Gas Recirculation (EGR) system – 181, 321
fault diagnosis – 172 to 176
Fuel Control system – 176
general information – 172, 317
oxygen sensor replacement – 52
Positive Crankcase Ventilation (PCV) system – 182
specifications – 51, 265
Thermostatically Controlled Air Cleaner (THERMAC) – 183
Transmission Converter Clutch (TCC) – 184
Trouble codes – 173, 317
vacuum hose routing – 171
Engine
camshaft
inspection and bearing replacement – 115
removal and installation
L4 – 70
V6 – 88
V8 – 100
crankcase front cover
V6 – 86, 306
V8 – 100, 307
crankshaft and main bearings – 113, 117, 121
crankshaft rear main oil seal (V6 and V8) – 119, 305, 307
cylinder block – 113, 114
cylinder head
assembly – 111
inspection – 109
disassembly – 108
inspection – 109

removal and installation
L4 – 64, 304
V6 – 82, 305
V8 – 98
disassembly – 108
engine mounts
L4 – 71
V6 – 89
V8 – 101
engine start-up after major repairs or overhaul – 124
exhaust manifolds
L4 – 63, 304
V6 – 81
V8 – 97
fault diagnosis – 107
flywheel/driveplate and rear main bearing oil seal (L4) – 71
general information
L4 – 62, 302
V6 – 78, 304
V8 – 94, 306
hydraulic lifters
L4 – 65, 302
V6 – 80
V8 – 97, 307
intake manifold
L4 – 63, 303
V6 – 79, 305
V8 – 94, 306
oil pan
L4 – 67, 302
V6 – 84
V8 – 98
oil pump
L4 – 69
V6 – 84
V8 – 99
oil pump driveshaft (L4 only) – 66
pistons and connecting rods – 111, 117, 121
piston rings – 118
Positive Crankcase Ventilation (PCV) system – 182
pushrod cover (L4) – 62
reassembly – 23
removal and installation
L4 – 72
V6 – 89
V8 – 101
repair operations possible with engine in vehicle – 106
rocker arm covers
L4 – 62
V6 – 78, 305
V8 – 95, 306
specifications
L4 – 61, 295
V6 – 74, 295, 296
V8 – 94, 295, 296
timing chain and sprockets
V6 – 88, 306
V8 – 100
timing gear cover (L4) – 69, 303
valve adjustment
V6 – 81
V8 – 97
valve servicing – 111
valve train component replacement
L4 – 62
V6 – 78
V8 – 95
Exhaust system
catalytic converter – 157
Exhaust Gas Recirculation (EGR) system – 181, 321
fault diagnosis – 37
general information – 157
removal and installation – 157

F

Front suspension – see Suspension and steering
Fuel pressure connection assembly (EFI) – 311
Fuel rail (EFI) – 311, 315, 316
Fuel system
carburetor check – 38
cold start valve (EFI) – 313, 316
Crossfire Injection System (CFI) – 156
Electronic Fuel Injection system (EFI) – 152, 308
emission control components – 169
engine idle speed – 38
exhaust system – 157
fuel and vapor lines – 136, 317
fuel filter replacement – 41, 302
fuel injection pressure relief procedure – 153
fuel injectors – 311, 315, 316
fuel pressure regulator – 154, 315, 316
fuel pump – 135, 153, 309
general information – 135
fuel tank – 136
Idle Air Control (IAC) – 153, 309, 316
injector removal and installation – 154, 311, 315, 316
mounting torque check – 44
overhaul
E2SE – 137
E4ME and E4MC – 152
removal and installation – 136
specifications – 132, 299
system check – 41
Throttle Body Injection (TBI) 220 – 308
Throttle Body Injection (TBI) 300 – 154, 156
Throttle Body Injection (TBI) 400 – 156
throttle body – replacement – 312
throttle linkage check – 43
TPS (EFI) – 317

H

Heater core – 130

I

Idle Air Control (IAC) valve (EFI) – 153, 309, 316
Ignition system
check – 159
distributor removal, servicing and installation – 159
general information and precautions – 158
ignition coil removal and installation – 161
ignition timing – 55
module replacement – 161
pick-up coil replacement – 161
spark plug replacement – 56
spark plug wires, distributor cap and rotor – 58
specifications – 27
Idle speed (minimum – TPI and MPFI systems) – 317

J

Jacking and towing – 16

M

Manual transmission
checking the transmission rear mount – 185
extension housing oil seal – 187
general information – 185
linkage (4-speed) – 186
shift control lever (5-speed) – 187
specifications – 185, 299

speedometer gear seal – 188
transmission removal and installation
4-speed – 187
5-speed – 187
Mass Air Flow (MAF) sensor (EFI) – 308, 312
Metric conversion tables – 338
Multi-Port Fuel Injection (MPFI) system – 308

P

Plenum (EFI) – 310, 313, 316

R

Rear axle
axle assembly – 203
axleshaft bearing – 202
axleshaft oil seal – 202
axleshaft removal and installation – 202
general information and identification – 201, 299, 321
specifications – 194
Routine maintenance – 26, 295, 301

S

Suspension and steering
cancelling cam – 279
checking – 37
front stabilizer bar – 269
front suspension balljoints – 270
front suspension coil springs – 270
front suspension control arms – 270
front suspension lubrication and inspection – 36
front wheel bearings – 47
general information
front suspension – 267
power steering – 278
rear suspension – 270
steering system – 275
ignition lock cylinder – 280
ignition key warning (buzzer) switch – 280
power steering gear – 278
power steering pump – 279
power steering system bleeding – 278
rear shock absorbers – 272
rear stabilizer bar – 274
rear suspension coil spring – 272
rear suspension lower control arm bushing – 274
rear suspension lower control arms – 273
specifications – 265, 301
steering column switches – 279
steering knuckle – 277
steering wheel – 278
strut assembly – 269
tie-rods – 276
torque arm – 274
track bar – 273
track bar brace – 273
turn signal switch – 279

T

Throttle Body Injection (TBI) – 308, 309, 317
Throttle cable (EFI) – 317
Tire rotation – 45, 302
Tools and working facilities – 9
Tuned-Port Injection (TPI) system – 308, 311, 316

V

Vehicle identification numbers – 7

W

Wiring diagrams – 245 to 264, 325 to 337

HAYNES AUTOMOTIVE MANUALS

NOTE: New manuals are added to this list on a periodic basis. If you do not see a listing for your vehicle, consult your local Haynes dealer for the latest product information.

ACURA

1776 **Integra & Legend** '86 thru '90

AMC

Jeep CJ – *see JEEP (412)*
694 **Mid-size models,** Concord, Hornet, Gremlin & Spirit '70 thru '83
934 **(Renault) Alliance & Encore** all models '83 thru '87

AUDI

615 **4000** all models '80 thru '87
428 **5000** all models '77 thru '83
1117 **5000** all models '84 thru '88

AUSTIN

Healey Sprite – *see MG Midget Roadster (265)*

BMW

276 **320i** all 4 cyl models '75 thru '83
632 **528i & 530i** all models '75 thru '80
240 **1500 thru 2002** all models except Turbo '59 thru '77
348 **2500, 2800, 3.0 & Bavaria** '69 thru '76

BUICK

Century (front wheel drive) – *see GENERAL MOTORS A-Cars (829)*
*1627 **Buick, Oldsmobile & Pontiac Full-size (Front wheel drive)** all models '85 thru '93
Buick Electra, LeSabre and Park Avenue; **Oldsmobile** Delta 88 Royale, Ninety Eight and Regency; **Pontiac** Bonneville
*1551 **Buick Oldsmobile & Pontiac Full-size (Rear wheel drive)**
Buick Electra '70 thru '84, Estate '70 thru '90, LeSabre '70 thru '79
Oldsmobile Custom Cruiser '70 thru '90, Delta 88 '70 thru '85, Ninety-eight '70 thru '84
Pontiac Bonneville '70 thru '81, Catalina '70 thru '81, Grandville '70 thru '75, Parisienne '84 thu '86
627 **Mid-size** all rear-drive **Regal & Century** models with V6, V8 and Turbo '74 thru '87
Regal – *see GENERAL MOTORS (1671)*
Skyhawk – *see GENERAL MOTORS J-Cars (766)*
552 **Skylark** all X-car models '80 thru '85

CADILLAC

*751 **Cadillac Rear Wheel Drive** all gasoline models '70 thru '90
Cimarron – *see GENERAL MOTORS J-Cars (766)*

CAPRI

296 **2000 MK I Coupe** all models '71 thru '75
205 **2600 & 2800** V6 Coupe '71 thru '75
375 **2800 Mk II** V6 Coupe '75 thru '78
Mercury Capri – *see FORD Mustang (654)*

CHEVROLET

*1477 **Astro & GMC Safari Mini-vans** all models '85 thru '91
554 **Camaro** V8 all models '70 thru '81
*866 **Camaro** all models '82 thru '91
Cavalier – *see GENERAL MOTORS J-Cars (766)*
Celebrity – *see GENERAL MOTORS A-Cars (829)*
625 **Chevelle, Malibu & El Camino** all V6 & V8 models '69 thru '87
449 **Chevette & Pontiac T1000** all models '76 thru '87
550 **Citation** all models '80 thru '85
*1628 **Corsica/Beretta** all models '87 thru '92
274 **Corvette** all V8 models '68 thru '82
*1336 **Corvette** all models '84 thru '91
704 **Full-size Sedans** Caprice, Impala, Biscayne, Bel Air & Wagons, all V6 & V8 models '69 thru '90
Lumina – *see GENERAL MOTORS (1671)*
Lumina APV – *see GENERAL MOTORS (2035)*
319 **Luv Pick-up** all 2WD & 4WD models '72 thru '82
626 **Monte Carlo** all V6, V8 & Turbo models '70 thru '88
241 **Nova** all V8 models '69 thru '79
*1642 **Nova and Geo Prizm** all front wheel drive models, '85 thru '90
*420 **Pick-ups '67 thru '87 – Chevrolet & GMC,** all full-size models '67 thru '87; Suburban, Blazer & Jimmy '67 thru '91
*1664 **Pick-ups '88 thru '92 – Chevrolet & GMC** all full-size (C and K) models, '88 thru '92
*1727 **Sprint & Geo Metro** '85 thru '91
*831 **S-10 & GMC S-15 Pick-ups** all models '82 thru '92
*345 **Vans – Chevrolet & GMC,** V8 & in-line 6 cyl models '68 thru '92

CHRYSLER

*1337 **Chrysler & Plymouth Mid-size** front wheel drive '82 thru '89
K-Cars – *see DODGE Aries (723)*
Laser – *see DODGE Daytona (1140)*

DATSUN

402 **200SX** all models '77 thru '79
647 **200SX** all models '80 thru '83
228 **B-210** all models '73 thru '78
525 **210** all models '78 thru '82
206 **240Z, 260Z & 280Z** Coupe & 2+2 '70 thru '78
563 **280ZX** Coupe & 2+2 '79 thru '83
300ZX – *see NISSAN (1137)*
679 **310** all models '78 thru '82
123 **510 & PL521 Pick-up** '68 thru '73
430 **510** all models '78 thru '81
372 **610** all models '72 thru '76
277 **620 Series Pick-up** all models '73 thru '79
720 Series Pick-up – *see NISSAN Pick-ups (771)*
376 **810/Maxima** all gasoline models '77 thru '84
124 **1200** all models '70 thru '73
368 **F10** all models '76 thru '79
Pulsar – *see NISSAN (876)*
Sentra – *see NISSAN (982)*
Stanza – *see NISSAN (981)*

DODGE

*723 **Aries & Plymouth Reliant** all models '81 thru '89
*1231 **Caravan & Plymouth Voyager Mini-Vans** all models '84 thru '91
699 **Challenger & Plymouth Saporro** all models '78 thru '83
236 **Colt** all models '71 thru '77
610 **Colt & Plymouth Champ (front wheel drive)** all models '78 thru '87
*556 **D50/Ram 50/Plymouth Arrow Pick-ups & Raider** '79 thru '91
*1668 **Dakota Pick-up** all models '87 thru '90
234 **Dart & Plymouth Valiant** all 6 cyl models '67 thru '76
*1140 **Daytona & Chrysler Laser** all models '84 thru '89
*545 **Omni & Plymouth Horizon** all models '78 thru '90
*912 **Pick-ups** all full-size models '74 thru '91
*1726 **Shadow & Plymouth Sundance** '87 thru '91
*1779 **Spirit & Plymouth Acclaim** '89 thru '92
*349 **Vans – Dodge & Plymouth** V8 & 6 cyl models '71 thru '91

FIAT

094 **124 Sport Coupe & Spider** '68 thru '78
479 **Strada** all models '79 thru '82
273 **X1/9** all models '74 thru '80

FORD

*1476 **Aerostar Mini-vans** all models '86 thru '92
788 **Bronco and Pick-ups** '73 thru '79
*880 **Bronco and Pick-ups** '80 thru '91
268 **Courier Pick-up** all models '72 thru '82
789 **Escort & Mercury Lynx** all models '81 thru '90
*2046 **Escort & Mercury Tracer** all models '91 thru '93
*2021 **Explorer & Mazda Navajo** '91 thru '92
560 **Fairmont & Mercury Zephyr** all in-line & V8 models '78 thru '83
334 **Fiesta** all models '77 thru '80
754 **Ford & Mercury Full-size,** Ford LTD & Mercury Marquis ('75 thru '82); Ford Custom 500, Country Squire, Crown Victoria & Mercury Colony Park ('75 thru '87); Ford LTD Crown Victoria & Mercury Gran Marquis ('83 thru '87)
359 **Granada & Mercury Monarch** all in-line, 6 cyl & V8 models '75 thru '80
773 **Ford & Mercury Mid-size,** Ford Thunderbird & Mercury Cougar ('75 thru '82); Ford LTD & Mercury Marquis ('83 thru '86); Ford Torino, Gran Torino, Elite, Ranchero pick-up, LTD II, Mercury Montego, Comet, XR-7 & Lincoln Versailles ('75 thru '86)
*654 **Mustang & Mercury Capri** all models including Turbo '79 thru '92
357 **Mustang V8** all models '64-1/2 thru '73
231 **Mustang II** all 4 cyl, V6 & V8 models '74 thru '78
649 **Pinto & Mercury Bobcat** all models '75 thru '80
*1670 **Probe** all models '89 thru '92
*1026 **Ranger & Bronco II** all gasoline models '83 thru '92
*1421 **Taurus & Mercury Sable** '86 thru '92
*1418 **Tempo & Mercury Topaz** all gasoline models '84 thru '91
1338 **Thunderbird & Mercury Cougar/XR7** '83 thru '88
*1725 **Thunderbird & Mercury Cougar** '89 and '90
*344 **Vans** all V8 Econoline models '69 thru '91

GENERAL MOTORS

*829 **A-Cars** – Chevrolet Celebrity, Buick Century, Pontiac 6000 & Oldsmobile Cutlass Ciera all models '82 thru '90
*766 **J-Cars** – Chevrolet Cavalier, Pontiac J-2000, Oldsmobile Firenza, Buick Skyhawk & Cadillac Cimarron all models '82 thru '92
*1420 **N-Cars** – Buick Somerset '85 thru '87; Pontiac Grand Am and Oldsmobile Calais '85 thru '91; Buick Skylark '86 thru '91
*1671 **GM: Buick** Regal, **Chevrolet** Lumina, **Oldsmobile** Cutlass Supreme, **Pontiac** Grand Prix, all front wheel drive models '88 thru '90
*2035 **GM: Chevrolet Lumina APV, Oldsmobile Silhouette, Pontiac Trans Sport** '90 thru '92

GEO

Metro – *see CHEVROLET Sprint (1727)*
Prizm – *see CHEVROLET Nova (1642)*
Tracker – *see SUZUKI Samurai (1626)*

GMC

Safari – *see CHEVROLET ASTRO (1477)*
Vans & Pick-ups – *see CHEVROLET (420, 831, 345, 1664)*

(continued on next page)

** Listings shown with an asterisk (*) indicate model coverage as of this printing. These titles will be periodically updated to include later model years – consult your Haynes dealer for more information.*

Haynes North America, Inc., 861 Lawrence Drive, Newbury Park, CA 91320 • (805) 498-6703

HAYNES AUTOMOTIVE MANUALS

(continued from previous page)

NOTE: New manuals are added to this list on a periodic basis. If you do not see a listing for your vehicle, consult your local Haynes dealer for the latest product information.

HONDA

351 **Accord CVCC** all models '76 thru '83
*1221 **Accord** all models '84 thru '89
160 **Civic 1200** all models '73 thru '79
633 **Civic 1300 & 1500 CVCC** all models '80 thru '83
297 **Civic 1500 CVCC** all models '75 thru '79
*1227 **Civic** all models '84 thru '91
*601 **Prelude CVCC** all models '79 thru '89

HYUNDAI

*1552 **Excel** all models '86 thru '91

ISUZU

*1641 **Trooper & Pick-up**, all gasoline models '81 thru '91

JAGUAR

*242 **XJ6** all 6 cyl models '68 thru '86
*478 **XJ12 & XJS** all 12 cyl models '72 thru '85

JEEP

*1553 **Cherokee, Comanche & Wagoneer Limited** all models '84 thru '91
412 **CJ** all models '49 thru '86
*1777 **Wrangler** all models '87 thru '92

LADA

*413 **1200, 1300. 1500 & 1600** all models including Riva '74 thru '86

MAZDA

648 **626** Sedan & Coupe (rear wheel drive) all models '79 thru '82
1082 **626 & MX-6 (front wheel drive)** all models '83 thru '91
370 **GLC Hatchback (rear wheel drive)** all models '77 thru '83
757 **GLC (front wheel drive)** all models '81 thru '86
*2047 **MPV** '89 thru '93
Navajo – *see FORD Explorer (2021)*
*267 **Pick-ups** '72 thru '92
460 **RX-7** all models '79 thru '85
*1419 **RX-7** all models '86 thru '91

MERCEDES-BENZ

*1643 **190 Series** all four-cylinder gasoline models, '84 thru '88
346 **230, 250 & 280** Sedan, Coupe & Roadster all 6 cyl sohc models '68 thru '72
983 **280 123 Series** all gasoline models '77 thru '81
698 **350 & 450** Sedan, Coupe & Roadster all models '71 thru '80
697 **Diesel 123 Series** 200D, 220D, 240D, 240TD, 300D, 300CD, 300TD, 4- & 5-cyl incl. Turbo '76 thru '85

MERCURY

For all PLYMOUTH titles see FORD Listing

MG

111 **MGB** Roadster & GT Coupe all models '62 thru '80
265 **MG Midget & Austin Healey Sprite Roadster** '58 thru '80

MITSUBISHI

*1669 **Cordia, Tredia, Galant, Precis & Mirage** '83 thru '90
*2022 **Pick-ups & Montero** '83 thru '91

MORRIS

074 **(Austin) Marina 1.8** all models '71 thru '80
024 **Minor 1000** sedan & wagon '56 thru '71

NISSAN

1137 **300ZX** all Turbo & non-Turbo models '84 thru '89
*1341 **Maxima** all models '85 thru '91
*771 **Pick-ups/Pathfinder** gas models '80 thru '91
*876 **Pulsar** all models '83 thru '86
*982 **Sentra** all models '82 thru '90
*981 **Stanza** all models '82 thru '90

OLDSMOBILE

Custom Cruiser – *see BUICK Full-size (1551)*
658 **Cutlass** all standard gasoline V6 & V8 models '74 thru '88
Cutlass Ciera – *see GENERAL MOTORS A-Cars (829)*
Cutlass Supreme – *see GENERAL MOTORS (1671)*
Firenza – *see GENERAL MOTORS J-Cars (766)*
Ninety-eight – *see BUICK Full-size (1551)*
Omega – *see PONTIAC Phoenix & Omega (551)*
Silhouette – *see GENERAL MOTORS (2035)*

PEUGEOT

663 **504** all diesel models '74 thru '83

PLYMOUTH

For all PLYMOUTH titles, see DODGE listing.

PONTIAC

T1000 – *see CHEVROLET Chevette (449)*
J-2000 – *see GENERAL MOTORS J-Cars (766)*
6000 – *see GENERAL MOTORS A-Cars (829)*
1232 **Fiero** all models '84 thru '88
555 **Firebird** all V8 models except Turbo '70 thru '81
*867 **Firebird** all models '82 thru '91
Full-size Rear Wheel Drive – *see Buick, Oldsmobile, Pontiac Full-size (1551)*
Grand Prix – *see GENERAL MOTORS (1671)*
551 **Phoenix & Oldsmobile Omega** all X-car models '80 thru '84
Trans Sport – *see GENERAL MOTORS (2035)*

PORSCHE

*264 **911** all Coupe & Targa models except Turbo & Carrera 4 '65 thru '89
239 **914** all 4 cyl models '69 thru '76
397 **924** all models including Turbo '76 thru '82
*1027 **944** all models including Turbo '83 thru '89

RENAULT

141 **5 Le Car** all models '76 thru '83
079 **8 & 10** all models with 58.4 cu in engines '62 thru '72
097 **12 Saloon & Estate** all models 1289 cc engines '70 thru '80
768 **15 & 17** all models '73 thru '79
081 **16** all models 89.7 cu in & 95.5 cu in engines '65 thru '72
Alliance & Encore – *see AMC (934)*

SAAB

247 **99** all models including Turbo '69 thru '80
*980 **900** all models including Turbo '79 thru '88

SUBARU

237 **1100, 1300, 1400 & 1600** all models '71 thru '79
*681 **1600 & 1800** 2WD & 4WD all models '80 thru '89

SUZUKI

*1626 **Samurai/Sidekick and Geo Tracker** all models '86 thru '91

TOYOTA

*1023 **Camry** all models '83 thru '91
150 **Carina Sedan** all models '71 thru '74
*2038 **Celica** Front Wheel Drive '86 thru '92
935 **Celica** Rear Wheel Drive '71 thru '85
*1139 **Celica Supra** '79 thru '92
361 **Corolla** all models '75 thru '79
961 **Corolla** all models (rear wheel drive) '80 thru '87
*1025 **Corolla** all models (front wheel drive) '84 thru '91
*636 **Corolla Tercel** all models '80 thru '82
230 **Corona & MK II** all 4 cyl sohc models '69 thru '74
360 **Corona** all models '74 thru '82
*532 **Cressida** all models '78 thru '82
313 **Land Cruiser** all models '68 thru '82
200 **MK II** all 6 cyl models '72 thru '76
*1339 **MR2** all models '85 thru '87
304 **Pick-up** all models '69 thru '78
*656 **Pick-up** all models '79 thru '92

TRIUMPH

112 **GT6 & Vitesse** all models '62 thru '74
113 **Spitfire** all models '62 thru '81
322 **TR7** all models '75 thru '81

VW

159 **Beetle & Karmann Ghia** all models '54 thru '79
238 **Dasher** all gasoline models '74 thru '81
*884 **Rabbit, Jetta, Scirocco, & Pick-up** all gasoline models '74 thru '91 & **Convertible** '80 thru '91
451 **Rabbit, Jetta & Pick-up** all diesel models '77 thru '84
082 **Transporter 1600** all models '68 thru '79
226 **Transporter 1700, 1800 & 2000** all models '72 thru '79
084 **Type 3 1500 & 1600** all models '63 thru '73
1029 **Vanagon** all air-cooled models '80 thru '83

VOLVO

203 **120, 130 Series & 1800 Sports** '61 thru '73
129 **140 Series** all models '66 thru '74
*270 **240 Series** all models '74 thru '90
400 **260 Series** all models '75 thru '82
*1550 **740 & 760 Series** all models '82 thru '88

SPECIAL MANUALS

1479 **Automotive Body Repair & Painting Manual**
1654 **Automotive Electrical Manual**
1480 **Automotive Heating & Air Conditioning Manual**
1762 **Chevrolet Engine Overhaul Manual**
1736 **Diesel Engine Repair Manual**
1667 **Emission Control Manual**
1763 **Ford Engine Overhaul Manual**
482 **Fuel Injection Manual**
1666 **Small Engine Repair Manual**
299 **SU Carburetors** thru '88
393 **Weber Carburetors** thru '79
300 **Zenith/Stromberg CD Carburetors** thru '76

See your dealer for other available titles

* *Listings shown with an asterisk (*) indicate model coverage as of this printing. These titles will be periodically updated to include later model years – consult your Haynes dealer for more information.*

Over 100 Haynes motorcycle manuals also available

Haynes North America, Inc., 861 Lawrence Drive, Newbury Park, CA 91320 • (805) 498-6703